INDEX OF APPLICATIONS

FINITE
MATHEMATICS
AN APPLIED APPROACH
SECOND EDITION

FINITE
MATHEMATICS
AN APPLIED APPROACH
SECOND EDITION

Paul E. Long
University of Arkansas

Jay Graening
University of Arkansas

▲▲ **ADDISON-WESLEY**

An imprint of Addison Wesley Longman, Inc.

Reading, Massachusetts • Menlo Park, California • New York • Harlow, England
Don Mills, Ontario • Sydney • Mexico City • Madrid • Amsterdam

Sponsoring Editor: Kevin M. Connors/Greg Tobin
Developmental Editor: Sandi Goldstein
Project Editor: Anne Schmidt Ryan
Design Administrator: Jess Schaal
Art Development Editor: Vita Jay
Text and Cover Design: Lesiak/Crampton Design Inc; Cindy Crampton
Photo Researcher: Corrine Johns
Production Administrator: Randee Wire
Compositor: Interactive Composition Corporation
Printer and Binder: R. R. Donnelley & Sons Company
Cover Printer: Phoenix Color Corp.
Cover Image: Liaison International, © Ralph Mercer

Finite Mathematics, An Applied Approach, Second Edition

Library of Congress Cataloging-in-Publication Data

Long, Paul E.
    Finite mathematics: an applied approach/by Paul E. Long,
Jay Graening—2nd ed.
      p. cm.
    Includes index.
    ISBN 0-673-99600-X
    1. Mathematics.   I. Graening, Jay.   II. Title.
QA39.2.L66   1996   96-23302
510—dc20   CIP

2 3 4 5 6 7 8 9 10 – VH – 99 98 97

CONTENTS

PREFACE

Finite Mathematics: An Applied Approach, Second Edition, is designed to provide the noncalculus mathematics needed for students in business, economics, social sciences, life sciences, and agricultural sciences. In this text, we show, through applications, the relevance of mathematics to both real life and future courses in the student's discipline. A key goal of this text is to prepare students for mathematically oriented courses in their discipline, particularly courses in statistics, economics, accounting, and general business. This text also provides a general mathematics background for students who need a terminal mathematics course. To promote students' understanding of the material, we have used a personal, conversational writing style that minimizes symbolism. We believe that students will find it easy to read and unintimidating.

Course Flexibility

The first four chapters form a block of linear material designed to be covered in consecutive order, although well-prepared students might bypass Chapter 1. Chapters 5, 6, 7, and 8 form a second block of material covering probability and statistics. These chapters are also designed to be covered in consecutive order, although Chapter 7 could be omitted if desired. The order in which the two blocks are covered is immaterial. Chapter 9 has Chapters 3 and 6 as prerequisites. Chapter 10 has Chapters 3, 4, and 6 as prerequisites. Chapters 9 and 10 are independent of each other. Chapter 11 is independent of all others. The diagram on the following page shows the interdependence among the chapters.

Features New to This Edition

Several features designed to assist students in the learning process have been added or improved in this edition:

Exercises are more carefully graded in level of difficulty.

The Problems for Exploration at the end of each section are no longer marked as "individual" or "group."

Crown's pivoting rules for nonstandard linear programming problems have been improved.

The section on the dual problem in linear programming has been rewritten to reflect the true meaning of the term *dual.*

Probability is now covered in two chapters: a new chapter on basic probability that includes discrete random variables and expected value, plus a chapter covering advanced topics.

The solution to mixed strategy games now makes use of linear programming techniques.

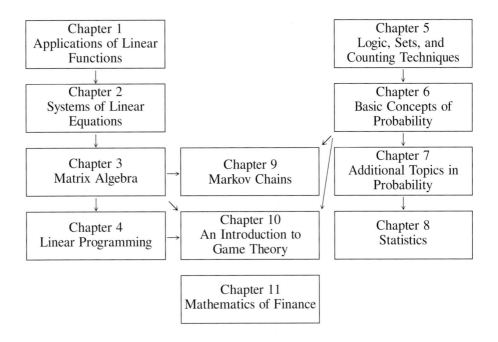

Distinguishing Features

Applications

Applications are found throughout the text in examples and exercises and use real data that demonstrate the relevance of mathematics to the real world. Each chapter begins with an overview of the content, an application problem, and an illustration. This opening application is then referred to and solved later in the chapter. Applications found in the exercise sets are drawn from a multitude of disciplines reflecting the diversity of the students who take this course.

Exercises

We include large numbers of graded exercises at the end of each section to allow the instructor to assign nonrepetitive homework for two semesters or two to three quarters during the academic year. This feature also permits "back-cycling" of homework assignments for constant review of concepts already covered in the course. The use of computer/calculator technology is encouraged throughout the text, and each section features some specially marked exercises for which computer programs or calculator capabilities will prove especially helpful.

Exercises taken from actuarial exams are included in appropriate exercise sets throughout the text. These exercises show students how the mathematics topics in the text are used by professionals.

Each exercise set ends with a suggested set of **Problems for Exploration.** These projects are intended to enrich and extend the text discussion. Some projects require gathering and analyzing data; some explore theoretical aspects of the section material; and some require students to explore a new topic that relates to the section.

Chapter Reflections and Other Final Features

Each chapter ends with two special features. The first is **Chapter Reflections,** a unique review that integrates chapter topics according to special themes:

(a) **Concepts** summarize briefly the main ideas of the chapter.

(b) **Computation/Calculation** provides basic drill exercises.

(c) **Connections** relate the chapter to other parts of mathematics or to other subjects through exercises—for example, using a matrix and its inverse to code and decode messages.

(d) **Communication** asks students to conceptualize and internalize the meaning of important topics by explaining them in their own words.

(e) **Cooperative Learning** has challenging extension problems for teams of students to explore.

(f) **Critical Thinking** requires students to analyze a situation, choose a position, and support their answer.

(g) **Computer/Calculator** discusses the computer programs and graphing calculator programs available for the chapter.

(h) **Cumulative Review** gives several pertinent review problems from previous chapters.

The second, **Sample Test Items,** provide a good study guide for the student upon completion and review of the chapter material.

And finally, many chapters conclude with an additional special feature: a **Career Uses of Mathematics** box that describes how real people use the mathematics of the chapter in their work.

Focus on Technology

We intend for students to use computers or graphing calculators to assist them in learning the course material. However, this book *does not require* this technology and can be used successfully without it. As appropriate, we make generic references to graphing calculators throughout the text. Our premise is that the mathematics should be supported, whenever feasible, by appropriate technology after students are exposed to the content topics of the pertinent sections, but not at the expense of the mathematics.

The computer programs included on the interactive tutorial software MasterIt® were developed with student input to make sure the programs are user friendly. The programs are DOS-based. These programs not only work problems for students but are also interactive and require thought and input. Consequently, most of the MasterIt® programs are designed as a tutorial for obtaining hands-on concept clarification, practice, and enrichment, as well as for printing out homework.

Supplement Package to Accompany This Text

For the Instructor

The **Instructor's Solution Manual** contains complete, worked-out solutions for all of the even-numbered exercises in the text. The **Printed Test Bank** includes test items for each chapter, with all answers supplied in a convenient format.

   Addison-Wesley TestGen-EQ with QuizMaster-EQ. This software is available in Windows and Macintosh versions and is fully networkable. TestGen-EQ's friendly graphical interface enables instructors to easily view, edit, and add questions, transfer questions to tests, and print tests in a variety of fonts and forms. Search and sort features let the instructor quickly locate questions and arrange them in a preferred order. Six question formats are available, including short-answer, true-false, multiple-choice, essay, matching, and bimodal formats. A built-in question editor gives the user power to create graphs, import graphics, insert mathematical symbols and templates, and insert variable numbers or text. **QuizMaster-EQ** enables instructors to create and save tests and quizzes using TestGen-EQ so students can take them on a computer network. Instructors can set preferences for how and when tests are administered. QuizMaster-EQ automatically grades the exams and allows the instructor to view or print a variety of reports for individual students, classes, or courses.

For the Student

The **Student's Solution Manual** includes complete, worked-out solutions to the odd-numbered problems in the text. To order, use ISBN 0-673-98334-X.

Interactive Mathematics Tutorial Software with Management System

This innovative software is available in DOS, Windows, and Macintosh versions and runs on commonly used LANs. As with the Test Generator/Editor, this software is algorithm driven so that the constants automatically regenerate and numbers rarely repeat in a problem type when students revisit any particular lesson. Each tutorial is based on textbook objectives, is self-paced, and provides unlimited opportunities to review concepts and practice problem-solving. If students give a wrong answer, they can ask to see the problem worked out and get a textbook page reference. Most problems include "smart hints" that respond to specific errors for the first incorrect responses. Tools such as an online glossary and Quick Reviews provide definitions and examples, and an online calculator aids students in computation. Section quizzes and chapter tests help students pinpoint their weak skills and prepare for classroom examination. The program is menu-driven for ease of use, and on-screen help can be obtained at any time with a single keystroke. Students' scores are calculated at the end of each lesson and can be printed out as a permanent record. The optional **Management System** lets instructors record student scores on disk and print diagnostic reports for individual students or classes. This software may also be purchased by students for home use. Student versions include record-keeping and practice tests.

   The **Graphing Calculator Lessons for Finite Mathematics** by Paula Young contain activities for using the TI-82 graphing calculator in the course. To order, use ISBN 0-06-501-330-1.

   Spreadsheet Topics in Finite Mathematics by Samuel W. Spero is a workbook that introduces students to the electronic spreadsheet, an important resource in business. To order, use ISBN 0-06-500300-4.

The interactive tutorial software **MasterIt**® by Richard Jetton was developed for this text. Available in DOS format, it consists of programs that relate directly to text discussions. Each program assists students with problem solving or demonstrates essential concepts.

Acknowledgments

Special thanks to Professor Arthur M. Hobbs of Texas A&M University for sharing his acute insight into linear programming, especially in improving the pivoting rules for nonstandard problems.

Our thanks go to Paula Young and Richard Jetton, our calculator and computer experts, who recognized the importance of technology in the classroom and implemented many changes to make student-friendly software a reality.

The following individuals reviewed the manuscript and offered many helpful comments and suggestions:

Mary Kay Abbey, Montgomery College at Takoma Park

Faiz Al-Rubaee, University of North Florida

Polly Amstutz, University of Nebraska at Kearney

Yvonne Brown, Pima Community College—West Campus

Sue Cluxton, Midlands Technical College

Thomas Davis, Sam Houston State University

Suhrit K. Dey, Eastern Illinois University

Michael Divinia, San Jose City College

Morteza Ebneshahrashoob, California State University at Long Beach

Frances Gulick, University of Maryland

Yvette Hester, Texas A&M University

Arthur M. Hobbs, Texas A&M University

Robert L. Horvath, El Camino College

Karla Karstens, University of Vermont

Edwin M. Klein, University of Wisconsin at Whitewater

Donald E. Myers, University of Arizona

Robert A. Nowlan, Southern Connecticut State University

C. T. Pan, Northern Illinois University

Thomas Ralley, The Ohio State University

John A. Roberts, University of Louisville

Jane M. Rood, Eastern Illinois University

Arnold L. Shroeder, Long Beach City College

June Shanholtzer, Eastern Illinois University

Lowell Stultz, Kalamazoo Valley Community College

Steve Wilson, Sonoma State University

James Wooland, Florida State University

We also wish to express our appreciation to the individuals who meticulously checked the answers for accuracy: Mark Carpenter, Sam Houston State University, Arthur M. Hobbs, Texas A&M University, and Steve Wilson, Sonoma State University.

At Addison Wesley Longman, we extend our sincere thanks to Sandi Goldstein, developmental editor, for her kind and efficient direction in helping us to the production stage. We also deeply appreciate the encouragement and advice of acquisitions editor Kevin Connors and marketing manager George Duda. And finally, our thanks to Anne Schmidt Ryan, assistant project editor, for her caring attention to details during the production process.

We welcome feedback from both students and instructors. Please send your comments and suggestions to us at our school address.

Paul E. Long
Jay Graening
University of Arkansas
Peabody Hall, #107B
Fayetteville, AR 72701

CHAPTER **1**

Applications of Linear Functions

If this trend continues, how many pounds of waste will every man, woman, and child be generating each day in the year 2003? (See Example 5, Section 1.5.)

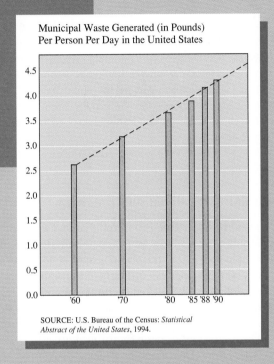

Municipal Waste Generated (in Pounds) Per Person Per Day in the United States

SOURCE: U.S. Bureau of the Census: *Statistical Abstract of the United States*, 1994.

CHAPTER PREVIEW

The central theme of this chapter is the modeling of various phenomena with linear functions. We begin by reviewing the rectangular coordinate system and the graphing of functions. Then the point-slope, slope-intercept, and general form equations of a straight line are developed. This leads to the concept of a linear function, whose graph is a straight line, and its use to model sales, costs, the book value of an asset depreciated by the straight-line method, the decline of fish population versus pollution, the growth of municipal solid waste, and so on. The break-even point and the equilibrium supply and demand quantities are then studied as applications of a pair of linear functions. Finally, the concept of a least squares regression line is introduced as an application of solving a system of two equations in two unknowns and is shown to occur in numerous fields of study.

CALCULATOR/COMPUTER OPPORTUNITIES FOR THIS CHAPTER

Graphing calculators may be used to graph functions used in this chapter. They also may be used to find the coordinates of the point of intersection between two lines, either by estimation with the "trace" function or more precisely with the touch of a "calculation" key found on most models. In Section 1.5 you will find the preprogrammed capability of a graphing calculator particularly useful in calculating the slope, *y*-intercept, and correlation coefficient of a least squares regression line. The supplement "Graphing Calculator Lessons for Finite Mathematics" that accompanies this book will aid in calculator uses for this course.

Certain computer programs popular in the mathematics community will also display and find the coordinates of the point of intersection between two lines as well as give solutions to more general systems of linear equations. These programs, along with most popular spreadsheets and some popular word-processing programs, will also automatically calculate the slope, *y*-intercept, and correlation coefficient of a least squares regression line. Finally, the computer disk MasterIt that accompanies this book contains a program called "LinReg," which does these same least squares calculations.

SECTION 1.1   The Cartesian Plane and Graphing

The **real number line** is a graphic model for the real number system. It consists of a horizontal line with an arbitrary point labeled with the number 0, called the **origin.** A point to the right of the origin is marked to correspond to the number 1, with positive numbers then scaled accordingly to the right of the origin, whereas negative numbers are scaled to the left of the origin (see Figure 1). Each real number then corresponds to precisely one point on the real number line, and, conversely, each point on the real number line corresponds to precisely one real number.

The **rectangular coordinate system,** also called the **Cartesian plane** after its inventor, René Descartes, provides a convenient way to graphically represent points in the plane as *ordered pairs of real numbers.* Such a coordinate system is formed by two real number lines intersecting at right angles, as shown in Figure 2. The horizontal number line is usually called the *x*-axis, while the vertical number line is usually called the *y*-axis. The point of intersection of the two lines is called the **origin,** and the axes separate the plane into four regions, called **quadrants.**

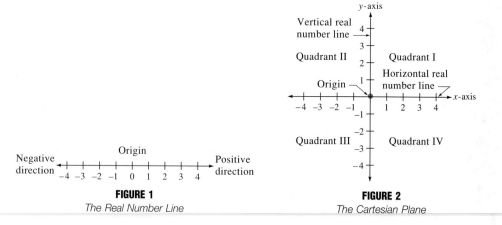

FIGURE 1
The Real Number Line

FIGURE 2
The Cartesian Plane

Each point in the plane can now be matched with exactly one **ordered pair** of real numbers (x, y) called **coordinates of the point;** conversely, each ordered pair of real numbers (x, y) corresponds to exactly one point in the plane. The **x-coordinate** represents the directed distance from the y-axis to the point, and the **y-coordinate** represents the directed distance from the x-axis to the point.

EXAMPLE 1
Plotting Points in the Cartesian Plane

Plot (locate) the points $(1, 3)$, $(-2, 4)$, $(4, 0)$, and $(-3, -2)$ in the Cartesian plane.

Solution To plot the point $(1, 3)$ envision a vertical line through 1 on the x-axis and a horizontal line through 3 on the y-axis. The intersection of these two lines represents the point $(1, 3)$, as shown in Figure 3. The other points are plotted in a similar way.

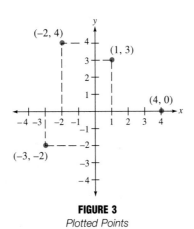

FIGURE 3
Plotted Points

The beauty of Descartes' rectangular coordinate system is that it allows us to visualize relationships between two variables. See Figure 4(b) of Example 2.

EXAMPLE 2
An Application of the Rectangular Coordinate System

Data collected by the U.S. National Highway Traffic Safety Administration show the following number of deaths (in thousands) per year resulting from motor vehicle accidents for the years 1980 through 1992.

Year	1980	1982	1984	1986	1988	1990	1992
Deaths	51.1	43.9	44.3	46.1	47.1	44.6	39.2

SOURCE: U.S. National Highway Traffic Safety Administration Data, U.S. Bureau of the Census: *Statistical Abstract of the United States, 1994.*

Show these data on a rectangular coordinate system.

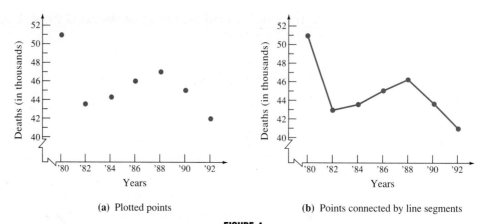

(a) Plotted points   (b) Points connected by line segments

FIGURE 4

Deaths from Motor Vehicle Accidents

Solution   Let x represent the years 1980 through 1988, and let y represent the number of deaths (in thousands). Because there are many years between 0 and 1980 for which no data are shown, it is customary to indicate these omissions graphically by showing a break in the x-axis, as in Figure 4(a). The same applies to omissions on the y-axis. Plotted points are sometimes connected by line segments, as in Figure 4(b), to more clearly depict any trends that are present. ■

The sharp early downward trend in the data reflects the imposition of a nationwide speed limit of 55 miles per hour. As of 1995, this speed limit was no longer in effect nationwide.

The relationship between two variables may sometimes be stated by an **equation.** Some examples of equations are:

$$y = 3x + 2$$
$$x^2 + y^2 = 9$$
$$C = 24x + 1800 \qquad \text{\small C is the cost, in dollars, of making x clock radios.}$$
$$R = 48x \qquad \text{\small R is the revenue gained by selling x ladies' blouses.}$$

Consider the first of these equations, $y = 3x + 2$. If $x = 1$, for instance, then $y = 3(1) + 2 = 5$. The point $(1, 5)$ is called a **solution** to the equation and may be plotted in the rectangular coordinate system. Another solution is $(0, 2)$ because the substitution of $x = 0$ into the equation results in $y = 2$. There are an infinite number of such solutions, and when they are displayed in the rectangular coordinate system, they constitute the **graph** of the equation. This is the link between *algebra* (the equation) and *geometry* (the representation of the solution points in the rectangular coordinate system) that Descartes created with his analytic geometry.

The Graph of an Equation

The **graph** of an equation is the geometric representation in the rectangular coordinate system of all solutions to the equation.

One method of graphing an equation is called the **plotting method.** It consists of the three steps stated next.

The Plotting Method of Graphing an Equation

The **plotting method** of graphing an equation consists of these three steps:

1. Make a representative table of solutions to the equation.

2. Plot the solutions as ordered pairs in the rectangular coordinate system.

3. Use a smooth curve to connect the points and also to portray solutions not shown in the table of representative solutions.

EXAMPLE 3
Graphing an Equation

Graph the equation $y = 3x + 2$.

Solution   Representative solutions can be efficiently displayed in a table with the headings x, y, and (x, y), as shown below. To find a solution (x, y), choose any number x and then substitute that chosen value into the equation $y = 3x + 2$ to find the corresponding value of y. A good sample of solutions will be found by choosing some positive and some negative values of x. Once the table is constructed, we are ready to plot these solutions in the rectangular coordinate system. The points appear as shown in Figure 5. Finally, we look for a pattern among the points just plotted to determine how the smooth curve through them will appear. In this case, the points appear to be in a straight line, so we connect them accordingly. ■

x	y	(x, y)
-3	-7	$(-3, -7)$
-2	-4	$(-2, -4)$
-1	-1	$(-1, -1)$
0	2	$(0, 2)$
1	5	$(1, 5)$
2	8	$(2, 8)$

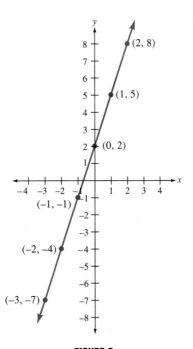

FIGURE 5
Graph of $y = 3x + 2$

The next section (Section 1.2) shows how to identify equations that have straight lines as their graphs. Once they are identified, we need only find *two solutions* (points on the line) to graph the equation.

EXAMPLE 4
Graphing an Equation

The Brite Lite Company hired an economist to study the relationship between the number of flashlights it could expect to sell and the price it charges for each. The economist discovered that sales S (in hundreds of flashlights per month) could be approximated by the equation

$$S = -\left(\frac{1}{2}\right)p^2 + 32$$

where p (in dollars) is the price charged for each flashlight. Graph this equation to get a visual picture of the relationship between sales and price.

Solution   For this equation, it is convenient to let p play the role of x and S the role of y. Because p (price) cannot be negative, we start seeking solutions to the equations with $p = 0$, even though this is actually somewhat unrealistic. Some values of p for which $S = -(\frac{1}{2})p^2 + 32$ can be readily calculated and are shown in the table to the left of Figure 6. We find that S drops off rather fast as p increases until $p = 8$ gives $S = 0$. In other words, when the flashlights are priced at \$8, no sales will be made. (Again, this may be somewhat unrealistic. Formulas such as these are always suspect at the endpoints.) We may *not* use p values greater than 8 because this would give negative sales. A sketch of the graph is also shown in Figure 6.   ▬

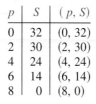

p	S	(p, S)
0	32	(0, 32)
2	30	(2, 30)
4	24	(4, 24)
6	14	(6, 14)
8	0	(8, 0)

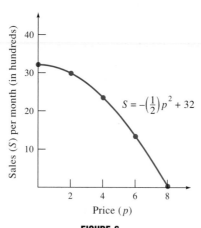

FIGURE 6
Flashlight Sales

In Example 4, the equation $S = -(\frac{1}{2})p^2 + 32$ tells us that the number of sales *depends on* the price charged for each flashlight. In mathematical language, we would say that the sales are a **function** of the price and would denote that fact by the equation $S = f(p)$, read "S equals f of p." In particular, the relationship between sales and the price may be written as

$$f(p) = -\left(\frac{1}{2}\right)p^2 + 32.$$

Such functional notation provides a convenient way to denote sales for various values of p without actually calculating the right side of the equation. For example, $f(4)$

denotes the sales (in hundreds) when $p = 4$; $f(7)$ denotes the sales (in hundreds) when $p = 7$, and so on. The interval $[0, 8]$ of possible values of p is called the **domain** of the function $S = f(p)$. The formal definition of a function using the variables x and y of the Cartesian plane is given next.

Defining a Function

A **function** is a rule f that assigns to each value of a variable x one—and only one—value $f(x)$, read as "f of x." The set of permissible values of x to which the rule f applies is called the **domain** of the function. The **graph** of a function consists of all ordered pairs $(x, f(x))$ in the rectangular coordinate system.

To graph a function (rule) f that is given in equation form, such as $f(x) = 2x - 5$, it is often more convenient to set $y = f(x) = 2x - 5$ and graph the equation $y = 2x - 5$ rather than use the functional notation exclusively. When considering $y = f(x)$, x is called the **independent variable,** and y is called the **dependent variable** since its value depends upon the choice of x.

EXAMPLE 5
Using Functional Notation

Sixteen grams of table salt are stirred into a container of hot water to dissolve. After x minutes the amount of salt that remains undissolved is given by the function $f(x) = (x - 4)^2$.

(a) How many grams of salt remain undissolved after 1 minute? After 2 minutes?

Solution The answers are found by computing

$$f(1) = (1 - 4)^2 = (-3)^2 = 9 \text{ grams}$$

and

$$f(2) = (2 - 4)^2 = (-2)^2 = 4 \text{ grams.}$$

(b) Find the domain of the function f. In other words, what values of time x are we permitted to use in $f(x) = (x - 4)^2$ to obtain grams of undissolved salt?

Solution Time starts at $x = 0$, the moment salt is put into the container. At that time the amount of undissolved salt is $f(x) = (0 - 4)^2 = (-4)^2 = 16$ grams. The values of x then increase until x becomes 4 at which time $f(4) = (4 - 4)^2 = 0$ grams of salt undissolved. That is, 4 minutes later all of the salt has been dissolved. Therefore, a realistic domain for this function is the interval $[0, 4]$.

(c) Graph the function $f(x) = (x - 4)^2$ for the values of x in its domain.

Solution We start by setting $y = (x - 4)^2$ and construct a table of solutions to the given equation using values of x in the interval $[0, 4]$ only. Selected values of x and the corresponding values of y are shown in the table to the left of Figure 7. ■■

x	$y = f(x)$	(x, y)
0	16	(0, 16)
1	9	(1, 9)
2	4	(2, 4)
3	1	(3, 1)
4	0	(4, 0)

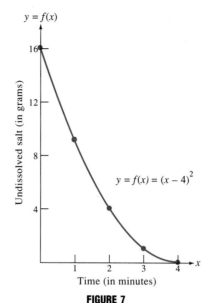

FIGURE 7
Relationship of Time and Undissolved Salt

We note that more suggestive letters than f may be used to denote functions. In Example 5, the notation $y = U(x)$ or $y = S(x)$, for instance, might have been used to denote the undissolved amount of salt. Furthermore, we might want to use t for time, rather than x. To avoid confusion, be sure to differentiate between the letter denoting the function or rule, say f, and the functional value, $f(x)$.

As demonstrated in Figures 6 and 7, it is not necessary to use the same scale on the horizontal and vertical axes when graphing functions.

EXERCISES   1.1

In Exercises 1 through 8, plot the given points on the same set of axes in the rectangular coordinate system.

1. (3, 5), (6, 1), (−2, −1)

2. (4, 4), (0, 6), (0, −2)

3. (5, 0), (−7, 1), (4, −3)

4. (6, 0), (0, 6), (−2, 5), (4, 2)

5. $(-\frac{1}{2}, 5)$, (4.5, 2.5), (−1, −4), (0, .5)

6. (3, −5), $(\frac{3}{4}, 4)$, (−3, −2), (0, 5)

7. (4, 0), $(4, \frac{1}{2})$, (0, 0), (4, −.5)

8. (−3.5, .5), (−2, −3), (0, 4.5), (3, −4)

In Exercises 9 and 10, write the coordinates of each point shown.

9.
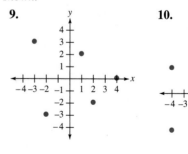

10.
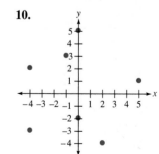

In Exercises 11 through 16, the data given were taken from a U.S. Census Bureau publication.* Plot the data in the rectangular coordinate system, and connect the points with line segments. Some estimation of the y-coordinates will be necessary.

11. Exports Dollar amounts (in millions) of artwork and antiques exported from the United States:

Year	Dollars (millions)
1990	2282
1991	1239
1992	1076
1993	952

12. Labor Force The number of women (over age 16) in the civilian labor force in the United States (in thousands):

Year	Women employed (thousands)	Year	Women employed (thousands)
1983	48,500	1988	54,742
1984	49,709	1989	56,030
1985	51,050	1990	56,554
1986	52,413	1991	56,893
1987	53,658	1992	57,798
		1993	58,407

13. Income The median income per U.S. household:

Year	Median income (dollars)	Year	Median income (dollars)
1982	20,171	1987	26,061
1983	21,018	1988	27,225
1984	22,415	1989	28,906
1985	23,618	1990	29,943
1986	24,897	1991	30,126
		1992	30,786

14. Newspapers The number of daily newspapers published in the United States:

Year	Newspapers	Year	Newspapers
1980	1744	1989	1773
1985	1701	1990	1788
1986	1651	1991	1781
1987	1646	1992	1755
1988	1745	1993	1850
		1994	1831

15. Smoking The percent of the U.S. population 18 years old and older who smoke cigarettes:

Year	Smokers (percent)	Year	Smokers (percent)
1965	42.2	1985	30.1
1974	37.1	1987	28.8
1979	33.5	1988	28.1
1983	32.1	1990	25.5
		1991	25.6

16. College Enrollment Percent of U.S. high school graduates who enrolled in college:

Year	Enrollees (percent)	Year	Enrollees (percent)
1960	23.8	1988	37.6
1970	33.3	1989	38.5
1975	33.1	1990	39.6
1980	32.3	1991	41.4
1985	34.3	1992	42.3

In Exercises 17 through 24, use the plotting method to graph each of the given equations for a straight line.

17. $y = 2x - 3$ 18. $y = x - 4$
19. $y = -2x$ 20. $y = 5x$
21. $2x - y = 6$ 22. $2x + y = 7$
23. $x + 2y = 4$ 24. $-x + y = 5$

In Exercises 25 through 30, use the plotting method to graph each of the given functions.

25. $f(x) = 3x - 1$ 26. $f(x) = -2x$
27. $f(t) = -t + 4$ 28. $P(t) = t^2 - 3$
29. $Q(p) = 2p - 4$ 30. $Z(p) = p - 1$

 In Exercises 31 through 36, use a graphing calculator to graph each of the given functions.

31. $y = x^2$ 32. $y = x^2 + 1$
33. $y = x^2 - 2$ 34. $y = x - 6$
35. $y = x^3$ 36. $y = (x - 1)^3$

37. A certain pollutant is being removed from a small lake. Let t be the number of days from the beginning of the removal process and $f(t) = 50\left(1 - \dfrac{t}{120}\right)$ the number of parts per million of this pollutant in the water. Find the domain for this function and sketch its graph throughout the domain.

*SOURCE (for Tables in Exercises 11–16): U.S. Bureau of the Census: *Statistical Abstract of the United States*, 1994.

38. Let x be the number of miles, in thousands, that a particular tire is driven and $f(x) = 2\left(1 - \dfrac{x}{40}\right)$ the tread depth in centimeters. Find the domain of this function and sketch its graph throughout its domain.

39. Let p be the price, in dollars, of a particular cassette tape and $S(p) = 25 - p^2$ the number sold, in thousands, each month by the ABC Department Store. Find the domain of this function and sketch its graph throughout its domain.

40. Let p be the price, in hundreds of dollars, of a particular brand of stereo and $S(p) = 30\left(16 - \dfrac{p^2}{4}\right)$ the number sold each year by Schriner's Department Store. Find the domain of this function and sketch its graph throughout its domain.

41. A particular state has a pollution control unit consisting of five inspectors who monitor municipal waste facilities. Each inspection costs the state $120 and requires one inspector to spend an entire day on an inspection. If x is the number of inspections during a given work week and $C(x) = 500 + 120x$ is the total cost to the state (in dollars) of those inspections, find the domain of this function and sketch its graph throughout its domain.

42. The Omega Yogurt store employs eight students, each of whom is on call to work a four-hour shift each of the five weekdays. If x is the number of student shifts needed each week and $C(x) = 80 + 40x$ is the store owner's cost (in dollars) for those x shifts, find the domain of this function and sketch its graph throughout its domain.

Problems for Exploration

1. 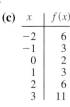 Use the plotting method to graph the equation $x^2 + y^2 = 1$.

2. Use a table of solutions (and a graph, if necessary) to discover a formula for each function.

(a)

x	$f(x)$
-3	-1
-2	1
-1	3
0	5
1	7
2	9
3	11

(b)

x	$f(x)$
-4	-3
-2	-2
0	-1
2	0
4	1

(c)

x	$f(x)$
-2	6
-1	3
0	2
1	3
2	6
3	11

SECTION 1.2 **Equations of Straight Lines**

In this section, we exhibit the various forms of an equation for a line. Such equations are known as **linear equations.** The term *line* will always mean *straight line.*

All lines, except those parallel to the y-axis (vertical lines), have a **slope,** which is a numerical value that measures the "steepness" of the line. To obtain the slope of a given nonvertical line, let $P(x_1, y_1)$ and $Q(x_2, y_2)$ be two distinct points on the line. (See Figure 8.) As we move along the line from P to Q, the change in the x-coordinate is given by

$$\text{change in } x\text{-coordinate} = x_2 - x_1 = \Delta x \qquad \Delta x \text{ is read "delta } x\text{"}$$

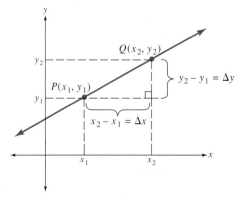

FIGURE 8
Coordinate Difference Between Points

and the corresponding change in the y-coordinate is

$$\text{change in } y\text{-coordinate} = y_2 - y_1 = \Delta y.$$

The slope of the line is then defined to be the ratio $\dfrac{\Delta y}{\Delta x}$ of these changes.

The Slope of a Line

The slope m of the nonvertical line passing through the points (x_1, y_1) and (x_2, y_2) is given by the equation

$$m = \frac{y_2 - y_1}{x_2 - x_1} = \frac{\Delta y}{\Delta x}.$$

(For a nonvertical line, $x_2 - x_1 \neq 0$.)

A feature that distinguishes a line from all other graphs in the plane is that the ratio $\dfrac{\Delta y}{\Delta x}$ is always the same number m. This occurs no matter which point is labeled P or Q or where the points are selected on the line. This implies that Δx and Δy may be thought of as directed changes in their respective variables: Δx positive to the right, negative to the left; Δy positive upward, negative downward. (See Figure 9.) The only note of caution needed in calculating the slope from given coordinates is that the order of subtraction in the numerator must be the same as that in the denominator:

$$m = \frac{y_2 - y_1}{x_2 - x_1} = \frac{\Delta y}{\Delta x}; \qquad m = \frac{y_1 - y_2}{x_1 - x_2} = \frac{-\Delta y}{-\Delta x}; \qquad m \neq \frac{y_2 - y_1}{x_1 - x_2}.$$

Since $m = \dfrac{\Delta y}{\Delta x}$, we see that if $\Delta x = +1$, then $m = \Delta y$. In other words, if the change in x is one unit to the right, then the change in y is precisely the slope of the line. In light of this, the slope may be defined as the **change in y per unit change in x.**

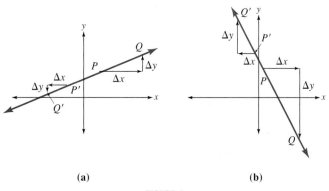

(a)   (b)

FIGURE 9

For a Particular Line, the Ratio $\dfrac{\Delta y}{\Delta x}$ is Always the Same.

The Slope as Unit Change

For a particular line with slope m and any point P on that line, if we move to another point on the line for which $\Delta x = 1$, then $\Delta y = m$. In words, **the slope is the change in y per unit change in x.**

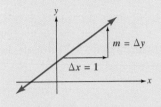

EXAMPLE 1

Finding the Slope of a Line Through Two Points

(a) Find the slope of the line through the points $(2, 4)$ and $(5, 6)$.

Solution Using the definition of the slope, we find that

$$m = \frac{\Delta y}{\Delta x} = \frac{6 - 4}{5 - 2} = \frac{2}{3}.$$

Observe in Figure 10 that the line rises when viewed from left to right. This is a characteristic of all lines with **positive slope.**

(b) Find the slope of the line through the points $(2, 5)$ and $(6, 3)$.

Solution The definition of the slope tells us that

$$m = \frac{\Delta y}{\Delta x} = \frac{3 - 5}{6 - 2} = \frac{-2}{4} = -\frac{1}{2}.$$

Notice in Figure 11 how the line falls when viewed from left to right. This is a characteristic of all lines with **negative slope.**

(c) Find the slope of the line through the points $(3, 4)$ and $(6, 4)$.

Solution The slope is

$$m = \frac{\Delta y}{\Delta x} = \frac{4 - 4}{6 - 3} = 0.$$

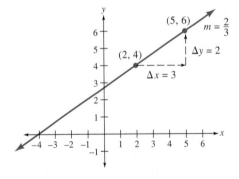

FIGURE 10

A Line Rises when m > 0.

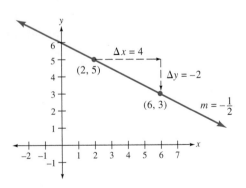

FIGURE 11

A Line Falls when m < 0.

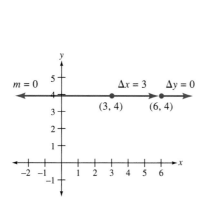

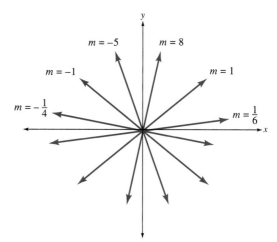

FIGURE 12
Horizontal Lines have Slope 0.

FIGURE 13
Slopes Indicate Steepness of Lines.

All horizontal lines have *slope* 0, as suggested in Figure 12. ■

The slope of a line measures the steepness of the line when viewed from left to right. As shown in Figure 13, the greater the **absolute value** (i.e., its magnitude, regardless of whether it is positive or negative) of the slope, the steeper the line.

Two points on a vertical line have the same *x*-coordinate, so the calculation of *m* is not defined, because division by 0 is not defined.

Vertical and Horizontal Lines

The slope of a **vertical line** is undefined. The slope of a **horizontal line** is 0.

The **point-slope form** of the equation of a line may be found if a fixed point (x_1, y_1) on the line is given and the line has a slope m that is known. The equation is found by considering any point (x, y) on the line other than (x_1, y_1) and noting that

$$\frac{y - y_1}{x - x_1} = m \quad \text{or} \quad y - y_1 = m(x - x_1).$$

The Point-Slope Form of a Line

If a line has slope m and passes through the point (x_1, y_1), then the **point-slope form** of the equation of the line is $y - y_1 = m(x - x_1)$.

EXAMPLE 2
Using the Point-Slope
Form of a Line

Find the equation of the line through (3, 5), with slope $m = \frac{3}{4}$.

Solution   The information needed for the point-slope form is given. Substituting $x_1 = 3$, $y_1 = 5$ and $m = \frac{3}{4}$ into this equation gives

$$y - 5 = \frac{3}{4}(x - 3)$$

$$4y - 20 = 3(x - 3)$$

$$4y - 20 = 3x - 9$$

$$-3x + 4y = 11. \quad ■$$

EXAMPLE 3

Using the Point-Slope
Form of a Line

Write the equation of the line through the points (2, 3) and (4, 7).

Solution   First, we find the slope:

$$m = \frac{7 - 3}{4 - 2} = \frac{4}{2} = 2.$$

Second, choose *either* of the given points since both will give the same equation. Using (2, 3), the point-slope form of a line gives:

$$y - 3 = 2(x - 2)$$
$$y - 3 = 2x - 4$$
$$y = 2x - 1.$$

The reader should verify that using the point (4, 7) gives the same equation. See Figure 14. ▇▇

The intended use of the equation of a line often dictates its final appearance. Sometimes it might be more advantageous to write the final form solved for y, as was done in Example 3: $y = 2x - 1$. At other times, the preferred appearance might be to put the terms involving x and y on the left and the constant on the right, as was done in Example 2: $-3x + 4y = 11$. Often, the final choice is a matter of taste. Nonetheless, different forms of an equation are equivalent to each other, and each is a special case of the **general form** $Ax + By = C$ where A and B are not both zero. The equations $-3x + 4y = 11$ and $3x - 4y = -11$ are both in general form. This shows that the general form is not unique.

The General Form of the Equation of a Line

An equation of the form

$$Ax + By = C$$

in which A, B, and C are constants and A and B are not both zero is called a **general form** of the equation of a line.

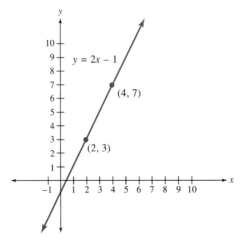

FIGURE 14
The Equation of a Line Through Two Points

If the graph of a function intersects the y-axis at a point $(0, b)$, the number b is called the **y-intercept.** Similarly, if the graph of a function intersects the x-axis at a point $(a, 0)$, the number a is known as the **x-intercept.** The **slope-intercept form** of a line may be found if we know the slope m of the line and the y-intercept b of the graph of the line. This form of the equation may be derived using the point-slope form since a point $(0, b)$ on the line is known:

$$y - b = m(x - 0)$$
$$y - b = mx$$
$$y = mx + b. \qquad \text{The slope-intercept form}$$

The Slope-Intercept Form of a Line

If the slope of a line is m and the y-intercept is b, then the **slope-intercept form** of the line is $y = mx + b$.

EXAMPLE 4
Using the
Slope-Intercept
Form of a Line

Find the equation of the line with y-intercept 3 and slope $m = 5$, and draw the graph of this line.

Solution The given information allows us to quickly write the equation in the slope-intercept form:

$$y = 5x + 3.$$

When $x = 0$, the equation shows that $y = 3$, which verifies that the point $(0, 3)$ is on the graph. At the point $(0, 3)$, if the x-coordinate is changed 1 unit to the right, the y-coordinate will change 5 units upward to return to the line because the slope is 5. These two facts may be used to construct the graph of $y = 5x + 3$, as shown in Figure 15. ■

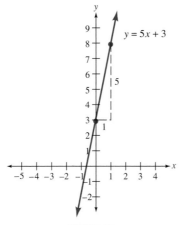

FIGURE 15
*Graphing by Use of the Slope and
the y-Intercept*

EXAMPLE 5

Finding the Slope and
y-Intercept of a Line

Find the slope and y-intercept of the line whose equation is $2x + 5y = 7$.

Solution   Solving this equation for y will give the slope-intercept form of the line:

$$5y = -2x + 7$$

$$y = -\frac{2}{5}x + \frac{7}{5}.$$

Now we can see that the slope is $-\frac{2}{5}$ and the y-intercept is $\frac{7}{5}$. ◼

EXAMPLE 6

Finding the x- and
y-Intercepts of a Line

Find the x- and y-intercepts of the line whose equation is $4x - 3y = 6$.

Solution   The line will intersect the y-axis at the point on its graph whose x-coordinate is 0. Substituting $x = 0$ into the given equation results in $-3y = 6$ or $y = -2$. This tells us that the y-intercept of the line is -2, as shown in Figure 16.

The line will intersect the x-axis at the point on its graph whose y-coordinate is 0. If we now let $y = 0$ in the given equation, the result is $4x = 6$ or $x = \frac{6}{4} = \frac{3}{2}$. It follows that the x-intercept is $\frac{3}{2}$, as shown in Figure 16. ◼

Summarizing the process used to find the intercepts, we see that:

to find the y-intercept, let $x = 0$ and solve for y, and

to find the x-intercept, let $y = 0$ and solve for x.

Two lines in the same plane are parallel if they never meet, however far extended. The definition of the slope of a line then implies that the two lines are parallel only when they have the same slope or when both have no slope. Conversely, if two distinct lines have the same slope, they are parallel.

EXAMPLE 7

Parallel Lines

Find the equation of the line through the point $(4, 6)$ and parallel to the line whose equation is $x - 2y = -4$. Write the answer in general form.

Solution   The line we seek must have the same slope as the line given by $x - 2y = -4$. Solving this equation for y gives

$$-2y = -x - 4 \quad \text{or} \quad y = \frac{1}{2}x + 2,$$

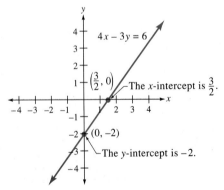

FIGURE 16
Finding the Intercepts of a Line

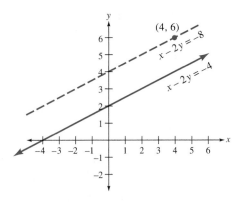

FIGURE 17
The Equation of Line Parallel to a Given Line

from which we conclude that the slope is $\frac{1}{2}$. The point-slope form may now be used to write the desired equation:

$$y - 6 = \frac{1}{2}(x - 4)$$
$$2y - 12 = x - 4$$
$$-x + 2y = 8 \quad \text{or} \quad x - 2y = -8.$$

See Figure 17 above.   ■■

We have already seen that a horizontal line has slope 0. If such a line passes through the point (h, k), its equation is

$$y - k = 0(x - h)$$
$$y - k = 0$$
$$y = k.$$

In other words, the defining property of a horizontal line is that every point on the line has the same y-coordinate.

Even though the slope of a vertical line is undefined, every vertical line has an equation. A vertical line passing through the point (h, k) has the defining property that every point on that line has x-coordinate h. Therefore the line can be described by the equation $x = h$.

EXAMPLE 8
Equations of Horizontal
and Vertical Lines

(a) Find the equation of the horizontal line through the point $(6, -4)$.

Solution   The equation is $y = -4$. See Figure 18(a).

(b) Find the equation of the vertical line through the point $(7, 4)$.

Solution   The equation is $x = 7$. See Figure 18(b).   ■■

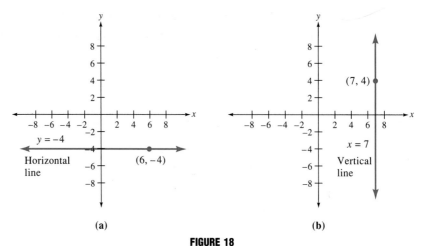

FIGURE 18
Equations of Horizontal and Vertical Lines

A summary of the different forms for the equation of a line is given next.

Various Forms of Linear Equations

Description	Equation
Point-slope form	$y - y_1 = m(x - x_1)$
Slope-intercept form	$y = mx + b$
General form	$Ax + By = C$
Horizontal line through (h, k)	$y = k$
Vertical line through (h, k)	$x = h$

The equation of a line that is not vertical can be expressed in any one of three forms: point-slope, slope-intercept, and general. In particular, the slope-intercept form $y = mx + b$ expresses y as a function of x. Such functions are called **linear.**

Linear Functions

A function is **linear** if it can be written in the form $f(x) = mx + b$ where m and b are real numbers.

EXAMPLE 9
A Linear Cost Function

Suppose that the cost (in dollars) of making x ladies' blouses of a certain brand is given by the linear function $C(x) = 15x + 1800$.

(a) What is the cost of making 80 of these blouses?

Solution The cost is found by replacing x with 80 in the equation $C(x) = 15x + 1800$. This gives

$$C(80) = 15(80) + 1800$$
$$= 1200 + 1800$$
$$= 3000 \quad \text{or} \quad \$3000.$$

(b) If $7800 has been spent making these blouses, how many blouses were made?

Solution   We are given that $C(x) = 7800$, and we are to find x. Therefore,

$$7800 = 15x + 1800$$
$$6000 = 15x$$
$$x = 400.$$

(c) Graph the function $C(x)$.

Solution   We may visualize the graph as ordered pairs in the form $(x, C(x))$ or simply set $y = C(x)$ and consider the graph as the usual ordered pairs (x, y) where $y = 15x + 1800$. In any event, we need only two such pairs to determine the graph that is a line, as shown in Figure 19. An easy one to find is by letting $x = 0$, which gives the pair $(0, 1800)$. Another $(80, 3000)$ was found in Example 9(a). Note that x cannot be negative. ▬▬

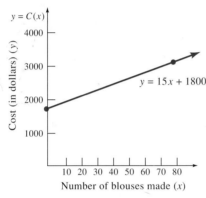

FIGURE 19
Costs for Making Ladies' Blouses

Linear functions are studied more extensively in the remainder of this chapter.

EXERCISES   1.2

 In Exercises 1 through 10:
  (a) Find the slope, if it exists, of the line through the given points.
(b) Check to see that your numerical calculation in part (a) is reasonable by using a graphing calculator to draw the line segment between the two points.

1. $(4, 5), (6, -1)$
2. $(3, -1), (-2, -1)$
3. $(\frac{1}{3}, 5), (-1, 5)$
4. $(0, 0), (-2, -1)$
5. $(\frac{7}{8}, 3.7), (1, 5.8)$
6. $(1.6, 250), (2.9, 123)$
7. $(3, -2), (3, 5)$
8. $(6, 3), (7, 8)$
9. $(1, 1), (3, 3)$
10. $(-4, 2), (6, 2)$

In Exercise 11 through 21, do the following with the given properties:
(a) Find both the slope-intercept and a general form of the equation of the line.
(b) Graph the line.

11. Through the point $(2, 5)$ with slope -2
12. Through the point $(-4, 6)$ with slope 4
13. Through the point $(0, 5.7)$ with slope $\frac{2}{3}$
14. Through the point $(4, 0)$ with slope -12
15. Through the points $(3, 4)$ and $(6, -1)$
16. Through the points $(\frac{1}{2}, 4)$ and $(-3, 2)$

17. Through the points $(.3, 1)$ and $(-2, 3)$

18. Through the points $(10, 500)$ and $(20, 700)$

19. Through the points $(17, 300)$ and $(23, 150)$

20. Having x-intercept 3 and y-intercept -2

21. Having x-intercept 1.6 and y-intercept 4.3

In Exercises 22 through 25, find the equations of both the vertical and the horizontal lines through the given point.

22. $(3, -6)$ **23.** $(-\frac{1}{2}, 4)$

24. $(-5.7, 200)$ **25.** $(1.2, -8.6)$

In Exercises 26 through 31 write the equations of three different lines, each of which is parallel to the given line, and graph all four lines on the same set of axes.

26. $y = 5x$ **27.** $4x - 7y = 6$

28. $y = -3x + 7$ **29.** $x = 5$

30. $y = 7$ **31.** $-2x + 4y = -3$

In each of the Exercises 32 through 35, find an equation of the line that passes through the given point and is parallel to the line for which the equation is given.

32. $(3, 5); x - 2y = 6$

33. $(1.3, 2); y = -3x - 7$

34. $(\frac{1}{2}, -3); 2x - 4y = 7$

35. $(0, -3.5); 2x + 3y = 6$

In Exercises 36 through 41, do the following:
(a) Write the equations of three different lines, each of which has the same y-intercept as the given line.
(b) Write the equations of three lines, each of which has the same x-intercept as the given line.

36. $y = 2x + 2$ **37.** $y = -x - 4$

38. $3x + 2y = 4$ **39.** $x - 2y = 5$

40. $y = 1.3x - .6$ **41.** $7x + 6y = 0$

For each of the Exercises 42 through 48, answer the following questions. If (x, y) is a point on the graph of the line and:
(a) $\Delta x = 1$, find Δy.
(b) If $\Delta x = 2$, find Δy.
(c) If $\Delta x = 5$, find Δy.

42. $y = 5x + 1$ **43.** $y = 3x - 4$

44. $y = -6x + 8$ **45.** $y = -2x - 5$

46. $x + 2y = 8$ **47.** $2x + 5y = 7$

48. $x - 2y = 4$

49. *Number of Subscribers*   During its first few weeks in operation, a new "advertisements only" weekly paper described its number of subscribers with the equation $S(t) = 800t + 6000$, where t represents the number of weeks in operation.

(a) How many subscribers did they have after five weeks of operation? 20 weeks?

(b) Are subscriptions increasing each week? If so, by how many?

(c) State how the slope and y-intercept of the graph of $y = S(t)$ is related to the number of subscribers.

(d) Graph the given subscriber function $y = S(t)$ to visually show the number of subscribers from the start of operations forward.

(e) How many weeks will it take for subscriptions to reach 18,000?

50. *Sales*   Sales by the Rockhard Concrete Company are described by $S(t) = 120t + 600$, where t is time in years, measured from $t = 0$ as the year 1985, and sales are in thousands of dollars.

(a) What were sales in 1985? In 1990? What is the estimate for sales in 2002?

(b) Are sales increasing each year? If so, by how much?

(c) State how the slope and the y-intercept of the graph of $y = S(t)$ are related to sales.

(d) Graph the given sales function $y = S(t)$ to visually show sales from 1985 forward.

(e) In what year does the company expect sales to reach $3 million?

51. *Population*   A county agent takes samples of the grasshopper population in late summer and concludes that the population is approximated by $P(t) = -2t + 100$, where t is measured in days, with $t = 0$ being August 31, and $P(t)$ represents thousands of grasshoppers per acre.

(a) What is the predicted grasshopper population on September 15? September 20?

(b) Is the grasshopper population increasing or decreasing each day? By how many?

(c) State how the slope and the y-intercept of the graph of $y = P(t)$ are related to the number of grasshoppers.

(d) Graph the function $y = P(t)$ for all applicable values of t.

(e) How many days will it take for the grasshopper population to reach 20,000 per acre?

52. *Monitoring Pollution*   The periodic monitoring of a particular water pollutant in Beaver Lake led to the equation $P(t) = -800t + 18,000$, where t is time, measured from $t = 0$ as the year 1987, and the pollutant is measured in parts per million.

(a) How many parts per million of this pollutant were in the water in 1990? In 1992? What is the estimate for 1996?

(b) Is the pollutant decreasing each year? If so, by how much?

(c) State how the slope and the y-intercept of the graph of $y = P(t)$ is related to the pollutant.

(d) Graph the function $y = P(t)$ for all applicable values of t.

(e) According to this model, when will this particular pollutant be completely removed from the water?

53. Temperature Scales Fahrenheit and Celsius temperature scales are related by the equation $C = (\frac{5}{9})F - (\frac{160}{9})$, where Celsius temperatures are given as a function of Fahrenheit temperatures.

(a) For each rise of 1 degree in Fahrenheit temperature, what is the corresponding rise in the Celsius temperature?

(b) If the Fahrenheit temperature is 80 degrees, what is the Celsius temperature?

(c) Solve the give equation for F to obtain an equation that gives Fahrenheit temperature as a function of the Celsius temperature.

(d) For each rise of 1 degree in the Celsius temperature, what is the corresponding rise in the Fahrenheit temperature?

54. Tire Tread Depth The maker of brand X tires claims that for well-maintained cars, the tread wear on brand X tires is a function of the number of miles driven and that the function is $T(x) = \dfrac{-x}{80{,}000} + \dfrac{1}{2}$, where x is the number of miles the tire has been driven and $T(x)$ is given in inches.

(a) Graph the function $y = T(x)$ for all applicable values of x.

(b) What is the tread depth on a new tire?

(c) How many miles of wear will eliminate the tread on one of these tires?

55. Determining Height Anthropologists have found that the height of a human can be determined by a linear function if the length x of the bone from the elbow to the shoulder is known in centimeters. The linear functions are $M(x) = 2.9x + 70.6$ for a male and $W(x) = 2.8x + 71.5$ for a female.

(a) Graph $y = M(x)$.

(b) If the bone in question measures 46 centimeters for a male, what is the height of that person?

(c) If a female is known to be 180 centimeters tall, what is the length of the bone in question?

56. Age From 1970 through 1986, the average age of men at their first marriage in a particular town is given by the linear function $A(x) = .25x + 21.6$, where x is the number of years beyond 1970.

(a) Graph the given function.

(b) What was the average age of first-married men in 1982?

(c) When will the average age of first-married men reach 30 years? 40 years? Do you think that this is a realistic model over a long period of time?

57. Sales A study of records by the Ace Cola Company determined that the number of cases of a particular soft drink (in thousands) that it sells could be estimated by the linear function $S(x) = .91x + 28.96$, where x is the number of years past 1985.

(a) Estimate sales for 1996.

(b) In what year should sales be 50,000 cases if this trend is continued?

(c) Graph the given function.

58. Depreciating a Car The "blue book" value of a particular make of car 2 years old or older is approximated by the linear function $V(x) = -1333x + 12{,}666$, where $V(x)$ is the value in dollars when x is the age of the car in years for $x \geq 2$.

(a) Graph this function.

(b) If you bought a 4-year-old car of this make, how much should you expect to pay?

(c) How old will this make of car be before its value is $8000?

Problems for Exploration

1. Discuss the relationship between how a line "leans" from the vertical and its slope (i.e., describe the direction and slope of the line in relation to a vertical line). Include how large slopes and small slopes affect the amount of "lean." Give some examples.

2. Compare vertical lines with horizontal lines. Can you use a general form, the point-slope form, or the slope-intercept form of a line to arrive at the equation of a vertical line? A horizontal line?

3. Suppose that the x-intercept of a line is a and the y-intercept is b. Develop a way to take this information and immediately write the equation of the line. Give some examples.

4. Examine your city's records to find out the number of gallons of water used per person in 1970 and for the most recent year available. Let $x = 0$ represent the x-coordinate of the 1970 data, and take the number of intervening years to the latest record to be the x-coordinate of that data. Use these two points to find the equation of the line that represents a linear trend over this period of time. Compare some actual data in the intervening years, and discuss the validity of your line.

SECTION 1.3 Linear Modeling

When we can describe a phenomenon (e.g., in the physical, business, social, or educational worlds) with a function, we say that we have constructed a **mathematical model** of that phenomenon. Sometimes, these models are exact, whereas at other times, they are rather crude approximations. There are many real-world situations in which the function is linear; in those cases, we call the model a **linear model.**

EXAMPLE 1
Modeling an Ecological Problem

A study of the fish population within a mile of a chemical plant on Spring River showed that when 10 tons per day of pollutants were dumped into the river, the fish population was 30,000. A few years later, when 20 tons per day were being dumped into the river, the fish population had declined to 24,000. Assume a linear relationship between the number of tons of pollutants and the number of fish when solving the problems presented in (a) and (b).

(a) Find an equation that will predict the fish population for any number of tons of pollutants being dumped into the river.

Solution   The assumption of a **linear relationship** means that the graph of the equation relating the variables will be a line. For such a relationship, a good rule is to let y be the variable to be predicted—in this case, the number of fish; x will represent the number of tons of pollutants. With this notation, the given information then tells us two points in the rectangular coordinate system: $(10, 30{,}000)$ and $(20, 24{,}000)$. The equation asked for will be that of the line through these two points. First, the slope of this line is

$$m = \frac{24{,}000 - 30{,}000}{20 - 10} = -600.$$

The slope represents the *average decline in number* of fish *per ton* of pollutant introduced into the river as indicated in Figure 20. Now choosing the point

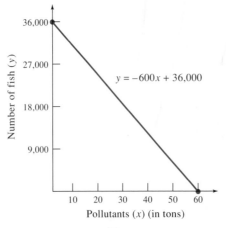

FIGURE 20
Relationship of Pollutants to Fish Population

(10, 30,000) and using $m = -600$, the point-slope formula gives the relationship we seek upon solving for y:

$$y - 30,000 = -600(x - 10)$$
$$y - 30,000 = -600x + 6,000$$
$$y = -600x + 36,000.$$

The equation $y = -600x + 36,000$ gives y (fish population) as a linear function of x (tons of pollutants).

(b) Use the function $y = -600x + 36,000$ to predict the fish population if 25 tons of pollutants per day are dumped into the river.

Solution   The answer is

$$y = -600(25) + 36,000 = 21,000 \text{ fish.}$$ ▬▬

EXAMPLE 2
The Price of
Single-Family Homes

According to U.S. Census Bureau data, the median price of a new privately owned, single-family home in the United States was $23,400 in 1970 and has been increasing rather linearly to $126,500 in 1993. See Figure 21.

(a) Use this data to find the average increase in median price per year from 1970 to 1993.

Solution   The average is found by calculating the difference in the two median prices and then dividing by the number of years between 1970 and 1993:

$$\frac{126,500 - 23,400}{1993 - 1970} = \frac{103,100}{23} \approx \$4483.$$

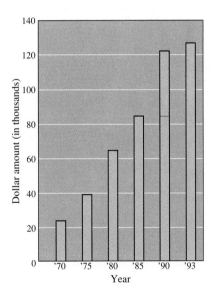

SOURCE: U.S. Bureau of the Census: *Statistical Abstract of the United States*, 1994.

FIGURE 21
Median Sales Price for New Private Homes

(b) Find a linear function that will approximate the median price of a new home throughout this period.

Solution Let x be the number of years from 1970 forward, with $x = 0$ representing 1970. Let y be the median cost of housing. Since slope measures the change in y (housing costs) per unit (year) change in x, part (a) tells us that the slope of our linear function is \$4483. Furthermore, when $x = 0$, $y = \$23,400$ (the cost in 1970). The slope-intercept form of a line can then be applied to give the linear function

$$y = 4483x + 23,400. \quad \blacksquare$$

In the short run, costs involved in manufacturing a product can be broken down into **fixed costs** and **variable costs.** As the name implies, fixed costs do not change and might include such things as design, production-machinery, and building costs. The firm pays these costs even if it produces no output. Variable costs, on the other hand, are costs that vary according to the production level. They include such things as labor, raw-material, or maintenance costs. The short-run **total costs** are then

$$\text{Total Costs} = \text{Variable Costs} + \text{Fixed Costs.}$$

In the long run, all costs are variable; fixed costs are zero.

For certain relatively simple situations, short-run variable costs may be calculated by multiplying the **direct cost per item produced,** considered to be the cost of labor and raw materials, by the number of items produced. For example, if the direct costs per item were \$50 and x items were produced, the variable costs would be $50x$ dollars. We confine our attention to variable costs such as these because they lead to a linear cost function:

$$\text{Total Costs} = (\text{Direct Cost per item}) \cdot x + \text{Fixed Costs.}$$

EXAMPLE 3
A Linear Cost Function

A new graphing calculator is being readied for production. The company has fixed costs of \$80,000, and company officials estimate that direct costs will be \$60 per calculator. Find a linear cost function that will give the total costs for producing x calculators.

Solution The company has an initial investment (fixed cost) of \$80,000 before production starts. Each calculator made adds \$60 to the company costs, so that upon making x calculators, a variable cost of $60x$ dollars is added to the fixed cost the company has incurred. If $C(x)$ represents the total production cost to the company for making x calculators, then $C(x) = 60x + 80,000$ dollars. $\blacksquare$

At any given production level, knowledge of the extra production costs incurred when changing that level is vital in making management decisions. As a measure of those extra costs, economists define the **marginal cost** as the *change* in total costs associated with producing one additional unit. In Example 3 for instance, if 100 calculators have been made, the marginal cost is \$60 because the change in total costs that result from making the 101st calculator is

$$C(101) - C(100) = 80,000 + 60(101) - [80,000 + 60(1000)]$$
$$= 80,000 + 6060 - 80,000 - 6000$$
$$= 60 \text{ dollars.}$$

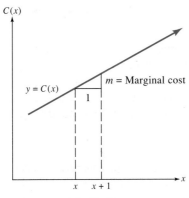

FIGURE 22
Additional Cost of Making Another Item

This result should come as no surprise. The slope of the graph to any linear function gives the vertical change in the graph per unit change in x; thus the value of the slope is always the marginal cost. Figure 22 gives a visual picture:

Linear Cost Functions

A **linear cost function** has the form

$$C(x) = mx + b$$

where the slope m is also known as the **marginal cost** as well as the **direct cost** per item, mx is the **variable cost,** and b is the **fixed cost.**

EXAMPLE 4
Marginal Cost Versus
Average Cost

A car rental firm charges $50 per day plus 30 cents per mile to rent a particular make of automobile.

(a) Find a linear cost function that will give the total rental cost per day as a function of the number of miles driven.

Solution   The fixed costs are $50, while the direct costs are 30 cents per mile. If we let x be the number of miles driven during the day, the cost per day for renting this car is $C(x) = .30x + 50$ dollars.

(b) If the car has been driven 100 miles, what is the total rental cost?

Solution   The total cost is $C(100) = .30(100) + 50 = \$80$.

(c) If 1 more mile is driven, what is the additional rental cost?

Solution   The additional cost is the marginal cost, 30 cents.

(d) If the car has been driven 150 miles, what is the additional rental cost for driving 1 more mile?

Solution   Again, the answer is 30 cents, the marginal cost.

(e) What is the average rental cost per mile for driving the car 100 miles?

Solution   The average rental cost is the total rental cost of driving 100 miles, divided by 100. That is, $\dfrac{C(100)}{100} = \dfrac{.30(100) + 50}{100} = \dfrac{80}{100} = \$0.80$ per mile.   ■

> ### The Average Cost
>
> If $C(x)$ is a linear cost function, the **average cost** per item for x items is given by
>
> $$\frac{C(x)}{x}.$$

EXAMPLE 5

Finding a Linear Cost Function, Based on Limited Information

Suppose that a particular manufacturing process has fixed costs of $2000, and it is known that 50 items cost a total of $6000. Find the linear cost function that models the cost for making x items.

Solution   Because the fixed cost is $2000 and the function must be linear, it must have the appearance $C(x) = mx + 2000$, where m is yet to be found. However, we are given that $C(50) = \$6000$, so, upon substituting 50 for x in the equation $C(x) = mx + 2000$, we get

$$C(50) = m(50) + 2000$$
$$6000 = 50m + 2000,$$

so that $50m = 4000$ or $m = 80$. Therefore, the function we seek is $C(x) = 80x + 2000$.

To check the validity of the function, substitute $x = 0$ and $x = 50$ into the equation to see whether 2000 and 6000, respectively, are obtained. This is indeed the case, so we can have confidence that our function is correct.   ■

An analysis similar to that shown in Example 5 may be done to find the fixed cost when the marginal cost is given instead of the fixed cost.

As machines and equipment wear out or become obsolete, businesses take into account the value lost each year over the useful lives of these items. This lost value is called **depreciation.** Depreciation may be calculated in several ways, in accordance with standard accounting procedures and tax laws. One widely used method, and perhaps the simplest, is that of **straight-line depreciation.** Straight-line depreciation assumes that the amount depreciated each year is always the same (i.e., a graph of the value of the object being depreciated would be a straight line).

EXAMPLE 6

Straight-Line Depreciation

Meek's cabinet shop bought a new saw for $6000. The shop estimates that the useful life of the saw will be 10 years, at which time its **salvage value** will be $1000. The difference between the purchase price and the salvage value, called the **net cost,** is the *amount to be depreciated:*

$$\$6000 - \$1000 = \$5000.$$

The straight-line method of depreciation assumes that over a 10-year period, the net cost will be depreciated the *same amount* each year. Therefore the **annual depreciation** is

$$\left(\frac{1}{10}\right)\$5000 = \$500.$$

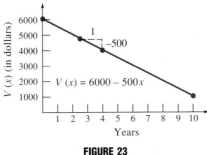

FIGURE 23
Straight-Line Depreciation

If x represents the number of years after purchase, the **book value** (the purchase price minus the total depreciation) of the saw is

$$V(x) = \$6000 - \$500x.$$

Figure 23 above illustrates this straight-line depreciation. Notice that the line stops at the point $(10, 1000)$. ▬▬

Straight-Line Depreciation

The **salvage value** of an item is the value, if any, at the end of its useful life. The **net cost** of an item is the difference between the purchase price and the salvage value. If N is the net cost of an item, the item's useful life is n years, and D is the item's annual depreciation, then $D = \dfrac{N}{n}$ is depreciated *each year* under the **straight-line method.** If P is the purchase price, then after x years, the value (V) of the item is given by the equation $V(x) = P - \left(\dfrac{N}{n}\right)x$.

Economists assert that, all other things being equal, the *supply* of a given item or commodity is a function of its price. In particular, it is assumed that as the price increases, producers will be willing to supply more of the item, so that the supply increases as well. Just the opposite is true of *demand:* As the price increases, fewer people will want to buy the item. In this text, we are particularly interested in situations in which both the supply and the demand are modeled by functions that are linear.

EXAMPLE 7
Linear Supply and
Demand Functions

The marketing department of Z-Mart stores estimates that the supply (in dozens) over the next year of a particular line of ladies' belts is given by

$$S(p) = 4p,$$

where p is the price per belt in dollars. The demand (D) (in dozens) is estimated to be

$$D(p) = 140 - 10p.$$

(a) What is the demand if the belts are priced at \$6 each?

Solution   The answer is $D(6) = 140 - 10(6) = 80$ dozen.

(b) At what price will the demand become zero?

Solution   We set $0 = 140 - 10p$ and solve for p: $p = \$14$. (In practice, the reliability of such models is usually suspect near the extremes. That is, even when priced at $14 per belt, some belts will probably be sold.)

(c) At what price will the supply equal the demand?

Solution   We set $S(p) = D(p)$ and solve for p:

$$4p = 140 - 10p$$
$$14p = 140$$
$$p = 10 \text{ dollars.}$$

When the demand for a commodity exceeds the supply, a shortage occurs; if the supply exceeds the demand, a surplus occurs. Of great interest to economists is the point at which the supply equals the demand; that is, there is neither a shortage nor a surplus of the commodity. When this occurs, the system is said to be in **equilibrium.** The **equilibrium point** is the point where the graphs of $y = S(p)$ and $y = D(p)$ intersect. The p-coordinate of that point is known as the **equilibrium price,** and the y-coordinate is known as both the **equilibrium supply** and the **equilibrium demand.** The point is found by first setting $D(p) = S(p)$, then solving for p, and then substituting back into either $D(p)$ or $S(p)$ to find the equilibrium demand (or supply). From Example 7(c), this procedure gives $p = \$10$, which is the equilibrium price. Next, let $p = 10$ in the supply function. Then $S(10) = 4(10) = 40$ dozen belts, which is the equilibrium supply (and demand). The equilibrium point is therefore $(10, 40)$. Figure 24 shows both $S(p)$ and $D(p)$ of Example 7, graphed on the same set of axes; it also indicates the equilibrium point.

The Equilibrium Point

If $D(p)$ is a demand function and $S(p)$ is a supply function for a given commodity in terms of the price p, then the **equilibrium price** is that value p_0 of p for which

$$D(p_0) = S(p_0).$$

The **equilibrium demand,** also known as the **equilibrium supply,** is the value $D(p_0)$. The **equilibrium point** is the ordered pair $(p_0, D(p_0))$.

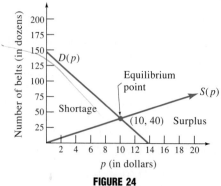

FIGURE 24
Supply and Demand Functions

EXERCISES 1.3

In Exercises 1 through 6, consider the starting number to represent the number of yogurt outlets a firm owns when starting a business, followed by the resulting expansion or contraction of outlets for a number of years thereafter. Decide which of these can be represented by a linear function. For those that are linear, find the function.

1. Start with ten, and double in number for each year thereafter.

2. Start with 200, and decline ten in number each year thereafter.

3. Start with 500, and increase 20 in number each year thereafter.

4. Start with 100, and decrease by half each year thereafter.

5. Start with 60, and show no change thereafter.

6. Start with ten, and increase by 20% each year thereafter.

In Exercises 7 through 12, a cost function is given. For each, do the following:
(a) Find the marginal cost.
(b) Find the fixed cost.
(c) Find the total cost for 20 items.
(d) Find the average cost per item for producing 20 items, 100 items, and 200 items.

7. $C(x) = 3x + 20$
8. $C(x) = 50x + 3000$
9. $C(x) = 3.2x + 1680$
10. $C(x) = 5.8x + 2000$
11. $C(x) = 1.6x + 5000$
12. $C(x) = 200x + 10,000$

13. For the cost function $C(x) = 8x + 75$, sketch a graph of $y = \frac{C(x)}{x}$ and $y = $ (marginal cost) on the same set of axes, and make a comparison of the average cost per item and the marginal cost as production increases.

14. For the cost function $C(x) = 14x + 125$, sketch a graph of $y = \frac{C(x)}{x}$ and $y = $ (marginal cost) on the same set of axes, and make a comparison of the average cost per item and the marginal cost as production increases.

In Exercises 15 through 18, find a linear cost function using the given information.

15. With fixed costs of $4000, twenty items cost a total of $10,000.

16. With fixed costs of $3500, twenty-two items cost a total of $5000.

17. With a marginal cost of $25, forty items cost a total of $4000.

18. With a marginal cost of $80, thirty items cost a total of $6000.

In Exercises 19 through 23, use the given data to find a linear cost function that will give the total cost of producing x units of the item described.

19. Ten refrigerators cost $2000; 15 refrigerators cost $2700.

20. Twenty radios cost $600; 35 radios cost $900.

21. Fifty ladies' coats cost $1000; 60 coats cost $1200.

22. Twelve microwave ovens cost $1400; 20 ovens cost $2250.

23. Fifteen sweepers cost $900; 30 sweepers cost $1560.

24. **Water Usage** Data from the U.S. Census Bureau* indicated that the average person in the United States in 1990 used 195 gallons of water each day. Using this fact, find a linear function that will give the total number of gallons y of water used by x people.

25. **Skipping Breakfast** If 25% of all Americans regularly skip breakfast, find a linear function that will indicate the number of people y who skip breakfast from a group of x people.

26. **Inventory Reduction** A camera store has 150 cameras in stock on January 1 and sells them at the rate of four per day. (Do not consider the possible arrival of new shipments of cameras.)
    (a) Find a linear function that gives the number of cameras in stock on January 1 and thereafter.
    (b) Graph the function found in Exercise 26(a).
    (c) State how the slope and the y-intercept of your graph are related to the camera inventory.
    (d) How many days will it take to reduce the stock to 90 cameras?

27. **Enrollment Growth** State University had an enrollment of 12,000 in 1985 and has experienced a growth of 600 students per year ever since.
    (a) Find a linear function that gives the enrollment at State University for 1985 and beyond.
    (b) Graph the function found in Exercise 27(a).
    (c) State how the slope and y-intercept of your graph are related to enrollment.
    (d) How many years will it take for the enrollment to reach 20,000 at this rate?

SOURCE: U.S. Bureau of the Census: Statistical Abstract of the United States, 1994.

28. *Advertising Revenue* Advertising revenue for a local newspaper was $500,000 in 1990, when the city's population was 60,000. The paper estimates that advertising revenue rises $250 for each population increase of 1000.

(a) Find a linear function that will give advertising revenue as a function of the population of the city.

(b) State how the slope and the y-intercept of the graph of the function found in Exercise 28(a) are related to advertising revenue.

(c) If the population increases by 3500 from its 1990 figure, what will the expected advertising revenue be?

29. *Automobile Depreciation* A new car initially costs $12,000 and loses value at the rate of $1000 each year.

(a) Write the linear function that gives the value of this car in terms of the years after its initial purchase.

(b) State how the slope and the y-intercept of the function found in Exercise 29(a) are related to the value of the car.

(c) How many years will it be before the value of the car will be $5000? $1500?

(d) Graph the function found in Exercise 29(a).

30. *Voter Decline* In 1985, a small town had 22,000 voters. Voter apathy has created a decline of 500 voters in each election since then.

(a) Write the linear function that will give the number of voters in terms of the number of elections since 1985.

(b) How many voters were there for the sixth election after 1985?

(c) According to these data, how many elections will be held before the number of voters declines to 12,000?

31. *Cost of Manufacturing* The Sleepmor Manufacturing Co. makes clock radios. A study of short-run costs indicates that fixed costs are $2000, and the direct cost is $20 per clock radio.

(a) What is the linear cost function that gives the total cost for manufacturing x clock radios? Graph this function.

(b) What is the total cost for manufacturing 150 clock radios?

(c) If 200 clock radios have been made, what is the additional cost of making the 201st clock radio?

(d) What is the average cost per clock radio for making 200 clock radios?

(e) What is the average cost per clock radio for making 500 clock radios? For making 1000? For making 2000?

32. *Cost of Manufacturing* A maker of fax/modems has short-run fixed costs of $75,000 and the direct cost of $80 for each fax/modem.

(a) What is the linear cost function that gives the total cost for manufacturing x fax/modems? Graph this function.

(b) What is the total cost for making 150 fax/modems?

(c) If 150 fax/modems have been made, what is the additional cost for making the 151st fax/modem?

(d) What is the average cost per fax/modem for making 300 fax/modems? For 500 fax/modems?

33. *Monitoring Pollution* A state pollution-control office estimates that the fixed costs incurred in monitoring waste treatment plants are $25,000 each year, and the cost of each on-site inspection is $1500.

(a) What is the linear cost function in terms of the number x of on-site inspections made each year?

(b) What is the total cost of 50 on-site inspections?

(c) What is the additional cost of making the 51st on-site inspection?

(d) What is the average cost per inspection of making 50 on-site inspections? 100 on-site inspections? 200 on-site inspections?

(e) Graph the function found in Exercise 33(a).

34. *Income Tax Brackets*

(a) What does it mean when your income puts you "in the 28% federal income tax bracket"? Suppose that your taxable income is $25,000 in a given year in which the first $17,850 is taxed at the rate of 15%, and the amount over $17,500 but less than $43,150 is taxed at the rate of 28%. Accounts would then say that you are "in the 28% tax bracket" and would refer to the 28% figure as a "marginal tax rate."[†]

(b) Compute the total tax owed on the $25,000 income. Note that it is quite different from 28% of $25,000.

35. *Sales Income* Assume that you are offered a job paying $1500 per month plus 4% of gross sales.

(a) Find a linear function that will give your monthly income in terms of sales.

(b) Graph the function found in Exercise 35(a).

(c) If you make sales of $20,000 during a given month, what will your income be for that month?

(d) If you make sales of $10,000 at some point during the month, how much additional income will you gain if you make another sale for $1?

(e) What dollar amount in sales will yield a monthly income of $3000?

[†] This means that above a certain level, 28 cents of each taxable dollar is paid as income tax.

36. Sales Related to Unemployment The Sharper View TV store found that the number of TV sets it sold was a function of the unemployment rate in its community. The store determined that when the unemployment rate was 2%, it could sell about 400 sets per month, and that for each 1% gain in the unemployment rate, sales fell about 50 sets per month.

(a) Find a linear equation that will give the number of sets sold each month as a function of the unemployment rate.

(b) Graph this function.

37. Apartment Occupancy An apartment building manager estimates that if the rental rate is $400 per month per apartment, all 80 apartments in the building will be rented. However, if the rent is raised to $500 per month, only 60 of the apartments will be rented. Assume that there is a linear relationship between rent per month and occupancy (i.e., number of apartments rented).

(a) Find a linear function that will predict the occupancy for a given monthly rent, and graph this function.

(b) Check your answer in Exercise 37(a) for accuracy.

(c) If the rent is established at $460 per month, use the function found in Exercise 37(a) to estimate the occupancy.

38. Parking Fees When a parking lot charged $3 per car per day, the attendants usually parked 60 cars. When the manager raised the fee to $4 per car, they parked only 50 cars per day.

(a) Find a linear function that gives the number of cars parked each day as a function of the charge per car.

(b) Use the function found in Exercise 38(a) to estimate the number of cars that will be parked if the fee is set at $3.75 per car.

(c) Use the function found in 38(a) to determine what should be charged per car if the operators of the lot want to have 80 cars parked each day.

(d) According to the function found in 38(a), what is the minimum amount that can be charged per car so that no (zero) cars will be parked there? Do you think this is realistic?

(e) If parking were free in this lot, how many cars could be parked there, according to the function found in 38(a)? Do you think this is realistic?

39. Sales Trends In 1986, Z-Mart retail stores had sales of $2.7 million, and in 1988, sales were $3.6 million. Assume that the trend in sales is linear.

(a) Find the equation of a line that will estimate sales (in millions of dollars) in succeeding years.

(b) Using the line found in Exercise 39(a), predict sales in 1995.

(c) Based on the line found in 39(a), when will sales reach the $6 million mark?

(d) Do you think that the line found in Exercise 39(a) will give a good estimate of sales in the year 2000? In the year 2025?

40. Mortality Rates Health statistics* show that infant mortality rates (per thousand live births) in the United States were 12.6 in 1980, 10.4 in 1986, and 8.9 in 1991.

(a) Assuming a linear decline in infant mortality rates from 1980 through 1991, find the equation of a line that will approximate the mortality rate in a given year during this period. Graph the equation of the line found.

(b) Tell how the slope of the line found in Exercise 40(a) relates to the mortality rate.

(c) Check the accuracy of your line for 1986. What do you predict that the rate will be in 2000?

41. Resource Recovery According to U.S. Census Bureau data,* the amount of paper and paperboard recovered (in millions of tons) in the United States in 1960 was 5.4, and this amount increased in a rather linear fashion to 20.9 in 1990.

(a) Assuming the growth in recovery to be linear, find the equation of a line that will approximate the amount recovered for any given year from 1960 to 1990.

(b) Tell how the slope of the line found in Exercise 41(a) relates to the recovery of paper and paperboard.

(c) The amount recovered in 1986 is listed as 14.8 (million tons). How accurate is your equation at this point? What is your estimate for the amount in the year 2000?

42. Dental Costs According to Social Security Administration statistics,* dental costs per person in the United States were $22 per year in 1970 and have increased rather linearly to $141 in 1991.

(a) Assuming linear growth during this period, find the equation of a line that will approximate the cost of dental service for any year during this period. Graph this line.

(b) The actual figure for 1985 was $94. Check your equation for accuracy at this point.

(c) Relate the slope of the line found in Exercise 42(a) to the cost of dental services.

*SOURCE: U.S. Bureau of the Census: *Statistical Abstract of the United States,* 1994.

43. Solid Waste Increase The following data from the U.S. Census Bureau* show (in millions of tons) how the generation of municipal solid waste has increased.

Year	1960	1970	1980	1985	1989	1990
Gross waste	87.8	121.9	151.5	164.4	191.4	195.7

(a) Assuming linear growth between 1960 and 1990, find the average growth per year.
(b) Find a function that will approximate the waste (in millions of tons) generated in any given year on or beyond 1960. Check the accuracy of your function for the year 1985.
(c) Use the function found in Exercise 43(b) to approximate the tons of waste that will be generated in 2005.
(d) At this rate, how many years will it take before 250 millions of tons of waste per year will be generated?

44. Aging Population The following data from the U.S. Census Bureau* show (in millions) the number of Americans at least 85 years old.

Year	1970	1980	1985	1988	1990	1991	1992
Americans age 85 and over	1.4	2.2	2.7	2.9	3.0	3.2	3.3

(a) Assuming linear growth between 1970 and 1992, find the average growth per year (in millions) of persons age 85 and over.
(b) Use the data for the years 1970 and 1992 to obtain a linear growth function for persons age 85 and over. Check for accuracy in 1985 and 1990.
(c) Assuming the linear trend in Exercise 44(b) continues into the future, what year will the number of persons age 85 and over reach 6 million?

45. Depreciating a Building A building is purchased for $100,000 and is to be totally depreciated by the straight-line method over a 20-year period. (Assume no salvage value.)

(a) What is the amount to be depreciated each year?
(b) Find a linear function that expresses the book value of the building as a function of years from date of purchase.
(c) Graph the function found in Exercise 45(b).

46. Depreciating a Machine A copying machine is purchased for $12,500 and is to be totally depreciated by the straight-line method over a five-year period. (Assume no salvage value.)

(a) What is the amount to be depreciated each year?
(b) Find a linear function that expresses the book value of this machine as a function of years from date of purchase.
(c) Graph the function found in Exercise 46(b).

47. Depreciating a Machine A particular machine costs $80,000 when purchased new and has a salvage value of $10,000 twenty years later. Straight-line depreciation is applied to the net value of the machine over a 20-year period.

(a) How much is depreciated each year?
(b) Write the linear function that represents the book value of the machine during its lifetime.
(c) Graph the function found in Exercise 47(b).

48. Depreciating a Building An apartment building is purchased for $160,000. Of this amount, $30,000 is considered the cost of the land on which the building sits. Excluding land cost, the purchase is to be depreciated over a 17-year period by the straight-line method.

(a) What is the amount to be depreciated each year?
(b) Find a linear function that gives the book value of the purchase as a function of the number of years after purchase.
(c) Check the accuracy of the function found in Exercise 48(b).
(d) Graph the function found in Exercise 48(b).

49. Depreciating a Building An apartment building is bought for $550,000, of which $50,000 is estimated to be the land cost. The total cost minus the land cost is fully depreciated by the straight-line method over a period of 25 years.

(a) How much is depreciated each year?
(b) Write the linear function that represents the book value of this building during this 25-year period.
(c) Graph the function found in Exercise 49(b).

50. Depreciating a Computer The Ziegler Bank and Trust Co. buys a computer for $90,000 and plans to use it for eight years, at which time the company estimates the salvage value to be $7500. On the accounting books, the company is depreciating the computer by the straight-line method over the eight-year period.

(a) What is the amount to be depreciated each year?
(b) Find a linear function that expresses the book value of the computer as a function of the number of years after purchase.
(c) Check the accuracy of your function by computing the book value at the time of purchase and again after eight years.
(d) Graph the function.

* SOURCE: U.S. Bureau of the Census: *Statistical Abstract of the United States*, 1994.

For Exercises 51 and 52, answer these questions about the given supply and demand functions for each exercise's set of equations:

(a) *What is the demand when p = 20? Is there a surplus or a shortage at this price?*

(b) *What is the price when the demand is 100? Is there a surplus or a shortage at this price?*

(c) *Graph S(p) and D(p) on the same set of axes, and identify the regions where shortages and surpluses exist.*

51. $S(p) = 6p$
     $D(p) = 160 - 6p$

52. $S(p) = 4p + 1$
     $D(p) = 148 - 3p$

53. *Demand Versus Price* When the price of soybeans is $4 per bushel, the demand is 12 million bushels. When the price is $6 per bushel, the demand is 9 million bushels.

(a) Use this data to find a linear demand function for soybeans (in millions of bushels).

Use the model found in Exercise 53(a) to answer the remaining parts.

(b) What is the demand when the price is $5.25 per bushel?

(c) If the demand for soybeans is 7 million bushels, what is the price?

(d) If soybeans are given away, what does the model in Exercise 53(a) say that the demand will be? Do you think that this is realistic?

54. *Demand Versus Price* A market research firm determines that the demand for a particular product, in terms of purchases per thousand population, is 15 when the product is priced at $p = \$20$, but only 5 when the product is priced at $60.

(a) Find a linear demand function for this product in terms of the price p.

Use the model found in Exercise 54(a) to answer the remaining parts.

(b) If the product were priced at $30, what would be the demand?

(c) At what price is the demand 0?

(d) According to your linear demand function, if the product were given away, what would the demand likely be?

Problems for Exploration

1. Make a list of several situations in your everyday life that are (a) linear and (b) not linear. Include utility bills, tuition and fees, car payments, and so on.

2. Write a paragraph comparing the marginal cost to the average cost per item for a linear cost function throughout the domain of both functions. Include graphs of each. Will the average cost per item ever equal the marginal cost?

3. Go to a local auto rental firm and report (complete with graphs) on the cost of renting a car for one day and driving it x miles. Include both marginal and average costs in your analysis. In particular, is the cost function linear?

SECTION 1.4   Two Lines: Relating the Geometry to the Equations

We have seen that every linear equation of the general form $Ax + By = C$ has as its graph a line in the plane. Any ordered pair of real numbers (x, y) that is a simultaneous solution to two (or more) linear equations is a **point of intersection** of the lines forming their graphs. For a pair of lines in the plane, one—and only one—of the following three statements is true:

 I. The two lines intersect in exactly one point.

 II. The two lines do not intersect at all (i.e., they are parallel).

 III. The two lines are actually the same line (i.e., they have the same graph).

The objectives of this section are (a) to discover from the equations of the lines which of these three statements apply, and, if the lines intersect, (b) to find the coordinates of the point or points of intersection. This process is referred to as "solving a system of two equations in two unknowns."

The following three pairs of linear equations are examples showing how the slopes of each respective pair of lines determine their classification according to the preceding statements I, II, or III.

$$x + 2y = 5 \qquad\qquad 2x - y = 3 \qquad\qquad 3x + 2y = 4$$
$$3x + y = 5 \qquad\qquad 4x - 2y = -9 \qquad\qquad 6x + 4y = 8$$

Example Equations (a)   Example Equations (b)   Example Equations (c)

EXAMPLE EQUATIONS (a)

I. If the slopes of the two lines are not equal, the lines must intersect in exactly one point. Example Equations (a) fall into this classification because the slope of the first line ($y = -\frac{1}{2}x + \frac{5}{2}$) is $-\frac{1}{2}$ and the slope of the second line ($y = -3x + 5$) is -3. See Figure 25.

EXAMPLE EQUATIONS (b)

II. If the slopes of the two lines are equal but the y-intercepts are different, then the two lines are parallel but distinct. Thus, there are no points of intersection. Example Equations (b) fall into this classification because when each is put into slope intercept form, $y = 2x - 3$ and $y = 2x + \frac{9}{2}$, the slope of both lines represented is 2 but the y-intercept of the first is -3 while that of the second is $\frac{9}{2}$. See Figure 26.

EXAMPLE EQUATIONS (c)

III. If the slopes of the two lines are equal, and they both have the same y-intercept, then the two lines are identical. Thus, every point on one line is also on the other, so there are infinitely many points of intersection. Example Equations (c) fall into this classification because the slopes of both lines represented ($y = -\frac{3}{2}x + 2$) are $-\frac{3}{2}$, and the y-intercepts are both 2. See Figure 27.

The general problem arising from these three cases is as follows:

> Given two equations of straight lines, find the coordinates of all points that lie on both lines, or determine that there are no points common to both lines.

The two methods of solution considered are the **method of substitution** and the **method of elimination.** Both are demonstrated on each of the pairs in Example Equations (a), (b), and (c), to show how they apply to the three classifications I, II, and III.

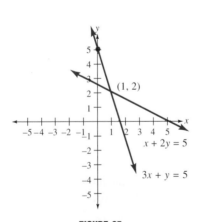

FIGURE 25
Lines that Intersect in One Point

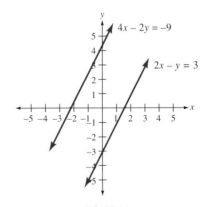

FIGURE 26
Lines that Have No Point in Common
(Parallel)

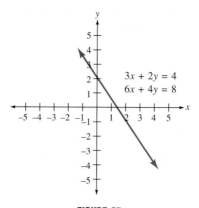

FIGURE 27
Lines that Have Every Point in Common

The Method of Substitution

This method involves solving for one of the variables in either equation and then substituting its value into the other equation to obtain one equation with only one unknown (often described as an "equation in one unknown").

The Method of Elimination

This method involves multiplying one of two original equations by a nonzero number to produce a third equation, which, when added to the other original equation, produces a final equation in one unknown.

EXAMPLE 1
Pairs of Lines in
Classification I (One
Point in Common)

Apply the two methods to the Example Equations (a):

$$x + 2y = 5 \qquad \textbf{(a.1)}$$
$$3x + y = 5. \qquad \textbf{(a.2)}$$

We already know that the two lines represented have exactly one point in common.

Solution

(a) First, we apply the *method of substitution* to this system of equations to obtain the point of intersection. In the given system, it is easy to solve for x in the first equation, (a.1):

$$x + 2y = 5 \qquad \textbf{(a.1)}$$
$$x = 5 - 2y. \qquad \textbf{(a.3)}$$

Now substitute this value into Equation (a.2) to obtain:

$$3(5 - 2y) + y = 5$$
$$15 - 6y + y = 5$$
$$-5y = -10$$
$$y = 2.$$

Finally, we substitute $y = 2$ into either of the original equations or into Equation (a.3) to find the corresponding value of x:

$$x = 5 - 2y \tag{a.3}$$
$$x = 5 - 2(2) = 1.$$

Therefore, the coordinates of the point of intersection are $x = 1$ and $y = 2$ (see Figure 25).

Solution

(b) The same point of intersection may be obtained by applying the *method of elimination* to this system. One option is to multiply Equation (a.1) by -3 so that when the resulting equation is added to the second equation, x is eliminated, leaving one equation in the unknown y:

$$x + 2y = 5 \tag{a.1}$$
$$3x + y = 5 \tag{a.2}$$

−3(Equation a.1)
$$-3x - 6y = -15$$
$$\underline{3x + y = 5} \tag{a.2}$$
$$-5y = -10$$
$$y = 2.$$

When $y = 2$ is substituted into Equation (a.1) (either original equation will do), we get $x = 1$. Again, the point of intersection is $(1, 2)$.

An alternative process would have been to multiply Equation (a.2) by -2, then add the result to Equation (a.1) to eliminate y and obtain one equation in the variable x. ■

We now examine how these two solution methods apply to a system in Classification II (i.e., it has no solution).

EXAMPLE 2
Pairs of Lines in
Classification II
(Parallel Lines)

Apply the two methods to Example Equations (b):

$$2x - y = 3 \tag{b.1}$$
$$4x - 2y = -9. \tag{b.2}$$

As mentioned previously, we already know that the two lines represented have no point in common.

Solution

(a) The *method of substitution* is applied first. An examination of these two equations leads to the conclusion that it is easiest to solve Equation (b.1) for y. Taking that approach we get

$$2x - y = 3 \tag{b.1}$$

or
$$y = 2x - 3.$$

Upon substituting this value for y into Equation (b.2), we get

$$4x - 2y = -9 \tag{b.2}$$
$$4x - 2(2x - 3) = -9$$
$$4x - 4x + 6 = -9$$

or
$$6 = -9.$$

The last statement is false (a contradiction), which is the signal that no point of intersection exists between the two lines (see Figure 26). Such a system is called **inconsistent.**

Solution

(b) Second, we apply the *method of elimination* to these two equations. If we multiply Equation (b.1) by -2 and add the results to Equation (b.2), we get

$$2x - y = 3 \qquad \text{(b.1)}$$
$$4x - 2y = -9 \qquad \text{(b.2)}$$
$$-2(\text{Equation b.1}): \quad -4x + 2y = -6$$
$$\underline{4x - 2y = -9} \qquad \text{(b.2)}$$
$$0 = -15.$$

Again, the last statement is a contradiction, which means there is no point of intersection between the two lines. ■

EXAMPLE 3
Pairs of Lines in
Classification III
(Same Line)

Apply the two methods to Example Equations (c):

$$3x + 2y = 4 \qquad \text{(c.1)}$$
$$6x + 4y = 8. \qquad \text{(c.2)}$$

From the previous discussion, we already know that these two equations represent the same line.

Solution

(a) First, we apply the *method of substitution.* Solving Equation (c.1) for y gives

$$3x + 2y = 4 \qquad \text{(c.1)}$$
$$2y = -3x + 4$$
$$y = \left(-\frac{3}{2}\right)x + 2.$$

Then, upon substitution into Equation (c.2), we get

$$6x + 4\left[\left(-\frac{3}{2}\right)x + 2\right] = 8$$
$$6x - 6x + 8 = 8$$
$$8 = 8.$$

The resulting equation, $8 = 8$, called an identity, is a perfectly correct mathematical statement. It does not tell us what the solutions are, but it does signal that one of the equations is unnecessary and does not contribute to the solution. (The graph in Figure 27 is the geometric justification for this statement.) Such a system of equations is known as a **dependent system;** in the case of two equations and two unknowns, this happens only when one equation is a multiple of the other. This means that when graphed, the two resulting lines coincide; hence, all points on one line are also on the other. As a consequence, only one of the equations is needed to determine the form of how the infinitely many solutions appear. (Again, see Figure 27.) A common method of denoting these points of intersection is to take one of the equations, say the first,

$$3x + 2y = 4, \qquad \text{(c.1)}$$

and then solve it for one of the variables, say y, to get

$$y = \left(-\frac{3}{2}\right)x + 2.$$

Then write the solution as

$$x = \text{any number},$$

$$y = \left(-\frac{3}{2}\right)x + 2.$$

If x is assigned any numerical value, we can solve for a corresponding y value so that (x, y) is a solution to both of the original equations and, hence, is a point of intersection. For instance, if $x = 2$, then $y = -1$, so that $(2, -1)$ represents a point on both lines. If $x = 0$, then $y = 2$, so $(0, 2)$ represents another point of intersection, and so on. In mathematical language, the variable in the solution that can be any number (x in this case) is called a **free variable** (or a **parameter**).

Solution

(b) Second, we apply the *method of elimination* to this pair of equations. If we multiply Equation (c.1) by -2 and add the results to Equation (c.2), the resulting system becomes

$$
\begin{array}{rll}
3x + 2y = 4 & \quad & \textbf{(c.1)}\\
6x + 4y = 8 & \quad & \textbf{(c.2)}
\end{array}
$$

$$
\begin{array}{rll}
-2(\text{Equation c.1}): \quad -6x - 4y = -8 & & \\
6x + 4y = 8 & \quad & \textbf{(c.2)}\\
\hline
0 = 0. & &
\end{array}
$$

This again shows that the system is dependent. This time, if the first equation is solved for x rather then y, the solutions take on the appearance

$$y = \text{any number}$$

$$x = \left(-\frac{2}{3}\right)y + \frac{4}{3}.$$

It is a matter of taste or convenience as to which variable is to be considered free. ■

Eliminating a variable between two linear equations may always be done by multiplying one of the original equations by a particular nonzero number and then adding the resulting equation to the other original equation. This was done in Examples 1(b), 2(b), and 3(b) where the respective multipliers were not difficult to find. Example 4 shows how this may be done for any pair of linear equations.

EXAMPLE 4
Using the Elimination Method

Use elimination to solve:

$$
\begin{array}{rl}
2x - 3y = -16 & \qquad \textbf{(1)}\\
5x + 4y = 29. & \qquad \textbf{(2)}
\end{array}
$$

Solution   After some study of these equations, we conclude that the variable x can be eliminated by multiplying Equation (1) by $-\frac{5}{2}$ and the result added to Equation (2):

$$(-5/2) \text{ Equation (1):} \qquad -5x + \frac{15}{2}y = 40$$

$$\text{Equation (2):} \qquad \underline{5x + \ 4y = 29}$$

$$\frac{23}{2}y = 69.$$

To ease the burden of selection for the multiplier in cases such as this, we now show how this selection may be done in a two stage process. First, multiply Equation (1) by $\frac{1}{2}$ to make the coefficient on x a 1:

$$(1/2) \text{ Equation (1):} \qquad x - \frac{3}{2}y = -8 \qquad \textbf{(1a)}$$

$$\text{Equation (2):} \qquad 5x + 4y = 29. \qquad \textbf{(2)}$$

Now compare the coefficients of x in Equations (1a) and (2): Equation (1a) should be multiplied by -5, then added to Equation (2) to eliminate x in Equation (2):

$$(-5) \text{ Equation (1a):} \qquad -5x + \frac{15}{2}y = 40$$

$$\text{Equation (2):} \qquad \underline{5x + \ 4y = 29}$$

$$\frac{23}{2}y = 69.$$

In either case, the value of y is then found to be $y = 69(\frac{2}{23}) = 3 \cdot 2 = 6$. The substitution of $y = 6$ into Equation (1), or a similar elimination of y between Equations (1) and (2), gives $x = 1$. ■

The accuracy of points of intersection obtained by either of the preceding methods should always be checked by substituting the coordinates back into the *original equations* of the lines. If the coordinates actually represent a point on both lines, an equality should result from both equations. This applies even for equations in Classification III.

Certain models of graphing calculators may be used to display not only the graphs of a pair of equations, but also any points of intersection to several decimal accuracies. Shown in Figure 28 is such a display for the linear equations:

$$3x - 4y = -14 \quad \text{or} \quad y = \left(\frac{3}{4}\right)x + \frac{14}{4} \qquad \text{and}$$

$$7x + 2y = -24 \qquad y = \left(-\frac{7}{2}\right)x + 12.$$

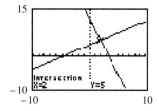

FIGURE 28

Solving a System Using a Graphing Utility

A knowledge of the three classifications of solutions we have discussed is essential in interpreting calculator screens and computer printouts, especially in cases in which no solution exists—or infinitely many solutions exist.

Now let's look at a real-world situation in which the point of intersection of two lines has a significant impact on business decisions. When starting any new manufacturing processes, the initial costs are almost always more than the revenue derived from initial sales of the product being manufactured. An example might be the introduction of a new automobile. Of great interest in such processes is the number n_0 that needs to be made so that the total revenue from sales finally matches the total costs incurred.

Accordingly, if a number x less than n_0 is made, the total cost $C(x)$ will be greater than the total revenue $R(x)$, so a *loss* results. On the other hand, if a number x greater than n_0 is made, then the total cost $C(x)$ is less than the total revenue $R(x)$, so a *profit* results. When n_0 are made, $R(n_0) = C(n_0)$, so that neither a profit nor a loss results; that is, we **break even.** The number n_0 is known as the **break-even quantity,** while the point where the graph of $y = C(x)$ and $y = R(x)$ intersect is known as the **break-even point.**

The Break-Even Point

For a cost function $C(x)$ and revenue function $R(x)$, the number n_0 for which $R(n_0) = C(n_0)$ is known as the **break-even quantity,** and the point of intersection of the graphs of $y = R(x)$ and $y = C(x)$ is known as the **break-even point.**

The **profit** from a manufacturing process is related to the revenue and the cost in the following way.

Profit

Profit is defined as the revenue minus the cost. In equation form, if $R(x)$ represents the revenue function, $C(x)$ represents the cost function, and $P(x)$ represents the profit function, then

$$P(x) = R(x) - C(x).$$

The profit is always zero at the break-even quantity, n_0. The reason is that this value makes the revenue equal to the cost. Therefore,

$$P(n_0) = R(n_0) - C(n_0) = 0.$$

EXAMPLE 5
Break-Even Analysis

The Ace Manufacturing Company makes clock radios. The company has determined that the short-run fixed costs are $2000 and the direct cost for each clock radio is $20. Suppose that x clock radios are made, and each is sold for $30.

(a) Find the total cost function.

Solution The total cost function is $C(x) = 20x + 2000$.

(b) Find the revenue function.

Solution If x clock radios are sold for $30 each, then $R(x) = 30x$.

(c) Find the profit function.

Solution The profit is the revenue minus the cost. Therefore,

$$P(x) = R(x) - C(x) = 30x - [20x + 2000]$$
$$= 30x - 20x - 2000$$
$$= 10x - 2000.$$

(d) Find the break-even quantity and the break-even point.

Solution The break-even quantity is found by setting $R(x) = C(x)$ and solving for x:

$$30x = 20x + 2000$$
$$10x = 2000$$
$$x = 200.$$

We lack only the y-coordinate for the break-even point. Substituting $x = 200$ into $y = R(x) = 30x$ gives $y = 30(200) = \$6000$. The break-even point is then $(200, 6000)$. (See Figure 29.)

(e) Graph $y = C(x)$, $y = R(x)$, and $y = P(x)$ on the same set of axes.

Solution The graphs are shown in Figure 29. Notice that at $x = 0$, there is a loss of \$2000. This is the fixed cost in our clock radio operation. Also, at the break-even quantity $x = 200$, the profit is \$0 as we expect. For $0 < x < 200$, the profit is negative (a loss occurs), and for $x > 200$, a positive profit is recorded. ■

One of the future uses for finding a point of intersection of two lines in this text occurs when the coordinates of "corner points" of a region bounded by lines are needed. The next example illustrates this.

EXAMPLE 6
An Application of
Intersection Points

Find the coordinates of the corner points of the shaded region shown in Figure 30.

Solution Of the four corner points, the three located on the coordinate axes are easy to find. Counterclockwise from the upper left, they are $(0, 3)$, $(0, 0)$, and $(2, 0)$. That leaves only the one marked A to be found. To do that, we first need the equations of the lines that intersect at point A. The line through $(0, 3)$ and $(5, 0)$ has slope

$$m = \frac{0 - 3}{5 - 0} = -\frac{3}{5}$$

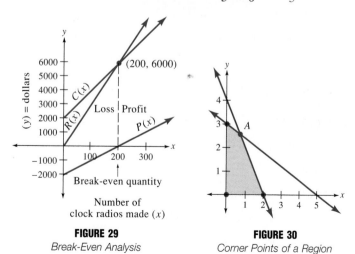

FIGURE 29
Break-Even Analysis

FIGURE 30
Corner Points of a Region

and since the y-intercept is 3, the slope-intercept form of a line may be used to obtain its equation

$$y = -\frac{3}{5}x + 3. \tag{1}$$

The line through $(0, 4)$ and $(2, 0)$ has slope

$$m = \frac{0-4}{2-0} = -2$$

and since the y-intercept is 4, the slope-intercept form of a line may again be used to obtain the equation of this line

$$y = -2x + 4. \tag{2}$$

Substituting Equation 1 into Equation 2 results in the equation

$$-\frac{3}{5}x + 3 = -2x + 4$$

$$2x - \frac{3}{5}x = 4 - 3$$

$$\frac{7}{5}x = 1$$

$$x = \frac{5}{7}.$$

Finally, substituting this value into Equation 2 shows that $y = -2(\frac{5}{7}) + 4$ or that $y = \frac{18}{7}$. Therefore, the point A has coordinates $(\frac{5}{7}, \frac{18}{7})$. ■

Even though there are other methods for solving systems of two linear equations in two unknowns (finding the point[s] of intersection), the two demonstrated in this section are among the easiest to use. Furthermore, either method may be used to solve any such system.

EXERCISES   1.4

In each of the Exercises 1 through 6, decide by examining the slope and the y-intercept of each pair of equations which of the classifications the resulting lines fall into: one point in common (I), no point in common (II), or all points in common (III). Do not solve these pairs of equations, though.

1. $4x + 6y = 9$
   $x - 3y = 1$

2. $x + 3y = 5$
   $2x + 6y = 10$

3. $x - 5y = 8$
   $3x - 15y = 3$

4. $2x - y = 4$
   $4x - 2y = 8$

5. $-x + 5y = 8$
   $x + 5y = 3$

6. $-x + 5y = 8$
   $-x + 5y = 3$

In each of the Exercises 7 through 24, do the following:
(a) First, determine how many solutions the system has, if any.
(b) Then find the point or points of intersection, if any exist, by an appropriate method.

7. $2x + y = 3$
   $x - y = 2$

8. $x + 2y = 4$
   $2x - 2y = 3$

9. $2x + y = 3$
   $x + .5y = 1.5$

10. $x + 3y = 2$
    $2x + 6y = 4$

11. $x + 2y = 0$
    $x - y = 0$

12. $2x - y = 0$
    $x + y = 0$

13. $x + 2y = 0$
    $2x + 4y = 0$

14. $x + 2y = 0$
    $2x + 4y = 1$

15. $x + y = 3$
$2x + 2y = 4$

16. $x - y = 5$
$2x - 2y = 1$

17. $.5x = 6$
$x + 3y = 4$

18. $y = 5$
$2x - y = 6$

19. $.5x + 2y = 3$
$x - y = 1$

20. $x + .6y = 0$
$x - 2y = 1$

21. $.2x - y = 0$
$x + .6y = 0$

22. $.4x - .3y = 6$
$x + y = 0$

23. $x - 2y = 3$
$2x - y = 6$

24. $.2x + .7y = 0$
$x - y = 0$

 In Exercises 25 through 28, use a graphing calculator or computer to help solve the given systems.

25. $2x + 3y = 150$
$1.8x + 3.5y = 167$

26. $y = 3x$
$y = 3x + .1$

27. $.3x - .4y = .8$
$6x - 8y = 16$

28. $.3x - .4y = 16$
$x + .5y = 21$

In each of the Exercises 29 through 34, do the following:
(a) Find the break-even quantity.
(b) Find the break-even point.
(c) Graph $R(x)$ and $C(x)$ on the same set of axes.

29. $R(x) = 50x$
$C(x) = 20x + 900$

30. $R(x) = .5x$
$C(x) = .3x + 20$

31. $R(x) = 20x$
$C(x) = 10x + 2500$

32. $R(x) = 6x + 10$
$C(x) = 2x + 90$

33. $R(x) = 15x$
$C(x) = 8x + 490$

34. $R(x) = 5x + 12$
$C(x) = 3x + 116$

In each of the Exercises 35 through 40, do the following:
(a) Find the break-even quantity. X
(b) Find the break-even point. a coord.
(c) Find the profit function. diff. R-C
(d) Graph the revenue, cost, and profit functions on the same set of axes.

35. $R(x) = 3x$
$C(x) = 2x + 12$

36. $R(x) = 50x$
$C(x) = 30x + 2000$

37. $R(x) = 20x$
$C(x) = 12x + 490$

38. $R(x) = 60x$
$C(x) = 30x + 1800$

39. $R(x) = 30x + 60$
$C(x) = 10x + 2800$

40. $R(x) = 16x + 88$
$C(x) = 5x + 231$

 In Exercises 41 through 44 use a graphing calculator or computer to find:
(a) The break-even quantity.
(b) The total cost spent to achieve the break-even quantity.

41. $R(x) = 8.45x$
$C(x) = 5.73x + 3760$

42. $R(x) = 14.80x$
$C(x) = 8.76x + 1475$

43. $R(x) = 7.35x + 40$
$C(x) = 5.13x + 185$

44. $R(x) = 22.45x + 128$
$C(x) = 15.60x + 1575$

45. *Revenue and Costs* The ABC Cassette Tape Shop sells only cassette tapes for stereo systems. The shop has a monthly overhead of $8000 (fixed costs) and an average direct cost of $8.36 per tape. Each tape sells for an average of $10.80. Let x be the number of tapes sold each month.

(a) If x tapes are sold during the month, find the linear functions for the cost, revenue, and profit.
(b) Find the number of tapes that must be sold each month in order to break even.
(c) Graph $C(x)$, $R(x)$, and $P(x)$ on the same set of axes.
(d) If 800 tapes are sold during the month, will a profit or a loss be incurred? What if 2000 are sold?

46. *Revenue and Costs* The Beta Beta Beta Fraternity is planning a picnic. The committee members in charge estimate that they will spend a total of $400 on pavilion rent, decorations, and so on. They also estimate that they will spend $7.50 per person for food and drinks. They plan to charge each person who attends $10. Let x be the number of people who attend the picnic.

(a) What is the cost function in terms of x?
(b) What is the revenue function in terms of x?
(c) How many people must attend for the committee to break even?
(d) What is the profit function in terms of x?
(e) Graph the cost, revenue, and profit functions on the same set of axes.

47. *Buying and Renting* A car rental firm buys a new car for $20,000 and estimates the cost for maintenance, taxes, insurance, and depreciation at 40 cents per mile. The firm charges $50 per day plus 50 cents per mile to rent the car. Let x be the number of miles the car is rented during the first year, and assume that the car is rented for 160 days.

(a) What is the cost function for owning and renting this car in terms of x?
(b) What is the revenue function in terms of x?
(c) How many miles must the car be driven for the company to break even during the first year?
(d) Graph the cost and revenue functions on the same set of axes for $0 \le x \le 150,000$ during the first year.

48. *Tool Rental* A rental firm has purchased a new ditchdigger for $36,000 and expects that maintenance costs will average 75 cents per actual hour of use. It plans to rent the ditchdigger for $200 per day plus $5 per hour of actual use. Assume that the ditchdigger will be rented for 150 days during the first year. Let x be the number of hours of actual use the ditchdigger gets during that first year.

(a) What is the cost function for purchasing and renting this machine in terms of x?

(b) What is the revenue function in terms of x?

(c) How many hours must this machine be used for the rental firm to break even during the first year?

(d) Graph the cost and revenue functions on the same set of axes.

49. *Revenue and Costs*   The Springdale Bottling Co. is planning to introduce a new soft-drink container that holds 3 liters. The company estimates the first-year fixed costs for setting up the new production line at $50,000, and the direct costs for each bottle will be $1.50. The sales department estimates that 60,000 bottles can be sold during the first year at $2 per bottle.

(a) Find the linear cost function $C(x)$ that will give the total costs for selling x bottles of the drink. Also find the revenue function $R(x)$.

(b) Find the profit function, and graph the cost, revenue, and profit functions on the same set of axes.

(c) If the sales department is correct in having estimated sales of 60,000 bottles during the first year, will a profit or a loss occur for the year?

(d) How many bottles must be sold to break even?

(e) What are the total cost and the average cost per 3-liter bottle if 20,000 bottles are produced? What are they if 40,000 bottles are produced?

(f) If 20,000 bottles have been produced, what is the additional cost of making the 20,001st bottle?

50. *Manufacturing Revenue and Costs*   The Hi-Tech Computer Company is planning to introduce a new graphics calculator. For the first year, it estimates that the fixed costs to set up the production line will be $100,000. Hi-Tech estimates the direct costs for making each calculator will be $50 each and expects to sell the calculators for $75 each.

(a) Find the linear cost function that gives the total cost of producing x calculators.

(b) Find the revenue function, and use the concept of "marginal" (previously applied to "marginal cost") to state the marginal revenue.

(c) Find the profit function, and graph the cost, revenue, and profit functions on the same set of axes.

(d) How many calculators must be made and sold for the company or break even?

(e) What is the total cost of making 25,000 calculators? If 25,000 have been made, what is the additional cost of making the 25,001st calculator?

(f) What is the average cost per calculator of making 25,000 calculators?

51. *Printing Costs*   The New Font Printing Company has two methods of printing the orders it receives. One is a computer method, which costs $300 to set up and then costs $25 to print 1000 copies. The other is an offset process, which costs $500 to set up and then costs $20 to print 1000 copies.

(a) Let x be the number to be printed in thousands. Find the linear cost function for each of the two methods.

(b) Graph both of these functions on the same set of axes.

(c) Find the quantity for which the costs are the same (the break-even cost).

(d) Assuming equal quality, which method would be most advantageous for the company to use to print an order of 30,000 copies? Of 70,000 copies?

52. *Car Revenue and Costs*   A new-car dealer has overhead costs (fixed costs) of $60,000 per month and average direct costs of $12,000 per car sold by the dealer (including the cost of the car). New cars are sold for an average price of $15,000 per car. Let x be the number of new cars sold each month by the dealer.

(a) Find the linear cost function $C(x)$ that represents the total cost of selling x new cars; also find the revenue function and the profit function for selling x new cars.

(b) Find the number of new cars the dealer must have sold each month to break even.

(c) Graph $C(x)$, $R(x)$, and $P(x)$ on the same set of axes.

(d) If 100 new cars are sold during the month, will a profit or a loss be incurred by the dealer? What if 200 cars are sold?

53. *Comparing Service Costs*   A new stereo of brand A costs $500 and comes with a five-year "no cost to you" guarantee at no additional cost. Brand B stereo of similar quality cost $350 but comes with an optional service contract, which costs $35 per year and will cover all repair costs.

(a) Let x be the number of years after purchase of one of these stereos, and assume that the service contract is purchased on brand B. Find the linear function that gives the cost of ownership for each stereo.

(b) Graph these two cost functions on the same set of axes.

(c) Find the number of years required for the two costs to be equal.

(d) Based on ownership of about five years, how would you decide which stereo has the lowest cost?

54. *Banking Costs* The cost of a student account at the Ozark Bank is $6 per month, with no fees for automatic teller machine (ATM) transactions. The cost for a similar account at the Ray Kelley Bank is $3 per month, and each ATM transaction costs 50 cents.

   (a) Let x be the number of ATM transactions per month. Find the linear cost function for having an account at each of these banks in terms of the number of ATM transactions each month.

   (b) Graph these cost functions on the same set of axes.

   (c) Find the number of ATM transactions for which the account costs are the same (the break-even cost).

   (d) Based on ATM usage, tell how you would decide which account would give lower overall account costs.

55. *Equal Sales* In 1980, A-Mart had sales of $36.3 million and has had a constant growth of $1.1 million in sales each year since. On the other hand, B-Mart had sales of $16.7 million in 1980 and has had a constant growth of $2.3 million in sales each year since. If these trends continue, when will B-Mart Sales overtake those of A-Mart?

56. *Equal Populations* In 1980, Harrisonville had a population of 6254 and has been gaining population at an average rate of 420 persons per year ever since. On the other hand, Smithville had a population of 17,213 in 1980 and has been losing population at an average rate of 330 persons per year ever since. If these trends continue, in what year will these two towns have the same population?

57. *Recycling* In 1980 the city of Rumford recycled 80 tons of waste paper and has been increasing this amount at an average rate of 2 tons per year ever since. On the other hand, Galeton recycled only 38 tons of waste paper in 1980, but has been increasing this amount at the rate of 3 tons per year ever since. If these trends continue, what year will the two cities recycle the same amount of paper?

In Exercises 58 through 63, find the equation of each line that forms a boundary of the shaded region. Then find the coordinates of each corner point of the shaded region.

58.

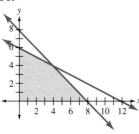

59.

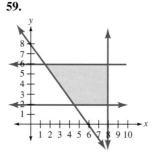

60.

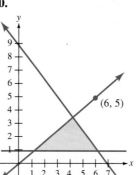

61.

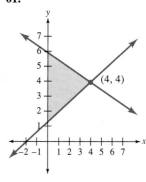

62.

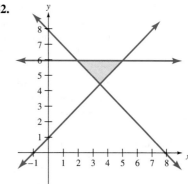

63.

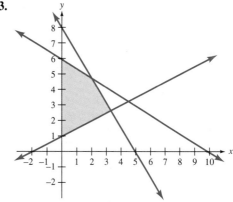

Problems for Exploration

1. Consider the equations

$$ax + by = c$$
$$dx + ey = f$$

where $a, b, c, d, e,$ and f are constants, a and b are not both zero simultaneously, and d and e are not both zero simultaneously. Find relationships between $a, b, c, d, e,$ and f that will result in a system of equations with:

   (a) Exactly one solution.

   (b) No solution.

   (c) Infinitely many solutions.

2. Check statistics from your city, county, or state regarding the amount of waste generated per person in 1980 and during several intervening years through the most recent year available. Then try to select two of these years that will, in your judgment, best "linearize" the data you have collected. Find the equation of this trend line, using these two years. In view of the data collected, rate how your line describes these data on a scale from 1 to 10, with 10 being very good. Once this is done, use the same process for the amount of waste that is recycled, again researching relevant data from 1980 to the most recent year available. Will these two trend lines ever intersect?

3. Find information on Cramer's Rule, a method of solution for systems of linear equations. Write a short explanation on how this method may be used to find the point (or points) of intersection of two lines or to detect when no point of intersection exists.

SECTION 1.5 **Regression and Correlation**

In Section 1.3, we found equations of lines passing through two points in which the two points represented observed or estimated data. We then saw how those lines could be used, within limits, for prediction purposes. Realistically, however, we would be reluctant to assume, based on just two observations, that the trend of the actual data is linear. If realistic predictions were desired, we would need more evidence of such a trend. Therefore, we usually collect data (observations) over as wide a range as is feasible to begin an investigation of just how, if at all, the two variables are interrelated. Once collected, these data are then plotted in the rectangular coordinate system to obtain a **scatter diagram.** Sometimes, the scatter diagram shows a linear tendency, as in Figure 31.

EXAMPLE 1
A Scatter Diagram
Showing a Linear
Tendency

An industrial psychologist gave an aptitude test to 15 new employees in a manufacturing plant and a few weeks later checked on their productivity. The results of the test scores and the corresponding production units for each of these employees were as follows.

x = aptitude score, y = production per day

x	y	x	y	x	y	x	y
25	60	60	80	30	62	73	88
80	102	55	76	38	71	40	73
68	85	70	86	50	73	40	68
35	70	45	71	58	78		

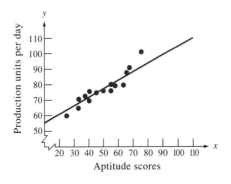

FIGURE 31
A Scatter Diagram Showing a Linear Tendency

When these points are plotted in the plane, the scatter diagram of Figure 31 shows that a straight line can be positioned to provide a reasonable representation of the given data. In such a case, we say that a **linear correlation** between the x and y variables exists. ■

At other times, the scatter diagram of collected data may show a pattern that is not linear or may even show no pattern at all. Figure 32 shows two nonlinear relationships: (a) quadratic correlation, and (b) no correlation.

The general study of how two or more variables are interrelated is called a **correlation** problem; the word *correlation* means "related together."

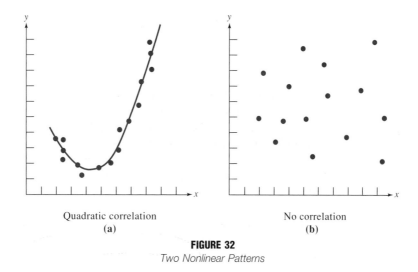

Quadratic correlation
(a)

No correlation
(b)

FIGURE 32

Two Nonlinear Patterns

After looking at a scatter diagram, we ask two questions: (1) Can we find a mathematical relationship between x and y, and (2) if so, how strong is the relationship? The first question leads us to try to find an equation that will express the relationship between the two variables. We can choose from many equations, but the simplest is the linear equation, the graph of which is a straight line, and this is where we concentrate our efforts. Once we decide to find a linear function that will represent the relationship between our two variables, we next encounter the problem of how to get the "best" one—that is, the equation that best describes the relationship.

The most widely accepted way to get a best fit to the data is by using the **least squares method.** The method works on this principle: A given data point (x_k, y_k), where x_k and y_k are constants, will differ in the y-coordinate from the point on a line with x-coordinate x_k by distance $d_k = y_k - f(x_k)$, and we say the best fit occurs when the sum of the squares of these distances is as small as possible (a minimum).

To be more specific, consider Figure 33, on the next page, which shows four points that represent collected data. As m and b vary, the line $y = mx + b$ changes, and therefore the vertical distances $d_1, d_2, d_3,$ and d_4 will change. This means that the sum $S = d_1^2 + d_2^2 + d_3^2 + d_4^2$ will change. The method of least squares gets its name from the fact that we want to find the particular values of m and b so that *the sum S of the squared distances is the least (smallest) possible.* Advanced mathematics is needed to show that the smallest value of S occurs at the solution point to a certain set of two equations whose variables are m and b and whose coefficients depend upon the x and

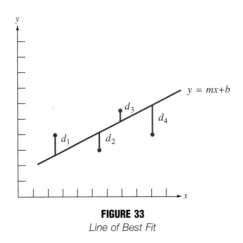

FIGURE 33
Line of Best Fit

y coordinates of the collected data. Those equations are shown next and use a mathematical shorthand symbol, Σ, which means "sum" or "add."

The Least Squares Regression Line

A linear function $y = mx + b$ will be the **least squares regression line** for the data points

$$(x_1, y_1), (x_2, y_2), \ldots (x_n, y_n)$$

in a scatter diagram if m and b are solutions to the system of equations

$$nb + (\Sigma x)m = \Sigma y \tag{1}$$
$$(\Sigma x)b + (\Sigma x^2)m = \Sigma xy \tag{2}$$

where Σx = sum of all x values in data points,
Σy = sum of all y values in data points,
Σxy = sum of products of both coordinates,
Σx^2 = sum of squares of all x values,
n = number of data points.

Once Equations 1 and 2 have been solved for m and b, the line $y = mx + b$ so obtained will be known as the **least squares regression line.** The word *regression* enters the picture to indicate the fact that once the equation describing the scatter diagram is known (see Figures 34 and 35), it may be used to predict the y-values by "regressing" or literally "going back to" known relationships between the x and y values.

EXAMPLE 2
Finding a Least
Squares Regression
Line

Find the least squares regression line for these data: (1, 6), (2, 4) (3, 3) (4, 1).

Solution   The scatter diagram in Figure 34 suggests that a line will express the relationship between the x and y values rather well. To find m and b in the equation $y = mx + b$, first use the data to organize a table such as this:

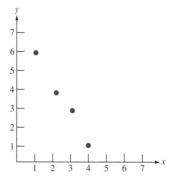

FIGURE 34

Scatter Diagram for Linear Regression

x	y	x^2	xy
1	6	1	6
2	4	4	8
3	3	9	9
4	1	16	4
sum 10	14	30	27

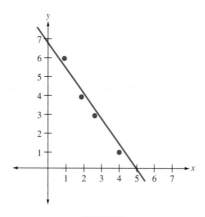

FIGURE 35

Regression Line and Data Points

Thus $\Sigma\, x = 10$, $\Sigma\, y = 14$, $\Sigma\, x^2 = 30$, and $\Sigma\, xy = 27$.

Then, because $n = 4$, Equations 1 and 2 become $4b + 10m = 14$ and $10b + 30m = 27$, respectively. Using one of the methods discussed in Section 1.4 of this chapter (i.e., substitution or elimination) now shows that the solution to this system is $b = 7.5$ and $m = -1.6$, so that the desired equation is

$$y = -1.6x + 7.5.$$

Figure 35 shows the scatter diagram with the regression line. Note that the regression line does not pass through any of the points in the scatter diagram; this is usually the case. ■

EXAMPLE 3
Finding a Regression Line for Prediction

Ms. Nugent is head of advertising sales at a TV station. As part of her sales pitch to other car dealers, she has collected the following data relating advertising spots per week and the number of sales Car Mart made during the week.

x = Number of ads run in a week, y = number of cars sold that week

x	y	x	y
5	15	20	32
15	20	0	10
12	19	9	18
		18	25

Organizing the given data as shown in the following table is helpful in finding the two desired equations in m and b.

x	y	x^2	xy
5	15	25	75
15	20	225	300
12	19	144	228
20	32	400	640
0	10	0	0
9	18	81	162
18	25	324	450
Sum 79	139	1199	1855

In this case, $n = 7$, so the resulting equations are

$$7b + 79m = 139$$
$$79b + 1199m = 1855,$$

and the solution (using two-decimal accuracy) is $m = .93$ and $b = 9.35$. The linear regression line may now be written as $y = .93x + 9.35$. This line, along with the scatter diagram, is shown in Figure 36.

(a) According to this regression line, how many car sales could be expected if the ad were run 30 times during the week?

Solution The answer would be $y = .93(30) + 9.35 \approx 37.25$, or about 37 cars. (See Figure 36.)

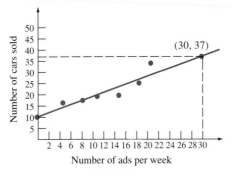

FIGURE 36
Predicting with a Regression Line

(b) If an advertising car agency sells 40 cars during the week, how many times did the sales ad run during the week, as predicted by the regression line?

Solution The answer is found by solving $40 = .93x + 9.35$ for x. The result is $x = 32.93$, or about 33 times. ■

CAUTION When using a regression line for prediction purposes, it usually is not wise to stray too far outside the limits of the data gathered.

EXAMPLE 4
Finding a Least
Squares Regression
Line

The accompanying graph (Figure 37) shows how the average number of members per household has been decreasing since 1850. The actual data are as follows:

Year	Members per household	Year	Members per household	Year	Members per household
1850	5.59	1900	4.76	1950	3.37
1860	5.28	1910	4.54	1960	3.33
1870	5.09	1920	4.34	1970	3.14
1880	5.04	1930	4.11	1980	2.76
1890	4.93	1940	3.67	1990	2.63
				1993	2.64

Both the graph and the actual data show that the decline is linear enough to be reasonably approximated by a linear function. The least squares fit of these data can be most easily obtained by letting $t = 0$ represent the year 1850, $t = 10$ the year 1860, and so on. The resulting regression line (found by the computer program LinReg) is $y = -.021t + 5.66$, where $0 \le t \le 222$. Its graph, along with the scatter diagram, is shown in Figure 38. (The reason for the upper limit on t is that the number of members per household cannot go below 1.)

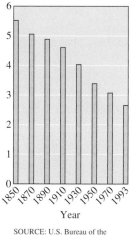

SOURCE: U.S. Bureau of the Census: *Statistical Abstract of the United States*, 1990 and 1994.

FIGURE 37

Average Number of Persons per Household for Selected Years

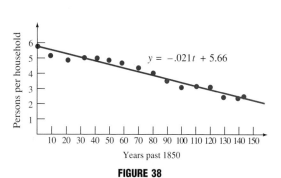

FIGURE 38

Average Number of People in a Household

This model of the data gives 5.66 members per household in 1850 and assumes a constant decline of 0.021 members each year thereafter until 1993, when the number is $y = -.021(143) + 5.66 = 2.66$. ■

If a least squares regression line has been graphed on the same set of axes as the scatter diagram, we can visually judge the "goodness" of fit of the line to the data. There is, however, a widely accepted numerical way to judge the goodness of fit, whether we have such a graph or not. It involves calculating a number, denoted by r, called the **correlation coefficient.** The correlation coefficient is *always* a number between -1 and 1, inclusive. Once calculated, the correlation coefficient works this way: The closer r is to $+1$ or -1, the better the least squares line fits the data, whereas values closer to 0 (zero) indicate a poor fit. As shown in Figure 39, if $r = +1$ or -1, there is a perfect linear relationship between the x and y values; that is, all points lie on the least squares line.

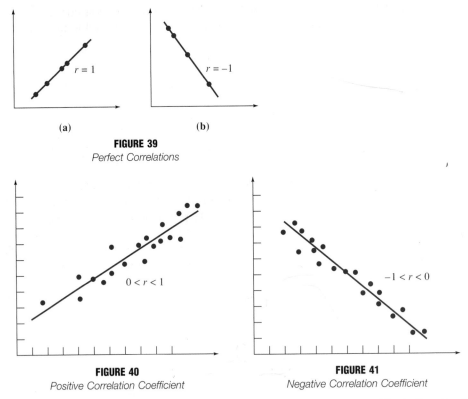

FIGURE 39
Perfect Correlations

FIGURE 40
Positive Correlation Coefficient

FIGURE 41
Negative Correlation Coefficient

As shown in Figure 40, for *r* **between 0 and 1** (excluding 0 and including 1), the *x* and *y* values have a *positive* correlation. Small *x* values are associated with small *y* values, and large *x* values are associated with large *y* values.

As shown in Figure 41, for *r* **between −1 and 0** (including −1 and excluding 0), the *x* and *y* values have a *negative* correlation. Small *x* values are associated with large *y* values, and large *x* values are associated with small *y* values.

If $r = 0$, there will be no apparent linear relationship between the *x* and *y* values in the scatter diagram. Figure 42 shows such a diagram.

The interpretation of *r* will vary, depending on the field of application. In the physical sciences, where experiments are closely controlled, *r* values with absolute values between .90 and 1 indicate a high correlation; in the social sciences and educational fields, *r* values with absolute values between .70 and 1 indicate a high to

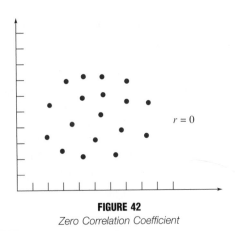

FIGURE 42
Zero Correlation Coefficient

very high correlation. Thus, the correlation coefficient must be judged within the context of the application. The formula used to find r follows.

The Correlation Coefficient

If $y = mx + b$ is the linear regression line for the points (x_1, y_1), $(x_2, y_2), \ldots (x_n, y_n)$, then a measure of how well it fits is given by the **correlation coefficient**

$$r = \frac{n(\Sigma\, xy) - (\Sigma\, x)(\Sigma\, y)}{\sqrt{n(\Sigma\, x^2) - (\Sigma\, x)^2}\,\sqrt{n(\Sigma\, y^2) - (\Sigma\, y)^2}}$$

where $\Sigma\, x$ = sum of all x values in data points,
$\Sigma\, y$ = sum of all y values in data points, and so on.

The correlation coefficient r for the line found in Example 3 is approximately .94, and the value of r for the line in Example 4 is approximately $-.99$. (These were found with the computer program LinReg.)

If r is to be calculated by hand, a table such as the one in Example 3 can be used effectively. Only one column showing the values of y^2 needs to be adjoined to that table to obtain the entries needed in the formula for r.

x	y	x^2	xy	y^2
5	15	25	75	225
15	20	225	300	400
12	19	144	228	361
20	32	400	640	1024
0	10	0	0	100
9	18	81	162	324
18	25	324	450	625
Sum 79	139	1199	1855	3059

Substituting the appropriate values from the table into the formula, the calculation of r becomes

$$r = \frac{7(1855) - (79)(139)}{\sqrt{7(1199) - (79)^2}\,\sqrt{7(3059) - (139)^2}}$$

$$\approx .94.$$

We encourage the use of technology to ease the calculation pain of finding the slope and y-intercept of a least squares regression line as well as the correlation coefficient r. We have pointed out in Example 4 and just after the defining formula for r that the computer program LinReg accompanying this text will compute all of these values upon entry of the data and the touch of a key. But much more is available. Popular spreadsheets and even some word processors perform these same calculations. In addition, graphing calculators are preprogrammed to do the same. Figure 43 shows a calculator screen after the data of Example 3 have been entered and the LinReg key pressed. Notice that the slope is labeled a rather than m. These calculations agree (to two-decimal accuracy) with our earlier hand calculations.

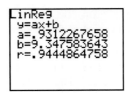

FIGURE 43
*Finding a Best-Fit Line
Electronically*

We conclude with Example 5, which answers the question posed in the graph at the outset of this chapter.

EXAMPLE 5
Modeling Municipal
Waste Generated per
Person Since 1960

Use a least squares regression line to model the following data, and use this model to predict the number of pounds of waste each person will generate per day in the year 2003.

Year	Pounds of waste per person
1960	2.66
1970	3.27
1980	3.65
1985	3.77
1988	4.12
1990	4.30

SOURCE: U.S. Bureau of the Census: *Statistical Abstract of the United States*, 1994.

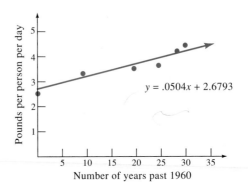

FIGURE 44
Waste Generated

Solution   If we let $x = 0$ represent the year 1960, then the data expressed as ordered pairs take on this appearance: $(0, 2.66)$, $(10, 3.27)$, $(20, 3.65)$, $(25, 3.77)$, $(28, 4.12)$, and $(30, 4.30)$. The least squares regression line for these points is $y = .054x + 2.6793$, and the correlation coefficient is $r \approx 0.986$. (See Figure 44.) (The values of m, b, and r were found by using a graphing calculator.)

To predict the amount of waste generated per person in the year 2003, first note that the year 2003 is represented by $x = 2003 - 1960 = 43$, and then substitute this value into the regression equation:

$$y = .054(43) + 2.6793 \approx 4.847 \text{ pounds.}$$

According to our model, almost 5 pounds of waste per person per day will be generated in 2003. ▬

We encourage the use of technology to find the slope, y-intercept, and correlation coefficients of the regression lines asked for in the exercises of this section.

EXERCISES   1.5

In Exercises 1 through 8, do the following:
(a) Plot the scatter diagram.
(b) Find the least squares regression line by setting up and solving the system of two equations and two unknowns that give m and b.
(c) Graph the line on the same axes as the scatter diagram. Pay particular attention to Problems 1–4; draw a conclusion about the least squares fit of data containing only two points.

1. $(1, 2), (3, 7)$ **2.** $(-1, 2), (6, 5)$

3. $(-1, 3), (2, -1)$ **4.** $(3, 5), (4, 8)$

5. $(1, 1), (2, 4), (3, 8)$

6. $(-1, 2), (1, 0), (5, -5)$

7. $(1, 1), (2, 3), (3, 6), (4, 5)$

8. $(0, 0), (1, -1), (2, -3), (5, -6)$

In Exercises 9 through 14, do the following:
(a) Plot the scatter diagram.
(b) Find and plot the least squares regression line.
(c) Find the correlation coefficient.

9. $(2, 6), (1, 3), (3, 9)$

10. $(2, 6), (4, 4), (5, 3)$

11. $(3, 8), (5, 3), (6, 7), (8, 2), (9, 4)$

12. $(1, 1), (4, 1), (2, 3), (6, 2), (5, 6), (7, 9)$

13. $(0, 4), (5, 0), (9, 5), (3, 7), (1, 1)$

14. $(3, 10), (4, 2), (6, 8), (7, 3), (8, 5)$

In Exercises 15 through 20, determine whether you think there would be a rather high or rather low correlation between the variables. For each exercise, estimate what you think the correlation coefficient would be.

15. The number of automobile accidents and the number of automobiles on the highways.

16. The heights and the ages of elementary school–age children.

17. The heights of person having bank accounts and the amount of money in their bank accounts.

18. The I.Q. of college students and their grade-point averages.

19. The ages of cars and their monetary value.

20. Student enrollment in universities and the number of football games won at those universities in 1990.

21. *Pollutants*   An experimental study of the environmental impact on black bass fish in large lakes was conducted. The number of fish (in thousands) within a $\frac{1}{2}$ mile of the source of pollution was estimated when various amounts of a particular pollutant were introduced into the water. The following data were collected:

Number of tons of pollutant	20	30	40	50	60	70
Number of fish	40	33	30	26	20	16

(a) Plot the scatter diagram, and find the least squares regression line for these data, using tons of pollutants as the independent variable x.

(b) Find the correlation coefficient.

(c) According to the linear model found in Exercise 21(a), how many fewer fish can be expected for each ton of the pollutant introduced into the water?

(d) Use the line found in 21(a) to estimate the black bass population if 65 tons of the pollutant were introduced into the lake.

(e) If the black bass population is estimated at 35,000, how many tons of the pollutant can be assumed to have been put into the lake?

22. *Estimating Height*   Anthropologists measured the length of the bone from the elbow to the shoulder on 10 women found in the burial grounds of an ancient civilization, and then measured their respective heights (all in centimeters). The results were as follows:

Bone length (cm)	Height (cm)
44	194.7
42.3	190
41.2	187
41.8	188.6
40.2	184.1
43.6	192.8
40.6	185.3
38.1	177.2
42.5	190
45	197.1

(a) Plot the scatter diagram for these data, using bone length as the independent variable x.

(b) Find the least squares regression line and the correlation coefficient.

23. *Reading Readiness* A reading-readiness test was administered to a random sample of ten first-grade students and compared with their I.Q. scores. The results were as follows:

Person	x = I.Q. score	y = Reading readiness score
A	102	45
B	110	50
C	102	48
D	98	38
E	100	40
F	101	45
G	96	42
H	100	45
I	104	39
J	94	38

(a) Plot the scatter diagram for these data.
(b) Find the least squares regression line, and plot it on the scatter diagram.
(c) Find the correlation coefficient.

24. *Aptitude Versus Performance* An accounting firm gave aptitude tests to all new employees during a given year and a year later compared these test results to the employees' performance ratings. The results for 20 randomly selected employees are as follows:

x = aptitude test score; y = performance rating

x	y	x	y	x	y	x	y
80	85	60	70	72	80	92	97
86	71	83	74	80	91	91	95
87	90	93	85	79	82	83	90
84	75	82	85	81	90	89	86
91	89	87	90	82	91	85	80

(a) Plot the scatter diagram for these data.
(b) Find the least squares regression line and the correlation coefficient.
(c) From your findings, do you think that the firm can relatively accurately predict job ratings from the aptitude test scores?

25. *Lumber in a Tree* Foresters have developed a rule to predict how many board feet of lumber are in a tree. To predict this, they cut several trees with known diameters and then measure the number of board feet of lumber in each tree. The following is a sample of their findings:

x = diameter (in inches):
17 19 21 23 25 28 32 38 39 41

y = volume (hundreds of board feet):
19 25 33 57 71 113 123 252 260 293

(a) Find the least squares regression line and the correlation coefficient.

(b) According to the linear model found in Exercise 25(a), what is the increase in board feet per inch of increase in diameter?
(c) How many board feet of lumber would be produced from a tree that had a diameter of 45 inches?
(d) If a tree produced 25,000 board feet of lumber, what could we assume about its diameter?

26. *Wages Versus Sales Volume* A department store chain collected data from each of its stores for a month. The data on total wages paid to sales personnel and on sales volume are shown below:

Store	Sales wages (thousands)	Sales volume (hundreds of thousands)
A	30.2	3.7
B	52.1	4.8
C	23.6	2.7
D	38.2	3.5
E	45.8	4.7
F	59.2	5.1
G	41.7	3.9
H	62.1	5.4

(a) Plot the scatter diagram, using wages as the independent variable x.
(b) Find the least squares regression line and the correlation coefficient.
(c) Using the regression line found in Exercise 26(b), determine how many thousands of dollars will be spent in sales wages to support a sales volume of $700,000.

27. *Earnings* Average hourly earnings in the U.S. retail trade industry (in current dollars and constant dollars) are shown in the table:*

Year	1980	1985	1990	1992	1993
Current dollars	4.88	5.94	6.75	7.13	7.29
Constant dollars	5.70	5.39	5.07	5.00	4.97

(a) Find the least squares regression line that approximates the average hourly earnings in both current dollars and constant dollars for this industry. Find the correlation coefficient in both cases.
(b) Plot both of the regression lines you found in Exercise 27(a) on the same set of axes.
(c) Use the regression lines found in Exercise 27(a) to estimate the difference in current dollar and constant dollar average hourly earnings in the year 2005.

*SOURCE: U. S. Bureau of the Census: *Statistical Abstract of the United States,* 1994.

28. *Cricket Chirps Versus Temperature* Biologists have studied the relationship between the temperature and the number of chirps made by a cricket. Generally, the warmer the weather is, the more often the cricket chirps. Here are some experimental data collected:

Temperature (in degrees Fahrenheit):
45 52 55 58 60 62 65 68 70 75 82 83

Number of seconds to count 50 chirps:
94 59 42 36 32 32 26 24 21 20 17 16

(a) Plot the scatter diagram.
(b) Find the least squares regression line and the correlation coefficient.
(c) If the temperature is 0 degrees Fahrenheit, how many seconds will it take to count 50 chirps, according to the linear model of Exercise 28(b)? Do you believe it?
(d) Use the linear model to estimate the temperature if it takes 1 second to count 50 chirps.

29. *Advertising* Dollars (in millions) spent advertising drugs and remedies in magazines is shown in the following table for select years.*

Year	1980	1985	1989	1990	1992
Dollars spent	79	135	135	163	279

(a) Find the least squares regression line that will estimate the number of dollars spent on magazine advertising for drugs and remedies. Also find the correlation coefficient.
(b) Assuming this trend continues, how many dollars will be spent for this type of advertising in the year 2005?

30. *Lead Emissions* The following table shows lead emissions in thousands of metric tons for selected years in the United States:*

Year	Lead (thousands of metric tons)	Year	Lead (thousands of metric tons)
1980	68.0	1986	6.6
1981	53.4	1987	6.2
1982	52.3	1988	5.9
1983	44.7	1989	5.5
1984	38.3	1990	5.1
1985	18.3	1991	4.6
		1992	4.7

(The abrupt drops from 1984 to 1986 reflect the unleaded gasoline phase-in, when leaded gasoline was no longer being sold at the pump.)

(a) Find the least squares regression line that will estimate the lead emissions over the period shown in the table. Also find the correlation coefficient.
(b) Find the least squares regression line that will estimate the lead emissions from 1986 through 1992. Also find the correlation coefficient.
(c) Using the lines found in Exercise 30(a) and (b), determine the year all lead emissions will be eliminated. Which do you judge to be the most accurate calculation?

31. *Recycling* The following table shows U.S. Census Bureau data* regarding the gross waste generated per year (in millions of tons) in the United States and the amount recovered (e.g., recycled, recovered for energy conversion):

Year	1960	1970	1980	1985	1987	1990
Gross waste generated	87.8	121.9	151.5	164.4	178.1	195.7
Materials recovered	5.9	8.6	14.5	16.4	20.1	33.4

(a) Find the least squares regression lines that approximate the waste generated and the amount recovered. Give the correlation coefficient in each case.
(b) At the rates shown, will we ever recover even 50% of the waste generated?
(c) Use the functions found in Exercise 31(a) to predict the waste generated and recovered in the year 2005.

Problems for Exploration

1. Find the height and shoe size of either male friends or female friends. Plot a scatter diagram of these data, with shoe size on the *x*-axis and height on the *y*-axis. Find the least squares regression line and the correlation coefficient. Do you think that your data indicate that the height of a person can be predicted with any degree of accuracy by his or her shoe size? Explain why your correlation coefficient might be considerably

*SOURCE: U.S. Bureau of the Census: *Statistical Abstract of the United States,* 1994.

different from that of another classmate doing the same experiment.

2. Measure the height and arm length (from shoulder to fingertip) of ten friends. Then plot a scatter diagram of these data, with length of arm on the x-axis and height on the y-axis. Find the equation of the regression line and the correlation coefficient. Do you think that your data indicate that a person's height can be predicted with any degree of accuracy by the length of his or her arm? Explain why your correlation coefficient might be considerably different from that of another classmate doing the same experiment.

3. Examine the records of your city for the past several years to find out the annual population. Does the trend appear linear? If so, use the data to obtain a least squares regression line, and use this line to predict the population ten years from now. Compare your figures with predictions made by the city. Are the city planners taking into account factors that your linear model does not?

4. Find a source that will show how temperatures cool as the height above the earth increases. Decide whether these data indicate a linear correlation. If so, find the least squares regression line, and use it to estimate temperatures for heights beyond your data.

CHAPTER 1 REFLECTIONS*

CONCEPTS

Descartes' *rectangular coordinate system* provides a convenient graphic scheme for representing ordered pairs of real numbers. The *graph* of every linear function is a straight line; conversely, every nonvertical straight line represents a linear function, the equation of which can be given by the *point-slope form,* $y - y_1 = m(x - x_1)$, or the *slope-intercept form,* $y = mx + b$. The *general form,* $Ax + By = C$, may be used to represent any line. A pair of lines in the plane may have one point in common, no points in common, or all points in common. To determine which is the case, either the *substitution* or *elimination* methods may be used, and either of these methods may also be used to find the point or points of intersection, if any. *Linear functions* are useful in modeling costs, revenue, depreciation, ecology concerns, and many other real-life applications. *Least squares regression lines* model data are collected in various fields and may be used for prediction purposes. The degree of the "goodness" of fit between the data and a linear regression line is given by the correlation coefficient.

COMPUTATION/CALCULATION

1. Calculate the x-intercept and the y-intercept for the line $3x - 5y = 30$.
2. Find the marginal cost and the fixed cost for the function $C(x) = 7x + 430$.
3. Determine the equations of the horizontal line and vertical line that pass through the point $(-3, 8)$.

4. Calculate the slope of the line $5x + 2y = 13$.
5. If (x, y) is a point on the line with equation $y = 3x + 4$, calculate the y-coordinate of the point on this line having x-coordinate $x + 2$.
6. Compute $f(-5)$ for the function $f(x) = 1 + \dfrac{2x}{3}$.
7. Write the equation of the line that passes through the origin and is parallel to the line $2y = 5x + 10$.
8. Use the substitution method to find the point or points, if any, that the lines represented by $2x + y = 5$ and $x + y = 7$ have in common.
9. Use the elimination method to find the point or points, if any, that the lines represented by $3x + 6y = 13$ and $5x - 6y = 11$ have in common.

Solve each of the following systems of equations.

10. $5x + 3y = 9$
    $10x + 6y = 10$

11. $5x + 3y = 9$
    $10x + 6y = 18$

12. Find the least squares regression line for the points $(2, 8)$, $(3, 6)$, $(4, 5)$, and $(5, 9)$. Would the correlation coefficient be nearer 1, -1, or 0?

CONNECTIONS

A manufacturing company has a total cost function given by $C(x) = 7.9x + 2520$.

13. Find the total cost of 60 items.
14. Find the average cost per item for 60 items.
15. Find the additional cost of producing the 61st item.

---

*The Chapter Reflections include a summary and various types of exercises for a thorough review of chapter topics. The Preface gives more information about the exercise subsections.

A discount store had sales of $1.8 million in 1990 and $2.7 million in 1992. Assume that the store's sales trend is linear.

16. Find an equation that will estimate sales in future years.

17. Predict the year that sales could be expected to reach $20 million.

An appliance company has revenue and cost functions given by $R(x) = 65x$ and $C(x) = 25x + 4000$.

18. Find the profit function.

19. Find the break-even point.

20. Graph the revenue, cost, and profit functions on the same set of axes.

COMMUNICATION

21. Using slopes and y-intercepts for systems of two linear equations in a plane, explain how to differentiate among systems that have one point, no points, or all points in common.

22. Identify and discuss all the similarities and differences you can find for the functions $f(x) = 7.2$, $y = -2$, and $f(x) = 39$.

23. Describe the limitations and potential errors that might occur when using linear functions as models of real-world situations.

24. Describe all of the advantages you can imagine for using linear functions to model real-world situations.

COOPERATIVE LEARNING

25. Create a system of two linear equations in two variables that would be very conveniently solved by using the substitution method. Solve the system of equations, and explain why this method is desirable for your example. Then do the same for the elimination method. Finally, write a summary comparing the advantages and disadvantages of these two methods.

CRITICAL THINKING

26. Compare and contrast the representation of vertical, horizontal, and other lines, using the point-slope, the slope-intercept, and the standard forms for a linear equation. Summarize your results or discoveries.

COMPUTER/CALCULATOR

27. Use the computer program LinReg from the disk MasterIt (or another computer program you are familiar with) to find the following for the points (2, 5), (4, 8), (5, 14), (6, 15), (3, 6), (4, 9), (7, 17), (9, 21), (5, 13), (12, 25), (15, 33), and (9, 19):

   (a) The slope m and the y-intercept b of the least squares regression line.

   (b) The correlation coefficient r.

   (c) The scatter diagram and graph of the regression line.

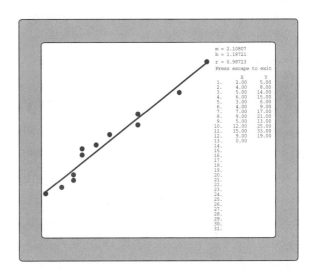

As each successive data point is entered (from the second point onward), new calculations of m, b, r, the scatter diagram, and the regression line appear on the screen. You may delete a data point (or points) at any stage and observe the effect on all of the calculations. The final screen of LinReg will appear as shown in the diagram.

28. Use a linear regression program designed for a graphics calculator to perform the calculations, and visually show the scatter plot requested in Exercise 27.

CHAPTER 1 SAMPLE TEST ITEMS

1. Plot the points $(2, 7)$, $(3, -2)$, $(-5, 0)$, $(-4, -1)$, and $(-2, 6)$ on the same set of axes in the rectangular coordinate system.

2. Graph the functions $y = x - 2$, $y = 2x - 4$, and $y = 3x - 6$ on the same set of axes.

3. Write the equation of a linear function for which the change in y per unit change in x is 3.7 and the y-intercept is -2.8.

4. Write a linear cost function that will represent the total cost of making x items if the total cost is $650 for 50 items and is $714 for 70 items.

5. A small town with a population of 2300 has been growing at a rate of about 250 per year. If this trend continues, find a linear function that will give the approximate population x years from now.

6. Find the equation of the vertical line that passes through the point $(-2, 7)$.

7. Find the slope, the x-intercept, and the y-intercept for the line having equation $x - 2y = 9$.

8. Write an equation of a line that has x-intercept -2 and y-intercept 6.4.

9. A restaurant sold 78 luncheon specials when the price was $4.75. When they raised the price to $5.50, they sold 63 specials. Find a linear function that will predict the number of specials sold for a given price.

10. A rancher buys a tractor for $5600 and plans to depreciate it over 10 years, at which time the value will be $2100. Find a linear function that gives the value of the tractor as a function of the years from the date of purchase.

11. By examining the slopes and y-intercepts, decide whether the equations $2x + 5y = 7$ and $2x - 5y = 7$ determine lines having one point, no points, or all points in common.

12. Solve the system $x + y = 5$, and $2x - y = 12$ by using the substitution method and then the elimination method.

13. Solve the system $.2x + .8y = 3$, and $x + 4y = 15$ by any appropriate method.

14. Find the profit function and the break-even point, given $R(x) = 75x$ and $C(x) = 45x + 1200$.

15. A graphing calculator company has fixed costs of $40,000 per month and direct sales costs of $73 per graphing calculator. If the calculators sell for an average of $98, what is the break-even point?

The Vestmore Company has a mutual fund that returned 8%, 9%, 12%, and 14% over the first four years.

16. Plot the scatter diagram for these four years.

17. Find the least squares regression line for these data.

18. Find the correlation coefficient.

19. Use the least squares regression line to predict the return on investments for the fifth year.

20. Predict how many years after the fund started that the return will be 20%.

Career Uses of Mathematics

As a systems engineer for IBM, I spend a great deal of time with my customers in the area of application development. This process involves designing and writing computer applications in one of the various computer programming languages. A large part of the programming function involves a heavy use of mathematics—from simple linear functions to much more complex computations.

I am frequently called upon to evaluate the performance of computers and the applications that run on them. This involves timing the execution of the application and then presenting the results in an easy-to-read format. I use mathematics in this function to calculate the results in a graphical form using the principles of mean, median, and standard deviation.

To be a successful computer analyst, one has to have a strong background in mathematical principles and be able to use them correctly to achieve the desired results.

Lindsay M. Hart, IBM

2

Systems of Linear Equations

A retailer has warehouses in Lima and Canton, from which two stores—one in Tiffin and one in Danville—place orders for bicycles. The store in Tiffin orders 38, and the store in Danville orders 46. Each warehouse has enough bicycles on hand to supply all that have been ordered. Economic conditions dictate that twice as many bicycles be sent from Lima to Danville as from Canton to Tiffin. Find a system of equations whose solutions will give possible shipping schedules from the warehouses to the two stores. (See Example 4, Section 2.1 and Example 5, Section 2.3.)

CHAPTER PREVIEW

Sometimes more than one equation is needed to create a mathematical model. In such cases, the model consists of a "system" of equations. Whenever possible, linear equations are used in such systems because of their simplicity. In this chapter, we first learn how to create mathematical models that involve two or more linear equations, each of which may involve several unknowns such as x, y, or z. Our attention then turns to finding the solutions of such systems. We first write the system in matrix form and then "pivot" on certain elements in the matrix through the use of the Gauss-Jordan elimination method. This process provides a method for recognizing when a given system has one or many solutions—or none at all. The introduction of the pivoting process in this chapter paves the way for its continued use in Chapters 3 and 4.

SECTION 2.1 Linear Systems as Mathematical Models

Up to this point, linear equations involving only two variables have been considered, and their straight-line graphs have provided visual help for many of the concepts to which they have been applied. The concept of a linear equation may, however, be generalized to any number of variables by following the pattern set forth by those of two variables, $ax + by = c$: The left side of the equation consists of the sum of terms, each of which is a constant multiplied by a variable, and the right side is a constant. For example, $3x + 2y - 6z = 7$ is a linear equation in the three variables x, y, and z. The equation $w - 4x + 5y - 7z = 9$ is a linear equation in the four variables w, x, y, and z. In general, a linear equation has the form

$$a_1x_1 + a_2x_2 + a_3x_3 + \cdots + a_nx_n = c$$

where $a_1, a_2, a_3, \ldots, a_n$ and c are constants, and $x_1, x_2, x_3, \ldots, x_n$ are variables. Linear equations in three variables have graphs that are planes in a three-dimensional Cartesian coordinate system, but those with four or more variables are too complex to visualize. Even so, algebraic methods exist for finding possible solutions for systems of linear equations regardless of the number of variables involved. One rather straightforward algebraic method is discussed in Section 2.2. Before concentrating on the solutions, however, we need to become proficient at setting up systems that model various situations. Actually, this is just a process of translating English sentences into equations, using mathematical symbols. Sometimes, this can be done with little effort, while at other times, more organization of the facts may be necessary. A method of organization is set forth in the next few examples that will serve us in working with systems of linear equations, as well as with systems of linear inequalities, which we will encounter in a later chapter.

EXAMPLE 1
A Mathematical Model of a Production Process

A firm produces bargain and deluxe TV sets by buying the components, assembling them, and then testing the sets before shipping. The bargain set requires 3 hours to assemble and 1 hour to test. The deluxe set requires 4 hours to assemble and 2 hours to test. The firm has enough employees so that 390 work hours are available for assembly, and 170 work hours are available for testing each week. Use a system of linear equations to model the number of each type of TV set that the company can produce each week while using all of their available labor.

Solution   Typically, the last sentence of the problem asks *how many* of each type of product (in this case, TV sets) should be produced. To start organizing the data presented, put the products "Bargain set" and "Deluxe set" as headings for two columns and at the bottom an *x* and a *y* to represent the number of each that will be produced (see Figure 1(a)). Then list the operations necessary to produce these products. In this case, there are two—assembly and testing—which are written in the left column as row headings (see Figure 1(b)). Now reread the problem and fill in the appropriate number for each box, including the numbers that limit each operation. The completed table should look like the one in Figure 2.

Bargain set	Deluxe set
x	*y*

	Bargain set	Deluxe set
Assembly		
Testing		
	x	*y*

       (a)   (b)

FIGURE 1

	Bargain set	Deluxe set	Limits on operations
Assembly	3	4	390
Testing	1	2	170
Number of each type of set produced	*x*	*y*	

FIGURE 2

     The equations modeling our production process can now be *read as rows* from Figure 2: $3x + 4y = 390$, $x + 2y = 170$. The reason is that each bargain set requires 3 hours of assembly time, so that to make *x* units, $3x$ work hours are needed. Similarly, each deluxe set requires 4 hours of assembly time, so that to make *y* units, $4y$ work hours are needed. Therefore, a total of $3x + 4y$ work hours are needed to produce *x* bargain sets and *y* deluxe sets, and we want that to be exactly the available 390 work hours. The reasoning for the second equation, representing testing work hours, is the same. The resulting system of equations is then

$$3x + 4y = 390$$
$$x + 2y = 170,$$

where *x* and *y* are nonnegative. ◼

EXAMPLE 2
Modeling a Mixture Problem

A dietitian is to combine a total of 5 servings of chicken soup, tuna, and green beans, among other ingredients, in making a casserole. Each serving of soup has 15 calories and 1 gram of protein, each serving of tuna has 160 calories and 12 grams of protein, and each serving of green beans has 20 calories and 1 gram of protein. If these three foods are to furnish 370 calories and 27 grams of protein in the casserole, how many servings of each should be used?

Solution This time the foods soup, tuna, and beans head the columns as "products," and the restricted items going across the rows are the total number of servings, the number of calories, and the number of grams of protein. See Figure 3. Because the total number of servings of these three foods is to be 5, it follows that $x + y + z = 5$. Notice how this fact is reflected in the first row of Figure 3. A reading of the calorie and protein requirements shows how the next two rows of Figure 3 are created.

	Soup	Tuna	Beans	Restrictions
Total Number of Servings	1	1	1	5
Calories per Serving	15	160	20	370
Grams Protein per Serving	1	12	1	27
Number Servings of Each Food	x	y	z	

FIGURE 3

Again, each row with a restriction represents an equation to be satisfied:

$$x + y + z = 5$$
$$15x + 160y + 20z = 370$$
$$x + 12y + z = 27. \quad \blacksquare$$

EXAMPLE 3
Modeling a Production
Process

Meek's Woodshop produces dollhouses and magazine racks. Dollhouses require 20 minutes of sawing time, 30 minutes' assembly time, and 10 minutes' sanding time. Magazine racks require 18 minutes of sawing time, 15 minutes' assembly time, and 12 minutes' sanding time. Due to limitations on machinery and work hours, 4 hours per day are devoted to sawing, 5 hours to assembly, and 3 hours to sanding. Find linear equations that will model the number of dollhouses and magazine racks that can be produced each day while fully utilizing the available labor.

Solution There are two products—dollhouses and magazine racks—and three operations—sawing, assembly, and sanding— each restricted by time. As usual, put the products on top as column headers and the restricted operations down the left side, to name the rows. Keeping time in minutes, fill in the boxes by reading the problem.

	Dollhouses	Magazine racks	Limitations (in minutes)
Sawing time	20	18	4(60) = 240
Assembly time	30	15	5(60) = 300
Sanding time	10	12	3(60) = 180
Number produced	x	y	

FIGURE 4

The rows in Figure 4 give the desired equations:

$$20x + 18y = 240$$
$$30x + 15y = 300$$
$$10x + 12y = 180. \quad \blacksquare$$

The next example contains a slightly different kind of restriction, but the same format as before may be used to obtain the equations.

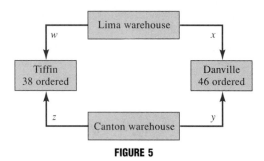

FIGURE 5

EXAMPLE 4
Modeling a Shipping
Problem

The problem stated at the outset of this chapter described shipping bicycles from two warehouses, one in Lima and one in Canton, to a store in Tiffin, which ordered 38, and a store in Danville, which ordered 46. Each warehouse has enough to supply all that have been ordered, but twice as many are to be shipped from Lima to Danville as from Canton to Tiffin. What equations will describe the possible shipping schedules?

Solution Let w, x, y, and z be the number of bicycles shipped from the respective warehouses to the stores, as shown in Figure 5. Because 38 are to be shipped to Tiffin,

$$w + z = 38,$$

and because 46 are to be shipped to Danville,

$$x + y = 46.$$

The requirement that twice as many are to be shipped from Lima to Danville as from Canton to Tiffin means that the x-variable is to be twice the z-variable; hence the equation between x and z is

$$x = 2z \quad \text{or} \quad x - 2z = 0.$$

This is why the third row in Figure 6 is as shown.

	w	x	y	z	
To Tiffin	1			1	38
To Danville		1	1		46
Relate x, z		1		-2	0

FIGURE 6

This time, the equations can be found rather easily without the aid of a table, even though one is shown. The complete system is

$$
\begin{aligned}
w \quad\quad + \quad z &= 38 \\
x + y \quad\quad &= 46 \\
x \quad\quad - 2z &= 0. \quad \blacksquare
\end{aligned}
$$

Once again, the tables suggested will always organize the data for any problem that can be modeled by a linear system of equations, so that the resulting rows give the desired equations. You may find it useful to describe some of the restrictions while the others can be put directly into equation form.

The configuration of numbers obtained in Figure 6, from which the equations were written, is an array of numbers consisting of three **rows** (the rows go horizontally) and five **columns** (the columns go vertically). If we insert a zero where each blank space appears and enclose the array with brackets, we obtain the table shown below. This table with brackets enclosing its entries is called a **matrix.**

$$\text{Row 2} \rightarrow \begin{bmatrix} 1 & 0 & 0 & 1 & 38 \\ 0 & 1 & 1 & 0 & 46 \\ 0 & 1 & 0 & -2 & 0 \end{bmatrix}$$
$$\uparrow$$
$$\text{Column 4}$$

Among the many applications of matrices is the convenient, useful, and natural representation of *any* system of linear equations, as suggested above: each row represents an equation, and each column, except the last, is used to keep the coefficients of a particular variable aligned. The last column consists of the constants to the right of the *equal* signs. Often, a vertical line is used in place of the equal signs to remind us that the last column is indeed the column of constants to the right of the equal signs.

$$
\begin{aligned}
w \quad\;\; + \;\; z &= 38 \\
x + y \qquad &= 46 \\
x \qquad - 2z &= 0
\end{aligned}
\quad \rightarrow \quad
\begin{array}{c}
\begin{matrix} w & x & y & z & \end{matrix} \\
\left[\begin{array}{cccc|c} 1 & 0 & 0 & 1 & 38 \\ 0 & 1 & 1 & 0 & 46 \\ 0 & 1 & 0 & -2 & 0 \end{array} \right]
\end{array}
$$

The matrix to the left of the vertical line is known as the **coefficient matrix** of the original system, for evident reasons: The coefficient matrix is the matrix consisting of the coefficients of the variables. The entire matrix is known as the **augmented matrix** of the original system. It is the coefficient matrix, augmented by the column of constants. Notice that *all* of the coefficients and constants, even those that are zero, are displayed in the augmented matrix.

EXAMPLE 5
Displaying the
Coefficient and
Augmented Matrices

For the linear system of equations

$$
\begin{aligned}
x + 3y + z &= 10 \\
-x \quad\;\; + z &= 5 \\
3x + \;\; y \quad &= 0,
\end{aligned}
$$

the coefficient matrix is

$$\begin{bmatrix} 1 & 3 & 1 \\ -1 & 0 & 1 \\ 3 & 1 & 0 \end{bmatrix},$$

and the augmented matrix is

$$\left[\begin{array}{ccc|c} 1 & 3 & 1 & 10 \\ -1 & 0 & 1 & 5 \\ 3 & 1 & 0 & 0 \end{array} \right]. \quad \blacksquare$$

Each entry (element, number) of a matrix is located where some row and some column intersect, and this leads naturally to a row-column address. In the augmented matrix of Example 5, for instance, the entry -1 is located in the second-row, first-column position; the entry 10 is in the first-row, fourth-column position; and the entry 0 is in the second-row, second-column position. In mathematics, it is customary to give the row number first and the column number second when stating the address of a given number in a matrix.

First row $\rightarrow$
$$\begin{bmatrix} 1 & 3 & 1 & \boxed{10} \\ -1 & 0 & 1 & 5 \\ 3 & 1 & 0 & 0 \end{bmatrix}$$

The number 10 is located in the first-row, fourth-column position.

$\uparrow$
Fourth column

General Notation

A general element in a matrix is denoted by a_{ij} where i is the row number and j is the column number in which the element is located. This may be shortened to say that a_{ij} is the element in the ith row and jth column in the matrix.

EXERCISES   2.1

In Exercises 1 through 8, write both the coefficient matrix and the augmented matrix for each system.

1. $x + 3y = 5$
$2x - y = -4$

2. $-3x + y = 0$
$x + 5y = 6$

3. $x + 3y = 0$
$2x + y - z = 0$
$x - y + z = 0$

4. $2x + 2y - z = 3$
$x - y = 0$
$5x + 3y + z = 5$

5. $w + 2x + 3y + z = 5$
$w + x - y = 3$
$3w + 5x + 2z = -2$

6. $x_1 + x_6 = 2$
$x_2 + x_3 - x_6 = 5$
$x_1 + x_3 + x_4 = -2$
$x_1 + x_2 + x_5 = 7$

7. $5x + 3y + 2z = 0$
$x - 2y = 0$
$3x + y + z = 1$
$x + 4z = 2$

8. $2a + 4b - c = 8$
$a + 3b + c = 6$
$a + b + c = 3$

In Exercises 9 through 14, find the system of equations that model the problem given by first making a table like the one suggested in the examples. Do not attempt to solve.

9. *Investing* Danielle bought two stocks, Datafix selling for $30 per share and Rocktite selling for $20 per share, for a total of $4000. The dividend from Datafix is $2 per share, and the dividend from Rocktite is $1 per share, each year. Danielle expects to receive a total of $220 in dividends from these two stocks during a given year. How many shares of each stock did she buy?

10. *Training Police Officers* A particular police department employs two grades of police personnel: rookies and sergeants. A person at the grade of rookie is to spend 20 hours training and 20 hours on patrol duty each week. A person at the grade of sergeant is to spend five hours training and 30 hours on patrol duty each week. The training center can effectively handle 240 person-hours each week, while the department needs at least 440 person-hours each week for patrol duty for these two grades of personnel. How many persons at each of these grades does the department have, assuming that the training center operates at full capacity and that the minimum requirements for patrol duty are met?

11. *Meeting a Pollution Allowance* An electrical plant uses coal and gas to generate electricity. Coal produces 3 megawatt-hours of electricity per ton, while gas produces 4.50 megawatt-hours per thousand cubic feet. Each ton of coal produces 60 pounds of sulfur dioxide per hour, and each thousand cubic feet of gas produces only 1 pound of sulfur dioxide per hour. The plant wants to produce electricity at the rate of 183 megawatt-hours while not exceeding the allowable sulfur dioxide pollution rate of 990 pounds per hour. How many tons of coal and how many thousand cubic feet of gas should the plant burn each hour to produce the desired amount of electricity while exactly meeting the maximum pollution allowance?

12. *Advertising* A firm wants to start an advertising campaign in newspapers, on radio, and on television to run for one month. It wants to run a total of 34 ads in these three media. Each time a newspaper ad is run, it costs $100; a radio ad, $80; and a television ad, $400. The company has budgeted $4840 for the entire campaign. Each newspaper ad has an effective rating of 8 points; each radio ad, 3 points; and each television ad, 12 points. The company wants a total of 206 effective rating points from its campaign. Assuming that the entire ad budget is spent and that the effective rating points are met, how many times should each ad be run?

13. *Manufacturing* A ballpoint-pen maker produces pens made of wood, silver, and gold. A wood pen requires 1 minute in a grinder and 2 minutes in a bonder, a silver pen requires $\frac{1}{2}$ minute in a grinder and 2 minutes in a bonder, and a gold pen requires 3 minutes in a grinder and $2\frac{1}{2}$ minutes in a bonder. If there are 200 hours of grinder time and 160 hours of bonder time available each week, how many pens of each type can be produced, assuming that all grinder and bonder time available is used?

14. *Manufacturing* A firm makes standard, deluxe, and super deluxe model microwave ovens. The assembly times for these ovens are, respectively, 20 minutes, 30 minutes, and 40 minutes. The painting times for these ovens are, respectively, 10 minutes, 10 minutes, and 12 minutes. The total production of deluxe and super deluxe models is to equal that of standard models. If the company has 100 hours of assembly time and 80 hours of painting time each day and these times are fully utilized, how many of each type of oven can be made on a daily basis?

For each of the remaining exercises, find a set of equations that models the given problem.

15. *Hiring and Training Workers* A particular department in a factory hires skilled workers, semiskilled workers, and supervisors. Each skilled worker is paid $12 per hour; each semiskilled worker is paid $9 per hour, and each supervisor is paid $15 per hour. The department is allowed an hourly payroll of $1560 for these three types of workers. The department requires that each

worker spend some time in training and safety schooling each week: skilled workers, 3 hours; semiskilled workers, 5 hours; and supervisors, 1 hour. The training center can handle a maximum of 588 person-hours each week, and the department needs 140 skilled and semiskilled workers to meet production schedules. Assuming that the allowable payroll is met, the training center is fully utilized, and production needs are met, how many of each type of worker does the department need?

16. *Combining Foods* A dietitian wants to combine spinach and lettuce to make a 5-pound salad mixture. The caloric content per pound of spinach and lettuce is, respectively, 80 and 30 calories. The units of vitamin A per pound for each are, respectively, 40 and 20 units. If the dietitian wants the total caloric content to be 250 and the total vitamin A units to be 140, how many pounds of each type of food should be used?

17. *Making Dog Food* A dog-food factory makes dog food from four ingredients: A, B, C, and D. They have, respectively, 2, 1, 3, and 1 units of vitamin A content and, respectively, 4, 6, 2, and 2 units of protein per pound. The company mixes the dog food in 200-pound batches, with 300 units of vitamin A per batch and 600 units of protein per batch. How many pounds of each of the four ingredients—A, B, C, and D—should be used to exactly meet the vitamin and protein requirements per batch?

18. *Manufacturing* A kitchen appliance manufacturer makes can openers and dough cutters and uses two machines: a press and a riveter. Each can opener requires 0.2 minutes in the press and 0.4 minutes in the riveter. A dough cutter requires 0.5 minutes in the press and 0.3 minutes in the riveter. The press can be operated only 3 hours per day and the riveter only 2.5 hours per day. If these two machines are fully utilized, how many can openers and dough cutters can be produced each day?

19. *Exporting* In a particular developing country, agricultural planners hope to have exports of soybeans, wheat, corn, and barley that will net $10 million from the coming year's crops. They estimate that the net profit per acre of these four commodities will be $25, $50, $40, and $30, respectively. They also estimate that the labor needed per acre to plant, harvest, and transport these four commodities will be 5, 6, 8, and 3 persons, respectively. Their country has 50,000 workers available for these operations. Assuming that the estimates are correct and that all 50,000 workers are used, how many acres of each of these four commodities should be planted to meet their goal?

20. Woodworking Jim's woodshop produces trivets, cutting boards, and bases for pen sets. The only operations involved are sawing and sanding. A trivet requires 10 minutes' sawing time and 5 minutes' sanding time; cutting boards require 5 minutes' sawing time and 5 minutes' sanding time; and pen bases require 10 minutes' sawing time and 15 minutes' sanding time. The per-day limit on sawing is 4 hours and on sanding 5 hours. If the sawing and sanding times are fully utilized, how many of each of these three items can be made each day?

21. Hiring Workers Brenda's Drive-In is located in a college town and has rush-hour business during the 5 o'clock hour and again during the 10 o'clock hour. To run her business efficiently, Brenda hires three shifts of workers, one to work from 10:00 A.M. to 6:00 P.M., another from 5:00 to 11:00 P.M., and a third from 10:00 P.M. to 1:00 A.M. Brenda knows from experience that she needs ten workers during the first- and second-shift overlap and eight workers during the second- and third-shift overlap. Can Brenda staff her worker needs for each shift if the first and third shifts are to have the same number of workers?

22. Shipping A discount store paid $413.28 for a shipment of two kinds of candy bars: Nutty Delite and Chewmoor. The Nutty Delite bars are sold for 50 cents each and the Chewmoor bars for 60 cents each, for a total of $508.40. The Nutty Delite bars cost 42 cents each and the Chewmoor bars cost 52 cents each. How many of each type of bar were in the shipment?

23. Highway Budget A state highway department has $120 million to spend on highway construction in three districts: Alma, Bedford, and Cooksville. The cost per mile for construction in the Alma district is $2 million, in the Bedford district $500,000, and in the Cooksville district $1 million. For political reasons, they need to build twice as many miles in the Alma district as in the Bedford district. How many miles of highway can be built in each of the three districts while meeting all these restrictions and using all $120 million?

24. City Budget The city of Aurora gets a matching grant of $1 million (making a total of $2 million) to spend in these three categories: streets, sewers, and parks. The city council decides that the amount spent on parks should equal the total amount spent on streets and sewers and that twice as much should be spent on parks as on sewers. How much money is to be allotted to each of these categories?

25. Buying Tires A tire store manager is given $28,500 to buy 600 new tires for the store. Two types of tires are to be purchased: four-ply nylon costing $40 each and steel-belted radials costing $50 each. The manager wants to buy three times as many steel-belted radial tires as four-ply nylon tires. Assuming that all $28,500 is spent, find a system of linear equations that would give the number of each type of tire bought.

26. Advertising A company is to run ads in newspapers, on radio, and on television. It wants to run a total of 20 ads in these three media. Each time a newspaper ad is run, it costs $30; a radio ad, $60; and a television ad, $200. The company has budgeted a total of $2400 for these ads. The company wants to run five times as many newspaper ads as television ads. How many times should each ad run to meet these conditions and utilize all of the ad budget?

27. Mixing Gasohol A gasoline supplier has two large gasohol tanks, one containing 8% alcohol and the other containing 13% alcohol. If 2000 gallons of gasohol with 10% alcohol are needed, how many gallons should be taken from each tank to provide the proper mixture?

28. Mixing Nuts A store is filling an order for 60 pounds of a mixture of walnuts, pecans, and peanuts, which will sell for 90 cents per pound. The walnuts normally sell for $1 per pound, the pecans for $1.10 per pound, and the peanuts 60 cents per pound. The order stipulates that 60% of the nuts are to be walnuts or pecans. If the income from selling the mixture should be the same as that for selling the nuts separately, how many pounds of each type of nut should be used in filling the order?

29. Mixing Acids A chemistry department wants to make 2 liters of an 18% acid solution by mixing a 21% solution with a 14% solution. How many liters of each type of acid solution should be used to produce the 18% solution?

30. Election Outcome A polling firm predicts that 35% of the "yes" vote and 40% of the "no" vote in an upcoming bond election will come from urban areas. A week before the election, supporters of the bond issue estimate that 100,000 votes will be cast in the election, with 37,000 votes coming from urban areas. Assuming that these predictions and estimates are correct, how will the bond issue fare in the election?

Problems for Exploration

1. Make up a word problem for which the mathematical translation is the following system of equations:

$$x + z = 14$$
$$x - z = 8.$$

2. Make up a word problem for which the mathematical translation is the following system of equations:

$$x + y = 8$$
$$x - 2y = 0.$$

3. Make up a word problem for which the mathematical translation is the following system of equations:

$$x + y + z = 30$$
$$2x \quad\ \ - z = 0$$
$$x + y \quad\ \ = 12.$$

Linear Systems Having One or No Solutions

In Section 1.4 of Chapter 1, we saw how two different methods could be used to find points in common with a pair of lines or to determine that no such points existed. There, we focused on the geometry and the signals that indicated the classification of a solution rather than on viewing these as just mechanical solutions to systems of two equations and two unknowns. That set the stage for our present work. As it turns out, *every* system of linear equations, regardless of size, falls into one of the same classifications:

    I. There is exactly one solution.

    II. There is no solution.

    III. There are an infinite number of solutions.

> **HISTORICAL NOTE**
> **Carl Friedrich Gauss (1777–1855)   Wilhelm Jordan (1842–1899)**
>
> The German mathematician Carl Friedrich Gauss (Gous) is recognized as the greatest mathematician of modern times, the last "complete mathematician." He was connected in some way with nearly every aspect of mathematics during his career but held number theory in high esteem. Wilhelm Jordan (1842–1899) was a German geodesist who presented the elimination method as a way to solve particular surveying problems.

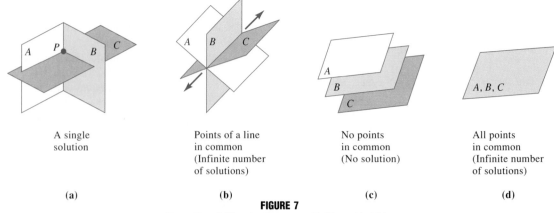

A single solution	Points of a line in common (Infinite number of solutions)	No points in common (No solution)	All points in common (Infinite number of solutions)
(a)	**(b)**	**(c)**	**(d)**

FIGURE 7
Some Possibilities for Systems with Three Variables

However, for larger systems of equations, it will not be as easy to tell by inspection which classification prevails, and the geometry will generally not be available for visual support. So we need a systematic approach that allows us to recognize the relationships between the equations in the system, which, in turn, will tell us why the system falls into a particular classification and why the solutions (if any) take on the appearance they do. Only when the nature of solutions to linear systems of equations is understood can we turn the complete task of solution over to computer programs. Then we can interpret the solution printout correctly, recognize when particular programs will not solve particular types of systems of linear equations, and know exactly what to seek in a more powerful program or how to proceed manually with a solution if all else fails. Undoubtedly, the most efficient approach to fulfilling the preceding objectives is to use a technique known as the **Gauss-Jordan elimination method,** which is actually an extension of the elimination method shown in Section 1.4 of Chapter 1. The objective is to systematically eliminate enough coefficients so that the solutions, if any, may be found with little or no effort. The Gauss-Jordan method makes use of three basic operations that may be applied to any system of linear equations, each of which *leaves unchanged* any solutions of the original system.

> **Basic Operations on Any Linear System**
> **That Leave the Solutions, If Any, Unchanged**
>
> **1.** Interchange any two equations.
>
> **2.** Replace any equation by a nonzero multiple of that equation.
>
> **3.** Replace any equation by that equation plus a multiple of another equation.

A new-appearing system obtained from the application of any one of these operations is called **equivalent** to the original system in the sense that the solution, if any, remains the same. The next example shows how these operations may be used to find the solution to a system of equations. Observe carefully the pattern that evolves to reach the final equivalent system, and note that the third operation is the one used to "eliminate" a variable.

EXAMPLE 1
Illustrating the
Gauss-Jordan Method
on a System of
Equations

Solve the following system of equations by the Gauss-Jordan method, and write the augmented matrix for each equivalent system of equations obtained in the process:

$$\begin{aligned} 2x + y - z &= -2 \\ y + 2z &= 2 \\ x - y + z &= 5. \end{aligned} \qquad \begin{bmatrix} 2 & 1 & -1 & | & -2 \\ 0 & 1 & 2 & | & 2 \\ 1 & -1 & 1 & | & 5 \end{bmatrix}$$

Solution   First, we find an equivalent system that has a 1 as the leading coefficient of the first equation. This may be done by replacing the first equation by $\frac{1}{2}$ times itself or by interchanging the first and last equations to avoid fractions. We choose the latter, giving this equivalent system:

$$\begin{aligned} x - y + z &= 5 \\ y + 2z &= 2 \\ 2x + y - z &= -2. \end{aligned} \qquad \begin{bmatrix} 1 & -1 & 1 & | & 5 \\ 0 & 1 & 2 & | & 2 \\ 2 & 1 & -1 & | & -2 \end{bmatrix}$$

Next, we replace the last equation by itself plus -2 times the first equation; this will eliminate the x-variable in the third equation:

$$\begin{aligned} x - y + z &= 5 \\ y + 2z &= 2 \\ 3y + 3z &= -12. \end{aligned} \qquad \begin{bmatrix} 1 & -1 & 1 & | & 5 \\ 0 & 1 & 2 & | & 2 \\ 0 & 3 & -3 & | & -12 \end{bmatrix}$$

Now we replace the first equation with itself plus 1 times the second equation and then replace the last equation by itself plus -3 times the second equation; this eliminates the y-variables in the first and third equations:

$$\begin{aligned} x \quad + 3z &= 7 \\ y + 2z &= 2 \\ -9z &= -18. \end{aligned} \qquad \begin{bmatrix} 1 & 0 & 3 & | & 7 \\ 0 & 1 & 2 & | & 2 \\ 0 & 0 & -9 & | & -18 \end{bmatrix}$$

The last equation is now replaced by $-\frac{1}{9}$ times itself; this puts a 1 in the third-row, third-column position:

$$\begin{aligned} x \quad + 3z &= 7 \\ y + 2z &= 2 \\ z &= 2. \end{aligned} \qquad \begin{bmatrix} 1 & 0 & 3 & | & 7 \\ 0 & 1 & 2 & | & 2 \\ 0 & 0 & 1 & | & 2 \end{bmatrix}$$

Finally, the second equation is replaced by itself plus -2 times the last equation and the first equation is replaced by itself plus -3 times the last equation; this eliminates the z-variables in the first and second equations:

$$\begin{aligned} x \quad &= 1 \\ y \quad &= -2 \\ z &= 2. \end{aligned} \qquad \begin{bmatrix} 1 & 0 & 0 & | & 1 \\ 0 & 1 & 0 & | & -2 \\ 0 & 0 & 1 & | & 2 \end{bmatrix}$$

This last equivalent system vividly displays the solution. ■

A study of the augmented matrices in Example 1 suggests that **basic row operations on a matrix,** analogous to those of the basic operations on equations, could be defined and used to arrive at the same solution. Such row operations are defined next, using the shorthand notation R_i to represent row i, an arrow $\rightarrow$ to be read as "replaces" and with the understanding that a "multiple of a row" means to multiply every number in that row by the same constant.

Basic Row Operations on a Matrix

1. Interchange any two rows. $(R_i \longleftrightarrow R_k)$

2. Replace any row by a nonzero multiple of itself. $(cR_i \rightarrow R_i)$

3. Replace any row by itself plus a multiple of another row. $(aR_i + R_k \rightarrow R_k)$

Note: The arrow $\rightarrow$ is read as "replaces."

From a matrix point of view, the use of these three operations would give a series of **row-equivalent** matrices, in the sense that each matrix so obtained has rows that represent a different appearing system of equations but which *have the same solution as the original system.* At some point, we want to arrive at a matrix that will enable us to easily tell the classification of the system (exactly one solution, no solution, or infinitely many solutions) and allow us to write the solution(s) with minimum effort. We suggest working toward the nice pattern of 0s and 1s exhibited in the last augmented matrix of Example 1. An efficient way to reach this goal is to work *column by column from left to right:*

1. First use row operations 1 and 2, if necessary, to put a 1 in the first-row, first-column position and *then* use row operation 3 to rewrite the remaining rows, if necessary, to eliminate all other nonzero entries in the first column.

2. Next, use row operations 1 and 2, if necessary, to put a 1 in the second-row, second-column position and *then* use row operation 3 to rewrite the remaining rows, if necessary, to eliminate all other nonzero entries in the second column.

3. Continue the pattern as shown below.

$$
\begin{bmatrix} 1 & \cdots \\ 0 & \cdots \\ 0 & \cdots \\ 0 & \cdots \end{bmatrix} \rightarrow
\begin{bmatrix} 1 & 0 & \cdots \\ 0 & 1 & \cdots \\ 0 & 0 & \cdots \\ 0 & 0 & \cdots \end{bmatrix} \rightarrow
\begin{bmatrix} 1 & 0 & 0 & \cdots \\ 0 & 1 & 0 & \cdots \\ 0 & 0 & 1 & \cdots \\ 0 & 0 & 0 & \cdots \end{bmatrix} \rightarrow
\begin{bmatrix} 1 & 0 & 0 & 0 & \cdots \\ 0 & 1 & 0 & 0 & \cdots \\ 0 & 0 & 1 & 0 & \cdots \\ 0 & 0 & 0 & 1 & \cdots \end{bmatrix}
$$

In general, a pivot operation consists of the following: the use of row operation 2 to turn any nonzero element in a matrix into a 1, followed by multiple uses of row operation 3 to introduce zeros into all remaining entries of the column in which that element resides. As a memory device, remember that a pivot operation consists of a "one" step followed by several "zero" steps.

The Pivot Operation

To pivot on a nonzero number a in a matrix means to do the following two steps. The number a is called the **pivot element,** the row in which a resides is called the **pivot row,** and the column in which a resides is called the **pivot column.**

1. First, if $a \neq 1$, change a into a 1 by multiplying the pivot row by $\frac{1}{a}$. If $a = 1$, proceed to the next step.

2. Then, if necessary, use row operation 3 repeatedly to change each nonzero number in the pivot column into a zero by replacing the row in which that nonzero number resides with itself plus an appropriate multiple of the pivot row.

NOTE   Once the first step has been completed, the pivot row does not change in appearance throughout the remainder of the pivot operation; only those rows where zeros are introduced change in appearance.

EXAMPLE 2
Details on How to
Perform a Pivot
Operation

Pivot on the number located in the first-row, first-column position of this matrix:

$$\left[\begin{array}{ccc|c} -2 & 1 & 4 & 0 \\ 4 & 1 & 8 & 3 \\ -3 & 2 & 1 & 4 \end{array}\right]$$

Solution According to the previous definitions, row 1 is the pivot row, column 1 is the pivot column, and -2 is the pivot element.

$$\text{Pivot row} \rightarrow \left[\begin{array}{ccc|c} \boxed{-2} & 1 & 4 & 0 \\ 4 & 1 & 8 & 3 \\ -3 & 2 & 1 & 4 \end{array}\right] \quad \text{The pivot element is } -2.$$

$$\underset{\text{Pivot column}}{\uparrow}$$

Since the pivot element is not a 1, replace row 1 by $-\frac{1}{2}$ times itself; stated differently, we are to multiply every number in row 1 by $-\frac{1}{2}$ and use the resulting numbers as the first row of a new matrix, leaving the second and third rows the same as before:

$$-\tfrac{1}{2}R_1 \rightarrow R_1 \left[\begin{array}{ccc|c} \boxed{1} & -\frac{1}{2} & -2 & 0 \\ 4 & 1 & 8 & 3 \\ -3 & 2 & 1 & 4 \end{array}\right]$$

Now use the new-appearing pivot row that has a 1 in the pivot position, along with row operation 3, to change the second and third rows so that each has a zero in the first-column position. To accomplish this, the multiplier -4 needs to be applied to the first row so that when the resulting numbers are added to the second row, a zero is obtained in the first-column position, and, separately, the multiplier 3 needs to be applied to the first row so that when these resulting numbers are added to the third row, the first-column position also becomes a zero. (This is the beauty of having a 1 in the pivot position; to turn a nonzero number k into zero, multiply the pivot row by $-k$ and add.)

$$\begin{array}{c} -4R_1 + R_2 \rightarrow R_2 \\ 3R_1 + R_3 \rightarrow R_3 \end{array} \left[\begin{array}{ccc|c} 1 & -\frac{1}{2} & -2 & 0 \\ 0 & 3 & 16 & 3 \\ 0 & \frac{1}{2} & -5 & 4 \end{array}\right]$$

The pivot operation is now complete. ■

Example 3 demonstrates the selection of pivot elements and the details leading to the solution of a particular system of linear equations.

EXAMPLE 3
Solving a System of
Equations by Pivoting

Solve the following system:

$$\begin{aligned} x \qquad\quad + z &= 4 \\ 2y + z &= 7 \\ 2x + y - z &= 1. \end{aligned}$$

Solution   The first step is to form the augmented matrix for this system:

$$\begin{array}{ccc} x & y & z \end{array}$$
$$\left[\begin{array}{ccc|c} \textcircled{1} & 0 & 1 & 4 \\ 0 & 2 & 1 & 7 \\ 2 & 1 & -1 & 1 \end{array}\right]$$

Then, working with the first column, we want the number in the first-row, first-column position, the circled 1, to be our first pivot element. Since that number is already a $+1$, we use appropriate multiples of row 1 to eliminate any other nonzero number in the first column:

$$\begin{array}{ccc} x & y & z \end{array}$$
$$-2R_1 + R_3 \rightarrow R_3 \quad \left[\begin{array}{ccc|c} 1 & 0 & 1 & 4 \\ 0 & \textcircled{2} & 1 & 7 \\ 0 & 1 & -3 & -7 \end{array}\right]$$

The first pivot operation is now complete. Next, working with the second column, the pivot element to be selected is in the second-row, second-column position, which is the circled 2 of the preceding matrix. Since that number is not a $+1$, the pivot operation is a two-stage process: We first replace the second row by $\frac{1}{2}$ times itself, then use appropriate multiples of row 2 to eliminate any nonzero numbers in the second column:

$$\frac{1}{2}R_2 \rightarrow R_2 \quad \begin{array}{ccc} x & y & z \end{array} \\ \left[\begin{array}{ccc|c} 1 & 0 & 1 & 4 \\ 0 & \textcircled{1} & \frac{1}{2} & \frac{7}{2} \\ 0 & 1 & -3 & -7 \end{array}\right] \quad -1R_2 + R_3 \rightarrow R_3 \quad \begin{array}{ccc} x & y & z \end{array} \\ \left[\begin{array}{ccc|c} 1 & 0 & 1 & 4 \\ 0 & 1 & \frac{1}{2} & \frac{7}{2} \\ 0 & 0 & \textcircled{-\frac{7}{2}} & -\frac{21}{2} \end{array}\right]$$

The second pivot operation is now complete. Finally, working with the third column, the pivot element to be selected is the circled $-\frac{7}{2}$ in the third-row, third-column position of the last matrix. Since that number is not $+1$, the pivot operation is again a two-stage process:

$$-\frac{2}{7}R_3 \rightarrow R_3 \quad \begin{array}{ccc} x & y & z \end{array} \\ \left[\begin{array}{ccc|c} 1 & 0 & 1 & 4 \\ 0 & 1 & \frac{1}{2} & \frac{7}{2} \\ 0 & 0 & \textcircled{1} & 3 \end{array}\right] \quad \begin{array}{c} -1R_3 + R_1 \rightarrow R_1 \\ -\frac{1}{2}R_3 + R_2 \rightarrow R_2 \end{array} \quad \begin{array}{ccc} x & y & z \end{array} \\ \left[\begin{array}{ccc|c} 1 & 0 & 0 & 1 \\ 0 & 1 & 0 & 2 \\ 0 & 0 & 1 & 3 \end{array}\right]$$

Writing the equations represented in the last matrix gives the solution:

$$x = 1, \qquad y = 2, \qquad z = 3. \quad \blacksquare$$

EXAMPLE 4
Solving a System of
Equations by Pivoting

Solve the following system:

$$\begin{aligned} x \phantom{{}+y} + z &= 2 \\ y - z &= 4 \\ 2x + y \phantom{{}- z} &= 5 \\ x + y \phantom{{}- z} &= 6. \end{aligned}$$

Solution   The first step is to form the augmented matrix for this system:

$$\left[\begin{array}{ccc|c} \textcircled{1} & 0 & 1 & 2 \\ 0 & 1 & -1 & 4 \\ 2 & 1 & 0 & 5 \\ 1 & 1 & 0 & 6 \end{array}\right]$$

Then, working with the first column, we want the first-row, first-column element to be our first pivot element. In this case, that element is already a 1. The remaining steps are shown next:

$$\begin{array}{c} \\ \\ -2R_1 + R_3 \to R_3 \\ -1R_1 + R_4 \to R_4 \end{array} \left[\begin{array}{ccc|c} 1 & 0 & 1 & 2 \\ 0 & \textcircled{1} & -1 & 4 \\ 0 & 1 & -2 & 1 \\ 0 & 1 & -1 & 4 \end{array}\right]$$

Next, we pivot on the element in the second-row, second-column position in the preceding matrix. Here is the result:

$$\begin{array}{c} \\ \\ -1R_2 + R_3 \to R_3 \\ -1R_2 + R_4 \to R_4 \end{array} \left[\begin{array}{ccc|c} 1 & 0 & 1 & 2 \\ 0 & 1 & -1 & 4 \\ 0 & 0 & \textcircled{-1} & -3 \\ 0 & 0 & 0 & 0 \end{array}\right]$$

The row of zeros in the fourth row is no cause for alarm. It is a signal that the last equation was a combination of the others and contributes nothing to the solution. Such a system is called **dependent.** Finally, we pivot on the -1 in the third-row, third-column position. Because this element is not a $+1$, the pivoting operation will be completed in *two* steps: First multiply the third row by -1; then turn all other entries in the column into zeros. Here are the details:

$$-1R_3 \to R_3 \left[\begin{array}{ccc|c} 1 & 0 & 1 & 2 \\ 0 & 1 & -1 & 4 \\ 0 & 0 & \textcircled{1} & 3 \\ 0 & 0 & 0 & 0 \end{array}\right] \begin{array}{c} -1R_3 + R_1 \to R_1 \\ 1R_3 + R_2 \to R_2 \\ \\ \end{array} \left[\begin{array}{ccc|c} 1 & 0 & 0 & -1 \\ 0 & 1 & 0 & 7 \\ 0 & 0 & 1 & 3 \\ 0 & 0 & 0 & 0 \end{array}\right]$$

The solution to the original system may now be read from the nonzero rows of the last matrix after recalling that the first column represents coefficients of x, the second column coefficients of y, and the third column coefficients of z:

$$x = -1, \qquad y = 7, \qquad z = 3. \quad \blacksquare$$

Dependent Systems of Equations

A system of n linear equations is called **dependent** if fewer than n of these equations will give the same solution as the original set of equations. Otherwise the system is **independent.**

Because dependent systems of linear equations in more than two variables are usually difficult or impossible to detect at the outset, one advantage of the pivoting approach is that each equation not needed to determine the solution is turned into a row of zeros in the augmented matrix (see Example 4) and effectively eliminated from consideration. We will see by the end of this chapter that the pivoting approach gives a systematic way to solve *any* system of linear equations.

Sometimes, the orderly progression of pivoting successively down the diagonal can be done but only after an interchange of rows is performed. For instance, if an augmented matrix has this appearance at some point

$$\left[\begin{array}{ccc|c} 1 & 3 & 4 & 5 \\ 0 & 0 & 2 & 1 \\ 0 & 1 & 3 & -2 \end{array}\right],$$

the next pivot element should be the one in the second-row, second-column position; but there is a zero there, making a pivot operation impossible. However, if the second and third rows are interchanged, a 1 will be in that position and not only can a pivot operation take place, but the interchange also won't destroy the integrity of the first column. The integrity of columns will be preserved on any interchange with rows *below* the desired pivot row.

At other times it may be impossible to select successive pivot elements down the diagonal and obtain the "ideal" arrangements of 1s and 0s exhibited in the coefficient matrices of Examples 1 and 3. For example, in the matrix

$$\left[\begin{array}{ccc|c} 1 & -4 & 3 & 5 \\ 0 & 0 & 2 & 3 \\ 0 & 0 & 6 & 3 \end{array}\right],$$

it is impossible to obtain a nonzero entry in the second-row, second-column position upon which to pivot. In such a case we do the next best thing: seek a pivot element in the second-row, *third-column position*. In this case we could pivot on the 2 in that position and proceed with the solution to the original system of equations. The goal is to get as close as possible to the "ideal" arrangement of 1s and 0s in the coefficient matrix. When this is done the coefficient matrix is said to be in **reduced row-echelon form.**

The Reduced Row-Echelon Form of a Matrix

A matrix is in **reduced row-echelon form** if the following three conditions are satisfied:

1. For each nonzero row, the first nonzero element is a 1.

2. A row with fewer leading zeros is above a row with more leading zeros.

3. The first nonzero element in a row is the only nonzero entry in its column.

EXAMPLE 5
The Reduced
Row-Echelon Form
of a Matrix

These matrices are all in row-echelon form:

$$\left[\begin{array}{ccc} 1 & 0 & 0 \\ 0 & 1 & 0 \\ 0 & 0 & 1 \end{array}\right] \left[\begin{array}{ccc} 1 & 0 & 0 \\ 0 & 1 & 3 \\ 0 & 0 & 0 \end{array}\right] \left[\begin{array}{ccccc} 1 & 0 & 0 & 0 & 8 \\ 0 & 0 & 1 & 6 & 7 \end{array}\right] \left[\begin{array}{cccc} 1 & 0 & 0 & 4 \\ 0 & 1 & 0 & 9 \\ 0 & 0 & 1 & 0 \end{array}\right] \left[\begin{array}{cccc} 1 & 0 & 0 & 8 \\ 0 & 0 & 1 & 0 \\ 0 & 0 & 0 & 0 \end{array}\right]$$

None of these matrices are in reduced row-echelon form.

$$\begin{bmatrix} 1 & 0 & 0 \\ 0 & 2 & 3 \\ 0 & 0 & 0 \end{bmatrix} \begin{bmatrix} 1 & 0 & 0 & 4 \\ 0 & 1 & 1 & 9 \\ 0 & 0 & 1 & 0 \end{bmatrix} \begin{bmatrix} 1 & 0 & 0 & 8 \\ 0 & 0 & 0 & 0 \\ 0 & 0 & 1 & 0 \end{bmatrix} \quad \blacksquare$$

The reduced row-echelon form of the coefficient matrix in Example 5 happens to lead to the conclusion that no solution to the system exists.

EXAMPLE 6

A System of Equations with No Solution

Solve the following system:

$$2x + y + 4z + 2w = 6$$
$$2z + w = 4$$
$$4z + 2w = 3.$$

Solution

$$\begin{bmatrix} \textcircled{2} & 1 & 4 & 2 & | & 6 \\ 0 & 0 & 2 & 1 & | & 4 \\ 0 & 0 & 4 & 2 & | & 3 \end{bmatrix} \qquad \frac{1}{2}R_1 \to R_1 \begin{bmatrix} 1 & \frac{1}{2} & 2 & 1 & | & 3 \\ 0 & 0 & \textcircled{2} & 1 & | & 4 \\ 0 & 0 & 4 & 2 & | & 3 \end{bmatrix}$$

$$\frac{1}{2}R_2 \to R_2 \begin{bmatrix} 1 & \frac{1}{2} & 2 & 1 & | & 3 \\ 0 & 0 & \textcircled{1} & \frac{1}{2} & | & 2 \\ 0 & 0 & 4 & 2 & | & 3 \end{bmatrix}$$

$$\begin{aligned} -2R_2 + R_1 \to R_1 \\ -4R_2 + R_3 \to R_3 \end{aligned} \begin{bmatrix} 1 & \frac{1}{2} & 0 & 0 & | & -1 \\ 0 & 0 & 1 & \frac{1}{2} & | & 2 \\ 0 & 0 & 0 & 0 & | & -5 \end{bmatrix}$$

The last row corresponds to the equation $0x + 0y + 0z + 0w = -5$, which has no solution. This means that the entire system has no solution. $\blacksquare$

Detecting When a System Has No Solution

At any stage in the pivoting process, if a row contains all zeros to the left of the vertical bar and a nonzero number to its right, then the system has **no solution.**

We should point out that there are several variations to the Gauss-Jordan elimination process that we used as a pivoting process. One variation is to (1) introduce zeros only below the main diagonal of the augmented matrix, (2) find the value of the last variable, and (3) then substitute back to find the values of the other variables. Another variation is to turn the pivot element into some convenient number (not necessarily 1) to avoid fractions. Even so, the objective is the same: Find an equivalent system of equations from which the solutions, if any, can easily be found.

Now that we know the appearance of single solutions, know about dependent systems, and recognize the signal when a system has no solution, we are in a position to take advantage of available technology to do this work for us. Certain models of graphing calculators are preprogrammed to give these solutions. All graphing calculators have the row operations used in the Gauss-Jordan elimination preprogrammed into them to emulate the hand calculations shown in this section. Derive is one of many computer programs that will solve any system of linear equations. Figure 8 shows the Derive screen that results from solving the four equations of Example 4. Line #1 was obtained first from the Author command, then by entering the equations, each separated by a comma. All equations are surrounded by square brackets. Line #2 and the remaining screen is the result of using the soLve command.

Figure 9 shows the outcome of asking Derive to solve the equations of Example 6. Notice the phrase "No solutions found" in the lower left portion of the screen.

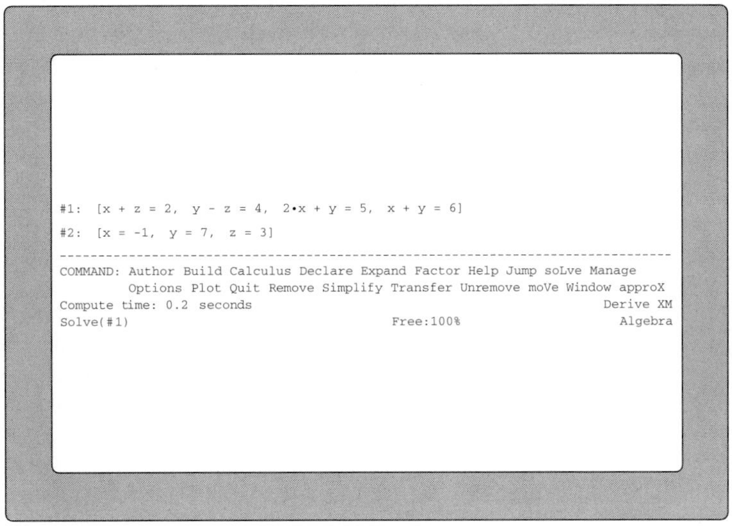

```
#1:   [x + z = 2,  y - z = 4,  2•x + y = 5,  x + y = 6]
#2:   [x = -1,  y = 7,  z = 3]
-----------------------------------------------------------------------------
COMMAND: Author Build Calculus Declare Expand Factor Help Jump soLve Manage
          Options Plot Quit Remove Simplify Transfer Unremove moVe Window approX
Compute time: 0.2 seconds                                        Derive XM
Solve(#1)                              Free:100%                    Algebra
```

FIGURE 8

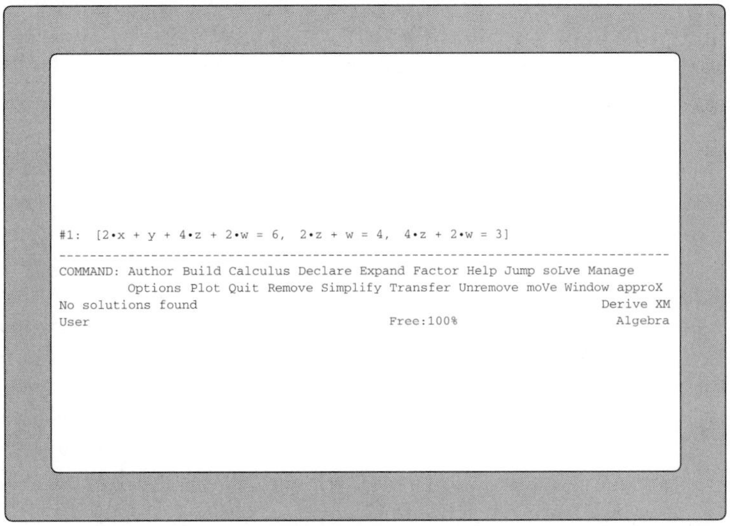

```
#1:   [2•x + y + 4•z + 2•w = 6,  2•z + w = 4,  4•z + 2•w = 3]
-----------------------------------------------------------------------------
COMMAND: Author Build Calculus Declare Expand Factor Help Jump soLve Manage
          Options Plot Quit Remove Simplify Transfer Unremove moVe Window approX
No solutions found                                              Derive XM
User                                   Free:100%                   Algebra
```

FIGURE 9

EXERCISES   2.2

In Exercises 1 through 18, find the solution, if one exists, by use of the pivoting process.

1.
$$x + y + z = 4$$
$$y - z = -1$$
$$x + 3y + z = 6$$

2.
$$2x + y + 2z = -5$$
$$x - y + z = -5$$
$$x + y = 2$$

3.
$$y - 3z = -11$$
$$x - y + z = 5$$
$$3x + z = 10$$

4.
$$y + 5z = 17$$
$$x + 2y - z = 2$$
$$2x - y + 3z = 9$$

5.
$$2x + 4y + 2z = 5$$
$$x + y + z = \tfrac{3}{2}$$
$$2x + y + 4z = 4$$

6.
$$3x + 3y + z = \tfrac{8}{3}$$
$$x - y + z = 1$$
$$3x + z = -1$$

7.
$$x + 2y + z = 3$$
$$3x + 3y + 3z = 7$$
$$2x + y + 2z = 1$$

8.
$$3x - z = 1$$
$$2x - y + z = 4$$
$$x + y - 2z = 3$$

9.
$$x + 3y = 8$$
$$2x - y = 4$$

10.
$$3x - y = 5$$
$$x + 2y = 0$$

11.
$$3x + y = -1$$
$$x - 2y = 3$$
$$4x - y = 2$$
$$8x + 5y = -6$$

12.
$$4x - 3y = -3$$
$$x + 2y = 1$$
$$5x - y = -2$$
$$9x - 4y = 3$$

13.
$$x + z = -2$$
$$x + y + z = -2$$
$$3x + 2y + 2z = -3$$
$$2x + y = 2$$

14.
$$3x + y = 9$$
$$x - y + z = 4$$
$$3x + z = 11$$
$$4x - y + 2z = 15$$

15.
$$x + y - 3z = 2$$
$$2x - z = 1$$
$$3x - 2y = 0$$
$$6x - y - 3z = 5$$

16.
$$y + 4z = 7$$
$$2x + y - z = 3$$
$$x - y + z = 1$$
$$3x + y + 5z = 0$$

17.
$$x + y + 3z = 0$$
$$\tfrac{1}{2}x - y - \tfrac{7}{2}z = 0$$
$$x + 2y = 0$$

18.
$$3x - y + z = 0$$
$$x + 3z = 0$$
$$2x + 3y + z = 0$$

In Exercises 19 through 24, use a calculator or computer to find the solution, if one exists.

19.
$$w + x + y - z = 1$$
$$2w + \tfrac{1}{2}x + \tfrac{2}{3}z = \tfrac{5}{12}$$
$$-w - x + \tfrac{1}{5}y = \tfrac{7}{20}$$
$$w + \tfrac{3}{4}x + y + 4z = \tfrac{17}{8}$$

20.
$$\tfrac{2}{5}w + \tfrac{1}{5}y - z = -\tfrac{34}{15}$$
$$-\tfrac{283}{100}w + 2x + y + z = 0$$
$$\tfrac{21}{20}w + \tfrac{2}{5}x + \tfrac{25}{10}y = 2$$
$$\tfrac{3}{10}w + \tfrac{1}{3}x + \tfrac{2}{5}y - z = -\tfrac{28}{15}$$

21.
$$.2x - .82y + .03z = -1.96$$
$$.02x + .14y - .6z = -2.82$$
$$.47x + .22y + .62z = 3.92$$

22.
$$2x + .3y + .82z = 7.72$$
$$.82x + 3y - .02z = 11.44$$
$$.57x - y + z = 4$$
$$3.39x + 2y - z = 5$$

23.
$$w - x + 2y + z = 9$$
$$.3w + .52y - z = 1.62$$
$$.6w + 2x - y = 3$$
$$.2w + .43x + .6y + z = 2.15$$

24.
$$2w + .6x + .3y - z = 2.80$$
$$.8w + .2x + 4y + z = 21.13$$
$$.36w + .4x - 3y = -12.41$$
$$w - x + y + .8z = 3.50$$

In the remaining exercises, first set up the system of equations involved, and then solve by using the pivoting process. Some of these exercises you have set up in Section 2.1.

25. *Meeting a Pollution Allowance* An electrical plant uses coal and gas to generate electricity. Coal produces 3 megawatt-hours of electricity per ton, while gas produces 4.50 megawatt-hours per thousand cubic feet. Each ton of coal produces 60 pounds of sulfur dioxide per hour, and each thousand cubic feet of gas produces 1 pound. The plant wants to produce electricity at the rate of 183 megawatt-hours while not exceeding the allowable sulfur dioxide pollution rate of 990 pounds per hour. How many tons of coal and how many thousand cubic feet of gas should the plant burn each hour to produce the desired amount of electricity while exactly meeting the maximum pollution allowance?

26. *Hiring and Training Workers* A particular department in a factory hires skilled workers, semiskilled workers, and supervisors. Each skilled worker is paid $12 per hour; each semiskilled worker, $9; and each supervisor, $15. The department is allowed an hourly payroll of $1560 for these three types of workers. The department requires that each worker spend some time in training and safety schooling each week: skilled workers, 3 hours; semiskilled workers, 5 hours; and supervisors, 1 hour. The training center can handle a maximum of 588 person-hours each week, and the department needs 140 skilled and semiskilled workers to meet production schedules. Assuming that the allowable payroll is met, that the training center is fully utilized, and that production needs are met, how many of each type of worker does the department need?

27. *Computer Printers* A large firm needs to purchase several new computer printers to meet its printing needs. The firm is considering buying the Q6, Q10, and Q12 models of a certain brand. The

Q6 prints at the rate of 6 pages per minute, has 2 megabytes of memory, and costs $800; the Q10 prints at the rate of 10 pages per minute, has 4 megabytes of memory, and costs $1500; and the Q12 prints at the rate of 12 pages per minute, has 12 megabytes of memory, and costs $3000. Various affected departments in the firm turned in a sum of 3720 pages per hour that they considered adequate for their printing needs, and the treasurer of the firm has set a budget of $11,400 to purchase printers with. The firm's engineers have decided that printing needs demand a total of 38 megabytes of memory. Annual maintenance costs are estimated at $20, $40, and $120, respectively, for the three models, and the firm has an annual maintenance budget of $380. Assuming the budgets, printing needs, and memory needs are exactly met, how many of each of these printers should be purchased?

28. **Manufacturing**   A kitchen appliance manufacturer makes can openers and dough cutters using two machines: a press and a riveter. Each can opener requires 0.2 minutes in the press and 0.4 minutes in the riveter. A dough cutter requires 0.5 minutes in the press and 0.3 minutes in the riveter. The press can be operated only 3 hours per day and the riveter only 2.5 hours per day. If these two machines are fully utilized, how many can openers and dough cutters can be produced each day?

29. **Advertising**   Members of a local charity want to advertise a forthcoming event that they are sponsoring by placing advertisements in the newspapers, on the radio, and on TV. They want to run a total of 24 ads, with three times as many radio ads as TV ads. The newspaper ads have an effective rating of 7; radio ads, 8; and TV ads, 9. The charity wants a total of 188 effective rating points. How many ads should be run in each medium to meet these objectives?

30. **Training Personnel**   The Vicky Valerin Secretarial Training Service trains secretaries for three types of office positions: Industrial plant (IP), small business office (SB), and executive secretarial (ES). Each student goes through training in three departments: Typing (T), telephone and miscellaneous (M), and ethics (E). The following table shows the number of units of training each type of secretary receives.

	IP	SB	ES
T	1	1	1
M	0.5	0.5	1.5
E	0.2	0.4	0.8

Each month, typing can deliver 42 units of training; telephone and miscellaneous, 29 units; and ethics, 16 units. For the service members to fully utilize their training capacity, how many of each type of secretary should they admit to their training service each month?

31. **Hiring Workers**   Brenda's Drive-In is located in a college town and has rush-hour business during the 5 o'clock hour and again during the 10 o'clock hour. To run her business efficiently, Brenda hires three shifts of workers: one from 10:00 A.M. to 6:00 P.M., another from 5:00 to 11:00 P.M. and a third from 10:00 P.M. to 1:00 A.M. Brenda knows from experience that she needs ten workers during the first- and second-shift overlap and eight workers during the second- and third-shift overlap. Can Brenda staff her worker needs for each shift if the first and third shifts have the same number of workers?

32. **Truck Leasing**   One day the Ace Delivery Company leased three trucks, with drivers, to make deliveries. The trucks were driven a total of 6940 miles. Truck A got 12 miles per gallon, and its driver was paid 18 cents per mile. Truck B got 14 miles per gallon, and its driver was paid 22 cents per mile. Truck C got 15 miles per gallon, and its driver was paid 24 cents per mile. The trucks used a total of 500 gallons of gasoline, and the drivers were paid a total of $1519.60. How many gallons did each truck consume during the day?

33. **Buying Tires**   A tire store manager is given $28,500 to buy 600 new tires for the store. Two types of tires are to be purchased, four-ply nylon costing $40 each and steel-belted radials costing $50 each. The manager wants to buy three times as many steel-belted radial tires as four-ply nylon tires. Assuming that all $28,500 is spent, find a system of linear equations that would give the number of each type of tire bought.

34. **Investing**   A bank trust department is to invest $100,000 for a client in three areas: certificates of deposit earning 8%, junk bonds earning 12%, and blue-chip stocks earning 6%. The trust is to earn $8000 each year on these investments, and the client stipulated that three times as much be invested in certificates of deposit and blue-chip stocks as in junk bonds. How much is to be invested in each of the three areas to meet the specified conditions?

35. **Dividing an Object**   Suppose you are to cut a piece of string 160 inches long into two pieces so that one piece is four times as long as the other. How long is each piece of string?

36. **Copy Machines**   The Maxi Printing Service is planning on purchasing several new copy machines to meet its customer needs. The company is considering a Model A machine that costs $2000 and prints at the rate of 30 pages per minute, a Model B machine that costs $4000 and prints at the rate of 45 pages per minute, and a Model C machine that costs $8000 and prints at the rate of 60 pages per minute. The firm has budgeted a total of $78,000 for the purchase of these

machines and wants to be able to print a total of 45,900 pages per hour. For these models, it would want to purchase three times as many Model A machines as Model B machines. Assuming that the firm's budget and printing needs are exactly met, how many of each model should be purchased?

37. *Manufacturing*   A particular company makes three models of chain saws: A, B, and C. Three operations are involved in the manufacturing process: assembly, painting, and testing. In the assembly operation, model A takes 1 hour, B takes $1\frac{1}{2}$ hours, and C takes 2 hours. In the painting department, each saw takes $\frac{1}{2}$ hour, and in the testing department, models A and C take $\frac{1}{5}$ of an hour each, and model B takes $\frac{1}{4}$ of an hour. The company has available each week 200 hours for assembly, 65 hours for painting, and 26 hours for testing. If all available labor hours are fully utilized, how many of each type of saw can be made each week?

38. *Mixing Acids*   A chemistry department wants to make 2 liters of an 18% acid solution by mixing a 21% solution with a 14% solution. How many liters of each type of acid solution should be used to produce the 18% solution?

39. *Consolidated Income*   Company A owns 20% of company B and 10% of company C. Company B owns 30% of company C, and company C owns 5% of company A and 40% of company B. The consolidated income of each company consists of its own net income plus its share of the net income of any other company in which it owns an interest. Given that the net income before intercompany adjustments of each company is

   A: $1.23 million

   B: $4.6 million

   C: $3.8 million,

find a system of linear equations that will model the consolidated income of each company and then solve the system.

40. *Consolidated Income*   The Nugent Company owns 30% of the Scott Company and 15% of the Ziegler Company. The Scott Company owns 5% of the Nugent Company and 8% of the Wickliff Company. The Wickliff Company owns 12% of the Nugent Company and 2% of the Scott Company, while the Ziegler Company owns 6% of the Nugent Company and 25% of the Scott Company. The consolidated income of each company consists of its own net income plus its share of the net income from other company ownership. Given that the net incomes before intercompany adjustments of these companies during a given year are

   Nugent Company: $5.4 million

   Scott Company: $4.8 million

   Wickliff Company: $8.3 million

   Ziegler Company: $4.0 million,

find a system of linear equations that will model the consolidated income for each of these companies, and then solve the system.

Problems for Exploration

1. Make up a system of four linear equations in four unknowns that has no solution.

2. A linear equation in three unknowns has as its graph a plane in a three-dimensional space. If a system of three equations in three unknowns has no solution, describe the possible geometric relationships between the planes represented.

3. Devise a way to perform pivoting operations so that the pivot element will not necessarily be made into a 1 and no fractions will ever occur in the pivoting process. Give some examples of your method.

SECTION 2.3   **Linear Systems Having Many Solutions**

In the previous section, we concentrated on systems that had exactly one solution or no solution at all. However, just as in the case of two equations and two unknowns discussed in Chapter 1, larger systems may also have an infinite number of solutions; this is the focus of our attention in this section. With the geometry available in Chapter 1, it was easy to tell at the outset when an infinite number of solutions was to occur. However, in three-variable cases, this will usually not be the case, so we rely on the pivoting process to signal when an infinite number of solutions results, and then use the final matrix to help us display the solutions.

EXAMPLE 1

A System Having an
Infinite Number of
Solutions

Find the solutions to this system:

$$x + 3y + z = 10$$
$$2x + 7y - z = 21.$$

Solution   The pivoting process applied to the augmented matrix gives this sequence:

$$\begin{bmatrix} ① & 3 & 1 & | & 10 \\ 2 & 7 & -1 & | & 21 \end{bmatrix}$$
$$-2R_1 + R_2 \rightarrow R_2 \begin{bmatrix} 1 & 3 & 1 & | & 10 \\ 0 & ① & -3 & | & 1 \end{bmatrix}$$
$$-3R_2 + R_1 \rightarrow R_1 \begin{bmatrix} 1 & 0 & 10 & | & 7 \\ 0 & 1 & -3 & | & 1 \end{bmatrix}$$

The pivoting process has now been carried as far as possible. Because the solution is not evident from this matrix, what do we do now? First, write the equivalent system of equations:

$$x + \quad 10z = 7$$
$$y - \quad 3z = 1.$$

This system of *two* equations and *three* unknowns has what we may call an "extra" unknown, and that is a signal that there will either be an infinite number of solutions or no solution at all. In this case, it is easiest to consider z to be the "extra" unknown and to solve the first equation for x and the second for y, thereby getting solutions for each in terms of the unknown z:

$$x = 7 - 10z$$
$$y = 1 + \quad 3z.$$

When the equations are written this way, we can see that x and y depend upon z. That is, if z is assigned a number, then x and y can be computed so that these three numbers are a solution to the system. *All* solutions may then be stated in the following form:

$$z = \text{any number}$$
$$x = 7 - 10z$$
$$y = 1 + 3z.$$

In particular, if $z = 0$, then $x = 7$, and $y = 1$, so that $x = 7$, $y = 1$, $z = 0$ is a solution. If $z = 2$, then $x = -13$ and $y = 7$ give $x = -13$, $y = 7$, $z = 2$ as a solution, and so on.   ■

The fact that we are free to assign z any number that we please in the solutions in Example 1 suggests that z may be called a **free variable** or, as is sometimes done, a **parameter.** We point out that it is not necessary for z to be the free variable; we could just as well have used x as the free variable and solved for y and z in terms of x, or let y be the free variable and solved for x and z in terms of y. Each of these will give exactly the same set of solutions, even though the *appearance* of the solution equations is not the same.

Solutions involving one or more free variables, like the one in Example 1, may be checked for accuracy just like any other solution. Just substitute the expression found for x—that is, $x = 7 - 10z$—and the expression found for y—$y = 1 + 3z$—into each of the original equations and see whether an equality results:

(1) $(7 - 10z) + 3(1 + 3z) + z = ?\ 10$

(2) $2(7 - 10z) + 7(1 + 3z) - z = ?\ 21$

(1) $7 - 10z + 3 + 9z + z = ?\ 10$

(2) $14 - 20z + 7 + 21z - z = ?\ 21$

(1) $10 = 10.$

(2) $21 = 21.$

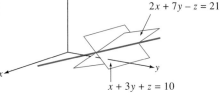

Line of intersection:
$x = 7 - 10z$
$y = 1 + 3z$
$z =$ any number

FIGURE 10
Planes Intersecting in a Line

Each of the equations in Example 1 has a plane in the three-dimensional rectangular coordinate system for its graph, and the solution (points of intersection) is a line, as shown in Figure 10.

EXAMPLE 2
Finding a Parametric Solution

Find the solution to this system of equations:

$$x - 2y + z = 3$$
$$2x + y + z = 1$$
$$4x - 3y + 3z = 7.$$

Solution   The pivoting scheme used on the augmented matrix for this system results in the following outcomes:

$$\begin{bmatrix} ① & -2 & 1 & 3 \\ 2 & 1 & 1 & 1 \\ 4 & -3 & 3 & 7 \end{bmatrix} \quad \begin{matrix} \\ -2R_1 + R_2 \to R_2 \\ -4R_1 + R_3 \to R_3 \end{matrix} \begin{bmatrix} 1 & -2 & 1 & 3 \\ 0 & ⑤ & -1 & -5 \\ 0 & 5 & -1 & -5 \end{bmatrix}$$

$$\tfrac{1}{5}R_2 \to R_2 \begin{bmatrix} 1 & -2 & 1 & 3 \\ 0 & ① & -\tfrac{1}{5} & -1 \\ 0 & 5 & -1 & -5 \end{bmatrix}$$

$$\begin{matrix} 2R_2 + R_1 \to R_1 \\ \\ -5R_2 + R_3 \to R_3 \end{matrix} \begin{bmatrix} 1 & 0 & \tfrac{3}{5} & 1 \\ 0 & 1 & -\tfrac{1}{5} & -1 \\ 0 & 0 & 0 & 0 \end{bmatrix}$$

Again, we have taken the pivoting process as far as possible. Because there are *two* nonzero rows and *three* variables, there is *one* free variable or parameter. Writing the equivalent equations from the last matrix gives

$$x + \frac{3}{5}z = 1$$

$$y - \frac{1}{5}z = -1,$$

from which it is easiest to solve for x and y in terms of z, so we let z be the free variable and arrive at this parametric solution:

$$x = \quad 1 - \frac{3}{5}z$$

$$y = -1 + \frac{1}{5}z$$

$$z = \text{any number.}$$

Geometrically, the solution to this set of equations is a line that is the intersection of the first two planes in the given system. The last row of zeros in the final matrix means that this line was contained in the plane represented by the third equation; hence, this last plane was not needed to determine the line. The system is dependent. ▄▄▄

The Gauss-Jordan elimination method, in conjunction with the pivoting process, is very helpful in ultimately revealing the relationships among linear equations that lead to the various types of solutions or in showing that no solution exists. By examining Examples 1 and 2 of this section, can you tell when an infinite number of solutions exist?

> **When an Infinite Number of Solutions Occur**
>
> If the reduced row-echelon form of the augmented matrix has the same number of nonzero rows in the coefficient matrix as in the augmented matrix, and that number is less than the number of variables, then the system has **an infinite number of solutions.**

A system of linear equations in which the constant term of every equation in the system equals zero is called **homogeneous.** The system in Example 3 is homogeneous. A homogeneous system of linear equations always has at least one solution. Do you see what that solution is?

EXAMPLE 3
Solving a
Homogeneous System
of Equations

Solve this homogeneous system of equations:

$$x + \qquad z = 0$$
$$x + y + \quad z = 0$$
$$2x + \qquad 2z = 0.$$

Solution   For this particular system, notice that the last equation is twice the first; hence, it does not contribute to the solutions and may be discarded. Solving the first equation for x, we get $x = -z$. Now substitute this value of x into the second equation and obtain $y = 0$. This shows that the solutions are

$$x = -z$$
$$y = 0$$
$$z = \text{any number.}$$

One particular solution for this system, and for any homogeneous system, is all variables equal zero. ▄▄▄

The previous example shows that once we understand the nature of the solutions, it is sometimes possible to obtain them without using the pivoting process, even on problems of substantial size. However, we should stress that the pivoting process applies equally well to *all* systems of linear equations and will guide us to the solution in every case.

A notational scheme for variables that is sometimes used, especially in large systems, is that of "subscripted" variables. The next example shows this notation, along with the fact that sometimes more than one parameter is needed in a solution.

EXAMPLE 4
Subscripted Notation;
More Than One
Parameter

Find the solution to this system of linear equations:

$$x_1 + x_2 - 2x_3 + 4x_4 = 6$$
$$3x_1 + x_2 + 2x_3 \qquad = 5.$$

Solution   The pivoting process can be used to advantage here.

$$\begin{bmatrix} ① & 1 & -2 & 4 & | & 6 \\ 3 & 1 & 2 & 0 & | & 5 \end{bmatrix} \quad -3R_1 + R_2 \rightarrow R_2 \quad \begin{bmatrix} 1 & 1 & -2 & 4 & | & 6 \\ 0 & ⓧ{-2} & 8 & -12 & | & -13 \end{bmatrix}$$

$$-\tfrac{1}{2}R_2 \rightarrow R_2 \quad \begin{bmatrix} 1 & 1 & -2 & 4 & | & 6 \\ 0 & ① & -4 & 6 & | & \tfrac{13}{2} \end{bmatrix}$$

$$-1R_2 + R_1 \rightarrow R_1 \quad \begin{bmatrix} 1 & 0 & 2 & -2 & | & -\tfrac{1}{2} \\ 0 & 1 & -4 & 6 & | & \tfrac{13}{2} \end{bmatrix}$$

The system of equations represented in the preceding matrix is

$$x_1 + \qquad 2x_3 - 2x_4 = -\frac{1}{2}$$

$$x_2 - 4x_3 + 6x_4 = \frac{13}{2}.$$

This time, there are two equations in four unknowns, resulting in *two* extra variables. It is easiest to solve for x_1 and x_2 in terms of x_3 and x_4, thus making x_3 and x_4 the free variables or parameters:

$$x_1 = -\frac{1}{2} - 2x_3 + 2x_4$$

$$x_2 = \frac{13}{2} + 4x_3 - 6x_4$$

$$x_3 = \text{any number}$$

$$x_4 = \text{any number}.$$

If we let $x_3 = 0$ and $x_4 = 1$, for example, then

$$x_1 = -\frac{1}{2} - 2(0) + 2(1) = \frac{3}{2}$$

and
$$x_2 = \frac{13}{2} + 4(0) - 6(1) = \frac{1}{2}.$$

This means that $x_1 = \frac{3}{2}, x_2 = \frac{1}{2}, x_3 = 0, x_4 = 1$ is a particular solution to the original system of equations. Each numerical choice of x_3 and x_4 gives a particular solution to the system. ▪

EXAMPLE 5
Solving the Shipping
Problem Given at the
Outset of this Chapter

The problem of shipping bicycles from two warehouses to two stores, as stated at the beginning of this chapter, is described by the equations found in Example 4 of Section 2.1 of this chapter. The equations found in Example 4 are

$$w \qquad + \quad z = 38$$
$$x + y \qquad = 46$$
$$x \qquad - 2z = 0.$$

If the pivoting process is applied directly to the augmented matrix of this system of equations, the final equivalent matrix in echelon form appears as

$$\begin{bmatrix} 1 & 0 & 0 & 1 & | & 38 \\ 0 & 1 & 0 & -2 & | & 0 \\ 0 & 0 & 1 & 2 & | & 46 \end{bmatrix}.$$

(Check your pivoting skills at this point by taking the initial augmented matrix to arrive at the preceding one.) The solutions can now be obtained by converting the last matrix to equation form,

$$w \quad + \quad z = 38$$
$$x - 2z = 0$$
$$y + 2z = 46,$$

and then considering z to be the parameter, so that the solutions become

$$w = 38 - \quad z$$
$$x = \qquad 2z$$
$$y = 46 - 2z$$
$$z = \text{any number.}$$

In reality, the variables in this particular problem represent numbers of bicycles and cannot be negative. This means that $z \geq 0$ but cannot be so large as to render any of the other variables negative. The first line of the solutions tells us that z cannot be larger than 38, while the third line tells us that z cannot be larger than 23. As a result, z must be *an integer from* 0 *through* 23. Now, our problem is completely solved, and the solutions give all possible shipping possibilities:

$$w = 38 - \quad z$$
$$x = \qquad 2z$$
$$y = 46 - 2z$$
$$z = \text{any integer from 0 through 23.} \quad \blacksquare$$

Later, we will encounter very special systems of equations having an infinite number of solutions, of which the following is a typical example:

$$x + 2y + u \qquad = 6$$
$$3x + \quad y \qquad + v = 8.$$

When dealing with such systems, we are not interested in all possible solutions, just in particular ones that may be found by inspection. In this system, if we consider x and y to be free variables (parameters) and set both equal to zero, then the first equation becomes $u = 6$ and the second $v = 8$. Therefore, one solution to this system is $x = 0$,

$y = 0$, $u = 6$, and $v = 8$. Suppose that we pivot on the element in the first-row, first-column position of the augmented matrix of this system to obtain

$$\begin{bmatrix} 1 & 2 & 1 & 0 & 6 \\ 0 & -5 & -3 & 1 & -10 \end{bmatrix}.$$

This matrix represents the equivalent system of equations,

$$x + 2y + \quad u \quad \ \ = 6$$
$$-5y - 3u + v = -10.$$

Can you find another particular solution from this matrix? The easiest way is to consider y and u free variables and to set both equal to zero, from which the first equation gives $x = 6$ and the second $v = -10$. A solution is then $y = 0$, $u = 0$, $x = 6$, and $v = -10$.

One final note is in order about the dependent systems encountered in Sections 2.2 and 2.3. An entire row of zeros in an augmented matrix means dependency only within the system. *This fact alone does not, in general, give any clue as to the type of solution the system has; any one of the three classifications might result.* Example 3 of Section 2.2 is a dependent system that has exactly one solution, Example 2 of this section is a dependent system that has an infinite number of solutions, and the following system is a dependent system that has no solution:

$$x + \quad y = 2$$
$$x + \quad y = 3$$
$$2x + 2y = 5.$$

Can you tell why just by visual inspection?

When using calculators or computers to solve systems of linear equations that have infinitely many solutions, be sure you know how to interpret the output. Realize also that in most cases the output is an *approximation* of the actual answer because of the round-off involved. Round-off can be so severe at times that a machine will give a solution to a system of equations that actually has no solution! Be sure to check such answers for accuracy. Figure 11 shows the screen output of Derive when used to solve the equations in Example 5.

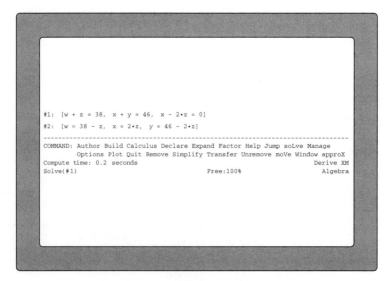

FIGURE 11

EXERCISES   2.3

In Exercises 1 through 16:
(a) Find the solution in parametric form.
(b) Check the answer.
(c) Write out three explicit solutions to the system.

1. $2x - y - 3z = 5$
$x - 2y + z = 3$

2. $3x + y - z = 0$
$2x + 3y + 2z = 1$

3. $2x + y - z = 0$
$x - 3y + z = 1$
$x + 4y - 2z = -1$

4. $2x - 3y + z = 1$
$x + y - 3z = 0$
$4x - y - 5z = 1$

5. $w + 2x + y - 3z = 4$
$-2w + x + y + z = 2$
$-w + 3x + 2y - 2z = 6$

6. $2x_1 + 3x_2 + 4x_3 - x_4 = -2$
$x_1 \qquad + 2x_3 \qquad = 0$
$x_1 - x_2 + x_3 + x_4 = 5$
$\qquad x_2 - 2x_3 \qquad = 3$

7. $x_1 + 4x_2 + 2x_3 - 3x_4 = 1$
$3x_1 + x_2 - 2x_3 \qquad = -11$
$x_1 - 2x_2 + x_3 + x_4 = 5$
$\qquad x_2 - \qquad 3x_4 = -8$

8. $\qquad 2w + x - 3y + z = 8$
$-w + 2x - y + 2z = 6$
$3w + 4x - 7y + 4z = 22$
$-2w + 4x - 2y + 4z = 12$

9. $x + y + z = 3$
$2x + 2y + 2z = 6$

10. $2x - y + z = 4$
$6x - 3y + 3z = 12$

11. $2x + 3y + z = 0$
$x - y + 2z = 0$
$4x + y + 5z = 0$

12. $3x + 3y - z = 0$
$4x + 5y - 3z = 0$
$x + 2y - 2z = 0$

13. $x \qquad - z = 0$
$\qquad y + z = 0$
$x + y \qquad = 0$

14. $x \qquad = 0$
$\qquad y + z = 0$
$\qquad 2y + 2z = 0$

15. $x \qquad + z = 0$
$\qquad y \qquad = 0$
$2x \qquad + 2z = 0$

16. $x + y \qquad = 0$
$\qquad z = 1$
$x + y + z = 1$

In Exercises 17 through 28, find the solutions, if any exist.

17. $x + 2y - z = 8$
$2x - y + z = 3$
$4x + 3y - z = 5$

18. $\qquad y + 2z5$
$x + 2y + z = 1$
$x + 3y + 3z = 2$

19. $\qquad w + 2x - 3y + z = 1$
$4w + 8x - 12y + 4z = 4$
$2w + 4x - 6y + 2z = 2$

20. $x + y - z = 1$
$2x + 2y - 2z = 2$
$3x + 3y - 3z = 3$

21. $2x - 3y + 2z = 3$
$x + 2y - 3z = 2$
$3x - y - z = 3$

22. $-w + x + y + 3z = 0$
$2w + 2x - y - z = 2$
$w + x + y - 4z = 1$
$w + 5x + 2y + z = 1$

23. $.3x + .5y + z = 4$
$5x - 2y + .8z = 7$
$5.6x - y + 2.8z = 15$

24. $6.2x + 3y + 2.7z = 2.3$
$1.8x - 2.3y - 1.9z = 5.6$
$9.8x - 1.6y - 1.1z = 13.5$

25. $.3x + 5.24y - 8.61z = 5.2$
$1.2x + 20.96y - 34.44z = 8$

26. $4.61w + 2x + 3.5y + z = 8$
$1.58w + .83x - 6.21y + 1.50z = 4$
$3w + .31x + 2y - z = 1.63$
$10.77w + 3.97x - 6.92y + 3z = 17.63$

27. $w + 2x - 3y + z = 6.3$
$2w - x + .54y + 2.8z = 5$
$2.83w + 5.72x + y - 3z = 2$
$12.66w + 15.44x - 2.92y + 2.6z = 32.9$

28. $w + x - y + 2z = 6.2$
$2w - .5x + y - .46z = 8.1$

In Exercises 29 through 32, solve these problems without using the pivoting process.

29. $x \qquad - 2z = 0$
$\qquad y \qquad = 0$
$2x \qquad - 4z = 0$

30. $3x - y \qquad = 0$
$\qquad z = 1$
$6x - 2y \qquad = 0$

31. $\qquad y + 2z = 0$
$x \qquad + z = 0$
$2x \qquad + 2z = 0$

32. $x + 4y \qquad = 0$
$\qquad y + z = 0$
$\qquad 3y + 3z = 0$

In Exercises 33 through 36:
(a) Find a specific solution by inspection.
(b) Pivot on the element in the second row and first column, and then find a second solution by inspection.

33. $2x + 2y + u \qquad = 10$
$x + y \qquad + v = 5$

34. $x + 3y + u = 4$
$x + y \qquad + v = 8$

35. $x - y - u \qquad = 6$
$2x + y \qquad + v = 4$

36. $2x + y + u \qquad = 7$
$3x - y \qquad - v = 10$

37. *City Budget* The city of Huntsville gets a grant for $1 million to spend on city projects. The city's board of directors decides to spend the money on three types of projects: streets, urban renewal, and parks. The board members also decide that the shares going to urban renewal and parks should equal the amount spent on streets. How can the money be allocated to these three projects?

38. *Highway Construction* A state highway department plans to build 100 miles of highways in three highway districts: A, B, and C. The cost per mile for building highways is $2 million in district A, $500,000 in district B, and $1 million in district C. If $80 million is allocated for construction, can 100 miles of highways be constructed at a total cost of $80 million so that some highway building takes place in each district?

39. *Advertising*   A new low-fat potato chip is being introduced in a small city through a series of advertisements in newspapers, on radio, and on television. A total of 30 advertisements are to run in these media. In addition, twice as many radio ads are to be run as television ads. Finally, the sum of the number of newspaper ads and two times the number of radio ads is to equal the number of ads run plus the number of television ads. How many of each type ad can be run?

40. *Construction*   A construction company is planning a development in which three types of housing— WoodIt, BrickIt, and StoneIt—are to be built. It estimates that the electrician hours required for WoodIt, BrickIt, and StoneIt houses are 30, 60, and 90 hours, respectively, and the plumber hours are 80, 100, and 120 hours, respectively. If subcontractors can furnish 1080 electrician hours and 1920 plumber hours per year, and these hours are fully utilized by the contractor, how many of each of these houses can be built in a year?

41. *Manufacturing*   A maker of computer printers has three models, the QT20, the QT40, and the QT60. Each QT20 model has 2 megabytes of memory; each QT40, 4 megabytes; and each QT60, 6 megabytes. To get the best prices on memory, the company has an agreement to buy a total of 500 megabytes of memory each week. It requires 3 hours to assemble each QT20 model, 4 hours to assemble each QT40 model, and 5 hours to assemble each QT60 model. The company has 560 worker assembly hours available each week. Each QT20 requires 5 minutes of test time; each QT40, 8 minutes; and each QT60, 11 minutes. The company has available 1060 minutes of testing time each week. Assuming that all available assembly and testing time is used each week, how many of each model printer can the company assemble?

42. *Investments*   An investor has $100,000 to invest in government bonds, mutual funds, and money market funds. The investor wants to invest as much in government bonds as the total in mutual funds and money market funds. If government bonds earn 7%, mutual funds earn 5%, and money market funds earn 5%, and the investor expects to earn $6000 in interest during the first year, how much should be invested in each of these three investments?

43. *Supplying Orders*   A company with two warehouses receives orders for 30 and 18 microwave ovens, respectively, from two stores, as shown in the following diagram. Twice as many ovens are to be shipped from Warehouse 1 to Store 1 as from Warehouse 2 to Store 2. Assuming that the warehouses have ample stock to supply the needs of both stores, find the equations that model the possibilities of meeting these orders, and then solve them. Give three possibilities for supplying the requested microwave ovens.

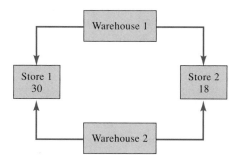

44. *Supplying Orders*   A company with two warehouses receives orders from three stores for TV sets, as indicated in the following diagram. Find the system of equations that model this supply problem, and solve it. Give three possibilities for supplying the stores with the requested TV sets.

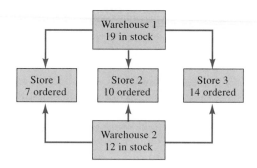

Problems for Exploration

1. Discuss what possibilities must occur in the pivoting process for a system of four linear equations in four unknowns to have an infinite number of solutions. Cover all possibilities for the parameters.

2. Suppose that the augmented matrix for a given system of linear equations has been reduced to the echelon form by pivoting. State rules involving the number of variables, the number of nonzero rows in the coefficient part of the matrix, and the number of nonzero rows in the entire augmented matrix that tell about the classification of the solution.

3. Show that for any homogeneous system of linear equations, the sum of two solutions to the system is also a solution to the system. Show also that any constant multiple of a solution is also a solution.

CONCEPTS

Linear systems of equations are useful mathematical models for many processes and situations. *Tables* can be helpful in organizing given information. *Augmented matrices,* the *pivoting process,* and the *row-echelon* form of a matrix together provide a convenient, systematic method for solving systems of equations. For systems having an infinite number of solutions, *free variables* or *parameters* can be used to give a general description for all of the solutions.

COMPUTATION/CALCULATION

Solve these exercises without using the pivoting process.

1.
$$3x \quad - \quad z = 0$$
$$y \qquad = 7$$
$$6x \quad - \quad 2z = 0$$

2.
$$x + y \qquad = 0$$
$$y - 3z = 0$$
$$5y - 15z = 0$$

Find three specific solutions by inspection.

3.
$$x + 2y + u \qquad = 12$$
$$x + 3y \qquad + v = 15$$

4.
$$x - y + u \qquad = 8$$
$$3x + y \qquad + v = 11$$

Find the solution, if one exists, using the pivoting process.

5.
$$x + 2y = 7$$
$$3x + y = 4$$

6.
$$\tfrac{1}{2}x - 3y = 5$$
$$2x - 12y = 18$$

7.
$$x + 2y \qquad = 9$$
$$4x - y + z = 7$$
$$2x - 5y + z = -11$$

8.
$$x + 2y + z = 0$$
$$3y + 5z = 1$$
$$5x - 4y - 10z = 12$$

9.
$$3x + 2y - z = -1$$
$$5x \quad + 7z = 5$$
$$-3y + 4z = 6$$
$$x - 6y + 3z = 13$$

10.
$$x - 2y = 13$$
$$3x + y = 4$$
$$2x - y = 11$$
$$4x + 2y = 2$$

11.
$$5x + 2y + z = 14$$
$$x + 2y + 3z = 6$$
$$2x \quad - z = 4$$

12.
$$x \quad + 4z = 7$$
$$y \quad = 5$$
$$2x \quad + 7z = 13$$
$$3x - y + 7z = 4$$

CONNECTIONS

13. Write a system of equations for which the given augmented matrix would be a model. Solve this system.

$$\begin{bmatrix} 3 & 0 & -2 & | & 7 \\ -1 & 8 & 5 & | & 0 \\ 0 & 2 & 1 & | & -4 \end{bmatrix}$$

14. A motorboat travels 33 miles in two hours downstream and 46 miles in four hours upstream. Find the rate of the stream and the speed of the boat.

15. Kevin agreed to train a horse for three months for $425 and a saddle. Instead, he trained the horse for five months and received $1025 and the saddle. What is the value of the saddle?

16. A student spends a total of 36 hours a week studying mathematics, English, psychology, and accounting. She spends as much time studying math and psychology together as she does studying English and accounting together. The amount of time she studies psychology and accounting together is five times as much time as she spends on studying English. The total time she spends on studying psychology and English is 16 hours less than the time she spends on math and accounting together. How much time during the week does she spend on each subject?

COMMUNICATION

17. Describe in your own words what it means for a matrix to be in reduced row-echelon form.

18. Explain the meaning of pivoting on a nonzero element of a matrix.

19. Describe how a system of equations having no solution can be detected using the pivoting process.

COOPERATIVE LEARNING

20. (a)
$$x = 7$$
$$y = -5$$
(b) No solution
(c) $x =$ any number
$$y = 3x + 4$$
(d)
$$x = 3z$$
$$y = 4$$
$$z =$$ any number
(e)
$$x = -2$$
$$y = 0$$
$$z = 3$$

For each solution given, determine three distinctly different systems of equations, each of which has the given solution as its solution. Give a geometric interpretation of each system. Write a paragraph describing your results and any noteworthy relationships.

CRITICAL THINKING

21. Two students were discussing the solution(s) to linear systems of equations having more equations than unknowns. Martina says that these systems always have an infinite number of solutions, while Johann says that these systems never have a solution. Is either correct? Explain and justify your response with supporting examples or analyses.

COMPUTER/CALCULATOR

Use the computer program Pivot from the disk MasterIt, or another suitable program, to solve the following systems of equations.

22. $\frac{8}{9}x + \frac{3}{4}y - \frac{7}{8}z = 9$ **23.** $x + .2y + 1.4z = 2$
$\frac{4}{5}x + \frac{5}{6}y + \frac{2}{3}z = \frac{12}{5}$ $.3x - 1.4y + .5z = 1.4$
$\frac{3}{4}x + \frac{2}{3}y - \frac{1}{4}z = 3$ $1.2x + .8y - .2z = 5$

The final screen for the Pivot program applied to Exercise 23 will appear as shown here.

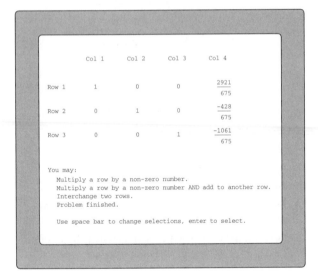

To solve linear systems, some calculators have built-in capabilities that will handle at least one of the three types of solutions that can occur. Others have programming capabilities that allow you to enter programs that will solve linear systems of equations. However, be sure you understand the capabilities of the program before you attempt to use it on just any system. Investigate the capabilities of some calculators you have access to. Most graphics calculators have built-in capabilities for doing the basic row operations involved in the pivoting process for matrices that are at least 6 × 6 in size. Most graphing calculators carry

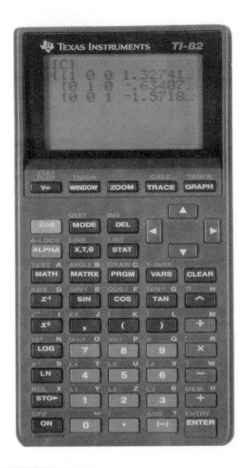

those operations out in decimal form, and the screen may be scrolled to view all of the matrix entries. Use an appropriate calculator to solve Exercises 22 and 23. The final screens will appear similar to the one above.

CUMULATIVE REVIEW

24. Calculate the value of the function f given by $f(x) = 2x - 7$, when $x = 29$.

25. Write the equation of a line having y-intercept 13, which is parallel to the line having equation $3x + 5y = 27$.

26. Find a total linear cost function having direct costs of $42 and fixed costs such that 30 items cost a total of $3000.

27. Solve the system $\begin{array}{l} 2x + 3y = 7 \\ 3x + 4y = 9 \end{array}$ using any appropriate method.

28. Find a function that will give the net value x years from now of a new printing press that costs $80,000, has scrap value of $12,000, and is depreciated by the straight-line method over a 15-year period.

1. Write the coefficient matrix for the system
$$
\begin{aligned}
x \quad\quad + 2z &= -7 \\
-3x + 5y - 8z &= 0 \\
x + y \quad\quad &= 43.
\end{aligned}
$$

2. Write the augmented matrix for the system
$$
\begin{aligned}
3a - \quad b &= 5 \\
-5a + 17b &= -1.2.
\end{aligned}
$$

Write a system of equations that models each of the next three problems.

3. A parking meter contained 213 coins in nickles, dimes, and quarters, having a total value of $21.15. If the number of dimes is twice the number of quarters, how many nickles were in the meter?

	Nickles	Dimes	Quarters	Total
Value in cents	5	10	25	2115
Relationship between y, z		1	-2	0
Total number of coins	1	1	1	213
Number of each coin	x	y	z	

4. A concert performance brings in receipts of $129,673 from the sale of 9073 tickets. If a student ticket costs $12.50 and an adult ticket costs $16.75, how many students and how many adults attended the concert?

5. Three strawberry pickers take turns working together in pairs. One pair working together can pick 62 quarts a day, another pair can pick 70 quarts a day, and the third pair can pick 58 quarts a day. If all three pickers worked together, how many quarts could they pick in a day?

6. Name *all* of the following matrices that are in reduced row-echelon form:

(a) $\begin{bmatrix} 1 & 0 & 0 \\ 0 & 1 & 0 \\ 0 & 1 & 2 \end{bmatrix}$ (b) $\begin{bmatrix} 1 & 0 & 0 \\ 0 & 0 & 1 \\ 0 & 1 & 0 \end{bmatrix}$ (c) $\begin{bmatrix} 0 & 1 & 0 \\ 0 & 0 & 1 \\ 0 & 0 & 0 \end{bmatrix}$

(d) $\begin{bmatrix} 1 & 0 & 0 \\ 0 & 0 & 0 \\ 0 & 0 & 1 \end{bmatrix}$ (e) $\begin{bmatrix} 1 & 0 & 3 \\ 0 & 1 & -7 \end{bmatrix}$

7. Complete the pivot operation on the circled element of the augmented matrix:

$$
\begin{bmatrix}
① & 1 & 3 & -6 \\
0 & 2 & 4 & 10 \\
5 & -2 & 1 & 0
\end{bmatrix}
$$

Find the solution, if one exists, using the pivoting process.

8.
$$
\begin{aligned}
x - 2y - z &= 1 \\
3x + y + z &= 10 \\
2x - y - z &= 0
\end{aligned}
$$

9.
$$
\begin{aligned}
6x - y &= -11 \\
x + 2y &= 9 \\
5x + y &= 0 \\
11x + 3y &= 4
\end{aligned}
$$

10. Use the pivoting process to solve the system of equations that models the problem given in Question 3 of this test.

11. List three explicit solutions to the following system:
$$
\begin{aligned}
x - 2y + z &= 5 \\
3x + 2y + z &= 7.
\end{aligned}
$$

12. Write the general solution in parametric form for this system:
$$
\begin{aligned}
x + y \quad\quad &= 5 \\
y + z &= 2 \\
x \quad\quad - z &= 3.
\end{aligned}
$$

13. Find the solution, if one exists:
$$
\begin{aligned}
x + 3y - 2z &= 5 \\
x + 2y + z &= 7 \\
y - 3z &= 2.
\end{aligned}
$$

14. Solve the system of equations that models the problem given in Question 4 of this test.

15. Solve the system of equations that models the problem given in Question 5 of this test.

Career Uses of Mathematics

From my early childhood, I knew that I would one day become a doctor. By high school, I had decided on radiology. I was told that it is a very technical specialty. I therefore emphasized both mathematics and physics in my curriculum during high school and college.

As a medical doctor specializing in radiology, I use x-rays, ultrasound, or large magnets to obtain diagnostic images of varying body parts. Even though my last finite math course was years ago, I am still (often to my surprise), in many phases of my medical practice, using the principles learned. For example, using magnetic resonance imaging (MRI) to evaluate a patient, we must have a working knowledge of matrices. The data we obtain vary by matrix size, depending on the body part to be imaged and the degree of resolution desired. For example, a large matrix (say, 256×256) will provide the high-resolution images needed for small structures like the pituitary gland. A smaller matrix (say, 256×192) might be used to image the entire brain.

Though radiology is not a primary care specialty and therefore does not involve a great degree of patient contact, it is still very rewarding. A disciplined, goal-oriented approach to my studies, which included mathematics, has brought me to this point.

Joseph S. Murphy, M.D.,
Radiology Consultants

3 *Matrix Algebra*

Pollsters find that intense campaigning in an upcoming bond issue election causes people to change the way they vote in the following way: At the beginning of a given week, 75% of those favoring the issue remain in favor a week later, while 25% change to opposing the issue; 60% of those opposing the issue still oppose the issue at the end of the week, while 40% change to supporting the issue. If a group of 100 voters is selected, 50 of whom support the issue and 50 of whom oppose it, predict how many of these voters will favor and how many will oppose the issue a week later, two weeks later, and three weeks later. (See Example 6, Section 3.2.)

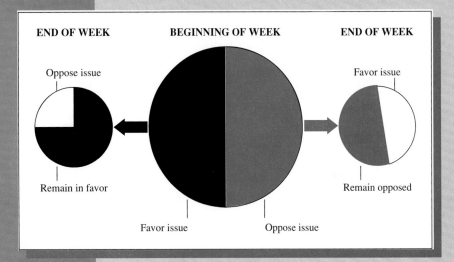

END OF WEEK   BEGINNING OF WEEK   END OF WEEK

Oppose issue

Remain in favor

Favor issue   Oppose issue

Favor issue

Remain opposed

CHAPTER PREVIEW

The notational convenience of using a matrix to represent a system of linear equations is only one of many applications in which matrices can be helpful. In this chapter, we define the addition, subtraction, and multiplication of matrices having compatible dimensions and the multiplication of matrices by a real number. The inverse of a matrix and how it may be found (when it exists) rounds out the operations needed to develop and explore a matrix algebra. The rules of operation are compared with those of the algebra of real numbers. The last section of this chapter gives three examples of how the inverse may be applied.

CALCULATOR/COMPUTER OPPORTUNITIES FOR THIS CHAPTER

Most graphing calculators are preprogrammed to do matrix operations including addition, multiplication, and finding the inverse. There are many computer programs that will do these same operations. Popular spreadsheets also have functions that will do matrix calculations.

The program MulTutor on the MasterIt disk, available through the publisher to accompany this text, provides a useful tutorial to help you learn matrix multiplication. The program Pivot on this disk may be used to find the inverse of a matrix or to determine when none exists.

SECTION 3.1   Matrix Addition and Applications

A **matrix** is a rectangular array of numbers. One useful way to classify matrices is by the number of rows and columns they have. If a matrix has m rows and n columns, then we say the **dimension** ("order," "size") of the matrix is $m \times n$, which is read "m by n." For example,

$$A = \begin{bmatrix} 2 & 3 & 5 \\ 1 & 0 & -2 \end{bmatrix}$$

has a dimension of 2×3, while

$$B = \begin{bmatrix} 2 & 1 \\ 4 & 6 \end{bmatrix}$$

has a dimension of 2×2. The matrix B is an example of a *square matrix*. The matrix A is not square.

> **The Dimension of a Matrix**
>
> A matrix with m rows and n columns has **dimension $m \times n$.** A matrix is **square** if the number of rows equals the number of columns.

One of the useful features of a matrix is that each entry is located at the intersection of some row and some column; this gives an "address" or "matrix position" for that entry in terms of a **row-column** designation. As already noted in Chapter 2, the general notation a_{ij} is used to denote the element in the ith row and jth column of a matrix. A general $m \times n$ matrix could then be written as

$$A = \begin{bmatrix} a_{11} & a_{12} & a_{13} & \cdots & a_{1j} & \cdots & a_{1n} \\ a_{21} & a_{22} & a_{23} & \cdots & a_{2j} & \cdots & a_{2n} \\ \vdots & & & & & & \\ a_{i1} & a_{i2} & a_{i3} & \cdots & a_{ij} & \cdots & a_{in} \\ \vdots & & & & & & \\ a_{m1} & a_{m2} & a_{m3} & \cdots & a_{mj} & \cdots & a_{mn} \end{bmatrix}$$

EXAMPLE 1
Organizing Data
in a Matrix

Suppose that food A contains 3 ounces of protein, 2 ounces of carbohydrate, and 6 ounces of fat per unit, while food B contains 4 ounces of protein, 5 ounces of carbohydrate, and 2 ounces of fat per unit. Using P, C, and F for protein, carbohydrate, and fat, respectively, this information may be displayed in one of the two ways shown, depending on preference or context:

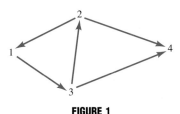

$$
\begin{array}{c}
 & \begin{array}{ccc} P & C & F \end{array} \\
\begin{array}{c} \text{Food A} \\ \text{Food B} \end{array} & \begin{bmatrix} 3 & 2 & 6 \\ 4 & 5 & 2 \end{bmatrix}
\end{array}
\quad \text{or} \quad
\begin{array}{c}
 & \begin{array}{cc} \text{Food A} & \text{Food B} \end{array} \\
\begin{array}{c} P \\ C \\ F \end{array} & \begin{bmatrix} 3 & 4 \\ 2 & 5 \\ 6 & 2 \end{bmatrix}
\end{array}
$$

EXAMPLE 2
Organizing Data in a
Matrix

Four stations are linked by communications lines, as shown in Figure 1. Direct communications may take place only in the direction of the arrows. Thus, 3 may communicate directly with 2, but 2 cannot communicate directly with 3.

FIGURE 1
Communication Lines

Assuming that each station communicates directly with itself, this communications network may be described by the matrix

$$
\begin{array}{c}
 & \begin{array}{cccc} 1 & 2 & 3 & 4 \end{array} \\
\begin{array}{c} 1 \\ 2 \\ 3 \\ 4 \end{array} & \begin{bmatrix} 1 & 0 & 1 & 0 \\ 1 & 1 & 0 & 1 \\ 0 & 1 & 1 & 1 \\ 0 & 0 & 0 & 1 \end{bmatrix}
\end{array}
$$

where a 1 in the ith row and jth column means that i may communicate directly with j, while a 0 means that it cannot.

EXAMPLE 3
Organizing Data in a
Matrix

A TV sales store sells brands W, X, Y, and Z, each with screen sizes of 20, 26, and 40 inches. A possible inventory matrix on January 1 might appear as follows:

$$
\begin{array}{c}
\\
\text{20-inch} \\
\text{26-inch} \\
\text{40-inch}
\end{array}
\begin{array}{cccc}
\text{W} & \text{X} & \text{Y} & \text{Z} \\
\end{array}
\begin{bmatrix}
8 & 6 & 2 & 0 \\
3 & 5 & 5 & 6 \\
1 & 3 & 2 & 4
\end{bmatrix} = A.
$$

How many 26-inch sets were on hand January 1?

Solution   There were 19.   ▄▄▄

To develop the full potential of matrices, we need to develop convenient ways of combining them and manipulating them as entities in their own right. The definitions and rules for doing this make up what is called **matrix algebra.** These rules are the familiar ones of regular algebra, with a few exceptions. To begin our study of matrix algebra, we must define when two matrices are equal.

> **Matrix Equality**
>
> Two matrices are equal if
>
> **1.** They have the same dimension.
>
> **2.** The entries in corresponding addresses are identical.

From this definition,

$$
\begin{bmatrix} 2 & 3 \\ 5 & 7 \end{bmatrix} \neq \begin{bmatrix} 1 & 2 & 3 \\ 0 & 4 & 5 \end{bmatrix}
$$

because they do not have the same dimension. Also,

$$
\begin{bmatrix} 3 & 5 \\ 2 & 1 \end{bmatrix} \neq \begin{bmatrix} 3 & 5 \\ 2 & 7 \end{bmatrix}
$$

because the entries in the second-row, second-column positions are not identical. The matrices

$$
\begin{bmatrix} 2 & x \\ 1 & 4 \end{bmatrix} \quad \text{and} \quad \begin{bmatrix} 2 & 3 \\ 1 & 4 \end{bmatrix}
$$

are equal if, and only if, $x = 3$.

> **Multiplying a Matrix by a Number**
>
> Let A be a matrix, and let c be any real number. Then the matrix cA is found by multiplying each entry in A by c. Multiplying a matrix by a number is known as **scalar multiplication.**

EXAMPLE 4
Multiplying a Matrix by
a Number

Let $A = \begin{bmatrix} 2 & 3 & 0 \\ 1 & -2 & 4 \end{bmatrix}$ and $c = 2$. Then

$$cA = 2A = \begin{bmatrix} 4 & 6 & 0 \\ 2 & -4 & 8 \end{bmatrix}.$$ ■

The concept of matrix addition offers no surprises.

> **The Sum of Two Matrices**
>
> For matrices A and B of the same dimension $m \times n$, $A + B$ denotes a matrix whose entry in each address is the sum of the entries in like addresses of A and B. Matrix addition is not defined for matrices of unequal dimension.

EXAMPLE 5
Adding Matrices

Let

$$A = \begin{bmatrix} 2 & 3 & 5 \\ 0 & 1 & 6 \end{bmatrix}, \quad B = \begin{bmatrix} 2 & 0 & -1 \\ 5 & 3 & 2 \end{bmatrix}, \quad \text{and} \quad C = \begin{bmatrix} 2 & 1 \\ 3 & -2 \end{bmatrix}.$$

Then

$$A + B = \begin{bmatrix} 2+2 & 3+0 & 5-1 \\ 0+5 & 1+3 & 6+2 \end{bmatrix} = \begin{bmatrix} 4 & 3 & 4 \\ 5 & 4 & 8 \end{bmatrix},$$

while $A + C$ and $B + C$ do not exist. To calculate $A + 3B$, we have

$$A + 3B = \begin{bmatrix} 2 & 3 & 5 \\ 0 & 1 & 6 \end{bmatrix} + \begin{bmatrix} 6 & 0 & -3 \\ 15 & 9 & 6 \end{bmatrix} = \begin{bmatrix} 8 & 3 & 2 \\ 15 & 10 & 12 \end{bmatrix}.$$

Note that $A + A = 2A$. This means that the two definitions involved yield results that appear like those of regular algebra. ■

EXAMPLE 6
An Application
Combining Matrix
Addition and
Multiplication of a
Matrix by a Number

An agricultural researcher is observing the growth of soybean, corn, and wheat plants under two fertilizer applications, N and P. Midway through the growing season, the average heights of the plants in each of the six plots were recorded in inches as follows:

$$\text{Fertilizer} \begin{array}{c} \\ N \\ P \end{array} \overset{\text{Soybeans} \quad \text{Corn} \quad \text{Wheat}}{\begin{bmatrix} 6 & 36 & 18 \\ 4 & 28 & 20 \end{bmatrix}} = A.$$

(a) Use a scalar multiplication to convert the average plant height in each plot into feet.

Solution   The multiplication of each entry in A by $\frac{1}{12}$ will convert inches to feet. The scalar multiplication is then

$$\frac{1}{12}A = \frac{1}{12}\begin{bmatrix} 6 & 36 & 18 \\ 4 & 28 & 20 \end{bmatrix} = \begin{bmatrix} \frac{1}{2} & 3 & \frac{3}{2} \\ \frac{1}{3} & \frac{7}{3} & \frac{5}{3} \end{bmatrix}.$$

(b) If the average plant heights in each plot are increasing at the rate of 5% per week, use a scalar multiplication to calculate the growth in inches during the week past midseason.

Solution The height increase during the week, in inches, is found by the scalar multiplication $0.05A$:

$$0.05A = 0.05\begin{bmatrix} 6 & 36 & 18 \\ 4 & 28 & 20 \end{bmatrix} = \begin{bmatrix} 0.3 & 1.8 & 0.9 \\ 0.2 & 1.4 & 1.0 \end{bmatrix}.$$

(c) Find a matrix expression that will give the average plant growth in inches for each plot during the second week past midseason.

Solution The growth (in inches) during the first week past midseason was found by $0.05A$. If we add these growths to the midseason heights found in A, the results will give plant heights at the end of one week past midseason: $A + 0.05A$. Now, applying the 5% growth rate to these heights gives

$$0.05(A + 0.05A) = 0.05A + (0.05)^2A = (0.05 + (0.05)^2)A = 0.0525A$$

for the growth during the second week past midseason. ▄▄▄

Matrix subtraction is a special case of matrix addition.

The Difference of Two Matrices

For matrices A and B of the same dimension, $A - B = A + (-1)B$. This means that the entry in each address of $A - B$ is the difference of the entries in the like addresses of A and B.

EXAMPLE 7
Matrix Subtraction

An investor bought Class A and Class B stock in each of three chemical companies, with dollar values as shown.

$$\begin{array}{cc} & \text{Class A} \quad \text{Class B} \\ \begin{array}{c} \text{Ion} \\ \text{Beaker} \\ \text{Atom} \end{array} & \begin{bmatrix} 1500 & 2000 \\ 800 & 1000 \\ 5000 & 3500 \end{bmatrix} = A. \end{array}$$

A bear market during the following year saw all of these stocks decline in value 10%. Find a matrix that will give the value of these stocks one year after purchase.

Solution The loss in value may be expressed as $0.10A$, so the actual value of the stocks would be given by $A - 0.10A$:

$$\begin{bmatrix} 1500 & 2000 \\ 800 & 1000 \\ 5000 & 3500 \end{bmatrix} - \begin{bmatrix} 150 & 200 \\ 80 & 100 \\ 500 & 350 \end{bmatrix} = \begin{bmatrix} 1350 & 1800 \\ 720 & 900 \\ 4500 & 3150 \end{bmatrix}. \quad ▄▄▄$$

The Zero Matrix

A matrix in which the element in every address is the number 0 is referred to as a **zero matrix** and is denoted by **0**.

For instance, $\mathbf{0} = \begin{bmatrix} 0 & 0 \\ 0 & 0 \\ 0 & 0 \end{bmatrix}$ is a 3×2 zero matrix. A zero matrix has the property that when added to any matrix A of the same size, the result is A. For example if $A = \begin{bmatrix} 2 & 5 & 6 \\ 1 & 9 & 3 \end{bmatrix}$ then, $\mathbf{0} + A = A$ because $\begin{bmatrix} 0 & 0 & 0 \\ 0 & 0 & 0 \end{bmatrix} + \begin{bmatrix} 2 & 5 & 6 \\ 1 & 9 & 3 \end{bmatrix} = \begin{bmatrix} 2 & 5 & 6 \\ 1 & 9 & 3 \end{bmatrix}$.

The word *compatible* is frequently used to describe matrices between which operations such as addition, subtraction, and so on are meaningful. If $A + B$ is written, for instance, we assume that the two matrices have the same dimension so that such symbolism actually represents a matrix, and we say A and B are **compatible** (with respect to addition). With this in mind, matrix addition satisfies the following properties.

Properties of Matrix Addition

Let A and B be $m \times n$ matrices. Then

1. Matrix addition is commutative:
$$A + B = B + A.$$

2. Matrix addition is associative:
$$(A + B) + C = A + (B + C).$$

3. There is an identity matrix $\mathbf{0}$ of dimension $m \times n$ for which
$$A + \mathbf{0} = \mathbf{0} + A = A.$$

4. For any numbers a and b,
$$(a + b)A = aA + bA.$$

These properties are similar to those for the algebra of real numbers.

EXAMPLE 8
Solving Matrix
Equations

(a) Solve this equation for X:

$$X + \begin{bmatrix} 2 & 1 & 3 \\ 5 & 0 & -1 \end{bmatrix} = \begin{bmatrix} 1 & -3 & 2 \\ 1 & 0 & 1 \end{bmatrix}.$$

Solution $X = \begin{bmatrix} 1 & -3 & 2 \\ 1 & 0 & 1 \end{bmatrix} - \begin{bmatrix} 2 & 1 & 3 \\ 5 & 0 & -1 \end{bmatrix} = \begin{bmatrix} -1 & -4 & -1 \\ -4 & 0 & 2 \end{bmatrix}.$

(b) Solve this equation for A:

$$2A - \begin{bmatrix} 1 & 2 \\ 0 & 1 \end{bmatrix} = \begin{bmatrix} 5 & 1 \\ 2 & 1 \end{bmatrix}.$$

Solution $2A = \begin{bmatrix} 5 & 1 \\ 2 & 1 \end{bmatrix} + \begin{bmatrix} 1 & 2 \\ 0 & 1 \end{bmatrix} = \begin{bmatrix} 6 & 3 \\ 2 & 2 \end{bmatrix}$ so that

$$A = \frac{1}{2}\begin{bmatrix} 6 & 3 \\ 2 & 2 \end{bmatrix} = \begin{bmatrix} 3 & \frac{3}{2} \\ 1 & 1 \end{bmatrix}.$$

(c) Solve this equation for X: $3X + 5A - B = 6C$.

Solution $3X = 6C + B - 5A$ so that $X = 2C + \frac{1}{3}B - \frac{5}{3}A$. ■

A typical use of matrices involves inventory tabulations. Usually, such information is stored in computers, perhaps in a spreadsheet, and on a much larger scale than for our problems here. The general rule of inventory status over a given period of time is as follows.

Inventory Rule

Let matrices be defined as follows: B is the beginning inventory, P is the purchases, S is the sales, and E is the ending inventory for a given time interval. Then $E = B + P - S$.

Whenever three of the four matrices in this equation are known, we may solve for the fourth. Let's look at an example of this rule.

EXAMPLE 9
An inventory
Application

A furniture dealer has four stores, A, B, C, and D, in a large city and keeps records about the sales of Early American and French provincial dining room sets in matrix form on a computer. Suppose that the records show

$$\text{Beginning inventory matrix} = \begin{array}{c} A \\ B \\ C \\ D \end{array} \begin{bmatrix} \overset{\text{E. Am.}}{8} & \overset{\text{Fr. Prov.}}{2} \\ 6 & 9 \\ 12 & 7 \\ 3 & 10 \end{bmatrix} = B,$$

$$\text{Purchase matrix} = \begin{array}{c} A \\ B \\ C \\ D \end{array} \begin{bmatrix} \overset{\text{E. Am.}}{8} & \overset{\text{Fr. Prov.}}{2} \\ 12 & 5 \\ 6 & 10 \\ 9 & 11 \end{bmatrix} = P,$$

$$\text{Ending inventory matrix} = \begin{array}{c} A \\ B \\ C \\ D \end{array} \begin{bmatrix} \overset{\text{E. Am.}}{7} & \overset{\text{Fr. Prov.}}{2} \\ 9 & 6 \\ 10 & 12 \\ 5 & 9 \end{bmatrix} = E.$$

Find the sales matrix.

Solution According to the general rule on inventory, $E = B + P - S$ so that $S = B + P - E$ or

$$
S = \begin{bmatrix} 8 & 2 \\ 6 & 9 \\ 12 & 7 \\ 3 & 10 \end{bmatrix} + \begin{bmatrix} 8 & 2 \\ 12 & 5 \\ 6 & 10 \\ 9 & 11 \end{bmatrix} - \begin{bmatrix} 7 & 2 \\ 9 & 6 \\ 10 & 12 \\ 5 & 9 \end{bmatrix} = \begin{bmatrix} 9 & 2 \\ 9 & 8 \\ 8 & 5 \\ 7 & 12 \end{bmatrix}.
$$

The last operation on a matrix to be mentioned in this section is the transpose of a matrix. We want to show how the transpose is found so that you are aware of this operation, but we delay its application until a later chapter.

The Transpose of a Matrix

The **transpose** of any matrix A, written A^T, is the matrix for which the columns are the respective rows of A.

EXAMPLE 10
Finding the Transpose
of a Matrix

Find the transpose of the matrix $A = \begin{bmatrix} 3 & 4 & -1 & 7 \\ 2 & 0 & 1 & 0 \\ 1 & 1 & 1 & 2 \end{bmatrix}$.

Solution According to the definition, we start with the first row of A and make it the first column of A^T, then take the second row of A and make it the second column of A^T, and so on. This gives

$$
A^T = \begin{bmatrix} 3 & 2 & 1 \\ 4 & 0 & 1 \\ -1 & 1 & 1 \\ 7 & 0 & 2 \end{bmatrix}.
$$

This section has begun the development of an algebra for matrices. The next two sections complete the basic algebra structure we need.

Graphing calculators have the preprogrammed ability to do matrix algebra, with certain makes and models having more size capacity than others. Figure 2 shows a graphing calculator screen indicating the calculation of $(2A + B)^T$ where

$$
A = \begin{bmatrix} 2 & 6 & 5 \\ 1 & 0 & 2 \end{bmatrix} \quad \text{and} \quad B = \begin{bmatrix} -2 & 3 & -5 \\ 1 & 3 & 4 \end{bmatrix}.
$$

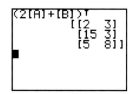

FIGURE 2
*Matrix Algebra
by Technology*

EXERCISES   3.1

In Exercises 1 through 4, assign the value to the variables so that a matrix equality results.

1. $\begin{bmatrix} t+1 & 5 & 2 \\ 7 & 6 & 0 \end{bmatrix} = \begin{bmatrix} 4 & 5 & 2 \\ 7 & x-3 & 0 \end{bmatrix}$

2. $\begin{bmatrix} 2 & x-4 \\ 3 & 0 \\ 7 & 0 \end{bmatrix} = \begin{bmatrix} 2 & 6 \\ t-1 & 0 \\ 1 & 0 \end{bmatrix}$

3. $\begin{bmatrix} 0 & 0 \\ x & -2 \end{bmatrix} = \begin{bmatrix} 0 & 0 \\ x & -2 \end{bmatrix}$

4. $\begin{bmatrix} x-1 & 4 & 5 \\ 8 & z & 0 \\ y-3 & 0 & 1 \end{bmatrix} = \begin{bmatrix} 2x+3 & 4 & t \\ 8 & z & 0 \\ 2y & 0 & 1 \end{bmatrix}$

In Exercises 5 through 12, use the given matrices A, B, and C to find the indicated expression, if possible. Let

$$A = \begin{bmatrix} 2 & 3 \\ 1 & -1 \end{bmatrix}, \quad B = \begin{bmatrix} 6 & -1 \\ 0 & 2 \end{bmatrix}, \quad \text{and}$$

$$C = \begin{bmatrix} 2 & -3 & 5 \\ 0 & 0 & -2 \end{bmatrix}.$$

5. $-3A + B$ **6.** $A + 4B$ **7.** $\frac{1}{3}(A + B)$

8. $\frac{1}{2}(B - A)$ **9.** $\frac{1}{4}(2A + C)$

10. $A + 2B - C$ **11.** $-5(C + 3C)$ **12.** $C - C$

In Exercises 13 and 14, use

$$A = \begin{bmatrix} a & b \\ c & d \end{bmatrix},$$

$$B = \begin{bmatrix} e & f \\ g & h \end{bmatrix},$$

and $C = \begin{bmatrix} i & j \\ k & l \end{bmatrix}$

in each expression to verify the equality.

13. $A + B = B + A$

14. $(A + B) + C = A + (B + C)$

15. *Seating Passengers*   An airline is buying two types of planes, P_1 and P_2. Plane P_1 will seat 30 first-class passengers, 50 tourist-class passengers, and 90 economy-class passengers. Plane P_2 will seat 50 first-class passengers, 60 tourist-class passengers, and 100 economy-class passengers.

(a) Display this information in a 2×3 matrix.
(b) Display this information in a 3×2 matrix.

16. *Advertising*   The John Cross firm encounters these costs in advertising its computers: $80 per minute per radio ad, $100 per column inch per newspaper ad, and $250 per minute per television ad.

(a) Display this information in a 1×3 matrix.
(b) Display this information in a 3×1 matrix.

17. *College Enrollment*   A particular small college has 300 freshmen, 287 sophomores, 250 juniors, and 240 seniors.

(a) Display this information in a 4×1 matrix.
(b) Display this information in a 1×4 matrix.

18. *Communications*   Three military outposts are linked by communication devices as shown in the graph below. Assume that each station communicates with itself and that communication with the other stations takes place only in the direction of the arrows. Write a matrix that displays this communications network.

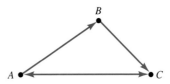

19. *Airline Flights*   Airline routes through cities A, B, C, D, and E are shown in the following diagram, and flights are available in the direction of the arrows only. For matrix considerations, assume that each city does not have a flight from itself to itself. Write a matrix that displays this flight network.

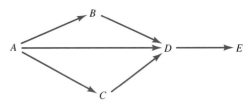

*A **dominance relation** expresses the fact that among members of a group, some members exert dominance over other members. Examples are the hierarchy of an office staff, the perceived status among students in a class, and an athletic team winning a sporting contest. A dominance matrix can be built that will be similar to the matrix in Example 2 of this section. For any two members x and y, put a 1 if x dominates y, otherwise put a 0, and assume that no persons dominate themselves.*

20. *Dominance Relations*   Among executives Smith, Jones, Barney, and Lemmons, Smith reports to Barney, Jones reports to Barney, and Barney reports to Lemmons. Make a dominance matrix showing these relationships.

21. Dominance Relations The Bears, the Wildcats, the Tigers, and the Cubs play in a basketball tournament. During the first two days of play, the Bears beat the Wildcats and the Tigers, the Tigers beat the Cubs, the Cubs beat the Bears and the Wildcats, and the Wildcats beat the Tigers. Construct a dominance matrix showing these tournament results.

In Exercises 22 through 27, solve each of the matrix equations for the indicated variable.

22. Solve for X: $X + 2B = C$

23. Solve for X: $\frac{3}{4}X - 5T = 3B$

24. Solve for A: $2A + 3B = \mathbf{0}$

25. Solve for B: $\frac{1}{2}A - 3B + C = 2D$

26. Solve for C: $2C + \frac{1}{2}B = 4C$

27. Solve for T: $A - T = 2C + 3T$

In Exercises 28 through 31, solve each matrix equation for X.

28. $X + \begin{bmatrix} 2 & 3 \\ 5 & 1 \end{bmatrix} = \begin{bmatrix} 5 & -1 \\ 2 & 6 \end{bmatrix}$

29. $\begin{bmatrix} 2 & -3 & 1 \\ 4 & 5 & 2 \end{bmatrix} + 2X = -3 \begin{bmatrix} -1 & 2 & -1 \\ 1 & 0 & 1 \end{bmatrix}$

30. $\begin{bmatrix} 2 & 1 & -3 \\ 5 & 0 & 0 \\ 1 & -2 & 3 \end{bmatrix} + \frac{1}{3}X = \begin{bmatrix} -1 & -2 & -3 \\ 0 & 1 & 1 \\ 5 & 0 & 2 \end{bmatrix}$

31. $2\begin{bmatrix} 1 & 1 \\ 2 & 0 \\ 3 & -2 \end{bmatrix} + 3X + \begin{bmatrix} 2 & 5 \\ -1 & 3 \\ -4 & -2 \end{bmatrix} = \begin{bmatrix} 0 & 0 \\ 0 & 0 \\ 0 & 0 \end{bmatrix}$

32. Is there a real number x that will solve this equation?

$$x\begin{bmatrix} 2 & 3 \\ 4 & 1 \end{bmatrix} + x\begin{bmatrix} 5 & 6 \\ 2 & 2 \end{bmatrix} = \begin{bmatrix} 14 & 18 \\ 12 & 9 \end{bmatrix}$$

33. Is there a real number x that will solve this equation?

$$x\begin{bmatrix} 2 & 1 & 1 \\ 0 & 1 & 1 \end{bmatrix} + x\begin{bmatrix} 1 & 2 & 2 \\ 3 & -1 & -1 \end{bmatrix} = \begin{bmatrix} 9 & 9 & 9 \\ 9 & 0 & 0 \end{bmatrix}$$

34. Multiplication by a Number If the matrix

$$A = \begin{bmatrix} 20.2 & 50.6 \\ 28.7 & 31.1 \end{bmatrix}$$

gives sales in thousands of dallars, perform a scalar multiplication that will result in a matrix that displays the sales in dollars.

35. Multiplication by a Number If the matrix

$$B = \begin{bmatrix} 0.06 & 0.02 \\ 0.03 & 0.05 \end{bmatrix}$$

gives times in hours, perform a scalar multiplication that will result in times given in minutes.

Exercises 36 and 37 refer to this inventory problem.

Inventory During a given week, a small convenience store has a beginning inventory of

	2% milk	Whole milk
Quarts	30	20
$\frac{1}{2}$ gal.	18	12

$= B,$

and sales inventory of

	2% milk	Whole milk
Quarts	21	18
$\frac{1}{2}$ gal.	22	15

$= S.$

36. If the ending inventory matrix is $\frac{1}{2}B$, what is the purchase matrix?

37. If the ending inventory matrix is

$$E = \begin{bmatrix} 10 & 8 \\ 12 & 10 \end{bmatrix},$$

what is the purchase matrix?

38. Pollution Control A state pollution-control office is monitoring two waste-treatment plants that discharge into the same watershed. Three of the quantities monitored from plant discharge are nitrates, lead, and oxygen content, measured in parts per million (ppm). At the beginning of the monitoring period, the measurements were

	Nitrates	Lead	Oxygen
Plant A	13	17	30
Plant B	20	22	50

$= B.$

At the end of the monitoring period six months later, the measurements were

	Nitrates	Lead	Oxygen
Plant A	10	15	30
Plant B	15	16	45

$= E.$

(a) Find the matrix $B - E$.

(b) What do the entries in $B - E$ represent?

(c) Does the matrix $B - E$ show which plant is making the best effort at reducing these pollutants in its discharge?

(d) If this trend continues, how many months will it take for plant A to entirely eliminate its nitrate discharge?

39. *Investing*   Mr. Jones has two types of IRA retirement accounts—certificates of deposit (CD) and municipal bonds (MB)—held in two banks, First National and Commercial National. The dollar amounts in each at the beginning of 1996 are given in the matrix:

$$\begin{array}{c} \\ \text{FN} \\ \text{CN} \end{array} \begin{array}{cc} \text{CD} & \text{MB} \\ \begin{bmatrix} 30{,}000 & 25{,}000 \\ 50{,}000 & 40{,}000 \end{bmatrix} \end{array} = A.$$

Assume an 8% appreciation rate on each of these accounts.

(a) Find a matrix expression representing the earnings of the various accounts during 1996.

(b) Compute the matrix expression found in (a).

(c) Find a matrix expression representing the value of each account at the end of 1996.

(d) Compute the matrix expression found in (c).

Exercises 40 through 44 refer to this problem.

College Costs   The Siffords are comparing the costs per year (in dollars) of sending their daughter to one of two universities. The costs are shown in the following matrix.

$$\begin{array}{c} \\ \text{Tuition, fees} \\ \text{Living expenses} \end{array} \begin{array}{cc} \text{State} & \text{Private} \\ \begin{bmatrix} 2500 & 6000 \\ 4800 & 5400 \end{bmatrix} \end{array} = C.$$

Assume an annual inflation rate of 5% in each of these costs.

40. Find a matrix expression that shows the increases in these costs for the second year at each university.

41. Find a matrix expression that represents the actual costs for the second year at each of the universities. Compute this expression.

42. Find a matrix expression that represents the actual costs for the third year at each of the universities. Compute this expression.

43. Find a matrix expression that represents the actual costs for the fourth year at each of the universities. Compute this expression.

44. Find a matrix expression that represents the actual costs of attending four years at each of the universities. Compute this expression.

45. *Psychology*   Three groups of people are tested for their ability to memorize and retain sequences of numbers. The first group had no training, the second group received a short course on memory, and the third group received intensive training on memory. The following table shows the average number of seconds the numbers were retained by each group:

Number of digits in sequence	Group I	Group II	Group III
1–8	14	16	18
9–14	8	11	15
15–18	6	10	13

Construct a matrix showing the percentage by which the performance of the second two groups surpassed the first.

In Exercises 46 through 49, find the transpose of each matrix.

46. $A = \begin{bmatrix} 2 & -3 & 1 \\ 0 & 5 & -2 \end{bmatrix}$ **47.** $B = \begin{bmatrix} 1 & 2 \\ 3 & -5 \end{bmatrix}$

48. $C = \begin{bmatrix} 2 \\ -1 \\ 3 \end{bmatrix}$ **49.** $D = \begin{bmatrix} 1 & 0 \\ 2 & -1 \\ -3 & 2 \\ 4 & 1 \end{bmatrix}$

Problems for Exploration

1. Investigate the "address system" for elements in two different spreadsheets, and write a short report on them and how they differ from the conventional mathematical system.

2. A matrix is called **symmetric** if $A = A^T$. If A is a 3×3 symmetric matrix, prove that $A + A^T$ is also symmetric. (Fill in the entries in A with letters a, b, c, ...)

3. A matrix is called **skew symmetric** if $A^T = -A$. If A is a 3×3 skew-symmetric matrix, prove that $A - A^T$ is also skew symmetric. (Fill in the entries in A with letters a, b, c, ...)

Matrix Multiplication and Applications

One of the simplest equations in algebra is $ax = b$, and we know that the solution is $x = a^{-1}b\left(a^{-1} = \dfrac{1}{a}\right)$, provided that $a \neq 0$. A motivation for matrix multiplication comes from the desire to convert a system of linear equations into a similar **matrix equation,** $AX = B$ and then solve for X (under some specific conditions) to obtain a solution to the system of the same form as that in algebra, $X = A^{-1}B$. To this end, our first concern is to define the multiplication of matrices A and X so that the product AX is the matrix B *and* represents an equivalent way of writing the original system of linear equations. An example points the way. It turns out that the process of multiplication is not as easily guessed as it was for addition.

Consider the system
$$2x + 3y = 5$$
$$4x + y = 6.$$

To arrive at the form $AX = B$, the appearance of the system suggests that B should be the 2×1 matrix $\begin{bmatrix} 5 \\ 6 \end{bmatrix}$. The matrix X plays the role of a variable, so it must somehow represent the variables of the system. To do this, define $X = \begin{bmatrix} x \\ y \end{bmatrix}$. The only thing left for A is the 2×2 coefficient matrix $A = \begin{bmatrix} 2 & 3 \\ 4 & 1 \end{bmatrix}$. Thus, if $AX = B$ or $\begin{bmatrix} 2 & 3 \\ 4 & 1 \end{bmatrix}\begin{bmatrix} x \\ y \end{bmatrix} = \begin{bmatrix} 5 \\ 6 \end{bmatrix}$ is to hold, then AX must be a 2×1 matrix; furthermore, to represent the given system of equations, it must look like this:

$$AX = \begin{bmatrix} 2x + 3y \\ 4x + y \end{bmatrix} = \begin{bmatrix} 5 \\ 6 \end{bmatrix} = B.$$

That is, $2x + 3y$ represents the single element in the first-row, first-column position and $4x + y$ represents the single element in the second-row, first-column position. Now notice how each row of A is combined with the column $\begin{bmatrix} x \\ y \end{bmatrix}$ to produce these results:

first row on the left, combined with the column on the right

second row on the left, combined with the column on the right

Do you see how they are combined? The rule of combination follows:

first element in row times first element in column, plus

second element in row times second element in column

This description of combining a row with a column to produce a single number may be thought of as a **row-column operation.** In mathematics, such an operation has historically been called the **dot product.** Based on the preceding pattern, we can now make a general rule for the dot product.

The Dot Product

Let A be an $m \times n$ matrix, and let B be an $n \times q$ matrix. Then for a typical

row in A, $[a_1\ a_2\ a_3\ \ldots\ a_n]$ and a typical column in B, $\begin{bmatrix} b_1 \\ b_2 \\ b_3 \\ \ldots \\ b_n \end{bmatrix}$, the **dot product,**

denoted by ■, is a **row-column operation** that gives this *number*:

$$[a_1\ a_2\ a_3\ \ldots\ a_n] \blacksquare \begin{bmatrix} b_1 \\ b_2 \\ b_3 \\ \ldots \\ b_n \end{bmatrix} = a_1b_1 + a_2b_2 + a_3b_3 + \ldots + a_nb_n.$$

The dot product can be calculated only when the rows of A have the same number of elements as the columns of B.

EXAMPLE 1
Demonstrating the Dot
Product

Let $A = \begin{bmatrix} 3 & 1 & 2 \\ 1 & -1 & 6 \\ 1 & 2 & 3 \end{bmatrix}$ and let $B = \begin{bmatrix} 5 & 1 \\ 2 & 0 \\ 3 & -1 \end{bmatrix}$.

(a) The dot product (row-column operation) applied to row 2 of A and column 1 of B would be

$$[1\ \ -1\ \ 6] \blacksquare \begin{bmatrix} 5 \\ 2 \\ 3 \end{bmatrix} = 1 \cdot 5 + (-1) \cdot 2 + 6 \cdot 3 = 21.$$

(b) Similarly, row 3 ■ column 2 $= [1\ \ 2\ \ 3] \blacksquare \begin{bmatrix} 1 \\ 0 \\ -1 \end{bmatrix}$

$$= 1 \cdot 1 + 2 \cdot 0 + 3 \cdot (-1) = -2. \quad \blacksquare$$

The multiplication of two matrices is now defined, using the dot product.

Defining Matrix Multiplication

Let A be an $m \times n$ matrix and let B be an $n \times q$ matrix. Then the **multiplication of A by B,** denoted by AB, is the $m \times q$ matrix whose entry in the ith row and jth column address is the dot product of the ith row of A with the jth column of B.

EXAMPLE 2
Illustrating Matrix
Multiplication

Let $A = \begin{bmatrix} 2 & 3 \\ 5 & 1 \end{bmatrix}$ and $B = \begin{bmatrix} 4 & 2 \\ -1 & 1 \end{bmatrix}$.

(a) Compute AB.

Solution $AB = \begin{bmatrix} 2 & 3 \\ 5 & 1 \end{bmatrix}\begin{bmatrix} 4 & 2 \\ -1 & 1 \end{bmatrix}$

$$= \begin{bmatrix} \text{row 1 of } A \ \blacksquare \ \text{column 1 of } B & \text{row 1 of } A \ \blacksquare \ \text{column 2 of } B \\ \text{row 2 of } A \ \blacksquare \ \text{column 1 of } B & \text{row 2 of } A \ \blacksquare \ \text{column 2 of } B \end{bmatrix}$$

$$= \begin{bmatrix} [2 \ \ 3] \ \blacksquare \ \begin{bmatrix} 4 \\ -1 \end{bmatrix} & [2 \ \ 3] \ \blacksquare \ \begin{bmatrix} 2 \\ 1 \end{bmatrix} \\ [5 \ \ 1] \ \blacksquare \ \begin{bmatrix} 4 \\ -1 \end{bmatrix} & [5 \ \ 1] \ \blacksquare \ \begin{bmatrix} 2 \\ 1 \end{bmatrix} \end{bmatrix} = \begin{bmatrix} 5 & 7 \\ 19 & 11 \end{bmatrix}.$$

(b) Compute BA.

Solution $BA = \begin{bmatrix} 4 & 2 \\ -1 & 1 \end{bmatrix}\begin{bmatrix} 2 & 3 \\ 5 & 1 \end{bmatrix}$

$$= \begin{bmatrix} \text{row 1 of } B \ \blacksquare \ \text{column 1 of } A & \text{row 1 of } B \ \blacksquare \ \text{column 2 of } A \\ \text{row 2 of } B \ \blacksquare \ \text{column 1 of } A & \text{row 2 of } B \ \blacksquare \ \text{column 2 of } A \end{bmatrix}$$

$$= \begin{bmatrix} [4 \ \ 2] \ \blacksquare \ \begin{bmatrix} 2 \\ 5 \end{bmatrix} & [4 \ \ 2] \ \blacksquare \ \begin{bmatrix} 3 \\ 1 \end{bmatrix} \\ [-1 \ \ 1] \ \blacksquare \ \begin{bmatrix} 2 \\ 5 \end{bmatrix} & [-1 \ \ 1] \ \blacksquare \ \begin{bmatrix} 3 \\ 1 \end{bmatrix} \end{bmatrix} = \begin{bmatrix} 18 & 14 \\ 3 & -2 \end{bmatrix}.$$

(c) Compute A^2. ($A^2 = AA$. It *does not* mean to square each element in A.)

Solution $A^2 = \begin{bmatrix} 2 & 3 \\ 5 & 1 \end{bmatrix}\begin{bmatrix} 2 & 3 \\ 5 & 1 \end{bmatrix} = \begin{bmatrix} 19 & 9 \\ 15 & 16 \end{bmatrix}.$ ■

Examples 2(a) and 2(b) show that matrix multiplication is *not* commutative.

    This computational suggestion might be helpful when forming the product of two matrices: Take the first row of the matrix on the left, and dot product it with *each* column of the matrix on the right. This will give the first row of the product. Next, take the second row of the matrix on the left, and dot product it with each column in the matrix on the right. This gives the second row of the product. Next, take the third row of the matrix on the left, and so on.

EXAMPLE 3
Illustrating Matrix
Multiplication

Let $A = \begin{bmatrix} 2 & 3 & 4 \\ -2 & 1 & 0 \end{bmatrix}$ and $B = \begin{bmatrix} 1 & 4 & -2 \\ 0 & 1 & 0 \\ 2 & 2 & 1 \end{bmatrix}$.

(a) The respective dimensions are A: 2×3 and B: 3×3. The number of columns, 3, in A is also the number of rows in B, and this must be the case in order for the dot product to be performed.

A: 2×3 B: 3×3

Must match for product

Dimension of product

Therefore, $AB = \begin{bmatrix} 2 & 3 & 4 \\ -2 & 1 & 0 \end{bmatrix} \begin{bmatrix} 1 & 4 & -2 \\ 0 & 1 & 0 \\ 2 & 2 & 1 \end{bmatrix} = \begin{bmatrix} 10 & 19 & 0 \\ -2 & -7 & 4 \end{bmatrix}$.

(b) The product BA does not exist. The reason is that when the order of B is considered first and the order A second, B: 3×3, A: 2×3, the middle numbers do not match. If a dot product were attempted, there would be a mismatch between the elements in any row of B with the elements of any column of A.

With matrix multiplication defined, we are now in a position to see how any linear system of equations can be written as an equivalent matrix equation in the form $AX = B$. We must wait until the next section, however, to arrive at the nice solution mentioned at the outset of this section.

EXAMPLE 4
A System Written as a
Matrix Equation
$AX = B$

The system

$$-2x + y - z = 3$$
$$x \qquad + z = 2$$
$$5x + y \qquad = -6$$
$$x + 2y + z = 3$$

may be written in the matrix equation form as

$$\begin{bmatrix} -2 & 1 & -1 \\ 1 & 0 & 1 \\ 5 & 1 & 0 \\ 1 & 2 & 1 \end{bmatrix} \begin{bmatrix} x \\ y \\ z \end{bmatrix} = \begin{bmatrix} 3 \\ 2 \\ -6 \\ 3 \end{bmatrix},$$

4×3 3×1 $4 \times 1.$

EXAMPLE 5
An Application of
Matrix Multiplication

Costs, in dollars, for radio (per minute), newspaper (per column inch), and TV ads (per minute) in two cities, Ocala and Hawthorne, are shown next.

	Radio	Newspaper	TV
Ocala	30	20	120
Hawthorne	25	18	140

(a) If ads are run five times in each of these media in Ocala and eight times in each of the media in Hawthorne, display a matrix product that will give the total amount spent on radio, newspaper, and TV ads.

Solution $[5 \quad 8] \begin{bmatrix} \overset{\textbf{R}}{30} & \overset{\textbf{N}}{20} & \overset{\textbf{TV}}{120} \\ 25 & 18 & 140 \end{bmatrix} = [\overset{\textbf{R}}{350} \quad \overset{\textbf{N}}{244} \quad \overset{\textbf{TV}}{1720}].$

This product matrix gives the totals requested.

(b) Suppose that the radio, newspaper, and TV ads are run 20, 15, and 30 times, respectively, in each of the two cities. Display a matrix product that will give the amount spent on ads in each of the cities.

Solution $\begin{matrix} \textbf{O} \\ \textbf{H} \end{matrix} \begin{bmatrix} 30 & 20 & 120 \\ 25 & 18 & 140 \end{bmatrix} \begin{bmatrix} 20 \\ 15 \\ 30 \end{bmatrix} = \begin{bmatrix} 4500 \\ 4970 \end{bmatrix} \begin{matrix} \textbf{O} \\ \textbf{H} \end{matrix}.$

The entries in the product matrix give the total costs in each city.

(c) Suppose that the radio, newspaper, and TV ads are run 30, 40, and 60 times, respectively, in each of the cities in January, and 20, 12, and 22 times, respectively, in February. Use a matrix multiplication to give the total amount spent in each city in January and February.

Solution $\begin{matrix} \textbf{O} \\ \textbf{H} \end{matrix} \begin{bmatrix} 30 & 20 & 120 \\ 25 & 18 & 140 \end{bmatrix} \begin{bmatrix} \overset{\textbf{J}}{30} & \overset{\textbf{F}}{20} \\ 40 & 12 \\ 60 & 22 \end{bmatrix} = \begin{bmatrix} \overset{\textbf{J}}{8900} & \overset{\textbf{F}}{3480} \\ 9870 & 3796 \end{bmatrix} \begin{matrix} \textbf{O} \\ \textbf{H} \end{matrix}.$

The first column in the product matrix gives the costs for ads in January in each of the cities, and the second column gives the costs for ads in February in the two cities.

In addition to graphing calculators and various other computer programs, spreadsheets offer not only a preprogrammed way to do matrix multiplication, but allow the labeling of rows and columns which is helpful in interpreting the meaning of the entries in the product matrix. Figure 3 on the next page shows how a spreadsheet might be used to calculate the matrix product in part (c) of Example 5. The product was calculated using the MMULT function operation of the spreadsheet.

EXAMPLE 6
Predicting Votes in an Election

The voting problem stated at the outset of this chapter tells how voter opinions change from week to week. Assuming that the stated trends hold, can we predict how these 100 voters will vote week by week? The answer is "yes." We first construct the matrix T whose first row records how the "for" voters at the beginning of the week will, as a percentage, take a "for" or "against" position by the end of the week; the second row records how the "against" voters at the beginning of the week will, as a percentage, take a "for" or "against" position by the end of the week.

$$\begin{matrix} & & & \text{End of week} \\ & & & \text{For} \quad\quad \text{Against} \\ \text{Beginning} & \text{For} & \begin{bmatrix} 0.75 & 0.25 \\ 0.40 & 0.60 \end{bmatrix} & = T. \\ \text{of week} & \text{Against} & & \end{matrix}$$

Then, we let A be the following matrix showing the initial position of the 100 voters:

$$A = \overset{\text{For} \quad \text{Against}}{[50 \quad\quad 50 \quad]}.$$

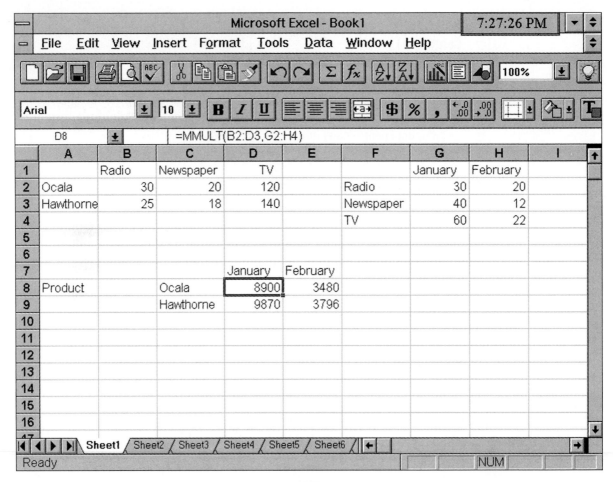

FIGURE 3

Matrix Multiplication Done on a Spreadsheet

Now, if we consider the product AT,

$$AT = \begin{bmatrix} 50 & 50 \end{bmatrix} \begin{bmatrix} 0.75 & 0.25 \\ 0.40 & 0.60 \end{bmatrix} = \begin{bmatrix} 57.5 & 42.5 \end{bmatrix},$$

the entries in the product will have these meanings:

$$50 \times 0.75 + 50 \times 0.40 = 57.5,$$

"For"   Still   "Against"   Change   "For" at end
     "For"   to "For"   of week

$$50 \times 0.25 + 50 \times 0.60 = 42.5.$$

"For"   Change to   "Against"   Still   "Against" at
     "Against"   "Against"   end of week

Therefore, about 58 voters will be for the issue and 42 will be against it at the end of the week. At the end of the second week, the figures will be

$$\begin{bmatrix} 57.5 & 42.5 \end{bmatrix} \begin{bmatrix} 0.75 & 0.25 \\ 0.40 & 0.60 \end{bmatrix} = \begin{bmatrix} 60.125 & 39.875 \end{bmatrix},$$

or about 60 of the voters in favor and about 40 against. At the end of the third week, we find that

$$[60.125 \quad 39.875]\begin{bmatrix} 0.75 & 0.25 \\ 0.40 & 0.60 \end{bmatrix} = [61.04 \quad 38.95],$$

or about 61 of the voters will be in favor of the issue, while about 39 will be against.

A matrix of special significance is an identity matrix.

The Identity Matrix

An **identity matrix** is any square matrix whose entries in the ith row, ith column addresses are all 1s and 0s elsewhere. An identity matrix is denoted by I regardless of dimension.

For example,

$$I = \begin{bmatrix} 1 & 0 \\ 0 & 1 \end{bmatrix}$$

has the dimension 2×2, and

$$I = \begin{bmatrix} 1 & 0 & 0 \\ 0 & 1 & 0 \\ 0 & 0 & 1 \end{bmatrix}$$

has the dimension 3×3. If

$$A = \begin{bmatrix} 3 & -5 \\ 2 & 8 \end{bmatrix},$$

observe that

$$AI = \begin{bmatrix} 3 & -5 \\ 2 & 8 \end{bmatrix}\begin{bmatrix} 1 & 0 \\ 0 & 1 \end{bmatrix} = \begin{bmatrix} 3 & -5 \\ 2 & 8 \end{bmatrix} = A$$

and

$$IA = \begin{bmatrix} 1 & 0 \\ 0 & 1 \end{bmatrix}\begin{bmatrix} 3 & -5 \\ 2 & 8 \end{bmatrix} = \begin{bmatrix} 3 & -5 \\ 2 & 8 \end{bmatrix} = A.$$

This illustrates the important fact that an identity matrix plays the *same role* for *square matrices* in matrix algebra as 1 does in ordinary algebra: $AI = IA = A$.

There are several exceptions to the usual rules of algebra when matrix multiplication is considered. For instance, we have already seen by example that matrix multiplication is not commutative. Enough properties do hold, however, so that we have a useful structure. The associative and distributive properties may not seem obvious, but we accept them for now until they can be verified by calculation in the exercises.

Properties of Matrix Multiplication

For matrices that are compatible under multiplication, matrix multiplication satisfies the following properties:

1. The associative law holds $(AB)C = A(BC)$.

2. If A is a square $n \times n$ matrix, $IA = AI = A$.

3. Matrix multiplication distributes over addition:

$$A(B + C) = AB + AC$$

and

$$(B + C)A = BA + CA.$$

The distributive property is particularly useful in matrix algebra because, as in ordinary algebra, it allows us to factor matrix expressions. For example,

$$AB + AC = A(B + C)$$
$$BA + A = (B + I)A.$$

Notice that the position of A in both factorizations is dictated by the fact that matrix multiplication is not commutative. Observe the identity matrix I in the second factorization; I in matrix algebra plays the role of one in ordinary arithmetic.

HINT   When factoring, multiply the result using the properties of matrix algebra, and make sure you get the original expression.

The commutative property of multiplication is not the only property of the algebra of real numbers that fails to carry over to matrix algebra.

Matrix Algebra Exceptions to Properties of Real Numbers

1. There are matrices A and B for which $AB \neq BA$.

2. There are matrices A, B, and C for which $AB = AC$, yet $B \neq C$. (That is, cancellation is not always permitted.)

3. There are matrices A and B, neither of which is the zero matrix, for which $AB = \mathbf{0}$.

EXAMPLE 7
Examples of Matrix
Algebra Exceptions

(a) Give an example of matrices A, B, and C for which $AB = AC$, but $B \neq C$.

Solution   Let $\quad A = \begin{bmatrix} 1 & 0 \\ 0 & 0 \end{bmatrix}$, $B = \begin{bmatrix} 2 & 6 \\ 8 & 7 \end{bmatrix}$ and $C = \begin{bmatrix} 2 & 6 \\ 4 & 3 \end{bmatrix}$.

Then $\quad AB = \begin{bmatrix} 2 & 6 \\ 0 & 0 \end{bmatrix}$ and $AC = \begin{bmatrix} 2 & 6 \\ 0 & 0 \end{bmatrix}$, but $B \neq C$.

(b) Give an example of matrices A and B for which $AB = \mathbf{0}$, but $A \neq \mathbf{0}$ and $B \neq \mathbf{0}$.

Solution Let $A = \begin{bmatrix} 2 & 0 \\ 0 & 0 \end{bmatrix}$ and $B = \begin{bmatrix} 0 & 0 \\ 5 & 1 \end{bmatrix}$. Then $AB = \begin{bmatrix} 0 & 0 \\ 0 & 0 \end{bmatrix} = \mathbf{0}$, but neither A nor B is the zero matrix. ∎

EXERCISES 3.2

In Exercises 1 through 4, write each linear system as a matrix equation of the form $AX = B$.

1. $\begin{aligned} x + 2y &= 7 \\ 3x - y &= 6 \end{aligned}$

2. $\begin{aligned} 3x + 2y - z &= 2 \\ x - y + z &= 4 \end{aligned}$

3. $\begin{aligned} x + y &= 2 \\ 3x - y &= 6 \\ x + 5y &= -1 \end{aligned}$

4. $\begin{aligned} 2x + 3y + z - 5w &= 1 \\ x - y + 8w &= 2 \\ -x + y + 2z &= 3 \end{aligned}$

In Exercises 5 through 8, write each matrix equation as a system of linear equations.

5. $\begin{bmatrix} 2 & 0 \\ 1 & 5 \\ 4 & 6 \end{bmatrix} \begin{bmatrix} x \\ y \end{bmatrix} = \begin{bmatrix} 3 \\ 5 \\ 2 \end{bmatrix}$

6. $\begin{bmatrix} 2 & 0 & 1 & -1 \\ 1 & 2 & 3 & 5 \end{bmatrix} \begin{bmatrix} x \\ y \\ z \\ w \end{bmatrix} = \begin{bmatrix} 0 \\ 0 \end{bmatrix}$

7. $\begin{bmatrix} 1 & 0 & 1 & 0 & 1 \\ 2 & 0 & 3 & 5 & 1 \\ -1 & 2 & 2 & 2 & 0 \end{bmatrix} \begin{bmatrix} x \\ y \\ z \\ w \\ v \end{bmatrix} = \begin{bmatrix} -2 \\ 3 \\ 6 \end{bmatrix}$

8. $\begin{bmatrix} 2 & 1 & 1 \\ 0 & 2 & 3 \\ 5 & 1 & -2 \end{bmatrix} \begin{bmatrix} x \\ y \\ z \end{bmatrix} = \begin{bmatrix} 1 \\ -1 \\ 3 \end{bmatrix}$

In Exercises 9 through 22, find each of the indicated matrices, if possible, for the given A, B, C, and D.

$$A = \begin{bmatrix} 2 & 3 \\ 1 & -2 \end{bmatrix}, B = \begin{bmatrix} 1 & 0 \\ 2 & 3 \end{bmatrix}, C = \begin{bmatrix} 2 & 1 \\ 1 & 0 \\ 3 & 2 \end{bmatrix},$$

$$and\ D = \begin{bmatrix} 1 & 3 & 5 & 9 \\ 2 & 0 & 1 & -3 \end{bmatrix}.$$

9. AB

10. CD

11. AC

12. DC

13. AD

14. CB

15. $2B + I$

16. $2I - BA$

17. $B^2 + 2I$

18. $I^2 - A^2$

19. A^2

20. $I - 2D$

21. $A^2 + 3B$

22. D^2

23. Is the matrix product $(A - B)(A + B) = A^2 - B^2$? Explain.

24. Use the rules for matrix algebra to find an expression for $(A + B)^2$. (**HINT** $(A + B)^2 = (A + B) \times (A + B)$, and use the distributive law.)

In Exercises 25 through 28 use

$$A = \begin{bmatrix} 3 & 1 \\ 2 & 5 \end{bmatrix}, B = \begin{bmatrix} 1 & 0 \\ 2 & 3 \end{bmatrix}, \quad and \quad C = \begin{bmatrix} 1 & 2 \\ 3 & 4 \end{bmatrix}$$

to verify the stated expressions.

25. $(A\ B)C = A(BC)$

26. $(B + C)A = BA + CA$

27. $A(B + C) = AB + AC$

28. $(AB)^T = B^T A^T$

29. Find matrices A, B, and C, each of the dimension 3×3, such that $AB = AC$, but $B \neq C$ and $A \neq \mathbf{0}$.

30. Find matrices A and B, each of the dimension 3×3, such that $AB = \mathbf{0}$, but $A \neq \mathbf{0}$ and $B \neq \mathbf{0}$.

In Exercises 31 through 33, use

$$A = \begin{bmatrix} 1 & 2 \\ 3 & 1 \end{bmatrix} \quad and \quad B = \begin{bmatrix} 2 & 3 \\ 4 & 1 \end{bmatrix}$$

to calculate the indicated quantities.

31. (a) $(2A)B$ (b) $A(2B)$ (c) $2(AB)$

32. (a) $(-3A)B$ (b) $A(-3B)$ (c) $-3(AB)$

33. (a) $(xA)B$ (b) $A(xB)$ (c) $x(AB)$

(d) What general conclusion is drawn from Exercises 31–33?

In Exercises 34 through 41, use the laws of matrix algebra, including the conclusion of Exercise 33 (d), to factor any common quantities from each expression.

34. $AX + BX$

35. $(AB)C + BC$

36. $AX + 2AY$

37. $AX - X$

38. $3AB + 4CB$

39. $AB + A$

40. $2AX - 3AY$

41. $2BC + 2C$

42. College Costs During Shawna's first year in college, she took 15 credits the first semester and 17 credits the second. She also worked a total of 160 hours the first semester and 130 hours the second. A matrix displaying this information is as follows:

$$
\begin{array}{c}
 \\
\text{Credits taken} \\
\text{Hours worked}
\end{array}
\begin{array}{cc}
\text{1st sem.} & \text{2nd sem.}
\end{array}
\left[\begin{array}{cc}
15 & 17 \\
160 & 130
\end{array}\right] = A.
$$

If tution costs $60 per credit and she received $4 per hour at work, find a matrix B so that a multiplication with A will show her income minus tuition each semester. Then do the matrix multiplication.

43. College Costs During her first two years at college, Sarah took course hours, plane trips home, and ski weekends, as indicated in this matrix:

$$
\begin{array}{c}
\text{Hours taken} \\
\text{Trips home} \\
\text{Ski trips}
\end{array}
\begin{array}{cccc}
\text{1st sem.} & \text{2nd sem.} & \text{3rd sem.} & \text{4th sem.}
\end{array}
\left[\begin{array}{cccc}
12 & 16 & 15 & 17 \\
2 & 3 & 1 & 2 \\
0 & 2 & 1 & 2
\end{array}\right] = A.
$$

Given that tuition costs $80 per hour, each round-trip plane fare home costs $260, and each ski weekend costs $120, find a matrix B so that a multiplication with the above matrix will give the total cost for each semester for these three activities. Then do the matrix multiplication.

44. Nutrition The amounts of fat (F), sodium (S), and protein (P), in units of grams, milligrams, and ounces, respectively, per serving of a soup, a meat, and a vegetable are as follows:

$$
\begin{array}{c}
\text{Soup} \\
\text{Meat} \\
\text{Vegetable}
\end{array}
\begin{array}{ccc}
F & S & P
\end{array}
\left[\begin{array}{ccc}
2 & 830 & 2 \\
5 & 30 & 8 \\
0 & 1 & 2
\end{array}\right] = A.
$$

(a) Given that a person eats 1 serving of soup, 1.5 servings of meat, and 0.73 servings of the vegetable, find a matrix B that, when multiplied with A, will give the total intake of fat, sodium, and protein. Then do the matrix multiplication.

(b) A person on a strict diet for health reasons gets -3 points for each gram of fat, -1.5 points for each milligram of sodium, and 2 points for each ounce of protein. Find a matrix C that, when multiplied with A, will give the total points accumulated for each food.

45. Manufacturing The times, in hours, for assembly, painting, and testing standard and deluxe lawn mowers are as shown:

$$
\begin{array}{c}
 \\
\text{Standard} \\
\text{Deluxe}
\end{array}
\begin{array}{ccc}
\text{Assembly} & \text{Painting} & \text{Testing}
\end{array}
\left[\begin{array}{ccc}
2 & 1 & .5 \\
2.5 & 1.2 & 1
\end{array}\right] = A.
$$

(a) If 80 standard and 100 deluxe mowers are made, what is a matrix B that, when multiplied with A, gives the total assembly, painting, and testing time required?

(b) If assembly time costs $5 per hour, painting costs $12 per hour, and testing costs $6 per hour, what is a matrix C that, when multiplied with A, gives the total cost for making each model of mower?

46. Advertising A particular product was advertised in a newspaper, on the radio, and on television, and the number of men and women influenced by the ads was recorded. The data were as follows:

$$
\begin{array}{c}
 \\
\text{Newspaper} \\
\text{Radio} \\
\text{Television}
\end{array}
\begin{array}{cc}
\text{Men} & \text{Women}
\end{array}
\left[\begin{array}{cc}
700 & 1000 \\
500 & 800 \\
1500 & 1200
\end{array}\right] = A.
$$

(a) Suppose that three, five, and two ads in the respective media were run in January, and two, four, and six ads in the respective media were run in June. Write this information in a 2 × 3 matrix B. Then compute BA, and interpret the entry in each address of the product matrix.

(b) Suppose that the ads were run once each month for three consecutive months. The first time the ads were run, men influenced by the ad spent an average of $10 each, and women influenced by the ad spent an average of $15 each. The second time the ads were run, the figures were $12 and $18, respectively. The third time the ads were run, the figures were $15 and $22, respectively. Record this information in a 2 × 3 matrix C. Then calculate the product AC, and interpret the entry in each address of the product matrix.

47. Transportation An airline purchases two types of airplanes, P_1 and P_2. Plane P_1 seats 30 first-class passengers, 50 tourist-class passengers, and 90 economy-class passengers. Plane P_2 seat 50 first-class passengers, 60 tourist-class passengers, and 100 economy-class passengers.

(a) Put this information in a 3 × 2 matrix labeled A.

(b) Suppose that five planes of type P_1 and eight planes of type P_2 make flights fully loaded. Put this information in a 2 × 1 matrix labeled B, and calculate the product AB.

(c) To fly between Kansas City and Seattle in January, each first-class passenger is charged $200, each tourist-class passenger is charged $180, and each economy-class passenger is charged $150. In August, these rates change to $240, $220 and $200, respectively. Put this information in a 2 × 3 matrix labeled C. Find the product CA, and interpret the entries in each address.

48. *Baking Costs* The Better Bread Bakery bakes whole wheat bread and oat bread, with mixing, baking, and packaging times, in hours, as shown.

$$\begin{array}{c} \\ \text{Whole wheat} \\ \text{Oat}\end{array} \begin{array}{ccc} \text{Mix} & \text{Bake} & \text{Package} \\ \left[\begin{array}{ccc} 0.03 & 0.05 & 0.01 \\ 0.02 & 0.03 & 0.01 \end{array}\right] \end{array} = A.$$

Given that an order is received for 200 loaves of whole wheat bread and 300 loaves of oat bread, and the cost of mixing, baking, and packaging is $12, $20, and $1, respectively, per hour, find matrices B and C so that the product BAC will give the total cost (excluding raw materials) for filling this order.

49. *Inventory* The Soft Sit Furniture Company has two stores that stock brands X and Y recliner chairs. The present inventory matrix is as follows:

$$\begin{array}{c} \\ \text{Store 1} \\ \text{Store 2}\end{array} \begin{array}{cc} \text{Brand X} & \text{Brand Y} \\ \left[\begin{array}{cc} 10 & 8 \\ 5 & 12 \end{array}\right] \end{array} = A.$$

Twenty percent of this inventory in Store 1 and 50% of this inventory in Store 2 is leather. Brand X leather chairs cost $400 each, and brand Y leather chairs cost $520 each. Find matrices B and C so that BAC gives the total cost of the leather chairs in inventory.

Use Example 6 in Section 3.2 as a guide in solving Exercises 50 and 51.

50. *Psychology* A mouse enters a T-shaped maze and can turn either left, where a shock and a piece of cheese awaits, or right, where just a piece of cheese awaits. Earlier experiments with this maze gave these results: Of those who go left on a particular day, 10% go to the left on the next day and 90% go to the right. Of those who go to the right on a particular day, 60% go to the right on the next day and 40% go to the left. Julie took 100 mice one day and found that 48 went to the left and 52 went to the right. Assuming that the preceding percentages hold, use a matrix product to determine how many of the mice should go left and how many right the following day. How many the third day? the fourth day?

51. *Social Research* A social science researcher studied children whose parents were classified as having high, middle, or low income and compared these classifications to the income levels the children eventually attained in life. The researcher found that of

children whose parents were in the high-income level, eventually 60%, 38%, and 2% were in the high-, middle-, and low-income levels, respectively. Of children whose parents were in the middle-income level, eventually 20%, 60%, and 20% were in the high, middle, and low levels, respectively. Of children whose parents were in the low-income level, eventually 5%, 40%, and 55% were in the high, middle, and low levels, respectively. According to this study, if 40 children from each income level were observed, use a matrix product to determine how many would be expected to eventually be in each income level.

52. *Labor Costs* Estimated hourly labor costs for the construction industry in several countries are shown in the following table:*

Germany's Sky-High Costs			
Estimated total hourly cost for construction workers, as charged by general contractors in the United States and several European countries, in U.S. dollars*			
COUNTRY	BASIC LABORER	CRAFTSMAN	DESIGN ENGINEER
Germany	**$35**	**$52**	**$83**
U.S.	31	48	70
Belgium	30	43	78
France	25	42	70
Britain	14	20	55
Spain	12	15	30
*Includes wages and other supplemental costs			
SOURCE: Hanscomb International Construction Consultants			

Put the information shown in this table in a 6 × 3 matrix denoted by A. Assume that construction productivity is equal in these various countries and that in building a 2000-square-foot house, basic labor accounts for 20%, craftspersons' work accounts for 75%, and design engineering accounts for 5% of the total labor.

(a) Construct a 3 × 1 matrix B using the proportions of basic labor, craftspersons' work, and design engineering, and calculate AB to find the weighted average labor cost for a 2000-square-foot house built in the various countries.

(b) Assume that it takes a total of 288 basic labor hours, 1080 craftsperson-hours, and/or 72 design engineer hours to produce our 2000-square-foot house. Put this information in a matrix C and use a matrix multiplication with A to find the total labor cost for such a house in the various countries.

*SOURCE: *The Wall Street Journal,* August 14, 1995.

53. *Executive Reporting* Executives A, B, C, D, and E in a particular corporation have reporting status as shown in this diagram

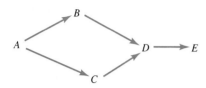

(a) Find a matrix A representing this reporting status diagram that will be similar to the one in Example 2 in Section 3.1, and assume, for purposes of constructing the matrix, that no executives report directly to themselves.

(b) Find the matrix A^2, and interpret the entries.

(c) Find the matrix A^3, and interpret the entries.

(d) Find $A + A^2 + A^3$, and interpret the entries.

54. *Airline Flights* A particular airline has flights between cities A, B, and C, with origin and destination as indicated on the following diagram.

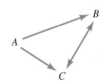

(a) Construct a matrix A representing these flights that is similar to Example 2 in Section 3.1, in which a 1 denotes a direct flight from a city on the left to a city on the top and we assume that there will be no direct flights from a city to the same city.

(b) Find A^2 and discuss how the entries give the number of two-stage flights between cities. Compare these results with the diagram.

(c) Find $A + A^2$, and interpret each entry in the result.

Problems for Exploration

1. A matrix is called **symmetric** if $A = A^T$. If A is a 3×3 symmetric matrix, prove that AA^T is also symmetric. (Fill in the entries of A with letters $a, b, c, \ldots$)

2. Let A and B be matrices for which $AB = BA$. Prove that $(AB)^2 = A^2B^2$.

3. A matrix is called **stochastic** if the sum of the elements in each row is 1. If A and B are both stochastic matrices, and AB exists, then prove that AB is also stochastic. (Fill in the entries of A and B with letters $a, b, c, \ldots$)

SECTION 3.3 **The Inverse of a Matrix**

We know that any linear system of equations can be written in the equivalent matrix equation form $AX = B$. Now if there were only some way of "dividing" both sides by A, the solution X for the system could be found. Even though *there is no division operation in matrix algebra,* the same result could be obtained by constructing a matrix A^{-1} having the property that $A^{-1}A = I = AA^{-1}$, *if such a matrix exists.* Then multiplication on both sides by A^{-1} on the left of $AX = B$ would give

$$A^{-1}(AX) = A^{-1}B$$

or
$$(A^{-1}A)X = A^{-1}B \qquad \text{Associative Law}$$

so that
$$IX = A^{-1}B,$$

and
$$X = A^{-1}B.$$

This would solve the problem if we knew when such a matrix A^{-1}, called the **inverse of A,** could be found for a given A and just how its construction took place. To investigate these concerns, we first should be aware that the inverse exists only for some specific square matrices.

The Inverse of a Matrix

Let A be a square $n \times n$ matrix. If there is a square $n \times n$ matrix, which when multiplied on either side by A, produces the identity matrix, then such a matrix is called the **inverse of** A and is denoted by A^{-1}. That is, A^{-1} must have this property

$$A^{-1}A = I \quad \text{and} \quad AA^{-1} = I.$$

In view of this definition, the solution to the previously mentioned matrix equation $AX = B$ will apply only when the coefficient matrix A is square and then only if the inverse of A exists.

Could it be possible for a matrix to have two or more inverses? The answer is "no"; if there is an inverse at all, there is only one. The reason is that if there were two inverses, A^{-1} and Q to a given matrix A, then $I = A^{-1}A = AA^{-1}$ and $I = QA = AQ$, so that

$$
\begin{aligned}
A^{-1} = IA^{-1} &= (QA)A^{-1} \\
&= Q(AA^{-1}) \qquad \text{Associative Law} \\
&= QI \qquad\qquad\;\; \text{Property of Inverse} \\
&= Q.
\end{aligned}
$$

Uniqueness of the Inverse

If a square matrix A has an inverse, then **the inverse is unique.**

EXAMPLE 1
Demonstrating the
Property of an Inverse

(a) Let $A = \begin{bmatrix} 2 & 3 \\ 1 & 1 \end{bmatrix}$. Then $A^{-1} = \begin{bmatrix} -1 & 3 \\ 1 & -2 \end{bmatrix}$ because

$$AA^{-1} = \begin{bmatrix} 2 & 3 \\ 1 & 1 \end{bmatrix} \begin{bmatrix} -1 & 3 \\ 1 & -2 \end{bmatrix} = \begin{bmatrix} 1 & 0 \\ 0 & 1 \end{bmatrix} = I$$

and

$$A^{-1}A = \begin{bmatrix} -1 & 3 \\ 1 & -2 \end{bmatrix} \begin{bmatrix} 2 & 3 \\ 1 & 1 \end{bmatrix} = \begin{bmatrix} 1 & 0 \\ 0 & 1 \end{bmatrix} = I.$$

(b) Let $A = \begin{bmatrix} 1 & -1 & -0 \\ 0 & 2 & 1 \\ 1 & 0 & 1 \end{bmatrix}$. Then $A^{-1} = \begin{bmatrix} 2 & 1 & -1 \\ 1 & 1 & -1 \\ -2 & -1 & 2 \end{bmatrix}$ because AA^{-1} and $A^{-1}A$ are both the 3×3 identity matrix. (You should verify these results.)

(c) The matrix $A = \begin{bmatrix} 0 & 0 \\ 3 & 0 \end{bmatrix}$ does not have an inverse. There is no 2×2 matrix $\begin{bmatrix} x & y \\ z & w \end{bmatrix}$ such that the product $\begin{bmatrix} 0 & 0 \\ 3 & 0 \end{bmatrix} \begin{bmatrix} x & y \\ z & w \end{bmatrix}$, which is $\begin{bmatrix} 0 & 0 \\ 3x & 3y \end{bmatrix}$, could equal the identity matrix $\begin{bmatrix} 1 & 0 \\ 0 & 1 \end{bmatrix}$. ∎

One straightforward, systematic way to calculate the inverse of a square matrix, or to detect its nonexistence, is to convert the problem into an equivalent system of equations and use the pivoting process to find the solution. To see why this may be done, consider the matrix $A = \begin{bmatrix} 2 & 3 \\ 1 & 1 \end{bmatrix}$ of Example 1(a). To construct the inverse of A, we need to find a matrix $A^{-1} = \begin{bmatrix} x & y \\ z & w \end{bmatrix}$ such that

$$AA^{-1} = \begin{bmatrix} 2 & 3 \\ 1 & 1 \end{bmatrix}\begin{bmatrix} x & y \\ z & w \end{bmatrix} = \begin{bmatrix} 1 & 0 \\ 0 & 1 \end{bmatrix}$$

or

$$\begin{bmatrix} 2x + 3z & 2y + 3w \\ x + z & y + w \end{bmatrix} = \begin{bmatrix} 1 & 0 \\ 0 & 1 \end{bmatrix}.$$

For the equality between the last two matrices to hold, it follows that

$$2x + 3z = 1$$
$$x + z = 0$$

and

$$2y + 3w = 0$$
$$y + w = 1.$$

The calculation of A^{-1} has now been reduced to solving two systems of linear equations, each in two unknowns. Notice what happens if each system were to be solved by the pivoting process:

$$\begin{bmatrix} 2 & 3 & | & 1 \\ 1 & 1 & | & 0 \end{bmatrix} \cdots \cdots \begin{bmatrix} 1 & 0 & | & x \\ 0 & 1 & | & z \end{bmatrix} \qquad (1)$$

$$\begin{bmatrix} 2 & 3 & | & 0 \\ 1 & 1 & | & 1 \end{bmatrix} \cdots \cdots \begin{bmatrix} 1 & 0 & | & y \\ 0 & 1 & | & w \end{bmatrix}. \qquad (2)$$

The coefficient matrix for both systems is exactly the same and is, in fact, the original matrix A. Therefore, the steps in the pivoting process for both systems to reach the final ideal matrix (the identity) would be identical, and furthermore, the values of x, y, z, and w will be in the indicated addresses shown in systems **(1)** and **(2)** above. Instead of doing the work twice, we can do the pivoting for both systems at once by beginning with this slightly adjusted augmented matrix:

$$\begin{bmatrix} 2 & 3 & | & 1 & 0 \\ 1 & 1 & | & 0 & 1 \end{bmatrix} = [A \,|\, I].$$

Constant column from system 1 ⎯↑ ↑⎯ Constant column from system 2

We then do the pivoting process *across entire rows* until the coefficient matrix becomes the identity, at which point both solutions appear on the right of the vertical bar, like this:

$$\begin{bmatrix} 1 & 0 & | & x & y \\ 0 & 1 & | & z & w \end{bmatrix} = [I \,|\, A^{-1}].$$

Solution column for system 1 ⎯↑ ↑⎯ Solution column from system 2

The actual steps to accomplish these solutions are shown next.

$$[A\,|\,I] = \begin{bmatrix} 2 & 3 & | & 1 & 0 \\ 1 & 1 & | & 0 & 1 \end{bmatrix}$$

$$R_1 \longleftrightarrow R_2 \quad \begin{bmatrix} 1 & 1 & | & 0 & 1 \\ 2 & 3 & | & 1 & 0 \end{bmatrix}$$

$$-2R_1 + R_2 \rightarrow R_2 \quad \begin{bmatrix} 1 & 1 & | & 0 & 1 \\ 0 & 1 & | & 1 & -2 \end{bmatrix}$$

$$-1R_2 + R_1 \rightarrow R_1 \quad \begin{bmatrix} 1 & 0 & | & -1 & 3 \\ 0 & 1 & | & 1 & -2 \end{bmatrix} = [I\,|\,\mathbf{A^{-1}}].$$

The part of the last matrix to the right of the vertical bar shows that the solution to System **(1)** is $x = -1$ and $z = 1$ (the first column), and the solution to System **(2)** is $y = 3$ and $w = -2$ (the second column), so that the inverse of A is

$$A^{-1} = \begin{bmatrix} -1 & 3 \\ 1 & -2 \end{bmatrix}.$$

It is easy to check the accuracy of A^{-1}; just make sure that $AA^{-1} = I$ or $A^{-1}A = I$. Either one will do, because it can be mathematically proven that if either of these equalities holds, then so will the other. A check of the inverse just found shows that

$$AA^{-1} = \begin{bmatrix} 2 & 3 \\ 1 & 1 \end{bmatrix}\begin{bmatrix} -1 & 3 \\ 1 & -2 \end{bmatrix} = \begin{bmatrix} 1 & 0 \\ 0 & 1 \end{bmatrix} = I.$$

The mechanics for finding the inverse of any 2×2 matrix are straightforward. Just form the 2×4 matrix $[A\,|\,I]$ and then use the pivoting process to transform A into I, if possible, and the resulting matrix will appear as $[I\,|\,A^{-1}]$. Remember that two systems of equations are being solved simultaneously and if either system fails to have a solution, then A will fail to have an inverse. Will this process work for 3×3 matrices and even higher-dimension square matrices? The answer is "yes," and for exactly the same reasons. For instance, to find the inverse of $A = \begin{bmatrix} 1 & -1 & 0 \\ 0 & 2 & 1 \\ 1 & 0 & 1 \end{bmatrix}$, the problem becomes one of making sure that

$$AA^{-1} = I \quad \text{or} \quad \begin{bmatrix} 1 & -1 & 0 \\ 0 & 2 & 1 \\ 1 & 0 & 1 \end{bmatrix}\begin{bmatrix} a & b & c \\ d & e & f \\ g & h & i \end{bmatrix} = \begin{bmatrix} 1 & 0 & 0 \\ 0 & 1 & 0 \\ 0 & 0 & 1 \end{bmatrix}.$$

Upon multiplying the two matrices on the left and equating addresses with the identity on the right, these three systems of equations result

$$\begin{aligned} a - d & = 1 \\ 2d + g & = 0 \\ a \quad\;\; + g & = 0. \end{aligned} \tag{1}$$

$$\begin{aligned} b - e & = 0 \\ 2e + h & = 1 \\ b \quad\;\; + h & = 0. \end{aligned} \tag{2}$$

$$\begin{aligned} c - f & = 0 \\ 2f + i & = 0 \\ c \quad\;\; + i & = 1. \end{aligned} \tag{3}$$

Because the coefficient matrix in each system is the same (namely the original matrix A), all three systems can be solved simultaneously, with the augmented matrix shown next on the left. Row operations will yield the last matrix.

$$[A \mid I] = \begin{bmatrix} 1 & -1 & 0 & 1 & 0 & 0 \\ 0 & 2 & 1 & 0 & 1 & 0 \\ 1 & 0 & 1 & 0 & 0 & 1 \end{bmatrix} \cdots \cdots \begin{bmatrix} 1 & 0 & 0 & 2 & 1 & -1 \\ 0 & 1 & 0 & 1 & 1 & -1 \\ 0 & 0 & 1 & -2 & -1 & 2 \end{bmatrix} = [I \mid A^{-1}].$$

The reason the inverse is to the right of the vertical bar is that the first column is the solution to System **(1)**, the second column is the solution to System **(2)**, and the third column is the solution to System **(3)**.

These last two examples show the pattern that develops for any size of square matrix.

Finding the Inverse of a Matrix

Let A be any $n \times n$ matrix. To determine whether A has an inverse,

1. Form the doublewide $n \times 2n$ matrix $[A \mid I]$, where I is the identiy matrix of the same dimension as A.

2. Then perform pivot operations on elements in the A side, considering rows to extend across the entire matrix $[A \mid I]$, until you obtain $[R \mid B]$ where R is in reduced row-echelon form.

3. If $R = I$, then $B = A^{-1}$. If $R \neq I$, then R will contain a row of zeros and A has no inverse.

EXAMPLE 2
Finding the Inverse
of a Matrix

Find the inverse of $A = \begin{bmatrix} 1 & 3 \\ 2 & 1 \end{bmatrix}$.

Solution

$$[A \mid I] = \begin{bmatrix} 1 & 3 & 1 & 0 \\ 2 & 1 & 0 & 1 \end{bmatrix}$$

$$-2R_1 + R_2 \to R_2 \quad \begin{bmatrix} 1 & 3 & 1 & 0 \\ 0 & -5 & -2 & 1 \end{bmatrix}$$

$$-\tfrac{1}{5}R_2 \rightarrow R_2 \quad \begin{bmatrix} 1 & 3 & | & 1 & 0 \\ 0 & 1 & | & \tfrac{2}{5} & -\tfrac{1}{5} \end{bmatrix}$$

$$-3R_2 + R_1 \rightarrow R_1 \quad \begin{bmatrix} 1 & 0 & | & -\tfrac{1}{5} & \tfrac{3}{5} \\ 0 & 1 & | & \tfrac{2}{5} & -\tfrac{1}{5} \end{bmatrix}$$

Therefore,

$$A^{-1} = \begin{bmatrix} -\tfrac{1}{5} & \tfrac{3}{5} \\ \tfrac{2}{5} & -\tfrac{1}{5} \end{bmatrix}.$$

Verify this result by calculating AA^{-1} and $A^{-1}A$ to see whether I is obtained. ▪

EXAMPLE 3
When the Inverse Does Not Exist

Find the inverse of $A = \begin{bmatrix} 1 & 2 \\ 2 & 4 \end{bmatrix}$.

Solution

$$[A \,|\, I] = \begin{bmatrix} 1 & 2 & | & 1 & 0 \\ 2 & 4 & | & 0 & 1 \end{bmatrix}$$

$$-2R_1 + R_2 \rightarrow R_2 \quad \begin{bmatrix} 1 & 2 & | & 1 & 0 \\ 0 & 0 & | & -2 & 1 \end{bmatrix}$$

The reduced row-echelon form of A (the matrix to the left of the vertical bar) is not the identity. Therefore, A has no inverse. From the viewpoint of solutions to systems of equations, the row of zeros to the *left* of the vertical bar means that no solution exists to either of the systems of equations represented. A row of zeros to the left of the vertical bar is *always* a signal that no inverse exists. ▪

An alternate method of finding the inverse of a 2×2 matrix is shown next. Its validity is shown by using the general method we have been discussing. (See Problem 7 in Problems for Exploration.)

Inverse Formula for a 2 × 2 Matrix

For the matrix $A = \begin{bmatrix} a & b \\ c & d \end{bmatrix}$, let $D = ad - bc$. If $D = 0$, then A^{-1} does not exist. If $D \neq 0$, then A^{-1} exists and is given by

$$A^{-1} = \frac{1}{D} \begin{bmatrix} d & -b \\ -c & a \end{bmatrix}.$$

To demonstrate the ease of use of this formula, find the inverse of $A = \begin{bmatrix} 3 & 2 \\ 5 & 4 \end{bmatrix}$. The value of D is $D = 3 \cdot 4 - 5 \cdot 2 = 2$. From the formula we find that:

$$A^{-1} = \frac{1}{2} \begin{bmatrix} 4 & -2 \\ -5 & 3 \end{bmatrix} = \begin{bmatrix} 2 & -1 \\ -\tfrac{5}{2} & \tfrac{3}{2} \end{bmatrix}.$$

At the outset of the section, we saw that if $AX = B$ is a matrix equation representing a system of linear equations and A has an inverse, then $X = A^{-1}B$. Example 4 shows how the product $A^{-1}B$ gives the solution to a specific system of equations.

EXAMPLE 4
Solving a System by
Using the Inverse
of a Matrix

Convert the system

$$2x + 3y = 1$$
$$x + y = 4$$

into a matrix equation. Then solve the system using the inverse of the coefficient matrix.

Solution   We first rewrite the given system as a matrix equation $AX = B$ and then use the fact that $X = A^{-1}B$ to find the solution. The matrix equation $AX = B$ is

$$\begin{bmatrix} 2 & 3 \\ 1 & 1 \end{bmatrix} \begin{bmatrix} x \\ y \end{bmatrix} = \begin{bmatrix} 1 \\ 4 \end{bmatrix}$$

with $A = \begin{bmatrix} 2 & 3 \\ 1 & 1 \end{bmatrix}$ and $B = \begin{bmatrix} 1 \\ 4 \end{bmatrix}$. Therefore, $X = A^{-1}B$ becomes

$$\begin{bmatrix} x \\ y \end{bmatrix} = \begin{bmatrix} 2 & 3 \\ 1 & 1 \end{bmatrix}^{-1} \begin{bmatrix} 1 \\ 4 \end{bmatrix}.$$

Omitting the details, $A^{-1} = \begin{bmatrix} 2 & 3 \\ 1 & 1 \end{bmatrix}^{-1} = \begin{bmatrix} -1 & 3 \\ 1 & -2 \end{bmatrix}$, so that

$$\begin{bmatrix} x \\ y \end{bmatrix} = \begin{bmatrix} -1 & 3 \\ 1 & -2 \end{bmatrix} \begin{bmatrix} 1 \\ 4 \end{bmatrix} = \begin{bmatrix} 11 \\ -7 \end{bmatrix}.$$

This means that the solution to the original system of equations is $x = 11$ and $y = -7$. Substituting these values for x and y into the original equations will verify the solution:

$$2(11) + 3(-7) = 1$$
$$(11) + (-7) = 4.$$

NOTE   We could also solve this system using techinques from Chapter 2, but we are using this example to illustrate the process and ideas involved when using a matrix inverse. This process is especially useful if technology is available to invert a matrix and perform matrix multiplication. ■■■

   Various computer programs, spreadsheets, and graphing calculators are preprogrammed to find the inverse of a matrix. Caution about round-off errors should be advised, however. Such errors may result in an inverse being found to a matrix that has no inverse. Figure 4 shows how a graphing calculator screen would look after the calculation of the inverse and the subsequent matrix multiplication used to solve the system of equations in Example 4.

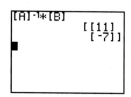

FIGURE 4
Calculator Solution to a
System of Equations

The real importance of the inverse of a matrix is in the structure of the matrix algebra. The inverse does in matrix algebra what the reciprocal (division) does in ordinary algebra with real numbers. However, in the real-number system, there is only one number, 0, that does not have an inverse (reciprocal). In matrix algebra, there are many matrices, in fact many square matrices, that do not have an inverse. One consequence of this difference is that for real numbers, if $ab = ac$ and $a \neq 0$, then we may cancel the a and obtain $b = c$, while for matrices, $AB = AC$ implies $B = C$ only when A has an inverse, and then the cancellation is done by multiplying both sides by A^{-1} *on the left*.

The inverse of a matrix has many applications other than those shown in this section. The next section shows three other applications from cryptography, economics, and manufacturing.

EXERCISES   3.3

In Exercises 1 through 4, decide whether matrix B is the inverse of A.

1. $A = \begin{bmatrix} 3 & 2 \\ 1 & 1 \end{bmatrix}$ $B = \begin{bmatrix} 1 & -2 \\ -1 & 3 \end{bmatrix}$

2. $A = \begin{bmatrix} 2 & 3 \\ 0 & 3 \end{bmatrix}$ $B = \begin{bmatrix} 2 & \frac{1}{2} \\ 3 & 1 \end{bmatrix}$

3. $A = \begin{bmatrix} 1 & -1 & 0 \\ 2 & 1 & 0 \\ 1 & 1 & 2 \end{bmatrix}$ $B = \begin{bmatrix} 1 & 0 & 0 \\ 2 & 1 & 1 \\ \frac{1}{2} & 3 & -1 \end{bmatrix}$

4. $A = \begin{bmatrix} 1 & 0 & 1 \\ 2 & 0 & 0 \\ 1 & 1 & 1 \end{bmatrix}$ $B = \begin{bmatrix} 0 & \frac{1}{2} & 0 \\ -1 & 0 & 1 \\ 1 & -\frac{1}{2} & 0 \end{bmatrix}$

In Exercises 5 and 6, write the systems of equations that are actually solved when the pivoting process is used to find the inverse of the given matrix.

5. $\begin{bmatrix} 3 & 5 \\ 7 & -1 \end{bmatrix}$

6. $\begin{bmatrix} 3 & 3 & 1 \\ 0 & 1 & 2 \\ 2 & 2 & 1 \end{bmatrix}$

In Exercises 7 through 22, find the inverse of each matrix, if it exists.

7. $\begin{bmatrix} 1 & 3 \\ 2 & 5 \end{bmatrix}$

8. $\begin{bmatrix} 2 & 4 \\ 3 & 2 \end{bmatrix}$

9. $\begin{bmatrix} 2 & 3 \\ 1 & \frac{3}{2} \end{bmatrix}$

10. $\begin{bmatrix} 2 & 0 \\ 0 & 1 \end{bmatrix}$

11. $\begin{bmatrix} 0 & 3 \\ 0 & 2 \end{bmatrix}$

12. $\begin{bmatrix} 2 & -1 \\ 0 & 1 \end{bmatrix}$

13. $\begin{bmatrix} -1 & 3 \\ 2 & -4 \end{bmatrix}$

14. $\begin{bmatrix} 1 & 5 \\ 2 & 10 \end{bmatrix}$

15. $\begin{bmatrix} 1 & 0 & 2 \\ 0 & 2 & 2 \\ 1 & 1 & 1 \end{bmatrix}$

16. $\begin{bmatrix} 1 & 2 & 3 \\ 0 & 1 & 1 \\ 2 & 2 & 2 \end{bmatrix}$

17. $\begin{bmatrix} 3 & 1 & 0 \\ 2 & 1 & 1 \\ 5 & 2 & 1 \end{bmatrix}$

18. $\begin{bmatrix} 1 & 1 & 1 \\ 2 & 2 & 2 \\ 3 & 3 & 4 \end{bmatrix}$

19. $\begin{bmatrix} 2 & 1 & 1 \\ 1 & 0 & 1 \\ 1 & 1 & 1 \end{bmatrix}$

20. $\begin{bmatrix} 1 & 0 & 1 \\ 0 & 2 & 1 \\ 2 & 2 & 2 \end{bmatrix}$

21. $\begin{bmatrix} 1 & 0 & 1 & 0 \\ 2 & 2 & 2 & 0 \\ 2 & 0 & 0 & 1 \\ 1 & 0 & 2 & 1 \end{bmatrix}$

22. $\begin{bmatrix} 2 & 1 & 1 & 1 \\ 1 & 1 & 2 & 1 \\ 1 & 2 & 1 & 1 \\ 1 & 1 & 1 & 1 \end{bmatrix}$

23. For $A = \begin{bmatrix} 2 & 3 \\ 1 & -4 \end{bmatrix}$, compute:

   (a) $I - A$
   (b) $(I - A)^{-1}$

24. For $A = \begin{bmatrix} 1 & 2 & 0 \\ 0 & 1 & 1 \\ 1 & 1 & 1 \end{bmatrix}$, compute:

   (a) $I - A$
   (b) $(I - A)^{-1}$

In Exercises 25 through 30, write each system as a matrix equation, and solve each if possible, using the inverse of the coefficient matrix.

25. $x + 3y = 7$
    $2x + 4y = -3$

26. $x + 5y = 3$
    $2x + 10y = 8$

27. $5x + 2y = 4$
    $4x + 2y = -3$

28. $2x + 3y = 8$
    $4x + 5y = -2$

29. $x \quad + 2z = 1$
    $\quad 2y + 2z = 3$
    $x + y + z = 5$

30. $x + 2y + 3z = 4$
    $\quad y + z = -2$
    $2x + 2y \quad = 3$

In Exercises 31 and 32, explain the failure of your calculator to aid in solving the system. Then after careful examination of the systems, proceed to find the infinitely many solutions.

31. $.5x + .3y - z = 5$
    $x - .4y + 2z = 4$
    $1.5x - .1y + z = 9$

32. $x - .4y + .2z = 4$
    $.5x + y - .3z = 6$
    $.5x - 1.4y + .5z = -2$

In Exercises 33 through 38, calculate the inverse of each diagonal matrix, and then draw a conclusion from the observed pattern. (A diagonal matrix is one whose only nonzero elements are in a_{ii} positions.)

33. $\begin{bmatrix} 3 & 0 \\ 0 & 1 \end{bmatrix}$

34. $\begin{bmatrix} -2 & 0 \\ 0 & 3 \end{bmatrix}$

35. $\begin{bmatrix} 3 & 0 & 0 \\ 0 & -1 & 0 \\ 0 & 0 & 5 \end{bmatrix}$

36. $\begin{bmatrix} -2 & 0 & 0 \\ 0 & 4 & 0 \\ 0 & 0 & 7 \end{bmatrix}$

37. $\begin{bmatrix} 3 & 0 & 0 & 0 \\ 0 & -6 & 0 & 0 \\ 0 & 0 & 2 & 0 \\ 0 & 0 & 0 & 5 \end{bmatrix}$

38. $\begin{bmatrix} 2 & 0 & 0 & 0 \\ 0 & 3 & 0 & 0 \\ 0 & 0 & 4 & 0 \\ 0 & 0 & 0 & 5 \end{bmatrix}$

In Exercises 39 through 50, use the rules for matrix algebra to solve each equation for the indicated variable. Assume that the orders are such that matrix multiplication and addition are possible and that inverses exist when needed.

39. Solve for A: $AB = C$

40. Solve for B: $AB = C$

41. Solve for X: $X - AX = B$

42. Solve for Z: $Z + AZ = B$

43. Solve for A: $AB + AC = C$

44. Solve for Z: $AZ + BZ = D$

45. Solve for T: $TA + B = C$

46. Solve for M: $AM - B = C$

47. Solve for A: $AX + AB = C - D$

48. Solve for A: $2AB + A = C$

49. Solve for T: $2T + BT = 3C$

50. Solve for X: $2X - 2AX = 5B$

In Exercises 51 through 54, use the following matrices to calculate the given expressions.

$$A = \begin{bmatrix} 2 & 3 \\ 1 & 2 \end{bmatrix} \quad and \quad B = \begin{bmatrix} 3 & 4 \\ 1 & 2 \end{bmatrix}$$

51. (a) $(AB)^{-1}$ (b) $B^{-1}A^{-1}$
    (c) After comparing the results of Exercise 51(a) and (b), state a general rule in words regarding what you have learned about these expressions.
    (d) Use your rule to rewrite $((AB)C)^{-1}$.

52. (a) $(A^2)^{-1}$ (b) $(A^{-1})^2$
    (c) After comparing the results of Exercise 52 (a) and (b), state a general rule in words regarding what you have learned about these expressions.

53. (a) $(A^T)^{-1}$ (b) $(A^{-1})^T$
    (c) After comparing the results of Exercise 53 (a) and (b), state a general rule in words regarding what you have discovered.

54. (a) $(A^{-1})^{-1}$
    (b) After calculating Exercise 54 (a), state a general rule in words regarding what you have discovered.

55. (a) For matrices A and B such that $AB = \mathbf{0}$, what condition must hold to conclude that $B = \mathbf{0}$?
    (b) If $AB = \mathbf{0}$, is it possible for both A and B to have inverses?
    (c) If both A and B have inverses, is it possible for $AB = \mathbf{0}$ to hold?
    (d) If $AB = \mathbf{0}$, is it possible that neither A nor B has an inverse?

This exercise appeared on an actuarial examination.

56. What is the inverse of the matrix $\begin{bmatrix} 1 & 1 & 1 \\ 0 & 1 & 1 \\ 0 & 0 & 1 \end{bmatrix}$?

(a) $\begin{bmatrix} 1 & 0 & 0 \\ 1 & 1 & 0 \\ 1 & 1 & 1 \end{bmatrix}$

(b) $\begin{bmatrix} 1 & 0 & 0 \\ -1 & 1 & 0 \\ 1 & -1 & 1 \end{bmatrix}$

(c) $\begin{bmatrix} 1 & -1 & 1 \\ 0 & 1 & -1 \\ 0 & 0 & 1 \end{bmatrix}$

(d) $\begin{bmatrix} 1 & -1 & 0 \\ 0 & 1 & -1 \\ 0 & 0 & 1 \end{bmatrix}$

(e) $\begin{bmatrix} 1 & -1 & 0 \\ 0 & 1 & 1 \\ 0 & 0 & 1 \end{bmatrix}$

Problems for Exploration

1. Write a short summary describing in your own words the concepts and rules of matrix algebra and how they differ from the algebra of real numbers studied in a regular algebra course.

2. Construct a 3×3 matrix in which every element is nonzero, but the matrix has no inverse. Can you give a general rule for constructing such a square matrix of any size?

3. Show that any square matrix with one row proportional to another row cannot have an inverse.

4. Prove that $(AB)^{-1} = B^{-1}A^{-1}$.
   (**HINT** Use the fact that when any matrix, including AB, is multiplied by its inverse, the identity matrix results.)

5. Let A be any square $n \times n$ matrix and let X be an $n \times 1$ column matrix. In terms of a system of equations, when will the matrix equation $AX = X$ hold? Give some examples.

6. Is it possible to have numbers a, b, c, and d so that either of the two systems of equations resulting from finding the inverse of $\begin{bmatrix} a & b \\ c & d \end{bmatrix}$ has an infinite number of solutions?

7. Use the pivoting method to find the inverse of $A = \begin{bmatrix} a & b \\ c & d \end{bmatrix}$.

SECTION 3.4 **More Applications of Inverses**

Elementary Cryptography

Cryptography is the study of sending and receiving coded messages. Such messages are **encoded** in such a way that even if intercepted, only the intended receiver should be able to **decode** them . There are many ways to encode messages. We consider one very simple example showing how an invertible $n \times n$ matrix A may be used to encode and decode a message. The sender uses A to encode the message, and the receiver uses A^{-1} to decode it. To encode a message, each letter of the alphabet and any other agreed-upon symbols are assigned a positive integer value known by both the sender and receiver. The message is then encoded using these integers and the matrix A and is sent as a string of integers. Finally, the receiver uses this string of integers and A^{-1} to retrieve the original integers, hence the message. Example 1 shows details.

EXAMPLE 1
A Cryptography
Application

Suppose that the sender and receiver agree to the numerical assignment $a = 1$, $b = 2, \ldots z = 26$, space $= 30$, period $= 40$, and apostrophe $= 60$. (These assignments could be reversed or otherwise scrambled in some way.) The sender has the encoding matrix

$$A = \begin{bmatrix} 1 & 0 & 1 \\ 2 & 0 & 0 \\ 1 & 1 & 1 \end{bmatrix},$$

and the receiver has the decoding inverse matrix.

$$A^{-1} = \begin{bmatrix} 0 & \frac{1}{2} & 0 \\ -1 & 0 & 1 \\ 1 & -\frac{1}{2} & 0 \end{bmatrix}.$$

Because A is 3×3, the message will be partitioned into sequences of three symbols each. Suppose that the message to be sent is "MEET ME AT SIX." The sequence breakdown, along with the agreed-upon assignment of numbers is then

M	E	E	T		M	E		A	T		S		I	X	.
13	5	5	20	30	13	5	30	1	20	30	19	9	24	40	

The code matrix is then constructed using each sequence as a row:

$$C = \begin{bmatrix} 13 & 5 & 5 \\ 20 & 30 & 13 \\ 5 & 30 & 1 \\ 20 & 30 & 19 \\ 9 & 24 & 40 \end{bmatrix}.$$

Now the message will be transmitted as the string of numbers in consecutive rows of CA.

$$CA = \begin{bmatrix} 13 & 5 & 5 \\ 20 & 30 & 13 \\ 5 & 30 & 1 \\ 20 & 30 & 19 \\ 9 & 24 & 40 \end{bmatrix} \begin{bmatrix} 1 & 0 & 1 \\ 2 & 0 & 0 \\ 1 & 1 & 1 \end{bmatrix} = \begin{bmatrix} 28 & 5 & 18 \\ 93 & 13 & 33 \\ 66 & 1 & 6 \\ 99 & 19 & 39 \\ 97 & 40 & 49 \end{bmatrix}.$$

That is, the message is sent as 28, 5, 18, 93, 13, 33, 66, 1, 6, 99, 19, 39, 97, 40, 49. The person receiving the message has the 3×3 matrix A^{-1} and knows to reassemble the message as a matrix having three columns. The receiver also knows that the message is represented by the matrix CA. How does the receiver recover the original matrix C? Just multiply CA on the right by A^{-1}: $(CA)A^{-1} = C$.

$$(CA)A^{-1} = \begin{bmatrix} 28 & 5 & 18 \\ 93 & 13 & 33 \\ 66 & 1 & 6 \\ 99 & 19 & 39 \\ 97 & 40 & 49 \end{bmatrix} \begin{bmatrix} 0 & \frac{1}{2} & 0 \\ -1 & 0 & 1 \\ 1 & -\frac{1}{2} & 0 \end{bmatrix} = \begin{bmatrix} 13 & 5 & 5 \\ 20 & 30 & 13 \\ 5 & 30 & 1 \\ 20 & 30 & 19 \\ 9 & 24 & 40 \end{bmatrix} = C.$$

Now, equating the numbers in C with the agreed-upon letters gives

$$\begin{bmatrix} M & E & E \\ T & & M \\ E & & A \\ T & & S \\ I & X & . \end{bmatrix}.\ \blacksquare$$

There are many applications of cryptography that touch our everyday lives. For example, a foreign embassy may receive messages from home in coded form. To gain access to your computer account, a password is needed; when the correct password is entered, the computer can decode it to allow access. A bank's automatic teller machine (ATM) must decode certain information we give it to allow access to our account. A computer encodes all information entered as binary numbers, does the internal work using those numbers, then returns the results in decoded form that we can understand. ASCII (American Standard Code for Information Interchange), a computer code, uses 33 for an exclamation point, 43 for +, and 48 for 0, for example.

An Economic Application

Another example of how matrices and their inverses are used is in **input-output analysis.** Briefly, input-output analysis is concerned with the interrelationship among various preselected sectors that make up an economy. Each sector usually requires **input** from the sectors of the economy, including itself. Stated differently, for each unit of output from a given sector, there is an **internal demand** within the system from the other sectors. The **output** of a sector refers to its production. For example, to produce a unit of steel requires machinery from the manufacturing sector output, various control devices from the electronics sector output, and so on. To be more specific, suppose that the economy is divided into the broad sectors of manufacturing (M), electronics (E), and agriculture (A). Then an **input-output matrix (technological matrix)** displaying the interrelations among these sectors might appear as

$$
\begin{array}{c}
\text{Output}\\
\begin{array}{ccc}
\text{1 of M} & \text{1 of E} & \text{1 of A}\\
\uparrow & \uparrow & \uparrow
\end{array}\\
\begin{array}{c}
\text{Input} \quad \text{M} \rightarrow\\
\text{E} \rightarrow\\
\text{A} \rightarrow
\end{array}
\begin{bmatrix}
0.01 & 0.02 & 0.001\\
0.10 & 0.01 & 0.02\\
0.12 & 0.15 & 0.02
\end{bmatrix} = T
\end{array}
$$

with this interpretation: For each unit output of M, 0.01 units of M itself will be required in the process; for each unit output of E, 0.02 units of M will be required; for each unit output of A, 0.001 units of M will be required. This same interpretation may be made for the inputs of E and A. Now let x be the total production of M, y the total production of E, and z the total production of A.

Then

$$
X = \begin{bmatrix} x \\ y \\ z \end{bmatrix}
$$

is called a **production matrix,** and the matrix product

$$
TX = \begin{bmatrix} 0.01 & 0.02 & 0.001\\ 0.10 & 0.01 & 0.02\\ 0.12 & 0.15 & 0.02 \end{bmatrix}\begin{bmatrix} x \\ y \\ z \end{bmatrix} = \begin{bmatrix} 0.01x + 0.02y + 0.001z\\ 0.10x + 0.01y + 0.02z\\ 0.12x + 0.15y + 0.02z \end{bmatrix} \begin{array}{l} \text{total M input}\\ \text{total E input}\\ \text{total A input} \end{array}
$$

is called an **internal demand** matrix because it gives the total input needed from each of the three sectors to produce x, y, and z units, respectively, of M, E, and A.

Now consumer demand enters into the picture. Others will want to purchase some of the output from the various sectors of the economy, and their wants are called the

final demand on the economy. That demad may be expressed in a matrix with one column:

$$D = \begin{bmatrix} \text{amount from M} \\ \text{amount from E} \\ \text{amount from A} \end{bmatrix}.$$

This economy may now be described by the equation

total production = internal consumption by sectors + final demand.

In symbols, this equation can be expressed as $X = TX + D$. In the United States, great sums of money are spent determining what and how much consumers will demand. Usually, a good estimate of the demand is known, so the trick is to find the production matrix X so that the equation $X = TX + D$ holds. Because both T and D are known, solving for X is a matrix algebra problem:

$$X = TX + D$$
$$X - TX = D$$
$$(I - T)X = D$$
$$X = (1 - T)^{-1}D \text{ (if the inverse exists).}$$

HISTORICAL NOTE
Wassily Leontief

In 1973, the Nobel Prize for Economics was awarded to Professor Wassily Leontief (lē-on´-tēf) of Harvard University. Professor Leontief's work focused on the interrelation of the various sectors of the American economy and made use of the so-called input-output matrices. In his original work, Professor Leontief divided the economy into 500 sectors.

The Output of an Economy

If T is the input–output matrix and D the final demand matrix for a given economy, then the total output for the economy is $X = (I - T)^{-1}D$, assuming that $(I - T)^{-1}$ exists.

The matrix $(I - T)^{-1}$ is known as the "Leontief Matrix for the Economy." It is so named in honor of Professor Wassily Leontief for his analysis of the U.S. economy, which is similar to the preceding discussion.

EXAMPLE 2
An Input-Output
Analysis

Given that the input matrix of an economy, divided into sectors A and B, is

$$\begin{array}{cc} & \text{Output} \\ & \begin{array}{cc} A & B \end{array} \\ \begin{array}{c} \text{Input: } A \\ B \end{array} & \begin{bmatrix} 0.02 & 0.15 \\ 0.18 & 0.10 \end{bmatrix}, \end{array}$$

and the final demand is

$$D = \begin{bmatrix} 320 \\ 580 \end{bmatrix} \begin{matrix} A \\ B \end{matrix},$$

find the total output matrix X.

Solution We know that $X = (I - T)^{-1}D$, so that

$$X = (I - T)^{-1}D = \left(\begin{bmatrix} 1 & 0 \\ 0 & 1 \end{bmatrix} - \begin{bmatrix} 0.02 & 0.15 \\ 0.18 & 0.10 \end{bmatrix} \right)^{-1} \begin{bmatrix} 320 \\ 580 \end{bmatrix}$$

$$= \begin{bmatrix} 0.98 & -0.15 \\ -0.18 & 0.90 \end{bmatrix}^{-1} \begin{bmatrix} 320 \\ 580 \end{bmatrix}$$

$$= \begin{bmatrix} 1.053 & 0.175 \\ 0.211 & 1.146 \end{bmatrix} \begin{bmatrix} 320 \\ 580 \end{bmatrix}$$

$$= \begin{bmatrix} 438.46 \\ 732.20 \end{bmatrix} \begin{matrix} A \\ B \end{matrix}$$

Note that once $(I - T)^{-1}$ is known, it is easy to recalculate the total production matrix X if the demand matrix D changes. ■

Manufacturing Application

Here is one final example of how the inverse of a matrix can be applied. Think of an assembly process in which a single part sometimes will require other parts for its subassembly. A simple parts diagram might appear as shown in Figure 5. The diagram is interpreted as follows: Part p_2 requires 2 of part p_1, part p_3 requires 1 of part p_1, part p_4 requires 3 of part p_1 and 1 of part p_2, and so on.

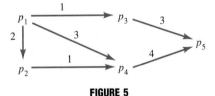

FIGURE 5
Diagram of Assembly Parts

This information can be put into a **parts matrix** as shown below. Each entry in this matrix represents the number of parts of the kind to the left of the element needed to assemble one part of the kind above the element. The assumption is now made that the total output of these parts is equal to the internal consumption of

$$
P = \begin{array}{c}
\begin{matrix} \text{Output:} & p_1 & p_2 & p_3 & p_4 & p_5 \end{matrix} \\
\begin{matrix} \text{Input:} & p_1 \\ p_2 \\ p_3 \\ p_4 \\ p_5 \end{matrix}
\begin{bmatrix} 0 & 2 & 1 & 3 & 0 \\ 0 & 0 & 0 & 1 & 0 \\ 0 & 0 & 0 & 0 & 3 \\ 0 & 0 & 0 & 0 & 4 \\ 0 & 0 & 0 & 0 & 0 \end{bmatrix}
\end{array}.
$$

the assembly process plus the demand outside the system. Therefore, if, as before, we let X be the total output column matrix and D the external demand matrix, then PX represents the number consumed internally in the assembly process, and $X = PX + D$. Again, D can usually be estimated rather closely so that the total production X is the sought-after matrix.

Therefore,

$$X = PX + D$$
$$X - PX = D$$
$$(I - P)X = D$$
$$X = (1 - P)^{-1}D \quad \text{(assuming that } (I - P)^{-1} \text{ exits).}$$

EXAMPLE 3
Using a Parts Matrix

For the assembly process decribed in Figure 5 and the resulting matrix P, let

$$D = \begin{bmatrix} 2000 \\ 2500 \\ 3000 \\ 2000 \\ 700 \end{bmatrix} \begin{matrix} p_1 \\ p_2 \\ p_3 \\ p_4 \\ p_5 \end{matrix}. \text{ Find the production matrix } X.$$

Solution   From the analysis above, $X = (I - P)^{-1}D$ so that for our situation

$$X = \left(\begin{bmatrix} 1 & 0 & 0 & 0 & 0 \\ 0 & 1 & 0 & 0 & 0 \\ 0 & 0 & 1 & 0 & 0 \\ 0 & 0 & 0 & 1 & 0 \\ 0 & 0 & 0 & 0 & 1 \end{bmatrix} - \begin{bmatrix} 0 & 2 & 1 & 3 & 0 \\ 0 & 0 & 0 & 1 & 0 \\ 0 & 0 & 0 & 0 & 3 \\ 0 & 0 & 0 & 0 & 4 \\ 0 & 0 & 0 & 0 & 0 \end{bmatrix} \right)^{-1} \begin{bmatrix} 2000 \\ 2500 \\ 3000 \\ 2000 \\ 700 \end{bmatrix}$$

$$= \begin{bmatrix} 1 & -2 & -1 & -3 & 0 \\ 0 & 1 & 0 & -1 & 0 \\ 0 & 0 & 1 & 0 & -3 \\ 0 & 0 & 0 & 1 & -4 \\ 0 & 0 & 0 & 0 & 1 \end{bmatrix}^{-1} \begin{bmatrix} 2000 \\ 2500 \\ 3000 \\ 2000 \\ 700 \end{bmatrix}$$

$$= \begin{bmatrix} 1 & 2 & 1 & 5 & 23 \\ 0 & 1 & 0 & 1 & 4 \\ 0 & 0 & 1 & 0 & 3 \\ 0 & 0 & 0 & 1 & 4 \\ 0 & 0 & 0 & 0 & 1 \end{bmatrix} \begin{bmatrix} 2000 \\ 2500 \\ 3000 \\ 2000 \\ 700 \end{bmatrix} = \begin{bmatrix} 36,100 \\ 7300 \\ 5100 \\ 4800 \\ 700 \end{bmatrix}. \quad \blacksquare$$

EXERCISES   3.4

In Exercises 1 through 4, use the given encoding matrix A and the given message to

(a) Construct the code matrix after assigning numbers to the letters and symbols in the message, as was done in Example 1.

(b) Compute the matrix the receiver gets.

(c) Find the decoding matrix A^{-1}, and use it to decode the message in a fashion similar to Example 1.

1. $A = \begin{bmatrix} 1 & 2 \\ -1 & 1 \end{bmatrix}$; message: I HAVE A JOB.

2. $A = \begin{bmatrix} 1 & 0 \\ 1 & 1 \end{bmatrix}$; message: GET LOST.

3. $A = \begin{bmatrix} 1 & 1 \\ 1 & 2 \end{bmatrix}$; message: NO WAY!

4. $A = \begin{bmatrix} 0 & 3 \\ 1 & 1 \end{bmatrix}$; message: I PASSED FINITE.

In Exercises 5 through 8, use matrix A and the encoded message to
(a) First find A^{-1}
(b) Then use A^{-1} to decode the message

5. $A = \begin{bmatrix} 1 & 0 & 1 \\ 2 & 0 & 3 \\ 1 & 2 & 1 \end{bmatrix}$; message: 48, 10, 66, 71, 60, 91, 26, 40, 27, 115, 60, 155

6. $A = \begin{bmatrix} 1 & -1 & 0 \\ 0 & 1 & 1 \\ 0 & 0 & 1 \end{bmatrix}$; message: 9, 51, 73, 30, −23, 25, 1, 3, 25, 1, 19, 29, 14, −7, 47

7. $A = \begin{bmatrix} 2 & 0 & 0 \\ 0 & 1 & 0 \\ 0 & 0 & 3 \end{bmatrix}$; message: 14, 12, 45, 4, 1, 36, 60, 23, 3, 36, 13, 27, 28, 7, 90, 6, 15, 42, 40, 9, 42, 42, 5, 57, 80, 30, 90

8. $A = \begin{bmatrix} 4 & 0 & 0 \\ 0 & 1 & 0 \\ 0 & 0 & 1 \end{bmatrix}$; message: 64, 15, 12, 48, 21, 20, 36, 15, 14, 120, 8, 1, 64, 16, 5, 56, 19, 40

In Exercises 9 through 14, answer these statements about each input-output matrix T and demand matrix D.
(a) Interpret the entries in each row of T.
(b) Compute $(I - T)^{-1}$.
(c) Compute the production matrix X.
(d) Find the internal consumption.

9. Input:
Output:
Steel Electronics
Steel $\begin{bmatrix} 0.02 & 0.10 \\ 0.15 & 0.01 \end{bmatrix}$ = T;
Electronics

$D = \begin{bmatrix} 500 \\ 800 \end{bmatrix}$

10. Input:
Output:
Agri. Mfg.
Agri. $\begin{bmatrix} 0.03 & 0.20 \\ 0.01 & 0.15 \end{bmatrix}$ = T; $D = \begin{bmatrix} 1000 \\ 1200 \end{bmatrix}$
Mfg.

11. Input:
Output:
Mfg. Electronics
Mfg. $\begin{bmatrix} 0.10 & 0.20 \\ 0.15 & 0.05 \end{bmatrix}$ = T;
Electronics

$D = \begin{bmatrix} 200 \\ 500 \end{bmatrix}$

12. Input:
Output:
Service Mfg.
Service $\begin{bmatrix} 0.20 & 0.50 \\ 0.10 & 0.15 \end{bmatrix}$ = T; $D = \begin{bmatrix} 300 \\ 800 \end{bmatrix}$
Mfg.

13. Input:
Output:
Agri. Mfg. Electronics
Agri. $\begin{bmatrix} 0.02 & 0.01 & 0.01 \\ 0.20 & 0.10 & 0.10 \\ 0.15 & 0.08 & 0.05 \end{bmatrix}$ = T;
Mfg.
Electronics

$D = \begin{bmatrix} 500 \\ 1000 \\ 2000 \end{bmatrix}$

14. Input:
Output:
Service Mfg. Electronics
Service $\begin{bmatrix} 0.40 & 0.20 & 0.10 \\ 0.01 & 0.10 & 0.20 \\ 0.30 & 0.20 & 0.10 \end{bmatrix}$ = T;
Mfg.
Electronics

$D = \begin{bmatrix} 300 \\ 400 \\ 500 \end{bmatrix}$

In Exercises 15 through 20:
(a) Construct the parts matrix P.
(b) Interpret the entries in each row.
(c) Find the production matrix based on P and the given demand matrix D.

15.

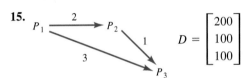

$D = \begin{bmatrix} 200 \\ 100 \\ 100 \end{bmatrix}$

16.

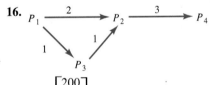

$D = \begin{bmatrix} 200 \\ 300 \\ 400 \\ 100 \end{bmatrix}$

17.

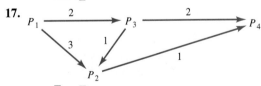

$D = \begin{bmatrix} 200 \\ 80 \\ 100 \\ 50 \end{bmatrix}$

18.

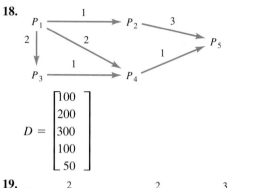

$$D = \begin{bmatrix} 100 \\ 200 \\ 300 \\ 100 \\ 50 \end{bmatrix}$$

19.

$$P_1 \xrightarrow{\quad 2 \quad} P_2 \xrightarrow{\quad 2 \quad} P_3 \xrightarrow{\quad 3 \quad} P_4$$

$$D = \begin{bmatrix} 250 \\ 300 \\ 150 \\ 60 \end{bmatrix}$$

20.

$$P_1 \xrightarrow{\quad 2 \quad} P_2 \xrightarrow{\quad 1 \quad} P_4 \xrightarrow{\quad 1 \quad} P_5$$

with $P_1 \xrightarrow{3} P_3 \xrightarrow{2} P_4$

$$D = \begin{bmatrix} 80 \\ 60 \\ 50 \\ 100 \\ 200 \end{bmatrix}$$

Each parts matrix represents the number of parts used internally to make other parts for subassembly in a production process. Construct the parts-arrow diagram (as shown in Exercises 15 through 20) for this subassembly process.

21.

	P_1	P_2	P_3	P_4
P_1	0	2	0	3
P_2	0	0	1	2
P_3	0	0	0	1
P_4	0	0	0	0

22.

	P_1	P_2	P_3	P_4	P_5
P_1	0	2	1	3	1
P_2	0	0	2	0	1
P_3	0	0	0	3	0
P_4	0	0	0	0	2
P_5	0	0	0	0	0

CHAPTER 3 REFLECTIONS

CONCEPTS

Matrices provide a convenient structure for organizing and interpreting data. Operations such as multiplying by a number, finding the transpose, determining the inverse, and squaring, and cubing can be performed on a single matrix. Operations such as *addition, subtraction,* and *multiplication* use a pair of compatible matrices. Matrix operations follow many of the rules of algebra, but *commutativity* and *cancellation for multiplication* are two notable exceptions. Matrices provide a powerful tool in a variety of areas of applications, including inventories, dominance relations, communication networks, cryptography, input-output analysis, and parts production.

COMPUTATION/CALCULATION

Let $A = \begin{bmatrix} 2 & 0 & 1 \\ 3 & -1 & 0 \end{bmatrix}$, $B = \begin{bmatrix} 1 & -2 \\ 0 & 4 \\ 3 & -1 \end{bmatrix}$, $C = \begin{bmatrix} 9 & 1 \\ 0 & -1 \end{bmatrix}$,

and $D = \begin{bmatrix} 1 & -2 \\ 3 & 0 \end{bmatrix}$. *Evaluate each expression, if possible.*

1. $C + 3D$ **2.** $\frac{1}{3}C - D$ **3.** B^T **4.** AB **5.** BA

6. BC **7.** CB **8.** D^2 **9.** A^2 **10.** B^{-1} **11.** C^{-1}

12. Solve the matrix equation $X + CX = D$ for X. (Assume that the matrices are compatible, and inverses exist where needed.)

13. Assign values to the variables so that

$$\begin{bmatrix} 27 & -w \\ 2t & 0 \end{bmatrix} = \begin{bmatrix} x^3 & 24 \\ 17 & -2v + 1 \end{bmatrix}.$$

14. Find the inverse of the matrix $\begin{bmatrix} 10 & 8 \\ 9 & 7 \end{bmatrix}$.

15. Find the inverse of the matrix $\begin{bmatrix} 2 & 0 & 1 \\ 1 & 3 & 2 \\ 1 & 1 & 0 \end{bmatrix}$.

16. Find matrices of order 3×3 with $A \neq \mathbf{0}$ and $B \neq \mathbf{0}$ such that $AB = \mathbf{0}$.

CONNECTIONS

17. Airline flights are available through the cities shown in the direction of the arrows only. If cities do not have flights from themselves to themselves, write a matrix that displays this flight network.

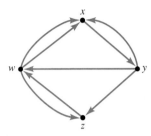

Use the given coding matrix A and the given message to solve Problems 18 through 21.

$$A = \begin{bmatrix} 0 & 1 \\ 1 & 3 \end{bmatrix} \quad \text{Message: NICE GOING!}$$

18. Construct the code matrix after assigning numbers to the letters and symbols in the message.

19. Compute the matrix the receiver gets.

20. Find the decoding matrix A^{-1}.

21. Decode the message using A^{-1}.

COMMUNICATION

22. What distinguishes a square matrix from other kinds of matrices?

23. Describe in your own words the conditions necessary for two matrices to be equal.

24. Identify the critical requirement for the matrix product AB to be defined.

25. Explain what it means for the matrix A^{-1} to be the inverse of the matrix A.

COOPERATIVE LEARNING

26. Write and solve an application word problem for which each given matrix equation would be a model.

(a) $\begin{bmatrix} 3 & 6 & 10 \end{bmatrix} \begin{bmatrix} 50 & 20 \\ 40 & 70 \\ 60 & 80 \end{bmatrix} = \begin{bmatrix} x & y \end{bmatrix}$

(b) $\begin{bmatrix} 2 & -1 & -1 \\ 1 & 1 & -2 \\ 1 & -1 & -1 \end{bmatrix} \begin{bmatrix} x \\ y \\ z \end{bmatrix} = \begin{bmatrix} 10 \\ 2 \\ 3 \end{bmatrix}$

(c) $\begin{bmatrix} 3 & -1 \\ 5 & -2 \\ 1 & 2 \\ 10 & -1 \end{bmatrix} \begin{bmatrix} x \\ y \end{bmatrix} = \begin{bmatrix} 1 \\ 0 \\ 12 \\ 15 \end{bmatrix}$

CRITICAL THINKING

27. Determine the conditions on matrices A and B for both matrix products AB and BA to be defined. Analyze several such products where both are defined, and estimate the likelihood that $AB = BA$.

COMPUTER/CALCULATOR

If you need help in mastering matrix multiplication, use the tutorial program MulTutor from the disk MasterIt. Entries must be in integer form, and you must give the correct order of the product before the program will accept these entries. After the matrices to be multiplied have been entered, you may move the row (in the left matrix) and column (in the right matrix), highlighting with the arrow keys. The computation of the corresponding row-column operation will then be displayed in the center of the screen, and the result of that computation will be highlighted in the proper address of the product matrix. The following is a typical screen output.

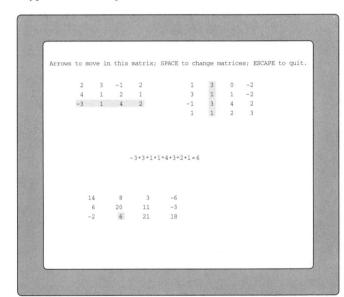

28. Use a program such as MulTutor to find the product AB if:

$$A = \begin{bmatrix} 2 & -1 \\ 3 & 5 \end{bmatrix}$$

and

$$B = \begin{bmatrix} 4 & 6 & 1 & 0 \\ 1 & 2 & 5 & 2 \end{bmatrix}.$$

29. Finding the inverse of a matrix may be viewed as solving systems of linear equations. Find the three systems of equations needed to find the inverse of

$$A = \begin{bmatrix} 2 & 0 & -1 \\ 1 & 1 & 3 \\ 3 & 0 & 2 \end{bmatrix},$$

then use MasterIt's program Pivot to solve all three simultaneously.

30. Use a program such as MatxUtil to
    (a) Find the inverse of the matrix given in Problem 29.
    (b) Find $A^2 + 3B$ where A is the matrix of Problem 29 and

$$B = \begin{bmatrix} -3 & 4 & 6 \\ 2 & 0 & 7 \\ 1 & 9 & 8 \end{bmatrix}.$$

Some hand-held calculators (especially graphing calculators) have a built-in capacity to find the inverse of a matrix at the stroke of a key upon entry of the matrix elements. All entries are to be in decimal form, and the inverse will also be given in decimal form.

31. Use a graphing calculator to find the inverse of

$$A = \begin{bmatrix} 2 & 4 \\ 1 & 5 \end{bmatrix}.$$

A screen output for the inverse is shown at the right.

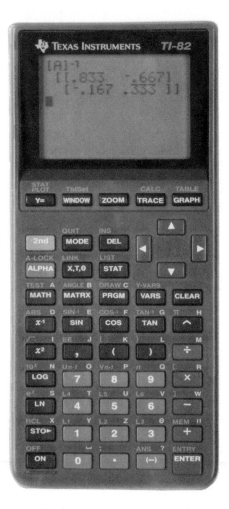

Cumulative Review

32. Find the slope, x-intercept, and y-intercept of the line having the equation $3x + 4y = 12$.

33. Write the equation of the horizontal line that contains the point $(-7, 0)$.

34. Solve the system

$$x + 3y = 7$$
$$3x - 2y = 9,$$

using the pivoting process.

35. Find the profit function and the break-even point for a company having a total cost function $C(x) = 60x + 2100$ and a revenue function $R(x) = 80x$.

36. Solve this system without using the pivoting process:

$$x + y \qquad = 0$$
$$y - 7z = 1$$
$$5y - 35z = 5.$$

37. Suppose you believe that you can make 72% on an upcoming Finite Math test with no further study but estimate that you can increase your test score by 5% for each additional hour of study. Find a linear function that will give your test score in terms of additional hours t of study. (Be sure to state the limits on t.)

CHAPTER **3** SAMPLE TEST ITEMS

1. Assign values to the variables so that the matrix
$\begin{bmatrix} 2 & 0 & x^2 \\ t+4 & 7 & 19 \end{bmatrix}$ equals the matrix
$\begin{bmatrix} y-7 & w & 9 \\ 10 & 7 & 3r+1 \end{bmatrix}$. Let $A = \begin{bmatrix} 2 & 0 & 3 \\ -1 & 1 & 0 \\ 0 & 0 & 4 \end{bmatrix}$,
$B = \begin{bmatrix} 1 & 2 & 3 \\ 4 & 5 & 6 \\ 7 & 8 & 9 \end{bmatrix}$, and $C = \begin{bmatrix} -2 & 1 & 4 \\ 6 & 3 & -1 \end{bmatrix}$.

Evaluate each expression, if possible.

2. $2A - B$ **3.** $\frac{1}{2}B + C$ **4.** A^T **5.** B^2 **6.** AB **7.** C^2

8. Three communication stations communicate directly only in the direction of the arrows. Write a matrix that displays this communication network.

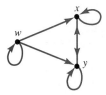

9. Write the linear system of equations corresponding to this matrix equation:

$$\begin{bmatrix} 2 & 0 & 1 \\ 3 & -2 & 5 \end{bmatrix} \begin{bmatrix} x \\ y \\ z \end{bmatrix} = \begin{bmatrix} 4 \\ 17 \end{bmatrix}.$$

10. The T-Mart Company has two stores that stock brands X and Y microwave ovens. The current inventory matrix is

$$\begin{array}{cc} & \text{Store 1} \quad \text{Store 2} \\ \begin{array}{c} \text{Brand X} \\ \text{Brand Y} \end{array} & \begin{bmatrix} 15 & 12 \\ 8 & 3 \end{bmatrix} = A \end{array}$$

Brand X microwave ovens cost \$175 each, and Brand Y ovens cost \$250 each. Find a matrix B and its product with A so that the resulting matrix gives the total cost of the microwave oven inventory in each store.

11. Decide whether $\begin{bmatrix} 3 & -1 \\ 2 & -1 \end{bmatrix}$ and $\begin{bmatrix} 1 & 1 \\ 2 & 3 \end{bmatrix}$ are inverses of each other.

12. Find the inverse of $\begin{bmatrix} 5 & 7 \\ 2 & 3 \end{bmatrix}$.

13. Find the inverse, if one exists, for the matrix

$$\begin{bmatrix} 1 & 4 & -3 \\ 0 & 1 & 7 \\ 0 & 3 & 20 \end{bmatrix}.$$

14. Solve for X in the matrix equation $XA + B = Y$.

15. Use the fact that

$$\begin{bmatrix} 3 & -1 & 0 \\ 1 & 0 & -1 \\ -1 & 1 & -1 \end{bmatrix}^{-1} = \begin{bmatrix} 1 & -1 & 1 \\ 2 & -3 & 3 \\ 1 & -2 & 1 \end{bmatrix}$$

to quickly solve the system

$$\begin{array}{rcl} 3x - y & = 2 \\ x \quad - z & = 0 \\ -x + y - z & = 3 \end{array}$$

by using matrix multiplication.

The arrow diagram below indicates the parts needed to assemble other parts. Use the diagram to answer Exercises 16 through 20.

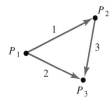

16. Construct the parts matrix P.

17. Interpret the entries in each row of P.

18. Compute $(I - P)^{-1}$.

19. If $D = \begin{bmatrix} 100 \\ 200 \\ 300 \end{bmatrix}$ is the demand matrix, calculate the production matrix X.

20. Find the number of each part consumed internally in the production process.

Career Uses of Mathematics

Our business manufactures rugs for the retail market. We have several hundred combinations of styles, sizes, and colors that are manufactured for shipment to several thousand retail stores and catalog distribution centers.

Our computer systems must prioritize all customer orders for a unique stock-keeping unit and then assign the available inventory or planned production for the unit. The order and inventory data are arranged in tables that are then matched to assign the appropriate number of pieces of inventory to the orders. The inventory table is prioritized by age, manufacturing lot, and number of pieces per location. If the inventory table is empty and there are additional orders to fill, we create a table of planned production and continue to assign promise dates to the order table in priority sequence. Once all orders are given inventory or a promise date, we use the tables to update the order, inventory, and product data bases. Processing against tables (or arrays) is the most practical way for us to handle all of the unique order and inventory conditions.

Understanding the elementary matrix operations introduced in finite mathematics is adequate for understanding our inventory tables and how we can combine them. Many other finite mathematics skills, such as linear programming and statistics, are used in other aspects of our company.

Paul L. Grafton, Burlington Industries

Linear Programming

Refrigerators are to be shipped from warehouses in Bentonville and Ashdown to two stores, one in Miami and one in Harrison, with orders and shipping cost per refrigerator as indicated. How many should be shipped from each warehouse to minimize shipping costs? (See Example 8, Section 4.6.)

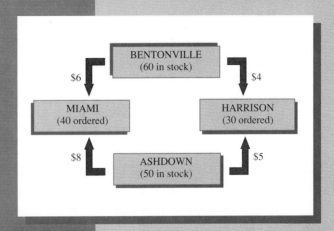

CHAPTER PREVIEW

Linear programming is one of the premier applications of modern mathematics. It is used extensively in business, health, agricultural, industrial, and military settings. Linear programming is a technique for finding the maximum or minimum value a linear function in several variables may attain; the variables in the function may assume values only in a region restricted by linear inequalities. After a section on how to mathematically model such problems, we show how a graphical solution may be attained for functions of two variables. Then we explore the simplex method (or algorithm) for standard maximum linear programming problems, which is an algebraic method of solution using pivoting that is adaptable to computer use for large problems. Next, we present a particularly simple pivoting scheme that leads to the solution of *any* linear programming problem. Finally, the dual problem is studied.

CALCULATOR/COMPUTER OPPORTUNITIES FOR THIS CHAPTER

The computer disk MasterIt has several programs to help you learn linear programming. LPGraph will graph the boundary lines for a feasible region and then shade in the feasible region. LinProg will help you learn the pivoting process needed to reach a solution by the simplex method (you must supply the brainpower) or the pivoting scheme required on nonstandard problems. Another program, LProg, will completely solve any large linear programming problem upon entering the data describing the problem.

The calculator manual available from the publisher to accompany this text has a calculator program that will quickly solve any linear programming problem (or indicate its nonsolvability) of up to 12 columns (including slack variable columns) to the left of the constants column and up to 6 rows (5 constraint and 1 objective function).

Computer programs such as Maple may also be used to solve linear programming problems.

SECTION 4.1   Modeling Linear Programming Problems

We resume our study of modeling with linear systems, but instead of exclusively using linear equalities, we now also allow **linear inequalities.** The problems modeled here fall under the classification of **linear programming problems.** Some examples illustrate the general nature of such problems and their mathematical formulation. We investigate how to find solutions later.

EXAMPLE 1
The Mathematical
Formulation of a Linear
Programming Problem

A company makes desks in two models: a student model and a secretarial model. Each student model requires 2 hours in woodworking and 3 hours in finishing. Each secretarial model requires 3 hours in woodworking and 5 hours in finishing. The company has a total of 240 work hours available in the woodworking department and a total of 390 work hours available in the finishing department each week. A profit of $20 is made on each student model and $50 on each secretarial one. Assuming that all desks are sold, how many of each type of desk should be made each week to maximize the company's profit from these items?   ■■■

We now develop the mathematical formulation for the problem in Example 1. The statement of Example 1 asserts that the woodworking department has a total of 240 work hours *available* each week. This means that the woodworking department may expend any number of hours *up to and including* 240 during the week. The mathematical symbol used to describe this limit is the **inequality sign** ≤, read, "less than or equal to." This restriction may then be stated as

[hours used in woodworking department] ≤ 240.

Similarly, the finishing department may use any number of hours *up to and including* 390 hours. This may be stated as

[hours used in finishing department] ≤ 390.

To describe how these inequalities relate to the production of the two types of desks, we use a table like those in Chapter 2 to organize the given data. The only difference now is that inequalities are involved and the profit is taken into account. Let x be the number of student models made and y the number of secretarial models. Figure 1 gives a systematic organization of all the data.

	Student desk	Secretarial desk	Limits
Woodworking	2	3	≤ 240
Finishing	3	5	≤ 390
Number made	x	y	
Profit	20	50	Maximize

FIGURE 1

Just as with equalities, we read the inequalities across the rows:

$$2x + 3y \leq 240$$
$$3x + 5y \leq 390.$$

The first of these inequalities states that if x student desks are made, $2x$ hours of woodworking will be required, and if y secretarial desks are made, $3y$ more hours of woodworking will be required for a total of $2x + 3y$ hours, and that the total must be less than or equal to 240. In mathematical symbols, this is precisely the description of the woodworking limitations. Similarly, the second inequality describes the limitations on the finishing department. Because of the linear form of the left side of these two inequalities, they are known as **linear inequalities** and are part of the **constraints** on the problem.

The statement of profit P is read from the last row of Figure 1 as $P = 20x + 50y$ (in dollars). The function P is known as the **objective function** for this problem. Notice that P is a function of *two* variables, x and y. To maximize P means to find values of x and y that obey all constraints and make P as large as possible.

One last consideration finishes our mathematical model of Example 1: Because x and y represent numbers of desks to be made, both must be at least as large as zero. Using the symbol $\geq$, which is read, "greater than or equal to," these conditions may be written $x \geq 0$ and $y \geq 0$ and become part of the system of constraints. In particular, these constraints are called **nonnegativity constraints.** The complete mathematical description of the problem in Example 1 may now be stated in the language of linear programming.

$$\text{Maximize } P = 20x + 50y, \qquad \leftarrow \text{Objective function}$$
$$\text{subject to } 2x + 3y \leq 240$$
$$3x + 5y \leq 390 \qquad \text{Constraints}$$
$$x \geq 0, y \geq 0$$

The next linear programming problem asks that the objective function be minimized (made as small as possible), subject to the given constraints.

EXAMPLE 2
The Mathematical Formulation of a Linear Programming Problem

A company makes two calculator models, one designed specifically for business use and the other designed for scientific use. The business model contains 10 microcircuits, and the scientific model contains 20. A contract with a microcircuit supplier requires the use of at least 3200 microcircuits each day. A contract with a supplier of the off-on switches used on both calculators requires the use of at least 300 such switches each day. The company would also like to produce at least 100 business models each day. If each business calculator requires 10 production steps and each scientific calculator requires 12 production steps, how many calculators of each type should be made to minimize production steps? Give the mathematical formulation of this problem.

Solution   The statement that "at least 3200 microcircuits" must be used each day means that 3200 *or more* are to be used. The mathematical symbol $\geq$, read "greater than or equal to," is used to describe this condition. Applying the restrictions as stated in the problem, we have

$$[\text{daily use of microcircuits}] \geq 3200$$
$$[\text{daily use of off-on switches}] \geq 300$$
$$[\text{daily production of business models}] \geq 100.$$

Let x be the number of business calculators made each day, and let y be the number of scientific calculators made each day. The data in the problem may then be organized as shown in Figure 2.

	Business calculators	Scientific calculators	Restrictions
Microcircuits	10	20	$\geq$ 3200
Off-on switches	1	1	$\geq$ 300
Number of business calculators	1	0	$\geq$ 100
Number made	x	y	
Production steps	10	12	Minimize

FIGURE 2

In this problem, the number of production steps S is to be minimized. Writing the total number of production steps first and the inequality restrictions next, the rows of Figure 2 give this mathematical formulation of the problem:

$$\text{Minimize } S = 10x + 12y, \qquad \leftarrow \text{Objective function}$$

$$\left.\begin{array}{rl} \text{subject to } 10x + 20y &\geq 3200 \\ x + y &\geq 300 \\ x &\geq 100 \\ x \geq 0, y &\geq 0 \end{array}\right\} \quad \text{Constraints}$$

(The nonnegativity constraint $x \geq 0$ is redundant because $x \geq 100$). ▬

Some linear programming problems have "mixed constraints," which means that some constraints will be inequalities of the form $\leq$, while others will be of the form $\geq$, and some may even be in the form of an equality.

EXAMPLE 3
The Mathematical
Formulation of a Linear
Programming Problem

A new line of snack cracker is being introduced by a food company, which has allotted a maximum of $50,000 for advertising in newspapers, on radio, and on television. Research shows that each newspaper ad costs $100 and will cause 200 people to try the new cracker, each radio ad costs $150 and will cause 300 people to try the new cracker, and each television ad costs \$300 and will cause 600 people to try the new cracker. The company wants to reach at least 10,000 people through these advertisements and has agreed to run at least twice as many radio ads as television ads. Company image ratings are 4 points for each newspaper ad, 2 for each radio ad, and 3 for each television ad. How can the company meet these objectives while maximizing the total number of company image points? Give the mathematical formulation of this problem.

Solution   Let x be the number of newspaper ads, y the number of radio ads, and z the number of television ads. The restriction on the total cost of ads may then be written as

$$100x + 150y + 300z \le 50{,}000.$$

The total number of people buying the new cracker is to be at least 10,000, so this restriction becomes

$$200x + 300y + 600z \ge 10{,}000.$$

The stipulation that "at least twice as many radio ads as television ads" are to be run means that the y number must be at least twice the z number, or that

$$y \ge 2z.$$

The company's total number of image points, which are to be maximized, may be written as $P = 4x + 2y + 3z$. The mathematical formulation of the problem may now be written.

$$
\begin{aligned}
\text{Maximize } P = 4x + 2y + 3z, &\quad \leftarrow \text{Objective function}\\
\text{subject to } 100x + 150y + 300z &\le 50{,}000\\
200x + 300y + 600z &\ge 10{,}000\\
y &\ge 2z\\
x \ge 0, \; y \ge 0, \; z &\ge 0
\end{aligned}
$$

Constraints

EXAMPLE 4
The Mathematical Formulation of a Linear Programming Problem

A firm makes three types of golf bags: economy, deluxe, and super deluxe. The economy bag requires 15 minutes of cutting time, 20 minutes of sewing, 30 minutes of trimming, and it sells for \$80. The deluxe bag requires 15 minutes of cutting time, 20 minutes of sewing, 50 minutes of trimming, and it sells for \$90. The super deluxe bag requires 20 minutes of cutting time, 30 minutes of sewing, 50 minutes of trimming, and it sells for \$100. The firm has available 80 work hours of cutting time, 90 work hours of sewing time, and 120 work hours of trimming time each day. A contract from a large discount store requires that the firm make at least 20 economy bags and 25 deluxe bags each day. The company also wants to make an equal number of deluxe and super deluxe bags. How many bags of each type should be made to meet the stated restrictions and to maximize revenue? Give the mathematical formulation of this problem.

Solution   Let x be the number of economy bags made, y the number of deluxe bags made, and z the number of super deluxe bags made. Taking care to convert hours to minutes, Figure 3 shows how these data may be organized. Note that the restriction $y = z$ has been rewritten as $y - z = 0$.

	Economy	Deluxe	Super deluxe	Restrictions
Cutting	15	15	20	$\leq 80(60)$
Sewing	20	20	30	$\leq 90(60)$
Trim	30	50	50	$\leq 120(60)$
Economy	1			≥ 20
Deluxe		1		≥ 25
Deluxe and super deluxe		1	-1	$= 0$
Number made	x	y	z	
Profit	80	90	100	Maximize

FIGURE 3

The mathematical formulation of this problem can be stated as follows:

$$\text{Maximize } P = 80x + 90y + 100z, \qquad \leftarrow \text{Objective function}$$

$$
\left.
\begin{aligned}
\text{subject to } 15x + 15y + 20z &\leq 4800 \\
20x + 20y + 30z &\leq 5400 \\
30x + 50y + 50z &\leq 7200 \\
x \qquad\qquad &\geq 20 \\
y \qquad &\geq 25 \\
y - \quad z &= 0 \\
x \geq 0, \, y \geq 0, \, z \geq 0
\end{aligned}
\right\} \quad \text{Constraints}
$$

These examples show the general nature of a linear programming problem: To maximize or minimize an objective function that is subject to particular constraints. The objective function is always a linear function of several variables, $x_1, x_2 \ldots x_n$, meaning that it has the form

$$f = a_1 x_1 + a_2 x_2 + \ldots + a_n x_n,$$

where $a_1, a_2, \ldots a_n$ are real numbers, not all of which are zero. All constraints (restrictions) are also linear and may be put into one of these forms:

$$b_1 x_1 + b_2 x_2 + \ldots + b_n x_n \leq c$$
$$b_1 x_1 + b_2 x_2 + \ldots + b_n x_n \geq c$$
$$b_1 x_1 + b_2 x_2 + \ldots + b_n x_n = c.$$

In addition, the nonnegativity conditions apply: $x_1 \geq 0, x_2 \geq 0, x_3 \geq 0, \ldots, x_n \geq 0$.

In the next two sections, we present a graphic solution method for linear programming problems with two variables. Then, in subsequent sections, an algebraic method of solution is presented.

EXERCISES 4.1

In Exercises 1 through 4:
(a) Organize the data in a diagram such as in Figures 1, 2, and 3.
(b) From this diagram, write the objective function and constraints that formulate the linear programming problem.

1. *Manufacturing* The Fine Line Pen Company makes two types of ballpoint pens: a silver model and a gold model. The silver model requires 1 minute in a grinder and 3 minutes in a bonder. The gold model requires 3 minutes in a grinder and 4 minutes in a bonder. Because of maintenance procedures, the grinder can

← convrct to min.

be operated no more than 30 hours per week and the bonder no more than 50 hours per week. The company makes $5 on each silver pen and $7 on each gold pen. How many of each type pen should be produced and sold each week to maximize profits?

Do 1st

2. **Inventory** An appliance dealer, William Young, wants to purchase for inventory a combined total of no more than 100 refrigerators and dishwashers. Refrigerators weigh 200 pounds each, and dishwashers weigh 100 pounds each. Suppose that the dealer is limited to a total of 12,000 pounds for these two items. If a profit of $35 on each refrigerator and $20 on each dishwasher is projected, how many of each should be purchased and sold to make the largest profit?

3. **Mixtures** A dietitian is to prepare two foods, A and B, to meet nutritional requirements. Each pound of Food A costs 20 cents and contains 100 units of vitamin C, 40 units of vitamin D, and 10 units of vitamin E. Each pound of Food B costs 15 cents and contains 10 units of vitamin C, 80 units of vitamin D, and 5 units of vitamin E. The mixture of the two foods is to contain at least 260 units of vitamin C, at least 320 units of vitamin D, and at least 50 units of vitamin E. How many pounds of each type of food should be used to minimize the cost?

4. **Medical** After a particular operation, a patient receives three types of painkillers. Type A rates a 2 in effectiveness, type B rates a 2.5, and type C rates a 3. For medical reasons, the patient can receive no more than 3 doses of type A, no more than 5 doses of types A and C combined, and no more than 7 doses of types B and C combined during a 24-hour period. How many doses of each type can be safely given to the patient in a 24-hour period with maximum painkilling benefits?

In the remaining exercises, write the mathematical formulation of the linear programming problem, and identify the objective function and all constraints.

5. **Health** A particular salad contains 4 units of vitamin A, 5 units of vitamin B-complex, and 2 mg of fat per serving. A nutritious soup contains 6 units of vitamin A, 2 units of vitamin B-complex, and 3 mg of fat per serving. If a lunch consisting of these two foods is to have at least 10 units of vitamin A and at least 10 units of vitamin B-complex, how many servings of each should be used to minimize the total milligrams of fat?

6. **Manufacturing** A tire company produces a four-ply, a radial, and an off-road tire. Each four-ply tire takes 2 minutes and costs $20 to produce; each radial tire, 6 minutes and $30; and each off-road tire, 10 minutes and $50. Orders the company receives dictate that it must make a total of at least 5000 tires each day and, furthermore, at least 1500 of the total must be radial

tires and at least 60 of the tires must be off-road tires. The union contract with the company stipulates that at least 400 worker-hours must be used each day on the production lines. How many of each type of tire should be made to meet these conditions and keep costs to a minimum?

7. **Radio Programming** A radio station is planning to air a new talk show, new entertainment show, new stock market show, and new national news show. Market research shows the expected number of listeners to be, respectively, 10,000, 2500, 800, and 600 each time the shows air. The production costs for each time these shows are aired are, respectively, $1,000, $800, $400, and $300. The weekly budget for production costs is not to exceed $20,000. Advertisers would like to run the talk show at least 5 times each week and the entertainment show at least 3 times each week. The station wants to limit the number of times each show is run during the week to 8, 5, 3, and 10 times, respectively. How many times each week should the station air each of these programs to maximize the total number of listeners?

8. **Manufacturing** A firm makes a standard and a deluxe model of electric skillet. Each standard model costs $25 and each deluxe model $40 to make. The firm must make at least 100 of the standard model and at least 75 of the deluxe. It sells each standard model for $30 and each deluxe model for $75, and it must have a total revenue of at least $2250 from sales of these two models. How many of each model should be made and sold to minimize costs?

9. **Mixtures** A road-paving firm has on hand three types of paving material. Each barrel of type A contains 2 gallons of carbon black and 2 gallons of thinning agent and costs $5. Each barrel of type B contains 3 gallons of carbon black and 1 gallon of thinning agent and costs $3. Each barrel of type C contains 3 gallons of carbon black and 1 gallon of thinning agent and costs $4. The firm needs to fill an order for which the final mixture must contain at least 12 gallons of carbon black and at least 6 gallons of thinning agent. How many barrels of each type of paving material should be used to fill this order at minimum expense?

10. **Agriculture** Members of the Ministry of Agriculture in a particular country are planning next year's rice, wheat, corn, and barley crops. They estimate that each hectare of rice planted will yield 20 cubic meters; each hectare of wheat, 18 cubic meters; each hectare of corn, 15 cubic meters; and each hectare of barley, 22 cubic meters. They would like to raise a total of at least 20 million cubic meters of these four grains. In addition, they would like to raise at least 4 million cubic meters of rice, 5 million of wheat, 6 million of

corn, and 2 million of barley. Soil conditions dictate that at least three times as many hectares of corn be planted as rice. The Ministry estimates that it has enough machinery to plant at least 1 million hectares of these four grains and that for planting, harvesting, and transportation, each hectare of these grains will require, respectively, 6, 3, 4, 3 number of workers. How many hectares of each of these grains should be planted to meet the requirements of the Ministry of Agriculture and to use the least number of workers?

11. **Investments** An investment club has at most $30,000 to invest in either junk bonds or premium-quality bonds. Each type of bond is bought in $1,000 denominations. The junk bonds have an average yield of 12%, and the premium-quality bonds yield 7%. The policy of the club is to invest at least twice the amount in premium-quality bonds as in junk bonds. How should the club invest its money in these bonds to receive the maximum return on its investment?

12. **Pollution** The Environmental Protection Agency (EPA) has issued new regulations that require the New Molecule Chemical Co. to incorporate a new process to reduce the pollution caused when a particular chemical is produced. The old process releases 2 grams of sulphur and 5 grams of lead for each liter of the chemical produced. The new process releases 1 gram of sulphur and 2.6 grams of lead for each liter of the chemical produced. The company makes a profit of 20 cents for each liter produced under the old method and 16 cents for each liter produced under the new process. The new EPA regulations do not allow for more than 10,000 grams of sulphur and 8000 grams of lead each day. How many liters of this chemical can be produced under the old method and how many under the new method to maximize profits and stay within EPA regulations?

13. **Agriculture** A farmer has 500 acres available on which to plant wheat or soybeans or both. The planting costs per acre are $200 for wheat and $350 for soybeans, and the farmer has a total of $25,000 to spend on planting costs. The anticipated income per acre from wheat is $350 and from soybeans $525. How many acres should be planted in wheat and how many in soybeans to maximize profit?

14. **Order Fulfillment** The Hi-Fi Record Store is about to place an order for cassette tapes and compact discs. The distributor it orders from requires that an order contain at least 250 items. The prices that Hi-Fi must pay are $12 for each cassette and $16 for each compact disc. The distributor also requires that at least 30% of any order be compact discs. How many cassettes and compact discs should Hi-Fi order so that its total ordering costs will be kept to a minimum?

15. **Transportation Purchases** A charter plane service contracts with a retirement group to provide flights from New York to Florida for at least 500 first-class, 1000 tourist-class, and 1500 economy-class passengers. The charter company has two types of planes. Type A costs $12,000 per flight and will carry 50 first-class, 60 tourist-class, and 100 economy-class passengers. Type B costs $10,000 per flight and will carry 40 first-class, 30 tourist-class, and 80 economy-class passengers. How many of each type of plane should be used to minimize flight costs?

16. **Manufacturing** A firm makes three models of microwave oven: a standard model, a deluxe model, and a super deluxe model. The firm has contracts that call for making at least 80 standard ovens, at least 50 deluxe ovens, at least 90 super deluxe ovens each week, and the total of all these models must be at least 300. The cost is $100 for making a standard model, $110 for a deluxe model, and $120 for a super deluxe model. How many ovens of each model should be made each week to minimize costs?

17. **Sales** An appliance store sells two brands of television sets: Nitebrite and Solar II. Each Nitebrite sells for $420, and each Solar II sells for $550. The store's warehouse capacity for television sets is 300, and new sets are delivered only once a month. Past records show that customers will buy at least 80 Nitebrite sets and at least 120 Solar II sets each month. How many of each type of set should the store buy each month to maximize revenue?

18. **Political Campaign** The campaign manager of a candidate for a local political office estimates that each 30-minute speech to a civic group will generate 20 votes, each hour spent on the telephone will generate 10, and each hour spent campaigning in shopping areas will generate 40. The candidate wants to make at least twice as many trips to shopping areas as speeches to civic groups and spend at least 5 hours on the telephone. The campaign manager thinks her candidate can win if he can generate a total of at least 1000 votes by these three methods. How can the candidate meet these goals and keep to a minimum his campaigning time?

19. **Diet** Jane is going on a low-carbohydrate diet that calls for her to eat two foods at each meal for a week. One unit of Food I contains 5 grams of carbohydrate, 80 calories, and 2 grams of fat. One unit of Food II contains 7 grams of carbohydrate, 110 calories, and 5 grams of fat. Jane wants each meal to provide at least 400 calories but no more than 105 grams of carbohydrate. In addition, at least three times as many units of Food I as Food II should be used in each meal for taste appeal. How many units of each food should Jane eat per meal to minimize her intake of fat?

20. *Refining*   A refinery is processing gasoline and fuel oil. The refining process demands that at least 3 gallons of gasoline to each gallon of fuel oil be refined. To meet the demand, at least 4.2 million gallons of fuel oil per day must be refined. The demand for gasoline is no more than 7.6 million gallons per day. If the refinery sells gasoline for 87 cents per gallon and fuel oil for $1.20 per gallon, how many gallons of each of these products should be refined each day to maximize revenues?

Problems for Exploration

1. Make up a linear programming word problem whose mathematical formulation is as follows:

$$\text{Maximize} \quad P = 30x + 50y,$$
$$\text{subject to} \quad x + y \le 80$$
$$2x + y \le 10$$
$$x \ge 0, y \ge 0.$$

2. Make up a linear-programming word problem whose mathematical formulation is as follows:

$$\text{Minimize} \quad S = 3x + 5y + 7z,$$
$$\text{subject to} \quad x + y + z \ge 200$$
$$x \qquad\qquad \ge 10$$
$$y + z \ge 40$$
$$x \quad - 2z \le 0$$
$$x \ge 0, y \ge 0, z \ge 0.$$

SECTION 4.2   ## Linear Inequalities in Two Variables

The inequality constraints of a linear programming problem in two variables form a **system of linear inequalities** whose solutions may be displayed as a shaded region in the plane. Learning how to find such a shaded region and its corner points is the goal of this section. Once this is done, we will have taken the first step toward the graphic solution to a linear programming problem.

A linear inequality in two variables will have one of the forms $ax + by < c$, $ax + by \le c, ax + by > c$, or $ax + by \ge c$, where a, b, and c are real numbers and where a and b are not both zero. For instance, $2x + y \le 8, x - 3y > 7$, and $x \ge 0$ are all examples of linear inequalities. Should we need them, the usual laws governing inequalities prevail.

Some Laws of Inequalities

1. If $a < b$ and c is any real number, then $a + c < b + c$.

2. If $a < b$ and c is any *positive* number, then $ac < bc$.

3. If $a < b$ and c is any *negative* number, then $ac > bc$.

EXAMPLE 1
Illustrating the Laws of Inequalities

Solve the inequality $x - 2y \le 6$ for y.

Solution   The following steps will isolate y on the left side of the inequality.

$$x - 2y \le 6$$
$$-x + (x - 2y) \le -x + 6 \qquad \text{Law (1)}$$
$$-2y \le -x + 6 \qquad \text{Algebra of real numbers}$$
$$\left(-\frac{1}{2}\right)(-2y) \ge \left(-\frac{1}{2}\right)(-x + 6) \qquad \text{Law (3)}$$
$$y \ge \frac{1}{2}x - 3 \qquad \text{Algebra of real numbers} \quad \blacksquare$$

Solutions to a Linear Inequality

The solution to a linear inequality in two variables consists of all ordered pairs (x, y), which, when substituted into the inequality, result in a true statement.

For the inequality $x - 2y \leq 6$ of Example 1, the ordered pair $(5, 5)$ is a solution because the substitution of $x = 5$ and $y = 5$ into this inequality results in

$$5 - 2(5) \leq 6,$$

or $-5 \leq 6$, a *true* statement.

The ordered pair $(10, 1)$ is not a solution because, upon substitution, we have

$$10 - 2(1) \leq 6$$

or $8 \leq 6$, a *false* statement.

How do we identify all such solutions? Actually, an infinite number of solutions exists, and they are best understood when graphically displayed in the coordinate plane. To see where such solutions lie, first replace $\leq$ with $=$ to obtain the equation $x - 2y = 6$. Now graph the line $x - 2y = 6$, as shown in Figure 4(a), by finding the x- and y-intercepts: When $y = 0$, $x = 6$, so that 6 is the x-intercept, and when $x = 0$, $y = -3$, which shows that -3 is the y-intercept. This line divides the plane into two **half-planes**—in this case an "upper half-plane" and a "lower half-plane"—and is the **boundary line** between them. Now, if a point (x_1, y_1) in the plane is substituted into the equation $x - 2y = 6$, which is equivalent to $y = \frac{1}{2}x - 3$, and the left side of the latter form is compared to the right side, three possibilities exist:

(a) $y_1 = \dfrac{1}{2}x_1 - 3;$ the point is on the line.

(b) $y_1 < \dfrac{1}{2}x_1 - 3;$ the point is below the line.

(c) $y_1 > \dfrac{1}{2}x_1 - 3;$ the point is above the line.

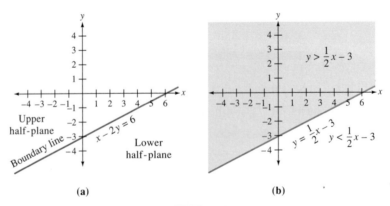

(a)   (b)

FIGURE 4

The Shaded Region Along with the Boundary Line Represents the Solution
to $x - 2y \leq 6$.

Because we know from Example 1 that the inequality $x - 2y \leq 6$ can be rewritten as $y \geq \frac{1}{2}x - 3$, it follows that all points making (a) or (c) true represent the part of the plane satisfying the inequality. Geometrically, this consists of all points in the plane that lie on or above the line $x - 2y = 6$, as shown in Figure 4(b).

The preceding analysis suggests that the graphic solution to an inequality *always* contains *one* of the *half-planes* determined by the boundary line but *not* the other. This may be proved mathematically. Some typical solutions to inequalities are shown in Figure 5. A solid line [as in (a) and (b)] indicates that the boundary line is part of the solution, while a dotted line [as in (c) and (d)] indicates that it is not. Note that for vertical lines, we have left and right half-planes. The fact that one of the half-planes is always part of the solution means that besides deciding whether to include the boundary line, we need is to determine which half-plane should be shaded for the graphic solution. The easiest way to do this is to substitute the coordinates of any point not on the boundary line into the original inequality and then to decide whether this point is a solution to the inequality.

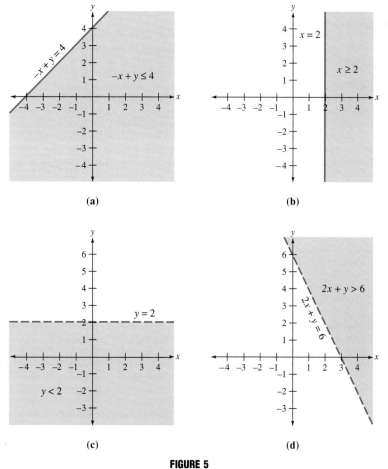

FIGURE 5
Typical Graphic Solutions to Inequalities

True - shade where pt is
False - Shade other Side of pt.

Pick (0,0) if available

The Graphic Solution to a Linear Inequality

1. Replace the inequality symbol with $=$ to obtain the boundary equation.

2. Graph the line represented by the boundary equation.

3. Select a test point *not* on the line just graphed, and substitute it into the original inequality. If the inequality is true, shade in the half-plane in which the test point lies. If the inequality is false, shade in the half-plane opposite where the test point lies.

4. If the inequality is of the form $\leq$ or $\geq$, do include the line in the set of solutions. If the inequality is of the form $<$ or $>$, use a dotted line.

NOTE   The quickest and easiest point to use as a test point is the origin $(0, 0)$, provided, of course, that it is not on the boundary line that divides the plane into two half-planes. If $(0, 0)$ is on the line, the next best test point to use is of the form $(a, 0)$ or $(0, b)$ on one of the axes.

EXAMPLE 2

Graphic Solution for a
Boundary Line
Containing the Origin

Shade in the graphic solution to the inequality $3x - y > 0$.

Solution   Replace the inequality with an equality to get the boundary line $3x - y = 0$. If we let $x = 0$, we find that $y = 0$ also. Therefore the boundary line passes through the origin, $(0, 0)$, and consequently the intercepts $x = 0$ and $y = 0$ determine only a single point. One more point is needed to graph the line. To determine such a point, select *any number not* 0 for either x or y, substitute into the equation, and solve for the other variable. Suppose we let $x = 1$. Then $3(1) - y = 0$ or $y = 3$. Therefore $(1, 3)$ is another point on the line. (See Figure 6.) Due to the strict inequality $>$, the line is not part of the feasible region and is dashed.

Since $(0, 0)$ is on the boundary line $3x - y > 0$, it may *not* be used as a test point. Suppose the point $(4, 0)$ is arbitrarily selected as a test point. The test of this point is shown next.

Test Point	Substituted in the Original Inequality
$(4, 0)$	$3(4) - 0 > 0$ or $12 > 0$, which is true.

As a consequence of this test, the half-plane determined by the line $3x - y = 0$ and containing $(4, 0)$ should be shaded. See Figure 6.

Knowing how to construct the graphic solution to a single inequality leads to the construction of the graphic solution to a system of two or more linear inequalities. In such systems, a solution will consist of points, each of which makes *all* of the inequalities in the system true simultaneously.

EXAMPLE 3

Finding Solutions to a
System of Linear
Inequalities

Find the graphic solution to this system of inequalities:

$$x + 3y \geq 6$$
$$x - y \leq 2$$
$$x \quad\quad \geq 0$$

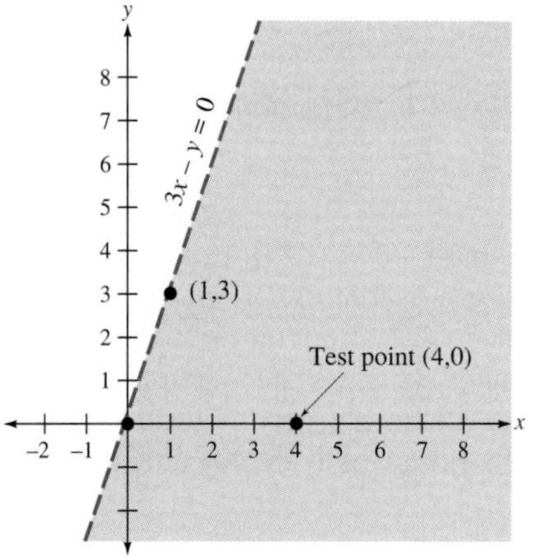

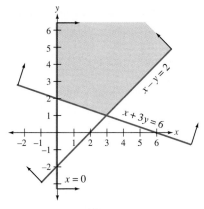

FIGURE 6
A Test Point That Is Not the Origin

Solution Replace each inequality with an equality to give the boundary equations:

$$\begin{array}{ll} \textit{Inequality} & \textit{Boundary Line} \\ x + 3y \ge 6 & x + 3y = 6 \\ x - y \le 2 & x - y = 2 \\ x \ge 0 & x = 0 \text{ (the } y\text{-axis)} \end{array}$$

Consider each of the inequalities, one at a time. First graph $x + 3y = 6$ by using the x-intercept 6 and the y-intercept 2. (See Figure 7.) Then substitute $(0, 0)$ as a test point in the inequality $x + 3y \ge 6$ because $(0, 0)$ is not a point on the line $x + 3y = 6$:

$$\begin{array}{ll} \textit{Test Point} & \textit{Substituted in Original Inequality} \\ (0, 0) & 0 + 3(0) \ge 6 \quad \text{or} \quad 0 \ge 6, \text{ which is false.} \end{array}$$

Therefore, the half-plane *not* containing the origin is to be shaded, as indicated by the arrows attached to the line $x + 3y = 6$ in Figure 7. Now the second line, $x - y = 2$,

FIGURE 7
Graphic Solution to a System of Linear Inequalities

is to be graphed. The x-intercept is 2 and the y-intercept is -2. Again, the test point $(0, 0)$ may be used:

Test point	Substituted in Original Inequality
$(0, 0)$	$0 - 0 \leq 2$ or $0 \leq 2$, which is true.

So the half-plane that *does* contain the origin is to be shaded, as indicated by the arrows attached to the line $x - y = 2$ in Figure 7.

Finally, the inequality $x \geq 0$ has as solutions all points in the plane that have a nonnegative x-coordinate (the points on or to the right of the y-axis), as shown in Figure 7. The solutions to the given system are those points in the plane that *simultaneously make all three* inequalities *true,* as indicated by the shaded region in Figure 7. This collection of points is known as the **feasible region** for the given system. ■

The Feasible Region

For any system of linear inequalities, the set of points that makes all inequalities in the system true simultaneously is called the **feasible region (set of feasible solutions)** for that system. In the plane, the **graph of the feasible region** is found by shading the set of points in the rectangular coordinate system that represents the feasible region.

When graphing the feasible region for a system of two or more inequalities in the plane, we suggest using the less confusing indicator arrows like those shown in Figure 7 to help locate the final graph rather than shading each individual half-plane involved. Because of the conceptual difficulties of drawing three-dimensional figures on paper, no graphs of feasible regions in three dimensions are attempted here.

The feasible region shown in Figure 7 has two **corner points.** Such points belong to the feasible region and are the intersection of two lines that bound the feasible region.

A Corner Point of a Feasible Region

For a feasible region in the plane defined by $\leq$ or $\geq$ constraints, a point of intersection of two boundary lines that is also part of the feasible region is called a **corner point (vertex) of the feasible region.**

A corner point is always the intersection of two boundary lines, but not all such intersections are corner points, as is demonstrated in Example 4.

EXAMPLE 4
Finding the Corner Points of a Feasible Region

Find the corner points of the feasible region described by

$$x + \quad y \leq 5$$
$$x + \quad 2y \leq 8$$
$$x \geq 0, \ y \geq 0.$$

Solution   The boundary lines for these inequalities are, respectively, $x + y = 5$, $x + 2y = 8$, $x = 0$, and $y = 0$. You should be able to mentally calculate the x- and y-intercepts for each of these lines. (Compare to Figure 8.) Inserting the test point $(0, 0)$ into each of the first two inequalities shows that we should shade on the origin side of both boundary lines. The nonnegativity constraints *always* tell us to shade in

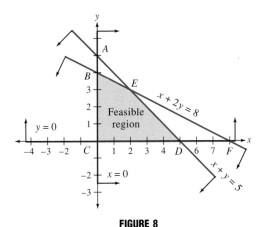

FIGURE 8
*B, C, D, and E Are Corner Points
of the Feasible Region.*

the first quadrant. The feasible region thus obtained is shown in Figure 8, with the intersection of each pair of boundary lines denoted by one of the letters A, B, C, D, E, or F. According to the previous definition, the point labeled B is a corner point of the feasible region because it is part of the feasible region and because it is the intersection of the lines $x = 0$ (the y-axis) and $x + 2y = 8$. The substitution of $x = 0$ into $x + 2y = 8$ gives $y = 4$, so the coordinates of B are $(0, 4)$. The point labeled C is also a corner point of the feasible region because it is the intersection of the lines $x = 0$ and $y = 0$. The coordinates of C are $(0, 0)$. The point labeled D is another corner point of the feasible region represented by the intersection of $y = 0$ and $x + y = 5$. Substitution of $y = 0$ into $x + y = 5$ gives $x = 5$. The coordinates of D are therefore $(5, 0)$. The corner point E is the intersection of the lines $x + 2y = 8$ and $x + y = 5$. The elimination method intoduced in Chapter 1 may be used to obtain the coordinates of E:

$$x + 2y = 8 \quad \textbf{(1)} \qquad\qquad\qquad x + 2y = 8$$
$$x + y = 5 \quad \textbf{(2)} \qquad -1(\text{Equation (2)}) \quad \underline{-x - y = -5}$$
$$y = 3$$

Substituting $y = 3$ into Equation (2) results in $x + 3 = 5$, so that $x = 2$. The coordinates of E are then $(2, 3)$. The points labeled A and F are the intersection of two boundary lines but are not part of the feasible region; therefore, they are not corner points of the feasible region. ■

NOTE When a corner point lies on one of the coordinate axes, its coordinates can most efficiently be obtained from the graph if such points are marked during the graphing process.

EXAMPLE 5
Finding the Corner
Points of a Feasible
Region

Find the feasible region and its corner points if the constraints are

$$-2x + y \le 1$$
$$x + y \le 4$$
$$y \ge 2.$$

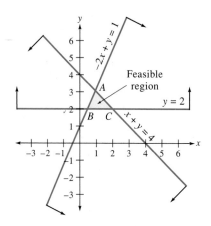

FIGURE 9

A, B, C Are Corner Points of the Feasible Region.

Solution The boundary lines are $-2x + y = 1$, $x + y = 4$, and $y = 2$. Compare your mental calculations of the x- and y-intercepts of each line with the intercepts shown in Figure 9. The test point $(0, 0)$ applied to the first two inequalities shows that we should shade on the origin side of both corresponding boundary lines. The third inequality tells us to shade above the line $y = 2$. The feasible region is shown in Figure 9.

The corner point A is the intersection of the boundary lines $-2x + y = 1$ and $x + y = 4$. Solving the first for y $(y = 2x + 1)$ and substituting the value of y into $x + y = 4$ gives $x + (2x + 1) = 4$, or $x = 1$. When $x = 1$ in $x + y = 4$, $y = 3$, which shows that the corner point A is $(1, 3)$. Corner point B is the intersection of the boundary lines $-2x + y = 1$ and $y = 2$. Substituting $y = 2$ into $-2x + y = 1$ gives $-2x + 2 = 1$, or $x = \frac{1}{2}$. The coordinates of B are $(\frac{1}{2}, 2)$. Corner point C is the intersection of $x + y = 4$ and $y = 2$. Substituting $y = 2$ into $x + y = 4$ gives $x + 2 = 4$, or $x = 2$. The coordinates of C are $(2, 2)$. ■

Given any linear programming problem in two variables, we are now in a position to graph the feasible region as defined by the constraints of the problem and to find the corner points of that region. As the next section shows, these are the first steps in the graphic solution of a linear programming problem.

EXERCISES 4.2

In each of the Exercises 1 through 12:
(a) Write the equation of the boundary line.
(b) Select an appropriate test point and use it to identify the half-plane in the feasible region.
(c) Shade in the feasible region.

1. $3x + y \leq 9$

2. $2x + 3y < 6$

3. $x + 2y \geq 6$

4. $2x + y \geq 10$

5. $-2x + y \geq 8$

6. $3x - 2y \leq 12$

7. $-x \geq 3$

8. $y < x$

9. $2x > 6$

10. $x + y \leq 0$

11. $3y \leq 6$

12. $x \geq 0$

In Exercises 13 thorough 20, do the following:

(a) Find the boundary line for each inequality.

(b) Identify the various half-planes represented by attaching arrows to the separating boundary lines; then shade in the feasible region.

(c) For each corner point of the feasible region, identify the two lines whose intersection is that point, then solve each system of equations to find the coordinates of the corner point.

13. $x + 2y \leq 6$
$\quad x \quad\quad \geq 0$
$\quad\quad\quad y \geq 0$

14. $x \quad\quad\quad \geq 3$
$\quad\quad\quad y \geq 4$
$\quad x + \quad y \leq 10$

15. $x \geq \quad y$
$\quad x \leq 2y$
$\quad x \leq 3$

16. $x - \quad y \geq 0$
$\quad x \quad\quad \leq 3$
$\quad\quad\quad y \geq 1$

17. $x + 3y \leq 9$
$\quad 3x + \quad y \leq 11$
$\quad x \quad\quad \geq 0$
$\quad\quad\quad y \geq 0$

18. $2x + 5y \leq 20$
$\quad x + 2y \leq 9$
$\quad x \quad\quad \geq 0$
$\quad\quad\quad y \geq 0$

19. $x + \quad y \leq 8$
$\quad 2x + \quad y \leq 10$
$\quad 4x + \quad y \leq 16$
$\quad x \quad\quad \geq 0$
$\quad\quad\quad y \geq 0$

20. $x + \quad y \leq 11$
$\quad x + 2y \leq 18$
$\quad x + 5y \leq 20$
$\quad x \quad\quad \geq 0$
$\quad\quad\quad y \geq 0$

In Exercises 21 through 30, graph the feasible region, find the coordinates of the corner points, and determine whether the points $(0, 0)$ and $(1, 3)$ are in the feasible region.

21. $2x - \quad y \leq 4$
$\quad 2x + 3y \leq 12$
$\quad x \geq 0, y \geq 0$

22. $2x + \quad y \leq 8$
$\quad x + \quad y \leq 6$
$\quad x \geq 0, y \geq 0$

23. $x + 3y \leq 9$
$\quad x + \quad y \leq 7$
$\quad x \geq 0, y \geq 0$

24. $x + 5y \leq 25$
$\quad x - 2y \leq 4$
$\quad x \geq 0, y \geq 0$

25. $x \quad\quad \geq 1$
$\quad x \quad\quad \leq 4$
$\quad\quad\quad y \geq 0$
$\quad\quad\quad y \leq 3$

26. $-2x + \quad y \leq 0$
$\quad x + \quad y \leq 4$
$\quad\quad\quad y \geq 0$

27. $-2x + \quad y \geq 0$
$\quad x + \quad\quad \geq 0$
$\quad\quad\quad y \leq 4$

28. $-3x + \quad y \geq 3$
$\quad x \quad\quad \geq 0$
$\quad\quad\quad y \leq 5$

29. $x - \quad y \geq 4$
$\quad x \quad\quad \leq 6$
$\quad\quad\quad y \geq 0$

30. $x - \quad y \leq 6$
$\quad x \quad\quad \geq 6$
$\quad\quad\quad y \geq 0$

In Exercises 31 through 42:

(a) Give the mathematical formulation of the linear programming problem.

(b) Graph the feasible region described by the constraints, and find the corner points. Do not attempt to solve.

31. *Sales* A salesman covers territory in two states, Iowa and Kansas. His daily travel expenses average $120 in Iowa and $100 in Kansas. His company provides an annual travel allowance of $18,000. His company also stipulates that he must spend at least 50 days in Iowa and 60 days in Kansas per year. If sales average $3000 per day in Iowa and $2500 per day in Kansas, how many days should be spent in each state to maximize sales?

32. *Tutoring* A college student, Karin Ann Sandberg, tutors her peers in finite math and calculus, limiting herself to no more than 12 hours per week tutoring these two courses. She also limits herself to no more than 8 hours per week for tutoring in finite math. If she charges $10 per hour to tutor in finite math and $12 per hour to tutor in calculus, how many hours should she spend tutoring each subject to maximize her income?

33. *Mixtures* A dietitian is blending two foods for a special diet. Food I has 30 units of vitamin A and 90 units of vitamin B-complex per kilogram. Food II has 40 units of vitamin A and 50 units of vitamin B-complex per kilogram. The mixture is to have at least 1200 units of vitamin A and at least 2970 units of vitamin B-complex. If Food I costs 80 cents per kg and Food II costs $1.10 per kg, how many kilograms of each should be used to minimize costs?

34. *Purchasing Mixtures* A concrete company needs at least 4 tons of #2 grade crushed rock and at least 12 tons of #6 grade crushed rock to mix an upcoming concrete order. The Hardrock Company ships carloads containing 1 ton of #2 rock and 6 tons of #6 rock. The Morstone Company ships carloads containing 2 tons of #2 rock and 4 tons of #6 rock. If each carload from the Hardrock Co. costs $2000 and each carload from the Morstone Co. costs $2300, how many carloads should be ordered from each to minimize costs?

35. *Shipping Parts* A particular part is shipped each week to assembly plants in Sacramento and Hobbs. The plant in Sacramento needs at least 1000 of these parts each week, and the plant in Hobbs needs at least 1200 of these parts each week. It costs $2 to ship each part to Sacramento and $3 to ship each part to Hobbs. The budget dictates that costs for shipping these parts are not to exceed $14,000 each week. The work hours required to ship a part to Sacramento are 0.2 hours and to ship a part to Hobbs are 0.25 hours. How many parts should be shipped to each location if the work hours are to be kept to a minimum?

36. *Manufacturing* A firm makes a standard model and a deluxe model of microwave oven. Each standard model requires 10 minutes of painting time, and each deluxe model requires 15 minutes of painting time. The firm has available a total of 6 hours each day for painting purposes. The firm must produce at least 15 standard units and at least 8 deluxe units each day. If each standard model sells for $120 and each deluxe model sells for $150, how many of each should be produced each day to maximize revenue?

37. *Manufacturing* The WriteMor Pen Company makes two types of ballpoint pens. A silver pen requires 1 minute in a grinder and 3 minutes in a bonder. A gold pen requires 3 minutes in a grinder and 4 minutes in a bonder. Because of cleaning, adjustments, and maintenance, the grinder can be operated no more than 30 hours per week, and the bonder can be operated no more than 50 hours per week. If each silver pen sells for $30 and each gold pen sells for $80, how many of each should be made each week to maximize revenue?

38. *Specialty Products* A gift company employs two workers to make gavels and plaques. To make a gavel, Ms. Jones performs some tasks that take $\frac{1}{2}$ hour; Mr. Smith performs the remaining tasks, which also take $\frac{1}{2}$ hour. To make a plaque, Ms. Jones's part takes $\frac{1}{2}$ hour, while Mr. Smith's part takes 1 hour. Mr. Smith can work no more than 25 hours per week, whereas Ms. Jones can work no more than 40 hours per week. If each gavel sells for $12 and each plaque for $20, how many of each should be made each week to maximize revenue?

39. *Manufacturing* A company makes two calculators— a business model and a scientific model. The business model contains 10 microcircuits, and the scientific model contains 20 microcircuits. The company has a contract that requires it to use at least 320 microcircuits each day and the manufacturing capacity for business calculators is 50 per day. The company also wants to make at least twice as many business calculators as scientific calculators. If each business model is sold for $50 and each scientific model for $60, how many of each should be made and sold each day to maximize revenue?

40. *Purchasing* The Psychology Department at State University buys mice and rats for experimental purposes. It buys at least three times as many mice as rats. Each mouse costs $2, each rat costs $5, and the departmental budget dictates that no more than $500 be spent for such purchases. If each mouse can be used for three experiments and each rat for two, how many

of each should be purchased to maximize the total number of experiments that can be run?

41. *Survey* A fashion magazine has commissioned the Psychology Department at State University to do a survey on the lifestyles of men who have graduated from college within the past five years and of men under age 40 years who have never attended college. The interviewing is done by professors and graduate students. A professor spends 30 minutes interviewing a college graduate and 35 minutes interviewing a person who never attended college. A graduate student spends 20 minutes interviewing a college graduate and 45 minutes interviewing a person who never attended college. During a semester, professors have at most 80 hours for interviewing, and graduate students have a total of 100 hours for interviewing. How many college graduates and how many men who never attended college should be interviewed during a semester to maximize the total number interviewed?

42. *Entertainment* During a week-long state fair, officials want to provide entertainment that will draw crowds to the fair. They find that a top star demands $10,000 for a performance and should draw a crowd of 8000 people. A slightly faded star demands $6000 for a performance and should draw a crowd of 3000 people. Fair officials have no more than $40,000 to spend on such entertainers and would like to draw a total of at least 25,000 people for their performances. Assume that each entertainer performs only once at the fair. If all top stars donate $500 and all slightly faded stars donate $450 of their fees back to community charities, how many top stars and how many slightly faded stars should be brought in so that donations to community charities are maximized?

Problems for Exploration

1. Construct a system of inequalities whose feasible region is the triangular region with vertices (0, 0), (2, 0), and (1, 1).

2. Construct a system of two inequalities so that there are no points in the feasible region.

3. Construct a system of inequalities so that the feasible region is the single point (2, 3). Is it possible to have a system of inequalities whose feasible region consists of exactly two points? Explain your answer.

Solving Linear Programming Problems Graphically

The constraints for a linear programming problem lead naturally to a feasible region for the problem. In the previous section, we learned how to graph and find the corner points for such regions in the coordinate plane. In this section, we use these concepts as the basis for completing the steps involved in graphically solving a linear programming problem.

A typical problem, such as the following one, shows how the graphic solution process unfolds.

$$\text{Maximize } f = 5x + 2y, \qquad \leftarrow \text{Objective function}$$

$$\text{subject to} \quad \left.\begin{array}{r} x + y \le 6 \\ 2x - y \le 6 \\ x \ge 0, y \ge 0. \end{array}\right\} \text{Constraints}$$

The problem is to select the point or points in the feasible region determined by the constraints that will make $f = 5x + 2y$ as large as possible. Because there are an infinite number of points in the feasible region, the task of finding this maximum value for f may seem hopeless; but it is not that difficult, as we shall see in this section. We begin by graphing the feasible region determined by the constraints and finding the coordinates of the corner points. See Figure 10.

We now experiment for just a moment. Suppose, for instance, that the point $(2, 0)$ is arbitrarily selected from the feasible region, and f is evaluated at this point. Then $f = 5(2) + 2(0) = 10$. However, that is not the only point in the feasible region that gives f a value of 10; there is an entire line, $5x + 2y = 10$, for which each point (x, y) on the line gives $f = 10$. (See Figure 11.) In particular, points on this line and also in the feasible set meet the constraints of the problem. What if the point $(3, 1)$ were chosen? Then $f = 5(3) + 2(1) = 17$ at that point. Again, any point (x, y) on the line $5x + 2y = 17$ gives f a value of 17. Notice the relationship between the graphs of

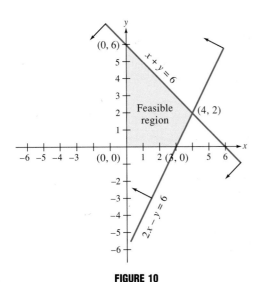

FIGURE 10

Graph and Corner Points of a Feasible Region

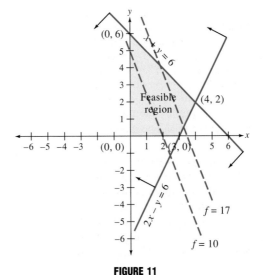

FIGURE 11

Objective Function Values and the Feasible Region

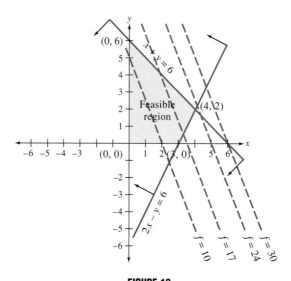

FIGURE 12

f Is Maximized at a Corner Point.

these lines. The slope-intercept form of each, both $y = -\frac{5}{2}x + \frac{10}{2}$ and $y = -\frac{5}{2}x + \frac{17}{2}$, shows that both have the same slope and that the second has the larger y-intercept (see Figure 11). Solving any line $ax + by = c$ in standard form for y, to obtain the point-slope form gives $-\dfrac{a}{b}$ as the slope and $\dfrac{c}{b}$ as the y-coordinate of the y-intercept. Therefore, all lines in standard form with left side $5x + 2y$ have the same slope, $-\frac{5}{2}$, and are parallel. Furthermore, if various values of c are assigned to f, then $f = 5x + 2y = c$; the larger c becomes, the higher the y-intercept on the y-axis. This is shown in Figure 12. These graphs show us that the larger the value of c, the farther upward and away from the origin the graph of $5x + 2y = c$ will be. Now the picture should be clear as to how we find the point in the feasible set that makes f as large as possible: The *last point of the feasible region* touched by the *upward-moving parallel lines* $5x + 2y = c$ is the point that *maximizes f*. In this case, that point is $(4, 2)$, and the value of f at that point is $f = 5(4) + 2(2) = 24$. Our linear programming problem is now solved.

This example suggests that if a linear objective function f has a maximum value on the feasible region bounded by linear constraints, then such a maximum will occur at a corner point of that region. Could this indeed always be the case—even for problems asking for the minimum value of f? Does every linear programming problem have a solution? The mathematical groundwork for investigating these questions has already been done. One aspect of the answer concerns *bounded feasible regions,* which may be described as those regions for which a positive number a can be selected so that a circle with its center at the origin and radius a completely contains the region. The other aspect concerns the concept of a *convex region,* which means that, if any two points in the region are connected by a line segment, then that segment lies entirely within the region. (See Figure 13.)

The fundamental theorem that is stated next is the cornerstone to finding solutions to linear programming problems.

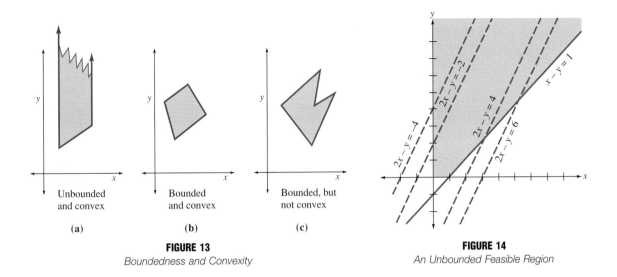

FIGURE 13
Boundedness and Convexity

FIGURE 14
An Unbounded Feasible Region

The Fundamental Theorem of Linear Programming

If the feasible region for any linear programming problem has at least one point and is convex and *if* the objective function has a maximum (or minimum) value within that set, then that maximum (or minimum) will always occur at a corner point in that region.

The fundamental theorem does not guarantee a solution to any linear programming problem. It just says that *if* the objective function has a maximum or minimum value, then it must occur at a corner point of the feasible region. We can assert from the discussion of Example 1 of this section, however, that if the feasible region has at least one point and is *bounded,* then *any* objective function will have both a maximum and minimum value at a corner point. On the other hand, if the feasible region is *unbounded,* then the existence of a solution depends upon both the nature of the objective function and the shape of the feasible region. To elaborate on this last statement, consider the unbounded region in Figure 14 described by $x - y \leq 1$, $x \geq 0$, and $y \geq 0$ and the three objective functions given in (a), (b), and (c).

(a) The function $f = 2x - y$ has no maximum or minimum value over this region. (See Figure 14 for a visual justification.) Along the boundary $x - y = 1$ $(y = x - 1)$, $f = 2x - y$ is reduced to $f = x + 1$ by substitution. This function has no maximum and therefore $f = 2x - y$ has no maximum over the entire region. On the other hand, along the y-axis $(x = 0)$, and $f = 2x - y$ becomes $f = -y$ from which we see that f has no minimum. As a consequence, f has no minimum over the entire region.

(b) The function $g = x + y$ has a minimum, but no maximum value over this region. The minimum value is 0 and occurs at the point $(0, 0)$. Inspection leads to the conclusion that no maximum exists.

(c) The function $h = x - y$ has a maximum, but no minimum value over this region. To see this, note that the graph of $h = x - y = 1$ is precisely that of the "lower-right" boundary of the feasible region. (See Figure 14.) Thus at any point along this boundary line, the value of h is 1 and that is the maximum value that h may attain. The reason is that if $k > 1$, the graph of $h = x - y = k$ does not intersect the feasible region at all while if $k < 1$, the graph of $h = x - y = k$ always intersects the feasible region; thus the more negative k becomes, the more negative the value of h becomes which shows that no minimum exists.

For an unbounded feasible region in the first quadrant and objective function $f = ax + by$, we can be assured of a minimum if a and b are both positive and a maximum if both a and b are negative. Otherwise, the shape of the feasible region and the signs of a and b will determine if a maximum or minimum exists and where it is located.

The Existence of Solutions

For a linear programming problem with feasible region R containing at least one point and having an objective function f,

1. If R is bounded, then f has both a maximum and minimum value at some corner point of R.

2. If R is unbounded and if f has a maximum or minimum value, that value will occur at a corner point. However, the nature of f and the shape of R will determine whether a maximum or minimum exists.

This background shows us how to graphically solve a linear programming problem. The details are listed next.

How to Graphically Solve a Linear Programming Problem

If a linear programming problem has a solution, these steps will locate the solution:

1. Graph the feasible region, as determined by the constraint set.

2. Find the coordinates of each corner point of the feasible region.

3. Evaluate the objective function f at each corner point, and select the point (or points) that gives the maximum or minimum, as required.

EXAMPLE 1
Graphically Solving a
Linear Programming
Problem

Maximize $f = 3x + 4y$,
subject to $2x + y \le 8$
$x + 3y \le 9$
$x \ge 0, y \ge 0$.

Solution   First, the feasible region must be graphed and the corner points found. Figure 15 shows the completed graph of the feasible region. The coordinates of all corner points except A may be found from the graphing process, where the intercepts

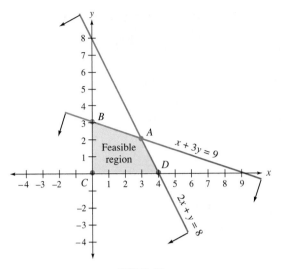

FIGURE 15

Feasible Region for a Linear Programming Problem

were recorded: $B(0, 3)$, $C(0, 0)$, and $D(4, 0)$. Point A is the intersection of the lines whose equations are

$$2x + y = 8$$
$$x + 3y = 9.$$

To find that point, solve the first equation for y; (that is, $y = -2x + 8$), and then substitute the result into the second:

$$x + 3(-2x + 8) = 9$$
$$x - 6x + 24 = 9$$
$$-5x = -15$$
$$x = 3.$$

Now substituting $x = 3$ into $y = -2x + 8$ gives

$$y = -2(3) + 8 = 2.$$

The coordinates of the point A are $(3, 2)$. Finally, evaluating f at each corner point of the feasible region in a tabular form, such as the one shown next, enables us to determine the maximum value of f.

Corner point	Value of $f = 3x + 4$
$A(3, 2)$	17—Maximum
$B(0, 3)$	12
$C(0, 0)$	0
$D(4, 0)$	12

These calculations reveal the solution to this problem: The maximum value of f is 17 and occurs when $x = 3$ and $y = 2$ (the corner point $(3, 2)$). ■

EXAMPLE 2

Graphically Solving a
Linear Programming
Problem

Maximize $f = x + 6y$ over the same set of constraints as those of Example 1.

Solution   We already know the corner points, so the tabulation of f at the corner points is all that remains to be done.

Corner point	Value of $f = x + 6y$
$A(3, 2)$	15
$B(0, 3)$	18—Maximum
$C(0, 0)$	0
$D(4, 0)$	4

The maximum value of f is now 18 and occurs when $x = 0$ and $y = 3$. ▬

EXAMPLE 3

Graphically Solving a
Linear Programming
Problem

Maximize $f = 2x + y$ over the same set of constraints as in Example 1.

Solution   Again, only the tabulation is necessary to find the maximum.

Corner point	Value of $f = 2x + y$
$A(3, 2)$	8—Maximum
$B(0, 3)$	3
$C(0, 0)$	0
$D(4, 0)$	8—Maximum

This time the solution is 8 and it occurs at two corner points, $(3, 2)$ and $(4, 0)$. In fact, this solution occurs at any point on the line segment that joins these two points. The reason is that when the objective function $f = 2x + y$ is set equal to any constant c, the slope of $2x + y = c$ is the same as the slope of the boundary line $2x + y = 8$. This means that the upward moving parallel lines $2x + y = c$ last touch the feasible region along the entire segment of $2x + y = 8$ between $(3, 2)$ and $(4, 0)$. The fact remains, nonetheless, that the solution still occurs at a corner point of the feasible region. ▬

EXAMPLE 4

Graphically Solving a
Linear Programming
Problem

Find the maximum and minimum value of $f = 2x + 5y$ subject to $\begin{array}{c} 2x + 3y \le 12 \\ -2x + y \le 0. \\ y \ge 2 \end{array}$

Solution   The feasible region is shown in Figure 16. The corner points are intersections of lines, as shown next.

A: $\begin{array}{c} -2x + y = 0 \\ y = 2 \end{array}$ **Solution** $\begin{array}{c} x = 1, \\ y = 2 \end{array}$ B: $\begin{array}{c} 2x + 3y = 12 \\ y = 2 \end{array}$ **Solution** $\begin{array}{c} x = 3, \\ y = 2 \end{array}$ C: $\begin{array}{c} -2x + y = 0 \\ 2x + 3y = 12 \end{array}$ **Solution** $\begin{array}{c} x = \dfrac{3}{2}, \\ y = 3 \end{array}$

The following tabulation shows where the minimum and maximum occur.

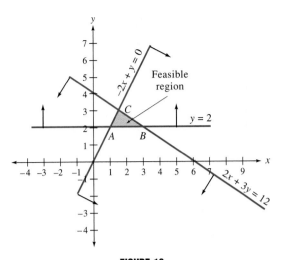

FIGURE 16
*Feasible Region Used to Find Maximum and
Minimum Values*

Corner point	Value of $f = 2x + 5y$
$A(1, 2)$	**12**—Minimum
$B(3, 2)$	16
$C(\frac{3}{2}, 3)$	**18**—Maximum

EXAMPLE 5
Graphically Solving a
Linear Programming
Problem

A dietitian is blending two foods for a special diet. Food I has 30 units of vitamin C and 90 units of vitamin D per kilogram. Food II has 40 units of vitamin C and 50 units of vitamin D per kilogram. The mixture is to have at least 1200 units of vitamin C and at least 2970 units of vitamin D. If Food I costs 30 cents per kilogram and Food II costs 20 cents per kilogram, how many kilograms of each food should be used to meet these requirements while keeping costs to a minimum?

Solution   The constraints may be written from the tabulation of the following data, with the condition that neither x nor y can be negative.

	Food I	Food II	Limits
Vitamin C	30	40	≥ 1200
Vitamin D	90	50	≥ 2970
Kilograms used	x	y	
Cost	30	20	

Now the problem may be formulated as follows:

$$\text{Minimize} \quad C = 30x + 20y,$$
$$\text{subject to} \quad 30x + 40y \geq 1200$$
$$90x + 50y \geq 2970$$
$$x \geq 0, y \geq 0$$

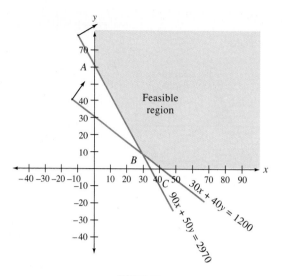

FIGURE 17
Feasible Region for Minimizing Costs of a Food Blend

The feasible region is shown in Figure 17, with the corner points labeled *A*, *B*, and *C*. Point *B* is intersection of

$$30x + 40y = 1200$$
$$90x + 50y = 2970$$

and the solution is $x = 28$ and $y = 9$. The other corner points, $A(0, 59.4)$ and $C(40, 0)$, are found from the graph of the constraint set. The tabulation of the cost function at the corner points is shown next.

Corner point	Value of $C = 30x + 20y$
$A(0, 59.4)$	1188
$B(28, 9)$	1020—Minimum
$C(40, 0)$	1200

This shows that the minimum cost is $10.20 when 28 kg of Food I and 9 kg of Food II are used.

The graphic approach to solving linear programming problems is limited to two or, perhaps for the energetic and visually perceptive, three variables. This is evidence enough that an algebraic approach is needed because most real-world problems involve many more variables than two or even three. We begin the development of such an approach in the next section. However, we keep the graphic theme intact to help us understand why particular techniques are needed.

EXERCISES **4.3**

Exercises 1 and 2 apply to the feasible region shown next.

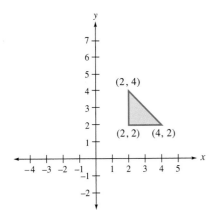

1. **(a)** If the objective function is $f = 3x + 2y$, graph the lines $3x + 2y = c$, where c takes on the values -6, 6, 12, 18, and 24 superimposed on the feasible region.
   (b) Based on these lines, where do the maximum and minimum values of f occur? Make a statement about the movement of the parallel lines $f = c$ versus the values of c.
   (c) If the objective function is $g = -f = -3x - 2y$, graph the lines $-3x - 2y = c$, where c takes on the values 6, -6, -12, -18, and -24 superimposed on the feasible region.
   (d) Based on these lines, where do the maximum and minimum values of $g = -f$ occur? Make a statement about the movement of the parallel lines $g = c$ versus the values of c.

2. **(a)** If the objective function is $f = 3x - 2y$, graph the lines $3x - 2y = c$, where c takes on the values of -6, 6, 12, 18, and 24 superimposed on the feasible region.
   (b) Based on these lines, where do the maximum and minimum values of f occur? Make a statement about the movement of the parallel lines $f = c$ versus the values of c.
   (c) If the objective function is $g = -f = -3x + 2y$, graph the lines $-3x + 2y = c$, where c takes on the values $-6, 6, 12, 18$, and 24 superimposed on the feasible region.
   (d) Based on these lines, where do the maximum and minimum values of $g = -f$ occur? Make a statement about the movement of the parallel lines $g = c$ versus the values of c.

Solve each of the linear programming problems in Exercises 3 through 15 by the graphic method:
(a) Graph the feasible region.
(b) Find the coordinates of the corner points.
(c) Evaluate f at each corner point, and find the extreme value requested.

3. Maximize $P = 4x + 4y$,
   subject to $x + 3y \leq 30$
   $2x + y \leq 20$
   $x \geq 0, y \geq 0$

4. Maximize $P = x + 5y$,
   subject to $2x + y \leq 12$
   $2x + 5y \leq 20$
   $x \geq 0, y \geq 0$

5. Maximize $f = x + 2y$,
   subject to $2x + y \geq 10$
   $x \leq 4$
   $y \leq 6$

6. Maximize $f = 6x + y$,
   subject to $x + y \geq 8$
   $-3x + y \geq 0$
   $y \leq 0$

7. Minimize $C = x - 2y$,
   subject to $x \geq 2, x \leq 4$
   $y \geq 1, y \leq 5$

8. Minimize $C = 4x + 2y$,
   subject to $-x + y \leq 0$
   $x + 2y \geq 6$
   $y \geq 0$

9. Minimize $C = 2x + 3y$,
   subject to $9x + 4y \geq 40$
   $x + 4y \geq 8$
   $x \geq 0, y \geq 0$

10. Minimize $C = 5x + 6y$,
    subject to $4x + y \geq 6$
    $x + 8y \geq 17$
    $x \geq 0, y \geq 0$

11. Minimize $C = 5x + 3y$,
    subject to $x + y \geq 6$
    $6x + y \geq 16$
    $x + 6y \geq 16$
    $x \geq 0, y \geq 0$

12. Maximize $P = 3x + 5y$,
    subject to $x + 2y \leq 8$
    $x + y \leq 6$
    $4x + 2y \leq 16$
    $x \geq 0, y \geq 0$

13. Maximize $P = 8x + y$,
    subject to
    $$x + y \leq 6$$
    $$x + 3y \leq 12$$
    $$3x + y \leq 12$$
    $$x \geq 0, y \geq 0$$

14. Maximize $f = 2x + y$,
    subject to
    $$x + y \leq 5$$
    $$x = 2$$
    $$y \geq 0$$

15. Minimize $f = 2x - 5y$,
    subject to
    $$x + y \leq 6$$
    $$x \geq 1$$
    $$y = 2$$

16. Consider the feasible region defined by the constraints $x \geq 1, y \geq 1, y \leq 3$.

    (a) Does the objective function $f = 2x + y$ attain a maximum at some point in the feasible region? If so, what is that point, and what is the maximum value of f?

    (b) Does the objective function $f = 2x + y$ attain a minimum at some point in the feasible region? If so, what is that point, and what is the minimum value of f?

17. Consider the feasible region defined by the constraints $x \geq 1, x \leq 4, y \geq 2$.

    (a) Does the objective function $P = x + 3y$ attain a maximum at some point in the feasible region? If so, what is that point, and what is the maximum value of f?

    (b) Does the objective function $P = x + 3y$ attain a minimum at some point in the feasible region? If so, what is that point, and what is the minimum value of f?

In the remaining exercises, first write the mathematical formulation of the problem, and then solve it by the graphic method. Some of these exercises you have already been asked to set up in Section 4.1.

18. **Manufacturing** The Fine Line Pen Company makes two types of ballpoint pens: a silver model and a gold model. The silver model requires 1 minute in a grinder and 3 minutes in a bonder. The gold model requires 3 minutes in a grinder and 4 minutes in a bonder. Because of maintenance procedures, the grinder can be operated no more than 30 hours per week and the bonder no more than 50 hours per week. The company makes $5 on each silver pen and $7 on each gold pen. How many of each type pen should be produced and sold each week to maximize profits?

19. **Inventory** The NewAge Technical Avenues store is accumulating an inventory of computers and fax ma-

chines. The store wants no more than 30 of these two items in the store. On the average, each computer it stocks costs $2000 and each fax machine $800, and it has set an inventory budget constraint of $40,800 for these two items. If the net profit on each computer is $200 and on each fax machine is $100, how many computers and how many fax machines should NewAge Technical Avenues have in inventory to maximize net profit?

20. **Mixing Foods** A dietitian is to prepare two foods to meet certain requirements. Each pound of Food I contains 100 units of vitamin C, 40 units of vitamin D, 10 units of vitamin E, and costs 20 cents. Each pound of Food II contains 10 units of vitamin C, 80 units of vitamin D, 5 units of vitamin E, and costs 15 cents. The mixture of the two foods is to contain at least 260 units of vitamin C, at least 320 units of vitamin D, and at least 50 units of vitamin E. How many pounds of each type of food should be used to minimize the cost?

21. **Manufacturing** The Traction Tire Company produces two types of tires, four-ply and radial. Each four-ply tire requires 2 minutes to make and costs $20. Each radial tire requires 6 minutes to make and costs $30. Orders received by the company dictate that they must make a total of at least 4000 tires each day of which at least 1600 must be radials. The union contract with Traction requires that they use at least 320 work-hours each day making these two types of tires. How many tires of each type should be made to meet the conditions and minimize costs?

22. **Radio Programming** A radio station is introducing a new talk show and a new gardening show to their programming. Their marketing consultant estimates that each time the talk show is aired, 20,000 people will listen while each time the gardening show is aired, 8000 people will listen. The station does not want to air these two shows a total of more than 12 times each week. Furthermore, they want to air the gardening show at least three times, but not more than five times each week. How many times each week should each show be aired to maximize the number of listeners each week?

23. **Mixtures** Two grains, barley and corn, are to be mixed for an animal food. Barley contains 1 unit of fat per pound, and corn contains 2 units of fat per pound. The total units of fat in the mixture are not to exceed 12 units. No more than 6 pounds of barley and no more than 5 pounds of corn are to be used in the mixture. If barley and corn each contain 1 unit of protein per pound, how many pounds of each grain should be used to maximize the number of units of protein in the mixture?

24. *Manufacturing*   A company makes two calculators—a business model and a scientific model. The business model contains 10 microcircuits, and the scientific model contains 20 microcircuits. The company has a contract that requires it to use at least 320 microcircuits each day. The company also wants to make at least twice as many business calculators as scientific calculators. If each business calculator requires 10 production steps and each scientific calculator requires 12 production steps, how many calculators of each type should be made to minimize production steps?

25. *Manufacturing*   A company makes desks in two models: a student model and a secretary model. They make $40 profit on each student model and $50 on each secretary model they sell. Each student model requires 2 hours in woodworking and 3 hours in finishing. Each secretary model requires 3 hours in woodworking and 5 hours in finishing. The company has a total of 240 work-hours available in the woodworking department and a total of 390 work-hours available in the finishing department. Due to new business orders the company needs to make at least 40 of the secretary model desks during the next month. How many desks of each model should the company make and sell during the next month to maximize their profit?

26. *Pollution*   The Environmental Protection Agency (EPA) has issued new regulations that require the New Molecule Chemical Co. to install a new process to help reduce pollution caused when producing a particular chemical. The old process releases 2 grams of sulphur and 5 grams of lead for each liter of the chemical produced. The new process releases 1 gram of sulphur and 2.6 grams of lead for each liter of the chemical produced. The company makes a profit of 20 cents for each liter produced under the old method and 16 cents for each liter produced under the new process. The new EPA regulations do not allow for more than 10,000 grams of sulphur and 8000 grams of lead to be released each day. How many liters of this chemical should be produced under the old method and how many liters should be produced under the new method to maximize profits and yet stay within EPA regulations?

27. *Transportation Purchases*   A charter-plane service contracts with a retirement group to provide flights from New York to Florida for at least 500 first-class, 1000 tourist-class, and 1500 economy-class passengers. The charter company has two types of planes. Type A will carry 50 first-class, 60 tourist-class and 100 economy-class passengers and costs $12,000 to make each flight. Type B will carry 40 first-class, 30 tourist-class and 80 economy-class passengers and costs $10,000 to make each flight. How many of each type plane should be used to minimize flight costs?

28. *Order Fulfillment*   The Hi-Fi Record Store is about to place an order for cassette tapes and compact discs. The distributor it orders from requires that an order contain at least 250 items. The prices Hi-Fi must pay are $12 for each cassette and $16 for each compact disc. The distributor also requires that at least 30% of any order be compact discs. How many cassettes and how many compact discs should Hi-Fi order so that its costs will be kept to a minimum?

29. *Manufacturing*   A firm makes a standard and a deluxe model of microwave oven. Each oven requires one plastic case, and the firm has 300 cases on hand. Each standard model requires 10 minutes to assemble, and each deluxe model requires 15 minutes to assemble. The orders received dictate that the firm produce at least 120 standard models and at least 150 deluxe models each day. To do this, it has eight employees in the assembly department, each working an eight-hour day. If the firm makes a profit of $25 on each standard model and $35 on each deluxe model, how many of each model should be made each day to maximize profit?

30. *Agriculture*   Agricultural planners in a certain country are planning next year's rice and wheat acreages. They estimate that each acre of rice planted will yield 30 bushels and each acre of wheat planted will yield 20 bushels. They want to raise at least 4 million bushels of rice and at least 5 million bushels of wheat. They have enough machinery to plant at least one million acres of these two crops combined. For planting, harvesting, and transporting the grains, the planners estimate that it will take six workers for each acre of rice planted and three workers for each acre of wheat planted. How can the planners meet the requirements set forth and use the minimum number of workers?

31. *Investments*   An investment club has at most $30,000 to invest in junk and premium-quality bonds. Each type of bond is bought in thousand-dollar denominations. The junk bonds have an average yield of 12% and the premium-quality bonds 7%. The policy of the club is to invest at least twice the amount in premium-quality bonds as in junk bonds. How many of each type bond should the club buy to maximize its investment return?

32. *Agriculture* The New Age Protein Company has 500 acres of land available on which to plant wheat and soybeans. The planting costs per acre are $120 for wheat and $200 for soybeans. The company has a budget of $76,000 for planting costs. Due to soil conditions on the land, no more than 400 acres may be planted in wheat, while exactly 100 acres must be planted in soybeans. If the anticipated income per acre from wheat is $350 and from soybeans is $520, how many acres of each grain should be planted to maximize income?

33. *Animal Experiments* A particular psychology department buys mice and rats for experimental purposes. It buys at least three times as many mice as rats. Each mouse costs $2, each rat costs $4, and the total departmental budget for such purchases is $500. Due to space limitations, the number of mice purchased cannot exceed 200, and the number of rats purchased needs to be exactly 25. If each mouse can be used in six experiments and each rat in four, how many of each should be purchased to maximize the number of experiments?

Problems for Exploration

1. Is it possible to have a linear programming problem where every point in the feasible region gives a maximum (or minimum) value of the objective function? Give an example, or explain why this cannot happen.

2. For an objective function $f = ax + by$ where a and b are constants and not both simultaneously zero, is it possible to have a feasible region in the first quadrant containing at least one point, where f will have neither a maximum nor a minimum?

SECTION 4.4 **Slack Variables and Pivoting**

Practical applications of linear programming problems usually involve many more than just two variables. In fact, problems with dozens of variables are not uncommon. An algebraic method to handle large-scale problems, called the **simplex method (simplex algorithm),** was developed by George Dantzig in 1947. The simplex algorithm has special significance when applied to a particular system of linear equations obtained from the constraints, and it is this aspect that we explore in this section. The final touches needed to actually use the simplex algorithm to solve linear programming problems are described in the next section.

The simplex algorithm is used to solve linear programming problems that already are, or that can be converted to, **standard maximum-type** problems. The following is an example of a standard maximum-type problem.

HISTORICAL NOTE
George Dantzig

The simplex method was developed in 1947 by George Dantzig, currently a professor emeritus at Stanford University. The method he developed helped the U.S. Air Force to allocate resources in the least expensive way.

EXAMPLE 1
A Standard Maximum-Type Problem

Maximize $f = 2x + y$, ← Objective function
subject to $2x + y \le 8$
$x + y \le 5$
$x \ge 0, y \ge 0$ ← Nonnegativity constraints

The Standard Maximum-Type Problem

A linear programming problem is a **standard maximum-type problem** if these conditions are met:

1. The objective functions is linear and is to be *maximized.*

2. The variables are all nonnegative.

3. Constraints, other than the nonnegativity constraints, are all of the form $ax + by + \ldots \leq c$, where $c \geq 0$.

The first step toward using the simplex algorithm, and the key to viewing a linear programming problem as an algebraic process, is the conversion of each constraint not of the nonnegativity type into an equality by the addition of a **slack variable** to the left side. To turn the constraint $2x + y \leq 8$ into an equality, for example, we add a variable, s_1, to the left side to create the equality $2x + y + s_1 = 8$. As x and y change, the variable s_1 continually "takes up the slack" to ensure that an equality exists, hence the name **slack variable.** Adding the slack variable s_2 to the second constraint of Example 1 completes a system of equations called **slack equations.**

$$\left. \begin{array}{l} 2x + y + s_1 \qquad = 8 \\ x + y \qquad\quad + s_2 = 5 \end{array} \right\} \qquad \text{Slack equations for Example 1}$$

NOTE Slack variables are never added to the nonnegativity constraints in any linear programming problem.

Now think of the first slack equation, $2x + y + s_1 = 8$, as one equation in three variables, whose solutions may be expressed as

$$s_1 = 8 - (2x + y)$$
$$x = \text{any number}$$
$$y = \text{any number},$$

where x and y are free variables. The last expression for s_1 tells us that

$$
\begin{array}{lll}
\text{if } 2x + y = 8, & \text{then} & s_1 = 0 \text{ (there is no slack)}, \\
\text{if } 2x + y < 8, & \text{then} & s_1 \text{ is positive, and} \\
\text{if } 2x + y > 8, & \text{then} & s_1 \text{ is negative.}
\end{array}
$$

That is, for points (x, y) in the first quadrant satisfying $2x + y \leq 8$, the value of the slack variable is either positive or zero. A similar situation prevails for the second slack equality, $x + y + s_2 = 5$. This shows us that for points in the feasible region where both of the inequalities involved are satisfied, each slack variable has a value that is either positive or zero, as shown in Figures 18 and 19. This conclusion is valid for all linear programming problems.

Property of Slack Variables

For any point in the feasible region of a linear programming problem, the value of each slack variable is either positive or zero.

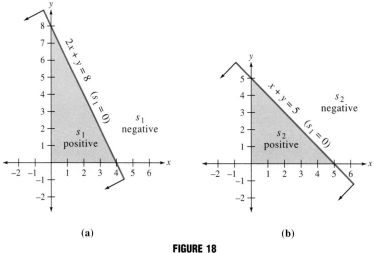

FIGURE 18
Values of a Slack Variable.

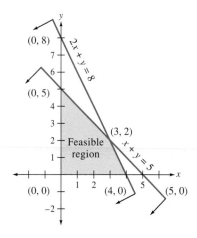

FIGURE 19
*Slack Variables Are Nonnegative
in the Feasible Region.*

We explore one last bit of geometry connected with the slack variables in the next example.

EXAMPLE 2
Coordinates of Points
Where Two Boundary
Lines Intersect

Find the values of x, y, s_1, and s_2 at each point of intersection for all pairs of boundary lines from Example 1.

Solution   The x and y coordinates of these points have already been demonstrated in Figure 19. Next substitute each of these points into the slack equations

$$2x + y + s_1 \qquad = 8$$
$$x + y \qquad + s_2 = 5.$$

The first equation gives the value of s_1 and the second equation the value of s_2.

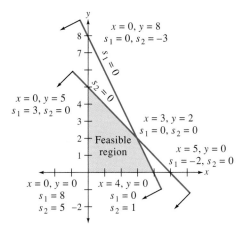

FIGURE 20

Two of the Four Variables Are Always Zero.

Point	First equation	Second equation
$(0, 8)$	$s_1 = 0$	$s_2 = -3$
$(0, 5)$	$s_1 = 3$	$s_2 = 0$
$(0, 0)$	$s_1 = 8$	$s_2 = 5$
$(4, 0)$	$s_1 = 0$	$s_2 = 1$
$(5, 0)$	$s_1 = -2$	$s_2 = 0$
$(3, 2)$	$s_1 = 0$	$s_2 = 0$

Notice that when a point of intersection lies outside of the feasible region, one of the slack variables is *negative,* as shown in Figure 20. ■

Now that we have the geometry of Example 1 in mind, we turn our attention to how particular solutions to the slack equations relate to that geometry. The system of slack equations

$$2x + y + s_1 \qquad = 8$$
$$x + y \qquad + s_2 = 5$$

is a system of two equations in four unknowns and, therefore, has an infinite number of solutions. Because we have four variables and only two equations in the system, we may choose *two* variables to be free and assign them any value we please. The choice that offers the easiest solution is to let x and y be the free variables, set both equal to zero, so that by inspection of the system, $s_1 = 8$ and $s_2 = 5$. Notice that this solution—$x = 0$, $y = 0$, $s_1 = 8$, and $s_2 = 5$—corresponds to the variables labeling the origin in Figure 20. Solutions following this pattern are not only the easiest to find, but will also be seen as the most useful in arriving at the final solution to a linear programming problem by the simplex algorithm. At any stage of the solution process, such a solution relies upon dividing the variables into two groups, each with a special name in the language of linear programming. In one group each variable is considered to be free and is set equal to zero; these are called **nonbasic variables.** The remaining variables, whose values are determined once the nonbasic variables are set equal to zero, are called **basic variables.** The solution is most easily read by first identifying the nonbasic variables and then "mentally blocking out" their columns (the equivalent of setting them equal to zero) leaving each row as an equation involving a basic variable only. Observe how this may be done from the augmented matrix for the above slack equations:

$$
\begin{array}{cccc}
x & y & s_1 & s_2 \\
\end{array}
$$
$$
\left[\begin{array}{cccc|c}
2 & 1 & 1 & 0 & 8 \\
1 & 1 & 0 & 1 & 5
\end{array}\right]
\tag{1}
$$

Nonbasic variables (set = 0)   Basic variables

To further investigate the nature of basic and nonbasic variables, suppose we take the augmented matrix **(1)** (rewritten below) and pivot on the *arbitrarily chosen* element that is circled in the second-row, second-column position, obtaining the matrix on the right.

$$
\begin{array}{cccc}
x & y & s_1 & s_2 \\
\end{array}
\qquad\qquad\qquad
\begin{array}{cccc}
x & y & s_1 & s_2 \\
\end{array}
$$
$$
\left[\begin{array}{cccc|c}
2 & 1 & 1 & 0 & 8 \\
1 & ① & 0 & 1 & 5
\end{array}\right]
\quad -1R_2 + R_1 \to R_1 \quad
\left[\begin{array}{cccc|c}
1 & 0 & 1 & -1 & 3 \\
1 & 1 & 0 & 1 & 5
\end{array}\right]
\tag{2}
$$

Nonbasic variable   Basic variables   Nonbasic variable

After the pivot a new-appearing system of two equations and four unknowns with infinitely many solutions is obtained. Which two variables should we now choose to set equal to zero (be free, nonbasic), and hence mentally "block out" those columns, so as to most easily read the values of the remaining variables? Let us choose x and s_2 as those nonbasic variables. Once set equal to zero, the first row represents the equation $s_1 = 3$, and the second row represents the equation $y = 5$. A solution is then

$$
x = 0, \quad s_2 = 0, \quad y = 5 \quad \text{and} \quad s_1 = 3.
$$

Relate this solution to the point $(0, 5)$ in Figure 20 and observe that two things have happened due to this pivot. First, the pivot operation, from the standpoint of the solutions obtained, has "moved" us from the corner point $(0, 0)$ to the adjacent corner point $(0, 5)$. Secondly, the classification of two variables changed: y changed from nonbasic to basic and s_2 changed from basic to nonbasic. Such a trade-off will occur every time a pivot operation is performed, which implies that the *number* of variables in each classification *never* changes. This, in turn, implies that there will always be two nonbasic variables whose values are zero, and that is precisely what is needed to identify the intersection of two boundary lines. (See Figure 20.)

The foregoing two augmented matrices show the pattern for identifying basic and nonbasic variables according to the distinctive appearance of their corresponding columns. Nonbasic variables, the free ones to be set equal to zero, are those whose columns have a "cluttered" look, while the basic variables are those whose columns contain a 1 somewhere and all other elements are zero.

Basic and Nonbasic Variables

For a system of slack equations represented in an augmented matrix, the **basic variables** are those whose column of coefficients contain exactly one positive 1 and all other entries 0. The remaining variables to the left of the vertical bar are called **nonbasic variables.**

Continuing our investigation of pivoting and the solutions obtained from classifying basic and nonbasic variables, suppose we now pivot on the arbitrarily chosen element circled in the first-row, fourth-column position of augmented matrix **(2)**. The result of this pivot is shown next.

$$
\begin{array}{cccc}
x & y & s_1 & s_2 \\
\end{array}
\left[\begin{array}{cccc|c}
1 & 0 & 1 & \boxed{-1} & 3 \\
1 & 1 & 0 & 1 & 5
\end{array}\right]
\quad \rightarrow \quad
\begin{array}{cccc}
x & y & s_1 & s_2 \\
\end{array}
\left[\begin{array}{cccc|c}
-1 & 0 & -1 & 1 & -3 \\
2 & 1 & 1 & 0 & 8
\end{array}\right]
\tag{3}
$$

Nonbasic variables ⟶   Basic variables

The basic variables are now y and s_2, while x and s_1 are nonbasic. Letting $x = 0$ and $s_1 = 0$, the first row shows that $s_2 = -3$ and the second that $y = 8$. This is the solution shown at the point $(0, 8)$ in Figure 20. Even though it is not a corner point of the feasible region, it is the intersection of two boundary lines, $x = 0$ and $s_1 = 0$ ($2x + y = 8$). This solution coincides with what we have already observed geometrically about any variable: If a slack variable has a negative value, then the point represented lies beyond the bounds of the feasible region.

> ### The Geometry of Pivoting
>
> If a pivot is performed on any nonzero element in the augmented matrix of slack equations, and if the nonbasic variables are set equal to zero in the resulting matrix, then the resulting solution is a point where two boundary lines to the feasible region intersect.

We have seen that sometimes when we pivot, the resulting solution obtained from the classification of basic and nonbasic variables gives another corner point of the feasible region and at other times it does not. To be useful in solving a linear programming problem, the solution obtained after *every* pivot operation must result in a corner point of the feasible region and not just the intersection of two of its boundary lines. There is a way to select the row in which the pivot resides to make this happen. We now develop a selection rule using the initial augmented matrix of slack equations for Example 1.

$$
\begin{array}{cccc}
x & y & s_1 & s_2 \\
\end{array}
\left[\begin{array}{cccc|c}
2 & 1 & 1 & 0 & 8 \\
1 & 1 & 0 & 1 & 5
\end{array}\right]
$$

Suppose that we have decided to arbitrarily pivot on an element in the x-column and we want to be sure that the resulting basic solution represents a corner point of the feasible region shown in Figure 19. The initial solution is the origin, and a pivot in the x-direction will take us either to the point $(4, 0)$ or $(5, 0)$ of Figure 19. These two points were determined from the graphing of the boundary lines *by dividing each coefficient of x into the corresponding constant term on the right*. Now look at the augmented matrix, and see how the same thing is accomplished:

$$
\begin{array}{c}
\text{2 into 8} \\
\left[\begin{array}{cccc|c}
2 & 1 & 1 & 0 & 8 \\
1 & 1 & 0 & 1 & 5
\end{array}\right]
\begin{array}{l}
\text{Quotient: 4} \\
\text{Quotient: 5}
\end{array} \\
\text{1 into 5}
\end{array}
$$

The *smaller* quotient, 4, identifies the point $(4, 0)$ and the larger quotient, 5, identifies the point $(5, 0)$ on the x-axis of Figure 19. We should choose the element in the x-column that gives the *smallest quotient*—the 2 in the first-row, first-column position—as the pivot element in order to remain in the feasible region. The actual pivot operation gives this result:

$$\begin{array}{cccc} x & y & s_1 & s_2 \end{array}$$
$$\begin{bmatrix} \textcircled{2} & 1 & 1 & 0 & | & 8 \\ 1 & 1 & 0 & 1 & | & 5 \end{bmatrix} \rightarrow \begin{array}{cccc} x & y & s_1 & s_2 \end{array} \begin{bmatrix} 1 & \frac{1}{2} & \frac{1}{2} & 0 & | & 4 \\ 0 & \frac{1}{2} & -\frac{1}{2} & 1 & | & 1 \end{bmatrix}$$

The solution given by the last augmented matrix is $y = 0$, $s_1 = 0$, $x = 4$, and $s_2 = 1$, and this is precisely the corner point we wanted. To reinforce the "smallest quotient" pivot selection, we next show that a pivot on the "1" in the first column *does not* give a corner point of the feasible region:

$$\begin{bmatrix} 2 & 1 & 1 & 0 & | & 8 \\ 1 & 1 & 0 & 1 & | & 5 \end{bmatrix} \rightarrow \begin{bmatrix} 0 & -1 & 1 & -2 & | & -2 \\ 1 & 1 & 0 & 1 & | & 5 \end{bmatrix}$$

The solution, $x = 5$, $y = 0$, $s_1 = -2$, and $s_2 = 0$, reflects the point $(5, 0)$ where two boundary lines intersect, but is not a corner point of the feasible region. (See Figure 20.)

Similarly, entries in the y-column give quotients of 8 and 5, from top to bottom. Can you use this information to tell in advance where a pivot on either element in the y-column will take the solution in a geometric sense?

The division of elements in the pivot column into the corresponding constants clearly holds the key to staying in the feasible region; the *smallest quotient* restricts the movement to the feasible region, while the larger quotient allows us to move beyond the feasible region to the next intersection of two boundary lines. This analysis points to a smallest-quotient rule. Before finalizing this, however, we look at one more example.

EXAMPLE 3
A Solution After
Pivoting

Consider the linear programming problem

$$\text{Maximize } f = 5x + 2y,$$
$$\text{subject to} \quad x + y \le 6$$
$$2x - y \le 6$$
$$x \ge 0, y \ge 0.$$

Inserting a slack variable for each appropriate constraint gives this system of slack equations:

$$x + y + s_1 \qquad = 6$$
$$2x - y \qquad + s_2 = 6.$$

Figure 21 shows the feasible region and the values of all variables where two boundary lines intersect.

The initial augmented matrix for the system of slack equations is

$$\begin{array}{cccc} x & y & s_1 & s_2 \end{array}$$
$$\begin{bmatrix} 1 & 1 & 1 & 0 & | & 6 \\ 2 & -1 & 0 & 1 & | & 6 \end{bmatrix}$$

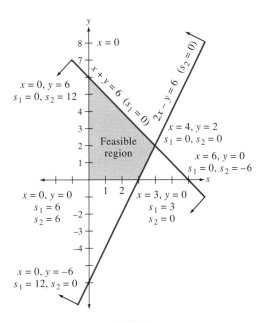

FIGURE 21
*Coordinates of Points Where Boundary
Lines Intersect*

Upon setting the nonbasic variables x and y equal to zero, this augmented matrix has a solution of $x = 0$, $y = 0$, $s_1 = 6$, and $s_2 = 6$. Suppose we decide to pivot on an element in the y-column. Computing the quotients as before gives

$$\begin{bmatrix} 1 & 1 & 1 & 0 & 6 \\ 2 & -1 & 0 & 1 & 6 \end{bmatrix}$$ Quotient: 6
Quotient: -6.

with "1 into 6" pointing into the first row and "-1 into 6" pointing into the second row.

We know in advance that if a pivot is performed on the 1 (which gave the quotient 6), the new solution will give the point $(0, 6)$:

$$\begin{bmatrix} 1 & 1 & 1 & 0 & 6 \\ 3 & 0 & 1 & 1 & 12 \end{bmatrix},$$ (Solution: $x = 0$, $y = 6$, $s_1 = 0$, and $s_2 = 12$; see Figure 21.)

while if a pivot is performed on the -1 (which gave the quotient -6), the new solution should give the point $(0, -6)$:

$$\begin{bmatrix} 3 & 0 & 1 & 1 & 12 \\ -2 & 1 & 0 & -1 & -6 \end{bmatrix}$$ (Solution: of $x = 0$, $y = -6$, $s_1 = 12$, and $s_2 = 0$; see Figure 21.)

Thus the pivot on a negative element has resulted in a point that is not in the feasible region. ▪

Example 3 illustrates that division by a negative number should be avoided when choosing a pivot row if the solution determined by setting the nonbasic variables equal to zero is to reflect a corner point of the feasible region after the pivot has been completed.

The following conditions set the stage for our pivoting rule that takes us from one corner point to another corner point.

Assuming that we have

(a) started with the augmented matrix to a system of slack equations obtained from *any standard maximum linear programming problem,*

(b) obtained a *solution* by setting the nonbasic variables equal to zero, and

(c) already made any number of pivot operations on the initial augmented matrix resulting in a matrix whose *solution reflects a corner point* of the feasible region, we then have the following rule:

The Smallest-Quotient Rule

Under conditions (a), (b) and (c) above:

1. Select any new pivot column.

2. Then divide each *positive* number in that column into the corresponding number in the rightmost column of the matrix.

3. Select the pivot row to be the one corresponding to the *smallest nonnegative quotient* so obtained. (Zero is included.)

Pivoting on the element where the pivot column and pivot row intersect will always result in a solution that occurs at a corner point of the feasible region. The solution at this point is called a **basic feasible solution** to the linear programming problem.

If the smallest-quotient rule is followed, it tells us that "once on a corner point, always on a corner point" of the feasible region.

EXAMPLE 4
Using the
Smallest-Quotient Rule

Pivot on the system of slack equations shown in Example 3, each time using the smallest-quotient rule, where the first pivot column is the second column. Then on the resulting matrix, choose the pivot column to be the first column, then on *that* resulting matrix, choose the pivot column to be the third column.

Solution   The system of slack equations has the following augmented matrix:

$$\begin{array}{cccc} x & y & s_1 & s_2 \\ \end{array}$$
$$\left[\begin{array}{cccc|c} 1 & 1 & 1 & 0 & 6 \\ 2 & -1 & 0 & 1 & 6 \end{array}\right]$$

Initially, s_1 and s_2 are basic variables, while x and y are nonbasic variables and equal zero. Letting $x = 0$ and $y = 0$, the first row reads $s_1 = 6$ and the second $s_2 = 6$. This solution represents the origin of Figure 22. Choose the y-column from which to select a pivot element. The appropriate quotients are 6 from the first row and -6 from the second row. The row with the smallest nonnegative quotient is the first row; hence, pivot on the 1 in the first-row, second-column position:

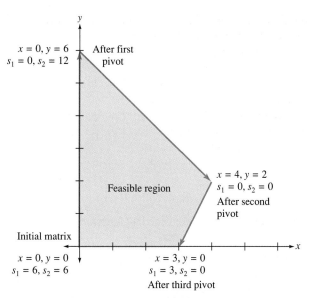

FIGURE 22

*Tracing the Movement of Solutions after
Successive Pivots*

$$\begin{array}{cccc} x & y & s_1 & s_2 \end{array}$$
$$\begin{bmatrix} 1 & ① & 1 & 0 & | & 6 \\ 2 & -1 & 0 & 1 & | & 6 \end{bmatrix} \begin{array}{l} \text{Quotient: } 6 \quad \leftarrow \text{Pivot row} \\ \text{Quotient: } -6 \end{array}$$
$$\underset{\text{Pivot column}}{\big|}$$

$$1R_1 + R_2 \rightarrow R_2 \begin{bmatrix} 1 & 1 & 1 & 0 & | & 6 \\ 3 & 0 & 1 & 1 & | & 12 \end{bmatrix}$$

The basic feasible solution for the last augmented matrix is $x = 0$, $s_1 = 0$, $y = 6$, and $s_2 = 12$. (See Figure 22.)

From the last matrix, we choose the x-column from which to select a pivot element. The selection process is shown next, along with the result of the pivot.

$$\begin{array}{cccc} x & y & s_1 & s_2 \end{array}$$
$$\begin{bmatrix} 1 & 1 & 1 & 0 & | & 6 \\ ③ & 0 & 1 & 1 & | & 12 \end{bmatrix} \begin{array}{l} \text{Quotient: } 6 \\ \text{Quotient: } 4 \leftarrow \text{Pivot row} \end{array}$$
$$\underset{\text{Pivot column}}{\big\uparrow}$$

$$\tfrac{1}{3}R_2 \rightarrow R_2 \begin{array}{cccc} x & y & s_1 & s_2 \\ \begin{bmatrix} 1 & 1 & 1 & 0 & | & 6 \\ 1 & 0 & \tfrac{1}{3} & \tfrac{1}{3} & | & 4 \end{bmatrix} \end{array}$$

$$-1R_2 + R_1 \rightarrow R_1 \begin{array}{cccc} x & y & s_1 & s_2 \\ \begin{bmatrix} 0 & 1 & \tfrac{2}{3} & -\tfrac{1}{3} & | & 2 \\ 1 & 0 & \tfrac{1}{3} & \tfrac{1}{3} & | & 4 \end{bmatrix} \end{array}$$

The last augmented matrix gives the basic feasible solution $s_1 = 0$, $s_2 = 0$, $y = 2$, and $x = 4$. (See Figure 22.)

Finally, on the last matrix, we pivot on an element in the s_1 column. The pivot selection and resulting pivot operation are shown next.

$$
\begin{array}{cccc}
x & y & s_1 & s_2 \\
\end{array}
$$

$$
\begin{bmatrix}
0 & 1 & \boxed{\tfrac{2}{3}} & -\tfrac{1}{3} & \bigg| & 2 \\
1 & 0 & \tfrac{1}{3} & \tfrac{1}{3} & \bigg| & 4
\end{bmatrix}
\begin{array}{l}
\text{Quotient: } 3 \quad \leftarrow \text{Pivot row} \\
\text{Quotient: } 12
\end{array}
$$

$$\uparrow \text{Pivot column}$$

$$
\tfrac{3}{2}R_1 \rightarrow R_1 \quad
\begin{array}{cccc}
x & y & s_1 & s_2 \\
\end{array}
\begin{bmatrix}
0 & \tfrac{3}{2} & 1 & -\tfrac{1}{2} & \bigg| & 3 \\
1 & 0 & \tfrac{1}{3} & \tfrac{1}{3} & \bigg| & 4
\end{bmatrix}
$$

$$
-\tfrac{1}{3}R_1 + R_2 \rightarrow R_2 \quad
\begin{array}{cccc}
x & y & s_1 & s_2 \\
\end{array}
\begin{bmatrix}
0 & \tfrac{3}{2} & 1 & -\tfrac{1}{2} & \bigg| & 3 \\
1 & -\tfrac{1}{2} & 0 & \tfrac{1}{3} & \bigg| & 3
\end{bmatrix}
$$

The basic feasible solution to the last augmented matrix is $y = 0$, $s_2 = 0$, $x = 3$, and $s_1 = 3$. (See Figure 22.) ▄▄

We are now in a position to complete the rules for the simplex algorithm in the next section.

In Exercises 1 through 8, determine which constraints are not in the form for a standard maximum-type problem.

1. $x + y \leq 3$ **2.** $3x + 4y \leq -4$

3. $3x - 2y \leq 6$ **4.** $x - y \geq 0$

5. $4x - 5y \geq 6$ **6.** $-2x + y \leq 6$

7. $7x + 5y \geq -4$ **8.** $3y \geq 7$

For the inequality $x + 3y \leq 9$, insert a slack variable s_1, and answer the questions in Exercises 9 through 13 about s_1.

9. What is the value of s_1 if $x = 3$ and $y = 2$?

10. What is the value of s_1 if $x = -2$ and $y = 5$?

11. What is the value of s_1 if $x = 0$ and $y = 0$?

12. What is the range of values of s_1 if $x \geq 0$, $y \geq 0$, and $x + 3y \leq 9$?

13. If s_1 is negative, where must the points (x, y) lie in the plane, relative to the boundary line $x + 3y = 9$?

For the inequality $2x + y \leq 8$, insert a slack variable s_2, and answer the questions in Exercises 14 through 18 about s_2.

14. What is the value of s_2 if $x = 2$ and $y = 1$?

15. What is the value of s_2 if $x = -2$ and $y = 7$?

16. What is the value of s_2 if $x = 0$ and $y = 0$?

17. What is the range of values of s_2 if $x \geq 0$, $y \geq 0$, and $2x + y \leq 8$?

18. If s_2 is negative, where must the points (x, y) lie in the plane, in relation to the boundary line $2x + y = 8$?

Exercises 19 through 24 apply to this linear programming problem:

$$
\begin{aligned}
\text{Maximize} \quad & P = 3x + 2y, \\
\text{subject to} \quad & x + y \leq 6 \\
& 2x + y \leq 8 \\
& x \geq 0, y \geq 0
\end{aligned}
$$

19. Insert slack variables where appropriate in the constraints, and form the system of slack equations.

20. Graph the feasible region, and label the intersection of each pair of boundary lines with four coordinates.

21. Form the initial augmented matrix from the slack equations, and find the pivot element in the x-column so that when the pivot operation is performed on that element, a basic feasible solution will result. Pivot on the other element in the x-column to confirm your answer.

22. Form the initial augmented matrix from the slack equations, and find the pivot element in the y-column so that when the pivot operation is performed on that element, a basic feasible solution will result. Pivot on the other element in the y-column to confirm your answer.

23. Pivot on an element in the *x*-column of the initial augmented matrix of slack equations so that a basic feasible solution results. Then apply the smallest-quotient rule to the *y*-column of the resulting augmented matrix to find a pivot element in the *y*-column so that, upon pivoting, another basic feasible solution results.

24. Pivot on an element in the *y*-column of the initial augmented matrix of slack equations so that a basic feasible solution results. Then apply the smallest-quotient rule to the *x*-column of the resulting augmented matrix to find a pivot element in the *x*-column so that, upon pivoting, another basic feasible solution results.

Exercises 25 through 30 apply to this linear programming problem:

$$\text{Maximize} \quad f = x + 3y,$$
$$\text{subject to} \quad x + y \le 9$$
$$-x + 2y \le 12$$
$$x \ge 0, y \ge 0$$

25. Insert slack variables where appropriate in the constraints, and form the system of slack equations.

26. Graph the feasible region, and label the intersection of each pair of half-plane lines with four coordinates.

27. Form the initial augmented matrix from the slack equations, and find the pivot element in the *x*-column, so that when the pivot operation is performed on that element, a basic feasible solution results. Pivot on the other element in the *x*-column to confirm your answer.

28. Form the initial augmented matrix from the slack equations, and find the pivot element in the *y*-column so that if the pivot operation were performed on that element, a basic feasible solution results. Pivot on the other element in the *y*-column to confirm your answer.

29. Pivot on an element in the *x*-column of the initial augmented matrix of slack equations so that a basic feasible solution results. Then apply the smallest-quotient rule to the *y*-column of the resulting augmented matrix to find a pivot element in the *y*-column so that, upon pivoting, another basic feasible solution results.

30. Pivot on an element in the *y*-column of the initial augmented matrix of slack equations, so that a basic feasible solution results. Then apply the smallest-quotient rule to the *x*-column of the resulting augmented matrix, to find a pivot element in the *x*-column so that, upon pivoting, another basic feasible solution results.

Exercises 31 through 36 apply to this linear programming problem:

$$\text{Maximize} \quad f = 5x + 2y,$$
$$\text{subject to} \quad x + y \le 10$$
$$2x - y \le 8$$
$$x \ge 0, y \ge 0$$

31. Insert slack variables where appropriate in the constraints, and form the system of slack equations.

32. Graph the feasible region, and label the intersection of each pair of half-plane lines with four coordinates.

33. Form the initial augmented matrix from the slack equations, and find the pivot element in the *x*-column so that when the pivot operation is performed on that element, a basic feasible solution results. Pivot on the other element in the *x*-column to confirm your result.

34. Form the initial augmented matrix from the slack equations, and find the pivot element in the *y*-column so that when the pivot operation is performed on that element, a basic feasible solution results. Pivot on the other element in the *y*-column to confirm your result.

35. Pivot on an element in the *x*-column of the initial augmented matrix of slack equations so that a basic feasible solution results. Then apply the smallest-quotient rule to the *y*-column of the resulting augmented matrix to find a pivot element in the *y*-column so that, upon pivoting, another basic feasible solution results.

36. Pivot on an element in the *y*-column of the initial augmented matrix of slack equations so that a basic feasible solution results. Then apply the smallest-quotient rule to the *x*-column of the resulting augmented matrix, to find a pivot element in the *x*-column so that, upon pivoting, another basic feasible solution results.

Exercises 37 through 42 refer to this linear programming problem:

$$\text{Maximize} \quad P = x + 4y,$$
$$\text{subject to} \quad x + y \le 8$$
$$x + 2y \le 10$$
$$x \ge 0, y \ge 0$$

37. Insert slack variables where appropriate in the constraints, and form the system of slack equations.

38. Graph the feasible region, and label the intersection of each pair of half-plane lines with four coordinates.

39. Form the initial augmented matrix from the slack equations, and find the pivot element in the *x*-column so that when the pivot operation is performed on that element, a basic feasible solution will result. Pivot on the other element in the *x*-column to confirm your result.

40. Form the initial augmented matrix from the slack equations, and find the pivot element in the y-column so that when the pivot operation is performed on that element, a basic feasible solution will result. Pivot on the other element in the y-column to confirm your result.

41. Pivot on an element in the x-column of the initial augmented matrix of slack equations so that a basic feasible solution results. Then apply the smallest-quotient rule to the y-column of the resulting augmented matrix to find a pivot element in the y-column so that, upon pivoting, another basic feasible solution results.

42. Pivot on an element in the y-column of the initial augmented matrix of slack equations so that a basic feasible solution results. Then apply the smallest-quotient rule to the x-column of the resulting augmented matrix to find a pivot element in the x-column so that, upon pivoting, another basic feasible solution results.

In each of the Exercises 43 and 44, a set of inequalities describing the feasible region to a standard maximum-type linear programming problem is given. Insert the appropriate slack variables, and sketch the feasible region. Starting at the origin, select successive pivot elements in the resulting augmented matrices so that the respective solutions, with nonbasic variables set equal to zero, will occur at successive corner points going completely around the feasible region with a clockwise motion and return to the origin. Then do the same thing with a counter-clockwise motion.

43. $x + y \le 8$
$\quad\ 3x + y \le 12$
$\quad\ x \ge 0, y \ge 0$

44. $2x - y \le 8$
$\quad\ 4x + y \le 22$
$\quad\ x \ge 0, y \ge 0$

In Exercises 45 through 50, consider each augmented matrix to be the result of one or more pivots from an initial augmented matrix representing the slack equations of a linear programming problem in standard form. Find the solution in which the nonbasic variables are zero, and decide whether the solution is a corner point of the feasible region for the problem.

45.
$$\begin{array}{cccc} x & y & s_1 & s_2 \\ \end{array}$$
$$\left[\begin{array}{cccc|c} 2 & 1 & 0 & 3 & 2 \\ 1 & 0 & 1 & 5 & 7 \end{array}\right]$$

46.
$$\begin{array}{cccc} x & y & s_1 & s_2 \\ \end{array}$$
$$\left[\begin{array}{cccc|c} 1 & -3 & 5 & 0 & 6 \\ 0 & 2 & -1 & 1 & 1 \end{array}\right]$$

47.
$$\begin{array}{cccc} x & y & s_1 & s_2 \\ \end{array}$$
$$\left[\begin{array}{cccc|c} 2 & 0 & -4 & 1 & -2 \\ 0 & 1 & 2 & 0 & 5 \end{array}\right]$$

48.
$$\begin{array}{cccc} x & y & s_1 & s_2 \\ \end{array}$$
$$\left[\begin{array}{cccc|c} 1 & 0 & 0 & 2 & -2 \\ 0 & -3 & 1 & -1 & 5 \end{array}\right]$$

49.
$$\begin{array}{cccccc} x & y & z & s_1 & s_2 & s_3 \\ \end{array}$$
$$\left[\begin{array}{cccccc|c} 2 & 1 & 0 & 1 & 4 & 0 & 2 \\ -1 & 0 & 1 & 1 & 2 & 0 & 1 \\ 3 & 0 & 0 & 3 & 0 & 1 & 5 \end{array}\right]$$

50.
$$\begin{array}{cccccc} x & y & z & s_1 & s_2 & s_3 \\ \end{array}$$
$$\left[\begin{array}{cccccc|c} 0 & 3 & -1 & 1 & -2 & 0 & 1 \\ 1 & 0 & 1 & 0 & -3 & 0 & 5 \\ 0 & 1 & 0 & 0 & 4 & 1 & 2 \end{array}\right]$$

51. The initial augmented matrix shown represents the slack equations for a standard maximum-type linear programming problem.

$$\begin{array}{cccc} x & y & s_1 & s_2 \\ \end{array}$$
$$\left[\begin{array}{cccc|c} 1 & 1 & 1 & 0 & 6 \\ 7 & 3 & 0 & 1 & 21 \end{array}\right]$$

(a) Write the set of all constraints for the problem.
(b) Graph the feasible region.
(c) Perform one pivot on the given matrix that will give a solution representing the point $(3, 0)$.
(d) Perform one pivot on the given matrix that will give a solution representing the point $(0, 6)$.
(e) Perform two successive pivots on the given matrix so that the last resulting matrix will give the solution that represents the point $(\frac{3}{4}, \frac{21}{4})$.

52. The initial augmented matrix shown represents the slack equations of a standard maximum-type linear programming problem.

$$\begin{array}{cccc} x & y & s_1 & s_2 \\ \end{array}$$
$$\left[\begin{array}{cccc|c} 1 & 1 & 1 & 0 & 4 \\ 1 & -1 & 0 & 1 & 2 \end{array}\right]$$

(a) Write the set of all constraints for the problem.
(b) Graph the feasible region.
(c) Perform one pivot on the given matrix that will give a solution representing the point $(2, 0)$.
(d) Perform one pivot on the given matrix that will give a solution representing the point $(0, -2)$.
(e) Perform two successive pivots on the given matrix so that the last resulting matrix will give a solution that represents the point $(3, 1)$.

Problems for Exploration

1. It is possible that two columns in the augmented matrix of slack equations are identical, but one represents a basic variable, while the other represents a nonbasic variable. Give an example of such a situation, and explain the meaning of this condition geometrically.

2. Graph the feasible region given by these inequalities:
$$-2x + y \leq 0$$
$$x - y \leq 0$$
$$x + y \leq 6$$
$$x \geq 0, y \geq 0.$$

Use this example to explain why the smallest-quotient rule is not stated this way: Divide each positive element in the pivot column into the corresponding element in the column of constants, and select the pivot row to be the one giving the smallest *positive* quotient so formed.

The Simplex Algorithm

The simplex algorithm (method) is an algebraic technique that solves standard maximum-type linear programming problems. In view of the smallest-quotient rule that was established in Section 4.4, all that remains to be done is to involve the objective function in the pivoting procedure on the set of slack equations so that we can actually find the maximum value. To the augmented matrix representing the slack equations, we add a new row at the bottom that represents the objective function in such a way that its value becomes part of any basic feasible solution read from this matrix. Then, finally, we discover the signal that tells us when the maximum has been reached. We now consider a typical problem with two variables and view the associated geometry.

$$\text{Maximize} \quad f = 5x + 3y, \quad \leftarrow \text{Objective function}$$
$$\text{subject to} \quad 2x + 3y \leq 18$$
$$2x + y \leq 10$$
$$x \geq 0, y \geq 0 \quad \leftarrow \text{Nonnegativity constraints}$$

The feasible region for this problem, shown in Figure 23, has each corner point labeled with its coordinates and its value of f.

The first step in the simplex algorithm (already noted in the previous section) is the formation of the slack equations.

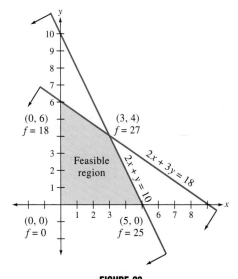

FIGURE 23
*Relating the Simplex Algorithm to the
Feasible Region*

Step 1 in the Simplex Algorithm

Insert a slack variable into each of the constraints, except into the nonnegativity constraints.

The result is this system of slack equations:

$$2x + 3y + s_1 \quad\quad = 18$$
$$2x + \ y \quad\quad + s_2 = 10.$$

Now rewrite the objective function as $-5x - 3y + f = 0$ to match the format of the slack equations, and adjoin it to the bottom of the slack equations in this fashion:

$$2x + 3y + s_1 \quad\quad\quad\quad = 18$$
$$2x + \ y \quad\quad + s_2 \quad\quad = 10$$
$$-5x - 3y \quad\quad\quad\quad + f = 0$$

Step 2 in the Simplex Algorithm

Rewrite the objective function to match the format of the slack equations, and adjoin it to the bottom of the slack equation system.

This creates a new system of three equations and five unknowns, x, y, s_1, s_2, and f. Traditionally, the last equation is separated from the slack equations in the augmented matrix by a horizontal bar, to remind us of the special role this equation plays in the maximization process that unfolds.

Step 3 in the Simplex Algorithm

Write the augmented matrix for this new system. This matrix is called the **initial simplex tableau.** Each number in the last row to the left of the vertical bar is called an **indicator.**

In our present problem, the initial simplex tableau is

$$
\begin{array}{ccccc}
x & y & s_1 & s_2 & f \\
\end{array}
$$
$$
\left[
\begin{array}{ccccc|c}
2 & 3 & 1 & 0 & 0 & 18 \\
2 & 1 & 0 & 1 & 0 & 10 \\
\hline
-5 & -3 & 0 & 0 & 1 & 0
\end{array}
\right]
\begin{array}{l}
\left.\begin{array}{l}\\ \\\end{array}\right\}\text{Slack equations} \\
\leftarrow\text{Objective function}
\end{array}
$$

Indicators

The constraints of a standard maximum-type linear programming problem assure us that the origin (all original variables equal zero) will always be in the feasible region; in fact, it will always be the **basic feasible solution** to the initial simplex tableau: $x = 0$, $y = 0$, $s_1 = 18$, and $s_2 = 10$. Continuing to read this solution from the last row, we see that $f = 0$ (see Figure 23).

The simplex algorithm provides an orderly scheme for selecting pivot elements. The pivot column is selected in such a way that the column's variable will cause f to increase faster per unit change than with any other column variable. The last equation of our present simplex tableau (the initial one) indicates that $-5x - 3y + f = 0$ or $f = 5x + 3y$, which shows that a unit increase in x will produce an increase of 5 in f, while a unit increase in y will produce an increase of 3 in f. Therefore, the pivot column will be the x-column. Notice that this choice is displayed in the simplex tableau as *the most negative indicator* in the last row. We now move up to the slack equations and use the smallest-quotient rule to select the pivot row.

$$\begin{array}{ccccc} x & y & s_1 & s_2 & f \\ \end{array}$$

$$\left[\begin{array}{ccccc|c} 2 & 3 & 1 & 0 & 0 & 18 \\ ② & 1 & 0 & 1 & 0 & 10 \\ \hline -5 & -3 & 0 & 0 & 1 & 0 \end{array}\right] \begin{array}{l} \text{Quotient: 9} \\ \text{Quotient: 5} \leftarrow \text{Pivot row} \\ \end{array}$$

↑
└──── Pivot column

The circled element where the two arrows meet is the pivot element.

Step 4 in the Simplex Algorithm

1. The most negative indicator in the last row of the tableau determines the pivot column.

2. The pivot row is selected from among the slack equations by using the smallest-quotient rule.

3. The intersection of the pivot column and the pivot row determines the pivot element.

Next, we pivot on the circled element by making that element a 1 and *all* other elements in that column a 0, including the -5 is the last row.

Step 5 in the Simplex Algorithm

Perform the pivot operation on the selected pivot element.

$$\begin{array}{ccccc} & x & y & s_1 & s_2 & f \\ \end{array}$$

$$\tfrac{1}{2}R_2 \leftarrow R_2 \quad \left[\begin{array}{ccccc|c} 2 & 3 & 1 & 0 & 0 & 18 \\ 1 & \tfrac{1}{2} & 0 & \tfrac{1}{2} & 0 & 5 \\ \hline -5 & -3 & 0 & 0 & 1 & 0 \end{array}\right]$$

$$\begin{array}{ccccc} & x & y & s_1 & s_2 & f \\ \end{array}$$

$$\begin{array}{l} -2R_2 + R_1 \rightarrow R_1 \\ \\ 5R_2 + R_3 \rightarrow R_3 \end{array} \left[\begin{array}{ccccc|c} 0 & 2 & 1 & -1 & 0 & 8 \\ 1 & \tfrac{1}{2} & 0 & \tfrac{1}{2} & 0 & 5 \\ \hline 0 & -\tfrac{1}{2} & 0 & \tfrac{5}{2} & 1 & 25 \end{array}\right]$$

The basic feasible solution in now $y = 0$, $s_2 = 0$, $x = 5$, $s_1 = 8$, and $f = 25$. Note the corner point with these x, y, and f values in Figure 23. This pivot operation has moved the basic feasible solution from the origin to the point $(5, 0)$ and simultaneously gives us the value of $f = 25$ at this point!

We know from Figure 23 that the maximum value of f has not yet been reached. How do we know this from the last matrix? Note that the last row, $-\tfrac{1}{2}y + \tfrac{5}{2}s_2 + f = 25$, can be rewritten as $f = 25 + \tfrac{1}{2}y - \tfrac{5}{2}s_2$. Because both y and s_2 are zero in the present solution, and neither are permitted to be negative, if we are to stay in the feasible region, any positive change in s_2 would *decrease* the value of f from its present value of 12. On the other hand, any positive change in y will further *increase* the value of f because y has a positive coefficient. However, this positive coefficient *is displayed as negative* in the format of the augmented matrix. This implies that, from the matrix point of view, *any negative indicator* indicates that further growth in f is possible. Furthermore, the most negative of all such indicators identifies the variable giving f the fastest growth per unit change of any variable. Thus, our next pivot column will be the y-column and, again, the smallest-quotient rule applied to those coefficients in the slack variable equations will determine the pivot row.

$$
\begin{array}{ccccc}
x & y & s_1 & s_2 & f \\
\end{array}
$$

$$
\left[\begin{array}{ccccc|c}
0 & \textcircled{2} & 1 & -1 & 0 & 8 \\
1 & \frac{1}{2} & 0 & \frac{1}{2} & 0 & 5 \\
0 & -\frac{1}{2} & 0 & \frac{5}{2} & 1 & 25
\end{array}\right]
\begin{array}{l}
\text{Quotient: } 4 \longleftarrow \text{Pivot row} \\
\text{Quotient: } 10
\end{array}
$$

Pivot column

Pivoting on the circled element requires two steps:
First,

$$
\frac{1}{2}R_1 \rightarrow R_1 \quad
\begin{array}{ccccc}
x & y & s_1 & s_2 & f \\
\end{array}
$$

$$
\left[\begin{array}{ccccc|c}
0 & 1 & \frac{1}{2} & -\frac{1}{2} & 0 & 4 \\
1 & \frac{1}{2} & 0 & \frac{1}{2} & 0 & 5 \\
0 & -\frac{1}{2} & 0 & \frac{5}{2} & 1 & 25
\end{array}\right]
$$

Second,

$$
\begin{array}{ccccc}
x & y & s_1 & s_2 & f \\
\end{array}
$$

$$
\begin{array}{l}
\\
-\frac{1}{2}R_1 + R_2 \rightarrow R_2 \\
\frac{1}{2}R_1 + R_3 \rightarrow R_3
\end{array}
\left[\begin{array}{ccccc|c}
0 & 1 & \frac{1}{2} & -\frac{1}{2} & 0 & 4 \\
1 & 0 & -\frac{1}{4} & \frac{3}{4} & 0 & 3 \\
0 & 0 & \frac{1}{4} & \frac{9}{4} & 1 & 27
\end{array}\right]
$$

This time, the basic feasible solution is $s_1 = 0$, $s_2 = 0$, $x = 3$, $y = 4$, and $f = 27$. Note the point in Figure 23 with these x, y, and f values. This pivot has moved the solution from the point $(5, 0)$ to the point $(3, 4)$ and has simultaneously calculated the new value of f at the point, 27.

Geometrically, we know that the maximum value of f for points in the feasible region has been reached. However, can we tell this from the augmented matrix? The answer is yes, and the proof is found by examining the last row. This time, the equation in the last row is $\frac{1}{4}s_1 + \frac{9}{4}s_2 + f = 27$, which can be rewritten as $f = 27 - \frac{1}{4}s_1 - \frac{9}{4}s_2$. At present, both s_1 and s_2 are 0; but if we pivot further, one will necessarily change from a nonbasic variable to a basic variable. This will result in a positive value for that variable, which will *decrease* the value of f from its present value of 27. Therefore, no further increase in f is possible. The coefficients of these variables, when viewed as indicators in the last simplex tableau, are *positive,* which leads to the conclusion that if no negative indicator appears in the last row of the matrix, then the maximum value of f has been reached.

Step 6 of the Simplex Algorithm

If a negative indicator is present, repeat steps 4 and 5. If no negative indicator is present, the maximum of the objective function has been reached.

Even though it is instructive and interesting to match the geometry with each basic feasible solution whenever possible, the beauty of the simplex algorithm is that it is algebraic in nature and can be performed on any standard maximum-type linear programming problem, regardless of size. Furthermore, the repetitive nature of the pivoting process means that we can think of the algorithm as a step-by-step procedure.

Flow Chart for the Simplex Algorithm

Step 1: Insert the slack variables, and find the slack equations.

Step 2: Rewrite the objective function to match the format of the slack equations, and adjoin at the bottom.

Step 3: Write the initial simplex tableau.

Step 4: Determine the pivot element.

Step 5: Perform the pivot operation.

Step 6: Determine whether there is a negative number in the last row to the left of the vertical bar.

yes   no

The maximum has been reached.

Just as with our developmental example, this step-by-step procedure always starts with the basic solution at the origin (all original problem variables equal zero), then successively pivots, giving basic feasible solutions at adjacent corner points on the feasible region until the corner that gives the maximum of the objective function is reached.

EXAMPLE 1
Using the Simplex Algorithm

Use the simplex algorithm to solve this problem:

$$\text{Maximize} \quad P = 3x + 4y + z,$$
$$\text{subject to} \quad x + 2y + z \le 6$$
$$2x \quad\quad + 2z \le 4$$
$$3x + y + z \le 9$$
$$x \ge 0, y \ge 0, z \ge 0.$$

Solution   Following in steps in the flow chart, the slack equations are formed first. They are

$$x + 2y + z + s_1 \quad\quad\quad = 6$$
$$2x \quad\quad + 2z \quad\quad + s_2 \quad\quad = 4$$
$$3x + y + z \quad\quad\quad\quad + s_3 = 9.$$

Now, rewriting the objective function as $-3x - 4y - z + P = 0$ and adjoining it to the bottom of the slack equations, we obtain a new system of four equations with seven unknowns, whose initial simplex tableau is as follows:

$$
\begin{array}{ccccccc|c}
x & y & z & s_1 & s_2 & s_3 & P & \\
1 & ② & 1 & 1 & 0 & 0 & 0 & 6 \\
2 & 0 & 2 & 0 & 1 & 0 & 0 & 4 \\
3 & 1 & 1 & 0 & 0 & 1 & 0 & 9 \\
\hline
-3 & -4 & -1 & 0 & 0 & 0 & 1 & 0
\end{array}
$$

} Stack equations

← Objective function

Indicators

Next, the most negative indicator in the last row is -4, showing that the pivot column is the y-column. Applying the smallest-quotient rule to the coefficients of y in the slack equations *only,* gives quotients of 3 (from the first row) and 9 (from the third row). We cannot divide by 0 in the second row. Because the smallest quotient, 3, was obtained from the first row, the first row will be the pivot row. The resulting pivot element will be the circled 2 in the initial simplex tableau. The completed pivot operation and the preparations for selecting the next pivot element are shown next.

$$
\begin{array}{ccccccc|c}
x & y & z & s_1 & s_2 & s_3 & P & \\
\tfrac{1}{2} & 1 & \tfrac{1}{2} & \tfrac{1}{2} & 0 & 0 & 0 & 3 \\
\boxed{2} & 0 & 2 & 0 & 1 & 0 & 0 & 4 \\
\tfrac{5}{2} & 0 & \tfrac{1}{2} & -\tfrac{1}{2} & 0 & 1 & 0 & 6 \\
\hline
-1 & 0 & 1 & 2 & 0 & 0 & 1 & 12
\end{array}
$$

Quotient: 6

Quotient: 2 ← Pivot row

Quotient: $\tfrac{12}{5}$

↑ Pivot column

The resulting pivot element is circled in the tableau. Pivoting on that element gives this tableau.

$$
\begin{array}{ccccccc|c}
x & y & z & s_1 & s_2 & s_3 & P & \\
0 & 1 & 0 & \tfrac{1}{2} & -\tfrac{1}{4} & 0 & 0 & 2 \\
1 & 0 & 1 & 0 & \tfrac{1}{2} & 0 & 0 & 2 \\
0 & 0 & -2 & -\tfrac{1}{2} & -\tfrac{5}{4} & 1 & 0 & 1 \\
\hline
0 & 0 & 2 & 2 & \tfrac{1}{2} & 0 & 1 & 14
\end{array}
$$

No negative indicators exist, which indicates that the maximum for P on the feasible region has been reached. The solution to the linear programming problem is then $P = 14$ when $z = 0$, $s_1 = 0$, $s_2 = 0$, $x = 2$, $y = 2$, and $s_3 = 1$. That is, the maximum value of P on the feasible region is 14 and occurs at the point $(2, 2, 0)$. ▨

EXAMPLE 2

Using the Simplex Algorithm

Use the simplex algorithm to solve this problem:

$$
\begin{aligned}
\text{Maximize} \quad & f = 3x + 2y, \\
\text{subject to} \quad & x + y \le 4 \\
& -2x + y \le 2 \\
& x \ge 0, \ y \ge 0.
\end{aligned}
$$

Solution   The corner points of the feasible region, each labeled with its x, y, and f values, are shown in Figure 24. Inserting slack variables into the first two inequalities and rewriting f as $-3x - 2y + f = 0$ gives this system of equations:

$$
\begin{aligned}
x + y + s_1 \qquad\quad & = 4 \\
-2x + y \quad + s_2 \quad & = 2 \\
-3x - 2y \qquad\quad + f & = 0.
\end{aligned}
$$

The initial simplex tableau and preparations for selecting the first pivot element are shown next.

$$
\begin{array}{ccccc|c}
x & y & s_1 & s_2 & f & \\
\boxed{1} & 1 & 1 & 0 & 0 & 4 \\
-2 & 1 & 0 & 1 & 0 & 2 \\
\hline
-3 & -2 & 0 & 0 & 1 & 0
\end{array}
$$

Quotient: 4 ← Pivot row

↑ Pivot column

Recall that the smallest-quotient rule allows division by *positive* numbers only. The first pivot element is the circled 1 in the first-row, first-column position. The result of this pivot appears next.

$$
\begin{array}{c}
\\
\\
3R_1 + R_2 \rightarrow R_2 \\
3R_1 + R_3 \rightarrow R_3
\end{array}
\begin{array}{cccccc}
x & y & s_1 & s_2 & f & \\
\end{array}
\left[
\begin{array}{ccccc|c}
1 & 1 & 1 & 0 & 0 & 4 \\
0 & 3 & 2 & 1 & 0 & 10 \\
0 & 1 & 3 & 0 & 1 & 12
\end{array}
\right]
$$

The fact that no negative indicators appear shows that the maximum value for f on the feasible region has been reached. That maximum value is 12 and occurs when $x = 4$ and $y = 0$. Check this result against Figure 24. ■

EXAMPLE 3
Using the Simplex Algorithm

Use the simplex algorithm to solve this linear programming problem:

$$\text{Maximize} \quad f = 2x + y,$$
$$\text{subject to} \quad x + y \le 6$$
$$y \le 2x$$
$$x \le y$$
$$x \ge 0, y \ge 0.$$

Solution   The graph of the feasible region is shown in Figure 25. Next, the algebra of inequalities can be used to rewrite the second and third constraints so that they satisfy the conditions set forth for a standard maximum type problem:

$$-2x + y \le 0$$
$$x - y \le 0.$$

Therefore, the constraint system now appears as

$$x + y \le 6$$
$$-2x + y \le 0$$
$$x - y \le 0.$$

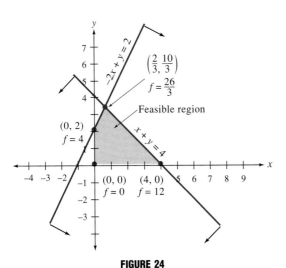

FIGURE 24
A Graphic Illustration of Basic Feasible Solutions

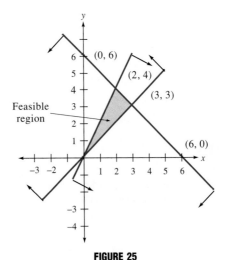

FIGURE 25
Illustrating Why 0 Is Allowed as a Smallest Quotient

Next, inserting the appropriate slack variables and rewriting f gives this system of equations:

$$
\begin{aligned}
x + y + s_1 &= 6 \\
-2x + y + s_2 &= 0 \\
x - y + s_3 &= 0 \\
-2x - y + f &= 0.
\end{aligned}
$$

The initial simplex tableau and preparations for selecting the first pivot element are shown next.

$$
\begin{array}{cccccc|c}
x & y & s_1 & s_2 & s_3 & f & \\
\left[\begin{array}{cccccc|c}
1 & 1 & 1 & 0 & 0 & 0 & 6 \\
-2 & 1 & 0 & 1 & 0 & 0 & 0 \\
\boxed{1} & -1 & 0 & 0 & 1 & 0 & 0 \\
-2 & -1 & 0 & 0 & 0 & 1 & 0
\end{array}\right]
\end{array}
$$

Quotient: 6

Quotient: 0 ← Pivot row

↑ Pivot column

The smallest-quotient rule of the simplex algorithm states that the smallest *nonnegative* quotient is to be used to select the pivot row. Because 0 is that number in this case, the third row is the pivot row. The result of this pivot is shown next.

$$
\begin{array}{cccccc|c}
x & y & s_1 & s_2 & s_3 & f & \\
\left[\begin{array}{cccccc|c}
0 & \boxed{2} & 1 & 0 & -1 & 0 & 6 \\
0 & -1 & 0 & 1 & 2 & 0 & 0 \\
1 & -1 & 0 & 0 & 1 & 0 & 0 \\
0 & -3 & 0 & 0 & 2 & 1 & 0
\end{array}\right]
\end{array}
$$

Quotient: 3 ← Pivot Row

↑ Pivot column

The negative indicator, -3, shows the need for another pivot; and, according to the smallest-quotient rule, that pivot element will be the circled 2 in the preceding augmented matrix. The pivot operation results in this tableau:

$$
\begin{array}{cccccc|c}
x & y & s_1 & s_2 & s_3 & f & \\
\left[\begin{array}{cccccc|c}
0 & 1 & \frac{1}{2} & 0 & -\frac{1}{2} & 0 & 3 \\
0 & 0 & \frac{1}{2} & 1 & \frac{3}{2} & 0 & 3 \\
1 & 0 & \frac{1}{2} & 0 & \frac{1}{2} & 0 & 3 \\
0 & 0 & \frac{3}{2} & 0 & \frac{1}{2} & 1 & 9
\end{array}\right]
\end{array}
$$

The last tableau shows the maximum value of f to be 9 when $x = 3$ and $y = 3$. ■

EXAMPLE 4
Using the Simplex
Algorithm

The Hi Tech Company is making three models of computer desks—an executive model, an office model, and a student model. Construction of the executive model requires 2 hours in the cabinet shop, 1 hour in the finishing department, and 1 hour in the crating department. The office model requires 1, 2, and 1 hour in these departments, respectively, while the student model requires 1, 1, and $\frac{1}{2}$ hour in these departments, respectively. On a daily basis, the cabinet shop has 16 hours available; the finishing department, 16; and the crating department, 10. If the company realizes a profit of \$150 on each executive model, \$125 on each office model, and \$50 on each student model, how many of each type of computer desk should be made and sold to maximize profits?

Solution Let x, y, and z be the number of executive, office, and student models, respectively, to be produced and sold. The problem may then be translated into a linear programming problem as follows:

$$\text{Maximize} \quad P = 150x + 125y + 50z,$$

$$
\begin{aligned}
\text{subject to} \quad 2x + \;\;\; y + \;\; z &\leq 16 \qquad \text{(Cabinet hours)}\\
x + 2y + \;\; z &\leq 16 \qquad \text{(Finishing hours)}\\
x + \;\;\; y + \tfrac{1}{2}z &\leq 10 \qquad \text{(Crating hours)}\\
x \geq 0, \; y \geq 0, z &\geq 0.
\end{aligned}
$$

After inserting a slack variable for each of the first three constraints and rewriting the objective function as $-150x - 125y - 50z + P = 0$, the initial simplex tableau, along with the notations for finding the first pivot element, may be written as shown next.

$$
\begin{array}{ccccccc}
x & y & z & s_1 & s_2 & s_3 & P \\
\end{array}
$$

$$
\left[
\begin{array}{ccccccc|c}
\textcircled{2} & 1 & 1 & 1 & 0 & 0 & 0 & 16 \\
1 & 2 & 1 & 0 & 1 & 0 & 0 & 16 \\
1 & 1 & \frac{1}{2} & 0 & 0 & 1 & 0 & 10 \\
\hline
-150 & -125 & -50 & 0 & 0 & 0 & 1 & 0
\end{array}
\right]
\begin{array}{l}
\text{Quotient: 8}\\
\text{Quotient: 16}\\
\text{Quotient: 10}
\end{array}
$$

↑ Pivot column

The smallest-quotient rule implies that the 2 in the first-row, first-column position should be the first pivot element. Upon completion of the pivoting process, the resulting augmented matrix, along with the preparations for finding the next pivot element, appear as follows.

$$
\begin{array}{ccccccc}
x & y & z & s_1 & s_2 & s_3 & P \\
\end{array}
$$

$$
\left[
\begin{array}{ccccccc|c}
1 & \frac{1}{2} & \frac{1}{2} & \frac{1}{2} & 0 & 0 & 0 & 8 \\
0 & \frac{3}{2} & \frac{1}{2} & -\frac{1}{2} & 1 & 0 & 0 & 8 \\
0 & \textcircled{$\frac{1}{2}$} & 0 & -\frac{1}{2} & 0 & 1 & 0 & 2 \\
\hline
0 & -50 & 25 & 75 & 0 & 0 & 1 & 1200
\end{array}
\right]
\begin{array}{l}
\text{Quotient: 16}\\
\text{Quotient: }\frac{16}{3}\\
\text{Quotient: 4}
\end{array}
$$

↑ Pivot column

The negative indicator shows that another pivot is required, and the smallest-quotient rule indicates that the new pivot element will be the $\frac{1}{2}$ in the third-row, second-column position. The augmented matrix resulting from the pivoting process is shown next.

$$
\left[
\begin{array}{ccccccc|c}
1 & 0 & \frac{1}{2} & 1 & 0 & -1 & 0 & 6 \\
0 & 0 & \frac{1}{2} & 1 & 1 & -3 & 0 & 2 \\
0 & 1 & 0 & -1 & 0 & 2 & 0 & 4 \\
\hline
0 & 0 & 25 & 25 & 0 & 100 & 1 & 1400
\end{array}
\right]
$$

Because no negative indicators are present, the maximum for P within the feasible region has been found. The maximum value of P is 1400 (dollars), and this occurs when $x = 6$, $y = 4$, and $z = 0$. That is, to maximize profits under the given constraints, the company should make six executive models, four office models, and no student models. ■

Important Facts About the Simplex Algorithm

A SURPLUS IN A GIVEN CONSTRAINT   The values of the slack variables in the solution to Example 4 are $s_1 = 0$, $s_2 = 2$, and $s_3 = 0$. Each of these variables is associated with a particular constraint of the problem: s_1, with the cabinet constraint; s_2, with the finishing constraint; and s_3, with the crating constraint. The fact that s_1 and s_3 are 0 means that there is no slack in those constraints. In other words, all of the available resources of the cabinet and crating departments are used when P is maximized. On the other hand, $s_2 = 2$ means that there are 2 hours of time in the finishing department that are unused or are **surplus** to the production process.

THE COLUMN REPRESENTING THE OBJECTIVE FUNCTION   Note that the column representing the function to be maximized never changes throughout the solution process, and some textbooks do not bother to put it in the tableau. If you choose not to put it in, just remember that the number in the lower right corner of the tableau is the current value of f.

WHAT IF THERE IS A "TIE" IN SELECTING THE PIVOT COLUMN OR PIVOT ROW?   What if there is a tie for the choice of pivot column or pivot row? Just use the column or row of your choice. In the rare event that your choice leads to looping within a fixed set of tableaus, go back and try the other choice.

THERE IS NEVER A NEGATIVE NUMBER IN THE RIGHTMOST COLUMN ABOVE THE HORIZONTAL BAR.   Due to the nature of a standard maximum-type linear programming problem, the initial simplex tableau can never have a negative number in the rightmost column above the horizontal bar. In addition, if the pivoting rules of this book are correctly followed, there can *never* be a negative number appearing in the rightmost column above the horizontal bar in any matrix obtained from successive pivots on the initial simplex tableau, regardless of how many pivot operations have been performed.

IS IT POSSIBLE FOR A NEGATIVE NUMBER TO APPEAR IN THE LOWER RIGHT-HAND CORNER?   Depending on the nature of the objective function, it is possible for a negative number to appear (even in the final solution matrix) in the lower right-hand corner of a matrix associated with the pivoting process. If a negative number appears in the final solution matrix, it means that the maximum value of the objective function at that point is negative. For example, if $f = -2x - 3y$ is to be maximized over a feasible region in the first quadrant that does not contain the origin, then the corner point at which the maximum occurs will have at least one coordinate that is not zero. Hence when f is evaluated there, a negative maximum value will result, and this value will appear in the lower right-hand corner of the final tableau.

WHAT IF NO SOLUTION EXISTS?   No solution exists when, at some point, we are unable to move forward in the selection of the next pivot element. As an example, suppose we ask the simplex algorithm to find the maximum value of $f = x + y$ subject to

$$x \le 1$$
$$x \ge 0, y \ge 0.$$

(The feasible region is a vertical strip along the y-axis, one unit wide but unbounded in the y-direction. We already know there is no maximum for this objective function over this region.) There is only one inequality, $x \le 1$ that requires a slack variable:

$$x + s_1 = 1.$$

Rewriting the objective function, we get

$$-x - y + f = 0.$$

The initial simplex tableau, the selection of the first pivot element, and the matrix resulting from that pivot is shown next.

$$
\begin{array}{cccc}
x & y & s_1 & f \\
\end{array}
$$

$$
\left[\begin{array}{cccc|c}
① & 0 & 1 & 0 & 1 \\
\hline
-1 & -1 & 0 & 1 & 0
\end{array}\right]
\quad 1R_1 + R_2 \to R_2
\quad
\left[\begin{array}{cccc|c}
1 & 0 & 1 & 0 & 1 \\
0 & -1 & 1 & 1 & 1
\end{array}\right]
$$

The negative indicator in the second matrix indicates that the maximum for f has not yet been reached and that at least one more pivot is required. The pivot column must be the y-column; however it is impossible to select a pivot row using the smallest-quotient rule. This is the signal that no maximum exists.

EXERCISES   4.5

In each of the Exercises 1 through 6, suppose that the matrix shown represents a simplex tableau after a sequence of pivot operations has been performed on an initial tableau. Answer these questions about each:

(a) *Does the tableau show that a maximum for the objective function has been reached?*

(b) *Find the value of f and the point at which that value occurs at this stage of the simplex algorithm.*

(c) *Does a surplus exist for any constraint? If so, which constraint and how much surplus?*

1.
$$
\begin{array}{ccccc}
x & y & s_1 & s_2 & f \\
\end{array}
$$
$$
\left[\begin{array}{ccccc|c}
1 & 0 & \frac{2}{3} & \frac{1}{3} & 0 & 5 \\
0 & 1 & 3 & 0 & 0 & 8 \\
0 & 0 & 5 & 6 & 1 & 20
\end{array}\right]
$$

2.
$$
\begin{array}{ccccc}
x & y & s_1 & s_2 & f \\
\end{array}
$$
$$
\left[\begin{array}{ccccc|c}
2 & 0 & 1 & 1 & 0 & 40 \\
1 & 1 & 0 & 3 & 0 & 60 \\
-1 & 0 & 0 & 5 & 1 & 38
\end{array}\right]
$$

3.
$$
\begin{array}{ccccc}
x & y & s_1 & s_2 & f \\
\end{array}
$$
$$
\left[\begin{array}{ccccc|c}
1 & \frac{2}{5} & \frac{1}{5} & 0 & 0 & \frac{8}{5} \\
0 & \frac{5}{6} & \frac{1}{3} & 1 & 0 & 16 \\
0 & \frac{8}{5} & \frac{6}{5} & 0 & 1 & \frac{80}{5}
\end{array}\right]
$$

4.
$$
\begin{array}{cccccc}
x & y & z & s_1 & s_2 & f \\
\end{array}
$$
$$
\left[\begin{array}{cccccc|c}
1 & 2 & 0 & 2 & 8 & 0 & 8 \\
0 & 3 & 1 & 3 & 1 & 0 & 20 \\
0 & 0 & 0 & -1 & 2 & 1 & 50
\end{array}\right]
$$

5.
$$
\begin{array}{cccccc}
x & y & s_1 & s_2 & s_3 & f \\
\end{array}
$$
$$
\left[\begin{array}{cccccc|c}
1 & 2 & 0 & 0 & \frac{5}{3} & 0 & 6 \\
0 & -1 & 1 & 0 & 3 & 0 & 8 \\
0 & 4 & 0 & 1 & \frac{1}{3} & 0 & 2 \\
0 & -\frac{2}{5} & 0 & 0 & \frac{2}{3} & 1 & 25
\end{array}\right]
$$

6.
$$
\begin{array}{cccccc}
x & y & s_1 & s_2 & s_3 & f \\
\end{array}
$$
$$
\left[\begin{array}{cccccc|c}
1 & 0 & -\frac{2}{5} & \frac{5}{2} & 0 & 0 & \frac{12}{5} \\
0 & 1 & 0 & -2 & 0 & 0 & 8 \\
0 & 0 & \frac{1}{6} & \frac{2}{5} & 1 & 0 & \frac{4}{5} \\
0 & 0 & \frac{5}{6} & \frac{7}{6} & 0 & 1 & \frac{17}{3}
\end{array}\right]
$$

In each of the Exercises 7 through 22:

(a) *Graph the feasible region.*

(b) *At each corner point of the feasible region, find the value of x, y, f, and each slack variable.*

(c) *Solve the linear programming problem by the simplex algorithm, and—after each pivot—identify the basic feasible solution with a corner point of the feasible region.*

7. Maximize $f = 8x + 6y$,
subject to $2x + y \le 10$
$$x + y \le 7$$
$$x \ge 0, y \ge 0$$

8. Maximize $f = 2x + 3y$,
subject to $x + y \le 8$
$$3x + 5y \le 30$$
$$x \ge 0, y \ge 0$$

9. Maximize $f = x + 4y$,
subject to $2x + 5y \le 20$
$$x + 4y \le 12$$
$$x \ge 0, y \ge 0$$

10. Maximize $f = 2x + y$,
subject to $x + 3y \le 14$
$$2x + y \le 11$$
$$x \ge 0, y \ge 0$$

11. Maximize $f = 5x + y$,
subject to $x + 2y \le 6$
$$4x + 3y \le 120$$
$$x \ge 0, y \ge 0$$

12. Maximize $f = 2x + 3y$,
   subject to $x + y \leq 2$
   $ x \leq 2$
   $x \geq 0, y \geq 0$

13. Maximize $f = 2x + y$,
   subject to $x + y \leq 5$
   $ y \leq 5$
   $x \geq 0, y \geq 0$

14. Maximize $f = 3x + 5y$,
   subject to $2x + y \leq 6$
   $ y \leq 4$
   $x \geq 0, y \geq 0$

15. Maximize $f = 3x + 3y$,
   subject to $x + y \leq 9$
   $-x + 2y \leq 10$
   $x \geq 0, y \geq 0$

16. Maximize $f = -x - 3y$,
   subject to $x + y \leq 5$
   $2x - y \leq 4$
   $x \geq 0, y \geq 0$

17. Maximize $f = -2x - y$,
   subject to $x - y \leq 6$
   $x + 2y \leq 8$
   $x \geq 0, y \geq 0$

18. Maximize $f = 2x + 2y$,
   subject to $-2x + 3y \leq 12$
   $2x + y \leq 6$
   $x \geq 0, y \geq 0$

19. Maximize $f = 3x + y$,
   subject to $y \leq x$
   $x \leq 2$
   $x \geq 0, y \geq 0$

20. Maximize $f = x + 4y$,
   subject to $y \leq x$
   $x \leq 2$
   $x \geq 0, y \geq 0$

21. Maximize $f = 5x + 2y$,
   subject to $x \leq y$
   $y \leq 3$
   $x \geq 0, y \geq 0$

22. Maximize $f = 3x + y$,
   subject to $x + y \leq 8$
   $x \leq 2y$
   $y \leq 6$
   $x \geq 0, y \geq 0$

Use the simplex algorithm to solve each of the linear pro-
gramming problems in Exercises 23 through 37.

23. Maximize $f = 3x + 2y$,
   subject to $x + y \leq 9$
   $2x + y \leq 12$
   $x + 4y \leq 24$
   $x \geq 0, y \geq 0$

24. Maximize $P = 2x + y$,
   subject to $4x + y \leq 20$
   $x + 2y \leq 16$
   $3x + 2y \leq 20$
   $x \geq 0, y \geq 0$

25. Maximize $f = 2x + y + 3z$,
   subject to $x + 2y + z \leq 25$
   $3x + 2y + 2z \leq 30$
   $x \geq 0, y \geq 0, z \geq 0$

26. Maximize $P = x + 2y + 4z$,
   subject to $x + 2z \leq 10$
   $3y + z \leq 24$
   $x \geq 0, y \geq 0, z \geq 0$

27. Maximize $f = 5x + 3y + 4z$,
   subject to $2x + y + 2z \leq 40$
   $2x + 2y + 3z \leq 50$
   $x + 3y + 2z \leq 20$
   $x \geq 0, y \geq 0, z \geq 0$

28. Maximize $f = 3x + 2y + z$,
   subject to $-x + 2y + z \leq 12$
   $2x + y - z \leq 6$
   $x + 2y + 2z \leq 8$
   $x \geq 0, y \geq 0, z \geq 0$

29. Maximize $P = 2x + 2y + z$,
   subject to $x - 2y + z \leq 4$
   $x - 4y + 3z \leq 12$
   $2x + 2y + z \leq 17$
   $x \geq 0, y \geq 0, z \geq 0$

30. Maximize $f = 2x + 3y + z$,
   subject to $x + z \leq 8$
   $y - z \leq 10$
   $x - y - z \leq 12$
   $x \geq 0, y \geq 0, z \geq 0$

31. Maximize $f = 2x + 3y + 3z$,
   subject to $x + 2z \leq 8$
   $-y + 3z \leq 10$
   $-x + y - z \leq 12$
   $x \geq 0, y \geq 0, z \geq 0$

32. Maximize $f = x + 3y + z$,
   subject to $x + y \leq 6$
   $y \leq 4$
   $x \leq 5$
   $x \geq 0, y \geq 0, z \geq 0$

33. Maximize $f = x + 2y + 4z$,
   subject to $y + z \leq 4$
   $y \leq 3$
   $z \leq 2$
   $x \geq 0, y \geq 0, z \geq 0$

34. Maximize $f = 4x + 3y + 2z + w$,

subject to
$$2x + y \quad\quad + 2w \leq 12$$
$$-x + 2y + z + 3w \leq 6$$
$$x + \quad\quad 2z + 3w \leq 15$$
$$x + y + z + w \leq 10$$
$$x \geq 0,\ y \geq 0,\ z \geq 0,\ w \geq 0$$

35. Maximize $f = 4x + 4y + 2z + w$,

subject to
$$x \quad\quad + z + 2w \leq 6$$
$$2x + y \quad\quad + w \leq 12$$
$$x + \quad\quad 2z + 3w \leq 18$$
$$x + y + 2z + w \leq 10$$
$$x \geq 0,\ y \geq 0,\ z \geq 0,\ w \geq 0$$

36. Maximize $f = 2x_1 + 3x_2 - 4x_3 - x_4$,

subject to
$$x_1 + 2x_2 + 3x_3 + x_4 \leq 6$$
$$x_2 + 2x_3 \quad\quad \leq 5$$
$$x_1 - x_2 + \quad\quad 2x_4 \leq 4$$
$$x_1 \geq 0,\ x_2 \geq 0,\ x_3 \geq 0,\ x_4 \geq 0$$

37. Maximize $f = 2x_1 - x_2 + x_3 - 2x_4 - 3x_5$,

subject to
$$2x_1 + x_2 + \quad\quad x_4 + 2x_5 \leq 2$$
$$4x_1 + \quad\quad 3x_3 + 2x_4 + x_5 \leq 8$$
$$x_1 + 2x_2 + 4x_3 + x_4 \quad\quad \leq 4$$
$$x_1 \geq 0,\ x_2 \geq 0,\ x_3 \geq 0,\ x_4 \geq 0,\ x_5 \geq 0$$

In the remainder of the exercises, find the objective function and the constraints, and then solve the problem by using the simplex method.

38. *Manufacturing* The Fine Line Company makes silver and gold ballpoint pens. A silver pen requires 1 minute in the grinder and 3 minutes in the bonder. A gold pen requires 3 minutes in the grinder and 4 minutes in the bonder. The grinder can be operated no more than 30 hours per week and the bonder no more than 50 hours per week. If the company makes 30 cents on each silver pen and 60 cents on each gold pen, how many of each type of pen should be made and sold each week to maximize profits?

39. *Manufacturing* A bicycle manufacturer makes a three-speed model and a ten-speed model with two operations, assembly, and painting. Each three-speed bike requires 1 hour to assemble and 2 hours to paint. Each ten-speed bike takes 1 hour to assemble and 1 hour to paint. The assembly operation has 80 work hours per week available, and the painting operation has 100 work hours each week available. If the company makes $80 on each three-speed bike and $60 on each ten-speed bike, how many bicycles should be made and sold each week to maximize profits?

40. *Advertising* An auto dealer's staffers can run a particular size newspaper ad in black and white for $200 per printing and in color for $300 per printing. They have budgeted a maximum of $6000 for such ads. They know that they can run each ad no more than 20 times before it loses its effectiveness. They estimate that each time the black-and-white version is run, it will be read by 3000 people; each time the color version is run, it will be read by 5000 people. How many times should each ad—the black and white and the color—be run to reach the maximum number of readers?

41. *Tests* Suppose that you take a finite math test that has some true-false questions and some problem-solving questions. The true-false questions are worth five points each and the problem-solving questions are worth eight points each. The rules of the test are that no more than 20 true-false questions are to be answered, and the total number of questions answered must be no more than 30. How many of each type of question should you answer (assume correctly) to maximize the number of points scored on the test?

42. *Manufacturing* A maker of microwave ovens is gearing up for a production run of deluxe and standard model ovens. A deluxe model requires 3 hours to assemble and a standard model requires 2 hours to assemble. There are 180 hours of assembly time available. The manufacturer has on hand 50 deluxe model cabinets and 30 standard model cabinets for its production run. The company makes a profit of $30 on each deluxe model and $25 on each standard model. How many of each type of oven should be made to maximize profits?

43. *Purchasing* The Journalism Department at State U wants to buy computers to serve two departmental purposes. For one purpose, brand X computers cost $2500 each; for the other purpose, brand Y computers cost $5000 each. The department has space for no more than 15 brand Y computers. If the department can spend no more than $100,000 for computers, how many of each brand of computer should be bought to maximize the total number bought?

44. *Manufacturing* A confectioner has 600 pounds of chocolate, 100 pounds of nuts, and 50 pounds of fruit in inventory with which to make three types of candy—Sweet Tooth, Sugar Dandy, and Dandy Delite. A box of Sweet Tooth uses 3 pounds of chocolate, 1 pound of nuts, 1 pound of fruit, and sells for $8. A box of Sugar Dandy requires 4 pounds of chocolate and $\frac{1}{2}$ pound of nuts and sells for $5. A box of Dandy Delite requires 5 pounds of chocolate, $\frac{3}{4}$ pound of nuts, 1 pound of fruit, and sells for $6. How many boxes of each type of candy should be made from the inventory available to maximize revenue?

45. Advertising An advertising agency has developed radio, newspaper, and television ads for a particular business. Each radio ad costs $200, each newspaper ad costs $100, and each television ad costs $500 to run. The business does not want the television ad to run more than 20 times, and the sum of the times the radio and newspaper ads can be run is to be no more than 110. The agency estimates that each airing of the radio ad will reach 1000 people, each printing of the newspaper ad will reach 800 people, and each airing of the television ad will reach 1500 people. If the total amount to be spent on ads is not to exceed $15,000, how many times should each type of ad be run so that the total number of people reached is a maximum?

46. Ecology A state fish and game commission is charged with management of an area designated as a wildlife refuge. Its members are interested in the populations of deer, quail, and turkeys. Studies indicate that each deer needs 1 unit of food group 1; 2 units of food group 2; and 2 units of food group 3. Each quail needs 1.2 units of food group 1; 1.8 units of food group 2; and 0.6 units of food group 3. Each turkey needs 2 units of food group 1; 0.8 units of food group 2; and 1 unit of food group 3. With its management practices, the commission estimates that there will be 600 units of food group 1 available, 900 units of food group 2, and 720 units of food group 3. If the total number of these three species of animals is to be a maximum, how many of each species will the refuge support?

47. **Manufacturing** A maker of toaster ovens has three models: a super deluxe, a deluxe, and a standard model. It uses three operations to produce these ovens: assembly, painting, and testing/packaging. Each super deluxe model requires 4 hours in assembly, 2 hours in painting, and 1 hour in testing/packaging. Each deluxe model requires 2 hours in assembly, 1 hour in painting, and 1 hour in testing/packaging. Each standard model requires 1 hour in assembly, 1 hour in painting, and $\frac{1}{2}$ hour in testing/packaging. The manufacturer has 160 work hours available in assembly, 100 in painting, and 60 in testing/packaging each week. If a profit of $20 is made on each super deluxe model, $16 on each deluxe model, and $12 on each standard model, how many should be made each week to maximize profits?

48. Psychology The psychology department at State University buys mice, rats, and snakes for experimental purposes. Each mouse costs $2, each rat costs $5, and each snake costs $6. The departmental budget requires that no more than $800 be spent on such purchases. Each mouse requires 1 square foot of living space, each rat requires 2 square feet of living space, and the department is limited to a maximum total of 90 square feet for these rodents. It wants to purchase no more than 10 snakes and no more than 30 rats. How many mice, rats, and snakes should be purchased by the department to accommodate the maximum total number? Will there be surplus money in the budget? Will there be a surplus of living space for the rodents?

49. Entertainment The directors of a state fair want to bring in entertainers to bolster attendance at the fair. They find that a top star demands $10,000, a faded star $6000, and high-quality local talent $3000 for each performance. The directors estimate that 8000 people will attend a top star performance, 3000 will attend a faded star performance, and 1200 will attend a high-quality local talent performance. The total amount to be spent for such entertainers is not to exceed $50,000 and they decide to contract for no more than six of them. The amount spent on advertising is not to exceed $4000, and the directors estimate that the advertising costs for each type of entertainer will be $5000, $400, and $250, respectively. Assuming one performance only for each entertainer, how many of each type should be contracted to maximize attendance? Will there be any surplus contracting or advertising funds left over? If so, how much?

50. Investing An investor has up to $150,000 to invest in any of three areas: stocks, bonds, or a money market fund. The annual rates of return are estimated to be 5%, 9%, and 8%, respectively. No more than $75,000 can be invested in any one of these areas. How much should be invested in each area to maximize annual income? Will there be any of the $150,000 that is not invested in one of these three areas?

51. Manufacturing A maker of computer printers has four models: the QP20, the QP40, the QP60, and the QP100. Each QP20 model has 2 megabytes of memory; each QP40, 4 megabytes; each QP60, 6 megabytes, and each QP100, 10 megabytes. In planning the next week's production, the company wants to buy no more than 820 megabytes of memory because of budget constraints, and furthermore, it wants to make at least twice as many QP20 models as QP100 models, at least as many QP40 models as QP60 models, and no more than 20 QP100 models. If the printers each sell for $500, $600, $1000, and $2000, respectively, how many of each model printer should the company assemble during the next week to maximize revenue?

52. Investments An investor has $100,000 to invest in government bonds, an aggressive growth mutual fund, a derivative-laden foreign bond fund, and a money market fund. The investor wants to invest at least as much in government bonds as the total in the aggressive growth mutual and the foreign bond fund. Furthermore, no more than \$20,000 is to be invested in the money market fund. If government bonds earn 7%, the aggressive growth mutual fund 5%, the foreign bond fund 8%, and the money market funds 5%, how much should be put into each of these investments to maximize annual returns?

Problems For Exploration

1. Write a paragraph or two explaining in your own words what the simplex algorithm does and how.

2. Explain why the simplex algorithm cannot be used on a real-world minimizing problem.

3. Standard maximum-type problems always have positive or zero constants to the right of the inequality signs; hence, positive or zero elements always appear in the rightmost column of the initial simplex tableau. Write an explanation of why the smallest-quotient rule, as we have formulated it, will never allow a negative number to appear in the right-most column of any succeeding tableau. Cover all cases. (One case will test your knowledge of the rules for inequalities.)

SECTION 4.6 **Nonstandard Problems**

There are many useful linear programming problems that do not meet the conditions of the standard maximum-type problem. Examples include any minimizing problem, maximizing problems in which one or more of the constraints (aside from nonnegativity) is of the $\geq$ type, problems with an equality constraint, and so forth. Such problems fall into the classification of **nonstandard problems.** In this section, we develop the necessary tools for solving all such problems, if a solution exists. First, we turn the problem into a **maximum-type** linear programming problem.

> ### A Maximum-Type Linear Programming Problem
>
> A maximum-type linear programming problem has this form:
>
> 1. The objective function is to be maximized.
>
> 2. The nonnegativity constraints are present. That is, all variables in the original problem are nonnegative.
>
> 3. All other constraints are of the form $ax + by + \ldots \leq c$ (c need not be positive).

Then, if any constant c on the right side of a constraint is negative, we consider an ingenious pivoting scheme that will eventually turn the initial tableau for the maximum-type problem into one on which the simplex algorithm can be applied to finish the solution process.

Notice that the only difference between a maximum-type problem and a standard maximum-type problem is that the requirement $c \geq 0$ has been dropped.

To accomplish the first step in our plan, we next examine the possible conversion procedures needed to change a nonstandard problem into a maximum-type problem.

Constraints of the $\geq$ type may be converted to the desired $\leq$ form by multiplying the inequality by -1 because we know that multiplication by any negative number reverses an inequality. For example, the inequality $2x + 3y \geq 5$ becomes $-2x - 3y \leq -5$, upon multiplying by -1; the inequality $5x - 3y + 7z \geq 6$ becomes $-5x + 3y - 7z \leq -6$, upon multiplying by -1.

Equality constraints may be eliminated by solving for one of the variables, then substituting to reduce the number of problem variables. Example 6 will show the details of how this may be done. An alternative approach is shown in the problems for exploration at the end of this section.

Finally, how do we change a problem whose objective function is to be minimized into an equivalent one whose objective function is to be maximized? The key lies in observing that the *minimum* of an objective function f occurs at *exactly the same point in the feasible region* that the *maximum* of $-f$ does. For example, consider $f = 2x + y$ and $g = -f = -2x - y$ over the arbitrary feasible region shown in Figure 26. Note that in that figure, the line $f = 2x + y = 8$ is the same line as $g = -2x - y = -8$. Figure 26 demonstrates the relationship between f and g: The further downward that f is moved, the *smaller* its value, while the further downward that g is moved, the *larger* its value. We conclude that the *minimum of f* over the feasible region occurs where the *maximum of g* occurs over the same feasible region.

EXAMPLE 1
Converting a
Minimizing Problem to
a Maximum-Type
Problem

Convert this problem to a maximum-type problem.

$$\text{Minimize} \quad f = 5x + 3y + z,$$
$$\text{subject to} \quad x + 2y + 5z \geq 6$$
$$2x - y + 6z \geq 8$$
$$x \geq 0, y \geq 0, z \geq 0$$

Solution   The constraints are converted to $\leq$ by multiplying by -1:

$$-x - 2y - 5z \leq -6$$
$$-2x + y - 6z \leq -8.$$

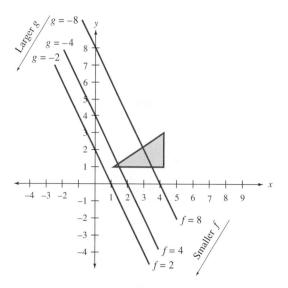

FIGURE 26
The Maximum for g Gives the Minimum for f.

As noted, the objective function part—minimize $f = 5x + 3y + z$—is replaced with maximize $g = -5x - 3y - z$. The maximum-type problem is written as

$$\begin{aligned}
\text{Maximize} \quad & g = -5x - 3y - z, \\
\text{subject to} \quad & -x - 2y - 5z \leq -6 \\
& -2x + y - 6z \leq -8 \\
& x \geq 0, y \geq 0, z \geq 0. \quad \blacksquare
\end{aligned}$$

Now that we see how to convert any linear programming problem to a maximum-type problem, why not just proceed with the simplex algorithm? The next example explains why.

EXAMPLE 2
A Maximizing Problem
with Mixed Constraints

$$\begin{aligned}
\text{Maximize} \quad & f = 2x + 3y, \\
\text{subject to} \quad & x + y \leq 8 \\
& 2x + y \geq 10 \\
& x \geq 0, y \geq 0.
\end{aligned}$$

When some of the inequalities are $\leq$ and some are $\geq$, as in this example, the problem is referred to as one with **mixed constraints.** The feasible region is shown in Figure 27, along with the coordinates of all points where two boundary lines intersect. The only thing that prevents our problem from being in the maximum-type format is the constraint $2x + y \geq 10$. Upon multiplying this inequality through by -1, the first two constraints now have the appearance

$$\begin{aligned}
x + y &\leq 8 \\
-2x - y &\leq -10
\end{aligned}$$

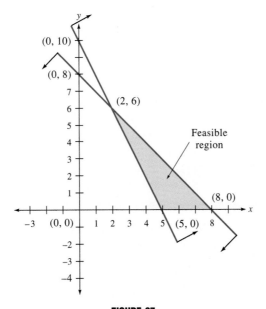

FIGURE 27
The Origin Is Not a Basic Feasible Solution.

and, hence, are in the desired form for a maximum-type problem. We proceed as usual to insert the slack variables and to rewrite the objective function to make it a part of the system of equations on which we pivot:

$$
\begin{aligned}
x + y + s_1 &= 8 \\
-2x - y \quad\quad + s_2 &= -10 \\
-2x - 3y \quad\quad\quad\quad + f &= 0.
\end{aligned}
$$

The initial simplex tableau may now be written

$$
\begin{array}{ccccc}
x & y & s_1 & s_2 & f \\
\end{array}
$$
$$
\left[
\begin{array}{ccccc|c}
1 & 1 & 1 & 0 & 0 & 8 \\
\hline
-2 & -1 & 0 & 1 & 0 & -10 \\
\hline
-2 & -3 & 0 & 0 & 1 & 0
\end{array}
\right]
$$

Up to this point, everything falls nicely into place. However, on reading the solution from the initial tableau in the usual manner ($x = 0$, $y = 0$, $s_1 = 8$, $s_2 = -10$), we realize that this solution does not represent a corner point of the feasible region because the value of one of the variables, s_2, is negative. (See Figure 27.) The fact that our first basic solution does not reflect a corner point of the feasible region means that we cannot employ the simplex algorithm to go from corner point to corner point of the feasible region until the maximum is reached. ▪

Since the initial simplex tableau for any linear programming problem has the problem variables as the first set of nonbasic variables, the first solution obtained by setting these variables equal to zero always gives the origin of the coordinate system as the first solution. Every standard maximum-type problem has the origin as a corner point of the feasible region, hence the initial tableau always starts with a solution at a corner point of the feasible region. However, when the origin is not in the feasible region, as in Example 2, such a first solution is of no value to us because we have no rules to guide us in the pivoting process. To deal with this problem, we first give an easy way to recognize when the solution to any simplex tableau does not reflect a corner point of the feasible region. This recognition is based on the fact that at least one variable will have a negative value. (See the tableau of Example 2.)

When a Tableau Does Not Represent a Corner Point

For any simplex tableau that has one or more *negative numbers* in the right-most column above the horizontal bar, the solution obtained by setting all nonbasic variables equal to zero does *not* represent a corner point of the feasible region.

Next, we need an organized pivoting scheme that will *eliminate these negative numbers,* thereby producing nonnegative entries in the rightmost column above the horizontal bar, and hence will give a *basic feasible solution* at that point in the solution process. Once this is accomplished, the appearance of the problem becomes that of a standard maximum-type problem and the rules of the simplex algorithm, developed in Section 4.5, may be used to complete the maximization process because those rules will continue to give tableaus whose solutions are basic feasible solutions to the problem. The pivoting scheme we prefer to use in eliminating these negative numbers is a set of rules first invented by Professor J. Conrad Crown in 1982, but slightly modified by Professor Arthur Hobbs to improve their efficiency. We refer to these rules as **Crown's rules.**

Crown's Rules: Eliminating Negative Numbers in the Rightmost Column Above the Horizontal Line in a Simplex Tableau

1. If there is a row above the horizontal line in the tableau whose only negative entry is in the rightmost column, then the problem has no solution; stop in such a case. Otherwise,

2. (a) To find the pivot column: Locate the first negative number from the top in the rightmost column of the tableau. Then select the most negative entry in that row to the left of the vertical bar. The column in which that negative number resides is the pivot column.*

    (b) To find the pivot row: Divide each *positive* entry in the pivot column into the corresponding entry in the rightmost column *only* if that entry is *nonnegative.* If such quotients exist, the row giving the smallest nonnegative quotient is the pivot row. If no such quotients exists, divide each *negative* entry in the pivot column into the corresponding entry in the rightmost column only if that entry is also *negative.* In this case, the row giving the *largest* positive quotient is the pivot row. (If no positive/positive-zero quotient exists, there will always be at least one negative/negative quotient.)

    (c) Pivot as usual, by first turning the pivot element into a one and all other elements in that column into zeros.

3. Repeat step 2 until there are no negative entries in the rightmost column above the objective function row.

The pivoting scheme just described assures us that the number of negative entries in the rightmost column above the horizontal bar can never go up; each time the scheme is applied, it either leaves the topmost negative entry in the rightmost column alone (if the pivot row has a zero as its rightmost entry) or makes it less negative (including making it, and possibly others, nonnegative) thereby giving a measure of progress toward a feasible solution. Once these negatives are eliminated, the usual simplex algorithm pivoting process is to be applied until there are no negative indicators in the objective function row.

EXAMPLE 3
Using the Crown Pivoting Rules

Complete the solution to the linear programming problem of Example 2.

Solution The initial tableau has already been constructed in Example 2 and appears as follows.

$$\begin{array}{ccccc} x & y & s_1 & s_2 & f \\ \left[\begin{array}{ccccc|c} \boxed{1} & 1 & 1 & 0 & 0 & 8 \\ -2 & -1 & 0 & 1 & 0 & -10 \\ \hline -2 & -3 & 0 & 0 & 1 & 0 \end{array}\right] \end{array}$$

---

*The selection of a pivot column corresponding to *any* negative number in that row to the left of the vertical bar is just as good. Sometimes one choice will be more efficient in removing negatives in the rightmost column than another, but there is no consistently "most efficient" choice. We choose the "most negative" simply to have a specific set of instructions. For pencil and paper calculations, investigate by inspection to gain efficiency.

The Crown pivoting rules are used first to eliminate the -10 in the rightmost column above the horizontal bar, which is the first, and in this case only, negative number in this area. (Ignore the indicators until all negatives in this area have been removed.) The most negative number to the left of the vertical bar in the row where -10 resides is -2. Since -2 is in the x-column, the x-column is the pivot column. In this column, above the horizontal bar, there is only one positive/nonnegative quotient, 1 divided into 8, so it is automatically the smallest such quotient. Since this quotient comes from the first row, the first row is the pivot row. Consequently, the 1 in the first-row, first-column position is the pivot element. The resulting tableau from pivoting on this element is shown next. Notice that this pivot eliminated the one and only negative number in the rightmost column above the bar. This means that a basic feasible solution has been found. ($f = 16$ when $x = 8$ and $y = 0$; see Figure 27.) Now we check the indicators to see if further pivoting is required to complete the maximization of the objective function.

$$
\begin{array}{ccccc}
x & y & s_1 & s_2 & f \\
\end{array}
$$
$$
\left[
\begin{array}{ccccc|c}
1 & 1 & 1 & 0 & 0 & 8 \\
0 & \textcircled{1} & 2 & 1 & 0 & 6 \\
\hline
0 & -1 & 2 & 0 & 1 & 16 \\
\end{array}
\right]
\quad
\begin{array}{l}
\text{Quotient: 8} \\
\text{Quotient: 6} \leftarrow \text{Pivot row} \\
\\
\end{array}
$$

Pivot column

Noting a negative indicator, the resulting quotients used in the smallest-quotient rule imply that the next pivot element is the circled 1 in the second-row, second-column position. The result of that pivot is the following tableau:

$$
\begin{array}{ccccc}
x & y & s_1 & s_2 & f \\
\end{array}
$$
$$
\left[
\begin{array}{ccccc|c}
1 & 0 & -1 & -1 & 0 & 2 \\
0 & 1 & 2 & 1 & 0 & 6 \\
\hline
0 & 0 & 4 & 1 & 1 & 22 \\
\end{array}
\right]
$$

Because no negative indicators now appear, the maximum of f has been reached and is 22. The maximum occurs when $x = 2$ and $y = 6$. (See Figure 27.)

In this particular problem, had we selected the first pivot column as the one determined by the least negative number in the row containing -10 (the y-column), the first pivot element would have been the 1 in the first-row, second-column position. The pivot operation on that element would have turned the -10 into a -2; progress toward a nonnegative entry there would have been made, but not as much as the pivot we made in this particular case. ▄▄▄

EXAMPLE 4
Minimizing an Objective
Function

Minimize $C = x + 2y + 4z,$

subject to $x + 2y + z \geq 5$

$2x - y \geq 8$

$x + y + 2z \leq 12$

$x \geq 0, y \geq 0, z \geq 0.$

Solution   As noted earlier, to turn this problem into a maximum-type problem, we must maximize $g = -C = -x - 2y - 4z$. Upon multiplying the first two constraints by -1, our problem can then be restated as:

$$\text{Maximize} \quad g = -x - 2y - 4z,$$
$$\text{subject to} \quad -x - 2y - z \le -5$$
$$-2x + y \qquad \le -8$$
$$x + y + 2z \le 12$$
$$x \ge 0,\ y \ge 0,\ z \ge 0.$$

After inserting the slack variables and rewriting g in the usual fashion, our initial tableau is

$$
\begin{array}{ccccccc|c}
x & y & z & s_1 & s_2 & s_3 & g & \\
\hline
-1 & -2 & -1 & 1 & 0 & 0 & 0 & -5 \\
-2 & 1 & 0 & 0 & 1 & 0 & 0 & -8 \\
1 & \textcircled{1} & 2 & 0 & 0 & 1 & 0 & 12 \\
\hline
1 & 2 & 4 & 0 & 0 & 0 & 1 & 0
\end{array}.
$$

The first negative number encountered in the rightmost column is -5 and the most negative number in that row is -2 in the y-column. Therefore, the y-column is our first pivot column. There is one positive/nonnegative quotient in that column, 1 divided into 12, establishing the third row as the pivot row. Therefore, the first pivot element is the circled one in the third-row, second-column position. The result of that pivot is shown next:

$$
\begin{array}{ccccccc|c}
x & y & z & s_1 & s_2 & s_3 & g & \\
\hline
1 & 0 & 3 & 1 & 0 & 2 & 0 & 19 \\
-3 & 0 & -2 & 0 & 1 & -1 & 0 & -20 \\
\textcircled{1} & 1 & 2 & 0 & 0 & 1 & 0 & 12 \\
\hline
-1 & 0 & 0 & 0 & 0 & -2 & 1 & -24
\end{array}
$$

The rightmost column now has only one negative number, -20, above the objective function, and the most negative number in that row is -3 in the x-column. This establishes the x-column as our next pivot column. In that column, there are two positive/nonnegative quotients, with 12 being the smallest. Therefore, the circled 1 in the third-row, first-column position is our next pivot element. The result of this pivot is

$$
\begin{array}{ccccccc|c}
x & y & z & s_1 & s_2 & s_3 & g & \\
\hline
0 & -1 & 1 & 1 & 0 & \textcircled{1} & 0 & 7 \\
0 & 3 & 4 & 0 & 1 & 2 & 0 & 16 \\
1 & 1 & 2 & 0 & 0 & 1 & 0 & 12 \\
\hline
0 & 1 & 2 & 0 & 0 & -1 & 1 & -12
\end{array}.
$$

Finally, the Crown pivoting rules have put us on a corner point of the feasible region. The negative indicator shows, however, that the maximum of g has not yet been reached. So the s_3-column is the pivot column and the smallest-quotient rule tells us that the circled one in the first-row, sixth-column position is the new pivot element. That pivot gives our final tableau:

$$
\begin{array}{ccccccc|c}
x & y & z & s_1 & s_2 & s_3 & g & \\
\hline
0 & -1 & 1 & 1 & 0 & 1 & 0 & 7 \\
0 & 5 & 2 & -2 & 1 & 0 & 0 & 2 \\
1 & 2 & 1 & -1 & 0 & 0 & 0 & 5 \\
\hline
0 & 0 & 3 & 1 & 0 & 0 & 1 & -5
\end{array}
$$

The maximum of g has now been reached, and that maximum value is -5. This means that the minimum of f, which is the negative of g, is **5**. This minimum occurs when $x = 5$, $y = 0$, and $z = 0$. ▆▆▆

EXAMPLE 5
When Fixed Costs Are Involved

A small subcontractor for a large shoe company does the cutting and gluing for walking shoes and jogging shoes. Each pair of walking shoes requires 2 minutes' cutting time and 4 minutes' gluing time. Each pair of jogging shoes requires 3 minutes' cutting time and 2 minutes' gluing time. The subcontractor has two employees who do the cutting and three employees who do the gluing and guarantees that each will work at least 30 hours per week. The cost of materials is $8 for each pair of walking shoes and $10 for each pair of jogging shoes, with an overhead cost of $200. How many pairs of each type of shoe should be made during the week to minimize total material costs?

Solution   Let x be the number of walking shoes and y be the number of jogging shoes to be made during the week. Then the constraints are

$$2x + 3y \geq 2(30)(60) = 3600 \text{ (minutes)}$$
$$4x + 2y \geq 3(30)(60) = 5400 \text{ (minutes)}$$

and the feasible region is shown in Figure 28. The objective function representing the total cost of materials becomes

$$\text{Minimize } C = 8x + 10y + 200.$$

This problem can then be stated as a maximum-type problem:

$$\text{Maximize } \quad g = -8x - 10y - 200 \text{ (in dollars)}$$

or
$$8x + 10y + g = -200,$$
$$\text{subject to } -2x - 3y \leq -3600$$
$$-4x - 2y \leq -5400$$
$$x \geq 0, y \geq 0.$$

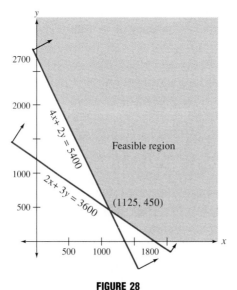

FIGURE 28
Feasible Region for Shoe Production

The initial simplex tableau then becomes

$$
\begin{array}{ccccc}
x & y & s_1 & s_2 & g
\end{array}
$$

$$
\left[
\begin{array}{ccccc|c}
-2 & -3 & 1 & 0 & 0 & -3600 \\
-4 & \boxed{-2} & 0 & 1 & 0 & -5400 \\
\hline
8 & 10 & 0 & 0 & 1 & -200
\end{array}
\right]
\begin{array}{l}
\text{Quotient: } 1200 \\
\text{Quotient: } 2700
\end{array}
$$

Since negative entries appear in the rightmost column, the Crown pivoting rules must be applied. The first one encountered is -3600 and the most negative number in that row to the left of the vertical bar is -3. This means that the y-column will be the pivot column. Because only negative/negative quotients exist in that column, the larger quotient determines the pivot row. The first pivot element is then the circled -2 located in the second-row, second-column. To get a 1 in that position, replace row 2 by $-\frac{1}{2}$ times itself:

$$
-\tfrac{1}{2}R_2 \to R_2 \quad
\begin{array}{ccccc}
x & y & s_1 & s_2 & g
\end{array}
$$

$$
\left[
\begin{array}{ccccc|c}
-2 & -3 & 1 & 0 & 0 & -3600 \\
2 & 1 & 0 & -\frac{1}{2} & 0 & 2700 \\
\hline
8 & 10 & 0 & 0 & 1 & -200
\end{array}
\right]
$$

Now introducing zeros above and below the 1 completes the pivot operation:

$$
\begin{array}{c}
3R_2 + R_1 \to R_1 \\
\\
-10R_2 + R_3 \to R_3
\end{array}
\quad
\begin{array}{ccccc}
x & y & s_1 & s_2 & g
\end{array}
$$

$$
\left[
\begin{array}{ccccc|c}
\boxed{4} & 0 & 1 & -\frac{3}{2} & 0 & 4500 \\
2 & 1 & 0 & -\frac{1}{2} & 0 & 2700 \\
\hline
-12 & 0 & 0 & 5 & 1 & -27{,}200
\end{array}
\right]
\begin{array}{l}
\text{Quotient: } 1125 \\
\text{Quotient: } 1350
\end{array}
$$

No negative entries appear in the rightmost column above the horizontal bar, which is the signal to return to the regular simplex algorithm rules. The pivot column will be the x-column and the smallest quotient indicates that the first row will be the pivot row. Pivoting on the circled 4 in the first-row, first-column position yields this matrix:

$$
\begin{array}{ccccc}
x & y & s_1 & s_2 & g
\end{array}
$$

$$
\left[
\begin{array}{ccccc|c}
1 & 0 & \frac{1}{4} & -\frac{3}{8} & 0 & 1125 \\
0 & 1 & -\frac{1}{2} & \frac{1}{4} & 0 & 450 \\
\hline
0 & 0 & 3 & \frac{1}{2} & 1 & -13{,}700
\end{array}
\right]
$$

The absence of negative indicators signals that the maximum of g has been reached and is $-13{,}700$. This means that the minimum cost C is \$13,700 and occurs when 1125 walking shoes and 450 jogging shoes are made during the week. ▄▄

Sometimes a maximizing or minimizing linear programming problem has an equality as a constraint. The simplest way to handle such a constraint is to solve for one of the variables in the equality and substitute for it throughout the problem, thereby reducing the number of variables by one. Example 6 will show the details.

EXAMPLE 6
A Constraint That Is an
Equality

$$
\begin{array}{ll}
\text{Maximize} & f = 3x + 2y + 5z, \\
\text{subject to} & x + y + z \le 8 \hspace{2cm} \text{(1)} \\
& x + y \phantom{{} + z} \le 4 \hspace{2cm} \text{(2)} \\
& \phantom{x + {}}2y + z = 6 \hspace{2cm} \text{(3)} \\
& x \ge 0,\, y \ge 0,
\end{array}
$$

Solution   Solving the equality in **(3)** for z, we have $z = 6 - 2y$. Substituting the expression for z into constraint **(1)** gives

$$x + y + (6 - 2y) \le 8$$

or

$$x - y \le 2. \qquad (1')$$

Constraint **(2)** does not contain z and therefore is not affected by the substitution process:

$$x + y \le 4. \qquad (2')$$

Because $z \ge 0$, we must have $z = 6 - 2y \ge 0$ or $2y - 6 \le 0$

or

$$y \le 3. \qquad (3')$$

Finally, rewrite the objective function using $z = 6 - 2y$:

$$f = 3x + 2y + 5z = 3x + 2y + 5(6 - 2y) = 3x - 8y + 30.$$

Rewriting this expression in suitable form for the initial simplex tableau becomes

$$-3x + 8y + f = 30.$$

The initial simplex tableau, using constraints **(1')**, **(2')**, and **(3')** is displayed next.

$$
\begin{array}{c}
x \quad\; y \quad s_1 \quad s_2 \quad s_3 \quad f \\
\left[\begin{array}{cccccc|c}
① & -1 & 1 & 0 & 0 & 0 & 2 \\
1 & 1 & 0 & 1 & 0 & 0 & 4 \\
0 & 1 & 0 & 0 & 1 & 0 & 3 \\
\hline
-3 & 8 & 0 & 0 & 0 & 1 & 30
\end{array}\right]
\end{array}
$$

There are no negative numbers in the rightmost column and therefore the Crown pivoting rules are not needed; the initial tableau has a solution that is a basic feasible solution. Consequently, we turn to the rules of the simplex algorithm. The most negative indicator is in the x-column, hence it is the pivot column; the smallest-quotient rule then tells us that the first row is the pivot row. After pivoting on the 1 in the first-row, first-column position, the resulting matrix looks like this:

$$
\begin{array}{c}
x \quad\;\; y \quad\;\; s_1 \quad\;\; s_2 \quad s_3 \quad f \\
\left[\begin{array}{cccccc|c}
1 & -1 & 1 & 0 & 0 & 0 & 2 \\
0 & 2 & -1 & 1 & 0 & 0 & 2 \\
0 & 1 & 0 & 0 & 1 & 0 & 3 \\
\hline
0 & 5 & 3 & 0 & 0 & 1 & 36
\end{array}\right]
\end{array}
$$

Since there are no negative indicators in the last matrix, the maximum has been reached. The maximum value of f is 36 and occurs when $x = 2$, $y = 0$, and $z = 6 - 2y = 6 - 2(0) = 6$.

IMPORTANT NOTE   When an equality constraint is present, a new constraint of the type **(3')** may be needed to ensure the nonnegativity of the variable which has been solved (in this case z). ■

EXAMPLE 7
A Problem Having
No Solution

Use the simplex algorithm to solve this linear programming problem:

$$\text{Maximum} \quad f = 2x + y,$$
$$\text{subject to} \quad x \geq 1$$
$$x \leq 2$$
$$y \geq 0.$$

Solution Rewriting the first inequality as $-x \leq -1$, the slack equations become

$$-x + s_1 \qquad = -1$$
$$x \qquad + s_2 = 2.$$

The initial simplex tableau is then

$$\begin{bmatrix} -1 & 0 & 1 & 0 & 0 & -1 \\ \textcircled{1} & 0 & 0 & 1 & 0 & 2 \\ -2 & -1 & 0 & 0 & 1 & 0 \end{bmatrix}$$

The Crown pivoting rules indicate that the circled one in the second row and first column is the first pivot element. On completing that pivot, the tableau becomes

$$\begin{bmatrix} 0 & 0 & 1 & 1 & 0 & 1 \\ 1 & 0 & 0 & 1 & 0 & 2 \\ 0 & -1 & 0 & 2 & 1 & 4 \end{bmatrix}$$

The negative indicator implies that a pivot is needed in the second column. However, the only choice for a pivot element is a zero, and this is an impossible choice. We conclude that the problem has no solution. Sketch the feasible region for this problem, and discover graphically why this is true. ◼

The problem posed at the outset of this chapter is solved in the next (and last) example of this section. This problem is on an extremely small scale when compared to most **transportation problems,** and, in fact, you may be able to solve it without the aid of linear programming. Nonetheless, the technique shown for setting up the constraints can be applied to problems of large scope.

EXAMPLE 8
Solving a
Transportation Problem

The problem stated at the outset of this chapter involves shipping refrigerators at minimum cost from two different warehouses to fill orders from two different stores. If we let x be the number of refrigerators to be shipped from Bentonville to Miami, then $40 - x$ is the number that must be shipped from Ashdown to Miami. Likewise, if we let y be the number of refrigerators to be shipped from Bentonville to Harrison, then $30 - y$ is the number that must be shipped from Ashdown to Harrison. Figure 29 describes the problem. The inequalities and objective function may now be constructed from the diagram: Minimize $f = 6x + 4y + 8(40 - x) + 5(30 - y) = -2x - y + 470$,

$$\text{subject to} \quad x \qquad \leq 40$$
$$y \leq 30$$
$$x + y \leq 60$$
$$x \geq 0, y \geq 0.$$
$$(40 - x) + (30 - y) \leq 50 \text{ or } -x - y \leq -20.$$

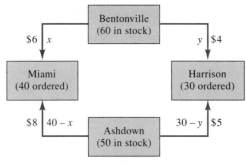

FIGURE 29

To minimize f, we need to maximize $g = -f$, or $g = 2x + y - 470$. The initial tableau and the subsequent pivots are shown next.

$$\begin{bmatrix} 1 & 0 & 1 & 0 & 0 & 0 & 0 & | & 40 \\ 0 & \textcircled{1} & 0 & 1 & 0 & 0 & 0 & | & 30 \\ 1 & 1 & 0 & 0 & 1 & 0 & 0 & | & 60 \\ -1 & -1 & 0 & 0 & 0 & 1 & 0 & | & -20 \\ \hline -2 & -1 & 0 & 0 & 0 & 0 & 1 & | & -470 \end{bmatrix} \rightarrow \begin{bmatrix} 1 & 0 & 1 & 0 & 0 & 0 & 0 & | & 40 \\ 0 & 1 & 0 & 1 & 0 & 0 & 0 & | & 30 \\ \textcircled{1} & 0 & 0 & -1 & 1 & 0 & 0 & | & 30 \\ -1 & 0 & 0 & 1 & 0 & 1 & 0 & | & 10 \\ \hline -2 & 0 & 0 & 1 & 0 & 0 & 1 & | & -440 \end{bmatrix}$$

$$\begin{bmatrix} 0 & 0 & 1 & \textcircled{1} & -1 & 0 & 0 & | & 10 \\ 0 & 1 & 0 & 1 & 0 & 0 & 0 & | & 30 \\ 1 & 0 & 0 & -1 & 1 & 0 & 0 & | & 30 \\ 0 & 0 & 0 & 0 & 1 & 1 & 0 & | & 40 \\ \hline 0 & 0 & 0 & -1 & 2 & 0 & 1 & | & -380 \end{bmatrix} \rightarrow \begin{bmatrix} 0 & 0 & 1 & 1 & -1 & 0 & 0 & | & 10 \\ 0 & 1 & -1 & 0 & 1 & 0 & 0 & | & 20 \\ 1 & 0 & 1 & 0 & 0 & 0 & 0 & | & 40 \\ 0 & 0 & 0 & 0 & 1 & 1 & 0 & | & 40 \\ \hline 0 & 0 & 1 & 0 & 1 & 0 & 1 & | & -370 \end{bmatrix}$$

The final tableau shows the solution to be $x = 40$ and $y = 20$, which means that the shipping schedule should look like this:

From Bentonville to Miami, 40

From Bentonville to Harrison, 20

From Ashdown to Miami, 0

From Ashdown to Harrison, 10.

The minimum shipping costs will be $370. ◼

EXERCISES 4.6

In each of the Exercises 1 through 16:

(a) Sketch the feasible region, and find the coordinates of the corner points.

(b) Solve the problem by using the simplex algorithm, using the Crown pivoting rules if necessary, and record the movement of the corner points for each tableau.

(c) Check the solution by computing the value of f at each corner point of the feasible region.

1. Maximize $f = 2x + y$,
subject to $x + y \le 6$
$4x + y \le 12$
$x \ge 0, y \ge 0$

2. Maximize $f = x + 2y$,
   subject to $x + y \geq 6$
   $3x + y \leq 9$
   $x \geq 0, y \geq 0$

3. Maximize $f = 4x + 2y$,
   subject to $x + y \geq 5$
   $3x + y \leq 12$
   $x \geq 0, y \geq 0$

4. Maximize $f = 4x + y$,
   subject to $x - y \geq 0$
   $x \leq 4$
   $y \geq 1$

5. Maximize $f = 2x + 3y$,
   subject to $2x - y \leq 0$
   $x \geq 1$
   $y \leq 6$

6. Maximize $f = 2x - 3y$,
   subject to $y \geq x$
   $y \leq 4$
   $y \geq 2$
   $x \geq 0$

7. Maximize $f = -4x + 5y$,
   subject to $x - y \geq 1$
   $x + y \leq 7$
   $x \geq 0, y \geq 0$

8. Maximize $f = 2x + 3y$,
   subject to $x + y \geq 6$
   $2x + y \geq 8$
   $y \leq 8$
   $x \leq 6$
   $x \geq 0, y \geq 0$

9. Minimize $f = x + 3y$,
   subject to $x + y \geq 6$
   $2x + y \geq 8$
   $x \geq 0, y \geq 0$

10. Minimize $f = 4x + 3y$,
   subject to $2x + y \geq 8$
   $x + 2y \geq 6$
   $x \geq 0, y \geq 0$

11. Minimize $f = x + y$,
   subject to $x + y \leq 6$
   $x \geq 2$
   $y \geq 2$

12. Minimize $C = 5x + y$,
   subject to $2x + y \geq 8$
   $2x + y \leq 12$
   $x \geq 1$
   $x \leq 2$

13. Minimize $C = x - 2y$,
   subject to $x + y \geq 6$
   $x \geq 2$
   $x \leq 8$
   $y \geq 2$
   $y \leq 6$

14. Minimize $C = -3x + 2y$,
   subject to $x - 2y \leq 0$
   $x \geq 2$
   $y \leq 7$

15. Minimize $f = 2x + y + 10$,
   subject to $y \leq x$
   $2x - y \geq 6$
   $x \geq 0, y \geq 0$

16. Minimize $f = 2x + y + 18$,
   subject to $x \geq 2$
   $y \geq x$
   $y \leq 2x$

Make use of the Crown pivoting rules to solve Exercises 17 through 30.

17. Minimize $C = 3x + 4y + z + 50$,
   subject to $x \leq y$
   $z \geq 2$
   $x \geq 4$
   $x + y + z \geq 6$
   $y \geq 0$

18. Minimize $C = 2x - 3y + z + 80$,
   subject to $x + 2y + z \geq 2$
   $z \leq 6$
   $x \geq y$
   $y \leq 4$
   $x \geq 0, y \geq 0, z \geq 0$

19. Maximize $f = 3x + 2y + 6z$,
   subject to $x + 2y + z \leq 6$
   $y + 2z \leq 4$
   $2x + z \geq 4$
   $x \geq 0, y \geq 0, z \geq 0$

20. Minimize $P = 4x + 5y + z$,
   subject to $x + 2y + 3z \leq 6$
   $x + y + 2z \geq 3$
   $3x + y \geq 5$
   $y \geq 1$
   $x \geq 0, y \geq 0, z \geq 0$

21. Maximize $P = 3x + 4y + z$,
   subject to $3x + 2y + 4z \leq 12$
   $x + z \geq 1$
   $y \geq 1$
   $z \leq 1$
   $x \geq 0, y \geq 0, z \geq 0$

22. Minimize $f = 2x - y + 2z$,

subject to
$$x + 2y \geq 40$$
$$x + 2y + 3z \geq 60$$
$$y + z \geq 30$$
$$x \geq 0, y \geq 0, z \geq 0$$

23. Minimize $C = 4x + 3y + 2z$,

subject to
$$x + 2y + z \geq 6$$
$$y + 2z \geq 1$$
$$2x + y + 2z \leq 4$$
$$x \geq 0, y \geq 0, z \geq 0$$

24. Minimize $f = x + 2y + 4z$,

subject to
$$x \geq 2$$
$$x + y + z \geq 6$$
$$z \geq 3$$
$$x \geq 0, y \geq 0$$

25. Maximize $f = x + 2y + z$,

subject to
$$x + y + 2z \leq 12$$
$$2x + y + z \leq 16$$
$$x + y = 3$$
$$x \geq 0, y \geq 0, z \geq 0$$

26. Maximize $P = 2x - 3y + z$,

subject to
$$x + y + 2z \leq 6$$
$$y \leq 2$$
$$x \geq z$$
$$x + y = 4$$
$$x \geq 0, y \geq 0, z \geq 0$$

27. Minimize $C = 3x + y - 2z$,

subject to
$$2x - y = 3$$
$$x + y - 2z \leq 4$$
$$x - z \geq 0$$
$$y \geq 0, z \geq 0$$

28. Minimize $f = 2x + 3y + z$,

subject to
$$x + y + z \geq 4$$
$$x + 3z \geq 6$$
$$x = 1$$
$$x \geq 0, y \geq 0, z \geq 0$$

29. Maximize $f = 3x_1 + 4x_2 + x_3$,

subject to
$$x_2 = 2$$
$$x_1 + x_2 + x_3 \leq 8$$
$$x_2 + x_3 \leq 4$$
$$x_1 \geq 0, x_2 \geq 0$$

30. Minimize $C = 2x_1 + 3x_2 - x_3 + 100$,

subject to
$$x_1 \leq x_2$$
$$x_3 \geq 2$$
$$x_2 \geq 4$$
$$x_1 + x_2 + x_3 \geq 8$$
$$x_1 \geq 0$$

For the remaining exercises, find the mathematical formulation of the problem, then solve by applying the Crown pivoting rules, if necessary, and then the simplex algorithm pivoting rules. Some of these exercises may be familiar to you from earlier sections.

31. *Sales* An appliance store sells two brands of television sets. Each Daybrite set sells for $420 and each Noglare set for $550. The store's warehouse capacity for television sets is 300 sets, and new sets are delivered only once each month. Records show that customers will buy at least 80 Daybrite sets and at least 120 Noglare sets each month. How many of each brand should the store stock and sell each month to maximize revenue?

32. *Making Golf Bags* A firm makes three types of golf bags, A, B, and C. Type A requires 15 minutes to cut, 30 minutes to sew, 30 minutes to trim, and sells for $80. Type B requires 15 minutes to cut, 20 minutes to sew, 50 minutes to trim, and sells for $90. Type C requires 20 minutes to cut, 30 minutes to sew, 50 minutes to trim, and sells for $100. The firm has available 80 work-hours for cutting, 90 work-hours for sewing, and 120 work-hours for the trimming each day. A contract from a large discount store requires that they make at least 20 bags of type A and 25 bags of type B each day. How many of each type bag should be produced to maximize revenue?

33. *Political Campaign* The campaign manager of a candidate for a local political office estimates that each 30-minute speech to a civic group will generate 20 votes, each hour spent on the phone will generate 10 votes and each hour spent campaigning in shopping areas will generate 40 votes. The candidate wants to make at least twice as many shopping area stops as speeches to civic groups and wants to spend at least three hours on the phone. The campaign manager thinks her candidate can win if he generates at least 1000 votes by these three methods. How can the candidate meet these goals with minimum time spent?

34. *Advertising* A new line of snack cracker is being introduced by a food company that has allotted no more than $50,000 for advertisements in newspapers, on the radio, and on television. Each newspaper ad costs $100 and will cause 200 people to try the new cracker; each radio ad costs $150 and will cause 300 people to try it; and each television ad costs $300 and will cause 600 people to try it. The company wants to reach at least 10,000 people through these advertisements and has agreed to run at least twice as many radio ads as television ads. How can the company meet these objectives at minimum cost?

35. **Medical** After a certain operation, a patient receives three types of painkiller, A, B, and C. A dose of A rates a 2 in effectiveness, a dose of B rates a 2.5 in effectiveness, and a dose of C rates a 3 in effectiveness. For medical reasons, the patient can receive no more than three doses of A, no more than five doses of A and C combined, and no more than seven doses of B and C combined during a 24-hour period. How many doses of each type should be given to the patient in a 24-hour period in order to maximize the pain-killing effect?

36. **Refining Gasoline** A refiner blends high- and low-octane gasoline into three grades for sale to the whole-saler: regular, premium, and super premium. The regular grade consists of 30% high-octane and 70% low-octane gasolines, the premium grade consists of 50% of each, and the super premium grade consists of 70% high-octane and 30% low-octane gasolines. The refiner can sell each gallon of regular for 60 cents, each gallon of premium for 65 cents, and each gallon of super premium for 70 cents. If there are currently in stock 150,000 gallons of high-octane gasoline and 130,000 gallons of low-octane gasoline, how many gallons of regular, premium, and super premium should be made if the refiner is to maximize revenues?

37. **Inventory** The InfoAge Communication Store stocks fax machines, multimedia computers, and portable CD players. Space restrictions dictate that it stock no more than a total of 100 of these three machines. Past sales patterns indicate that it should stock an equal number of fax machines and multimedia computers and at least 20 CD players. If each fax machine sells for $500, each multimedia computer for $1800, and each CD player for $1000, how many of each should be stocked and sold for maximum revenues?

38. **Mixing Foods** A dietitian is to prepare two foods to meet certain requirements. Each pound of Food I contains 100 units of vitamin C, 40 units of vitamin D, 10 units of vitamin E, and costs 20 cents. Each pound of Food II contains 10 units of vitamin C, 80 units of vitamin D, 5 units of vitamin E, and costs 15 cents. The mixture of the two foods is to contain at least 260 units of vitamin C, at least 320 units of vitamin D, and at least 50 units of vitamin E. If the fixed cost for handling and storage of these two foods is $10, how many pounds of each food should be used to minimize the total cost?

39. **Manufacturing** The Wide-Track Tire Company makes radial, mud and snow, and off-road tires. Each radial tire takes four minutes and costs $20 to pro-duce; each mud and snow tire takes 6 minutes and costs $30 to produce; each off-road tire takes 8 minutes and costs $40 to produce. Orders received by the company dictate that at least 1000 radial, at least 400 mud and snow, and at least 100 off-road tires be made each day. The union contract with Wide-Track requires the use of at least 500 work-hours each day. If the fixed costs for production of these tires is $2000, how many of each type tire should be made each day to minimize total costs?

40. **Radio Programming** Radio station KIX-104 is adding to its programming line-up a talk show, a sports call-in show, and a country and western show. The company's marketing consultant estimates that each time these shows are aired, they will attract, respectively, 5000, 6000, and 3000 listeners. Advertising revenues dictate that, together, the talk show and the sports call-in show can be aired a total of no more than eight times each week and the country and western show no more than six times each week. Furthermore, the station manager wants to run the country and western show twice as many times as the talk show. How many times should each show be aired each week to attract the maximum number of radio listeners?

41. **Mixture** Two grains, barley and corn, are to be mixed for an animal food. Barley contains 1 unit of fat per pound and corn contains 2 units of fat per pound. The total units of fat in the mixture should not exceed 12. No more than 6 pounds of barley and no more than 5 pound of corn are to be used in the mixture. If barley and corn each contain 1 unit of protein per pound, how many pounds of each grain should be used to maximize the number of units of protein in the food mixture?

42. **Diet** Jane is going on a low-carbohydrate diet that calls for her to eat two foods at each meal for a week, Food I and Food II. One unit of Food I contains 5 grams of carbohydrate, 80 calories, and 2 grams of fat. One unit of Food II contains 7 grams of carbohydrates, 110 calories, and 5 grams of fat. Jane wants each meal to provide at least 400 calories but no more than 105 grams of carbohydrate. In addition, at least three times as many units of Food I as Food II is to be used in each meal for taste appeal. How many units of each food should Jane use per meal to minimize the intake of fat?

43. *Shipping Costs*   Find the minimum cost for shipping TV sets from the warehouses to the stores in the following figure.

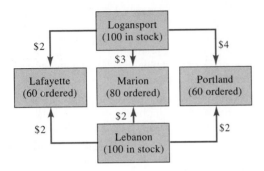

44. *Shipping Costs*   The Raw Wood Furniture Company gets an order from its store in Glasgow for 20 desks, an order from its store in Winifred for 12 desks, and an order from its store in Browning for 15 desks. The warehouse in Havre has 25 desks and the warehouse in Melville has 34 desks from which these orders are to be filled. The costs to the company for shipping desks from Havre to these three stores are $4, $6, and $5, respectively. The costs to the company for shipping desks from Melville to these three stores are $5, $4, and $5, respectively. How many desks should the company ship from each warehouse to these three stores so that shipping costs will be a minimum?

45. *Shipping Costs*   The ABC Book Company has boxes of a particular textbook in two warehouses: one in Mansfield with 28 boxes and one in Athens with 40 boxes. It gets orders from three bookstores for this text: 20 boxes from Weirton, 15 boxes from Delaware, and 10 boxes from Kenton. The shipping costs per box from Mansfield to these three locations are $1, $1, and $2, respectively. The shipping costs per box from Athens to these three locations are $2, $2, and $1, respectively. How many boxes of books should be shipped from each warehouse to these three locations so that shipping costs will be minimal?

Problems for Exploration

1. Consider this problem:

$$\text{Maximize } f = 2x + y,$$
$$\text{subject to} \quad y = 1$$
$$x + y \leq 2$$
$$x \geq 0, y \geq 0$$

(a) Graph the feasible region.
(b) Set up the initial tableau, and solve the problem by using the rules shown in this section. Keep track of the geometry of the basic solutions.
(c) Take the initial tableau of part (b), and attempt to solve this problem by using this pivot-selection rule:

Select the pivot column as usual, then in that column, divide each element above the horizontal line into the corresponding element in the rightmost column, and select the pivot row to be the one giving the smallest *positive* quotient. At any time that a negative element appears in the rightmost column, use the Crown pivoting rules to eliminate it. Keep track of the geometry of the basic solutions. What is happening after several pivots that tells us that this rule is not desirable?

2. Consider this problem:

$$\text{Maximize } f = 2x + 5y,$$
$$x = 3$$
$$x + y \leq 4$$
$$x \geq 0, y \geq 0$$

(a) Graph the feasible region.
(b) Use this problem to demonstrate that any problem with an equality constraint may be solved by replacing $=$ with two constraints, one using $\leq$ and the other using $\geq$.

Note: The reason this method might be preferable to substitution is that sometimes a problem produces a simplex tableau with many zeros. Such a problem is called "sparse." The substitution method can destroy the sparseness of such a problem, while the method suggested by this problem does not. There are solution techniques that take advantage of sparseness in pivoting and which are fast on sparse problems (but not on problems that are not sparse).

3. Locate a city government office, a state government office, or a local industry that uses linear programming to solve one of its problems. Write up an account of this problem, telling what it does, its size, and so forth.

SECTION 4.7 The Dual Problem

Every linear programming problem has associated with it another linear programming problem called its **dual.** If the original problem is one whose objective function is to be maximized, its dual problem is one whose objective function is to be minimized. On the other hand, if the original problem is one whose objective function is to be minimized, its dual problem is one whose objective function is to be maximized. As our exploration of the dual problem unfolds, we will see that the final tableau for a dual problem not only gives the solution to the dual problem, but also contains the solution to the original problem.

Here is how the relationship between the original problem and its dual works: For a maximizing problem, every constraint (except the nonegativity constraints) must first be written in the $\leq$ form. This may require multiplying certain of the inequalities by -1. The dual problem then becomes a minimizing problem with constraints in the $\geq$ form. Before finding the dual to a minimizing problem, every constraint must first be written in the $\geq$ form. Again, this may require multiplying certain of the inequalities by -1. The dual problem then becomes a maximizing problem with constraints in the $\leq$ form. In either case, once the constraints are written in the proper form, the problem is known as the **primal problem.** We summarize our comments with the following chart.

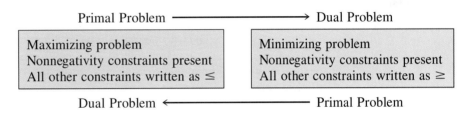

Primal Problem $\longrightarrow$ Dual Problem

| Maximizing problem
Nonnegativity constraints present
All other constraints written as $\leq$ | Minimizing problem
Nonnegativity constraints present
All other constraints written as $\geq$ |

Dual Problem $\longleftarrow$ Primal Problem

We use the example problem,

$$\text{Minimize} \quad f = x + 2y,$$
$$\text{subject to} \quad x + \ y \geq 6$$
$$x + 3y \geq 12$$
$$x \geq 0, y \geq 0$$

to explain how the dual problem is formed and how its final tableau gives the solution to our example problem. Notice that all constraints are already in the $\geq$ form; hence, the original problem is also the primal problem. Figure 30 shows the feasible region to our problem. A few calculations show that the minimum value of f is 9 and occurs when $x = 3$ and $y = 3$. With that in mind, here is how we proceed to create the dual problem associated with this example problem. First, put the constraints (without slack variables) and the coefficients of the objective function into a matrix, as follows:

$$P = \begin{bmatrix} 1 & 1 & | & 6 \\ 1 & 3 & | & 12 \\ \hline 1 & 2 & | & 0 \end{bmatrix}$$

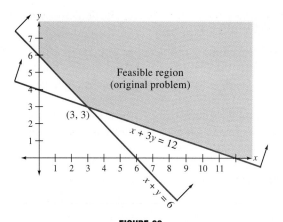

FIGURE 30

Feasible Region for Original Minimizing Problem

The matrix P is known as the **primal matrix** for this problem. Next, form the transpose of this matrix, P^T:

$$P^T = \begin{bmatrix} 1 & 1 & \big| & 1 \\ 1 & 3 & \big| & 2 \\ \hline 6 & 12 & \big| & 0 \end{bmatrix} = D.$$

Finally, from D, a standard maximum-type problem, called the **dual problem,** is created by considering the last row to be a new objective function that is to be maximized and the remaining rows as new constraints of the $\leq$ type. Thus, the dual problem becomes

$$\text{Maximize} \quad d = 6u + 12v,$$
$$\text{subject to} \quad u + v \leq 1$$
$$u + 3v \leq 2$$
$$u \geq 0, v \geq 0$$

Figure 31 shows the feasible region for the newly created dual problem. Calculations show that the *maximum* value of $d = 6u + 12v$ over this region is 9 and occurs at the point where $u = \frac{1}{2}$ and $v = \frac{1}{2}$. Note that this *maximum* value is the same as the *minimum* value of the original objective function in the primal problem.

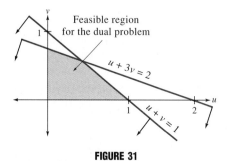

FIGURE 31

Feasible Region for the Dual Problem

Even though the original problem and its dual are quite different in appearance and have very different feasible regions, there is a link between the two. This link is described by a rather deep mathematical theorem known as the **duality theorem.** The assertions of that theorem are stated next.

The Duality Theorem

Suppose that a primal linear program, with objective function f, and its dual, with objective function d, both have solutions. Then

1. If the primal problem is a maximization problem, the dual will be a minimization problem and max f = min d.

2. If the primal problem is a minimization problem, the dual will be a maximization problem and min f = max d.

We have already observed in a geometric sense that our example problem agrees with the general statement of the theorem. But there is more; the duality theorem also tells us where to look in the final tableau of the dual problem for the corner points at which these equal optimal solutions occur.

The Duality Theorem, Continued

When solving the dual problem by the simplex algorithm (using Crown's rules first, if necessary), the coordinates of the point at which the optimal solution occurs are read from the final tableau in the usual way. In that same final taleau, the coordinates of the point at which the optimal solution to the primal problem occurs are found in the objective function row and the slack variable columns.

To demonstrate this fact, we solve our present example by using the simplex algorithm. First the slack variables are inserted and the objective function rewritten as usual:

$$
\begin{aligned}
u + \quad v + s_1 \qquad\qquad\quad &= 1 \\
u + \quad 3v \qquad + s_2 \qquad\quad &= 2 \\
-6u - 12v \qquad\qquad\quad + d &= 0.
\end{aligned}
$$

The initial simplex tableau is then found to be

$$
\begin{array}{ccccc}
u & v & s_1 & s_2 & d \\
\end{array}
$$
$$
\left[
\begin{array}{ccccc|c}
1 & 1 & 1 & 0 & 0 & 1 \\
1 & \textcircled{3} & 0 & 1 & 0 & 2 \\
\hline
-6 & -12 & 0 & 0 & 1 & 0
\end{array}
\right]
\quad
\begin{array}{l}
\text{Quotient: } 1 \\
\text{Quotient: } \frac{2}{3} \leftarrow \text{Pivot row} \\
\end{array}
$$

Pivot column (under -12)

The first pivot element is the circled 3 in the second-row, second-column position. Completion of that pivot results in this matrix:

$$
\begin{array}{ccccc}
u & v & s_1 & s_2 & d \\
\end{array}
$$
$$
\left[
\begin{array}{ccccc|c}
\textcircled{\frac{2}{3}} & 0 & 1 & -\frac{1}{3} & 0 & \frac{1}{3} \\
\frac{1}{3} & 1 & 0 & \frac{1}{3} & 0 & \frac{2}{3} \\
\hline
-2 & 0 & 0 & 4 & 1 & 8
\end{array}
\right]
\quad
\begin{array}{l}
\\
\text{Quotient: } \frac{1}{2} \leftarrow \text{Pivot row} \\
\text{Quotient: } 2 \\
\end{array}
$$

Pivot column (under $\frac{2}{3}$)

The negative indicator, -2, in the last row indicates that the maximum for d has not yet been reached, so another pivot is necessary on the circled $\frac{2}{3}$ in the first-row, first-column position. The result of this pivot is shown next:

$$
\begin{array}{ccccc}
u & v & s_1 & s_2 & d \\
\end{array}
$$
$$
\left[
\begin{array}{ccccc|c}
1 & 0 & \frac{3}{2} & -\frac{1}{2} & 0 & \frac{1}{2} \\
0 & 1 & -\frac{1}{2} & \frac{1}{2} & 0 & \frac{1}{2} \\
\hline
0 & 0 & 3 & 3 & 1 & 9
\end{array}
\right]
$$

The maximum value for d on the feasible region of the dual problem has now been reached. It is 9 and occurs when $u = \frac{1}{2}$, $v = \frac{1}{2}$, $s_1 = 0$, and $s_2 = 0$. As asserted in the duality theorem, the maximum value of d is the same as the minimum value of the original objective function f. How do we read these from the final tableau? Remember that the solution occurred when $x = 3$ and $y = 3$. As promised by the duality theorem, the bottom row of the final tableau of the dual problem has, in the slack-variable columns, the values for x and y, respectively: $x = 3$, and $y = 3$. Now our problem is completely solved from the final tableau of the dual problem: furthermore, this solution is confirmed by our geometry.

The example problem just completed shows one reason why we study the dual problem; if an original minimizing problem has all constraints of the form $\geq$, then the dual problem becomes a standard maximum-type problem that can be solved by using only the pivoting rules of the simplex algorithm thereby dispensing with the Crown pivoting rules. We show one more example of this type before turning to more general minimizing and maximizing problems.

EXAMPLE 1
Using the Dual to Solve a Minimizing Problem

A company makes three types of pencil sharpeners—NoDull, SharpenIt, and Pin-Point. Production costs for each are, respectively, $7, $12, and $10. Contracts for the upcoming month require that the number of NoDull models plus the number of SharpenIt models be at least 200, the number of SharpenIt models be at least 100, and the sum of all three models must be at least 500. How many of each model should be scheduled for production next month to minimize production costs?

Solution   Let x, y, and z represent the number of NoDull, SharpenIt, and PinPoint models, respectively, to be produced. The problem then can be stated as:

$$
\begin{aligned}
\text{Minimize} \quad & C = 7x + 12y + 10z, \\
\text{subject to} \quad & x + y \qquad\quad \geq 200 \\
& \qquad\; y \qquad\quad \geq 100 \\
& x + y + z \geq 500 \\
& x \geq 0,\, y \geq 0,\, z \geq 0
\end{aligned}
$$

The primal and dual matrices are derived next.

$$
P = \left[
\begin{array}{ccc|c}
1 & 1 & 0 & 200 \\
0 & 1 & 0 & 100 \\
1 & 1 & 1 & 500 \\
\hline
7 & 12 & 10 & 0
\end{array}
\right]
; P^T = D = \left[
\begin{array}{ccc|c}
1 & 0 & 1 & 7 \\
1 & 1 & 1 & 12 \\
0 & 0 & 1 & 10 \\
\hline
200 & 100 & 500 & 0
\end{array}
\right].
$$

The dual problem can be stated from D, as shown in the following:

$$\text{Maximize}\quad d = 200u + 100v + 500w,$$
$$\text{subject to}\quad u \qquad\quad + w \leq 7$$
$$u + v + w \leq 12$$
$$w \leq 10$$
$$u \geq 0,\ v \geq 0,\ w \geq 0.$$

Inserting the three required slack variables and rearranging the objective function to be maximized, the initial simplex tableau for our dual problem is as shown next:

$$\begin{array}{ccccccc|c}
u & v & w & s_1 & s_2 & s_3 & d & \\
\hline
1 & 0 & \textcircled{1} & 1 & 0 & 0 & 0 & 7 \\
1 & 1 & 1 & 0 & 1 & 0 & 0 & 12 \\
0 & 0 & 1 & 0 & 0 & 1 & 0 & 10 \\
\hline
-200 & -100 & -500 & 0 & 0 & 0 & 1 & 0
\end{array}$$

Our first pivot element is the circled 1 in the initial tableau. The result of that pivot operation and the next pivot element identified are shown next.

$$\begin{array}{ccccccc|c}
u & v & w & s_1 & s_2 & s_3 & d & \\
\hline
1 & 0 & 1 & 1 & 0 & 0 & 0 & 7 \\
0 & \textcircled{1} & 0 & -1 & 1 & 0 & 0 & 5 \\
-1 & 0 & 0 & -1 & 0 & 1 & 0 & 3 \\
\hline
300 & -100 & 0 & 500 & 0 & 0 & 1 & 3500
\end{array}$$

The negative indicator in the objective-function row tells us that at least one more pivot operation is needed. The next pivot element is circled above and the result of that pivot operation is as follows:

$$\begin{array}{ccccccc|c}
u & v & w & s_1 & s_2 & s_3 & d & \\
\hline
1 & 0 & 1 & 1 & 0 & 0 & 0 & 7 \\
0 & 1 & 0 & -1 & 1 & 0 & 0 & 5 \\
-1 & 0 & 0 & -1 & 0 & 1 & 0 & 3 \\
\hline
300 & 0 & 0 & 400 & 100 & 0 & 1 & 4000
\end{array}$$

The lack of negative indicators in this tableau shows the maximum for d has now been achieved and is 4000. Therefore, the minimum production costs will be $4000, and this will occur when 400 of the NoDull model, 100 of the SharpenIt model, and none of the PinPoint model are made. (These values are read from the objective-function row in the slack-variable columns.) ▬

Since we have the Crown pivoting rules, we can exploit the full nature of the dual problem. The next example is a much more general minimizing problem.

EXAMPLE 2
Using the Dual to Solve a Minimizing Problem

Solve this linear programming problem through the use of its dual problem:

$$\text{Minimize}\quad f = 2x - 3y,$$
$$\text{subject to}\quad y \leq 3$$
$$x \leq 4$$
$$x + y \geq 5$$
$$x \geq 0,\ y \geq 0$$

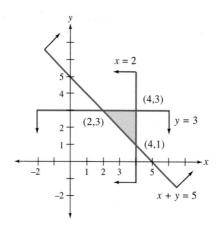

FIGURE 32
The Feasible Region for Example 2

Solution   The feasible region is shown in Figure 32. The minimum value of f is -5 at the point $x = 2$, $y = 3$.

To obtain the dual problem, we first rewrite the main constraints to the problem as

$$-y \geq -3$$
$$-x \qquad \geq -4$$
$$x + y \geq 5$$

so that the primal matrix becomes

$$P = \begin{bmatrix} 0 & -1 & -3 \\ -1 & 0 & -4 \\ 1 & 1 & 5 \\ \hline 2 & -3 & 0 \end{bmatrix}$$

and

$$P^T = \begin{bmatrix} 0 & -1 & 1 & 2 \\ -1 & 0 & 1 & -3 \\ \hline -3 & -4 & 5 & 0 \end{bmatrix} = D.$$

The dual problem may now be written as

$$\text{Maximize} \quad d = -3u + -4v + 5w,$$
$$\text{subject to} \qquad -v + w \leq 2$$
$$-u \qquad + w \leq -3$$
$$u \geq 0, v \geq 0, w \geq 0$$

The initial tableau for the dual problem is stated next.

u	v	w	s_1	s_2	d	
0	−1	1	1	0	0	2
⊝−1	0	1	0	1	0	−3
3	4	−5	0	0	1	0

The Crown pivoting rules are needed first to remove the -3 in the rightmost column. According to those rules, the first pivot element will be -1 in the second-row, first-column position. After multiplying the second row by -1 and pivoting on the resulting 1 in the second-row, first-column position, the resulting tableau is

$$
\begin{array}{cccccc}
u & v & w & s_1 & s_2 & d \\
\end{array}
$$
$$
\left[\begin{array}{cccccc|c}
0 & -1 & \boxed{1} & 1 & 0 & 0 & 2 \\
1 & 0 & -1 & 0 & -1 & 0 & 3 \\
\hline
0 & 4 & -2 & 0 & 3 & 1 & -9
\end{array}\right].
$$

The negative above the horizontal bar in the rightmost column has now been removed, so we return to the simplex algorithm rules. The most negative indicator is -2 and the only positive number in that column is the 1 in the first-row, third-column position. That becomes the new pivot position. Upon completion of that pivot, the next and final tableau becomes

$$
\begin{array}{cccccc}
u & v & w & s_1 & s_2 & d \\
\end{array}
$$
$$
\left[\begin{array}{cccccc|c}
0 & -1 & 1 & 1 & 0 & 0 & 2 \\
1 & -1 & 0 & 1 & -1 & 0 & 5 \\
\hline
0 & 2 & 0 & 2 & 3 & 1 & -5
\end{array}\right].
$$

The solution to the maximization of d has now been completed. The maximum value for d is -5 and that occurs when $u = 5, v = 0$, and $w = 2$. According to the Duality Theorem, the minimum value of f in our original problem is also -5 and occurs when $x = 2$ (from the bottom of the s_1 column) and $y = 3$ (from the bottom of the s_2 column). ■

We now turn to the dual of a problem whose objective function is to be maximized.

EXAMPLE 3
The Dual of a
Maximization Problem

Find and solve the dual problem to this linear programming problem:

$$
\begin{aligned}
\text{Maximize} \quad & f = 2x + 3y, \\
\text{subject to} \quad & x + y \le 6 \\
& 2x + y \le 8 \\
& x \ge 0, y \ge 0
\end{aligned}
$$

Solution Since the original problem has all main constraints in the $\le$ form, it is also the primal problem. The primal matrix may now be constructed as follows:

$$
P = \left[\begin{array}{cc|c}
1 & 1 & 6 \\
2 & 1 & 8 \\
\hline
2 & 3 & 0
\end{array}\right].
$$

The transpose of this matrix, the dual matrix, then becomes

$$
P^T = \left[\begin{array}{cc|c}
1 & 2 & 2 \\
1 & 1 & 3 \\
\hline
6 & 8 & 0
\end{array}\right] = D.
$$

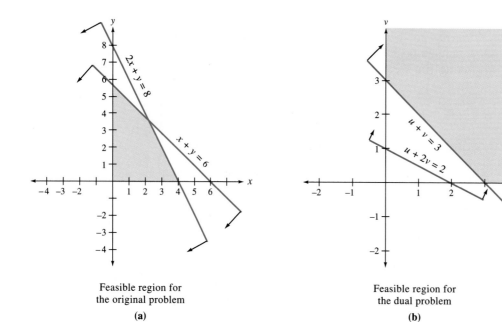

Feasible region for
the original problem

(a)

Feasible region for
the dual problem

(b)

FIGURE 33

From this matrix the dual problem may be stated as

$$\text{Minimize} \quad d = 6u + 8v,$$
$$\text{subject to} \quad u + 2v \geq 2$$
$$u + v \geq 3$$
$$u \geq 0, v \geq 0$$

The feasible region for the original and dual problems are shown in Figure 33. The maximum value of f is 18 and occurs at the point $x = 0$, $y = 6$. The minimum value of d is also 18 and occurs at the point $u = 3$, $v = 0$.

In order to solve the resulting minimum problem, we rewrite it as a maximum-type problem:

$$\text{Maximize} \quad g = -6u - 8v,$$
$$\text{subject to} \quad -u - 2v \leq -2$$
$$-u - v \leq -3$$
$$u \geq 0, v \geq 0$$

These inequalities lead to the initial tableau.

$$
\begin{array}{ccccc}
u & v & s_1 & s_2 & g \\
\end{array}
$$
$$
\left[
\begin{array}{ccccc|c}
-1 & -2 & 1 & 0 & 0 & -2 \\
-1 & \boxed{-1} & 0 & 1 & 0 & -3 \\
\hline
6 & 8 & 0 & 0 & 1 & 0 \\
\end{array}
\right]
$$

The Crown pivoting rules are needed first to remove the negative numbers in the rightmost column. Those rules imply that the circled -1 in the second-row, second-column position is the first pivot element. Upon completion of the pivot operation, the tableau has this appearance:

$$
\begin{array}{ccccc}
u & v & s_1 & s_2 & g \\
\end{array}
$$

$$
\left[
\begin{array}{ccccc|c}
1 & 0 & 1 & -2 & 0 & 4 \\
\boxed{1} & 1 & 0 & -1 & 0 & 3 \\
\hline
-2 & 0 & 0 & 8 & 1 & -24
\end{array}
\right]
$$

The negatives above the horizontal bar have both been removed, but there is a negative indicator in the last row indicating that at least one more pivot is needed. The smallest-quotient rule applied to the first column shows that the next pivot element is the 1 in the second-row, first-column position. Pivoting there gives this tableau:

$$
\begin{array}{ccccc}
u & v & s_1 & s_2 & g \\
\end{array}
$$

$$
\left[
\begin{array}{ccccc|c}
0 & -1 & 1 & -1 & 0 & 1 \\
1 & 1 & 0 & -1 & 0 & 3 \\
\hline
0 & 2 & 0 & 6 & 1 & -18
\end{array}
\right]
$$

No further pivots are needed. From this matrix we find that the maximum of g is -18; hence, the minimum of d is 18 and occurs when $u = 3$ and $v = 0$. Furthermore, the Duality Theorem tells us that the maximum value of f in the original problem is also 18 and occurs when $x = 0$ (from the bottom of the s_1 column) and $y = 6$ (from the bottom of the s_2 column). ■

For the particular problem in Example 3, there was nothing to be gained by using the dual to a maximization problem other than solving two linear programming problems at once. Indeed, we can solve any linear programming problem by using the Crown pivoting rules, if necessary, without the use of the dual problem. The one big computational edge the dual problem has to offer is that, in certain cases, it reduces the number of constraints upon which the pivoting rules must be applied to reach a solution when compared to the original problem. For instance, if a linear programming problem has 10 variables and 20 constraints, then the dual will have only 10 constraints.

EXERCISES   4.7

In each of the Exercises 1 through 8, complete the following steps:
(a) Find the primal matrix.
(b) Find the dual matrix.
(c) Formulate the dual problem.
(d) Graph the feasible sets for both the primal and the dual problems, and find their solutions graphically.
(e) Use the simplex method to solve the dual problem, and, from the final tableau, find the solution to the primal and dual problems and the points at which they occur.

1. Minimize $f = 2x + y$,
subject to $\quad x + y \geq 4$
$\qquad\qquad 3x + y \geq 8$
$\qquad\qquad x \geq 0, y \geq 0$

2. Minimize $f = 3x + 4y$,
subject to $\quad x + \ \ y \geq 6$
$\qquad\qquad x + 2y \geq 8$
$\qquad\qquad x \geq 0, y \geq 0$

3. Minimize $C = 3x + y$,
subject to $\qquad x + \ \ y \geq 10$
$\qquad\qquad 2x + 3y \geq 12$
$\qquad\qquad x \geq 0, y \geq 0$

4. Minimize $C = x + 2y$,
subject to $\quad 2x + \ \ y \geq 20$
$\qquad\qquad 3x + 4y \geq 60$
$\qquad\qquad x \geq 0, y \geq 0$

5. Minimize $f = 2x - 3y$,
subject to $\quad 4x + \ \ y \geq 18$
$\qquad\qquad x + 2y \leq 8$
$\qquad\qquad x \geq 0, y \geq 0$

6. Minimize $f = 3x + 4y$,
   subject to $3x + y \geq 9$
   $x + y \leq 6$
   $x \geq 0, y \geq 0$

7. Maximize $f = 2x + y$,
   subject to $x + y \leq 8$
   $x \leq 1$
   $x \geq 0, y \geq 0$

8. Maximize $f = x - 2y$,
   subject to $x + y \geq 6$
   $x + y \leq 8$
   $x \geq 0\ y \geq 0$

In Exercises 9 through 16, solve the original problem by using the dual problem. Write the point at which the minimum or maximum of the original problem occurs.

9. Minimize $f = x + 2y$,
   subject to $x + y \geq 8$
   $2x + y \geq 12$
   $x \geq 1, y \geq 0$

10. Minimize $f = 2x + 3y$,
    subject to $x + y \geq 10$
    $4x + y \leq 16$
    $x \geq 0, y \geq 0$

11. Minimize $f = 3x + y + 2z$,
    subject to $x + y \qquad \geq 6$
    $2x + \qquad z \geq 4$
    $y + 2z \geq 5$
    $x \geq 0, y \geq 0, z \geq 0$

12. Minimize $f = 2x + 5y$,
    subject to $2x + y \geq 9$
    $4x + 3y \leq 23$
    $x \qquad \leq 8$
    $x \geq 0, y \geq 0$

13. Minimize $f = x + y + 3z$,
    subject to $x + y + z \geq 8$
    $y \qquad \leq 4$
    $x \qquad + z \geq 6$
    $x \geq 0, y \geq 0, z \geq 0$

14. Minimize $C = 2x + 3y + z$,
    subject to $x + y + z \geq 10$
    $x + 2y \qquad \geq 4$
    $x \geq 0, y \geq 0, z \geq 0$

15. Maximize $f = 2x + y - z$,
    subject to $x + \qquad z \geq 6$
    $x + y \qquad \geq 4$
    $x + y + z \leq 10$
    $x \geq 0, y \geq 0, z \geq 0$

16. Maximize $f = x + 2y + z$,
    subject to $x + y + z \leq 8$
    $x \qquad \geq 1$
    $y + z \geq 3$
    $x \geq 0, y \geq 0, z \geq 0$

Use the dual problem to solve the remaining exercises. Some of the exercises you have been asked to set up or solve graphically in earlier sections.

17. **Health** A serving of Food A contains four units of vitamin A, five units of vitamin B-complex, and two mg of fat per serving. A serving of Food B contains six units of vitamin A, two units of vitamin B-complex, and three mg of fat per serving. If a mixture of these two foods is to have at least ten units of vitamin A and at least ten units of vitamin B-complex, how many servings of each food should be used to minimize the total mg of fat?

18. **Manufacturing** A firm makes a standard and deluxe model of electric skillet. Each standard model costs $25 to make and each deluxe model costs $40 to make. The firm must make at least 100 standard models and at least 75 deluxe models. The firm sells each standard model for $30 and each deluxe model for $75, and they must produce a total revenue of at least $2250 from sales of these two models of skillet. How many of each model should be made and sold to minimize costs?

19. **Road Paving** A road-paving firm has on hand three types of paving materials. Each barrel of type A contains two gallons of carbon-black and two gallons of thinning agent and costs $5. Each barrel of type B contains four gallons of carbon-black and one gallon of thinning agent and costs $3. Each barrel of type C contains three gallons of carbon-black and one gallon of thinning agent and costs $4. The firm needs to fill an order for a final mixture that must contain at least 12 gallons of carbon-black and at least 6 gallons of thinning agent. How many barrels of each type of paving material should be used to fill this order at minimum expense?

20. **Agriculture** Agricultural planners in a certain country are planning next year's rice and wheat acreages. They estimate that each acre of rice planted will yield 30 bushels and each acre of wheat planted will yield 20 bushels. They want to raise at least 4 million bushels of rice and at least 5 million bushels of wheat. They want to plant at least 1 million acres of these two crops combined. For planting, harvesting, and transporting the grains, the planners estimate that it will take six workers for each acre of rice planted and three workers for each acre of wheat planted. How can the planners meet the requirements set forth and use the minimum number of workers?

21. **Order Fulfillment** The Hi-Fi Record Store is about to place an order for cassette tapes and compact discs. The distributor it orders from requires that an order contain at least 250 items. The prices Hi-Fi must pay are $12 for each cassette and $16 for each compact disc. The distributor also requires that at least 30% of any order be compact discs. How many cassettes and how many compact discs should Hi-Fi order so that its ordering costs will be kept to a minimum?

22. *Plane Charter*   A charter plane service contracts with a retirement group to provide fights for New York to Florida for at least 500 first-class, 1000 tourist-class, and 1500 economy-class persons in their organization. The charter company has two types of planes. Type A will carry 50 first-class, 60 tourist-class, and 100 economy-class passengers, and costs $12,000 to make the flight. Type B will carry 40 first-class, 30 tourist-class, and 80 economy-class passengers, and costs $10,000 to make the flight. How many of each type of plane should be used to minimize flight costs?

23. *Manufacturing*   A firm makes three models of microwave oven: standard, deluxe, and super deluxe. It has contracts that call for making at least 80 standard models, at least 50 deluxe models, and at least 90 super deluxe models. In addition, the sum of all three models produced must be at least 300. If the cost is $100 for each standard model, $110 for each deluxe model, and $120 for each super deluxe model, how many of each should be made each week to minimize costs?

24. *Advertising*   An advertising agency tells Fun Foods that each television commercial will cost $150,000 and reach 8 million people, each series of radio commercials will cost $30,000 and reach 2 million people, and each series of newspaper ads will cost $26,000 and reach 1.8 million people. Fun Foods wants to run at least three television commercials, at least two series of radio commercials, and at least three series of newspaper ads. How many of each type of ad should the company run while minimizing advertising costs if they want to reach at least 80 million people?

Problem For Exploration

1. Make up a linear programming problem for which the solution to the dual problem occurs at a point whose coordinates are the same as the coordinates of the point that is the solution to the original problem.

C HAPTER   **4**   REFLECTIONS

CONCEPTS

The problem of finding the maximum or minimum value of a linear function subject to particular constraints occurs frequently in industrial, business, military, and agricultural settings. These *linear programming* problems can be solved *graphically* by using the *corner points* of the *feasible region* when only two variables are involved. The *simplex algorithm* provides a systematic, *algebraic method* (in any number of variables) for solving linear programming problems that can be reduced to the *standard maximum-type problem*. The *Crown pivoting rules,* along with the *smallest-quotient rule,* provide a convenient and efficient scheme for solving all types of standard and nonstandard linear programming problems. The *dual problem* can be efficiently used to solve particular minimization problems.

COMPUTATION/CALCULATION

1. Write the system of inequalities that would determine the given feasible region shown in the diagram. Use this information to solve Problem 2.

2. Find all the corner points for the given feasible region.

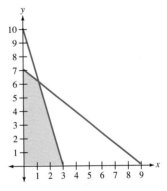

3. Graph the feasible region for the given system of inequalities.

$$\begin{cases} 5x + 2y \le 20 \\ x + 4y \le 10 \\ x \ge 0, y \ge 0 \end{cases}$$

4. Find each corner point of the feasible region determined by the given inequalities in the previous problem.

5. Graphically solve the linear programming problem:

$$\text{Minimize} \quad C = 3x + 2y,$$
$$\text{subject to} \quad 2x + 4y \geq 6$$
$$x + y \geq 2$$
$$x \geq 0, \ y \geq 0$$

Use the following problem as the basis for answering Exercises 6 through 9:

$$\text{Maximize} \quad P = 4x + 3y,$$
$$\text{subject to} \quad x + y \leq 5$$
$$3x + 2y \leq 12$$
$$x \geq 0, \ y \geq 0$$

6. For the given linear programming problem, insert appropriate slack variables in the structural constraints, and form the system of slack equations.

7. Graph the feasible region of this problem and label the intersection of each pair of boundary lines with four coordinates.

8. Form the initial augmented matrix, suggested by Exercise 6 from the slack equations, and find the pivot element in the x-column so that pivoting on this element will give a basic solution representing a corner point.

9. Solve the linear programming problem given by the simplex algorithm.

10. Use both the dual-problem method and the simplex algorithm to solve the problem:

$$\text{Minimize} \quad C = 3x + y,$$
$$\text{subject to} \quad x + y \geq 8$$
$$2x + y \geq 12$$
$$x \geq 0, \ y \geq 0.$$

11. Use the Crown pivoting rules to solve the problem:

$$\text{Maximize} \quad f = 2x - y + z,$$
$$\text{subject to} \quad x + 4y + z \leq 12$$
$$y + 4z \leq 8$$
$$4x + z \geq 4$$
$$x \geq 0, \ y \geq 0, \ z \geq 0$$

CONNECTIONS

12. A department store makes two types of dress shirts. The Gent brand costs $6 to make and sells for $12. The King brand costs $8 to make and sells for $18. The store needs to make at least 150 of the Gent brand and at least 100 of the King brand. If the store must have a total revenue of at least $9000, how many of each brand should be made and sold to minimize costs?

13. The Fibonacci Math Club is planning a fund-raising cookie sale. It plans to bake sugar cookies, peanut butter cookies, and pecan sandies in batches of three dozen cookies. Each batch of sugar cookies uses 2 eggs, 3 cups of flour, and $1\frac{1}{8}$ cups of sugar. Batches of peanut butter cookies use one egg, $1\frac{1}{4}$ cups of flour, and 1 cup of sugar, whereas pecan sandies use 2 cups of flour, $\frac{1}{3}$ cup of sugar, and no eggs. Ten dozen eggs, 300 cups of flour, and 200 cups of sugar have been donated to the club. All other ingredients are on hand in sufficient quantities. If a dozen sugar cookies sell for $1.50, peanut butter cookies for $1.75, and pecan sandies for $2.25, how many batches of each should the club bake to maximize their income?

COMMUNICATION

14. Explain the meaning of *feasible region* for a system of linear inequalities.

15. Describe the appropriate steps for graphically solving a linear programming problem.

16. Indicate what conditions would be required for a linear programming problem to be a *standard maximum-type* problem.

17. Describe the *smallest-quotient rule*.

18. List the major steps for applying the *simplex algorithm*.

19. Describe the rules to follow when using the Crown pivoting rules.

COOPERATIVE LEARNING

20. Write a complete word problem for a linear programming application that fits the following given feasible region.

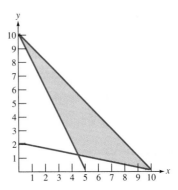

21. Write a complete word problem for a linear programming application that fits the following given system.

$$x + 2y + s_1 \qquad\qquad = 12$$
$$4x + 5y \quad + s_2 \qquad\quad = 20$$
$$5x + \quad y \qquad + s_3 \quad = 10$$
$$-3x - 4y \qquad\qquad + f = 0$$

CRITICAL THINKING

22. Make up an objective function f and a feasible region so that when minimizing f by use of the simplex algorithm, successive pivots always give a positive number in the lower right-hand corner of the tableau. Draw a conclusion about a positive number versus a negative number in that position when maximizing as well as minimizing objective functions.

COMPUTER/CALCULATOR

The program LPGraph, developed for this textbook, will help you graphically solve linear programming problems by shading in the feasible region and graphing the objective function ($f = c$) for various selections of c that you choose.

23. Solve this problem with the aid of LPGraph or a similar program:

$$\text{Minimize} \quad f = 2x + y,$$
$$\text{subject to} \quad x + y \geq 6$$
$$y \geq 1$$
$$y \leq 3$$
$$x \qquad \leq 6$$

After entering $f = 2x + y = c$ for $c = 4, 6, 7, 8, 10,$ and 12, the screen for LPGraph appears as follows:

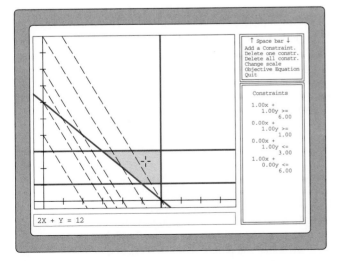

24. (a) Use a program such as LinProg to solve the following problem:

$$\text{Maximize} \quad f = 3x + 2y,$$
$$\text{subject to} \quad x + y \leq 8$$
$$3x + y \leq 12$$
$$x \geq 0, y \geq 0$$

(b) Use a program such as LProg to solve this same problem.

25. Use a program such as LProg and a linear programming program for a graphing calculator to solve Exercise 23 of Section 4.6.

Cumulative Review

26. The formula $F = \frac{9}{5}C + 32$ can be used to change temperatures given in degrees Celsius to degrees Fahrenheit. Find all the temperatures, if any, that give the same number on both temperature scales.

27. A message was coded using matrix

$$A = \begin{bmatrix} 1 & 2 \\ 0 & 2 \end{bmatrix}$$

and the numbers 1 through 26 for the letters a through z and the number 30 for spaces. Find A^{-1}, and use it to decode the message: 25, 80, 21, 102, 7, 24, 20, 100, 1, 30, 30, 62.

28. The given parts matrix represents the number of parts used internally to make another part for subassembly in a production process. Construct the parts arrow diagram for this subassembly process.

$$\begin{array}{c} \\ p_1 \\ p_2 \\ p_3 \\ p_4 \end{array} \begin{array}{cccc} p_1 & p_2 & p_3 & p_4 \\ \begin{bmatrix} 0 & 1 & 2 & 3 \\ 0 & 0 & 2 & 1 \\ 0 & 0 & 0 & 1 \\ 0 & 0 & 0 & 0 \end{bmatrix} \end{array}$$

Solve each system, using the pivoting process.

29.
$$3x \qquad - \quad z \qquad\qquad = 0$$
$$2y + \quad z \qquad\qquad = 0$$
$$x + \quad y + \quad z + \quad w = 6$$
$$2x + 3y + 5z - 10w = 15$$

30.
$$x + \quad y + \quad z = 0$$
$$3x \qquad - 2z = 0$$
$$5x + 2y \qquad = 0$$
$$3y + 5z = 0$$

CHAPTER 4   SAMPLE TEST ITEMS

1. Graph and shade the feasible region determined by the inequalities $x + 2y \leq 6$, $5x + 2y \leq 10$, $x \geq 0$, $y \geq 0$.

2. Find all the corner points for the feasible region determined by $x + 5y \leq 13$, $2x + y \leq 8$, $x \geq 0$, $y \geq 0$.

3. Write all the inequalities that describe the following data, and graph the feasible region: You are to select the number of each type of objective question you will answer for a test containing true-false questions worth three points each and multiple-choice questions worth five points each. You must answer at least ten true-false and at least six multiple-choice questions. Because of time constraints, you can answer at most 30 true-false and at most 20 multiple-choice questions. You are allowed to answer at most 36 questions in total. How many of each type of question should you answer to maximize your potential score? What is this maximum score?

4. If the objective function is $f = x + 2y$, graph the lines $x + 2y = c$ (for $c = 6, 11, 14, 18,$ and 22) superimposed on the given feasible region. Based on these lines, where do the maximum and minimum values of f occur?

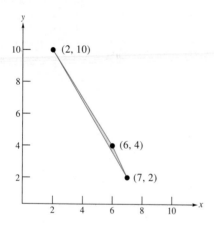

Solve each of the following linear programming problems graphically.

5. Maximize $f = 2x + 5y$,
   subject to $7x + y \leq 21$
   $2x + 9y \leq 18$
   $x \geq 0, y \geq 0$

6. Minimize $g = 4x - y$,
   subject to $3x + y \geq 6$
   $3x + 4y \geq 12$
   $x \geq 0, y \geq 0$

The initial augmented matrix shown represents the slack equations for a linear programming problem in standard form.

$$\begin{array}{cccc} x & y & s_1 & s_2 \\ \end{array}$$
$$\left[\begin{array}{cccc|c} 2 & 1 & 1 & 0 & 8 \\ 1 & 5 & 0 & 1 & 10 \end{array}\right]$$

7. Write the set of all constraints for this problem.

8. Perform one pivot on the given matrix above that will give a basic solution representing the point $(4, 0)$.

The matrix shown below represents a simplex tableau after a sequence of pivot operations.

$$\begin{array}{cccccc} x & y & s_1 & s_2 & s_3 & f \\ \end{array}$$
$$\left[\begin{array}{cccccc|c} 0 & 2 & 0 & 1 & 5 & 0 & 4 \\ 0 & -1 & 1 & 0 & 9 & 0 & 8 \\ 1 & 4 & 0 & 0 & 3 & 0 & 2 \\ 0 & -2 & 0 & 0 & 4 & 1 & 35 \end{array}\right]$$

9. Does this tableau show that a maximum for the objective function has been reached?

10. Find the value of f at this stage of the algorithm and the point at which this value occurs.

11. Name all the constraints for which a surplus exists, and give the amount of surplus for each.

For Exercises 12 and 13, a linear programming problem requires you to

$$\text{Maximize} \quad f = 2x + y,$$
$$\text{subject to} \quad x + y \leq 6$$
$$y \leq 4$$
$$x \geq 0, y \geq 0$$

12. Graph the feasible region, and find the values of x, y, and each slack variable at each corner point.

13. Solve this linear programming problem by using the simplex algorithm.

For Exercises 14 and 15, a linear programming problem requires you to

$$\text{Minimize} \quad f = 2x + 3y,$$
$$\text{subject to} \quad x + y \geq 3$$
$$x + 2y \geq 4$$
$$x \geq 1, y \geq 0$$

14. Formulate the dual problem.

15. Use the simplex algorithm to solve the dual problem, and find the solution to the primal and dual problems, including the points at which it occurs.

For Exercises 16 through 18, a linear programming problem requires you to

$$\text{Minimize} \quad f = 3x + y,$$
$$\text{subject to} \quad x + y \geq 7$$
$$2x + y \geq 10$$
$$x \geq 0, y \geq 0$$

16. Sketch the feasible region, and find the coordinates of the corner points.

17. Solve this problem by using the Crown pivoting rules.

18. Check the solution to this problem by evaluating f at each corner point of the feasible region.

19. Use the Crown pivoting rules to

$$\text{Maximize} \quad f = 2x + 3y + 5z,$$
$$\text{subject to} \quad x + y + 2z \leq 6$$
$$2y + z \geq 4$$
$$2x + z \leq 4$$
$$x \geq 0, y \geq 0, z \geq 0$$

20. To minimize costs, a college mathematics department is analyzing its faculty salaries for professors and lecturers. The average annual salary is $40,000 for professors and $15,000 for lecturers. Space allocations permit no more than 50 total faculty, but at least 25 must be professors, and no more than 15 can be lecturers. To teach the necessary classes requires at least 35 faculty members. How many professors and lecturers should be employed to minimize total salary costs?

Career Uses of Mathematics

Mathematics has a profound impact on many industries—from biotechnology to fast food. Consider, for example, the effect of a mathematical technique called linear programming on the airline industry. Linear programming can be used to determine patterns within very complex problems and to propose a number of solutions to the problem.

Airlines are required to limit their flight crew members' working time and impose other scheduling restraints. As a result, airlines frequently have to give flight crew employees "penalty" pay for working overtime (for instance, waiting for another plane that has been delayed). In an effort to cut the amount of penalty pay by improving scheduling, operations planners at American Airlines use linear programming software to make up their monthly flight schedules. This allows the airlines to reduce the amount of time crews spend waiting around in airports and to schedule the use of bigger planes for the most popular routes. With all the relevant data concerning flight crews' home cities and working time and route popularity at various times of the day, the linear programming software can assign crews and planes in such a way to minimize crew overtime and maximize route potential.

Mechanical failures and unpredictable weather still cause delays and cancellations, but linear programming has enabled American to reduce penalty payments from 15 percent of payroll in 1980 to 4 percent in 1990.

SOURCE: Jon Van, "Cutting Costs a Matter of Math: Advances in Problem-Solving Can Save Millions," *Chicago Tribune*, Monday, 30 July 1990, sec. N.

5

Logic, Sets, and Counting Techniques

The eight blood types are A⁺, A⁻, B⁺, B⁻, AB⁺, AB⁻, O⁺, and O⁻. In how many ways is it possible to get five units of blood from a blood bank so that all five have different blood types and each group of five has at least one of the two O types? (See Example 3, Section 5.7)

C HAPTER PREVIEW

The main objective of this chapter is to learn some specialized counting techniques that lay the foundation for later work in probability and statistics. These techniques include the multiplication principle, combinations, and permutations, along with a few others. The last exercise set contains a random collection of all types of counting problems discussed. To support these efforts, the chapter begins with a brief study of the logical use of the words *and, or,* and *not* in a mathematical sense, as well as what it means for two sentences to be logically equivalent. From there we move to the closely related subject of sets and then, finally, to the counting techniques.

CALCULATOR/COMPUTER OPPORTUNITIES FOR THIS CHAPTER

Calculators can be used effectively to compute factorials, combinations, and permutations in this chapter. Common calculator keys for these calculations are an exclamation point (!) for factorials, *nCr* for combinations, and *nPr* for permutations.

The computer disk MasterIt has programs named Count I and Count II, which will check your counting ability at both basic and more advanced levels.

SECTION 5.1 ## Logic

To comprehend successfully the concepts and exercises in our upcoming work in counting, probability, and statistics, you must be able to interpret the mathematical meanings of key words such as *and, or, not, at least,* and *at most.* For example, suppose a survey of college freshmen and sophomores gave this data:

	Taking English	Taking Geology
Freshmen	20	15
Sophomores	18	10

Do you know how many of these students are described by each of the two sentences that follow?

The student is a freshman *and* is taking geology. (There are 15.)

The student is a sophomore *or* is taking geology. (There are 43.)

These key words are also used in defining operations between sets and in rephrasing statements in a logically equivalent way to make them more useful. Who sets the ground rules for the use of these words? How do we know when the rephrasing of a statement is logically equivalent to the original statement? The answer to these questions is found in the subject of logic, and we will use that viewpoint to first obtain a basic understanding of the key words needed to carry us through counting techniques (and beyond). The subject of logic is a vast one, but we will confine our attention only to the ideas directly needed for our mathematical developments. We begin by exploring the concept of statements and how they are combined to form compound sentences.

By a **statement,** we mean a declarative sentence that is either true or false, and not both, or an open sentence that can be made true or false by substituting for the pronoun or variable in the sentence. Some examples of statements are

There are five positions on a basketball team. (True)

Every student wears glasses. (False)

December has 31 days. (True)

$2 + 3 = 7$. (False)

$x + 3 = 5$. (True for $x = 2$; false otherwise.)

He is the President of the United States. (True for one person; false for all others.)

From such statements, we may form compound sentences by the use of connectives. Two important and often-used connectives are *and* and *or*. To illustrate, let

p: The student is a freshman.

and

q: The student is taking geology.

be two statements. Then the **conjunction** of *p* and *q* is constructed as follows: The student is a freshman *and* is taking geology.

The special logical symbolism used for the conjunction is shown next.

Conjunction of *p* and *q*: $p \wedge q$ (*p* **and** *q*).

A compound sentence using the *or* connective is known as the **disjunction** of *p* and *q* and is formed in the following manner:

The student is a freshman *or* is taking geology.

The special logical symbolism for the disjunction is as follows.

Disjunction of *p* and *q*: $p \vee q$ (*p* **or** *q*).

The word *not* is also considered a connective and, for a statement *p*, is called the **negation** of *p*. It works just as you think it should:

p: The student is a freshman.

The negation of *p*: The student is *not* a freshman.

There is also a logical symbolism for the negation.

Negation of *p*: $\sim p$ (**not** *p*).

The three connectives just introduced—*and, or,* and *not*—serve as building blocks for most of our needs in this textbook, with the understanding that variations in the use of English may sometimes give alternative wording. For example, the use of "but . . . not" has the same meaning as "and . . . not." (See Example 2(d) below.) These connectives, and others, form a logical basis for a formal mathematical language.

EXAMPLE 1
Illustrating Connectives
and Symbols

Consider the statements

p: The shirt was made by the Ace Company.

q: The shirt has a flaw.

Then

$p \wedge q$: The shirt was made by Ace and has a flaw.

$p \vee q$: The shirt was made by Ace or has a flaw.

$\sim p$: The shirt was not made by Ace.

$\sim q$: The shirt does not have a flaw.

$p \wedge \sim q$: The shirt was made by Ace and does not have a flaw.

$\sim p \wedge \sim q$: The shirt was not made by Ace and does not have a flaw.

$\sim p \vee q$: The shirt was not made by Ace or has a flaw. ▄▄

EXAMPLE 2

Translating Sentences to Symbols

Consider the statements

 p: The computer has a CD-ROM drive.

 q: The Internet is accessible.

We now show each of the following sentences in symbolic form:

The computer has a CD-ROM drive and the Internet is accessible: $p \wedge q$.

The Internet is not accessible: $\sim q$.

The computer has a CD-ROM drive or the Internet is accessible: $p \vee q$.

The computer has a CD-ROM drive, but the Internet is not accessible: $p \wedge \sim q$.

Because we are dealing only with statements that are true or false and not both, each such statement has exactly two possible *truth values,* T or F, and of these, only one may describe the statement. A simple display of truth values is shown next.

Statement	True value
All Fords are red.	F
3 + 5 = 8	T
Some Fords are red.	T

A sentence using any two such statements p and q will always have four possible pairs of truth values to consider:

p	q
T	T
T	F
F	T
F	F

The truth value for a particular sentence involving p and q and involving the connectives we have discussed will then be assigned for each of the possible combinations just listed. The basic building blocks for such a process are the truth values assigned to $p \wedge q$, $p \vee q$, and $\sim p$. The last one would be assigned truth values just as you think it should: If p is true, then $\sim p$ is false, while if p is false, then $\sim p$ is true.

p	$\sim p$	Examples of Assigning Truth Values to $\sim p$
T	F	p: 3 + 5 = 8
		$\sim p$: 3 + 5 ≠ 8
F	T	p: All Fords are red.
		$\sim p$: Not all Fords are red. *or*
		$\sim p$: Some Fords are not red. *or*
		$\sim p$: There is at least one Ford that is not red.

The accepted way of assigning truth values to $p \wedge q$ is shown next along with an example of each case.

p	q	$p \wedge q$	Examples of Assigning Truth Values to $p \wedge q$
T	T	T	Chicago is in Illinois, and Dallas is in Texas.
T	F	F	Chicago is in Illinois, and Dallas is in Alaska.
F	T	F	Chicago is in Wyoming, and Dallas is in Texas.
F	F	F	Chicago is in Wyoming, and Dallas is in Alaska.

That is, the entire sentence $p \wedge q$ (p and q) is true when *both* p and q are *true;* otherwise, the entire sentence is false. In other words, for the sentence $p \wedge q$ to be true, both p and q must be true simultaneously.

The accepted way of treating the disjunction $p \vee q$ is shown next, with accompanying examples to help us grasp why the truth values are assigned as they are.

p	q	$p \vee q$	Examples of Assigning Truth Values to $p \vee q$
T	T	T	Two is an even number or 5 is odd.
T	F	T	Two is an even number or 5 is even.
F	T	T	Two is odd or 5 is odd.
F	F	F	Two is odd or 5 is even.

That is, the entire sentence $p \vee q$ (p or q) is true when *either* p is true or q is true or *both* are true; otherwise the entire sentence is false. Stated differently, $p \vee q$ is true when *at least* one of the two components is true; otherwise, $p \vee q$ is false. Still differently, p or q is false if both p and q are false; otherwise, the entire sentence is true.

NOTE Throughout the remainder of this book, and, indeed, throughout all of mathematics and statistics, the words *and* and *or* are used as logical connectives with the truth values as defined above.

With these building blocks, we are now ready to construct a truth table for any sentence using p and q and the connectives $\wedge$, $\vee$, and $\sim$.

EXAMPLE 3
Constructing a Truth Table

Make a truth table to exhibit the truth values of $(p \wedge q) \vee \sim p$. Some intermediate steps will prove useful in reaching our final goal.

p	q	$p \wedge q$	$\sim p$	$(p \wedge q) \vee \sim p$
T	T	T	F	T
T	F	F	F	F
F	T	F	T	T
F	F	F	T	T

The table shows that the given sentence is false when p is true and q is false, but is true for the other possible combinations of T and F. ■

An important application of truth tables is in using them to help us decide when two sentences are logically equivalent.

Logically Equivalent Sentences

Two sentences are **logically equivalent** if they have identical truth values. The notation $p \equiv q$ is used to denote the fact that p and q are logically equivalent.

The usefulness of this concept lies in the fact that we can often translate a given sentence into one that is easier to understand.

EXAMPLE 4
Showing the Equivalence of Sentences

The statements p and $\sim(\sim p)$ are equivalent, as shown by the first and third columns of the following table.

p	$\sim p$	$\sim(\sim p)$
T	F	T
F	T	F

■

EXAMPLE 5 Show that $\sim(p \wedge q) \equiv \sim p \vee \sim q$ by the use of a truth table.

Showing the
Equivalence of
Statements

p	q	$p \wedge q$	$\sim(p \wedge q)$	$\sim p$	$\sim q$	$\sim p \vee \sim q$
T	T	T	F	F	F	F
T	F	F	T	F	T	T
F	T	F	T	T	F	T
F	F	F	T	T	T	T

The fourth and last columns are identical and therefore show the desired equivalence. Note that columns 3, 5, and 6 are inserted solely for the purpose of making the two desired columns easier to complete. ◼

Example 5 actually demonstrates the equivalence of the first of two very important equivalences known as **DeMorgan's laws of logic.**

> **DeMorgan's Laws of Logic**
> **1.** $\sim(p \wedge q) \equiv \sim p \vee \sim q$
> **2.** $\sim(p \vee q) \equiv \sim p \wedge \sim q$

HISTORICAL NOTE
Augustus DeMorgan (1806–1871)

Augustus DeMorgan was an English mathematician. He was among the first mathematicians who became aware of the structure in algebra—the commutative laws of addition and multiplication, the associative laws, the distributive law, and so on. He is credited with posing this problem: "I was x years old in the year x^2. When was I born?"

EXAMPLE 6 Each of the following uses one of DeMorgan's laws of logic to translate a given sentence into a logically equivalent one.

Using DeMorgan's
Laws

(a) What is the negation of the "The fax machine is busy, and the copier is jammed."

Solution "The fax machine is not busy, or the copier is not jammed."

(b) What is the negation of "The first ball is red, or the second ball is not green"?

Solution "The first ball is not red, and the second ball is green."

(c) What is the negation of "The shirt has a flaw and did not come from Smith's Department Store"?

Solution "The shirt does not have a flaw or came from Smith's." ◼

In previous chapters, we occasionally encountered the term *at least* in rather straightforward numerical settings. For example, if we say "at least 20 TV sets are to be made," then we know immediately that any number that is 20 or larger will satisfy the statement. In some situations that we will encounter in the future, however, a translation of *at least* into a logically equivalent statement using *or* and *and* connectives will be very helpful. The next and last example of this section illustrate what we mean.

EXAMPLE 7
Translating an At Least Statement

Translate the following sentence into one using the connectives *or* and *and:* From a random sample of six calculators, at least four are in good working order.

Solution   The given sentence means that from among these six calculators, four or more are in good working order. This, in turn, can be restated in terms of *or* as

"Four are in good working order, or five are in good working order, or six are in good working order."

To be given more specific, one last sentence will tell the full story:

"(Four are in good working order, and two are not), or (five are in good working order, and one is not), or (all six are in good working order)."

The last translation is the most specific and is also the most useful in counting and probability problems. Notice that with this breakdown, no two of the component parts separated by an *or* can happen simultaneously—more about this later. ■

EXAMPLE 8
Translating an At Most Statement

Translate the following sentence into one using connectives *or* and *and:* Among five students, at most two are taking psychology.

Solution   The given sentence means that none, one, or two of these students are taking psychology. To be more precise,

"None of the five are taking psychology, or one is taking psychology and four are not, or two are taking psychology and three are not." ■

EXERCISES   5.1

For the statements,

> p: The student is a male.
>
> q: The student is a sophomore.

write the indicated sentence in Exercises 1 through 6.

1. $p \vee q$ **2.** $\sim p \vee q$

3. $p \wedge q$ **4.** $\sim p \wedge \sim q$

5. $\sim p$ **6.** $p \wedge \sim q$

For the statements,

> p: The blouse is red.
>
> q: The belt is black.

write the indicated sentences in Exercises 7 through 11, making use of logically equivalent forms to simplify your answers, if possible.

7. $p \wedge \sim q$ **8.** $\sim(p \wedge q)$

9. $\sim(p \wedge \sim q)$ **10.** $\sim(p \vee q)$

11. $\sim(\sim p \vee \sim q)$

For the statements,

> p: The die shows an odd number.
>
> q: The die does not show a five.

write the sentences indicated in Exercises 12 through 16, making use of logically equivalent forms to simplify your answers, if possible.

12. $\sim p$ **13.** $\sim(p \vee q)$

14. $\sim p \vee q$ **15.** $\sim(p \wedge \sim q)$

16. $\sim(p \wedge q)$

For the statements,

> p: Ninety percent of all cataract surgeries are successful.
>
> q: Eighty percent of all kidney transplants are successful.

write each of the sentences in Exercises 17 through 20 in symbolic form.

17. Ninety percent of all cataract surgeries and 80% of all kidney transplants are successful.

18. Ten percent of all cataract surgeries fail, or 80% of all kidney transplants are successful.

19. Ten percent of all cataract surgeries fail, and 20% of all kidney transplants fail.

20. Ninety percent of all cataract surgeries are successful, and 20% of all kidney transplants are unsuccessful.

For the statements,

   p: The first number drawn is a three.

   q: The sum of the two numbers drawn is five.

write each of the sentences in Exercises 21 through 24 in symbolic form.

21. The first number drawn is a three, and the sum is not five.

22. The first number drawn is not a three.

23. The first number drawn is not a three, or the sum is five.

24. The first number drawn is not a three, and the sum of the numbers is five.

Make a truth table for each of the Exercises 25 through 30.

25. $p \vee \sim q$

26. $p \vee (q \wedge \sim p)$

27. $\sim p \vee \sim q$

28. $\sim p \vee (q \wedge p)$

29. $\sim(\sim p \vee q)$

30. $\sim p \wedge (\sim q \vee q)$

Use a truth table to show the equivalences in Exercises 31 through 33.

31. $\sim(p \vee q) \equiv \sim p \wedge \sim q$

32. $p \vee q \equiv q \vee p$

33. $p \wedge (q \vee r) \equiv (p \wedge q) \vee (p \wedge r)$

In Exercises 34 through 37, use truth tables to decide which of the pairs of sentences are equivalent.

34. $p \vee \sim q$: $\sim p \wedge q$

35. $p \wedge q$: $\sim p \vee (q \wedge p)$

36. $\sim p \wedge q$: $\sim(p \vee \sim q)$

37. $q \wedge (\sim p \vee q)$: $\sim(\sim p \vee q)$

Decide what truth value (T or F) to assign to each of the sentences in Exercises 38 through 46.

38. All dogs are cocker spaniels.

39. Some tapes play country music.

40. $2 + 3 = 5$, and January has 35 days.

41. $2 + 3 = 5$, or January has 35 days.

42. $2 + 3 \neq 7$, or Social Security numbers have nine digits.

43. Some college students drink coffee, and all college students own automobiles.

44. Some college students drink coffee, or all college presidents are female.

45. All Hondas are red, or there have been at least two male presidents of the United States.

46. All heart transplants are successful, and Oprah Winfrey's television show is watched by millions.

Write the negation of each of the sentences in Exercises 47 through 56.

47. At least one of the two marbles is red.

48. At least one of the two students is majoring in history.

49. At least two of the three students graduated with honors.

50. At least two of the three cars have an automatic transmission.

51. At most three of the four marbles are green.

52. At most one of the three students is on the basketball team.

53. The car is red or has a CD-player.

54. The modem is fast, or the scanner will reproduce color.

55. The blouse is not blue, and the belt is black.

56. The I.D. number begins with the digit 2, and the employee is not computer literate.

Write an equivalent statement to each of those in Exercises 57 through 62, using the connectives "and" and "or," similar to the last translation of Example 7.

57. In a group of five students, at least four are taking a mathematics course.

58. Among a collection of eight calculus books, at least five have computer support.

59. In a lot of six shirts, at least four have flaws.

60. Among a set of 20 light bulbs, at least 19 are good.

61. In a bowl of 26 marbles, at most three are red.

62. In a group of six students, at most two are male.

Problems For Exploration

1. A **tautology** *t* is a sentence whose only possible truth value is T. Give an example of a tautology in symbolic form (e.g., using $p, q, \vee, \wedge, \sim$), and verify it through the use of a truth table.

2. A **contradiction** *c* is a sentence whose only possible truth value is F. Give an example of a contradiction in symbolic form (e.g., using $p, q, \vee, \wedge, \sim$), and verify it by using a truth table.

3. In the algebra of logic, the tautology t and contradiction c play the roles of zero and one in regular algebra. Which plays the role of which? Why?

4. A sentence of the form "if p, then q" is called a **conditional** and is denoted by the symbol $\rightarrow$. A conditional $p \rightarrow q$ is defined to be false if p is true and q is false, otherwise it is true. Make a truth table for $\sim(p \land q) \rightarrow \sim q$.

5. The **contrapositive** of $p \rightarrow q$ is defined to be the conditional $(\sim q) \rightarrow (\sim p)$.

    (a) Make a truth table showing that a conditional and its contrapositive are equivalent.

    (b) List three mathematical theorems, and state the contrapositive for each.

SECTION 5.2 Sets

The language of sets provides a common way to express ideas in a universal notation that permeates all fields of mathematics. In addition, sets will provide us with a link to logic and a visual aid in expressing the thoughts of counting and probability.

A **set** is a well-defined collection of distinct objects that are not necessarily related. The term *well defined* means that some property, rule, or description will allow us to determine whether a given object belongs to the set. Some examples of well-defined sets are

> The set of people in the United States who voted in the 1996 presidential election.
>
> The set of letters in the Greek alphabet.
>
> The integers between 0 and 10, inclusive.

These sets are not well defined:

> The set of small numbers.
>
> The set of young people in Boston.

The objects that make up a set are called **elements** of the set, and we write $a \in S$ to denote the fact that the element a belongs to the set S. If a is not an element of S, we write $a \notin S$. If the elements in a set can be listed, braces $\{\ \}$ are usually used to enclose the elements. Sometimes, a set may have no elements at all, in which case, we call the set **empty**. The symbol for an empty set is $\emptyset$ (with *no* braces around it) or braces $\{\ \}$ with no symbol within.

EXAMPLE 1
Set Notation

(a) Let A denote the set of integers from 0 to 5, inclusive. Then $A = \{0, 1, 2, 3, 4, 5\}$, where $3 \in A$, but $8 \notin A$.

(b) Let B represent the set of positive integers. Then $B = \{1, 2, 3, 4, \ldots\}$ is a common way of expressing a set that has a definite pattern but cannot be listed in its entirety. Another way is to write $B = \{n : n \text{ is a positive integer}\}$, where the colon is read "such that"; B is the set of all n such that n is a positive integer.

(c) Let C be the set of integers between $\frac{1}{2}$ and $\frac{3}{4}$. There are no integers in this range of numbers, so we write $C = \emptyset$. ■

If it happens that each element of one set, say A, is also an element of another set, B, then we write $A \subseteq B$ and say that A is a **subset** of B.

Definition of Subset

$A \subseteq B$ if, and only if, each element in A is also an element in B.

The concept of a subset now allows us to define the concept of **set equality.**

Set Equality

$A = B$ if, and only if, $A \subseteq B$ and $B \subseteq A$. In other words, two sets are equal if they contain exactly the same elements.

For example, $A = \{1, 2, 3\}$ is a subset of $B = \{0, 1, 2, 3, 4, 5\}$, so we could write $A \subseteq B$. In this case, B is not a subset of A because there is at least one element of B that is not an element of A, so $A \neq B$. As another example of a subset, the set consisting of the months whose names begin with a "J," January, June, and July, is a subset of the set of 12 months of the year. Yet another example is the subset of people who had successful kidney transplants from among all people who had such transplants in 1996.

It is generally convenient to regard all sets in a particular discussion to be subsets of a single larger set U called the **universal set.** For example, sets of numbers may come from the universal set of real numbers; sets of letters may come from the universal set called the English alphabet. The idea of a universal set allows us to create an algebra of sets that results in properties paralleling those of logic and, of course, many of those you have studied in algebra.

The Intersection of Sets

Let A and B be sets in a universal set U. Then the **intersection** of A and B, denoted by $A \cap B$, is the set of all elements in U that belong to *both* A and B. The logical connective *and* describes the intersection

$$A \cap B = \{x : x \in A \text{ and } x \in B\} = \{x : x \in A \land x \in B\}.$$

HISTORICAL NOTE
John Venn (1834–1923)

Venn diagrams are named after the English logician John Venn, who was a lecturer of moral sciences at Cambridge University. His diagrams for representing the relationships between sets became the accepted standard and are still used today. Such diagrams are also important in the study of logic.

A convenient way of pictorially describing the intersection (as well as other sets) is shown in Figure 1 by a **Venn diagram** on the next page. A Venn diagram is constructed by drawing a square to represent the universal set U and by drawing within that square two or more circles, which are usually overlapping, one for each set under discussion. The diagram is completed by shading in the part or parts representing the elements in the given set expression. For the present case of $A \cap B$, we shade the part that belongs to *both* circles.

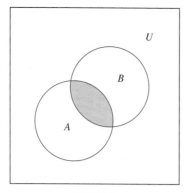

FIGURE 1
The Shaded Portion Represents
$A \cap B.$

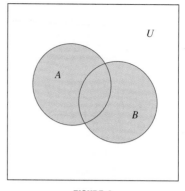

FIGURE 2
The Shaded Portion Represents
$A \cup B.$

The Union of Sets

Let A and B be sets in a universal set U. Then the **union** of A and B, denoted by $A \cup B$, is the set of all elements in U that belong to at least one of the sets A or B. The logical connective *or* describes the union

$$A \cup B = \{x : x \in A \text{ or } x \in B\} = \{x : x \in A \lor x \in B\}.$$

Figure 2 shows the Venn diagram where, according to the definition, the union consists of elements in the A circle or the B circle, or both.

EXAMPLE 2
Illustrating the
Intersection and Union

Let $A = \{0, 1, 2, 3\}$, $B = \{2, 3, 5\}$, and $C = \{5\}$ be subsets of $U = \{0, 1, 2, 3, 4, 5\}$. Then,

$$A \cap B = \{2, 3\}$$
$$A \cup B = \{0, 1, 2, 3, 5\}$$
$$A \cap C = \emptyset$$
$$B \cup C = \{2, 3, 5\} = B$$
$$B \cap C = \{5\} = C$$
$$A \cup C = \{0, 1, 2, 3, 5\}. \quad \blacksquare$$

Notice that the intersection of two sets will always be a subset of each of the sets being intersected ($\emptyset$ is a subset of every set), and the union will always contain each of the sets from which the union is formed. In fact, you can think of the union as taking one of the sets and then adjoining to it all the remaining elements of the other set. Observe that we do not list a given element in a set twice, and the order in which we list the elements of a set is immaterial.

The Complement of a Set

Let A be a subset of a universal set U. Then the **complement** of A, denoted by A', is the set of all elements in U that are not in A. That is,

$$A' = \{x : x \in U \text{ and } x \notin A\}.$$

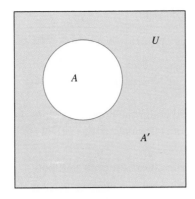

FIGURE 3
The Shaded Portion Represents A'.

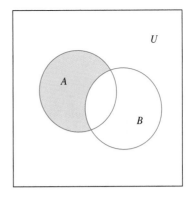

FIGURE 4
The Shaded Portion Represents
$A - B = A \cap B'.$

Figure 3 shows a Venn diagram of A'.

> **The Difference of Two Sets**
>
> Let A and B be two sets. Then the **difference** $A - B$ consists of all elements in A that are not in B. That is,
>
> $$A - B = \{x : x \in A \text{ and } x \notin B\} = A \cap B'.$$

Figure 4 depicts $A - B$ in a Venn diagram.

EXAMPLE 3
Illustrating Set
Operations

Let $A = \{1, 2\}$, $B = \{0, 1, 2, 3\}$, $C = \{4, 5\}$, and $U = \{0, 1, 2, 3, 4, 5, 6, 7, 8\}$. Then,

$$A' = \{0, 3, 4, 5, 6, 7, 8\}$$
$$A - B = \emptyset$$
$$B - A = \{0, 3\}$$
$$A \cap B = \{1, 2\} = A$$
$$A \cup B = \{0, 1, 2, 3\} = B$$
$$B \cup C' = \{0, 1, 2, 3, 6, 7, 8\}$$
$$B \cap C = \emptyset$$
$$C - B = \{4, 5\} = C$$
$$B - C = \{0, 1, 2, 3\} = B. \quad \blacksquare$$

You may have sensed by now that there is a strong connection between logic and sets as far as we have developed them. To elaborate, suppose that P is the set of elements in a universal set U that makes statement p true, and suppose that $Q \subseteq U$ is the set that makes statement q true. The sets P and Q are then known as **truth sets** for p and q, respectively. Sentences constructed using p, q, and the connectives we have studied then also have truth sets. For example, the set of elements in $P \cap Q$ would make $p \wedge q$ true. The following table now gives the translation between some logical connectives and the corresponding set operations.

Statement	Truth set
And: $p \land q$	Intersection: $P \cap Q$
Or: $P \lor q$	Union: $P \cup Q$
Negation: $\sim p$	Complement: P'
$p \land \sim q$	$P \cap Q' = P - Q$

EXAMPLE 4

Illustrating the Connection Between Logic and Sets

Roll a single six-sided die, and count the number of dots showing on the top side. The number of dots showing is an element of the set $\{1, 2, 3, 4, 5, 6\}$. Consider the statements

p: The die shows an even number.

q: The die shows a number greater than 4.

The truth set for p is $P = \{2, 4, 6\}$ and for q is $Q = \{5, 6\}$. Then, the conjunction of p and q,

$p \land q$: The die shows an even number and a number greater than 4.

has the truth set

$$P \cap Q = \{6\}.$$

The sentence describing the disjunction, *p or q*,

$p \lor q$: The die shows an even number or a number greater than 4.

has the truth set

$$P \cup Q = \{2, 4, 5, 6\}.$$

The negation

$$\sim p$$

has the truth set

$$P' = \{1, 3, 5\}.$$

The sentence

$\sim p \land q$: The die does not show an even number
and does show a number greater than 4.

has the truth set

$$P' \cap Q = \{5\}. \quad \blacksquare$$

A study of Example 4 illustrates how we are naturally and, quite correctly, led to use a logical sentence structure and its truth set representation interchangeably. In fact, most of the time, we just refer to translating a sentence into set form or vice versa without the truth set connection. This process is sometimes made easier through the use of a Venn diagram, as shown in the next two examples. As is often the case, the intersection or union of more than two sets will be involved.

General Unions and Intersections

For the sets $A_1, A_2, \ldots A_n$, the **intersection**

$$A_1 \cap A_2 \cap \ldots \cap A_n = \{x : x \text{ belongs to every set}\},$$

while the **union**

$$A_1 \cup A_2 \cup \ldots \cup A_n = \{x : x \text{ belongs to at least one of the sets}\}.$$

EXAMPLE 5
From Logic to Sets

A survey was made of all shoppers who checked out in a particular lane of Harps IGA store one day between the hours of 2:00 and 7:00 P.M. Among those shoppers, let C represent those who bought cinnamon, P those who bought pepper, and S those who bought salt. In each of the following, use a Venn diagram to identify the set of shoppers, and then use the symbols C, P, and S, along with $\cup$, $\cap$, $'$, and $-$ to express the set.

(a) Shoppers who bought cinnamon, but not pepper.

Solution   The Venn diagram shown next identifies the shoppers in question.

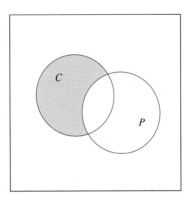

The shaded set may be expressed as

$$C - P = C \cap P'.$$

(b) Shoppers who bought cinnamon and pepper, but not salt.

Solution   The Venn diagram shown next identifies the shoppers in question.

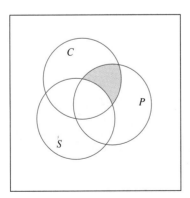

The shaded region may be expressed as

$$(C \cap P) - S = (C \cap P) \cap S'.$$

(c) Shoppers who bought at least one of these products.

Solution The Venn diagram shown next identifies the shoppers in question.

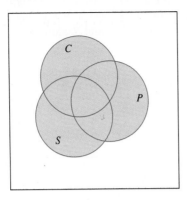

The shaded region may be expressed as

$$C \cup P \cup S.$$

(d) Shoppers who bought exactly one of these products.

Solution The Venn diagram shown next identifies the shoppers in question.

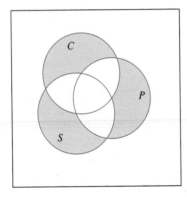

The shaded region may be expressed in more than one way:

$$[C \cap (P \cup S)'] \cup [P \cap (C \cup S)'] \cup [S \cap (C \cup P)']$$
$$= [C \cap (P' \cap S')] \cup [P \cap (C' \cap S')] \cup [S \cap (C' \cap P')]. \quad \blacksquare$$

EXAMPLE 6
From Sets to Logic

Two balls are drawn in succession from a box containing five red, seven green, and four purple balls. Describe in sentence form the meaning of each of the following set expressions. (R_1 means red ball on the first draw, G_2 means green ball on the second draw, and so on.)

(a) $R_1 \cap R_2$

Solution The first ball drawn was red and the second ball drawn was red.

(b) $(R_1 \cup G_2)'$

Solution The first ball drawn was not red and the second ball drawn was not green.

(c) $R_1' \cap P_2$

Solution The first ball drawn was not red and the second ball drawn was purple.

or

The first ball drawn was green or purple and the second ball drawn was purple. ▬

EXERCISES 5.2

For the sets $A = \{1, 2, 3\}$, $B = \{2, 3, 5\}$, $C = \{6, 7\}$, and $U = \{0, 1, 2, 3, 4, 5, 6, 7\}$, find the elements in the sets described in Exercises 1 through 10.

1. A' **2.** $A \cup B$ **3.** $\emptyset'$

4. $B \cap C$ **5.** $A' \cap B$ **6.** $A \cup B'$

7. $(A \cup C)'$ **8.** $A' \cup B'$ **9.** $A - B'$

10. $C - A$

For the sets $A = \{a, b, c\}$, $B = \{c, d\}$, $C = \{e\}$, and $U = \{a, b, c, d, e, f\}$, find the sets described in Exercises 11 through 20.

11. B' **12.** $A' \cup B$ **13.** $A \cap B$

14. $B \cap C$ **15.** U' **16.** $(A \cap B')'$

17. $B' \cup C$ **18.** $(A' \cup B')'$ **19.** $A - C$

20. $A - B$

In each of the Exercises 21 through 28, shade in the part of a Venn diagram representing the given set.

21. $A \cup B'$ **22.** $(A \cup B)'$

23. $A' \cap B'$ **24.** $(A \cup B) - C$

25. $(A \cap B) - C$ **26.** $A - (B \cap C)$

27. $A' \cap B$, if $A \subseteq B$

28. $A \cup (B' \cap C)$, if $A \subseteq B$

Answer Exercises 29 through 34 as true or false. If false, make up sets that demonstrate the falseness of the statement.

29. A is always a subset of $A \cup B$.

30. A is never a subset of $A \cap B$.

31. $A \cap B$ is always a subset of $A \cup B$.

32. $A \cup B$ is never a subset of A.

33. $A \cap \emptyset = \emptyset$ **34.** If $A \neq \emptyset$, $A \cup \emptyset = \emptyset$

Selecting Numbers Let $U = \{1, 2, 3, \ldots, 16\}$ be the universal set, and P be the set of those numbers that makes the following statement true:

p: The number is odd.

and let Q be the set of those numbers that makes this statement true:

q: The number is divisible by 3.

Write each of the Exercises 35 through 40 in sentence form.

35. $P \cap Q$ **36.** $(P \cap Q)'$

37. $P \cup Q'$ **38.** $P' \cap Q$

39. $P - Q$ **40.** $Q' - P$

Survey Let the universal set U be the set of people currently living in your hometown. Let G represent all of those people who make the following statement true:

p: The person wears glasses.

and let F represent all of those people who make the following statement true:

q: The person is over 40 years of age.

Write each of the Exercises 41 through 46 in sentence form.

41. F' **42.** $G' \cup F$

43. $G \cap F$ **44.** $G' - F$

45. $G \cap F'$ **46.** $(G \cup F)'$

Survey Let the universal set U be the set of households in Elkins, let G be the set of households in Elkins who subscribe to the *Gazette,* and let T be the set of households in Elkins who subscribe to the *Tribune.* In each of the Exercises 47 through 54,

(a) Use these set symbols to construct a Venn diagram identifying the requested households.

(b) Use these set symbols, along with $\cap$, $\cup$, $'$, or $-$, to write a set expression identifying these households.

47. Households in Elkins that do not subscribe to the *Gazette.*

48. Households in Elkins that subscribe to both the *Gazette* and the *Tribune.*

49. Households in Elkins that subscribe to the *Gazette* but not to the *Tribune.*

50. Households in Elkins that subscribe to neither of these two papers.

51. Households in Elkins that do not subscribe to the *Gazette* but do subscribe to the *Tribune.*

52. Households in Elkins that subscribe to exactly one of these two newspapers.

53. Households in Elkins that do not subscribe to the *Gazette* and do not subscribe to the *Tribune*.

54. Households in Elkins that subscribe to the *Gazette* or do not subscribe to the *Tribune*.

Survey Let the universal set U be the set of students enrolled in a given finite math class, let E be the set of those students currently enrolled in an English class, H those students currently enrolled in a history class, and G those students currently enrolled in a geology course. In each of the Exercises 55 through 62,

(a) Use these set symbols to construct a Venn diagram identifying the given students.

(b) Use these set symbols, along with $\cap$, $\cup$, $'$, and $-$, to write a set expression identifying these students.

55. Students who are currently enrolled in English and history but not geology.

56. Students who are currently enrolled in English but not in history or geology.

57. Students who are currently enrolled in geology or English but not history.

58. Students who are currently enrolled in all four of these courses.

59. Students who are currently enrolled in exactly one of these four courses.

60. Students who are currently enrolled in none of these four courses.

61. Students who are currently enrolled in at least one of these four courses.

62. Students who are currently enrolled in at least two of these four courses.

Problems for Exploration

1. Find an alternative set expression for $A \cup B'$.

2. Use truth tables to show that
$$A \cap (B \cup C) = (A \cap B) \cup (A \cap C).$$

3. Use a logical argument to prove that for any two sets A and B, $A \cap B \subseteq A$. Start by taking an element x in $A \cap B$.

4. (a) Use a logical argument to prove that
$$A \cup (B \cap C) \subseteq (A \cup B) \cap (A \cup C).$$
Start by taking an element in $A \cup (B \cap C)$.

(b) Use a logical argument to prove that
$$(A \cup B) \cap (A \cup C) \subseteq A \cup (B \cap C).$$
Start by taking an element in
$$(A \cup B) \cap (A \cup C).$$

(c) Conclude that
$$A \cup (B \cap C) = (A \cup B) \cap (A \cup C).$$

SECTION 5.3 Application of Venn Diagrams

Identifying a region in a Venn diagram to represent a particularly complicated set expression may prove to be taxing. A method that eliminates the need to cross-shade component parts needed to reach the final conclusion is shown in Example 1.

EXAMPLE 1
Constructing a Venn Diagram

Use a Venn diagram to represent $A \cap (B \cup C')$.

Solution   First, draw three overlapping circles in a universal set U. Next, label each separate region with a number, letter, or symbol, as shown in Figure 5(a). (We prefer to label with numbers.) Now, record the numbers that make up each of the regions A, B, and C' involved in the given expression $A \cap (B \cup C')$. Then record the numbers that identify regions in any component parts of the expression, such as $(B \cup C')$, that would be used as building blocks in constructing such an expression. (Our list is shown next.)

Set	Region labels
A	**1, 2, 3, 4**
B	**3, 4, 5, 6**
C'	**1, 4, 5, 8**
$B \cup C'$	**1, 3, 4, 5, 6, 8**

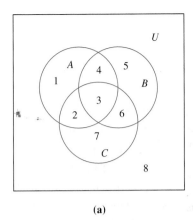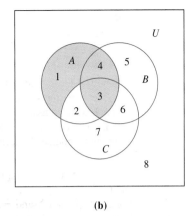

(a)   (b)

FIGURE 5

Construction of a Venn Diagram: (a) Labeling the Regions; (b) Shading the Regions

Finally, $A \cap (B \cup C')$: 1, 3, 4, which identifies the regions to be shaded, as shown in Figure 5(b). ▨

Venn diagrams provide a quick, easy way to indicate whether two set statements represent the same set.

EXAMPLE 2
Using Venn Diagrams to Show Set Equality

Use Venn diagrams to show that $(A \cap B)' = A' \cup B'$.

Solution   First, draw two Venn diagrams, one for each set expression on either side of the equality, and use identical labels to identify the separate regions in each diagram, as shown in Figure 6. Then, as was done in Example 1, find the final region to be shaded for each of the expressions in question.

$(A \cup B)'$

Set	Region labels
A	1, 2
B	2, 3
$A \cap B$	2
$(A \cap B)'$	**1, 3, 4**

$A' \cup B'$

Set	Region labels
A	1, 2
B	2, 3
A'	3, 4
B'	1, 4
$A' \cup B'$	**1, 3, 4**

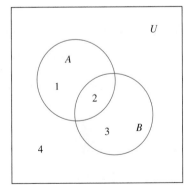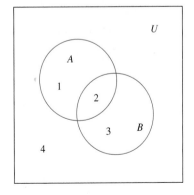

For $(A \cap B)'$ For $A' \cup B'$

FIGURE 6

Using Venn Diagrams to Show Set Equality

The last line in each column shows that the same region should be shaded, which demonstrates that $(A \cap B)' = A' \cup B'$. ▬▬

The equality shown in Example 2 is one of two that are known as **DeMorgan's laws of sets.** Notice how they parallel DeMorgan's laws of logic.

DeMorgan's Laws of Sets

Let A and B be sets in a universal set U. Then

$$(A \cup B)' = A' \cap B'$$

and $\qquad\qquad\qquad (A \cap B)' = A' \cup B'.$

These are but two of many equalities among set expressions. A partial list of some of the more prominent ones is shown next for sets A, B, and C in a universal set U. Note the similarity between these and the usual rules of algebra.

Some Properties of Set Algebra

Commutative laws:

1. $A \cap B = B \cap A$ $\qquad\qquad\qquad$ **2.** $A \cup B = B \cup A$

Associative laws:

1. $A \cap (B \cap C) = (A \cap B) \cap C$ $\qquad$ **2.** $A \cup (B \cup C) = (A \cup B) \cup C$

Distributive laws:

1. $A \cap (B \cup C) = (A \cap B) \cup (A \cap C)$

2. $A \cup (B \cap C) = (A \cup B) \cap (A \cup C)$

DeMorgan's laws of sets:

1. $(A \cap B)' = A' \cup B'$ $\qquad\qquad\qquad$ **2.** $(A \cup B)' = A' \cap B'$

Set difference: $\quad A - B = A \cap B'$
Double complementation: $\quad (A')' = A$

Venn diagrams are also used in counting problems. A counting problem is one that asks, "How many?" and the answer is always one of the whole numbers, 0, 1, 2, 3, ... For a finite set A, the notation $n(A)$ is used to denote the number of elements in the set. For instance,

If $A = \{a, b, c\}$, then $n(A) = 3$.

If $B = \{0, 1, 2, 3, 4, 5, 6\}$, then $n(B) = 7$.

If S is the set of cards in a standard deck of cards, then $n(S) = 52$.

For $\emptyset$, $n(\emptyset) = 0$.

The next example shows how a Venn diagram may be used to help in counting the number of elements in some set expressions.

EXAMPLE 3
Using a Venn Diagram
to Help with Counting

The admissions office of Jefferson County Memorial Hospital checked the records of 40 recently admitted patients and found that 24 had major medical coverage (M), 15 had catastrophic medical coverage (C), and 6 had both types of coverage. Use a Venn diagram to find the number of these patients who had neither of these types of coverage.

Solution We first draw a Venn diagram with two circles, one labeled M and the other labeled C. We now label each separate region of that diagram with a number that serves two purposes: (1) to identify the region, *and* (2) to tell how many of these patients have the particular type of insurance represented by that region. Start the labeling process by considering the intersection $M \cap C$. The data tell us that 6 patients had both types of coverage, which means that $n(M \cap C) = 6$. So we label the intersection with a 6, as shown in Figure 7(a). Now we look back at the data and read that 24 had major medical coverage, which means that a *total* of 24 should be represented in the M circle. There are already 6 in this circle so $24 - 6 = 18$ should be used to count the remainder, $M - (M \cap C)$, as shown in Figure 7(b). The total in the M circle is now 24. The data also tell us that a total of 15 should be represented in the C circle. There are already 6 in this circle, so $15 - 6 = 9$ should be used to count the region $C - (M \cap C)$ as shown in Figure 7(c). Finally, the total in our universal set is 40, which means that the remaining uncounted region outside the two circles should be labeled with $40 - (9 + 6 + 18) = 40 - 33 = 7$, as shown in Figure 7(d). We now can answer the question posed at the outset of this example and any other counting question about these data directly from the Venn diagram in Figure 7(d).

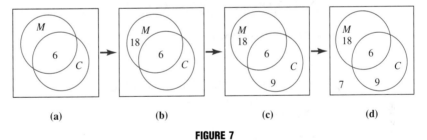

(a)   (b)   (c)   (d)

FIGURE 7
Sequence of Steps in Labeling a Venn Diagram for Counting Purposes

(a) How many of these patients had neither of the two types of insurance? The answer is the number in the region outside the two circles: 7.

(b) How many of these patients had major medical but not catastrophic insurance? The answer is the number in the region represented by $(M - C)$: 18.

(c) How many of these patients had major medical or catastrophic insurance? The answer is the number in $M \cup C$: $18 + 6 + 9 = 33$. ■

EXAMPLE 4
Counting with Venn
Diagrams

One hundred customers of a discount grocery store were asked about their purchases of candy, apples, and popcorn during the previous month. Thirty-two said they had purchased candy, 48 had purchased apples, 35 had purchased popcorn, 15 had purchased both candy and apples, 7 had purchased both candy and popcorn, 5 had purchased both apples and popcorn, and 2 had purchased all three of these foods. A Venn diagram representing these purchases may be constructed similar to that in

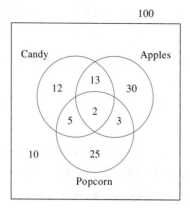

FIGURE 8
Grocery Store Purchases

Example 3. Here, we start with the intersection of all three circles and build outward to obtain the numbers shown in Figure 8.

(a) Among these customers, how many purchased candy or popcorn, but not apples?

Solution   In symbols, Question (a) asks $n[(C \cup P) - A]$, and the answer from Figure 8 is found to be $12 + 5 + 25 = 42$.

(b) For these customers, how many purchased none of these brands?

Solution   The answer is 10.

(c) For these customers, how many purchased exactly two of these brands?

Solution   The answer is $5 + 3 + 13 = 21$. ■

A useful counting formula arises for $n(A \cup B)$ that we can use now as well as in later sections. To justify this formula and its special case, consider the Venn diagrams in Figure 9. From Figure 9(a), $n(A) = 5$ and $n(B) = 7$. However, $n(A) + n(B) = 12$ **includes** the count of the two elements in the intersection *twice*. Therefore, **excluding** $n(A \cap B)$ once will leave each element in $A \cup B$ counted exactly once. It follows that

$$n(A \cup B) = n(A) + n(B) - n(A \cap B).$$

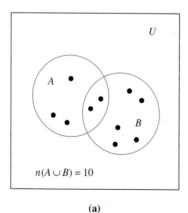

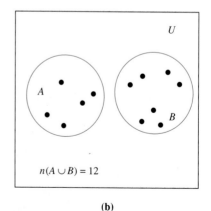

(a)   (b)

FIGURE 9
Developing a Counting Formula

Figure 9(b) shows the special case when $A \cap B = \emptyset$. Because $n(A \cap B) = 0$, the formula reduces to $n(A \cup B) = n(A) + n(B)$.

> **Important Counting Formula: The Inclusion-Exclusion Principle**
>
> For any two sets A and B,
>
> $$n(A \cup B) = n(A) + n(B) - n(A \cap B),$$
>
> or equivalently,
>
> $$n(A \text{ or } B) = n(A) + n(B) - n(A \text{ and } B).$$

The inclusion-exclusion principle may be restated in words as

"When counting, 'or' means 'to add,' but don't double-count any outcome."

We note that when using a Venn diagram for counting purposes, the construction we have demonstrated is specifically designed to eliminate any double-counting of elements, so there is no particular need for a formula there. However, as we move forward with other counting techniques where the word "or" is involved, the use of the inclusion-exclusion principle, or its restatement shown above, will preclude any double-count of outcomes.

EXAMPLE 5
Using the Inclusion-Exclusion Principle

If $n(A) = 26$, $n(B) = 32$, and $n(A \cup B) = 47$, what is $n(A \cap B)$?

Solution Anytime we know three of the four terms in the inclusion-exclusion principle, we can algebraically solve for the fourth. Direct substitution of the given data into

$$n(A \cup B) = n(A) + n(B) - n(A \cap B)$$

gives
$$47 = 26 + 32 - n(A \cap B)$$
so that
$$47 = 58 - n(A \cap B)$$
or
$$n(A \cap B) = 58 - 47 = 11. \quad \blacksquare$$

EXAMPLE 6
Using the Word "or" in Counting

A check of 90 light bulbs revealed the following data.

Bulb type	Defect	No defect
60W bulbs	3	20
100W bulbs	5	22
150W bulbs	7	33

(a) How many of these bulbs are 60W or defective?

Solution Consider one set to be the 60W bulbs and the other to be the defective bulbs. We are asked to find $n(60\text{W or defective})$. From the chart, the meaning of "or," and the fact that no bulb is to be counted twice, we have

$$n(60\text{W or defective}) = 3 + 20 + 5 + 7 = 35$$

without remembering a formula.

Of course, the inclusion-exclusion counting formula gives the same result when we realize that three bulbs lie in both sets:

$$n(60\text{W or defective}) = n(60\text{W}) + n(\text{defective}) - n(60\text{W and defective})$$
$$= 23 + 15 - 3 = 35.$$

(b) Find $n(60W$ or $150W)$.

Solution   Since no bulb can be both 60W and 150W,

$$n(60W \text{ or } 150W) = n(60W) + n(150W)$$
$$= 23 + 40 = 63. \quad \blacksquare\blacksquare$$

EXERCISES   5.3

Shade in a Venn diagram to represent each of the sets in Exercises 1 through 10.

1. $(A' \cup B) \cap C$

2. $(A \cap B) \cap C'$

3. $(A \cup B)' \cap C'$

4. $A' \cap (B \cup C')$

5. $(A \cap B')'$, if $A \cap B = \emptyset$

6. $A \cup (B' \cap C)$, if $A \subseteq B$

7. $(A \cap B) \cup C'$

8. $[(A \cup B)' \cap C] - B$

9. $(A - C) \cup (B \cap C')$

10. $B - [(A \cup B)' - C]$

In Exercises 11 through 14, use a Venn diagram to verify the set equalities.

11. $(A \cup B)' = A' \cap B'$

12. $A \cap (B \cup C) = (A \cap B) \cup (A \cap C)$

13. $[A' \cap (B \cup C)] - B = (A \cup B)' \cap C$

14. $[(A \cup B) - C] - B = A - (B \cup C)$

In Exercises 15 through 20, use a Venn diagram to decide which pairs of sets are equal.

15. $A \cap B'$; $A - B$

16. $A' \cap B$; $A \cup B'$

17. $(A \cup B)'$; $A' \cap B$

18. $(A \cup B) - C$; $(A \cap C') \cup (B \cap C')$

19. $C - (A \cup B)$; $(C \cap A) \cup (C \cap B)$

20. $B - (A \cup C)$; $(B \cap A') \cup (B \cap C')$

Use the counting formula $n(A \cup B) = n(A) + n(B) - n(A \cap B)$ to answer Exercises 21 through 26.

21. Find $n(A \cup B)$ if $n(A) = 8$, $n(B) = 12$, and $n(A \cap B) = 3$.

22. Find $n(A \cup B)$ if $n(A) = 6$, $n(B) = 3$, and $A \cap B = \emptyset$.

23. Find $n(A)$ if $n(A \cup B) = 8$, $n(B) = 5$, and $n(A \cap B) = 2$.

24. Find $n(B)$ if $n(A) = 6$, $n(A \cup B) = 8$, and $n(A \cap B) = 0$.

25. Find $n(A \cap B)$ if $n(A \cup B) = 12$, $n(A) = 7$, and $n(B) = 9$.

26. Find $n(A \cap B)$ if $n(A \cup B) = 20$, $n(A) = 12$, and $n(B) = 8$.

27. *Proficiency*   Thirty senior students at a state university were asked about their reading proficiency in French and German. Twelve said they could read French, five said they could read German, and two said they could read both of these languages.

(a) Draw and label a Venn diagram that numerically represents this survey.

In (b) through (f), rewrite each question in set notation, and then answer the question using the Venn diagram constructed in (a). How many of these students read

(b) French, but not German?

(c) French or German?

(d) At least one of these two languages?

(e) Exactly one of these two languages?

(f) Neither of these two languages?

(g) Rework Exercise 27(c) using the counting formula $n(A \cup B) = n(A) + n(B) - n(A \cap B)$.

28. *Survey*   A food store surveyed 50 of its shoppers as to whether they had purchased bacon or steak during the past week. The results were 30 had purchased bacon, 20 had purchased steak, and 8 had purchased both bacon and steak.

(a) Draw and label a Venn diagram to numerically represent this survey.

In (b) through (f), rewrite each question in set notation, and then answer the question using the Venn diagram constructed in (a). How many of these shoppers bought

(b) Bacon, but not steak?

(c) Bacon or steak?

(d) At least one of these two meats?

(e) Exactly one of these two meats?

(f) Neither of these two meats?

(g) Rework Exercise 28(c) using the counting formula $n(A \cup B) = n(A) + n(B) - n(A \cap B)$.

In each of the remaining exercises, use a Venn diagram to help answer the counting question.

29. Survey In a particular city, 80 households were asked these two questions:

Q1: Do you own a Chevrolet?

Q2: Do you own at least two cars?

Twenty-five households answered "yes" to Question 1, 52 answered "yes" to Question 2, and 18 answered "yes" to both questions. How many of these households

(a) Answered "yes" to at least one of these questions?
(b) Answered "yes" to exactly one of these questions?
(c) Answered "no" to both questions?
(d) Own a Chevrolet but no other car? What percentage of these households own only a Chevrolet car?

30. Survey In a particular school district, 90 families were asked these two questions:

Q1: Do you have children attending public kindergarten?

Q2: Do you have children in grades 1 through 5 attending public school?

Thirty answered "yes" to Question 1, 50 answered "yes" to Question 2, and 10 answered "yes" to both questions.

(a) How many answered "yes" to at least one of these questions?
(b) How many answered "no" to both questions?
(c) How many have no children in grades 1 through 5 attending public school?
(d) What percentage of these families had children in public kindergarten or in public school grades 1 through 5?

31. Survey There were 150 respondents to a survey about the ownership of radios and color TV sets. They were asked these questions:

Q1: Do you own a color TV set?

Q2: Do you own at least two radios?

There were 110 people who answered "yes" to Question 1, 90 who answered "yes" to Question 2, and 120 who answered "yes" to Question 1 or Question 2.

(a) How many of these people answered "yes" to both questions?
(b) How many of these people answered "no" to both questions?
(c) How many of these people answered "yes" to Question 1 but "no" to Question 2?
(d) What percentage of these people own at least two radios but no color TV set?

32. Survey In a history class, 30 students were asked about other courses they were enrolled in: 20 said they were in math, 22 in English, and 26 in either math or English.

(a) How many students are in both math and English?
(b) How many are in neither?
(c) How many are in math but not English?
(d) What percentage of these students are in math but not English?

33. Survey Ninety customers of a discount store were asked about their purchases during the past month. Twenty said they had purchased books, 45 had purchased film, 38 had purchased jewelry, 15 had purchased both books and film, 8 had purchased both books and jewelry, 6 had purchased both film and jewelry, and 3 had purchased all three articles. How many of these people had purchased

(a) Books or film but not jewelry?
(b) Books and film but not jewelry?
(c) Jewelry but not books or film?
(d) At least one of these three articles?
(e) Exactly one of the three articles?

34. Poll A poll of 100 executives showed that 45 subscribe to *The Wall Street Journal*, 25 to *Business Week*, 8 to both *The Wall Street Journal and Barron's*, 12 to *Barron's and Business Week*, 6 to *The Wall Street Journal and Business Week*, 2 to all three, and 22 to none of these publications. How many of these executives

(a) Subscribe to at least one of these publications?
(b) Subscribe to *Business Week* but not to *The Wall Street Journal* or *Barron's*?
(c) Do not subscribe to *Barron's*?
(d) Subscribe to exactly two of these publications?
(e) Subscribe to *Business Week* or *The Wall Street Journal* but not to *Barron's?*
(f) Subscribe to *Barron's* and *The Wall Street Journal* but not to *Business Week?*

35. Survey In a particular city, 150 residents were asked to answer the following questions:

Q1: Are you in favor of building the proposed East-West expressway?

Q2: Are you in favor of the loop expressway?

Q3: Are you in favor of the proposed city bond issue?

Sixty-three answered "yes" to Question 1, 82 answered "yes" to Question 3, 15 answered "yes" to Questions 1 and 2, 42 answered "yes" to Questions

1 and 3, 40 answered "yes" to Questions 2 and 3, 12 answered "yes" to all three questions, and 27 answered "no" to all three questions. From among these people, how many

(a) Answered "yes" to Question 2 or Question 3?
(b) Answered "yes" to at least one of the questions?
(c) Answered "yes" to exactly one of these questions?
(d) Favored the East-West expressway but did not favor the other two projects?

36. Survey The following questions were asked of a random sample of 1000 people:

Q1: Do you favor nationalized health care?

Q2: Do you favor higher Social Security taxes?

Q3: Do you favor more reliance on private retirement systems?

Of these people, 553 answered "yes" to Question 1, 224 answered "yes" to Question 2, 438 answered "yes" to Question 3, 87 answered "yes" to Questions 1 and 2, 102 answered "yes" to Questions 1 and 3, 93 answered "yes" to Questions 2 and 3, and 50 answered "yes" to all three questions. What percentage of these people

(a) Favor nationalized health care or higher Social Security taxes?
(b) Do not favor any of these three ideas?
(c) Favor nationalized health care but not increased Social Security taxes?

37. Psychological Study A psychologist took 20 monkeys and spent a week trying to teach them to ride a tricycle, another week teaching them to do chin-ups, and yet another week teaching them to shoot a basketball into a low hoop. The results were as follows:

2 could be taught to do all three activities
4 could be taught to ride a tricycle and shoot a basketball
3 could be taught to do chin-ups and shoot a basketball
5 could be taught to ride a tricycle and do chin-ups
9 could be taught to ride a tricycle
8 could be taught to shoot a basketball
5 could not be taught to do any of these activities

How many of these monkeys were taught to

(a) Do chin-ups but could not be taught to ride a tricycle?
(b) Do at least two of these activities?

How many could *not* be taught to

(c) Ride a tricycle or shoot a basketball?

38. Options A Chevrolet dealer has 160 new cars for sale on her lot. Among these cars:

50 have four-cylinder engines
80 have tilt wheel
30 have power windows
42 have four-cylinder engines and tilt wheel
18 have four-cylinder engines and power windows
15 have all three of these features
65 have none of these features

How many of these cars have

(a) Tilt wheel and power windows?
(b) At least one of these features?
(c) Exactly two of these features?

39. Survey A student was hired to stand near the spice section of a supermarket from 5:00 to 8:00 P.M. on a particular day to count the number of people who bought spices and to make special note of the number of people who bought cinnamon or pepper. The student reported that 60 people bought a spice of some kind, 50 bought cinnamon, 40 bought pepper, and 5 bought both cinnamon and pepper. Should the student be believed?

40. Survey A plant supervisor reported that of the 20 people being supervised, 16 could operate a lathe, 12 could operate a grinder, 8 could operate a bender, 2 could operate a lathe and grinder, 4 could operate a lathe and bender, 5 could operate a grinder and bender, and 1 could operate all three machines. Should the supervisor be believed?

41. Unemployment data for a small town in Texas are as follows.

	Employed	Unemployed
Male	60	10
Female	40	5

How many of these people are
(a) Male or unemployed?
(b) Female and employed?

42. The Bureau of the Census, *Statistical Abstract of the United States*, 1994, shows these amounts (in millions of dollars) spent on newspaper advertising:

Year	Nationally	Locally
1990	3867	28,414
1991	3685	26,724
1992	3602	27,135

According to these data, how much was spent (in millions of dollars) on advertising

(a) During 1992 or in local newspapers?
(b) During 1991 or in national newspapers?
(c) During 1991 or 1992?
(d) During 1992?

The following problem appeared on an actuarial examination.

43. Which of the following events is identical to $(B \cap C) \cup (A' \cap B \cap C')$?

$$\text{I.} \ \ B \cap (A' \cup C)$$
$$\text{II.} \ \ (A' \cap B) \cup (B \cap C)$$
$$\text{III.} \ \ (A' \cap C') \cup (B \cap C)$$

(a) I and II only
(b) I and III only
(c) II and III only
(d) I, II, and III
(e) The correct answer is not given by (a), (b), (c), or (d).

Problems for Exploration

1. Use the algebra of set rules to factor A from the expression $(A \cup B) \cap [(A \cup C) \cup (A \cup D)]$.

2. Use a Venn diagram to help find a simpler expression for $(A \cap B) \cup (B \cap C')$.

3. Use a Venn diagram to show that
$$p \wedge (q \vee r) \equiv (p \wedge q) \vee (p \wedge r).$$

4. Draw Venn diagrams (and simplify the results when possible) for $A \cup B, A \cap B, (A \cup B)'$, and $(A \cap B)'$ under each given condition:

(a) $A \subseteq B$ (b) $A = B$ (c) $A = \emptyset$
(d) A and B are disjoint $(A \cap B = \emptyset)$
(e) $A = U$

5. The *symmetric difference* of A and B is defined as $A \Delta B = (A \cup B) - (A \cap B)$.

(a) Check the commutative, associative, and distributive laws for Δ.
(b) Show that $A \Delta B = (A - B) \cup (B - A)$.
(c) Show that $A \cap B = (A \Delta B) \Delta (A \cup B)$.

6. Invent a Venn diagram for four sets. Remember that every possible intersection must have a corresponding region, and no intersection should have more than one corresponding region.

SECTION 5.4

The Multiplication Principle

The key to success in understanding the concepts of probability in Chapter 6 is knowing how to count. To *count* means to find the number of items in lists, some of which may be far too lengthy to write down. It is the lengthy lists that imply the need for counting techniques to answer such "How many?" questions. We have already seen that Venn diagrams may be used to answer a particular type of counting problem. Other types of problems, however, call for different techniques, and it is those that we explore in the remainder of this chapter. Regardless of the counting technique applied, it is the case that

1. A counting problem always asks, "How many?"

2. The answer is always found to be one of the *whole numbers,* 0, 1, 2, 3, ...

Here are some examples of counting problems that are unlike any we have seen up to this point, but that we should be able to answer after studying this section.

How many different Social Security numbers are possible?

How many different telephone numbers can be put on the area code 501?

In how many ways can a president and a vice president be selected from among 12 people?

How many ways can a sum of 8 be rolled on a pair of dice?

One way to solve these, or any other counting problems, is to make a list of all objects to be counted and then simply count them. However, as you can see from the preceding examples, this technique is limited in its application because of the extremely long lists that can occur. Even so, one helpful suggestion on any counting problem is to list a few of the items to be counted so that it is clear exactly what is, and what is not, to be counted. Beyond that, we need to know techniques that will give us a count of the number in the list without actually viewing the entire list. Some examples in which we can write down the entire list of items to be counted may help in setting the stage for the techniques of this section.

EXAMPLE 1
Viewing the List, Then
Finding a Formula

A box contains three slips of paper, on each of which is written the letter a, b, or c, respectively. A slip is drawn, the letter on it recorded, and the slip is *not* returned to the box. A second slip is then drawn and the letter recorded. Each sequence of two letters so selected is known as an **outcome** from this activity. How many different outcomes are there?

Solution   All possible outcomes may be listed as

$$(a, b), (a, c), (b, a), (b, c), (c, a), (c, b)$$

A **tree diagram** like the one shown in Figure 10 gives a visual description of the selection process leading to the six possible outcomes. From left to right, there are a total of six paths through the diagram, and each of these paths corresponds to the ordered pair listed at the end of the path. Since each ordered pair is an outcome, there are six possible outcomes.

The tree diagram helps explain how we may obtain the total number of outcomes in terms of the number of choices available for each of the two successive draws: There are three choices of slips for the first draw and two choices of slips for the second draw. (The first slip was not returned to the box.) Thus,

$$\underbrace{3}_{\substack{\text{Number of choices} \\ \text{on first draw}}} \times \underbrace{2}_{\substack{\text{Number of choices} \\ \text{on second draw}}} = 6$$

gives the correct result. Note that the product correctly counts the interchange of two letters in an outcome as a new and different outcome; for example, (a, b) and (b, a) are different outcomes. The *order* in which the letters appear is *important*. We say that each outcome is an **ordered arrangement** of the elements in the set. ∎

In Example 1, each selection comes from the same set $\{a, b, c\}$, but once a letter is removed, we are not allowed to use it again in future selections. This is known as selection **without replacement** or selection when **no repetitions are allowed.**

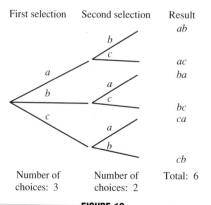

First selection   Second selection   Result

Number of   Number of   Total: 6
choices: 3   choices: 2

FIGURE 10
Tree Diagram

EXAMPLE 2
Viewing the List, Then
Finding a Formula

A coin is tossed three times. An **outcome** from this activity is an ordered sequence of three symbols: a T for tails or H for heads to represent the result of the first toss, a T or H to represent the result of the second toss, and a T or H to represent the result of the third toss. How many outcomes are possible?

Solution Recording the possibilities for successive tosses of the coin gives this list:

$$HHH, \ HHT, \ HTH, \ THH, \ TTH, \ THT, \ HTT, \ TTT$$

Therefore, there are a total of eight possible outcomes for this activity. As in Example 1, we may obtain this total by thinking in terms of the number of choices available for each of the three tosses making up an outcome: There are two choices (H or T) for the first toss, two choices for the second toss, and two for the third toss, from which the product

$$\underbrace{2}_{\substack{\text{Number of} \\ \text{choices on} \\ \text{first toss}}} \times \underbrace{2}_{\substack{\text{Number of} \\ \text{choices on} \\ \text{second toss}}} \times \underbrace{2}_{\substack{\text{Number of} \\ \text{choices on} \\ \text{third toss}}} = 8$$

gives the correct result. Note that the product correctly counts the interchange of two different letters in an outcome as a new and different outcome. ■

In Example 2, we may view the selections as coming from the set $\{H, T\}$ of two different objects, an H and a T, and, once a letter is selected, it is returned to the set for possible selection again in the same outcome. (For example, HHT or TTT.) We refer to this as **selection with replacement** or selections when **repetitions are allowed.**

EXAMPLE 3
Viewing the List, Then
Finding a Formula

A model of hair dryer has three blower-speed settings—B_1, B_2, and B_3—and two heat settings—H_1 and H_2. How many settings for styling purposes does this dryer have?

Solution The possible settings (outcomes) can easily be listed:

$$B_1; H_1 \qquad B_2; H_1 \qquad B_3; H_1$$
$$B_1; H_2 \qquad B_2; H_2 \qquad B_3; H_2$$

The list tells us that there are six settings for styling purposes. Why do we not put $H_1; B_1$ (and so on) in the list? Because it is *not* a different dryer setting. That is, changing the order in which the individual settings are selected does *not* result in a new dryer setting. Even so, the total number may be obtained by multiplying the respective number of choices, as was done in Examples 1 and 2:

$$\underbrace{3}_{\substack{\text{Number of} \\ \text{blower choices first}}} \times \underbrace{2}_{\substack{\text{Number of} \\ \text{heat choices second}}} = 6. \ ■$$

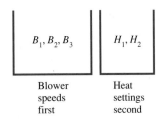

Blower speeds first

Heat settings second

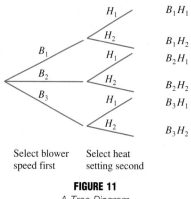

FIGURE 11
A Tree Diagram

Think of making selections from two *different sets:* the blower speeds in one set and the heat settings in another set. Select from the set of blower speeds first, then select from the set of heat settings second. The product $3 \cdot 2 = 6$ gives all possible settings for selections made in this order. A tree diagram showing this selection process is shown in Figure 11. Notice that the leading branches of the tree are labeled with blower settings *only,* reflecting a selection from only *one* of the two sets. The attached branches reflect selections made from a *different* set, namely the heat settings. A path along two consecutive branches results in the blower setting shown at the end of the second branch.

Our tree could have been drawn by depicting heat selections first and blower settings second, still giving $2 \cdot 3 = 6$ different paths or dryer settings.

The **multiplication principle** demonstrated in Examples 1, 2, and 3 is a very general and powerful counting technique. It counts *all possible distinct ordered outcomes,* each consisting of successive selections made under the following conditions:

1. Each outcome of successive selections is viewed from left to right in a row (or on a line).

2. If selections are made from the same set, either with or without replacement, the multiplication principle *counts the interchange of two distinguishable objects* (elements) in any outcome as a *new outcome.* The reason is that such an interchange gives another possible ordered outcome of objects from the same set. (See Examples 1 and 2.)

3. If selections are made from different sets A and B, then multiplying the number of possible selections from A by the number of possible selections from B gives the total number of possible ordered outcomes, where selections from A are made *first* and those from B are made *second.* (See Example 3.)

These conditions may be condensed into the **multiplication principle** stated next.

The Multiplication Principle

If an outcome consists of k successive selections where there are n_1 choices for the first selection, n_2 choices for second selection, . . . and n_k choices for the kth selection, the total number of distinct ordered outcomes possible is the product

$$n_1 \cdot n_2 \cdot n_3 \cdot \ldots \cdot n_k.$$

NOTE   Sets are always considered to have different appearing or **distinguishable elements** unless otherwise noted. The two *a*'s in {*a, a, b, c*} are considered **indistinguishable.** (More on this in the next two sections.) Direct application of the multiplication principle *will not* correctly count all ordered outcomes in sets with indistinguishable elements.

EXAMPLE 4
Using the Multiplication Principle

How many telephone numbers can be put on a 555-*xxxx* exchange?

Solution   We have the set of ten digits—0, 1, 2, 3, . . . , 9—from which to choose. Because a telephone number may have repeated digits, we assume that the selection process takes place with replacement; that is, once a digit is selected, we still may use it again. The multiplication principle tells us that the total number of four-digit numbers possible is

$$\underbrace{10}_{\substack{\text{Number of}\\\text{choices on}\\\text{first digit}}} \times \underbrace{10}_{\substack{\text{Number of}\\\text{choices on}\\\text{second digit}}} \times \underbrace{10}_{\substack{\text{Number of}\\\text{choices on}\\\text{third digit}}} \times \underbrace{10}_{\substack{\text{Number of}\\\text{choices on}\\\text{fourth digit}}} = 10{,}000. \quad\blacksquare$$

EXAMPLE 5
Using the Multiplication Principle

How many four-digit personal identification numbers (like those used for ATM cards) are possible if zero cannot be used as the first digit and no digit may be repeated?

Solution   Because no repetition of digits in the identification number is allowed, the selection process is one without replacement. Since the first digit cannot be zero, there are nine choices for that digit. For the second digit, we may now use zero or any digit not used in the first selection: There are nine choices for the second digit. For the third digit, there are eight choices; for the fourth digit, seven choices. The multiplication principle then tells us that the total number of different personal identification numbers is

$$\underbrace{9}_{\substack{\text{Number of}\\\text{choices on}\\\text{first digit}}} \times \underbrace{9}_{\substack{\text{Number of}\\\text{choices on}\\\text{second digit}}} \times \underbrace{8}_{\substack{\text{Number of}\\\text{choices on}\\\text{third digit}}} \times \underbrace{7}_{\substack{\text{Number of}\\\text{choices on}\\\text{fourth digit}}} = 4536. \quad\blacksquare$$

Up to this point, all of our selections in forming an outcome have rather naturally gone from left to right. The multiplication principle does not require that any particular order be followed, however. We are allowed to start in any digital slot that is most convenient.

EXAMPLE 6
Using the Multiplication Principle

Three different novels and two different technical manuals are to be arranged on a shelf. In how many ways can they be arranged if a novel is to occupy the middle position?

Solution   It is most convenient to select a novel for the middle position first, and there are three choices from which to do so. We now may select one from the four remaining books to occupy any one of the four remaining positions. Once selected and put on the shelf, a book cannot be used again in the selection process; we are selecting without replacement. Here are the number of choices involved.

$$\underbrace{4}_{\substack{\text{Second}\\\text{selection}}} \times \underbrace{3}_{\substack{\text{Third}\\\text{selection}}} \times \underbrace{3}_{\substack{\text{First}\\\text{selection}}} \times \underbrace{2}_{\substack{\text{Fourth}\\\text{selection}}} \times \underbrace{1}_{\substack{\text{Fifth}\\\text{selection}}} = 72. \quad\blacksquare$$

EXAMPLE 7

Using the Multiplication Principle

Three boys and two girls are to be seated in a row of chairs to have a picture taken. In how many ways can they be seated for the picture

(a) If there are no restrictions?

Solution   Think of five slots in a row, each representing a chair, and then start denoting the successive choices for the number of people available for the respective chairs. (We may start this process at any position; however, because we normally read from left to right, it seems logical to begin with the position on the left.) There are five choices for seating a person in the left chair. Once the person on the left is seated, there are four choices left for the second seat (once seated, a person cannot be seated again), and so forth. Therefore, there are

$$5 \cdot 4 \cdot 3 \cdot 2 \cdot 1 = 120$$

possible seating arrangements (outcomes).

(b) If boys are to sit in the end chairs?

Solution   There are three choices for seating a boy in the left seat, then two choices for seating a boy in the right end seat. The remaining three people may be seated with no restrictions. Thus, there are $3 \cdot 3 \cdot 2 \cdot 1 \cdot 2 = 36$ different ways. Figure 12 below may give some help.   ■

EXAMPLE 8

Using the Counting Principle with Inclusion-Exclusion

Three boys and two girls are to be seated in a row. In how many different ways can this be done if

(a) The boys are to sit next to each other and the girls are to sit next to each other?

Solution   To meet the seating conditions of the problem,

  A: The boys could be seated in the first three seats on the left, then followed by the girls on the right, *or*

  B: The girls could be seated in the first two seats on the left, then followed by the boys on the right.

Recall that the inclusion-exclusion formula for counting may be stated as "or means add, but don't double-count any outcome." In this case, no arrangement (outcome) in *A* is found in *B*, leading us to conclude there will be no double-count when adding the number of possibilities for *A* to those for *B*. In symbols $n(A \cup B) = n(A) + n(B)$. Therefore, the total number of arrangements (outcomes) is

$$(3 \cdot 2 \cdot 1 \cdot 2 \cdot 1) + (2 \cdot 1 \cdot 3 \cdot 2 \cdot 1) = 2(3 \cdot 2 \cdot 1 \cdot 2 \cdot 1) = 24.$$

Figure 13 shows how we arrive at this product.

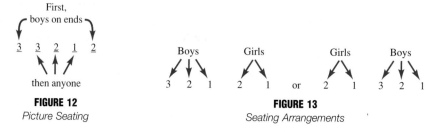

First,
boys on ends

$\underline{3}$ $\underline{3}$ $\underline{2}$ $\underline{1}$ $\underline{2}$

then anyone

FIGURE 12
Picture Seating

Boys   Girls   Girls   Boys

3 2 1   2 1   or   2 1   3 2 1

FIGURE 13
Seating Arrangements

(b) A boy is seated on the left end or a girl is seated on the right end?

Solution   In some seating arrangments, a boy is seated on the left and a girl on the right. Therefore

n(boy on left *or* girl on right)

$= n$(boy on left $+ n$(girl on right) $- n$(boy on left *and* girl on right)

$$\underset{\substack{\nwarrow\nwarrow\nearrow\nearrow \\ \text{then anyone}}}{= \underset{\substack{\text{Boy first} \\ \downarrow}}{\mathbf{3}} \cdot 4 \cdot 3 \cdot 2 \cdot 1} \quad + \quad \underset{\substack{\nwarrow\nwarrow\uparrow\nearrow \\ \text{then anyone}}}{4 \cdot 3 \cdot 2 \cdot 1 \cdot \underset{\substack{\text{Girl first} \\ \downarrow}}{\mathbf{2}}} \quad - \quad \underset{\substack{\nwarrow\uparrow\nearrow \\ \text{then anyone}}}{\underset{\substack{\text{Boy first,} \\ \downarrow}}{\mathbf{3}} \cdot 3 \cdot 2 \cdot 1 \cdot \underset{\substack{\text{girl second} \\ \downarrow}}{\mathbf{2}}}$$

$= 72 + 48 - 36 = 84.$ ▬

EXAMPLE 9
Using the Multiplication
Principle

An experiment in a psychology laboratory involves luring a rat through the following maze from A to D, with various enticements in each tunnel. How many different paths (outcomes) can the rat take through the maze as shown in Figure 14?

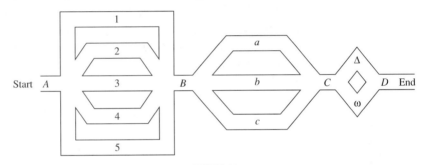

FIGURE 14
Paths Through a Maze

Solution   Think of the maze as consisting of three different sets of paths: one set from A to B, one set from B to C, and one set from C to D. We have five choices from the first set of paths, three choices from the second set of paths, and two choices from the third set of paths. The multiplication principle then tells us that there are a total of

$$5 \cdot 3 \cdot 2 = 30$$

different paths through the maze. Observe that this product does *not* count (and correctly so) the paths from left to right and those from right to left because they are the same paths. This is because we are making our selections from different sets. ▬

EXAMPLE 10
Using the Multiplication
Principle

Suppose that a license plate is made up of two letters followed by five digits. How many different such plates can be made?

Solution   Because no restrictions are given, we assume that license plates such as AA–00000 are permitted. We are to select letters first, then digits. Think of the 26 letters in the alphabet in a first box (set) and the ten digits 0, 1, 2, 3, 4, 5, 6, 7, 8, 9, in a second box.

There are $26 \cdot 26$ ways of selecting two letters, with replacement, from the first box. For each of these ways, there are $10 \cdot 10 \cdot 10 \cdot 10 \cdot 10$ ways of selecting five digits, with replacement, from the second box. The multiplication principle states that the total number of different license plates is the product of $26 \cdot 26$ and $10 \cdot 10 \cdot 10 \cdot 10 \cdot 10$ or

$$26 \cdot 26 \cdot 10 \cdot 10 \cdot 10 \cdot 10 \cdot 10 = 67,600,000. \quad \blacksquare$$

It is *very important* to recognize that the multiplication principle counts as a new outcome the interchange of two different elements in a given outcome when selections are made from the same set. This knowledge is used throughout the remainder of this chapter and carries forward into our study of probability.

EXERCISES 5.4

How many possible outcomes are there for each of the activities described in Exercises 1 through 4?

1. Roll a single die.

2. Draw one card from a standard deck of cards.

3. Record a score on a ten-point quiz.

4. Have a birthday in July.

In Exercises 5 through 11, write down the entire list to be counted, then count. Assume that there are no restrictions unless indicated.

5. *Coin Tossing* A coin is tossed four times. How many possible outcomes are there?

6. *Constructing Numbers* Two-digit numbers are to be made up from the digits 2, 3, 4. How many such numbers can be made if repetition of digits is allowed?

7. *Constructing Numbers* Two-digit numbers are to be made up from the digits 0, 1, 2, 3. How many such numbers can be made if 0 cannot be the first digit and repetition of digits is allowed?

8. *Sampling* Three slips of paper have the numbers 1, 2, and 3, respectively, written on them. A slip is drawn, the number recorded, and the slip laid aside. Now a second slip is drawn and the number recorded. How many such outcomes are there?

9. *Correspondence* A secretary types letters to three people and then types the corresponding envelopes. In how many ways can the letters be put in the envelopes so that exactly two are in the wrong envelopes?

10. *Routes* If there are three roads from Goshen to Gravette and two roads from Gravette to Tontitown, how many ways are there to go from Goshen to Tontitown through Gravette?

11. *Experiment* An agricultural experimenter wants to test three varieties of soybeans under two levels of fertilizer and two types of insecticides. How many experimental plots are needed?

In the remaining exercises, use any appropriate technique that has been presented to obtain the count requested.

12. *Elections* In how many ways can a president, a secretary, and a treasurer be elected from among ten people?

13. *Social Security Number* How many Social Security numbers are there if

(a) There are no restrictions?

(b) No number can begin with zero?

(c) Neither of the first two digits can be zero?

14. *Telephone Numbers* How many seven-digit telephone numbers can be put on the 501 area code if

(a) There are no restrictions?

(b) No telephone number can begin with zero?

(c) No telephone number can begin with zero, and the last digit must be odd?

15. *Arrangements* In how many ways can five boys and three girls be seated in a row if

(a) There are no restrictions?

(b) Boys and girls are seated alternately?

(c) Girls must be seated in both end seats?

16. *Arrangements* In how many ways can five boys and five girls be seated in a row if

(a) Boys and girls are seated alternately?

(b) Boys sit together and girls sit together?

(c) One of the girls, Sue, must be seated on the left end?

17. *Arrangements* In how many ways can eight books be arranged on a shelf if

(a) There are no restrictions?

(b) One of the books, *Of Mice and Men*, must be displayed on the left end?

18. *Constructing Numbers* How many three-digit numbers can be constructed from the digits 2, 3, 4, 5, 6, 7 if

(a) There are no restrictions?

(b) The numbers constructed must all be even?

(c) The numbers constructed must all be even, and no digit may be repeated?

19. **Constructing Numbers** How many four-digit numbers can be constructed if
    (a) There are no restrictions? (Assume that zero can be the first digit.)
    (b) Zero cannot be the first digit, and no digit can be repeated?
    (c) The first digit is even, no digit can be repeated, and each number constructed must be odd?

20. **Schedules** If three subjects are available for an 8 A.M. class and five other subjects for a 10 A.M. class, how many class schedules are possible for these two classes?

21. **Tossing Coins** A coin is tossed twice. How many possible outcomes are there? How many if the coin is tossed five times? Seven times?

22. **Serial Numbers** Serial numbers assigned to calculators by a manufacturer have as a first symbol J, H, or T to indicate the plant in which the calculator was made, followed by one of the numbers 01, 02, . . . 12 to indicate the month in which made, followed by four digits. How many different serial numbers are possible?

23. **I.D. Numbers** If a firm has 600 employees, what is the smallest number of digits that an employee I.D. number can have if each I.D. number has x digits? What is the smallest number if the firm has 12,000 employees?

24. **Arrangements** In how many ways can four of nine monkeys be arranged in a row for a genetics experiment?

25. **Quiz** In how many ways can a ten-question true-false quiz be answered?

26. **Test** In how many ways can a ten-question multiple-choice test be answered if there are four choices for each question?

27. **Selections** In how many ways can a couple select a bathtub if a tub is made of one of three materials and each material comes in two colors?

28. **Licenses** How many different car license plates can be made using three letters followed by three digits if
    (a) There are no restrictions?
    (b) No letter can be repeated?
    (c) No letter can be repeated, and zero cannot be used as the first digit?

29. **Licenses** How many different car license plates can be made using two letters followed by four digits if
    (a) There are no restrictions?
    (b) No letter can be repeated?
    (c) The last digit must be a 4, and no digit can be repeated?

30. **Call Letters** How many four-letter radio station call letters are there if
    (a) Each must begin with a K or a W?
    (b) Each must begin with a K or a W, and no letter can be repeated?
    (c) Each must begin with a K or a W, no letter may be repeated, and the last letter must be a Z?

31. **Call Letters** How many four-symbol radio station call signals are there that use either all letters or one letter followed by three digits if
    (a) The first letter must be a K or a W?
    (b) The first letter must be a K or a W, and no digit is repeated?
    (c) The first letter must be a K or a W, and no letter or digit may be repeated?

32. **Selections** A microwave oven dealer has in stock three models, each in five colors. In how many ways can a customer select a microwave oven to purchase?

33. **Winners** Thirty runners enter a race in which first-, second-, third-, and fourth-place winners are recognized. In how many ways can these runners place in these four positions?

34. **Code Words** How many three-letter codes are there if
    (a) There are no restrictions?
    (b) The letter z may not be used as the first letter?
    (c) The letter z may not be used as the first letter, and no letter may be repeated?

35. **Selections** In how many ways can a 40-member football team select a captain and co-captain?

36. **Batting Orders** The manager of an amateur baseball team has selected the nine players for an upcoming game. How many batting orders are possible if
    (a) There are no restrictions?
    (b) The pitcher must be last, the first baseman must bat sixth, and the catcher must bat second?

37. **Arrangements** In how many different ways can four of eight people be seated in four chairs?

38. **Arrangements** In how many different ways can five of nine people be seated in five chairs?

39. **Tossing Coins** A coin is tossed three times. How many of these outcomes have
    (a) A head on the first or third toss?
    (b) A head on the first toss or a tail on the second toss?

40. **Code Words** Three-letter code words are to be made from the letters a, b, c, and d, with no repetition of letters allowed. How many three-letter code words are possible if
    (a) The first letter is an a and the last letter a b?
    (b) The first letter is an a or the last letter a b?

41. *I.D. Numbers* Four-digit identification numbers are being made, with repetition of digits within the numbers allowed. How many such identification numbers are possible if

(a) Each must begin with a zero and end with an eight?

(b) Each must begin with a zero or end with an eight?

42. ***Computer Programming*** In certain programming languages, an identifier is a sequence of characters in which the first character must be a letter in the English alphabet and the remaining characters may be either a letter or a digit. For identifiers of length five, how many

(a) Begin with an h and end with a 6?

(b) Begin with an h or end with a 6?

43. *Sampling* From among eight computers, how many samples of three can be selected

(a) Without replacement, where the order of selection is important?

(b) With replacement, where the order of selection is important?

44. *Sampling* From among ten computer disks, how many samples of four can be selected

(a) Without replacement, where the order of selection is important?

(b) With replacement, where the order of selection is important?

Problems for Exploration

1. For each type of counting technique discussed in this section, give examples of how a business or a government agency might need that type to function efficiently.

2. In how many ways can one right and one left shoe be selected from ten pairs of shoes without obtaining a matching pair?

3. If six people are to speak at a convention, in how many orders can they do so with person B speaking immediately before person A?

SECTION 5.5   **Permutations**

Counting problems sometimes involve the product of consecutive positive integers, beginning with 1 and continuing to some fixed integer n. To save massive amounts of writing when n is large, the mathematical shorthand notation ***n!***, read ***"n factorial,"*** is used to denote such a product. For example, $4! = 1 \cdot 2 \cdot 3 \cdot 4 = 24$, and $6! = 1 \cdot 2 \cdot 3 \cdot 4 \cdot 5 \cdot 6 = 720$.

> ***n* Factorial**
>
> Let n be a positive integer. Then the product of the integers from 1 through n is denoted by $n!$, read "n factorial."
>
> $$1 \cdot 2 \cdot 3 \cdot \ldots \cdot n = n!$$
>
> By definition, $0! = 1$.

EXAMPLE 1
Using Factorial
Notation

In how many ways can eight children be seated in a row for a school picture?

Solution   According to the multiplication principle, the answer is

$$8 \cdot 7 \cdot 6 \cdot 5 \cdot 4 \cdot 3 \cdot 2 \cdot 1 = 8!$$
$$= 40,320.$$

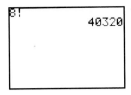

FIGURE 15

Figure 15 shows how a graphing calculator screen might appear upon calculating 8!. Many models of less sophisticated calculators also have a factorial key. ■■

In Example 1, when two children in the row exchange seats, the resulting new arrangement is said to be a **permutation** of the former arrangement. In that sense, our interpretation of the word "permutation" is the same as the dictionary meaning. To *permute* is to *interchange*. In general, then, we may think of successively selecting a first object, then a second object, then a third object, etc., as giving an **ordered outcome,** whereas the selection of these same objects in a different order gives a permutation of the first outcome. Stated differently, a **permutation** is any **ordered arrangement of objects on a line,** and any reordering of those objects on the line is a *new permutation.* Accordingly, a counting problem would be to count the possible number of different permutations of a given set of objects. In Example 1, for instance, there are 8! = 40,320 different permutations, or different orderings, of the eight children when all are seated in a row (line).

In carrying these thoughts further, we might ask about the number of permutations of the letters in a word, for example, "batter." Some of the permutations are batter, abttre, and tarebt. Notice that for any ordered arrangement, the interchange of the two t's *does not* give a new permutation. We say that the two t's are *indistinguishable.* This is in contrast to Example 1 involving children; people are, of course, *distinguishable.* Evidently, the counting process for determining the possible number of permutations in which indistinguishable objects are present will be different from that in which all objects are distinguishable. It seems logical to begin with the case in which all objects are distinguishable, because the multiplication principle will count the number of permutations as it did in Example 1. In fact, we take several aspects of Example 1 to begin our study of permutations.

Selection Rules for Permutations

1. All objects are selected from the *same set S.*

2. All objects in *S* are considered *distinguishable.*

3. Successive selections from *S* are made *without replacement.*

4. Successive selections are considered to be arranged on a line.

When selecting elements (objects) from a given set *S*, it is not necessary that *all* objects in the set be in the ordered arrangement that serves as a permutation. Example 2 illustrates this idea with a set of letters from which the permutations can actually be written down and then counted.

EXAMPLE 2
Counting Permutations

Find the number of permutations of the letters in the set $S = \{a, b, c, d\}$ when selected two at a time.

Solution   In this case the ordered arrangements on a line can be enumerated as follows:

$$
\begin{array}{cccccc}
ab & ba & ac & ca & ad & da \\
bc & cb & bd & db \\
cd & dc
\end{array}
$$

There are 12 such permutations, so we say: "There are 12 permutations of four things taken two at a time," and we use the notation $P(4, 2)$ to denote this fact.

The multiplication principle also gives this same result: There are four choices for the first selection, and, for each of these, there are three choices for the second selection. Therefore, $P(4, 2) = 4 \cdot 3 = 12$. ■

EXAMPLE 3
Computing the Number
of Permutations

In how many ways can the positions of nursing coordinator, insurance benefits coordinator, and records coordinator at Washington Regional Medical Center be filled from among six internal candidates, all of whom are qualified to do any of these three jobs?

Solution   The positions must be filled in some order. Suppose that the nursing coordinator is chosen first, then the insurance benefits coordinator and, finally, the records coordinator. The order in which people are selected for these positions is important in the sense that if Jane is selected for nursing and John is selected for insurance, then John for nursing and Jane for insurance is a new outcome of selections, or a permutation of the first. Of course, once a person is selected for a particular job, he or she cannot be put back into the pool of people eligible for the remaining jobs. The fact that we are to successively select three people, without replacement, from a set of six people in which the order of selection is important provides the conditions for a permutation. Therefore,

$$
\underbrace{6}_{\text{Choices for nursing}} \times \underbrace{5}_{\text{Choices for insurance}} \times \underbrace{4}_{\text{Choices for records}} = 120.
$$

This is the same as asking for the number of ordered arrangements of six people taken three at a time. In the language of permutations, we are saying that "the number of permutations of six things taken three at a time" is 120; this count is denoted by $P(6, 3)$. ■

Permutations

Let S be a set containing n distinguishable objects. Any *ordered arrangement* of r of these objects on a line, where $0 \le r \le n$, obtained by successive selections from S *without replacement* is called a **permutation.** In such a case, we say that we have a **permutation of n things taken r at a time.** The *total number* of such permutations is denoted by $P(n, r)$. (Calculators frequently use the notation $_nP_r$ or nPr.

Since permutations are reordering on a line, the term **linear permutations** is sometimes used to illuminate this fact.

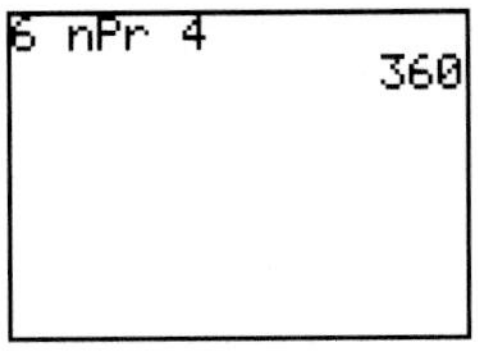

```
6 nPr 4
            360
```

FIGURE 16
*Calculator Screen
Showing a Permutation
Calculation*

EXAMPLE 4
Computing the
Number of
Permutations

Compute $P(6, 4)$.

Solution   We think of successively selecting (without replacement) four elements from a set of six elements. There are six choices for the first selection, five choices for the second, four for the third, and three for the fourth. That is,

$$6 \cdot 5 \cdot 4 \cdot 3 = 360 = P(6, 4).$$

Figure 16 shows how a graphing calculator screen would show a permutation calculation. ▬

EXAMPLE 5
Computing the Number
of Permutations

The order of administering five drugs to a recuperating patient is important. In how many ways can all five of the drugs be administered?

Solution   Think of a set of five drugs from which we are to successively select five without replacement. The total number of ways this may be done is

$$P(5, 5) = 5 \cdot 4 \cdot 3 \cdot 2 \cdot 1 = 5! = 120.$$

This is also the number of possible ordered arrangements of a set of five drugs. ▬

Even though we have seen that the multiplication principle will solve any linear permutation problem, there is a special "permutations formula" that will do the same thing. This formula may be obtained by generalizing the observation that

$$P(7, 3) = 7 \cdot 6 \cdot 5$$

$$= 7 \cdot 6 \cdot 5 \cdot \frac{4 \cdot 3 \cdot 2 \cdot 1}{4 \cdot 3 \cdot 2 \cdot 1} = \frac{7!}{4!} = \frac{7!}{(7-3)!}.$$

In general, the multiplication principle states that

$$P(n, r) = n \cdot (n-1) \cdot (n-2) \cdot \ldots \cdot (n-r+1),$$

which may be rewritten as

$$P(n, r) = n \cdot (n-1) \cdot (n-2) \cdot \ldots \cdot (n-r+1) \cdot \frac{(n-r) \cdot \ldots \cdot 2 \cdot 1}{(n-r) \cdot \ldots \cdot 2 \cdot 1}$$

$$= \frac{n!}{(n-r)!}.$$

> **The Number of Permutations**
>
> $$P(n, r) = n \cdot (n - 1) \cdot (n - 2) \cdot \ldots \cdot (n - r + 1)$$
>
> $$= \frac{n!}{(n - r)!}$$

Because 0! has been defined to be 1, this formula is also correct when $r = 0$ or $r = n$.

EXAMPLE 6
The Number of
Permutations
(By the Formula)

Calculate $P(9, 4)$ by using the preceding formula.

Solution

$$P(9, 4) = \frac{9!}{(9 - 4)!} = \frac{9!}{5!} = \frac{9 \cdot 8 \cdot 7 \cdot 6 \cdot 5 \cdot 4 \cdot 3 \cdot 2 \cdot 1}{5 \cdot 4 \cdot 3 \cdot 2 \cdot 1}$$
$$= 9 \cdot 8 \cdot 7 \cdot 6$$
$$= 3024. \quad \blacksquare$$

We summarize the link between the multiplication principle and permutations by recognizing that counting permutations of distinguishable objects may be done by a straightforward use of the multiplication principle, as well as with the formula we developed. However, not every multiplication principle problem is a permutation problem. (See Example 3 of Section 5.4, for instance.)

Suppose we now consider a set of n elements, not all of which are distinguishable, and ask about the number of permutations possible when selected n at a time.

EXAMPLE 7
Not All Objects to Be
Selected Are
Distinguishable

How many distinguishable linear permutations of the letters $\{a, a, a, b,$ and $c\}$ are there when taken five at a time?

Solution Consider any particular arrangement of these five letters, such as $a, b, a,$ a, c. Now, for the moment, consider the letters a to be distinguishable so they could be labeled as a_1, a_2, a_3. Then there are $3 \cdot 2 \cdot 1 = 3! = 6$ arrangements in the distinguishable case that are not in the distinguishable case:

Indistinguishable	Distinguishable
a, b, a, a, c	a_1, b, a_2, a_3, c
	a_1, b, a_3, a_2, c
	a_2, b, a_1, a_3, c
	a_2, b, a_3, a_1, c
	a_3, b, a_1, a_2, c
	a_3, b, a_2, a_1, c

Let x be the total number of permutations that may be obtained when considering the a's indistinguishable. We have just discovered that for each indistinguishable arrangement, there are 3! distinguishable arrangements. The multiplication principle now tells us that there must be a total of $3! \, x$ permutations when the a's are considered distinguishable. But we also know that there are a total of $5 \cdot 4 \cdot 3 \cdot 2 \cdot 1 = 5!$ arrangements of $\{a_1, a_2, a_3, b, c\}$ in the distinguishable case. Therefore

$$3! \, x = 5!$$

$$x = \frac{5!}{3!} = 20. \quad \blacksquare$$

After one more example, the general pattern for permutations with distinguishable objects will emerge.

EXAMPLE 8
Not All Objects to Be
Selected Are
Distinguishable

Find the number of distinguishable permutations of the letters $\{a, a, a, b, b, b, b, c, c, d\}$ when taken ten at a time.

Solution   Again, let x be the number of such permutations. Reasoning as we did in Example 7, there are $3! \, x$ permutations if we replace the a's by a_1, a_2, and a_3. Likewise, there are $(4!)(3!)x$ permutations if we then replace the b's by b_1, b_2, b_3, and b_4 corresponding to the number of ways in which b_1, b_2, b_3, b_4 can be arranged. Continuing, there are $(2!)(4!)(3!)x$ permutations if we also replace the c's by c_1 and c_2. But we know that there are $10!$ permutations of $\{a_1, a_2, a_3, b_1, b_2, b_3, b_4, c_1, c_2, d\}$. Therefore

$$(2!)(4!)(3!)x = 10!$$

or

$$x = \frac{10!}{2! \, 4! \, 3!}. \quad \blacksquare$$

The pattern emerging from these examples leads to the conclusion that whenever there are k indistinguishable objects in the set from which selections are to be made, a division by $k!$ needs to be made to count only the different permutations. We state this result in more specific terms next.

Permutations with Indistinguishable Objects

Let S be a set with n elements (objects).
Suppose that a subset of k_1 of these elements is indistinguishable, a different subset of k_2 elements is also indistinguishable, a different subset of k_3 elements is also indistinguishable, and so on, and, finally, a different subset of k_m elements is also indistinguishable. Then, the number of distinguishable permutations of these n elements taken n at a time is

$$\frac{n!}{(k_1!)(k_2!)(k_3!) \ldots (k_m!)}.$$

EXAMPLE 9
Permutations with
Indistinguishable
Objects

Find the number of permutations (ordered linear arrangements) of the letters in the word "TALLAHASSEE."

Solution   There are 11 letters in this word, and we consider arrangements of them 11 at a time. Following the above patterns, we get

$$\frac{11!}{3! \, 2! \, 2! \, 2! \, 1! \, 1!} = 831,600. \quad \blacksquare$$

EXAMPLE 10
Permutations with
Indistinguishable
Objects

A box contains five red indistinguishable marbles and three green indistinguishable marbles. The marbles are to be successively selected and arranged in a line. How many such distinguishable linear arrangements are there?

Solution   The answer is

$$\frac{8!}{5!\,3!} = 56. \quad \blacksquare$$

EXERCISES   5.5

Do the calculations required in Exercises 1 through 24, and simplify your answer to a counting number.

1. 6!   **2.** 4!   **3.** 8!

4. $\frac{6!}{3!}$ **5.** $\frac{8!}{4!}$ **6.** $\frac{6!}{4!\,2!}$

7. $\frac{12!}{7!\,5!}$ **8.** $\frac{51!}{48!\,3!}$ **9.** $\frac{52!}{47!\,5!}$

10. $(5!)^2$ **11.** $(2!)!$ **12.** $(3!)!$

13. $P(8,4)$ **14.** $P(12,3)$ **15.** $P(6,1)$

16. $P(10,5)$ **17.** $P(7,7)$ **18.** $P(5,0)$

19. $P(10,0)$ **20.** $P(12,12)$ **21.** $P(4,3)$

22. $P(4,1)$ **23.** $P(0,0)$ **24.** $(P(6,2))^2$

In Exercises 25 through 28:
(a) Restate each in permutation notation.
(b) Then solve the problem.

25. *Arrangements* In how many ways can eight books be arranged four at a time on a shelf?

26. *Elections* In how many ways can three of ten people be elected to the positions of president, secretary, and treasurer?

27. *Arrangements* In how many ways can seven monkeys be arranged in a row for a poster picture?

28. *Arrangements* How many arrangements are there of the letters in the word *computer?*

29. *Political Science* In how many ways can five candidates for an office be listed on a ballot?

30. *Political Science* A state political convention is selecting candidates for the office of Governor and Lieutenant Governor from among six candidates. A publicity firm has been hired to design campaign buttons for each possible outcome before the convention. How many different buttons must be designed?

31. *Elections* The eight-person board of directors of the Acme Corporation is to elect a president, a vice president, and a treasurer, and the remaining members will be a committee to study future expansion. In how many ways can these officers be elected?

32. *Appointments* The 12-person executive committee of the Flying Ace Corporation is to select an environmental compliance officer, a federal governmental relations officer, a state governmental relations officer, and a county governmental relations offiicer. In how many ways can these officers be selected?

33. *Music* The Music Department at State University will give a concert consisting of five pieces of music. In how many ways can these be listed in the program?

34. *Awards* Six members of the Bryan Fire Department are to receive awards for outstanding performance of duties during the past year. In how many ways can the awards ceremony be arranged?

35. *Arrangements* How many linear arrangements are there of the letters in the word *page?*

36. *Arrangements* How many linear arrangements are there of the letters in the word *beautiful?*

37. *Arrangements* How many permutations are there of the letters in the word *college?*

38. *Arrangements* How many permutations are there of the letters in the word *intelligible?*

39. *Psychology* For an experiment twelve psychology students are to be divided into two groups, one containing seven students and the other containing five students. In how many ways can this grouping be done?

40. *Spring Break* Fifteen students are going skiing on their spring break. They plan to travel in three vehicles—one seating seven, one seating five, and one seating three students. In how many ways can the students group themselves for their trip?

41. *Performer* A star musical performer is giving a concert at State University. At the concert, she will present 13 numbers, four of which are country and western, three of which are rock 'n' roll, and six of which are pop numbers. In how many ways can the star arrange her concert if songs within a category are to be considered indistinguishable?

42. *Selections* Six indistinguishable red balls and four indistinguishable green balls are arranged in a line. How many distinguishable arrangements of these balls are possible?

43. *Selections* Five indistinguishable red balls, three indistinguishable green balls, and two indistinguishable blue balls are arranged in a line. How many distinguishable arrangements of these balls are possible?

44. *Arrangements* Ten books are to be arranged on a shelf. Four of the books have red covers, three have green covers, and another three have gray covers. In how many ways can these books be arranged on a shelf if books of the same color must be arranged together?

45. *Constructing Numbers* How many two-digit or three-digit I.D. numbers can be constructed using the digits 1, 3, 4, 5, 6, 8, and 9 if no repetition of digits is allowed?

46. *Constructing Code Words* How many three-letter or four-letter code words can be made if no repetition of letters is allowed?

Problems for Exploration

1. A firm has 750 employees. Explain why at least two of the employees would have the same pair of initials for their first and last names.

2. In how many ways can three people be seated around a round table? In how many ways can *n* people be seated around a round table?

SECTION 5.6 **Combinations**

Another important counting problem arises when we ask, "How many five-card hands can be dealt from a standard deck of 52 cards?" The cards are selected in succession, without replacement, from a standard deck of 52. However, once the five cards have been dealt (selected), any linear reordering of those cards is *not* considered a new

---

Standard Deck of 52 Cards

4 suits →	Clubs(C)	Diamonds(D)	Hearts(H)	Spades(S)
13 denominations ↓	Ace of C	Ace of D	Ace of H	Ace of S
	King of C	King of D	King of H	King of S
	Queen of C	Queen of D	Queen of H	Queen of S
	Jack of C	Jack of D	Jack of H	Jack of S
	10 of C	10 of D	10 of H	10 of S
	9 of C	9 of D	9 of H	9 of S
	8 of C	8 of D	8 of H	8 of S
	. . .	. . .	. . .	. . .
	2 of C	2 of D	2 of H	2 of S
	↑ All are black	↑ All are red	↑ All are red	↑ All are black

(The kings, queens and jacks are *face cards*. Depending on the game, aces may be the highest or lowest card in the deck.)

---

hand. In the final analysis, the order of selection is unimportant; we are only interested in finding out how many *different subsets* of five cards can be created from the cards in the deck. In this sense, we might rephrase the original question as "How many subsets of five cards each are possible in a deck of 52 cards?" Here are two other examples of this type of counting.

Suppose the Internal Revenue Service has 50 personal income tax forms from which they are to select a sample of six to check for accuracy. In how many ways can such a sample be obtained? The forms will be randomly selected, in succession, without replacement, from the 50 distinguishable forms. Once selected, the interest is not in the order of their selection, but rather their accuracy. In other words, any reordering of the six forms does not give a new sample of six to be checked for accuracy. Once again, the question might have been rephrased as "How many sets of six tax forms can be created from among 50 forms?"

In how many ways can a committee of three be selected from among 12 people? Once again, a person selected is not returned to the pool of people from which the remaining committee members are chosen. Furthermore, once a committee is chosen, any rearrangement of its members *does not* result in a new committee.

Whenever objects are successively selected from a set of distinguishable objects and any rearrangement of that outcome does not give a new outcome, we say that we have a **combination** of those objects. Stated differently, a combination of objects is a set of objects; the order is not important. In general, counting problems having outcomes such as these are called **combination problems.** There are three central themes running through these examples and, hence, through all combination problems. They are stated next.

Selections Rules for Combinations

1. Objects are successively selected from a single set *without* replacement.

2. The objects in the set are considered *distinguishable.*

3. Any permutation (interchange, rearrangement) within a given outcome *does not* give a new outcome to be counted.

Combinations

Let S be a set of n distinguishable elements. The successive selection of r elements, where $0 \leq r \leq n$, *without regard to order*, is called a **combination of n things taken r at a time.** The total number of combinations of n things taken r at a time is denoted by $C(n, r)$.

NOTE   Other notations for the total number of combinations of n things taken r at a time are $_nC_r$ and $\binom{n}{r}$. The first is a common notation for calculators.

NOTE   Again, we assume that sets contain only distinguishable objects unless specifically noted. In particular, if a sample is to be taken from 100 light bulbs, think of each bulb as having a different number written on it. If a sample of balls is to be taken from a box containing 20 red balls, 8 green balls, and 3 purple balls, think of each of the 31 balls as having a different number written on it, and so on.

EXAMPLE 1
Comparing
Combinations with
Permutations

For selections from the set $S = \{a, b, c, d\}$, compare $C(4, 3)$ with $P(4, 3)$.

Solution   A table will enumerate the possible outcomes for $C(4, 3)$ and $P(4, 3)$.

Outcomes in $C(4, 3)$	Outcomes in $P(4, 3)$
abc	abc, acb, bac, bca, cab, cba
abd	abd, adb, bad, bda, dab, dba
acd	acd, adc, cad, cda, dac, dca
bcd	bcd, bdc, cbd, cdb, dbc, dcb
Total: 4	Total: 24

In Example 1, $C(4, 3)$ is smaller than $P(4, 3)$. In fact, for each combination, there are $3 \cdot 2 \cdot 1 = 3!$ permutations. Hence, $C(4, 3) \cdot 3! = P(4, 3)$. In general, $C(n, r)$ is smaller than $P(n, r)$ because, within each selection of r elements, there are $r!$ permutations that are counted in $P(n, r)$ but are not counted in $C(n, r)$. Therefore

$$C(n, r) \cdot r! = P(n, r)$$

or

$$C(n, r) = \frac{P(n, r)}{r!} = \frac{1}{r!} \cdot \frac{n!}{(n - r)!} = \frac{n!}{r! \, (n - r)!}.$$

The Number of Combinations

$$C(n, r) = \frac{n!}{r! \, (n - r)!}.$$

EXAMPLE 2
Calculating with the
Combination Formula

(a) $C(8, 5) = \dfrac{8!}{5! \cdot 3!} = \dfrac{8 \cdot 7 \cdot 6 \cdot 5 \cdot 4 \cdot 3 \cdot 2 \cdot 1}{5 \cdot 4 \cdot 3 \cdot 2 \cdot 1 \cdot 3 \cdot 2 \cdot 1} = 56.$

(b) $C(7, 2) = \dfrac{7!}{2! \cdot 5!} = \dfrac{7 \cdot 6 \cdot 5 \cdot 4 \cdot 3 \cdot 2 \cdot 1}{2 \cdot 1 \cdot 5 \cdot 4 \cdot 3 \cdot 2 \cdot 1} = 21.$

(c) $C(5, 5) = \dfrac{5!}{5! \cdot 0!} = \dfrac{5 \cdot 4 \cdot 3 \cdot 2 \cdot 1}{5 \cdot 4 \cdot 3 \cdot 2 \cdot 1 \cdot 1} = 1.$

(d) $C(4, 0) = \dfrac{4!}{0! \cdot 4!} = \dfrac{4 \cdot 3 \cdot 2 \cdot 1}{1 \cdot 4 \cdot 3 \cdot 2 \cdot 1} = 1.$

EXAMPLE 3
Computing the
Number of
Combinations

How many different five-card hands are there in a deck of 52 cards?

Solution   A card hand is considered to be the successive selection of five cards drawn without replacement from a standard deck of 52 cards in which any reordering within the five cards drawn does not give a new hand. Therefore the answer is

$$C(52, 5) = \frac{52!}{5! \cdot 47!} = 2{,}598{,}960.$$

Figure 17 shows how this same calculation might appear after being done on a graphing calculator.

EXAMPLE 4
A Sample of
Light Bulbs

The quality control unit in a light bulb factory is to test a sample of five light bulbs selected from among 60 bulbs. How many different samples, selected without replacement, are possible?

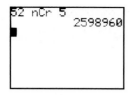

FIGURE 17

Solution   We assume that once five bulbs have been selected, any rearrangement of those bulbs does not give a new sample to be tested. Therefore, the answer is

$$C(60, 5) = \frac{60!}{5! \cdot 55!} = 5,461,512. \quad \blacksquare$$

EXAMPLE 5
Linking Combinations
with the Multiplication
Principle

Among 20 calculators, eight are defective. In how many ways can six calculators be selected from among the 20, without replacement and without regard to order, in which

(a) All are defective?

Solution   We are to select only from the eight defective calculators. Since calculators are selected without replacement and since the order of selection is immaterial, this question could be rephrased as "find the number of combinations of the 8 defective calculators taken 6 at a time." Therefore the answer is

$$C(8, 6) = \frac{8!}{6!\,2!} = 28.$$

(b) Exactly four in the sample will be defective?

Solution   This means that four of the calculators selected are to be defective and two are to have no defect, making a total of six selected. Think of the calculators being divided into two subsets (boxes), the first with the eight defective ones and the second with the 12 with no defect. Selecting from the front box, in which order is unimportant, we find that there are $C(8, 4)$ ways of selecting four defective calculators, and, for each of these ways, there are $C(12, 2)$ ways of selecting the two remaining calculators from the second box. Since we are selecting from different subsets, the multiplication principle tells us that these two numbers should be multiplied together to give the total number of ways of selecting four defective calculators, followed by two with no defect. However, once selected, any reordering is not considered to be a new selection of calculators. Therefore, the answer is given by the ordering in which the initial selections were made:

$$C(8, 4) \cdot C(12, 2) = \frac{8!}{4!\,4!} \cdot \frac{12!}{2!\,10!} = 4620. \quad \blacksquare$$

Defect   No Defect

3

EXAMPLE 6
An Application of
Combinations

State University, State Tech, and Hendrix College each send four representatives to a conference on student government. At that conference, a committee of five is to be selected from among these 12 representatives. In how many ways can this committee be selected if

(2 5)

(a) There are no restrictions?

Solution   Committee selection would be without replacement; once selected, any rearrangement of a committee does not give a new committee. (Contrast this with the selection of people for specific titles.) Therefore, all 12 representatives are eligible for selection, and the question asks, "How many combinations of 12 things taken five at a time are there?" The answer is

$$C(12, 5) = 792.$$

(b) Exactly two of the committee members must come from Hendrix College.

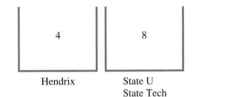

4 rep. 2

$$C(4, 2) \cdot C(8, 3)$$

4   8

Hendrix   State U
             State Tech

Solution   Think of the 12 representatives divided into two subsets, one containing the four Hendrix College students and the other containing the eight students from State University and State Tech. We are to select two students from the first-mentioned set and the remaining three students from the second-mentioned set; in each case, the order of selection is not important. The multiplication principle tells us that we should multiply the number of possible selections in each case together to get the total number of committees in which Hendrix students are selected first and the others second. But once a committee is selected, all reorderings give the same committee. Therefore the answer is

$$C(4, 2) \cdot C(8, 3) = 6 \cdot 56 = 336.$$

(c) Two students must come from State University, two from State Tech, and one from Hendrix College.

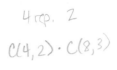

$$C(4, 2) \cdot C(4, 2) \cdot C(4, 1)$$

4   4   4

Hendrix   State U   State Tech

Solution   The number of possible committees under these conditions is

$$C(4, 2) \cdot C(4, 2) \cdot C(4, 1) = 6 \cdot 6 \cdot 4 = 144. \quad \blacksquare$$

EXAMPLE 7
An Application
of Combinations

Seven cards are drawn without replacement from a deck of 52 cards. In how many ways can this be done if

(a) All of them must be clubs?

Solution   There are 13 clubs, so there are $C(13, 7) = 1716$ different seven-card hands consisting of clubs only.

(b) All of them must come from the same suit?

Solution   First, select a suit. There are $C(4, 1) = 4$ ways to do this. Now for each selection of a suit there are $C(13, 7)$ ways of selecting the seven cards. The multiplication principle then gives a final answer of

$$4 \cdot C(13, 7) = 6864 \text{ ways.} \quad \blacksquare$$

Reflect on Examples 5, 6, and 7 to see common threads. All may be thought of as a deck of cards, for example, in which there may be more or less than four suits, and each suit need not have the same number of cards. In Example 5, think of a deck of cards with two "suits," *defective* with eight cards and *no defect* with 12 cards. In Example 6, think of a deck of cards with three "suits," State University, State Tech, and Hendrix College, each having four cards. So in a sense, any problem asking for the number of possible outcomes, selected without replacement and in which order is not important, can be thought of as a card hand if you wish.

EXAMPLE 8
Combinations and
Inclusion-Exclusion

A box contains five red, three green, and five blue balls.

(a) In how many ways is it possible to select a sample of three balls, without replacement, in which all of the balls are red or all of the balls are green?

Solution   By the inclusion-exclusion principle, n(all red or all green) $= n$(all red) $+ n$(all green) $- n$(all red and all green). The last term is zero, so the answer is n(all red) $+ n$(all green) $= C(5, 3) + C(3, 3) = 10 + 1 = 11$.

(b) In how many ways is it possible to select a sample of four balls, without replacement, in which at least three are red?

Solution   We translate "at least three are red" into the logically equivalent ("three are red, and one is not red") or ("all four are red"). Because it is impossible for the latter two statements to happen simultaneously, the inclusion-exclusion principle states that our answer is found by n (three are red, and one is not red) $+ n$ (all four are red) $=$

$$C(5, 3) \cdot C(8, 1) + C(5, 4) = 10 \cdot 8 + 5 = 85. \quad \blacksquare$$

EXAMPLE 9
An Application of
Combinations

How many different arrangements in a row can be made from three S's and five F's in which all letters are used each time?

Solution   Note that for any arrangement of these eight letters, such as

S F S F F F S F,

any interchange of elements among the S's (or the F's) does not give a different arrangement to be counted, because the S's are indistinguishable. First, find the number of ways to place three S's among the eight slots so that any interchange of any of the S's does not give a new arrangement. There are $C(8, 3)$ ways. How many ways are there of placing F's in the remaining slots so that any interchange does not give a new arrangement? The number is 1. The multiplication principle then gives $C(8, 3) \cdot 1 = 56 \cdot 1 = 56$ as the answer. (We could have selected slots for the five F's first, then the S's, and then obtained the answer of $C(8, 5)$. However $C(8, 5) = 56$ also.) $\quad \blacksquare$

We could have gotten the answer to Example 9 by using permutations of indistinguishable objects; consider both the S's and the F's to be indistinguishable:

$$\frac{8!}{3!\,5!} = 56.$$

But this is precisely the meaning of combinations: Objects are considered distinguishable during the *selection* process, but once selected, they are considered indistinguishable!.

EXERCISES 5.6

In Exercises 1 through 10, simplify each combination to a counting number.

1. $C(6, 2)$ **2.** $C(8, 5)$ **3.** $C(9, 3)$ **4.** $C(12, 4)$

5. $C(15, 2)$ **6.** $C(52, 3)$ **7.** $C(48, 4)$ **8.** $C(12, 6)$

9. $C(14, 7)$ **10.** $C(12, 0)$

11. (a) Calculate $C(5, 2)$ and $C(5, 3)$.
   (b) Calculate $C(9, 4)$ and $C(9, 5)$.
   (c) Calculate $C(8, 5)$ and $C(8, 3)$.
   (d) State a rule you discovered about combinations while working Exercise 11(a) through (c).

12. Calculate $C(5, r)$ for all possible values of r.

13. Calculate $C(6, r)$ for all possible values of r.

In the remaining exercises, assume that sets contain distinguishable elements (objects). Imagine they are numbered if you wish.

14. *Teams* How many tennis doubles teams can be formed from 12 players?

15. *Geometry* How many straight lines are determined by five points in the plane, no three of which are on the same line?

16. *Card Hands* How many four-card hands are possible from a standard deck of 52 cards?

17. *Card Hands* How many three-card hands are possible from a standard deck of 52 cards?

18. *Sampling* Among 18 computers, 12 are in working order and six are defective. How many samples of four are possible, selected without replacement and without regard to order, where all are defective?

19. *Sampling* Among 23 computer modems, 15 are in working order and eight are defective. How many samples of six are possible, selected without replacement and without regard to order, if all are in working order?

20. *Sampling* A warehouse contains 12 refrigerators, two of which are not in working order. In how many ways can a sample of five of these refrigerators be

selected, without replacement and without regard to order, if

(a) All are to be in working order?
(b) Exactly two are not in working order?
(c) All five are not to be in working order?

21. *Sampling* A box contains 20 computer disks, five of which are known to have bad sectors. In how many ways can three of these disks be selected, without replacement and without regard to order, so that

(a) None have bad sectors?
(b) All have bad sectors?
(c) Exactly two do not have bad sectors?

22. *Selections* A committee of four is to be selected from among eight graduate students and a professor to meet with the Dean about new classroom equipment. In how many ways can this committee be selected if

(a) The professor cannot be on the committee?
(b) The professor must be on the committee?
(c) There are no restrictions?

23. *Selections* Five people from among the mayor and seven other city officials are to be selected to make an inspection of solid-waste facilities. In how many ways can these five people be selected if

(a) The mayor must be included in the five?
(b) The mayor is not to be among the five selected?
(c) There are no restrictions?

24. *Selecting a Committee* Companies Alford *and* Baker each send three representatives to a conference. At the conference, a committee of four is to be selected from among these representatives. In how many ways can this be done if

(a) All are to come from Alford's?
(b) Exactly two are to come from Alford's?
(c) There are no restrictions?

25. *Selecting a Committee* Data Fix, Data Mix, and Data Six each send five representatives to a conference. At the conference, a committee of four from among these company representatives is selected. In how many ways can this be done if

(a) All are to come from Data Fix?
(b) None are to come from Data Fix?
(c) All are to come from Data Mix or Data Six?
(d) All are to come from Data Mix or all are to come from Data Six?

26. **Card Hands** In how many ways can five cards be drawn, without replacement, from a deck of 52 cards if

(a) All are to be clubs?
(b) Exactly two are clubs, two are diamonds, and one is a spade?
(c) All are to be clubs or diamonds?
(d) All are to be hearts or all are to be kings?

27. **Card Hands** Eight cards are to be drawn, without replacement, from a deck of 52 cards. In how many ways can this be done if

(a) All are to be spades?
(b) All are to be from the same suit?
(c) Exactly five are to be clubs?
(d) All are to be hearts or all are to be spades?

28. **Sampling** There are five red and seven green balls in a box, and a sample of three is to be chosen without replacement and without regard to order. In how many ways can this be done if

(a) None are to be green?
(b) At least two are to be green?

29. **Selecting a Committee** From among three conservatives and five liberals, a committee of three is to be selected. In how many ways can this be done if

(a) None are to be conservatives?
(b) Exactly one is to be a conservative?
(c) At least two are liberals?

30. **Card Hands** In how many ways can a three-card hand be dealt from a standard deck of 52 cards if

(a) All are to be red cards?
(b) All are to be face cards?
(c) At least one is to be a face card?

31. **Card Hands** In how many ways can a four-card hand be dealt from a standard deck of 52 cards if

(a) All are to be black cards?
(b) All are to be 8s?
(c) At least two are to be hearts?

32. **Sampling** A box contains four red, three green, and five blue marbles. In how many ways can three marbles be selected, without replacement and without regard to order, if

(a) All are to be blue?
(b) All are to be red or all are to be blue?
(c) Exactly one is red?

33. **Sampling** A box contains five red, six blue, and seven purple balls. In how many ways can four balls be selected, without replacement and without regard to order, if

(a) All are to be blue or all are to be purple?
(b) Exactly two are to be blue?
(c) Exactly one is to be purple?

34. **Selections** From a company of 15 soldiers, a squad of four is selected for patrol duty each night.

(a) For how many nights in a row could a squad go on duty without two of the squads being identical?
(b) In how many ways could the squad be chosen if Joe, a soldier in the company, must always be on the squad?
(c) In how many ways could the squad be chosen if Joe, a soldier in the company, can never be on the squad?

35. **Arrangements** How many different arrangements are there of four F's and seven S's in a row?

36. **Subcommittees** In how many ways can a committee of nine be divided into two subcommittees, one having five members and the other having four members?

37. **Sampling** In a lot of 100 light bulbs, six are defective. In how many ways can a sample of three be drawn from this lot, without replacement and without regard to order, if

(a) All are good?
(b) All are defective?
(c) Exactly two are defective?

38. **Card Hands** Six cards are to be drawn (or dealt), without replacement, from a deck of 52 cards. In how many ways can this be done if

(a) All are to be spades?
(b) All are to be kings?
(c) None are to be diamonds?
(d) All are to be clubs or all are to be kings?

39. **Sampling** A sample of five marbles is to be selected, without replacement and without regard to order, from a box that contains three red, four green, and six purple marbles. In how many ways can this be done if

(a) Exactly two are to be red, two are to be green, and one is to be purple?
(b) All are red or purple?
(c) All are to be red or all are to be purple?
(d) At least four are to be purple?

40. *Sampling*   A box contains two red, five green, six black, and four blue jellybeans. In how many ways can a sample of four jellybeans be selected from this box, without replacement and without regard to order, if

(a) All are to be the same color?
(b) Exactly two are to be the same color?

41. *Sampling*   A shipment of eight men's suits has three of size 38, three of size 40, and two of size 42. In how many ways can a sample of three suits be selected for a style show if

(a) All are the same size?
(b) Exactly two are the same size?

42. *Pizza*   A pizza chain advertises that it has ten toppings to choose from. How many ways are there of ordering a pizza from this firm with up to four toppings? (Assume a pizza has at least one topping.)

Problems for Exploration

1. Would this symbol make any sense: $C(2, 5)$? Why or why not?

2. Would this symbol make any sense: $(6, -3)$? Why or why not?

3. Write a summary of the counting techniques discussed in Sections 5.4, 5.5, and 5.6, focusing on the importance of order within a given outcome.

4. Do direct calculations with the symbols involved, to show that $C(n, r) = C(n, n - r)$.

5. Do direct calculations with the symbols involved, to show that $C(n, r - 1) + C(n, r) = C(n + 1, r)$.

6. Suppose there are k_1 indistinguishable objects of one kind and k_2 indistinguishable objects of another kind. Show that the number of distinguishable linear permutations of these $k_1 + k_2$ objects can be found by $C(k_1 + k_2, k_1) = C(k_1 + k_2, k_2)$.

7. Why should a "combination lock" be called a "permutation lock"?

SECTION	5.7

Other Counting Techniques: A Final Summary

In this section, we continue our study of counting methods and then summarize all of those we have studied. To help us recognize which method should be applied to a given problem, the exercises at the end of this section randomly use all of the methods discussed in this chapter.

Sometimes it is much easier to get a count of the outcomes that we do not want than of the ones we do. In such cases, we can get the count of outcomes we do want by subtracting the outcomes we do not want from the total number of outcomes possible. The next example shows how this may be done.

easier than to do a combt add them

EXAMPLE 1
Counting What We Want By Using a Count of What We Do Not Want

How many three-digit numbers from 000 to 999 have a 7 in them?

Solution   It is easy to count the numbers in this list that do not have a 7 in them: There are nine choices for the first digit, nine for the second digit, and nine for the third, so that there are $9 \cdot 9 \cdot 9 = 729$ such numbers. Because there are a total of $10 \cdot 10 \cdot 10 = 1000$ numbers from 000 to 999, the remainder, $1000 - 729 = 271$, must have a 7 in them somewhere.   ■

EXAMPLE 2
Counting What We Want By Using a Count of What We Do Not Want

In how many ways can a five-card hand be dealt so that it contains at least one spade?

Solution   There are $C(52, 5)$ possible five-card hands that may be dealt. Now, if we remove from these all five-card hands with *no* spades in them, $C(39, 5)$, then the remaining hands, $C(52, 5) - C(39, 5) = 2{,}023{,}203$, must have at least one spade.   ■

The counting technique employed in Example 2 can be used efficiently anytime a counting question asks for "at least one." The reason is that those outcomes that have "at least one" are logically equivalent to counting "the total number of outcomes − those that have none." Similarly, counting "at least two" is equivalent to counting "the total number − (none or one)."

We now examine the problem posed at the outset of this chapter.

EXAMPLE 3
Counting Blood Samples

The blood types are A^+, A^-, B^+, B^-, AB^+, AB^-, O^+, and O^-. How many different samples of five can be obtained from these eight blood types if each sample is to obtain *at least one* of the O types?

Solution The total number of ways of selecting different samples of five is $C(8, 5)$. Now, if we subtract from that number all of those that have no O type, $C(6, 5)$, the remaining samples must contain at least one of the O types. Therefore, the answer is $C(8, 5) - C(6, 5) = 56 - 6 = 50$.

Alternative Solution Translating "at least one O type" into "exactly one O type and the other four not" or "two O types and the other three not" will give

$$C(2, 1) \cdot C(6, 4) + C(2, 2) \cdot C(6, 3) = 2 \cdot 15 + 1 \cdot 20 = 50.$$

Seating people round a table represents a slightly different counting situation than that of arranging people in a row. For instance, in how many ways can five people be seated around a round table? Arrangements are determined by who sits on the right and left of a person. If the five people are seated, and they all get up and move one seat to the right, this is considered the same arrangement. To answer the question, first seat one of the five people. Then seat the remaining four people in a clockwise (or a counterclockwise) direction from the first seated person. This, in effect, changes our circular problem into a straight line problem. Thus, there are four choices for the first seat, then three for the next seat, then two and finally one for $4! = 24$ ways. In general, if there are n people to be seated around a round table, there are $(n - 1)!$ ways of doing so.

Chapter 6, on probability, poses counting problems of all the various types that we have studied. This, in turn, calls for knowing the proper method of counting to be applied to the situation at hand. A summary of these methods is given next.

Guide to Counting Techniques

Problems solved by Venn diagrams:
  The distinctive wording of Venn diagram problems points toward their use. Label separate regions as shown in Section 5.3 to remove any "double-counting."
Problems that involve the connective *or*:
  When *or* connects two sentences, use the inclusion-exclusion principle or a Venn diagram.
Problems that involve the words *at least*:
  Translate into a logically equivalent problem using *or* connectives, or use a count of what you do not want.
The multiplication principle:
  When making successive selections from the same set of distinct elements, with or without replacement, when the order of selection is important, apply this principle. (Any reordering within a given outcome gives a new outcome to be counted.) Apply the multiplication principle when making successive selections from different sets.

Permutations:

For distinguishable objects, permutations are a special case of the multiplication principle. There is also a special formula, $P(n, r)$. For indistinguishable objects, there is another special formula, $\dfrac{n!}{(k_1!)(k_2!)(k_3!) \ldots (k_m!)}$.

Combinations:

Use combinations when successive selections are made without replacement from the same set of distinct elements and when, once the selections are made, any reordering within an outcome does not give a new outcome to be counted. The formula $C(n, r)$ for combinations is important.

Count what you want by counting what you do not want:

May be used on any counting problem but is very useful on problems involving *at least one*.

Problems may call for combining one or more of these methods. The exercises at the end of this section test your ability to apply the proper techniques.

EXERCISES 5.7

The first five exercises in this set involve concepts discussed in this section. The remaining exercises involve all types of counting problems introduced in this chapter.

In Exercises 1 through 5, use a count of what you do not want to get a count of what you do want.

1. *Counting Numbers* How many numbers from 00 to 99 have a 3 in them?

2. *Counting Cards* How many nine-card hands have at least one king? one or 2, 3, etc. to 9

3. [icon] *Sampling* From among 80 light bulbs, ten are defective. How many samples of eight have at least one defective bulb?

4. *Card Hands* How many eight-card hands have at least two queens? 2 up to 8

5. *Selecting a Committee* At the Boy Scout camp council, there are two boys from each of the states Iowa, New Mexico, Florida, and Maine. In how many ways can a committee of four be formed from among these boys, in which there is at least one boy from Maine?

6. *Arrangements* In how many ways can five of eight books be arranged on a shelf?

7. *Elections* In how many ways can a president, secretary, and treasurer be elected from among ten people?

8. *Rolling a Die* A six-sided die is rolled once. How many ways are there of rolling

(a) An odd number?
(b) A number greater than three?
(c) A number less than three or greater than four?
(d) A number less than three and greater than four?

9. *Arrangements* In how many ways can eight people be seated around a round table?

10. *Forming a Committee* In how many ways can a committee of five be selected from among ten people?

11. *Cards* One card is drawn from a deck of 52 cards. In how many ways can this be done if it is to be

(a) A king? consider only kings to begin w/
(b) A king or a spade?
(c) A king and a spade?
(d) A king and a queen? not possible
(e) Not a king or not a queen?
(f) A king or not a spade?

12. [icon] *Experiment Categories* A psychology experiment observes groups of five individuals. In how many ways can the experiment take 15 people and group them into three groups of five?

13. *Zip Codes* For five-digit postal zip codes,

(a) How many are possible if there are no restrictions on the digits?
(b) How many are possible if zero is not allowed as a first digit?

14. *Arrangements* Four boys and two girls are arranged in a row. In how many ways is this possible if

(a) There are no restrictions?

(b) Boys and girls are seated alternately?

(c) The two girls must be seated side by side?

15. *Survey* A survey of 60 shoppers reveals that in the past week, 20 bought toothpaste, 15 bought deodorant, and eight bought both of these items. How many of these shoppers

(a) Bought toothpaste or deodorant last week?

(b) Bought exactly one of these two items?

16. *Arrangements* In how many ways can seven monkeys be arranged in a row for a poster picture?

17. *Sampling* In how many ways can three refrigerators be selected, without replacement and without regard to order, from among eight refrigerators?

18. *Geometry* Seven distinct points are marked on a circle. How many triangles can be drawn, using these points as vertices?

19. *Tossing Coins* How many possible outcomes are there if a coin is tossed eight times?

20. *Samples* A bag contains eight red apples and four yellow apples. In how many ways can a shopper select a sample of three apples, without replacement and without regard to order, if

(a) All are to be red?

(b) There are no restrictions?

(c) At least two must be yellow?

21. *Arrangements* In how many ways can the letters in the word *turkey* be arranged in a row if

(a) There are no restrictions?

(b) The first letter must be a *k*?

(c) The first letter must be *r* and the last letter *t*?

22. *Arrangements* In how many distinguishable ways can a row of two Ss and nine Fs be arranged?

23. *Survey* Fifty students were asked about their purchase of candy during the past week. Twenty said they bought gumdrops, 12 said they bought candy bars, five said they bought both gumdrops and orange slices, three said they bought both gumdrops and candy bars, two said they bought both candy bars and orange slices, two said they bought all three of these types of candy, and 12 said they bought none of these types of candy. How many of these students bought

(a) Exactly one of these types of candy?

(b) At least one of these types of candy?

(c) Orange slices or candy bars but not gumdrops?

24. *Teams* Ten students are going to play basketball. In how many ways can two teams of five be formed?

25. *License Plates* How many license plates are possible using three letters followed by two digits if

(a) No letter can be repeated?

(b) A plate must begin with an *A*, and no letter or digit can be repeated?

(c) The letters *A*, *B*, and *C* cannot be used, the digit 0 cannot be used, and no letter or digit can be repeated?

26. *Card Hands* In how many ways can a five-card hand be dealt if

(a) There are no restrictions?

(b) All must be kings?

(c) Exactly two must be queens?

(d) All are to be hearts or all are to be spades?

27. *Arrangements* In how many ways can seven people be arranged in a row?

28. *Ranking Teams* A sportswriter is asked to rank eight teams. How many rankings are possible if

(a) There are no restrictions?

(b) A particular team, the Mishaps, must be rated last?

29. *Group Projects* A group of ten mathematics students is to work on two group projects, one relatively difficult and one relatively easy. In how many ways can they be divided so that six will work on the more difficult project and four will work on the less difficult project?

30. *Selecting a Committee* From among five women and seven men, four are to be selected as sales representatives. In how many ways can this be done if

(a) All must be women?

(b) Exactly two must be men?

(c) At least three must be men?

(d) All are to be men or all are to be women?

31. *Constructing Numbers* How many four-digit numbers can be made from the digits 1, 2, 3, 4, 5, and 6 if

(a) There are no restrictions?

(b) The numbers formed must be odd?

(c) The numbers formed must be odd, and no digit can be repeated?

32. *Arrangements* How many possible arrangements of the letters in the word *football* are there?

33. *Testing* A professor has 12 questions about permutations and 15 questions about combinations in a computer test bank. If the computer is to randomly draw questions from this question bank for a five-question quiz, how many tests may be generated if

(a) All questions are about permutations?

(b) All questions are about combinations?

(c) There are no restrictions?

34. **Downsizing**   A firm plans to close four of its current 12 plants. In how many ways can the four be chosen if
   (a) The order of closing is not important?
   (b) The order of closing is important?

35. **Expansion**   A pizza chain plans to add a single new store in eight of 15 targeted cities. In how many ways can these eight cities be selected if
   (a) The order of expansion is not important?
   (b) The order of expansion is important?

36. **Defects**   Among 20 computers, four have defective screens, two have defective keyboards, and one has both of these defects. How many of these computers have neither of these defects?

37. **Sampling**   A sample of four textbooks is to be selected from among ten textbooks. How many samples are possible if
   (a) There is no replacement, and the order of selection is not important?
   (b) There is no replacement, and the order of selection is important?
   (c) There is replacement, and the order of selection is important?

38. **Pizza**   If eight toppings are available to you when ordering a pizza, in how many ways can you order a pizza with up to five toppings? (Assume that a pizza must have at least one topping.)

39. **Sampling**   A box contains four red and seven green marbles. Two marbles are drawn without replacement and without regard to order. In how many ways is this possible if
   (a) One is red and one is green?
   (b) At least one is red?
   (c) Both are to be red or both are to be green?

40. **Sampling**   A box contains two red, four green, six black, and three blue balls. In how many ways can a sample of three balls be selected, without replacement and without regard to order, if
   (a) All of the balls are to be red?
   (b) All are the same color?
   (c) Exactly two of the balls are the same color?

41. **Ratings**   An investment firm rates a group of 18 stocks. In how many ways could they rate three of them AAAA, six of them AAA, four of them AA, and five of them A?

42. **Selections**   Four cards are drawn, without replacement, from a deck of 52. In how many ways can this be done if
   (a) Exactly two are to be spades?
   (b) There are to be no spades?
   (c) All are to be kings or all are to be queens?

43. **Test**   In how many ways can an eight-question true-false test be answered?

44. **Test**   In how many ways can a multiple-choice test of six questions be answered if there are three choices for each answer?

45. **Sampling**   There are five red, three green, and eight blue marbles in a box. In how many ways can a sample of four be selected, without replacement and without regard to order, if
   (a) All are the same color?
   (b) Exactly two are the same color?

46. **Selecting a Committee**   A committee of four people is to be selected from among three Republicans and four Democrats. In how many ways can this be done if
   (a) All are to be Republicans?
   (b) At least one is to be a Republican?
   (c) All are from the same party?

47. **Selections**   Ten cards are to be selected from a deck of 52 cards. In how many ways can this be done if
   (a) At least two are to be spades?
   (b) At least three are to be spades?

48. **Selections**   A box contains 12 red, eight green, and 20 blue marbles. How many different samples of eight marbles may be selected, without replacement and without regard to order, if
   (a) At least three are red?
   (b) All are to be green?

49. **Sampling**   A sample of five pencils is to be selected from among ten pencils. In how many ways can this sample be selected if
   (a) There is no replacement, and order of selection is not important?
   (b) There is no replacement, and order of selection is important?
   (c) There is replacement, and order of selection is important?

50. **Sampling**   A sample of three teabags is to be selected from among 11 teabags. In how many ways can this sample be selected if
   (a) There is no replacement, and order of selection is not important?
   (b) There is no replacement, and order of selection is important?
   (c) There is replacement, and order of selection is important?

Problems for Exploration

1. Summarize, in your own words, the counting techniques that have been introduced in this chapter and how to use them.

2. A child is learning to count money with a penny, a nickel, a dime, a quarter, and a half-dollar. If the child selects one or more of these coins at a time, how many different sums can be formed?

3. In how many ways can two keys be arranged on a key ring? Three keys? Four keys? *n* keys?

CHAPTER **5** REFLECTIONS

CONCEPTS

Problems that ask "How many?" play an important role in computer science, statistics, and probability. These *counting problems* involve special key words of logic—such as *and, or,* and *not*—and their *set operation* counterparts of intersection, union, and complementation. *Venn diagrams* are a useful tool for exploring set relationships and for solving some types of counting problems. The *multiplication principle* is a basic computational procedure for counting the number of different sequences formed from successive choices of objects from the same set (called permutations), or of objects from different sets. Counting problems involving the number of sets that can be selected from a collection of objects are called *combinations*. In some situations, counting can be done most efficiently by counting what we do not want.

COMPUTATION/CALCULATION

1. Use truth tables to determine whether the statements $p \vee (q \wedge p)$ and $(p \vee q) \wedge p$ are equivalent.

2. Use DeMorgan's laws of logic to simplify $\sim(p \vee q) \wedge \sim q$.

3. Write the negation of "All of the marbles are green." For the sets $U = \{1, 2, 3, 4, 5, 6\}$, $A = \{1, 2\}$, $B = \{3, 5, 6\}$, and $C = \{2, 4, 5, 6\}$, compute the results of the given set operation(s).

4. $B \cap C$ **5.** $A - C$ **6.** B'

7. $A \cup C'$ **8.** $(C')'$ **9.** $(A \cup C)'$

10. $A \cap B$ **11.** $C \cap (A \cup B)$ **12.** $(C \cap A) \cup B$

13. Use set operations to describe the shaded set in the Venn diagram.

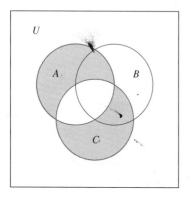

14. How many possible outcomes are there when a coin is tossed seven times?

15. In how many ways can three 6-sided dice come up?

16. How many five-digit numbers can be formed if zero cannot be the first digit?

 Calculate and simplify to a counting number.

17. $P(96, 5)$ **18.** $C(96, 5)$ **19.** $C(96, 91)$

20. $P(23, 7)$ **21.** $C(47, 38)$

CONNECTIONS

22. What logical connective(s) should be used between the constraints of a linear programming problem to completely determine the feasible region? (For example, $x + 2y \le 8, 4x + y \le 20, x \ge 0, y \ge 0$.)

23. Write the negation of "All Fords are red or blue."

24. Write the negation of "Some college students drink tea and coffee."

25. The Food-4-Less grocery store questioned 75 of its shoppers about their bread purchases during the past week. The results indicated that 43 bought white, 39 bought whole wheat, 16 bought rye, 19 bought both white and whole wheat, 6 bought both whole wheat and rye, 5 bought both white and rye, and 3 bought none of these. How many bought all 3 kinds of bread?

26. Eight students arrived at Cinema Ten to see a newly released movie. In how many ways could four of them line up to buy tickets?

27. An astronomy club has 20 members, including four officers. How many possible delegations of five persons could be selected to visit Chicago's Museum of Science and Industry if the president must go, but none of the other officers may go?

COMMUNICATION

28. Describe one of DeMorgan's laws of logic without using any logic symbols.

29. Explain what it means for two sentences to be logically equivalent.

30. Without using logic or set symbols, define the difference of two sets.

31. Explain in your own words how permutations and combinations are alike and how they are different.

COOPERATIVE LEARNING

32. Develop general formulas for determining the number of elements in the union of three sets and the number of elements in the union of four sets.

(**HINT** Use the formula for the number of elements in the union of *two* sets, along with the associative property, or explore the situations with the aid of Venn diagrams.)

CRITICAL THINKING

33. Analyze the following problem and the proposed solution. Make a case for why you think the proposed solution is correct or incorrect. Then, in either instance, find two other different and separate solutions to this problem.

Problem: In how many ways can three balls be selected without replacement from a bowl containing seven red, five green, and three yellow balls if at least one ball must be red?

Proposed solution: The first ball could be red, then anything, then anything; or the first ball could not be red, then red, then anything; or the first ball could not be red, then not red, then red. So there are $7 \cdot 14 \cdot 13 + 8 \cdot 7 \cdot 13 + 8 \cdot 7 \cdot 7$, or 2394, ways of selecting at least one red ball.

COMPUTER/CALCULATOR

34. Calculate $C(83, 61)$

Cumulative Review

35. Find the least squares regression line for the points (10, 21), (20, 27), (30, 43), (40, 56), and (50, 58).

36. Write a linear cost function for traveling x miles in a taxi cab, if 3 miles cost $3.70 and 10 miles cost $8.25.

37. Write the general solution in parametric form for this system:

$$5x + 2y - z = 6$$
$$-3x + y \quad\;\; = 2.$$

38. Calculate AB and BA for the matrices

$$A = \begin{bmatrix} 2 & 0 & -1 \\ 1 & -3 & 0 \end{bmatrix} \quad \text{and} \quad B = \begin{bmatrix} 0 & 2 \\ 5 & -3 \\ 1 & -1 \end{bmatrix}.$$

39. Use Crown pivoting rules to solve this linear programming problem:

$$\text{Maximize} \quad f = 3x + 4y,$$
$$\text{subject to} \quad 5x + 3y \le 30$$
$$x + 4y \ge 8$$
$$-x + y \le 6$$
$$x \ge 0, y \ge 0$$

CHAPTER 5 SAMPLE TEST ITEMS

Using the statements

   p: The first ball drawn is red.

   q: The second ball drawn is not blue.

write the indicated sentences. Make use of logically equivalent sentences where appropriate.

1. $\sim p \vee \sim q$ **2.** $\sim p \wedge q$ **3.** $\sim(\sim p \vee q)$

4. Make a truth table for $\sim p \vee (\sim q \wedge p)$

Let $U = \{\emptyset, 2, \{2\}, 3, 4, 5\}$, $A = \{\emptyset, 2, 3, 5\}$, $B = \{2, 4, 5\}$, *and* $C = \{\emptyset, \{2\}, 3, 5\}$. *Decide whether each statement is true (T) or false (F), and circle your choice.*

T F **5.** $2 \in C$ T F **6.** $\{2\} \subseteq C$

T F **7.** $\emptyset \subseteq B$ T F **8.** $2 \subseteq A$

T F **9.** $\emptyset \in C$ T F **10.** $A = C$

T F **11.** All students like mathematics, or some students drink tea.

Using sets U, A , B, and C, given for Questions 5 through 10, list the elements of each indicated set.

12. $(A \cup C)'$ **13.** $A \cap C'$ **14.** $A - B$

15. Shade the accompanying Venn diagram to represent $A \cap (B \cap C')$.

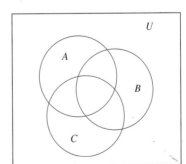

For Exercises 16 and 17: Fifty students were randomly surveyed from those that ate regularly at a university dining hall. This survey revealed that 38 ate breakfast, 25 ate lunch, 18 ate breakfast and lunch, 35 ate breakfast and dinner, 24 ate lunch and dinner, and 17 ate all three.

16. How many of these students ate lunch only?

17. How many of these students ate dinner?

18. How many three-digit numbers are there in our usual system if zero cannot be the first digit?

19. In how many ways can a five-card hand be drawn, without replacement, from a standard deck of 52 cards if exactly three are to be hearts?

20. In how many ways can the letters in the word *banana* be arranged?

For Exercises 21 and 22: Nine people are to travel to dinner in a five-passenger van and a four-passenger sports car.

21. How many different groups of five and four are possible for the trip?

22. How many seating arrangements are possible if two persons, Juanita and Kareem, are designated drivers and the others can sit in any of the remaining seats?

23. You have two different pairs of tennis shoes, three different pairs of shorts, and six different T-shirts. How many different outfits are possible?

24. In how many ways can four couples be seated in a row of eight seats at a theater if each couple is seated together?

25. How many ways can a ten-question true-false test be answered so that at least one answer is correct?

Career Uses of Mathematics

In my day-to-day work as a sole proprietor of a public accounting business, it is imperative for me to be able to use critical thinking. The application of statistics, probability, present value, and algebra in the areas of auditing and accounting not only involves the proper use of predetermined formulas, but also the knowledge of when and how to use them and with which elements of a given situation.

I had been out of school for eight years, teaching foreign languages in secondary schools, when I decided to get a degree in business administration. Finite math was my first course upon returning to school. The concepts learned in finite math were the groundwork for all of my courses in accounting, finance, and statistics. Taking the finite math course gave me a basis in logic, quantitative thinking, and mathematical procedures.

I liken my success in learning mathematical concepts to knowing how to study a foreign language. The key is repetition. Repetition of textbook problem solving in mathematics prepares the student to use the mathematical concepts in the real world.

Rosann P. Gonzalez

Rosann P. Gonzalez, CPA

Basic Concepts of Probability

A company sells one-year term policies designed for 40-year-old women. The face value of the policy is $10,000 and each policy sells for $300. If the company's mortality tables show that a woman of this age will live for another year with probability .98, what is the average earnings per policy the company can expect from the sale of such policies? (See Example 2, Section 6.3.)

CHAPTER PREVIEW

Suppose an action is to be taken that will produce one of several possibilities. (For instance, a state income tax form is to be checked and classified as correct, overpaid, or underpaid.) Before the action takes place, the concept of probability can be used to measure, usually with a fraction or decimal number, the likelihood that each of these possibilities will indeed happen. Sections 1 and 2 of this chapter investigate how these numbers are assigned to each of these several possibilities as well as certain subsets of these possibilities. In Section 3 the concept of "expected value" is explored as a useful application of probability to long-term trends. Liberal use will be made of the counting techniques discussed in Chapter 5.

CALCULATOR/COMPUTER OPPORTUNITIES FOR THIS CHAPTER

The computer disk MasterIt contains a computer program called "Probability I" that allows you to test your skill in answering basic probability questions. If your answer is wrong, you are given a hint and a second chance. If you are wrong again, the answer and how it is obtained will be shown. Another program on the MasterIt disk simulates the repeated toss of a coin or a die.

A calculator with a permutation key, a combination key, and a factorial key will be useful in calculations throughout the chapter.

SECTION 6.1

Equally Likely Outcomes

The subject of probability can be traced back to the seventeenth century, when it arose from the study of gambling problems. The range of applications today goes far beyond gambling problems, reaching into such areas as medical science, business decisions, plant experiments, and weather forecasting, to mention just a few. In fact, even though the term *probability* is not commonly used in everyday publications, we still come into contact with it in disguised ways. For example, a weather forecast telling of "an 80% chance of rain tomorrow" or a sportscaster mentioning that "the odds for the Eagles winning the Super Bowl are six to five" are actually ways of stating probability. Generally speaking, questions of probability arise when we see that there are a number of possibilities that can occur, and we do not know beforehand just which one will actually happen. We then attempt to assign a number to each possibility that reflects the proportional share of the time that possibility is expected to happen. If one possibility is more likely than another, then the more likely possibility should be assigned a larger fractional share than the less likely possibility. On the other hand, if all possibilities are **equally likely,** then each one should be assigned identical fractional shares of time. Since the equally likely case occurs so often in natural settings, we take it as our starting point in the study of probability.

HISTORICAL NOTE
Blaise Pascal (1623–1662) and Pierre de Fermat (1601–1665)

The rigorous mathematical foundations of modern probability theory were jointly laid by two French mathematicians Blaise Pascal and Pierre de Fermat (Fer-'mä). Probability arose from twin roots: games of chance and statistical data for such matters as mortality tables and insurance rates.

EXAMPLE 1
Tossing a Balanced Coin

Suppose a balanced coin is to be tossed. What is the probability that it will fall with heads up?

Solution We reason this way: Upon being tossed, there are only two ways the coin can fall, heads or tails, and since the coin is balanced, it is just as likely to fall heads as tails. Since there is a 1 in 2 chance heads will be showing, the number $\frac{1}{2}$ is used to predict the likelihood that this possibility will occur. In the language of probability, we say the **probability** that the coin will fall with heads up is $\frac{1}{2}$. (The probability of the coin falling with tails up is also $\frac{1}{2}$.) ■

Even though the outcome of any single trial of tossing a balanced coin is uncertain, the importance of the probabilities found in Example 1 is that they are *reliable predictors* of the *long-term* behavior of repeatedly tossing a single balanced coin many times: About half of the time a head will turn up and about half the time a tail will turn up. To give some experimental justification for our reasoning, three records of 100 tosses of a balanced coin, simulated by a programmable calculator that had a random number generator, are shown next.

	First trial	Second trial	Third trial
	49 heads	56 heads	50 heads
	51 tails	44 tails	50 tails

It is the long-term predictability feature that makes probability so useful. In the science of genetics, for example, it is uncertain whether an offspring will be male or female, but accurate long-term predictions can be made as to the percentage of males and females. An insurance company cannot predict which persons in the United states will die at the age of 60, but it can predict with great accuracy *how many* people will die at the age of 60.

Example 1 provides a framework for terminology used in the study of probability. An **experiment** is to be performed (tossing a coin) and there are two possible **outcomes,** heads (H) or tails (T). The set of all possible outcomes from an experiment gives a universal set within which all probability activity takes place and is given the special name **sample space.** For this example, the sample space is $S = \{H, T\}$.

Experiments

An **experiment** is an activity or procedure that produces distinct well-defined possibilities called **outcomes** that can be observed or measured, but that cannot be predicted with certainty. Each outcome is an element in the universal set of all possible outcomes for the experiment. This universal set is called the **sample space.**

EXAMPLE 2
Tossing a Balanced Coin Twice

A balanced coin is to be tossed twice and the result of each toss recorded. Find the probability of each outcome in the sample space.

Solution   The **experiment** is the act of tossing a balanced coin twice and recording the results of the first toss, then the second toss. The possible recorded **outcomes** are the ordered pairs

$$HH, \quad HT, \quad TH, \quad \text{and} \quad TT.$$

These four outcomes constitute the **sample space** for this experiment. Is there any reason to believe that any one of these outcomes is any more likely to happen than another? No; since the coin is balanced, we would agree they are equally likely. Since there are 4 possible outcomes, it seems natural to assign each outcome a probability of $\frac{1}{4}$.

Outcome	Probability
HH	$\frac{1}{4}$
HT	$\frac{1}{4}$
TH	$\frac{1}{4}$
TT	$\frac{1}{4}$
	Sum: 1

These probabilities do two things. First, they show the relative strength of each outcome given as a fractional part of the whole 1; thus the sum of probabilities of all outcomes in the sample space should be 1. Secondly, the predicted long-term effects of repeating this experiment over and over are for two heads to turn up about $\frac{1}{4}$ of the time, a head first and a tail second about $\frac{1}{4}$ of the time, a tail first and a head second about $\frac{1}{4}$ of the time, and two tails about $\frac{1}{4}$ of the time.   ■

NOTE   Unless stated to the contrary, we will henceforth always assume that coins are balanced.

EXAMPLE 3
Rolling a Single Die

A single balanced die is to be rolled and the number of dots showing on the top face recorded. Find the probability of each outcome in the sample space.

Solution   Figure 1 shows the possible outcomes for this experiment. The sample space may then be written as $S = \{1, 2, 3, 4, 5, 6\}$. Since the die is balanced, there is no reason to believe any one of these outcomes would occur more frequently than another. Hence it is reasonable to assume each outcome is equally likely, with probability $\frac{1}{6}$.

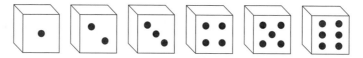

FIGURE 1
Sample Space for a Single Die

Outcome	Probability
1	$\frac{1}{6}$
2	$\frac{1}{6}$
3	$\frac{1}{6}$
4	$\frac{1}{6}$
5	$\frac{1}{6}$
6	$\frac{1}{6}$
	Sum: 1

The following table shows three different simulations of repeating this experiment 500 times on a programmable calculator with a random number generator. They give some experimental justification for our probability assignments since $\frac{1}{6} \approx 0.167$.

Dots showing	First Repetition of 500 Trials		Second Repetition of 500 Trials		Third Repetition of 500 Trials		Combined Repetition of 1500 Trials
	Number	Probability	Number	Probability	Number	Probability	Probability
1	84	.168	87	.174	85	.170	.171
2	96	.192	84	.168	103	.206	.189
3	82	.164	87	.174	71	.142	.160
4	71	.142	88	.176	73	.146	.155
5	88	.176	77	.154	91	.182	.171
6	79	.158	77	.154	77	.154	.156

NOTE   Henceforth, all dice will be considered balanced unless otherwise noted.

EXAMPLE 4
Rolling a Pair of Dice

A pair of dice is to be rolled and the number of dots on the upper face of each die is to be recorded. To aid in the thought process, think of one die as red and one as green. An outcome then consists of two numbers, one denoting the number of dots on the upper face of the red die and one denoting the number of dots on the upper face of the green die. Find the number of elements in the sample space and the probability of each outcome in that sample space.

Solution   There are six possible ways to record the dots on the top side of the red die, and, for each of these, there are six possible ways to record the dots on the top side of the green die. Applying the multiplication principle then gives $6 \cdot 6 = 36$ possible outcomes for this experiment. Figure 2 gives a visual picture of the sample space. Note that a red 2 and a green 3 is a different outcome than a red 3 and a green 2. A "double," such as red 4 and green 4, however, would be listed only once as an outcome.

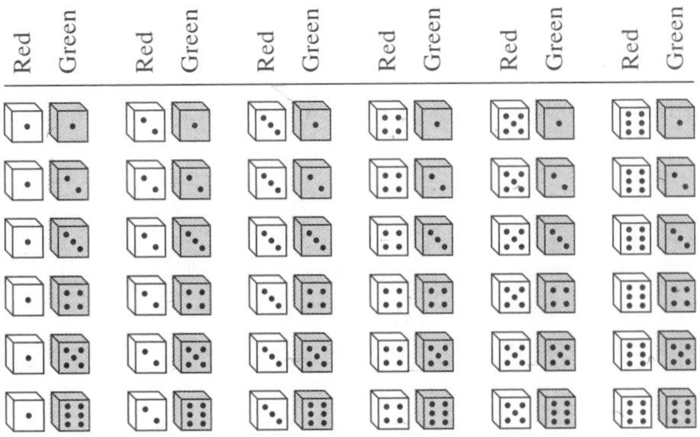

FIGURE 2
Sample Space for Two Dice

If we agree that each outcome is equally likely, then each should be assigned a probability of $\frac{1}{36}$. ■

As it was with counting problems, the simplest way to view any probability problem is to write out the list of outcomes as was done in Examples 1 through 4. After studying those examples you have probably sensed that for equally likely outcomes, the *number* of outcomes in the sample space is the key to assigning probability; if the number of outcomes is k, then each outcome is assigned the probability $\frac{1}{k}$. Knowing this, we can utilize the counting techniques of Chapter 5 to find the number of outcomes when it is not practical to write them all down. Example 5 will illustrate this idea.

EXAMPLE 5
Dealing a Card Hand

A five-card hand is dealt from a well-shuffled deck of cards. How many outcomes are there from such an activity and what is the probability of each outcome?

Solution   The experiment is that of selecting or dealing five cards from the deck, without replacement, and without regard to order. The number of such hands, hence the number of outcomes from the experiment, is found by computing the number of combinations of 52 things taken 5 at a time: $C(52, 5) = 2,598,960$. Since the cards were dealt from a well-shuffled deck, no one five-card hand can be assumed more likely to be dealt than another. This leads to the conclusion that each five-card hand (outcome) is equally likely and therefore has probability $\frac{1}{2,598,960}$. ■

EXAMPLE 6
Computer Passwords

To access a certain application in the ABC Company's computer network, a password consisting of three digits followed by two letters must be correctly entered. If an unauthorized person knows how the password is constructed, what is the probability of correctly guessing the password on the first try?

Solution   The experiment is the act of selecting and entering three digits followed by two letters into the computer. Each *ordered* sequence of three digits followed by two letters is an outcome for the experiment. The selections in the sequence are made with replacement (repetitions are allowed) and the order in which they are selected is important. The multiplication principle then tells us that there are

$$10 \cdot 10 \cdot 10 \cdot 26 \cdot 26 = 676,000$$

possible outcomes (passwords) to the experiment. Since the unauthorized person is guessing at the correct makeup of the password, we assume that each password is equally likely to be guessed. Therefore, the probability $\frac{1}{676,000}$ is assigned to each outcome. This means that the probability of correctly guessing the password with one try is $\frac{1}{676,000}$. ∎

Suppose we now return to Example 2 where a coin was to be tossed twice with a sample space of

$$S = \{HH, HT, TH, TT\}$$

and each outcome has probability $\frac{1}{4}$. The question "What is the probability that at least one head will turn up?" does not concern itself with just one outcome, but rather with a subset of three outcomes. In the language of probability, such a subset is called an **event.** In fact, the term "event" is used to describe *any* subset of a sample space including subsets consisting of a single outcome, the empty set, or the entire sample space. An event consisting of a single outcome is called a **simple event** while an event containing more than one outcome is called a **compound event.** If E denotes an event in a sample space S, the notation $P(E)$ is used to denote the probability of E. For the particular question at hand,

$$E = \{HH, HT, TH\}.$$

Figure 3 shows a Venn diagram of this event. The **event E is said to occur** if a trial of the experiment produces an outcome that belongs to E. For our present example,

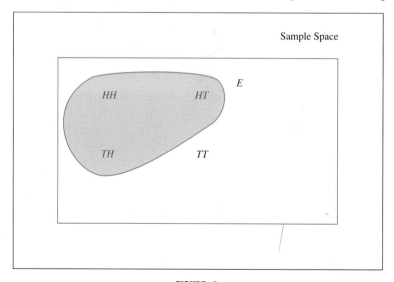

FIGURE 3
The Event "At Least One Head"

how likely is a trial of this experiment to produce one of the outcomes *HH*, *HT*, or *TH*? Since each outcome in the sample space is equally likely, three times out of four would sound reasonable. Accordingly, we write $P(E) = \frac{3}{4}$. Stated differently, the probability is the ratio of the number of outcomes in E to the total number of outcomes in the sample space S.

EXAMPLE 7
Rolling a Pair of Dice

A pair of dice is to be rolled. What is the probability that

(a) A sum of 5 is rolled? $(1,4)\ (4,1)(3,2)(2,3) \quad \frac{4}{36}$

Solution   Consider one of the pair to be red and one to be green as in Example 4. There are 36 outcomes in the sample space. Among these, "roll a sum of 5" describes an event E which, as a subset, can be listed with the number of dots on one of the die, say red, first, and the number of dots on the other (green) die second, as follows:

$$E = \{(2, 3), (3, 2), (4, 1), (1, 4)\}.$$

Since all outcomes are equally likely, and $n(E) = 4$, we reason that there are 4 chances out of 36 of getting one of the outcomes in E on a single trial of the experiment "roll a pair of dice." The probability of rolling a sum of 5 is then given by the ratio

$$P(E) = \frac{n(E)}{n(S)} = \frac{4}{36}.$$

(b) A sum of 8 is rolled? $(2,6)(6,8)(4,4) \quad (3,5)(5,3) \quad \frac{5}{36}$

Solution   This time the event described may be listed as

$$E = \{(4, 4), (5, 3), (3, 5), (6, 2), (2, 6)\}.$$

Because $n(E) = 5$, there are 5 chances out of 36 of getting one of the outcomes in E. Therefore, the probability of rolling a sum of 8 is given by

$$P(E) = \frac{5}{36}.$$ ■

In view of our development of events, any experiment that can be realistically modeled by assuming equally likely outcomes is, from a probability viewpoint, a counting problem: Count the number of elements (outcomes) in E and the number of elements (outcomes) in S, then divide.

The Probability of an Event

Let S be a sample space with $n(S) = k$ equally likely outcomes. Then the probability of each outcome is

$$\frac{1}{n(S)} = \frac{1}{k}.$$

If E is an event in this sample space with $n(E) = m$ outcomes, then the probability of E, denoted by $P(E)$, is

$$P(E) = \frac{n(E)}{n(S)} = \frac{m}{k}.$$

The smallest number of elements that E can have is 0 if E is the empty set. On the other hand, the largest number of elements E can have is k if $E = S$. This means that the smallest $P(E) = \dfrac{m}{k}$ could be is $\dfrac{0}{k} = 0$ while the largest $P(E) = \dfrac{m}{k}$ could be is $\dfrac{k}{k} = 1$. Therefore, the probability of any event is always between 0 and 1, inclusive.

EXAMPLE 8
Selecting a Card

One card is to be drawn from a deck of 52 cards and the denomination and suit recorded.

(a) Find the probability that a spade is drawn.

Solution   There are 52 possible cards that may be drawn, hence 52 outcomes in the the sample space. Since each card is just as likely to be drawn as another, we assign each outcome (card) a probability of $\frac{1}{52}$. The event E described by "the card is a spade" has 13 outcomes. Therefore, $P(\text{spade}) = \frac{13}{52} = \frac{1}{4}$.

(b) Find the probability that a queen is drawn.

Solution   The event "the card is a queen" contains 4 outcomes. It follows that $P(\text{queen}) = \frac{4}{52} = \frac{1}{13}$. ∎

EXAMPLE 9
Checking for Flaws

A maker of ladies' slacks selects a typical rack of 50 pairs of slacks for inspection by a buyer for a large discount chain. Assume that 2 of the 50 pairs will have a flaw. Suppose the buyer randomly selects a sample of 6 pairs of slacks from the rack, without replacement, and checks them for a flaw. (Random selection means no one pair is more likely to be selected than another pair.)

(a) What is the probability that none of the 6 pairs will have a flaw?

Solution   Because they are selected without replacement and the order in which they are inspected for a flaw is not important, the experiment may be considered as selecting a *set* of 6 pairs of slacks from 50. The number of possible outcomes (sets of 6 pairs of slacks) from such an experiment is $C(50, 6) = 15,890,700$.

   Among all possible outcomes, how many meet the criteria of the event E described by "none of the 6 pairs of slacks will have a flaw?" Only sets of 6 selected from the 48 having no flaw. The number of those is $C(48, 6) = 12,271,512$. Therefore, the probability of selecting 6 pairs of slacks in which none has a flaw is

$$P(E) = \frac{C(48, 6)}{C(50, 6)} = \frac{12,271,512}{15,890,700} \approx 0.772.$$

(b) What is the probability that exactly 1 of the 6 pairs has a flaw?

Solution   Among all possible sets of 6, this event asks about the number having 1 pair with a flaw and 5 that do not have a flaw. Since there are 2 pairs with a flaw and 48 without a flaw, that number is $C(2, 1) \cdot C(48, 5)$. Therefore, the probability we seek is

$$\frac{C(2, 1) \cdot C(48, 5)}{C(50, 6)} = \frac{2 \cdot (1,712,304)}{15,890,700} \approx 0.216. \quad ∎$$

EXAMPLE 10
Downsizing

A company has 13 manufacturing plants throughout the country. It plans to close three of these plants, and, the order in which they are closed is based on what least disrupts supply schedules. Assume that economic conditions are such that any one plant is just as likely to be closed as another. If the plants are numbered 1 through 13, what is the probability that plants 2, 6, and 7 will be closed in the order stated?

Solution   The experiment is the act of selecting 3 plants for closing from a total of 13, without replacement, in which order is important. The number of outcomes from this experiment is the number of permutations of 13 plants taken 3 at a time: $P(13, 3) = 1716$. From among these outcomes, the event E of interest to us is the single ordered selection 2, 6, and 7. It follows that

$$P(E) = \frac{1}{P(13, 3)} = \frac{1}{1716} \approx 0.000583. \quad \blacksquare$$

With our last example in Section 6.1, we want to reemphasize the importance of listing the outcomes in a sample space whenever possible. This will greatly aid in the counting process that yields probability.

EXAMPLE 11
Selecting Slips of Paper

Three slips of paper—each with the number 1, 2, or 3, respectively, written on it—are placed in a box. A slip is randomly selected, the number on it is recorded, and the slip is then returned to the box. Finally, a second slip is randomly selected and the number on it recorded.

(a) Write the sample space for this experiment.

Solution   Each outcome from this experiment may be viewed as an ordered pair of numbers, selected with replacement from the set $\{1, 2, 3\}$, where no pair is more likely to be selected than another. The totality of these ordered pairs (the sample space) is listed next:

$$\begin{array}{ccc} (1, 1) & (1, 2) & (1, 3) \\ (2, 1) & (2, 2) & (2, 3) \\ (3, 1) & (3, 2) & (3, 3) \end{array}$$

(b) What is the probability that the sum of the two numbers drawn is 4?

Solution   Of the nine possibilities, those in the event "the sum of the two numbers is 4" are (1, 3), (2, 2), and (3, 1). Therefore the probability we seek is $\frac{3}{9} = \frac{1}{3}$.

(c) What is the probability that the first number drawn is a 2 or that the sum of the two numbers drawn is 3?

Solution   The outcomes in the event described are

$$(2, 1), (2, 2), (2, 3), \text{ and } (1, 2).$$

As a consequence, the probability of the event described is $\frac{4}{9}$.

(d) What is the probability that the sum of the two numbers drawn is 8?

Solution   The event described is empty; there is no sum that equals 8. Since $n(\emptyset) = 0$, the probability of this event is $\frac{0}{9} = 0$. An impossible event always has probability 0. $\blacksquare$

EXERCISES 6.1

1. *Spinners* The arrow in the figure is spun and the letter representing the area in which the arrow points is then recorded.

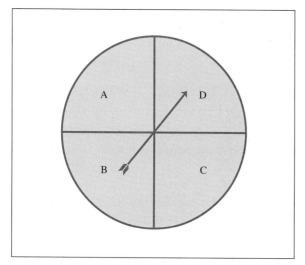

   (a) Write the outcomes in the sample space for this experiment.
   (b) Assign a probability to each outcome in the sample space.
   (c) Find the probability that the outcome is not a vowel.

2. *Spinners* The arrow in this figure is spun and the number representing the area in which the arrow points is then recorded.

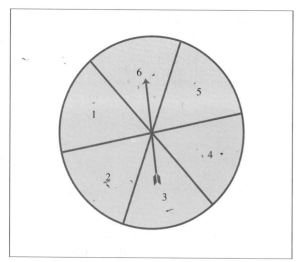

   (a) Write the outcomes in the sample space for this experiment.

   (b) Assign a probability to each outcome in the sample space.
   (c) What is the probability that the arrow points to an even number?
   (d) What is the probability that the arrow points to a number greater than 4?

3. *Two-child Families* A sociologist is studying two-child families. For probability considerations, she records the sex of the first and second child in a sample space and assumes each of these outcomes is as likely as another.

   (a) Write the possible outcomes in the sociologist's sample space.
   (b) Assign a probability to each outcome in that sample space.
   (c) What is the probability that such families will have at least one girl?
   (d) What is the probability that such families will have a child of each sex?

4. *Tossing a Coin* A coin is tossed three times. On each toss, a record is made of whether it lands with heads or tails turned up.

   (a) Write the outcomes of the sample space for this experiment.
   (b) Assign a probability to each outcome in the sample space.
   (c) What is the probability that at least one head turns up?
   (d) What is the probability that exactly two heads turn up?

5. *Three-child Families* A sociologist is studying three-child families. For probability considerations, he records the sex of the first and second and third child in a sample space and assumes that each of these possible outcomes is just as likely as another.

   (a) Write the outcomes in the sociologist's sample space.
   (b) Assign a probability to each outcome in that sample space.
   (c) What is the probability that such families will have at least one girl?
   (d) What is the probability that such families will have exactly two girls?

6. *Committees* A committee of three is selected from among Alice, Bob, Carl, and Desmond. Assume that the committee is selected in such a way that each person is just as likely as another to be a member of the committee.

   (a) Write the outcomes in the sample space for the experiment of committee selection.

(b) Assign a probability to each outcome.

(c) What is the probability that Alice will be on the committee?

(d) What is the probability that Bob or Desmond will be on the committee?

7. *Officers*   A president and secretary are selected from among Latonya, Yee, and Mary. Assume that the officers are selected in such a way that each slate of officers is equally likely to occur.

(a) Write the outcomes in the sample space for the experiment of officer selection.

(b) Assign a probability to each outcome.

(c) What is the probability that Latonya will hold one of these offices?

(d) What is the probability that Yee will be elected president?

8. *Seating*   Jacob, Lina, and Shawna are randomly seated in a row. (Each arrangement is just as likely as another to occur.)

(a) Write the outcomes in the sample space for the experiment of seating these people.

(b) Assign a probability to each outcome.

(c) What is the probability that Shawna will be seated on the left end of the row?

(d) What is the probability that Jacob or Lina will be seated in the middle?

9. *A Peculiar Die*   A four-sided pyramidal die is rolled, and the number of dots on the bottom side is counted. (Assume that one, two, three, or four dots, respectively, appear on each of the sides.)

(a) Write the outcomes in the sample space for this experiment.

(b) Assign a probability to each outcome.

(c) What is the probability of rolling an even number?

(d) What is the probability of rolling a number greater than 3?

10. *Peculiar Dice*   A pair of four-sided pyramidal dice, one red and one green, is rolled, and the number of dots on the bottom side of each die is recorded. (Assume that one, two, three, or four dots, respectively, appear on each of the sides.)

(a) Write the outcomes in the sample space for this experiment.

(b) Assign a probability to each outcome.

(c) What is the probability of rolling a sum of 6?

(d) What is the probability of rolling a sum greater than 5?

11. *Two Activities*   An experiment consists of tossing a coin and recording whether it lands with heads or tails turned up; then a single four-sided pyramidal die is rolled and the number of dots on the bottom face is recorded. (See Exercise 9.)

(a) Write the outcomes in the sample space for this experiment.

(b) Assign a probability to each outcome.

(c) What is the probability that the coin turns up heads and an even number of dots show on the bottom face of the die?

(d) What is the probability that the die shows a number greater than three on the bottom?

12. *Two Activities*   An experiment consists of tossing a coin and recording whether it lands heads or tails turned up; then a single standard die is rolled and the number of dots on the upper face is recorded.

(a) Write the outcomes in the sample space for this experiment.

(b) Assign a probability to each outcome.

(c) What is the probability that the coin lands heads up and an even number of dots shows on the upper face of the die?

(d) What is the probability that the die shows a number greater than 4?

13. *Sampling*   Nine balls numbered from 1 through 9, respectively, are in a box. One ball is to be randomly selected from the box and the number on it recorded.

(a) Write the outcomes in the sample space for this experiment.

(b) Assign a probability to each outcome.

(c) What is the probability that the number on the ball selected is greater than seven?

(d) What is the probability that the number of the ball selected is odd?

14. *Testing*   An economics professor gives a two-question, multiple-choice quiz in which each question has four possible answers—A, B, C, and D. Assume that a student guesses at the answers on this quiz.

(a) Write the outcomes in the sample space for the experiment of guessing the answers to this test.

(b) Assign a probability to each outcome.

(c) What is the probability that the student will answer both questions correctly?

(d) What is the probability that the student answers one question correctly?

15. *Sampling*   Three slips of paper—each with the number 1, 2, or 3, respectively, written on it—are placed in a box. A slip is to be drawn and the number on it recorded. That slip is laid aside, and a second slip drawn and the number on it recorded.

(a) Write the outcomes in the sample space for this experiment.

(b) Assign a probability to each outcome.

(c) Find the probability that the sum of the two numbers drawn is 5.

16. **Sampling** Four slips of paper—each with the number 1, 2, 3, or 4, respectively, written on it—are placed in a box. A slip is to be drawn and the number on it recorded. That slip is then returned to the box and a second slip drawn and the number on it recorded.

    (a) Write the outcomes in the sample space for this experiment.
    (b) Assign a probability to each outcome.
    (c) Find the probability that the sum of the two numbers drawn is 6.

17. **Rolling Dice** A pair of standard dice is rolled. Find the probability that

    (a) A sum of 7 is rolled.
    (b) A sum of 11 is rolled.
    (c) A sum greater than 8 is rolled.

18. **Rolling Dice** A pair of standard dice is to be rolled. Find the probability that

    (a) A sum of 10 is rolled.
    (b) The red die shows five dots on the top side.
    (c) The red die does not show five dots on the top side.

19. **Rolling Dice** A pair of standard dice is to be rolled. Find the probability that

    (a) A sum of 1 is rolled.
    (b) A sum of 2 is rolled.
    (c) The red die shows two dots on the top side.

20. **Rolling Dice** A pair of standard dice is to be rolled. Find the probability that

    (a) A sum greater than 1 is rolled.
    (b) A sum less than 5 is rolled.
    (c) A sum less than 3 or greater than 8 is rolled.
    (d) A sum less than 3 and greater than 8 is rolled.

21. **Rolling Dice** Three standard dice, one red, one green, and one blue, are to be rolled.

    (a) What is the probability of rolling a sum of 2?
    (b) What is the probability of rolling a sum of 4?
    (c) What is the probability of rolling a sum of 7?

22. **Rolling Dice** Three standard dice, one red, one green, and one blue, are to be rolled.

    (a) How many outcomes are in the sample space?
    (b) What probability is assigned to each outcome?
    (c) What is the probability of rolling a sum of 8?

23. **Drawing a Card** A single card is to be drawn from a standard deck of 52 cards.

    (a) How many outcomes are in the sample space for this experiment?
    (b) What is the probability of drawing a club?
    (c) What is the probability of drawing a king?

24. **Drawing a Card** A single card is to be drawn from a standard deck of 52 cards.

    (a) What is the probability that the card will have a 4 on it?
    (b) What is the probability that the card is red?
    (c) What is the probability that the card is a face card (a jack, queen, or king)?

25. **Card Hands** A five-card hand is to be dealt from a standard deck of 52 cards.

    (a) What is the probability that all five of the cards are clubs?
    (b) What is the probability that all five of the cards are from the same suit?
    (c) What is the probability that none of the five cards are hearts?

26. **Card Hands** A three-card hand is to be dealt from a standard deck of 52 cards.

    (a) What is the probability that all three of the cards are spades?
    (b) What is the probability that all three of the cards are from the same suit?
    (c) What is the probability that none of the three cards are hearts?

27. **Card Hands** A six-card hand is to be dealt from a standard deck of 52 cards.

    (a) What is the probability that all six of the cards are face cards?
    (b) What is the probability that all six of the cards are kings?
    (c) What is the probability that none of the six cards is a king?

28. **Card Hands** A two-card hand is to be dealt from a standard deck of 52 cards.

    (a) What is the probability that both of the cards are jacks?
    (b) What is the probability that both of the cards are clubs?
    (c) What is the probability that neither of the cards are hearts?

29. **Card Hands** A five-card hand is to be dealt from a standard deck of 52 cards.

    (a) What is the probability that exactly two of the cards are spades?
    (b) What is the probability that exactly two of the cards are queens?
    (c) What is the probability that exactly three of the five cards are not hearts?

30. Card Hands An eight-card hand is to be dealt from a standard deck of 52 cards.

(a) What is the probability that exactly five of the cards are clubs?

(b) What is the probability that exactly four of the cards are face cards?

(c) What is the probability that exactly four of the cards are aces?

31. Quality Control A buyer for a small chain store randomly selects ten shirts, without replacement, from a lot of 200 shirts to inspect them for a flaw. It is known that ten of the shirts among the 200 have a flaw. What is the probability that the sample selected by the buyer

(a) Contains no shirt with a flaw?

(b) Contains exactly two shirts with a flaw?

(c) Contains exactly five shirts with a flaw?

32. Quality Control A campus bookstore buys 100 calculators from which it assumes that two will have a defect of some kind. The mathematics department buys eight of these calculators from the bookstore. What is the probability that of those calculators bought by the mathematics department

(a) All are free of defects?

(b) Exactly one will have a defect?

(c) Exactly two will have a defect?

33. Seating People Three boys and two girls are to be randomly seated in a row. What is the probability that

(a) A boy will be seated in the middle seat?

(b) Girls will be seated on both ends?

(c) Boys and girls will be alternately seated?

34. Seating People Three boys and three girls are to be randomly seated in a row. What is the probability that

(a) Boys will be seated on both ends?

(b) Boys and girls will be alternately seated?

(c) A particular girl, Tonya, will be seated on the left end?

35. License Plates Suppose the license plates in your state are made with three letters followed by three digits. If you obtain a new license plate, what is the probability that it will

(a) Start with the letter A and end with the digit 8?

(b) Have no letter or digit repeated?

(c) Not contain the letter Z or the digit 0?

36. Identification Numbers Suppose a bank is to issue you an ATM card with a personal identification number (PIN) consisting of four digits randomly selected by a computer. What is the probability that your PIN will

(a) Begin with an 8?

(b) Have no digit repeated?

(c) Not contain the digit 7?

NOTE Recall that we consider all objects, such as light bulbs and colored balls, to be distinguishable unless otherwise noted.

37. Sampling Among 80 light bulbs, 20 are known to have defects. A sample of six of these bulbs is to be selected, without replacement, and without regard to order. What is the probability that the sample will

(a) Contain no defective bulbs?

(b) Contain exactly three defective bulbs?

(c) Contain all defective bulbs?

38. Sampling Among 30 microwave ovens, ten are known to have defects. A sample of five of these ovens is to be selected, without replacement, and without regard to order. What is the probability that the sample will

(a) Contain no defective ovens?

(b) Contain exactly two defective ovens?

(c) Contain all defective ovens?

39. Sampling From among 10 spark plugs, three are known to be defective. An ordered sample of four spark plugs is to be selected at random, with replacement, from these ten plugs and tested for the defect. What is the probability that

(a) None of the plugs will be defective?

(b) The first two plugs tested will be defective?

(c) All of the plugs tested will be defective?

40. Sampling From among four red and eight green balls, an ordered sample of five balls is to be randomly selected, with replacement. What is the probability that

(a) All of the balls will be green?

(b) The first two balls selected will be red?

(c) All of the balls will be red?

41. Sampling From among three red and five green balls, an ordered sample of three balls is to be randomly selected, without replacement. What is the probability that

(a) All of the balls will be green?

(b) All of the balls will be red?

(c) The first ball will be red and the remaining two will be green?

42. Sampling From a lot of ten ladies' blouses, four are known to have a flaw. An ordered sample of three blouses is to be randomly selected, without replacement, from these ten blouses. What is the probability that

(a) None will have a flaw?

(b) The first two will have a flaw and the third will not have a flaw?

(c) The first one selected will have a flaw and the remaining two will not have a flaw?

43. *Sampling* From among four red and six green balls, a sample of four balls is to be randomly drawn, without replacement and without regard to order. What is the probability that

(a) All of the balls will be green?
(b) Exactly two of the balls will be red?
(c) All of the balls will be red?

44. *Sampling* From among three red, four green, and three purple balls, a sample of three balls is to be randomly drawn, without replacement and without regard to order. What is the probability that

(a) All will be green?
(b) Exactly two will be red?
(c) Exactly one will be purple?

45. *Personnel* An executive committee of five people is to be selected from among ten managers, four of whom are men and six of whom are women. If the committee is selected by drawing lots, meaning that each committee is just as likely as another, what is the probability that

(a) All members of the committee will be women?
(b) Exactly two of the committee members will be men?
(c) Exactly three of the committee members will be women?

46. *Personnel* From among the mayor and five other city council members, a committee of three is to be selected for inspection of city waste management facilities. If this committee is selected by drawing lots,

meaning that each committee is just as likely as another, what is the probability that

(a) The mayor will not be on the committee?
(b) The mayor will be on the committee?

Problems for Exploration

The following problem appeared on an actuarial examination.

1. What is the probability that a hand of 5 cards chosen randomly and without replacement from a standard deck of 52 cards contains the king of spades, exactly 1 other king, and exactly 2 queens?

(a) $\dfrac{\binom{4}{2}\binom{4}{2}\binom{44}{1}}{\binom{52}{5}}$ (b) $\dfrac{\binom{3}{1}\binom{4}{2}\binom{44}{1}}{\binom{52}{5}}$

(c) $\dfrac{\binom{7}{3}\binom{44}{1}}{\binom{52}{5}}$ (d) $\dfrac{\binom{3}{1}\binom{4}{2}\binom{11}{1}}{\binom{52}{5}}$

(e) $\dfrac{\binom{3}{1}\binom{4}{2}}{\binom{52}{5}}$

2. At a bridge table, the four players are to be dealt 13 cards each. Find the probability that each hand will contain an ace.

SECTION 6.2   Outcomes with Unequal Probability; Odds

The structure of sample spaces will be complete when we consider experiments that do not have equally likely outcomes. Sometimes examples of such experiments arise in a basic sense, such as the spinner shown in Figure 4.

If the arrow is spun and the letter representing the area to which the arrow points is recorded, then the possible outcomes are A, B, C, and D and the sample space is

$$S = \{A, B, C, D\}.$$

From the appearance of the spinner, we cannot assign equal probabilities to each of these four outcomes; outcome D deserves a larger number than the other three. In fact, since the area labeled D is one-half of the total area of the circle, we would expect the arrow to point to D about half of the time and hence D should be assigned a probability of $\frac{1}{2}$. Similarly, the areas labeled A, B, and C each occupy $\frac{1}{3}$ of $\frac{1}{2}$, or $\frac{1}{6}$, of the total area of the circle which means that they should be assigned a probability of $\frac{1}{6}$ each.

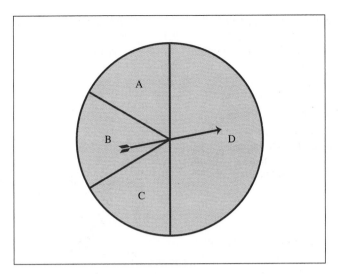

FIGURE 4

Outcome	Probability
A	$\frac{1}{6}$
B	$\frac{1}{6}$
C	$\frac{1}{6}$
D	$\frac{1}{2}$

Sum: 1

These probability assignments tell us that if the experiment is repeated over and over, then the long-term effects are that about $\frac{1}{6}$ of the time, the arrow will point to one of the regions A, B, or C, respectively, and about half of the time it will point to region D. Notice that the probabilities are assigned in such a way that their sum is 1, just as in the equally likely case of Section 6.1.

How would we find the probability of the event $E = \{C, D\}$, for example? Stated differently, "What proportionate share of the total should the region $C \cup D$ be assigned?" Would you agree that $\frac{1}{6} + \frac{1}{2} = \frac{4}{6} = \frac{2}{3}$ should be the answer? In other words, just add the probabilities of the various outcomes in the event E:

$$P(E) = P(\{C, D\}) = P(C) + P(D) = \frac{1}{6} + \frac{1}{2} = \frac{2}{3}.$$

Following this line of reasoning, the probabilities of some other events in this sample space are listed next:

$$P(\{A, B, D\}) = P(A) + P(B) + P(D) = \frac{1}{6} + \frac{1}{6} + \frac{1}{2} = \frac{5}{6}.$$

$$P(\{A, B\}) = P(A) + P(B) = \frac{1}{6} + \frac{1}{6} = \frac{2}{6} = \frac{1}{3}.$$

It seems reasonable to define the probability of an event to be the sum of the probabilities of all outcomes in the event. This agreement also includes the way events were handled in the equally likely case: If $n(S) = m$ and $n(E) = k$, then the probability of each outcome is $\frac{1}{m}$ and $P(E) = \frac{m}{k} = m\left(\frac{1}{k}\right) = \frac{1}{k} + \frac{1}{k} + \ldots + \frac{1}{k}$ (m factors).

> ### General Rules for Probability
>
> If a sample space S has a finite number of outcomes $O_1, O_2, \ldots, O_n$, then the assignment of probabilities, denoted by
>
> $$P(O_1) = p_1, \; P(O_2) = p_2, \ldots, P(O_n) = p_n,$$
>
> must obey these three rules:
>
> **1.** Each probability $p_1, p_2, \ldots, p_n$ is a number between 0 and 1, inclusive.
>
> **2.** $p_1 + p_2 + \ldots + p_n = 1$.
>
> **3.** $P(\emptyset) = 0$.
>
> The **probability of an event E** in S is found by adding the probabilities of all outcomes making up E.

Another way in which a sample space having outcomes that are not equally likely can arise is to realize that there are two components to an experiment: (1) an activity and (2) what is recorded from that activity. In essence then, there is no *one* sample space arising from a given activity. Example 1 shows how a familiar activity can result in a sample space that is different from our previous considerations.

EXAMPLE 1
Another Look at Tossing Coins

A coin is to be tossed twice and the number of times it lands heads up is recorded. Find the sample space and assign a probability to each outcome in this sample space.

Solution   We begin with the fundamental premise that outcomes from tossing a coin twice will result in four outcomes, each of which is an ordered pair,

$$HH, \quad HT, \quad TH, \quad TT$$

and where each of these outcomes has probability $\frac{1}{4}$. This creates an **underlying sample space** with equally likely outcomes. From this sample space, a new sample space is created whose outcomes reflect the number of heads that could possibly turn up: 0 (TT), 1 (HT or TH) or 2 (HH). Thus the outcomes in the new sample space, 0, 1, and 2, define *events* in the underlying space. For that reason, the term **event space** is sometimes applied to the new sample space. The probability assignments for the outcomes in the new sample space are those of corresponding events in the underlying sample space of equally likely outcomes. A list of those assignments is shown next.

Outcomes (Number of heads)	Defined event in underlying space	Probability
0	$\{TT\}$	$\frac{1}{4}$
1	$\{HT, TH\}$	$\frac{1}{4} + \frac{1}{4} = \frac{1}{2}$
2	$\{HH\}$	$\frac{1}{4}$
		Sum: 1

The first and last columns of this table show how the probability is distributed over the outcomes 0, 1, and 2 and is called the **probability distribution** for this sample space. (The middle column is for clarification of how the assignments are made). This

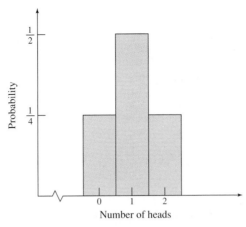

FIGURE 5

Probability Distribution Histogram for the
Number of Heads on Two Tosses
of a Coin

distribution can be pictured in a type of bar graph called a **histogram** as shown in Figure 5. ▪

EXAMPLE 2
Another Look at
Tossing a Die

A standard die has sides with one, two, and three dots colored red, two sides with four and five dots colored green, and a side with six dots colored blue. The die is tossed and the color of the top side is recorded. Find the sample space for this experiment and the probability of each outcome in that sample space.

Solution   The underlying sample space is the familiar one recording the number of dots on the top side having the six outcomes

$$1, 2, 3, 4, 5, 6$$

where each outcome has probability $\frac{1}{6}$. From this, a new sample space is constructed using the desired outcomes "red," "green," and "blue" whose probabilities are obtained from the events they define in the underlying sample space. The probability distribution of the new sample space is shown next.

Outcomes (The color is recorded)	Defined event in underlying space	Probability
Red	$\{1, 2, 3\}$	$\frac{1}{2}$
Green	$\{4, 5\}$	$\frac{1}{3}$
Blue	$\{6\}$	$\frac{1}{6}$
		Sum: 1

From the probability distribution of the new sample space, the probability that the die turns up green or blue is found to be $\frac{1}{3} + \frac{1}{6} = \frac{1}{2}$. ▪

NOTE   Recall that all objects such as colored balls and light bulbs are considered distinguishable unless otherwise noted.

EXAMPLE 3
Selecting a Ball

A box contains five red and three green balls. One ball is to be randomly selected from the box and its color recorded. Find the sample space for this experiment and the probability of each outcome in this sample space.

Solution Since the balls are distinguishable, visualize the red ones to be labeled with numerals 1 through 5, respectively, and the green ones to be labeled with the numerals 6 through 8, respectively. We consider the selection of any one ball to be as likely as any other ball. The outcome "red" then defines the event $\{1, 2, 3, 4, 5\}$ with probability $\frac{5}{8}$ and the outcome "green" defines the event $\{6, 7, 8\}$ with probability $\frac{3}{8}$. The probability assignments for these two outcomes are

Outcomes (color of ball)	Defined event in underlying space	Probability
Red	$\{1, 2, 3, 4, 5\}$	$\frac{5}{8}$
Green	$\{6, 7, 8\}$	$\frac{3}{8}$
		Sum: 1

EXAMPLE 4
Selecting a Card

Suppose that in a standard deck of cards, each ace is assigned the value 0, each face card is assigned the value 1, and the remaining cards are assigned the number on the card. One card is to be selected and the value of the card is noted.

(a) Find the sample space for this experiment and the probability of each outcome in that sample space.

Solution The underlying sample space consists of 52 outcomes (cards), each with probability $\frac{1}{52}$. The new sample space of card values has the eleven outcomes—0, 1, 2, 3, . . . , 10. The number 0 is assigned probability $\frac{4}{52} = \frac{1}{13}$ because there are 4 aces out of 52 cards in the underlying sample space; the outcome 1 is assigned $\frac{12}{52} = \frac{3}{13}$ because there are 12 face cards out of 52 cards in the underlying sample space; the outcome 2 is assigned probability $\frac{4}{52} = \frac{1}{13}$ because there are four 2s out of 52 cards in the underlying sample space, . . . , the outcome 10 is assigned probability $\frac{4}{52} = \frac{1}{13}$ because there are four 10s out of 52 cards in the underlying space.

Outcome:	0	1	2	3	4	5	6	7	8	9	10
Probability:	$\frac{1}{13}$	$\frac{3}{13}$	$\frac{1}{13}$	$\frac{1}{13}$	$\frac{1}{13}$	$\frac{1}{13}$	$\frac{1}{13}$	$\frac{1}{13}$	$\frac{1}{13}$	$\frac{1}{13}$	$\frac{1}{13}$

Note that the sum of these probabilities is 1.

(b) Find the probability that a card with value of at least 7 will be drawn.

Solution The event described consists of the outcomes

$$E = \{7, 8, 9, 10\}$$

and has probability

$$P(E) = P(7) + P(8) + P(9) + P(10)$$
$$= \frac{1}{13} + \frac{1}{13} + \frac{1}{13} + \frac{1}{13} = \frac{4}{13}.$$

(c) Find the probability that a card with value less than 3 will be drawn.

Solution The event described consists of the outcomes

$$E = \{0, 1, 2\}$$

and has probability

$$P(E) = P(0) + P(1) + P(2)$$
$$= \frac{1}{13} + \frac{3}{13} + \frac{1}{13} = \frac{5}{13}.$$

The probability assignments made throughout Section 6.1 and thus far in this section were **theoretical** in nature. That is, they were made by deductive reasoning alone. For example, when we said that the probability of a coin falling heads up was $\frac{1}{2}$, we did not require that a coin be tossed or that a coin even be at hand; we had a conceptual experiment and an idealized outcome. Nonetheless, we feel comfortable with such theoretical concepts because they seem logical and can be experimentally verified. On the other hand, suppose we had a certain coin slightly bent in such a way that upon 1000 tosses, it turned up heads 750 times. In this case, we would be much more inclined to rely on the experimental evidence, rather than an idealized concept, and assign heads a probability of $\frac{750}{1000} = 0.75$ based on the **relative frequency** with which heads turned up. Similarly, the probability or relative frequency that a tail would turn up would be $\frac{250}{1000} = 0.25$. Example 5 explores this concept further.

EXAMPLE 5
A Plant Experiment

From a large volume of pea pods, a plant experimenter broke open 20 pods and counted the number of peas in each pod. The results were as follows:

5 pods contained 4 peas each

9 pods contained 5 peas each

4 pods contained 6 peas each

2 pods contained 7 peas each

(a) Use this data to assign a probability to the number of peas that will be found in a typical pea pod.

Solution   The experiment is breaking open pea pods and recording the number of peas in each pod. The outcomes from this experiment are 4, 5, 6, or 7 peas. Based on this data, 5 of the 20 pods contained 4 peas each. Therefore it seems reasonable to assign the outcome 4 a probability of $\frac{5}{20}$. Stated differently, the **frequency** with which a pod contained 4 peas was 5; therefore, the **relative frequency** with which a pod contained 4 peas was $\frac{5}{20}$. Similarly, the relative frequency with which a pod contained 5 peas—hence the probability, is $\frac{9}{20}$. The probability (relative frequency) distribution for these pea pods is shown next.

Outcome of experiment (number of peas in pod)	Probability assignment (relative frequency)
4	$\frac{5}{20}$
5	$\frac{9}{20}$
6	$\frac{4}{20}$
7	$\frac{2}{20}$
	Sum: 1

A visual picture of this probability distribution is shown in the relative frequency histogram of Figure 6.

(b) Based on this data, what is the probability that a pea pod will contain fewer than 6 peas?

Solution   Stated differently, this question asks for the probability of the event $E = \{4, 5\}$. That probability is $\frac{5}{20} + \frac{9}{20} = \frac{14}{20}$.

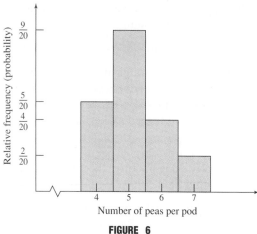

FIGURE 6
Relative Frequency Histogram

(c) Based on this data, what percentage of a large volume of pea pods will contain 6 or more peas? This question can be restated by asking for the probability of the event $E = \{6, 7\}$, then converting the probability to a percentage. The probability of E is $\frac{4}{20} + \frac{2}{20} = \frac{6}{20} = 0.30$, or about 30%. ▄▄

The assignment of probabilities in Example 5 is unlike the theoretical models in two ways. First, the sample size of 20 leads only to an approximation of the true probabilities for number of peas per pod. But as the sample size increases, it more accurately portrays the true proportion of pods containing 4, 5, 6, or 7 peas. Second, there is no realistic way to perform this experiment repeatedly under exactly the same conditions—such as with tossing a coin. So the best we can do is use the relative frequencies from a collected sample to assign probabilities and assume they are valid over the entire population even though *they are approximate* in that sense. Relative frequency assignments are also called **empirical probabilities** because they are based on empirical evidence.

EXAMPLE 6
Medical Application of
Empirical Probability

Medical authorities in a particular state examined randomly selected records of 250 residents who had been given flu shots in the fall and found that six of them had contracted the flu at some time during the following winter. The authorities then released these figures showing the chances of getting the flu even though a shot was given:

Outcome	Probability (relative frequency)
Had shot, will get flu	$\frac{6}{250} = .024$
Had shot, will not get flu	$\frac{244}{250} = .976$
	Sum: 1

Another way of expressing this probability would be to say that among the residents who had been given flu shots, 2.4% would get the flu, and 97.6% would not. ▄▄

EXAMPLE 7
Empirical Assignments
of Probability

An auditor of personal income tax forms submitted to a certain state gathered this information from a randomly selected sample of 300 forms:

Income Level	Correct	Understated tax	Overstated tax
$30,000–$50,000	125	20	10
$50,001–$100,000	110	30	5

Find the following empirical probabilities for a randomly selected individual income tax form submitted to this state.

(a) Find the probability that a person will submit a correct form.

Solution   We assume that any of the 300 forms are equally likely for selection. According to the data collected, $125 + 110 = 235$ of the forms were correct. Therefore, the probability assigned to this event is $\frac{235}{300}$.

(b) Find the probability that a person will be in the $30,000–$50,000 income level *and* will submit a tax form that understates his or her tax.

Solution   There are 20 tax forms meeting this criteria. This implies that the probability assigned to this event is $\frac{20}{300}$.

(c) Find the probability that a person will be in the $30,000–$50,000 income level *or* will submit a tax form that understates his or her tax.

Solution   There are $125 + 20 + 10 + 30 = 185$ tax forms meeting this criteria. Consequently the probability assignment to this event is $\frac{185}{300}$. ■

Another form of probability is in the statements of "odds" for or against some event. For example, "the odds are 7 to 5 that the Eagles will win over the Giants" or "the odds are 3 to 2 against having a recession next year" are actually statements about the likelihood of an event occurring and may therefore be put in the context of probability. Specifically, the statement "the odds are 7 to 5 that the Eagles will win over the Giants" means that if a 12-game series $(7 + 5 = 12)$ is played over and over between these two teams under identical conditions, the tendency would be for the Eagles to win 7 games and lose 5 games in each 12-game series. Accordingly, the odds statement implies that the probability of the Eagles winning against the Giants is $\frac{7}{12}$, while the probability that the Eagles will lose is $\frac{5}{12}$. Following this pattern, if the odds *for* an event E are stated as a to b, then

$$P(E) = \frac{a}{a + b} \quad \text{and} \quad P(E') = \frac{b}{a + b}.$$

This shows how an odds statement can be converted to a probability statement. It also shows that the odds statement "a to b" may be thought of as a comparison of the probability that E will occur to the probability that E will not occur.

EXAMPLE 8
Computing Probability
When the Odds
Are Known

If the odds for rain today are estimated to be 1 to 3, what is the probability of rain today?

Solution   From the meaning of odds, $P(\text{rain}) = \frac{1}{1 + 3} = \frac{1}{4}$. ■

Now suppose we were told that the probability is $\frac{7}{12}$ that the Eagles will win over the Giants. How do we find the odds of the Eagles winning over the Giants? Note that the probability for the Eagles to lose is $\frac{(12 - 7)}{12} = \frac{5}{12}$. The numerators provide the comparison of winning to losing: 7 to 5. This ratio can also be obtained by dividing the probability of the Eagles winning by the probability of the Eagles losing: $\frac{\left(\frac{7}{12}\right)}{\left(\frac{5}{12}\right)} = \frac{7}{5}$.

In general, suppose we know that $P(E) = \frac{m}{n}$. How can we find the odds for E? First, note that $P(E') = \frac{(n - m)}{n}$. Then think of odds as being a comparison of the probability of E occurring to the probability of E not occurring: m to $n - m$. This ratio can be obtained by computing

$$\frac{P(E)}{P(E')} = \frac{\dfrac{m}{n}}{\dfrac{n - m}{n}} = \frac{m}{n - m}.$$

Odds

Let E be an event in a sample space S where E is not all of S. Then the **odds for E** are found by reducing $\dfrac{P(E)}{P(E')}$ to lowest terms $\dfrac{a}{b}$ and writing "a to b" or "$a : b$."

Similarly, the **odds against E** are found by reducing $\dfrac{P(E')}{P(E)}$ to lowest terms.

EXAMPLE 9
Computing Odds

A single card is drawn from a deck of 52 cards, and the card is noted. What are the odds that the card will be an ace?

Solution   Let E be the event "an ace is drawn." Then $P(E) = \frac{4}{52}$ and $P(E') = \frac{48}{52}$. Therefore, the odds for E are

$$\frac{P(E)}{P(E')} = \frac{\dfrac{4}{52}}{\dfrac{48}{52}} = \frac{4}{48} = \frac{1}{12}$$

or "1 to 12" or 1 : 12. The odds *against* an ace being drawn are 12 : 1.   ■

EXERCISES   6.2

1. **Spinners**   The arrow on the spinner in the figure is spun and the letter representing the area the arrow points to is then recorded.

   (a) Find the sample space for this experiment and assign probabilities to each outcome in the sample space.
   (b) What is the probability that the arrow will point to the area labeled A or C?
   (c) What is the probability that the arrow will not point to the area labeled A?

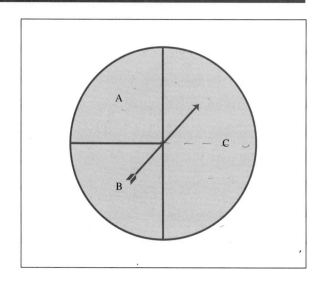

2. *Spinners* The arrow on the spinner in this figure is spun and the letter representing the area the arrow points to is then recorded.

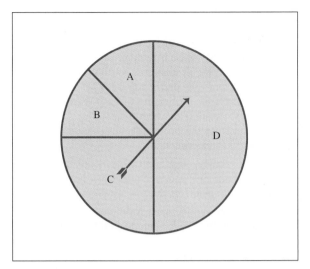

(a) Find the sample space for this experiment, and assign probabilities to each outcome in the sample space.

(b) What is the probability that the arrow will point to the area labeled A or D?

(c) What is the probability that the arrow will point to the area labeled B or C?

3. *Sampling* Four slips of paper—each with the letter *a*, *a*, *b*, or *c*, respectively, written on it—are placed in a box. One slip of paper is to be drawn from the box and the letter on the slip recorded.

(a) Find the sample space for this experiment and assign probabilities to each outcome in the sample space.

(b) What is the probability that a slip with the letter *b* on it is drawn?

(c) What is the probability that a slip with the letter *a* or *c* on it is drawn?

4. *Sampling* Six slips of paper—each with the letter *w*, *x*, *x*, *y*, *y*, or *z*, respectively, written on it—are placed in a box. One slip of paper is to be drawn from the box and the letter on the slip recorded.

(a) Find the sample space for this experiment and assign probabilities to each outcome in the sample space.

(b) What is the probability that a slip with the letter *x* or *y* on it is drawn?

(c) What is the probability that a slip with the letter *w* or *z* on it is drawn?

5. *Rolling a Die* A die has the face with 1 dot colored red, the face with two dots colored green, and the faces with three, four, five, and six dots colored blue. The die is to be rolled and the color of the upper side recorded.

(a) Find the sample space for this experiment and assign probabilities to each outcome in the sample space.

(b) What is the probability that the upper side will not be colored green?

(c) What is the probability that the color on the upper side will be red or blue?

6. *Rolling a Die* A die has the faces with one and two dots colored red, the face with three dots colored green, and the faces with four, five, and six dots colored blue. The die is to be rolled and the color of the upper side recorded.

(a) Find the sample space for this experiment and assign probabilities to each outcome in the sample space.

(b) What is the probability that the upper side will not be colored green?

(c) What is the probability that the color on the upper side will be red or blue?

7. *Rolling a Die* A die is to be rolled. If three dots show on the upper side, a "success" is recorded. Otherwise, a "failure" is recorded. Find the sample space for this experiment and assign probabilities to each outcome in the sample space.

8. *Colored Balls* Five red and nine green balls are in a box. One ball is to be randomly drawn from the box and the color recorded.

(a) Find the sample space for this experiment and assign probabilities to each outcome in the sample space.

(b) What is the probability that the ball will not be red?

9. *Colored Balls* Six red, two green, and seven blue balls are in a box. One ball is to be randomly drawn from the box and the color recorded.

(a) Find the sample space for this experiment and assign probabilities to each outcome in the sample space.

(b) What is the probability that the ball is red or green?

(c) What is the probability that the ball will not be red?

10. *Colored Balls* Eight red, five green, seven blue, and ten purple balls are in a box. One ball is to be randomly drawn from the box and the color recorded.

(a) Find the sample space for this experiment and assign probabilities to each outcome in the sample space.

(b) What is the probability that the ball is red or green?

(c) What is the probability that the ball will not be red?

11. Dice A pair of dice is to be rolled. If a sum of 7 or 8 is rolled, a "success" is recorded. Otherwise, a "failure" is recorded. Find the sample space for this experiment and assign probabilities to each outcome in the sample space.

12. Dice A pair of dice is to be rolled. If a sum of 2 or 10 is rolled, a "success" is recorded. Otherwise, a "failure" is recorded. Find the sample space for this experiment and assign probabilities to each outcome in the sample space.

13. Cards One card is to be drawn from a deck of 52 cards. If the card is a heart, a "success" is recorded. Otherwise a "failure" is recorded. Find the sample space for this experiment and assign probabilities to each outcome in the sample space.

14. Cards One card is to be drawn from a deck of 52 cards. If the card is a jack, a "success" is recorded. Otherwise a "failure" is recorded. Find the sample space for this experiment and assign probabilities to each outcome in the sample space.

15. Tossing Coins A coin is to be tossed twice. If two heads turn up, a "success" is recorded. Otherwise, a "failure" is recorded. Find the sample space for this experiment and assign probabilities to each outcome in the sample space.

16. Tossing Coins A coin is to be tossed twice. If one head and one tail turn up, a "success" is recorded. Otherwise, a "failure" is recorded. Find the sample space for this experiment and assign probabilities to each outcome in the sample space.

17. Tossing Coins A coin is to be tossed three times and the number of heads recorded.

(a) Find the sample space for this experiment and the probability of each outcome in that sample space.

(b) What is the probability that at least two heads turn up?

(c) What is the probability that two or fewer heads will turn up?

(d) Construct the probability distribution histogram for this experiment.

18. Tossing Coins A coin is to be tossed three times and the number of tails on the last two tosses is recorded.

(a) Find the sample space for this experiment and the probability of each outcome in that sample space.

(b) What is the probability that at least one tail turns up on the last two tosses?

(c) What is the probability that fewer than two tails will turn up on the last two tosses?

(d) Construct the probability distribution histogram for this experiment.

19. Dice A pair of dice is to be rolled, one red and one green. The number of dots on the top face of the red die is recorded.

(a) Find the sample space for this experiment and the probability of each outcome in that sample space.

(b) What is the probability that the red die shows at least four dots on the top face?

(c) What is the probability that the red die shows at most two dots on the top face?

(d) Construct the probability distribution histogram for this experiment.

20. Seating Three boys and two girls are to be seated randomly in a row and the number of boys who sit in the last two chairs of the row is recorded.

(a) Find the sample space for this experiment and the probability of each outcome in that sample space.

(b) Find the probability that at least one boy sits in the last two chairs.

(c) Construct the probability distribution histogram for this experiment.

21. Seating Three boys and two girls are to be seated randomly in a row and the number of girls who sit in the last two chairs of the row is recorded.

(a) Find the sample space for this experiment and the probability of each outcome in that sample space.

(b) Find the probability that at least one boy sits in the last two chairs.

(c) Construct the probability distribution histogram for this experiment.

22. I. D. Numbers A four-digit identification number is to be selected at random and the number of zeros in the last two digit positions is recorded.

(a) Find the sample space for this experiment and the probability of each outcome in that sample space.

(b) Find the probability that at most one zero appears in the last two digit positions.

(c) Construct the probability distribution histogram for this experiment.

23. I. D. Numbers A three-digit identification number is to be selected at random and the number of 2s in the last two digit positions is recorded.

(a) Find the sample space for this experiment and the probability of each outcome in that sample space.

(b) Find the probability that at most one of the last two digit positions contains the number 2.

(c) Construct the probability distribution histogram for this experiment.

24. **Unexpected Results**   A die is biased in such a way that the probability of a particular side turning up is proportional to the number of dots on that side. Such a die is to be rolled and the number of dots on the top side recorded.

  (a) Find the sample space for this experiment and the probability of each outcome in that sample space.
  (b) Find the probability that at least a four will be rolled.
  (c) Construct the probability distribution histogram for this experiment.

25. **Cards**   Two cards are to be dealt, without replacement, from a deck of 52 cards and the number of kings recorded.

  (a) Find the sample space for this experiment and the probability of each outcome in that sample space.
  (b) Find the probability that at least one king will be dealt.
  (c) Construct the probability distribution histogram for this experiment.

26. **Cards**   Two cards are to be dealt, without replacement, from a deck of 52 cards and the number of clubs recorded.

  (a) Find the sample space for this experiment and the probability of each outcome in that sample space.
  (b) Find the probability that at least one club will be dealt.
  (c) Construct the probability distribution histogram for this experiment.

27. **Sampling**   Among ten computers, three are known to have a defect. A sample of two computers is to be selected, without replacement and without regard to order, and the number of defective computers recorded.

  (a) Find the sample space for this experiment and the probability of each outcome in that sample space.
  (b) Find the probability that at least one of the computers will have a defect.
  (c) Construct the probability distribution histogram for this experiment.

28. **Sampling**   Among 12 microwave ovens, four are known to have a defect. A sample of two ovens is to be selected, without replacement and without regard to order, and the number of defective ovens recorded.

  (a) Find the sample space for this experiment and the probability of each outcome in that sample space.
  (b) Find the probability that at least one of the ovens will have a defect.
  (c) Construct the probability distribution histogram for this experiment.

29. **Survey**   A survey of 20 households asked how many radios they owned. The results were as follows:

  5 households answered 3
  8 households answered 4
  7 households answered 5

  (a) Construct the relative frequency (probability) distribution for this experiment.
  (b) Construct the relative frequency histogram.
  (c) Find the probability that a household will own at least four radios.

30. **Survey**   A survey of 25 households asked how many television sets they owned. The results were as follows:

  2 households answered 0
  8 households answered 1
  12 households answered 2
  3 households answered 3

  (a) Construct the relative frequency (probability) distribution for this experiment.
  (b) Construct the relative frequency histogram.
  (c) Find the probability that a household will own at least 2 television sets.

31. **Survey**   An agricultural worker broke open 18 soybean pods and counted the number of beans in each pod. The results were as follows:

  2 pods contained 2 beans
  5 pods contained 3 beans
  8 pods contained 4 beans
  3 pods contained 5 beans

  (a) Construct the relative frequency (probability) distribution for this experiment.
  (b) Construct the relative frequency histogram.
  (c) Find the probability that a bean pod will contain 3 or more beans.

32. **Survey**   On a certain day early in the growing season, a commercial tomato grower checked the number of blooms on each of 26 plants. The results were as follows:

  8 plants had 5 blooms
  12 plants had 8 blooms
  6 plants had 10 blooms

  (a) Construct the relative frequency (probability) distribution for this experiment.
  (b) Construct the relative frequency histogram.
  (c) Find the probability that a plant will have fewer than 8 blooms.

33. *Biased Coin* A coin is tossed twice and the number of heads recorded. This experiment was repeated 1000 times with these results:

Outcome recorded	Frequency of outcome
0 heads	290
1 head	510
2 heads	200

(a) Make an empirical probability distribution for this information.

(b) Find the probability that at least one head will occur in two tosses of this coin.

34. *Biased Die* A die was tossed 600 times and the number of dots on the top face recorded. The results were as follows:

Number of dots	Frequency of outcome
1	100
2	90
3	140
4	80
5	105
6	85

(a) Make an empirical probability distribution for this information. If this die is rolled, find the empirical probability that

(b) A 4 or greater is rolled.

(c) An even number is rolled.

(d) An odd number is rolled.

35. *Economic Trends* A survey of 120 economists shows that 64 predict a recession, 42 predict that the economy will show slow growth, and 14 predict sharply higher growth within the next year.

(a) Make an empirical probability distribution for this information.

(b) Based on this information, find the probability that there will be no recession within the next year.

36. *Politics* Jones, Brown, and Smith are in a three-way race for the Democratic Party's nomination for governor. A week before the election, 80 potential Democratic voters were surveyed, with 32 of them favoring Jones, 18 favoring Brown, 15 favoring Smith, and 15 undecided.

(a) Make an empirical probability distribution for this information.

(b) Assume that the undecided vote is split equally among the three candidates. Find the probability that a voter will favor Jones in the election.

37. *Product Reliability* A large number of computer chips was tested, and it was found that 92% were usable.

(a) Make an empirical probability distribution for this information.

(b) If one of these chips is randomly selected, what is the probability that it will be usable?

38. *Product Reliability* An inspection of 54 ladies' blouses checked for accuracy of sizing revealed that six were too small, three were too large, and 45 were within accepted tolerances.

(a) Make an empirical probability distribution for this information.

(b) Find the probability that a randomly selected blouse is not correctly sized.

(c) Find the probability that a randomly selected blouse is not too small.

Exercises 39 and 40 refer to these data: The maker of a certain brand of automobile gathered information on 450 persons who leased their automobile for two-year periods. The information they gathered is shown next.

Age	Number of times lease renewed			
	1	2	3	4
20–29	10	8	0	0
30–39	12	20	16	3
40–49	6	28	50	31
50–59	20	46	29	35
60 or over	8	42	36	50

39. *Market Analysis* Find the empirical probabilities for each of the following events.

(a) Being in the 30–39 age range and renewed a lease three times.

(b) Being in the 50–59 age range and renewed a lease at least twice.

(c) Being in the 50–59 age range or renewed a lease one time.

(d) Renewed a lease twice.

40. *Market Analysis* Find the empirical probabilities for each of the following events.

(a) Being in the 20–29 age range and renewed a lease three times.

(b) Being in the 60 or over age range and renewed a lease at most three times.

(c) Being in the 60 or over age range or renewed a lease at most three times.

(d) Being in the 40–49 age range and renewed a lease at least once.

Exercises 41 and 42 refer to these collected data: An audit of 250 billing statements from a large trucking firm for both short-haul and long-haul routes revealed the following information.

	Short-haul route	Long-haul route
Correct	100	82
Overbilled	30	20
Underbilled	10	8

41. *Market Analysis*   If a billing statement from this firm is randomly selected, find the empirical probability that

(a) The statement is correct.

(b) The statement is for a long-haul route.

(c) The statement overbills and is for a short-haul route.

(d) The statement underbills or is for a long-haul route.

42. *Market Analysis*   If a billing statement from this firm is randomly selected, find the empirical probability that

(a) The statement underbills.

(b) The statement is for a short-haul route.

(c) The statement is correct and is for a long-haul route.

(d) The statement overbills or is for a long-haul route.

Exercises 43 and 44 refer to these data: The library at a large state university asked 200 randomly selected students how many times they visited the main campus library each week to see whether there was a relationship between the number of visits and class standing. The number of students, by classification and number of library visits, is shown in the following table.

Class standing	Number of visits to the main library each week				
	0	1	2	3	Over 3
Freshman	3	4	5	3	1
Sophomore	8	7	8	4	3
Junior	4	12	20	18	12
Senior	1	25	30	22	10

43. *Library Habits*   If an undergraduate student attending this university is chosen at random, find the empirical probability that the student is

(a) A freshman who made three or more visits to the library in a week.

(b) A sophomore or junior who did not visit the library in a week.

(c) A junior and visited the library twice in a week.

(d) A senior or visited the library three times in a week.

44. *Library Habits*   If an undergraduate student attending this university is chosen at random, find the empirical probability that the student is

(a) A sophomore and visited the library more than three times in a week.

(b) A junior or senior who visited the library once in a week.

(c) A senior and visited the library once in a week.

(d) A senior or visited the library once in a week.

45. *Odds*   A single card is to be selected from a deck of 52 cards. Find the odds

(a) For the card being a king.

(b) For the card being a spade.

(c) For the card being red.

(d) Against the card being a club.

46. *Odds*   A pair of dice is to be rolled. Find the odds

(a) For the sum of the dots on the top sides to be 6.

(b) For the sum of the dots on the top sides to be 11.

(c) For the red die to show three dots on the top side.

(d) Against the dots' sum on the top side being 3.

47. *Odds*   A three-card hand is to be dealt. Find the odds

(a) For the hand to contain all spades.

(b) For the hand to contain all kings.

(c) Against the hand containing all jacks.

48. *Odds*   Three boys and four girls are to be seated in a row. Find the odds

(a) For a boy to be seated on either end.

(b) For a girl to be seated in the middle seat.

(c) Against a boy to be seated in the middle seat.

49. *Odds*   A survey of 20 households revealed that six households owned no chainsaws, ten households owned one chainsaw and four households owned two chainsaws. Find the odds

(a) For a household owning two chainsaws.

(b) For a household owning at least one chainsaw.

(c) Against a household owning two chainsaws.

50. *Odds*   The owner of an apple orchard took a sample of 30 apples. The check revealed that 20 apples contained no worms, six contained one, and four contained two. Find the odds

(a) For an apple containing no worm.

(b) For an apple containing at least one worm.

(c) Against an apple containing two worms.

Problems for Exploration

1. A recent *Reader's Digest* sweepstakes document stated that the odds for winning the grand prize and the extra prize are approximately 1 in 146,300,000. What is the approximate probability for winning this prize? What are the odds for not winning this prize?

2. To play one form of the Indiana state lottery, you are allowed to choose six different numbers from 1 to 44, inclusive. The Lottery Commission chooses six numbers randomly, and if your numbers match the six chosen by the commission, regardless of their order, then you win the Grand Prize. You win the second prize if five of your numbers match those chosen by the commission.

(a) Compute the probability of winning the grand prize.
(b) Compute the odds for winning the grand prize.
(c) Compute the probability of winning the second prize.
(d) Compute the odds for winning the second prize.

SECTION 6.3   **Discrete Random Variables and Expected Value**

The concept of probability allows us to compute what is "expected" in terms of an average for probability distributions that have numerical outcomes. Applications include games of chance and decision making in the business world.

What is "expected" is sometimes rather natural, and no formal mathematical formulation is needed. For example, suppose you consider playing this game with a friend: A coin is tossed. If it turns up heads, you give your friend $1; if it turns up tails, your friend gives you $1. If you were to play this game for a long period of time, how much would you *expect* to win? If you answered "nothing" then you have some feel for the terminology and realize that the laws of probability played a major role in your answer.

On the other hand, more involved situations require special mathematical tools to arrive at what is "expected." Consider this game offered by a game operator. A single die is rolled. If the number of dots on the top side is at least 5, then you win $9; otherwise, you give the game operator $5. As you play this game over and over, your "earnings" per game will be either a gain of $9 or a loss of $5 and might be recorded something like this: $9, −$5, −$5, −$5, $9, −$5, $9, $9, −$5, $9, −$5, −$5, −$5, −$5, $9, $9, −$5, $9, −$5, $9, −$5. Thus, as each game is played, your earnings **vary** between the two numbers 9 and −5, and furthermore, these earnings vary in a **random** fashion depending on the chance outcome of the die. For these reasons, the numbers $9 and −$5 are called *values* of a **random variable.** In addition, each of these random variable values has a **probability** that measures the likelihood of that value occurring relative to the other value. These probabilities come from the underlying sample space determined by rolling a single die: The value $9 is assigned a probability of $\frac{2}{6}$ because this value is attained when the event {5, 6} occurs and this event has probability $\frac{2}{6}$; the value −$5 is assigned a probability of $\frac{4}{6}$ because this value is attained when the event {1, 2, 3, 4} occurs and this event has probability $\frac{4}{6}$. The long-term implications of probability mean that if we played 1200 games, for example, we would *expect* the value $9 to appear $\frac{2}{6}$ of the time and the value −$5 to appear $\frac{4}{6}$ of the time in the above record of earnings. These expectations allow us to calculate the *average earnings per game played* by adding up all of the 1200 numbers in this record and then dividing by 1200. Since we expect $9 to appear $(\frac{2}{6})(1200)$ times and −$5 to appear $(\frac{4}{6})(1200)$ times in the list, the average should be

$$\frac{\$9\left(\dfrac{2}{6}\right)(1200) + (-\$5)\left(\dfrac{4}{6}\right)(1200)}{1200}$$

$$= \$9\left(\dfrac{2}{6}\right) + (-\$5)\left(\dfrac{4}{6}\right) \qquad \text{Upon cancelling the 1200s}$$

$$= \$\dfrac{18}{6} - \$\dfrac{20}{6} = -\$\dfrac{2}{6} = -\$\dfrac{1}{3}.$$

The number $-\$\frac{1}{3}$ is known as the **expected value** of this game; **expected value** means **long-term average.** That is, over the long term, you would expect to lose $\$\frac{1}{3}$ per game because the expected value is negative, and indeed, if you played 1200 games, you would expect to lose $\$(\frac{1}{3})(1200) = \$400$.

Notice that the calculation of expected value did not depend upon the number of games played (all of the 1200s cancelled out), but rather only on the values of the random variable and their respective probabilities as was shown in the second line:

$$\$9\left(\frac{2}{6}\right) + (-\$5)\left(\frac{4}{6}\right).$$

This is the key to finding expected value.

We give another example of a random variable and its expected value before formalizing these concepts.

EXAMPLE 1
Calculating Expected Value

Suppose you pay $10 to randomly draw one ball from a box containing three red, five green, and four blue distinguishable balls. If the ball is red, you win $14, if the ball is green, you win $7, while if the ball is blue, you win $11. What are your expected earnings from this game?

Solution   The underlying sample space consists of 12 equally likely outcomes. Because the color of the ball drawn dictates the exchange of money, we create a new sample space with the three outcomes, red, green, and blue with probabilities $\frac{3}{12}, \frac{5}{12}$, and $\frac{4}{12}$, respectively. As we play the game repeatedly, our "earnings" will vary among the numbers $4 (paid $10, won $14), −$3 (paid $10, won $7), and $1 (paid $10, won $11) depending on the chance occurrence of the respective outcomes red, green, or blue in the sample space:

$1, $1, −$3, $4, −$3, −$3, $4, $1, $4, −$3, $1, −$3, −$3, $4, . . .

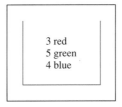

3 red
5 green
4 blue

The numbers $4, −$3, and $1 then become the random variable values. Each of these three numerical values is now assigned a probability equal to that of the corresponding color in the new sample space constructed above. A table showing the probability distribution for the random variable values may be constructed as follows:

Outcome in sample space	Random variable value	Probability of random variable value
Red	$4	$\frac{3}{12}$
Green	−$3	$\frac{5}{12}$
Blue	$1	$\frac{4}{12}$

This table shows that as the game is played over and over for a long period of time, we *expect* the value $4 to appear $\frac{3}{12}$ of the time, −$3 to appear $\frac{5}{12}$ of the time, and $1 to appear $\frac{4}{12}$ of the time. As we noted in our introductory example, it is the **long-term**

probability expectations that are the key ingredients to finding the expected value, not the actual (large) number of times the game is played:

$$\text{Expected Value} = \$4\left(\frac{3}{12}\right) + (-\$3)\left(\frac{5}{12}\right) + (\$1)\left(\frac{4}{12}\right)$$

$$= \$\frac{12}{12} - \$\frac{15}{12} + \$\frac{4}{12} = \$\frac{1}{12}.$$

As a consequence of this calculation, the average of all our earnings, in the long term, should be $\$\frac{1}{12} \approx \$0.083$ or about $0.08 per game.

If, for example, 2400 games were played, we could expect to win a total of $\$(\frac{1}{12})2400 = \$200$. ■

Our examples show that random variable values are numbers that are decided by the various outcomes in a sample space. In the introductory example of rolling a die, both of the outcomes 5 and 6 on a die were assigned to $9 while each of the outcomes 1, 2, 3, and 4 were assigned to −$5. In Example 1 the assignments were red to $4, green to −$3, and blue to $1. In each case, such assignments may be thought of as a function whose domain is a sample space and whose functional values are particular real numbers. Such a function is called a **random variable function** and is usually denoted by X rather than f or g, while the functional values, or random variable values, are denoted by $x_1, x_2, \ldots, x_n$ in the finite case. In the final analysis, it is the probability distribution of the random variables that allow us to find the expected value.

> ### Random Variables and Expected Value
>
> A **random variable function** X is a rule that assigns a numerical value to each outcome in a sample space. Each of these numerical values is known as a **random variable value.** If the random variable values are $x_1, x_2, x_3, \ldots,$ x_n with respective probabilities $p_1, p_2, p_3, \ldots, p_n$ then the **expected value of the random variable,** denoted by $E(X)$, is
>
> $$E(X) = x_1 p_1 + x_2 p_2 + x_3 p_3 + \ldots + x_n p_n.$$

Insurance companies rely on mortality tables prepared by actuaries to tell them the probability that certain aged persons will live through a specified time interval. Example 2 shows how an insurance company might use expected value.

EXAMPLE 2
Expected Value of a
Life Insurance Policy

An insurance company sells one-year term policies designed for 40-year-old women. The face value of the policy is $10,000, and each policy sells for $300. Suppose that the company's mortality tables show with probability .98 that a woman of this age will live for another year. What are the company's expected earnings from this type of policy?

Solution   The sale of these policies is a game of chance; we think of a sample space with two outcomes, "live" with probability .98, and "die" with probability .02. The company's "earnings" on each policy such as this depend on whether the buyer lives or dies during the year after purchase. If the buyer lives, then the company earns $300, whereas if the buyer dies, the company "earns" $300 − $10,000 = −$9700. It follows that company earnings per policy can be thought of as a random variable X with values $300 and $-$9700$. Assuming a large number of these policies are sold, the company would expect that .98 or 98% of these values would be $300 (the policy-holder lives) and .02 or 2% of these values would be $-$9700$ (the policyholder dies). The probability distribution for these random variable values is shown next.

Outcome in sample space	Random variable value	Probability of random variable value
Live	$300	.98
Die	$-$9700	.02

The expected value is

$$E(X) = \$300(.98) + (-\$9700)(.02) = \$100.$$

Knowing that expected value means "long-term average," we conclude that the company should profit an average of $100 per policy sold. ▬

EXAMPLE 3
Insuring Your
Computer

Suppose you plan to insure your new laptop computer, which you will be taking to campus, against theft for the amount of $3000. An insurance company claims that their records indicate 3% of such computers on college campuses are stolen within one year and offers to insure the computer for an annual premium of $100. What is *your* expected return per year from the insurance company on this policy?

Solution   You may think of this as a game of chance where you pay $100 (the premium) to play. If the computer is stolen during the year, you will gain $3000 $-$ $100 = $2900 with empirical probability .03. On the other hand, if the computer is not stolen during the year, you will lose the premium of $100 and gain nothing from the insurance company with empirical probability .97. Thus, $2900 and $-$100 are random variable values for a random variable X and we organize your earnings in the following table.

Outcome in sample space	Random variable value	Probability of random variable value
Stolen	$2900	.03
Not stolen	$-$100	.97

Your expected earnings from this policy are then

$$E(X) = (\$2900)(.03) + (-\$100)(.97) = \$87 - \$97 = -\$10.$$

This means that if you insure with this company over many years under the same circumstances, you will average a loss of $10 each year to the company. ▬

The next sample shows that expected value need not always involve money.

EXAMPLE 4
Dealing Cards

A two-card hand is to be dealt from a deck of 52 cards. Find the expected number of aces in the hand.

Solution   The number of aces in a two-card hand is a random variable X whose values are 0 (no aces), 1 (1 ace, 1 not an ace), and 2 (both cards are aces). The probability of such hands is shown next using the fact that there are 4 aces and 48 cards that are not aces:

$$P(\text{no aces}) = \frac{C(4, 0) \cdot C(48, 2)}{C(52, 2)} = \frac{1128}{1326}$$

$$P(1 \text{ ace}) = \frac{C(4, 1) \cdot C(48, 1)}{C(52, 2)} = \frac{192}{1326}$$

$$P(2 \text{ aces}) = \frac{C(4, 2)}{C(52, 2)} = \frac{6}{1326}.$$

The expected value for this random variable, or expected number of aces, is then

$$E(X) = 0\left(\frac{1128}{1326}\right) + 1\left(\frac{192}{1326}\right) + 2\left(\frac{6}{1326}\right) = \frac{204}{1326} \approx 0.154.$$

The number 0.154 is interpreted as follows: If two-card hands are dealt over and over for a long period of time and the number of aces counted in each hand, then the average of those numbers would be 0.154. ■

EXAMPLE 5
Outcomes May Be the
Random Variable
Values

Suppose a coin is tossed twice and the number of heads that turn up each time is recorded. Find the expected number of heads that turn up if this experiment is repeated over a long period of time.

Solution   The sample space for this experiment has outcomes 0 (no heads turned up), 1 (1 head, 1 tail turned up), or 2 (both turned up heads). The probability distribution for this experiment is shown in the following table.

Number of heads that turn up in two tosses of a coin	Probability
0	$\frac{1}{4}$
1	$\frac{1}{2}$
2	$\frac{1}{4}$

This time, the random variable values are precisely those of the outcomes in the experiment, 0, 1, and 2. The expected number of heads is then

$$0\left(\frac{1}{4}\right) + 1\left(\frac{1}{2}\right) + 2\left(\frac{1}{4}\right) = 0 + \frac{1}{2} + \frac{1}{2} = 1. \quad ■$$

EXAMPLE 6
An Agricultural
Application

An agricultural researcher broke open 20 pea pods and counted the number of peas in each pod. The results were as follows: 2, 4, 3, 2, 6, 4, 5, 4, 3, 3, 4, 6, 5, 2, 5, 4, 3, 3, 4, and 4. Find the expected number of peas per pod.

Solution   The number of peas per pod can be considered a random variable X with values 2, 3, 4, 5, and 6. These values and their associated probabilities can be organized in a table such as the one shown next.

Number of peas in pod	Frequency with which that number appears	Probability or relative frequency
2	3	$\frac{3}{20}$
3	5	$\frac{5}{20}$
4	7	$\frac{7}{20}$
5	3	$\frac{3}{20}$
6	2	$\frac{2}{20}$
	Sum: 20	Sum: 1

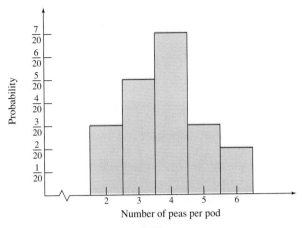

FIGURE 7

Histogram Showing Probability

Probability is assigned to each random variable value in accordance with the relative frequency with which it occurs as is shown in the last column of the table above. The expected number of peas per pod is now calculated as follows:

$$E(X) = 2\left(\frac{3}{20}\right) + 3\left(\frac{5}{20}\right) + 4\left(\frac{7}{20}\right) + 5\left(\frac{3}{20}\right) + 6\left(\frac{2}{20}\right) = \frac{76}{20} = 3.8 \text{ peas.}$$

We interpret 3.8 as the average number of peas per pod. In reality, it means that if this experiment could be repeated over and over, then in the long run, we would expect to find an average of 3.8 peas per pod. The probability distribution histogram of this random variable is shown in Figure 7.

EXAMPLE 7
A Business Decision

Two locations, A and B, are under consideration for a new fast-food franchise. Traffic counts reveal that location A can expect 50% of its customers in the morning, 20% in the afternoon, and 30% during the evening hours. On the other hand, location B can expect 25% of its customers in the morning, 20% in the afternoon, and 55% during the evening hours. Statistics from the parent company show that a typical customer will spend $2.50 in the morning, $2 in the afternoon, and $4.50 in the evening. Assuming that each location is projected to have about 1500 customers per day, which location will have the larger revenue?

Solution   We view this problem as one of finding the expected value of two experiments, one occurring at location A and the other at location B. The random variable values are the same in each location, $2.50, $2, $4.50, but the respective probabilities are different. A table will show the pertinent information.

Period of day	Random variable value (revenue per customer)	Percentage of volume stated as a probability Location A	Location B
Morning	$2.50	.50	.25
Afternoon	$2.00	.20	.20
Evening	$4.50	.30	.55

Expected value for location A: $2.50(.50) + $2(.20) + $4.50(.30) = $3.00.
Expected value for location B: $2.50(.25) + $2(.20) + $4.50(.55) = $3.50.

These calculations show that location B has the better expected value; each customer will spend an average of $3.50 there but only $3 at location A. For the 1500 projected customers, the expected revenue per day from location B will be 1500($3.50) = $5250 while only 1500($3) = $4500 at location A. ▬▬

EXAMPLE 8
A Game of Chance

Roulette wheels in Nevada have 38 equally spaced slots numbered 00, 0, 1, 2, 3, . . . , 36. Half of the slots numbered 1 through 36 are red, the other half are black, and the slots numbered 00 and 0 are green. Among the various types of betting on such a wheel is that of betting on a color. For instance, if a player places a sum of money on black and the winning number is black, then the player wins a sum of money equal to the bet placed. Otherwise, the bet is lost. Find the expected value of winnings if a $20 bet is placed on black.

Solution   The winnings from such betting may be considered to be a random variable X with values $20 and −$20. These values, along with their probabilities are organized in the following table:

Random variable values	Probability
$20	$\frac{18}{38}$
−$20	$\frac{20}{38}$

The expected value of winnings is then calculated to be

$$E(X) = \$20\left(\frac{18}{38}\right) - \$20\left(\frac{20}{38}\right) = \$\frac{360}{38} - \$\frac{400}{38} = -\$\frac{40}{38} \approx -\$1.05.$$

This calculation shows that if a player places a $20 bet over and over on black, then the player will average a loss of $1.05 *per game.* ▬▬

Suppose an unbiased coin is tossed. If it comes up heads, you give your friend $1 while if it comes up tails, your friend gives you $1. Your expected winnings from such a game are 1\left(\frac{1}{2}\right)$ − 1\left(\frac{1}{2}\right)$ = 0. When the expected value of the winnings from a game are zero, the game is called **fair.** In a fair game, neither player has an advantage.

A Fair Game

A game is called **fair** if the expected value of a player's winnings are zero. That is, if X is a random variable whose values represent a player's winnings from a game, then that game is fair if

$$E(X) = 0.$$

In general, random variables fall into one of two classifications, **discrete** or **continuous.** A discrete random variable is one that either has a finite number of values or whose values can be arranged in a sequence. All of the examples in this section have had a finite number of random variable values and hence fall into the discrete classification. Continuous random variables will be introduced in Chapter 8.

EXERCISES 6.3

To find the requested expected values of the exercises in this section, you may find it helpful to construct a table that will clearly identify the random variable values and their respective probabilities, as was done in several of the examples.

1. Suppose a coin is to be tossed. If the coin turns up heads, you win $1, while if the coin turns up tails, you lose $2.

   (a) What are your expected earnings from such a game?

   (b) If you played this game 1500 times, how much money would you expect to gain or lose?

2. Suppose a die is to be tossed. If the number of dots on the top face is at least five, you win $10. Otherwise, you lose $5.

   (a) What are your expected earnings from this game?

   (b) If you played this game 3000 times, how much money would you expect to gain or lose?

3. To move from your current position on a game board, you draw one ball from a container that has five red and six green balls. If the ball is red, you advance four places, while if the ball is green you move back three places.

   (a) What is your expected movement in this game?

   (b) In 44 turns at rolling the die, where would you expect to be located on the game board relative to your present position?

4. *Advertising* A firm allocates $\frac{1}{6}$ of its advertising dollar to newspaper ads, $\frac{1}{3}$ to radio ads, and $\frac{1}{2}$ to television ads. It is estimated that the return for that portion of a dollar spent on each of these three media is, respectively, $4, $3, and $6.

   (a) What is the firm's expected value from these advertising sources?

   (b) If $5000 is spent for these ads in the proportion indicated, what is the firm's expected dollar return?

5. Assume that a boy is just as likely as a girl at each birth. (Exclude multiple births.) In a two-child family, what is the expected number of boys?

6. Assume that a boy is just as likely as a girl at each birth. (Exclude multiple births.) In a three-child family, what is the expected number of girls?

7. A box contains four slips of paper, on each of which is written the number 4, 6, 9, or 18, respectively. One slip of paper is randomly drawn, and the number on the slip is considered to be a value of the random variable. Find the expected value of this random variable.

8. A box contains five slips of paper, on each of which is written the number 3, 4, 8, 9, or 20, respectively. One slip of paper is randomly drawn, and the number on the slip is considered to be a value of the random variable. Find the expected value of this random variable.

9. Suppose that you pay $5 to play this game. From a box containing six red and eight green marbles, you are allowed to select one marble after being blindfolded. If the marble is red, you win $10, but if the marble is green, you win nothing. What are your expected earnings from this game?

10. Suppose that you pay $25 to play this game. After being blindfolded, you draw one ball from a box containing five red, three green, and six white balls. If the ball is red, you win $5; if the ball is green, you win $100; and if the ball is white, you win nothing. What are your expected earnings from this game?

11. What are your expected earnings in this game? You draw one card from a deck of 52 cards. If the card is a spade, you win $5, but if the card is not a spade you lose $2.

12. What are your expected earnings from this game? You draw one card from a deck of 52 cards. If the card is a king or a queen, you win $50; otherwise, you win nothing.

13. Suppose that you pay $2 to play this game. A pair of dice is rolled. If the sum of the dots showing on top is 6, 7 or 8, you win $5. Otherwise, you win nothing. What are your expected earnings from this game?

14. What are your expected earnings from this game played with a friend? A pair of dice is rolled. If the sum of the dots on top is 6 or less, your friend gives you $10. For any other sum rolled, you give your friend $7.

15. You match pennies with a friend. If both coins show heads or both show tails, you win $1. Otherwise, you give your friend $1. What are your expected earnings from this game?

16. If you bet $1 on any three-digit number from 000 to 999 and your number is drawn, you win $500. What are your expected earnings from this game?

17. *Gambling* Suppose that you pay $1 to play this carnival game. The operator has a box containing 80 balls, each marked with a number from 1 to 80, respectively. You select a number from 1 through 80. Then the game operator randomly selects 20 balls. If your number is among the 20 on the balls selected, you get $3. What are your expected earnings?

18. *Gambling*   A single die is rolled and the amount of money you lose or win is the number of dots on the top side minus four. (A negative number means loss; a positive number means gain.) What are your expected earnings?

19. *Insurance*   You are considering insuring against theft the new $300 CD-player you just installed in your automobile. Your insurance agent tells you that such a policy option could be added to your present policy for $12 per year. The agent further states that the probability of theft is .2 in a given year. If you take this insurance, what is your expected return per year?

20. *Insurance*   You are considering an insurance policy for the new automobile you just purchased. A policy option is for towing insurance which costs $5 per year. The insurance agent tells you that company statistics show you will need towing during the year with probability .1 and that an average tow charge within your city is $60. If you take this insurance, what is your expected return per year?

21. *Insurance*   An insurance company is going to sell one-year life insurance policies with a face value of $100,000 to 25-year-old women for $1100. Their mortality tables show that such women will live for one year with probability .99. Consider the company's earnings from such policies to be values of a random variable.

 (a) Find the company's expected earnings per policy.
 (b) If the company hopes to sells 2000 of these policies, how much income could they expect from this source?

22. *Insurance*   An insurance company is going to sell one-year life insurance policies with a face value of $50,000 to 25-year-old men for $5500. Their mortality tables show that such men will live for one year with probability .9. Consider the company's earnings from such policies to be values of a random variable.

 (a) Find the company's expected earnings per policy.
 (b) If the company hopes to sell 3000 of these policies, how much income could they expect from this source?

23. *Insurance*   A life insurance company plans to sell one-year term life insurance policies with a face value of $50,000 to 50-year-old men. Their mortality tables show that such men will live for one year with probability .92. How should the company price such policies so that their average earnings per policy will be $500?

24. *Insurance*   A company plans to sell one-year term life insurance policies with a face value of $25,000 to

35-year-old women. Their mortality tables show that such women will live for one year with a probability of .97. How should the company price these policies to have average earnings of $50 per policy?

25. *Testing*   An upcoming test in a history class will have four true-false questions and eight multiple-choice questions, each with five answer choices. If a student guesses at the answer on every question on this test, what is the expected number of questions she will get correct?

26. *Testing*   An upcoming test in a mathematics class will have four true-false questions, five standard multiple-choice questions with four choices each, and 12 ten-response questions that require the selection of one of ten possible answers to the question. If a student guesses at the answer on every question on this test, what is the expected number of questions he will get correct?

27. *Roulette*   Roulette wheels in Nevada have 38 equally spaced slots numbered 00, 0, 1, 2, 3, . . . , 36. To play roulette, a player bets $1 on one of the 38 numbers. If the ball comes to rest on that number, the player wins $35, plus the $1 that was bet. Otherwise, the $1 bet is lost. What is the expected value for a player of this game?

28. *Business Decision*   A chain of video rental stores is considering City A and City B for the location of a new store. City A has 30% of its population 20 years of age or under, 30% from 21 through 30, and 40% 31 or older. City B has 25% of its population 20 years of age or under, 20% from 21 through 30, and 55% 31 or older. The company estimates that the 20-and-under age group rents an average of five videos per week, the 21-through-30 age group rents two videos per week, and the 31-and-over age group rents one video per week. In which of these cities should the chain locate its new store to rent the most videos per week?

29. *Business Decision*   A person is comparing how a copy shop would fare at a campus location versus a location in the heart of the business district. Estimates are that the campus location would get 20% of its business in the morning, 30% of its business in the afternoon, and 50% of its business during the evening hours. On the other hand, the business district location would get 40% of its business in the morning, 50% of its business in the afternoon, and 10% of its business during the evening hours. If it is projected that copy jobs average $5 in the morning, $10 in the afternoon and $3 in the evening, and that each location will get about 2000 customers each day, which location has the largest expected earnings?

30. **Survey** Twenty bean pods were broken open and the number of beans in each pod counted. The record looked like this:

Number of beans in pod	Frequency
2	5
3	8
4	4
5	2
6	1

Consider the number of beans in each pod to be a random variable. Find the expected number of beans per pod.

31. **Survey** Sixty people were asked about the number of years they had been in college. The responses are as tabulated. Find the expected number of years of college attended.

Number of years in college	Frequency
0	5
1	7
2	8
3	10
4	20
5	6
6	2
7	2

32. If Zal bet his best friend $25 that Zal's favorite team would win the Super Bowl and the announced odds for his team winning are 9 to 7, what were his expected earnings?

33. If Karla bet her friend $5 that Karla's favorite basketball team would win the National Collegiate Athletic Association (NCAA) tournament and the announced odds for her team winning were 89 to 10, what were her expected earnings?

34. **Lottery Tickets** A local civic organization sells 10,000 lottery tickets for $1 each. Prizes are awarded by a random drawing of three ticket numbers. The holder of the first number drawn wins a new portable TV set worth $300, the holder of the second number drawn wins a $200 shopping spree at a local supermarket, and the holder of the third number drawn wins $50 in cash. If you buy one ticket, what are your expected earnings?

35. If the odds for winning a $500 lottery prize are 1 to 62,586 and you spend $2 for the ticket and $1 for gasoline to go make the ticket purchase, what are your expected earnings for buying one ticket?

36. **Sampling** Among a rack of 30 ladies' blouses, four have a flaw of some kind. Suppose you buy two of these blouses and consider the number having a flaw to be a random variable.

(a) Make a probability distribution table for this random variable.

(b) Construct a probability histogram for this distribution.

(c) Of the two you bought, what is the expected number with a flaw?

37. **Sampling** Two balls are drawn, without replacement and without regard to order, from a box containing three red and five black balls. The number of red balls drawn is considered to be a random variable.

(a) Make a probability distribution table for this random variable.

(b) Construct a probability histogram for this distribution.

(c) Find the expected number of red balls in the sample drawn.

38. **Committees** A committee of three is selected from among five men and four women, and the number of women on the committee is considered to be a random variable.

(a) Make a probability distribution table for this random variable.

(b) Construct a probability histogram for this distribution.

(c) Find the expected number of women on the committee.

39. Two cards are drawn, without replacement and without regard to order, from a standard deck of 52 cards. What is the expected number of kings?

40. Three cards are drawn, without replacement and without regard to order, from a standard deck of 52 cards. What is the expected number of spades?

41. A box contains five red and three green marbles. Two marbles are randomly selected from this box, without replacement and without regard to order. What is the expected number of red marbles?

42. A bag contains three red, two green, and five black jellybeans. If you reach into the bag and randomly select three jellybeans, without replacement and without regard to order, what is the expected number of red jellybeans? Of green jellybeans?

43. A box contains one red, two green, four blue, and five black balls. If a sample of two balls is drawn from this box, without replacement and without regard to order, what is the expected number of red balls? Of green balls?

44. A two-card hand is dealt from a deck of 52 cards. Define a random variable X to have values that are the number of 8s in the hand. Find the expected number of 8s.

45. A two-card hand is dealt from a deck of 52 cards. Define a random variable X to have values that are the number of hearts in the hand. Find the expected number of hearts in the hand.

46. On each of three slips of paper in a box is written the number 5, 8, or 11, respectively. Two slips of paper are drawn, in succession, with replacement. Define the random variable X to have values that are one-half the sum of the two numbers drawn. Find the expected value of this random variable.

47. On each of three slips of paper in a box is written the number 2, 4, or 9, respectively. Two slips of paper are drawn, without replacement. Define the random variable X to have values that are three times the sum of the two numbers drawn. Find the expected value of this random variable.

48. A pair of dice is rolled. Let X be the random variable defined to have the value eight if a sum of 2, 3, or 4 is rolled and -5 otherwise. Find the expected value of this random variable.

49. A pair of dice is rolled. Let X be the random variable defined to have values equal to the sum of the dots showing on the top faces. Find the expected value of this random variable.

50. One card is drawn from a deck of 52 cards. Define the random variable X to have a value equal to the number on the card if the card has a number on it and 15 if the card does not have a number on it. Find the expected value of this random variable.

Problems for Exploration

1. Give two examples of games that are "fair".

2. A pair of dice is to be tossed. Let X be the random variable whose value is the larger of the two numbers on the top face. Find the expected value of X.

CHAPTER 6 REFLECTIONS

CONCEPTS

Probability numerically expresses the likelihood of particular outcomes occuring relative to the other possibilities in an *experiment*. *Events* are subsets of the sample space and have probabilities of at least zero but never more than one. *Simple events* consist of a single outcome whereas *compound events* contain more than one outcome. Probabilities can be assigned *theoretically* by deductive reasoning, or they can be assigned *empirically* based on relative frequencies. The *odds in favor* of an event is the ratio of the number of elements in the sample space that are in the event to those that are not; the *odds against* an event is the reverse ratio. A *random variable* assigns a number to each outcome in a sample space. The long-term average of these numerical values is called the *expected value* of the random variable.

COMPUTATION/CALCULATION

Find the number of outcomes in the sample space for the following experiments.

1. Drawing a card from a standard deck of 52 cards and then rolling a die.

2. Constructing a license plate having any three letters, followed by any four nonzero digits.

3. What are the odds in favor of getting a queen or a king when drawing one card from a standard deck?

4. What are the odds against an event that has a probability of $\frac{23}{79}$ of occurring?

A finite math class has 17 women and 12 men. A group of five is randomly selected to work on a cooperative learning project.   fraction % if is going to happen

5. Find the probability that the group has exactly two women.

6. Find the probability that three or more women are in the group of five.

A pair of dice is rolled. If a sum of 6 or more is rolled, a "success" is recorded. Otherwise, a "failure" is recorded.

7. Find the sample space for this experiment.

8. Determine the probability of a success.

9. Consider the number of girls in a three-child family to be a random variable. Find the expected value.

Fifty people were randomly surveyed to determine the number of videotapes they rented each month, as shown in the table. From this data:

Number of videotapes	Frequency
0	5
1	7
2	6
3	8
4	11
5 or more	13

10. Find the empirical probability that a person rents three or more videotapes per month.

11. Determine the expected number of videotapes rented per person each month.

CONNECTIONS

12. A lottery commission sells 1,000,000 tickets for $1 each. There are ten third prizes of $500 each, five second prizes of $5000 each, and one grand prize of $50,000. What is Corey's expected value if he buys one ticket?

13. An insurance company sells one-year term life insurance policies with a face value of $50,000 to 40-year-old women for $1800. Their mortality tables show that such women will live for one year with probability .98. Find the company's expected earnings per policy.

COMMUNICATION

14. Describe the characteristics of an event E if $P(E) = 0$. Give an example of such an event.

15. Describe the characteristics of an event E if $P(E) = 1$. Give an example of such an event.

16. Define *random variable* in your own words.

17. Define *expected value* in your own words.

COOPERATIVE LEARNING

18. A pair of dice is rolled. Find the probability that the sum of the numbers showing on the top sides is not 7 or that the product is not 12.

CRITICAL THINKING

19. Consider the experiment of drawing a single card from a standard deck of 52 cards. For this experiment, the probability that the card is not a heart or not an ace is *not* the same as the probability that the card is not a heart *plus* the probability that the card is not an ace. Explain why this is so, determine the correct probability, and discuss your method of solution.

COMPUTER/CALCULATOR

20. Use a computer or a calculator program to simulate the roll of a single die 120 times. Read the number of 1s, 2s, . . . , 6s obtained, and compute the relative frequency of each of the outcomes. Compare these numbers to the *theoretical relative* frequency of each outcome and the *theoretical* number of 1s, 2s, . . . , 6s that would be obtained.

Cumulative Review

21. The number of tornado deaths in Arkansas by decade for the 1970s and 1980s is 31 and 42, respectively. Using these data points, find a function to model this situation. Then use your function to predict the number of tornado deaths in Arkansas for the 1990s.

$$A = \begin{bmatrix} 1 & 0 & -3 \\ 0 & 2 & 4 \\ 1 & 0 & -4 \end{bmatrix}$$

22. Calculate the inverse of matrix A.

23. Compute and simplify $A^2 - 3A$ for the given matrix A.

24. Solve this linear programming problem:

$$\text{Minimize} \quad f = 2x + 5y,$$
$$\text{subject to} \quad 2x + 5y \geq 20$$
$$3x + 4y \geq 24$$
$$6x + y \geq 12$$
$$x \geq 0, y \geq 0$$

25. In a family of four children, which outcome is most likely? All four are of the same sex; three are of one sex; two are of each sex. Explain your answer.

CHAPTER 6 SAMPLE TEST ITEMS

1. Find the number of outcomes in the sample space for the experiment of selecting a sample of three marbles from a bag containing eight marbles—five blue and three orange.

2. Two boys and two girls randomly sit in a row of four seats. Find the probability that both boys and both girls sit next to each other.

3. What are the *odds* in favor of two heads and two tails coming up when a coin is tossed four times?

4. If the odds against an event *E* occurring are 5 to 9, then find the probability that *E* occurs.

Two standard dice, one red and one green, are to be rolled. Find the probability that

5. A sum of six will be rolled.

6. Both dice will show four dots on the top side.

7. The green die will not show two dots on the top side.

A three-card hand is to be dealt from a standard deck of 52 cards. Find the probability that

8. None of the cards will be aces.

9. Exactly two of the cards will be diamonds.

10. All three cards will be from the same suit.

Among eight graphing calculators, three are known to have a defect. A sample of two calculators is to be selected, without replacement and without regard to order, and the number of defective calculators is recorded.

11. Find the probability that exactly two calculators will be defective.

12. Find the probability that at least one calculator will be defective.

13. Construct the probability distribution histogram for this experiment.

In a five-child family, excluding multiple births and assuming that girls and boys are equally likely at birth, what is the

14. Probability of having two girls?

15. Probability of having at most two boys?

16. Expected number of girls?

17. Two cards are drawn, without replacement, from a standard deck of 52 cards. What is the expected number of jacks?

18. An upcoming sociology test will have 60 true-false questions and 40 multiple-choice questions with five options per multiple-choice item. What is the expected number of questions that can be correctly answered by a student who guesses on every question?

Career Uses of Mathematics

Did you receive any so-called junk mail last week? Some of it was probably tempting to you as a student, and you may have responded. The rest of the junk mail was not tempting and probably made you wonder why it was sent. Companies sending this mail have considered these different reactions before the mailing. Probability of response was a primary factor in their decision.

I use my M.S. in statistics to identify the types of households with the greatest probability of responding to direct mail. Business is great. Direct marketing is a fast-growing industry. You also get to work with some very creative people. I encourage you to continue your studies of mathematical tools like probability analysis.

Scott D. Wisner

Scott Wisner
Acxiom Corporation

Additional Topics in Probability

A particular rocket assembly depends on two key components. If either component is working properly, then the assembly will function properly, but, if both fail, the assembly will not function properly. The probability of failure for one component is .002 and the other is .01. If the ability or the failure to function of either component has no effect on the other, what are the odds that both will fail? (See Example 9, Section 7.3.)

CHAPTER PREVIEW

The set operations of union, intersection, and complement can be used to create new events from given events in a sample space. How to find the probability of a union and an intersection will be among the highlights of this chapter. In particular, an addition formula will be given for the union of two or more events. This formula reduces to an even simpler addition formula for events that are "mutually exclusive." Similarly, a multiplication formula will be given for the intersection of two or more events, and this formula becomes simpler when the events are "independent." The idea of conditional probability is introduced, not only for its own value and natural display on probability tree diagrams, but as a transitional vehicle to the probability of the intersection of events. Bayes' Theorem is shown to be an important application of conditional probability. Finally, the useful concept of binominal experiments is explored.

CALCULATOR/COMPUTER OPPORTUNITIES FOR THIS CHAPTER

The computer disk MasterIt contains two programs, ProbabilityI and ProbabilityII, which allow you to hone your skills in finding probability at two levels of difficulty. For each question, an incorrect answer will prompt a hint and another try. If the second try is incorrect, a correct answer and how it is obtained will appear on the screen. Another program on this disk will allow the simulation of coin tosses and throws of a single die.

A calculator with a permutation key, combination key, and factorial key will be useful in calculations throughout the chapter.

SECTION 7.1 Addition Rules for Probability; Mutually Exclusive Events

Set operations on two or more events (subsets) in a sample space will produce a new event. We will be interested in determining the probability of this new event. For instance, if A and B are two events in a sample space, then $A \cup B$, $A \cap B$, and $A \cap B'$ are also events in that sample space and each has a probability that is strongly dependent on the meaning of the set symbol or symbols involved. To review those symbols, recall that

the union symbol, $\cup$, is the set symbol for the logical connective "or,"

the intersection symbol, $\cap$, is the set symbol for the logical connective "and," and

the complement symbol, $'$, is the set symbol for the logical connective "not."

The primary focus of this section will be on the union of events and the logical connective "or." To set the stage for this work, recall the "inclusion-exclusion" principle which gave a rule for counting the number of elements in the union of two finite sets:

$$n(A \cup B) = n(A) + n(B) - n(A \cap B).$$

EXAMPLE 1
Venn Diagrams and Probability

Let S be a sample space in which the events A and B have the properties that $P(A) = .5$, $P(B) = .6$, and $P(A \cap B) = .2$.

(a) Illustrate these probabilities with a Venn Diagram.

Solution The construction will parallel that of a Venn diagram for counting problems. First draw two overlapping circles for A and B and write .2 in the intersection to denote the sum of the probabilities of all outcomes in $A \cap B$ as shown in Figure 1(a). Now, since $P(A) = .5$, the sum of probabilities of all outcomes in A is .5. But .2 of that total has already been listed in $A \cap B$, which is part of A (See Figure 1(a)); this means that $.5 - .2 = .3$ should be written in the part of A that is not in B as shown in Figure 1(b). Now we are assured that the probability of each outcome in A has been counted exactly once. Next, $P(B) = .6$, hence the part of B that is not in A should be labeled with $.6 - .2 = .4$ as shown in Figure 1(c). Finally, the sum of probabilities of all outcomes in S should be 1; thus far the total as shown in Figure 1(c)

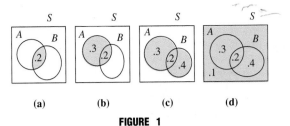

FIGURE 1

Showing Probability on a Venn Diagram

is $.3 + .2 + .4 = .9$; this means that outcomes whose probabilities sum to $.1$ have yet to be recorded in the diagram and those outcomes are not in either A or B. This is the reason for the $.1$ in Figure 1(d) and finishes the construction of the Venn diagram in such a way that the probability of each outcome in S has been counted only once.

(b) Use the Venn diagram in Figure 1(d) to find the probability of the event $A \cup B$ and compare the answer with $P(A) + P(B)$.

Solution   To find the probability of $A \cup B$, we are to add the probabilities of all outcomes in $A \cup B$. Figure 1(d) shows that sum displayed in nonoverlapping regions where the probability of no outcome is counted twice. Therefore, $P(A \cup B) = .3 + .2 + .4 = .9$.

The calculation $P(A \cup B) = .9$ is not the same as $P(A) + P(B) = .5 + .6 = 1.1$. A check of Figure 1(d) shows why: $P(A \cap B) = .2$ is counted once in $P(A) = .3 + .2$ and again in $P(B) = .2 + .4$, hence *twice* in $P(A) + P(B)$. Knowing this, we can assert that

$$P(A) + P(B) - P(A \cap B) = (.3 + .2) + (.2 + .4) - .2$$
$$= .3 + .2 + .4$$
$$= P(A \cup B).$$

(c) Use the Venn diagram in Figure 1(d) to find $P(A \cap B')$.

Solution   Locating the region depicting outcomes that are in A and not in B, we find the sum of those outcomes to be $.3$.   ■

EXAMPLE 2

Probability of a Union of Two Events

The following data were collected from a finite math class at State University:

	Have a scholarship	No scholarship
Freshman	8	5
Sophomore	5	7
Junior	3	6

A student is selected at random from these 34 students where each is considered equally likely to be selected with probability $1/34$.

(a) What is the probability that the student chosen is a freshman *or* a sophomore?

Solution   The event described by "the student is a freshman or a sophomore" may be thought of as the union of two events

A: (The student is a freshman.)   Students in the first row.

B: (The student is a sophomore.)   Students in the second row.

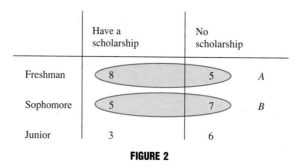

	Have a scholarship	No scholarship	
Freshman	8	5	A
Sophomore	5	7	B
Junior	3	6	

FIGURE 2

See Figure 2. We seek the probability of the event $A \cup B$. Since outcomes in the sample space are equally likely, the most straightforward way to find that probability is to count the number of outcomes in $A \cup B$ and then divide by 34, the total number of outcomes in the entire sample space. The first two rows in the table provide the count for $A \cup B$:

$$P(A \text{ or } B) = P(A \cup B) = \frac{(8 + 5) + (5 + 7)}{34} = \frac{8 + 5}{34} + \frac{5 + 7}{34}$$

$$= P(A) + P(B) = \frac{25}{34}.$$

Since A and B have no outcome in common, we find that $P(A \cup B) = P(A) + P(B)$. Events A and B for which $A \cap B = \emptyset$ are called **mutually exclusive.**

(b) What is the probability that the student chosen is a junior *and* has no scholarship?

Solution   From the given table, there are six students who meet the conditions of the event "the student is a junior and has no scholarship." (See Figure 3). Therefore,

$$P(A \text{ and } B) = P(A \cap B) = \frac{6}{34}.$$

(c) What is the probability that the student chosen is a junior *or* has no scholarship?

Solution   The event described by "the student chosen is a junior *or* has no scholarship" may be thought of as the union of the two events

A: (The student is a junior.)   Students in the third row.

B: (The student has no scholarship.)   Students in the second column.

	Have a scholarship	No scholarship	
Freshman	8	5	B
Sophomore	5	7	
Junior	3	6	
	A		

FIGURE 3

The probability of the event $A \cup B$ that we seek can again be found by counting the number of outcomes in $A \cup B$, being careful not to double-count any students, and then dividing by 34:

$$P(A \text{ or } B) = P(A \cup B) = \frac{3 + 6 + 7 + 5}{34} = \frac{21}{34}.$$

Figure 3 also suggests that this same probability could be found by adding $P(A)$ and $P(B)$, then removing the double-count of probability for the six outcomes in the intersection:

$$P(A \cup B) = P(A) + P(B) - P(A \cap B) = \frac{9}{34} + \frac{18}{34} - \frac{6}{34} = \frac{21}{34}. \quad \blacksquare$$

Examples 1 and 2 show that there is more than one way to approach the probability of a union of two events: Sometimes the visual impact of a Venn diagram will be helpful; for equally likely outcomes, count the number of outcomes in the event $A \cup B$ and then divide by the number of outcomes in the sample space; then use the formula $P(A \cup B) = P(A) + P(B)$ when A and B are mutually exclusive or $P(A \cup B) = P(A) + P(B) - P(A \cap B)$ when A and B are not mutually exclusive. It turns out that these formulas, which parallel the inclusion-exclusion principle for counting the number of elements in the union of two sets, are valid for *any* sample space.

When A and B are mutually exclusive, as shown in Figure 4, then $P(A \cup B)$ is found by adding together the probabilities of all outcomes in the event $A \cup B$: $P(A) + P(B)$. It follows that $P(A \cup B) = P(A) + P(B)$. On the other hand, if A and B are *not* mutually exclusive, then there must be at least one outcome that belongs to both A and B as suggested by Figure 5. As a consequence, $P(A) + P(B)$ would contain a double-count of all probabilities in $A \cap B$. Thus, $P(A) + P(B) - P(A \cap B)$ will give a single count to the probability of each outcome in $A \cup B$. It follows that $P(A \cup B) = P(A) + P(B) - P(A \cap B)$.

An Addition Formula for Probability: The Union of Two Events

For any two events A and B in any sample space, $P(A \text{ or } B) = P(A \cup B) = P(A) + P(B) - P(A \cap B)$. If $A \cap B = \phi$, then A and B are **mutually exclusive** and $P(A \text{ or } B) = P(A \cup B) = P(A) + P(B)$.

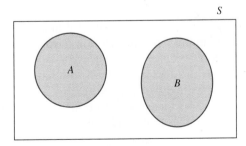

FIGURE 4

The Events A and B Are Mutually Exclusive (A and B Have No Outcome in Common). So, $P(A \cup B) = P(A) + P(B)$.

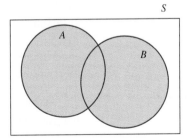

FIGURE 5

A and B Are Not Mutually Exclusive (A and B Have at Least One Outcome in Common). So, $P(A \cup B) = P(A) + P(B) - P(A \cap B)$

These formulas show that for probability considerations we may think like this: "or means add, but don't double count any of the probabilities."

EXAMPLE 3
Rolling Dice

A pair of dice is rolled.

(a) What is the probability of rolling a sum of 3 or a sum of 8?

Solution   As usual, consider one die to be red and the other green. The experiment of rolling a pair of dice has 36 equally likely outcomes, each with probability $\frac{1}{36}$. The event described by "roll a sum of 3 or a sum of 8" may be thought of as the union of the events

$$A \text{ (Roll a sum of 3)} = \{(2, 1), (1, 2)\}$$
$$B \text{ (Roll a sum of 8)} = \{(4, 4), (6, 2), (2, 6), (5, 3), (3, 5)\}$$

These events have no outcome in common and are therefore mutually exclusive. Since $n(A) = 2$ and $n(B) = 5$, we know that $P(A \cup B) = \frac{2+5}{36} = \frac{7}{36}$. In terms of the Addition Formula for the Union of Two Events,

$$P(A \cup B) = P(A) + P(B) = \frac{2}{36} + \frac{5}{36} = \frac{7}{36}.$$

(b) What is the probability of rolling a double or a sum of 4?

Solution   The event described by "roll a double or a sum of 4" may be thought of as the union of the events

$$A \text{ (Roll a double)} = \{(1, 1), (2, 2), (3, 3), (4, 4), (5, 5), (6, 6)\}$$
$$B \text{ (Roll a sum of 4)} = \{(2, 2), (3, 1), (1, 3)\}$$

These two events are not mutually exclusive; the outcome (2, 2) is common to both. Counting the number of outcomes in $A \cup B$, and remembering not to double count any outcome, shows that $n(A \cup B) = 8$. Therefore, $P(A \cup B) = \frac{8}{36}$. Of course the Addition Formula for the Union of Two Events will give the same result: Observing that $P(A) = \frac{6}{36}$, $P(B) = \frac{3}{36}$ and $P(A \cap B) = \frac{1}{36}$, the evaluation of that formula gives

$$P(A \cup B) = P(A) + P(B) - P(A \cap B) = \frac{6}{36} + \frac{3}{36} - \frac{1}{36} = \frac{8}{36}. \quad \blacksquare$$

EXAMPLE 4
A Sample of Automotive CD-Players

Among 20 automotive CD-players, four are known to have a defect. A sample of three of these players is selected, without replacement and without regard to order, to be checked for a defect. What is the probability that all three will have a defect or all three will not have a defect?

Solution   Think of the 20 CD-players as divided into two groups, one with the four that are defective and another with the 16 that have no defect. An outcome to this experiment is an unordered set of three CD-players. Is it possible for such an outcome to contain all players with a defect *and* all without a defect? The answer is no. Therefore, the events described by "all are defective" and "all do not have a defect" are mutually exclusive. The fact that selection is done without replacement and without regard to order implies that combinations is the counting technique needed.

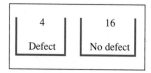

P (all defective or all without a defect)

$$= P \text{ (all defective)} + P \text{ (all without a defect)}$$

$$= \frac{C(4, 3)}{C(20, 3)} + \frac{C(16, 3)}{C(20, 3)} = \frac{4}{1140} + \frac{560}{1140} = \frac{564}{1140}.$$ ▬

EXAMPLE 5
Seating People

Three boys and two girls are to be randomly seated in a row. What is the probability that a boy is seated on the left end or a girl is seated on the right end?

Solution   An outcome in this experiment is an ordered arrangement of five people in a row. Is it possible for one of these outcomes to have a boy on the left end *and* a girl on the right end? The answer is yes. As a consequence, the events described by "a boy is seated on the left end" and "a girl is seated on the right end" are not mutually exclusive. Since seating is without replacement, but with regard to order, the counting technique involved is the multiplication principle.

P(boy on the left or girl on right)

$$= P \text{(boy on left)} + P \text{(girl on right)} - P \text{(boy on left and girl on right)}$$

$$= \frac{3 \cdot 4 \cdot 3 \cdot 2 \cdot 1}{5 \cdot 4 \cdot 3 \cdot 2 \cdot 1} + \frac{4 \cdot 3 \cdot 2 \cdot 1 \cdot 2}{5 \cdot 4 \cdot 3 \cdot 2 \cdot 1} - \frac{3 \cdot 3 \cdot 2 \cdot 1 \cdot 2}{5 \cdot 4 \cdot 3 \cdot 2 \cdot 1}$$

$$= \frac{3}{5} + \frac{4}{10} - \frac{3}{10} = \frac{7}{10}.$$ ▬

EXAMPLE 6
Another Look at Venn Diagrams

A check of 50 calculators just off the assembly line reveals that 5 have faulty cases, 8 have faulty batteries, and 3 have both of these defects. Let

C: Calculator has a faulty case.

B: Calculator has a faulty battery.

(a) What is the probability that a randomly selected calculator from this collection will have a faulty case or a faulty battery?

Solution   In symbols, the question asks $P(C \cup B)$? From the data given,

$$P(C \cup B) = P(C) + P(B) - P(C \cap B) = \frac{5}{50} + \frac{8}{50} - \frac{3}{50} = \frac{10}{50} = \frac{1}{5}.$$

Alternatively, the Venn diagram shown in Figure 6 will efficiently answer the question:

$$P(C \cup B) = \frac{n(C \cup B)}{n(S)} = \frac{2 + 3 + 5}{50} = \frac{10}{50} = \frac{1}{5}.$$

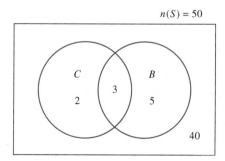

FIGURE 6
The Outcomes for Testing Calculators

(b) What is the probability that one of these calculators will have a faulty case but not a faulty battery?

Solution In symbols, the question asks $P(C \cap B')$? From looking at the Venn diagram we can determine that the answer is $\frac{2}{50} = \frac{1}{25}$.

(c) What is the probability that one of the calculators will have exactly one of these defects?

Solution In symbols, the question asks

$$P[(C \cap B') \cup (B \cap C')]?$$

Identifying the sets just described on the Venn diagram gives $\frac{2}{50} + \frac{5}{50} = \frac{7}{50}$. ■

There is a useful connection between the probability of any event E and its complement E'. Because $E \cap E' = \phi$, E and E' are mutually exclusive events. Furthermore, the outcomes in E along with those in E' make up all outcomes in the sample space S. See Figure 7. As a consequence,

$$P(E \cup E') = 1$$
or
$$P(E) + P(E') = 1$$
so that
$$P(E) = 1 - P(E').$$

This result is known as the **complement theorem.**

The Complement Theorem

Let S be any sample space and let E be any event in S. Then the **complement theorem** states that

$$P(E) = 1 - P(E').$$

EXAMPLE 7
Using the Complement Theorem

(a) The probability of rolling a sum of 5 with a pair of dice is $\frac{4}{36}$. Find the probability of not rolling a sum of 5.

Solution The probability of rolling a sum of 5 is $\frac{4}{36}$. According to the Complement Theorem, the probability of not rolling a 5 is $1 - \frac{4}{36} = \frac{32}{36}$.

(b) Find the probability that a 5-card hand dealt from a deck of 52 cards will contain all red cards as well as the probability that such a hand will not contain all red cards.

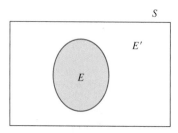

FIGURE 7
E and E' Are Mutually Exclusive

Solution   There are 26 red cards in the deck. The probability that a 5-card hand will contain all red cards is

$$\frac{C(26,\ 5)}{C(52,\ 5)} = \frac{65{,}780}{2{,}598{,}960} \approx 0.0253.$$

The probability that 5-card hand will not contain all red cards is then

$$1 - 0.0253 = 0.9747.$$

(Contrast this with the probability that the hand will contain no red card (all black cards): $\frac{C(26,\ 5)}{C(52,\ 5)} = 0.0253$.)   ▬

When more than two events are involved, formulas for the probability of their union become increasingly difficult except when the events are mutually exclusive. A collection of events is **mutually exclusive** if the intersection of every pair of the events is empty. In such cases, the probability of the union is just the sum of the probabilities of each event making up the union. This is expressed in the following extension of the addition theorem.

Extension of the Addition Formula for Mutually Exclusive Events

For mutually exclusive events $A_1, A_2, \ldots, A_n$,

$$P(A_1 \cup A_2 \cup \ldots \cup A_n) = P(A_1) + P(A_2) + \ldots + P(A_n)$$

The complement theorem can sometimes shorten calculation efforts in probability problems, especially in conjunction with the addition theorem for mutually exclusive events. This use arises naturally in Example 7. It also arises in the probability of an event described by "at least *one*," which is logically equivalent to finding $1 - P(none)$, or in the probability of an event described by "at least *two*," which is logically equivalent to $1 - P(none\ or\ one)$, and so on. Examples 8 and 9 demonstrate this.

EXAMPLE 8
Using the Extended
Addition Formula

The manager of Hill's Department Store wanted to know what sums of money shoppers who entered the store were spending. She took a typical day's receipts and broke them down in this way:

Sum spent in store (dollars)	Number of shoppers
$10.00 or less	50
$10.01 to $25.00	70
$25.01 to $50.00	40
$50.01 to $100.00	30
More than $100.00	10
	Total: 200

Based on this information,

(a) What is the probability that a shopper who comes into the store will spend $50.00 or less?

Solution   The mutually exclusive breakdown of amounts spent leads to the answer: $\frac{50 + 70 + 40}{200} = \frac{4}{5}$. Stated differently, 80% of the shoppers who enter the store will spend $50.00 or less. The complement theorem could also be used to answer this question:

$$P(\text{spend } \$50 \text{ or less}) = 1 - P(\text{spend more than } \$50)$$

$$= 1 - \frac{30 + 10}{200} = 1 - \frac{1}{5} = \frac{4}{5}$$

(b) What is the probability that a Hill's shopper will spend more than $10.00?

Solution   The answer can be found by adding the last four amounts in the table and dividing by 200, obtaining $\frac{150}{200}$. An easier way is to use the complement theorem: $P(\text{spend more than } \$10.00) = 1 - P(\text{spend more than } \$10.00)' = 1 - P(\text{spend } \$10.00 \text{ or less}) = 1 - \frac{50}{200} = \frac{150}{200} = \frac{3}{4}$. Stated differently, 75% of the shoppers who come into the store spend more than $10.00. ■

EXAMPLE 9
Using the Extended
Addition Formula

A pair of dice is rolled. Find the probability that

(a) A sum of 2 or 3 is rolled.

Solution   Because a sum of 2 and a sum of 3 cannot be rolled at the same time, the addition theorem for mutually exclusive events applies. $P(\text{roll a sum of 2 or a sum of } 3) = P(\text{roll a sum of 2}) + P(\text{roll a sum of 3}) = \frac{1}{36} + \frac{2}{36} = \frac{3}{36}$.

(b) A sum of at least 9 is rolled.

Solution   The event "a sum of at least 9 is rolled" means that a 9, 10, 11 or 12 is rolled. It is clear that if rolling each of these sums is an event, these sums (events) are mutually exclusive. Therefore, $P(\text{at least a 9 is rolled}) = P(\text{roll a sum of 9 or 10 or 11 or 12}) = P(\text{roll a sum of 9}) + P(\text{roll a sum of 10}) + P(\text{roll a sum of 11}) + P(\text{roll a sum of 12}) = \frac{4}{36} + \frac{3}{36} + \frac{2}{36} + \frac{1}{36} = \frac{10}{36}$.

(c) A sum greater than 4 is rolled.

Solution   The complement theorem comes in handy here:

$$P(\text{sum greater than } 4) = 1 - P[(\text{sum greater than } 4)']$$
$$= 1 - P(\text{sum of 2 or 3 or 4})$$
$$= 1 - [P(2) + P(3) + P(4)]$$
$$= 1 - \left(\frac{1}{36} + \frac{2}{36} + \frac{3}{36}\right) = \frac{5}{6}. ■$$

The uses of *or* shown in the various contexts of the examples in this section reappear from time to time throughout the remainder of this chapter.

EXAMPLE 10
How "At Least" is
Related to Mutually
Exclusive Events

A 3-card hand is to be dealt, without replacement, from a deck of 52 cards. What is the probability that at least 1 of the cards dealt will be a king?

Solution   As is always the case, "at least" must be translated into a more workable statement. A moment's reflecting about what a 3-card hand that contains at least 1 king could look like leads to this logical rephrasing: "exactly 1 king and 2 nonkings" or "exactly 2 kings and 1 nonking" or "all 3 kings." Because the three events involved are mutually exclusive, our problem may be written as: P(at least 1 king) $= P$(exactly 1 king and 2 nonkings) $+ P$(exactly 2 kings and 1 nonking $+ P$(all 3 are kings) $=$

$$\frac{C(4, 1) \cdot C(48, 2)}{C(52, 3)} + \frac{C(4, 2) \cdot C(48, 1)}{C(52, 3)} + \frac{C(4, 3)}{C(52, 3)} \approx$$

$$.20416 + .01303 + .00018 = .21737.$$

Another translation of "at least 1 king" uses the complement theorem:

$$P(\text{at least 1 king}) = 1 - P(\text{at least 1 king})' =$$

$$1 - P(\text{no kings}) = 1 - \frac{C(48, 3)}{C(52, 3)} \approx 1 - .78262 = .21738. \quad \blacksquare$$

EXERCISES   7.1

In Exercises 1 through 7, construct a Venn diagram to find the desired probabilities.

1. If $P(A) = .6$, $P(B) = .3$, and $P(A \cap B) = .1$, find:

(a) $P(A \cup B)$
(b) $P(A \cap B')$
(c) $P(A' \cup B)$

2. If $P(A) = .4$, $P(B) = .7$, and $P(A \cap B) = .3$, find: (a) $P(A \cup B)$
(b) $P(A \cap B')$
(c) $P(A' \cup B)$

3. If $P(A) = .6$, $P(B) = .6$, and $P(A \cap B) = .3$, find:

(a) $P(A \cup B)$
(b) $P(A' \cap B)$
(c) $P(A \cup B')$

4. If $P(A) = .7$, $P(B) = .5$, and $P(A \cap B) = .4$, find:

(a) $P(A \cup B)$
(b) $P(A' \cap B)$
(c) $P(A \cup B')$

5. If $P(A \cup B) = .8$, $P(A) = .4$, and $P(B) = .7$, find:

(a) $P(A \cap B)$
(b) $P(A' \cap B')$
(c) $P(A' \cup B')$

6. If $P(A \cup B) = .6$, $P(A) = .6$, and $P(B) = .5$, find:

(a) $P(A \cap B)$
(b) $P(A' \cap B')$
(c) $P(A' \cup B')$

7. If $P(A - B) = .4$, $P(B - A) = .3$, and $P(A \cup B)' = .1$, find:

(a) $P(A)$
(b) $P(A \cup B)$
(c) $P(A \cap B)'$

In Exercises 8 through 14, use a Venn diagram to find the desired probability.

8. A survey of married couples taken before an election shows that the husband will vote with a probability of .4, the wife will vote with a probability of .6, and both will vote with a probability of .3. What is the probability that either the husband or the wife will vote in the upcoming election?

9. A survey revealed that 30% of the people get their news from newspapers, 50% from television, and 20% get news through both media. What is the probability that a person will get the news through either a newspaper or a television?

10. Of the shirts sold in a clothing store, 5% have either faulty seams or a missing button, 2% have faulty seams, and 0.5% have both flaws. What is the probability that a shirt will have a missing button?

11. Eighty college students were asked about their eating habits during the last week. Sixty of them ate in a school cafeteria or at a fast-food place, 30 ate only in a school cafeteria, while 20 ate in both a school cafeteria and a fast-food place. What is the probability that a student ate in a fast-food place?

12. A survey of 130 households revealed that during the past year, 50 bought a car, 40 bought a TV set, 30 bought a camcorder, 18 bought a car and a TV set, 10 bought a car and a camcorder, 8 bought a TV and a camcorder, and 3 bought all three of these items. What is the probability that one of these households

(a) Bought a car or camcorder during the past year?
(b) Bought a car or TV set, but not a camcorder during the past year?
(c) Bought exactly one of these three items during the past year?

13. Among 180 households surveyed, 50 have a bicycle, 60 have a skateboard, 30 have a baseball bat, eight have a bicycle and a skateboard, seven have a bicycle and a baseball bat, four have a skateboard and baseball bat, and two have all three products. What is the probability that a household will

    (a) Have a skateboard or baseball bat?

    (b) Have a skateboard or bicycle, but not a baseball bat?

    (c) Have exactly two of these items?

14. **Product Inspections** On inspecting a lot of 120 ladies' blouses, it was learned that eight had faulty seams, five had improperly sewn buttons, and two had both of these defects. If a blouse is selected from this lot, what is the probability that

    (a) It will have at least one of these flaws? $\frac{15}{120}$

    (b) It will have neither of these flaws? $\frac{120-15}{120}$

    (c) It will have exactly one of these flaws?

15. (a) Is it possible for two mutually exclusive events A and B to have $P(A) = .2$ and $P(A \cup B) = .5$?

    (b) Is it possible for two events A and B to have $P(A \cup B) = .8$, $P(A) = .2$, and $P(A \cap B) = .5$?

    (c) Is it possible for two events A and B to have the property that $P(A \cup B) = .6$, $P(B) = .3$, and $P(A \cap B) = .8$?

16. (a) Is it possible for two mutually exclusive events A and B to have $P(A) = .6$ and $P(A \cup B) = .5$?

    (b) Is it possible to have two events A and B with the conditions that $P(A) = .6$ and $P(A \cup B) = .5$?

Employment Data Exercises 17 through 20 refer to the number of people in a small town who are either employed or unemployed.

	Male	Female
Employed	90	50
Unemployed	10	15

If a person in this town is selected at random, what is the probability that

17. The person is employed or female?

18. The person is employed and female?

19. The person is employed?

20. The person is male or unemployed?

Survey Exercises 21 through 24 refer to the data below regarding a survey of 100 young people in an upscale community.

	Drives only a Corvette	Drives only a Miata	Drives neither
Male	28	40	6
Female	12	10	4

Based on this survey, what is the probability that a young person will

21. Be female or drive a Corvette?

22. Drive a Miata?

23. Drive neither of the two makes of cars?

24. Be male or drive a Miata?

Sampling Exercises 25 and 26 apply to these data: A college student bought a bag of peanuts in the shell. He took each of the 20 shells in the bag, broke them open, and counted the number of peanuts in each shell. His record looked like this:

Number of peanuts in shell	Number of shells
1	3
2	6
3	10
4	1

Based on these results, if you buy a bag of peanuts and break open one of the shells, what is the probability that the shell will have

25. Three or more peanuts in it?

26. At least two peanuts in it?

27. **Cards** One card is drawn from a deck of 52 cards. What is the probability that the card will be

    (a) A 5 or an 8?

    (b) A jack or a club?

    (c) A jack and a club?

    (d) An ace or a spade?

28. **Cards** One card is drawn from a deck of 52 cards and the card noted. What is the probability that the card is

    (a) A king or a jack?

    (b) A king or a heart?

    (c) A king or a queen?

    (d) A card with a 2, 3, 4, or 5 on it?

29. **Seating** Four boys and three girls are randomly seated in a row. What is the probability that

    (a) A boy will be seated in the left end seat?

    (b) A boy will be seated in the left end seat, or a girl will be seated in the right end seat?

    (c) A girl will be seated in the middle seat, or a boy will be seated in the left end seat?

30. **Seating** Two boys and three girls are randomly seated in a row. What is the probability that

    (a) A girl will be seated in the middle seat?

    (b) A girl will be seated in the left end seat, or a boy will be seated in the right end seat?

    (c) A girl will be seated in the left end seat, or a girl will be seated in the right end seat?

31. *Seating* Three boys and three girls are randomly seated in a row. What is the probability that

    (a) A girl will be seated in the left end seat, and a girl will be seated in the right end seat?

    (b) A girl will be seated in the left end seat, or a girl will be seated in the right end seat?

    (c) Boys and girls are seated alternately, or a girl is seated in the left end seat?

    (d) Boys will be seated together and girls will be seated together, or girls will be seated in the two middle seats?

32. *Seating* Four boys and four girls are randomly seated in a row. What is the probability that

    (a) A girl will be seated in the left end seat, and a girl will be seated in the right end seat?

    (b) A girl will be seated in the left end seat, or a girl will be seated in the right end seat?

    (c) Boys and girls are seated alternately, or a girl is seated in the left end seat?

    (d) Boys will be seated together and girls will be seated together, or girls will be seated in the two middle seats?

33. *Arrangements* Seven balls, four of which are red and three of which are green, are randomly arranged in a row. What is the probability that

    (a) A red ball will be on the left end of the row?

    (b) A red ball will be on the left end of the row, or a green ball will be on the right end of the row?

    (c) The red balls are arranged together and the green balls are arranged together, or the first two positions on the left end of the row have different-colored balls?

    (d) The colors of the balls alternate, or the first two balls on the left end are red?

34. *Arrangements* Six balls, three of which are red and three of which are green, are randomly arranged in a row. What is the probability that

    (a) A red ball will be on the left end of the row?

    (b) A red ball will be on the left end of the row, or a green ball will be on the right end of the row?

    (c) The red balls are arranged together and the green balls are arranged together, or the two balls in the middle are both red?

    (d) The colors of the balls alternate, or the first two balls on the left end are red?

35. *Arrangements* Five balls, four of which are red and one of which is green, are randomly arranged in a row. What is the probability that

    (a) A red ball will be on the left end of the row?

    (b) A red ball will be on the left end of the row, or a green ball will be on the right end of the row?

    (c) The red balls are arranged together, or the first two positions on the left end of the row have different-colored balls?

    (d) The colors of the balls alternate, or the first two balls on the left end are red?

Work each of the Exercises 36 through 42 in two ways: (a) by restating each in a logically equivalent form, using the union of mutually exclusive events, and (b) by restating each, using the complement theorem.

36. A coin is tossed twice. What is the possibility that at least one lands heads up?

37. A coin is tossed three times. What is the probability that at least one lands heads up? 7/8

38. A pair of dice is rolled. What is the probability of rolling at least a sum of 3?

39. A pair of dice is rolled. What is the probability of rolling at least a sum of 4?

40. Three cards are dealt from a deck of 52 cards. What is the probability that at least one of the cards is a king?

41. Four cards are dealt from a deck of 52 cards. What is the probability that at least one of the cards is a spade?

42. A box contains six slips of paper, on each of which is written the number 1, 2, 3, 4, 5, or 6, respectively. One slip is randomly drawn. What is the probability of drawing a number greater than 2?

43. *Rolling Dice* A pair of four-sided pyramidal dice—one red and one green—is rolled, and the dots on the downward-facing sides are noted. What is the probability that the sum of the dots will be

    (a) A number less than 4 and greater than 6?

    (b) A number less than 4 or greater than 6?

    (c) What is the probability that the red die shows a 2?

44. *Rolling Dice* A pair of dice is rolled, and the sum of the dots showing on the upper faces is noted. What is the probability that

    (a) A sum less than 5 is rolled?

    (b) A sum less than 5 or greater than 10 is rolled?

    (c) A sum greater than 5 or less than 8 is rolled?

    (d) A sum greater than 2 is rolled?

    (e) A sum greater than 3 is rolled?

    (f) The sum of 1 is rolled?

45. *Rolling Dice* A pair of standard dice—one red and one green—is rolled. What is the probability that

    (a) The sum of the dots on the top faces is 6 or 8?

    (b) The sum of the dots on the top faces is 4, or a "double" is rolled?

    (c) The red die shows 4 dots on the top face, or the green die shows a 5 on the top face?

46. Rolling Dice A pair of standard dice—one red and one green—is rolled. What is the probability that

(a) The sum of the dots on the top faces is 2 or 9?

(b) The sum of the dots on the top faces is 7 or the sum of the dots on the top faces is greater than 5?

(c) The red die shows 4 dots on the top face, or the sum of the dots on the top faces is greater than 6?

47. **Cards** Five cards are drawn from a standard deck of 52 cards. What is the probability that

(a) All are hearts or all are spades?

(b) At least two are spades?

(c) All are kings or all are queens?

48. Cards Three cards are dealt from a standard deck of 52 cards. What is the probability that

(a) At least one of the cards is a club?

(b) All are kings or all are queens? ————

(c) All are kings or all are face cards?

49. Cards Seven cards are dealt from a standard deck of 52 cards. What is the probability that

(a) All are hearts or all are diamonds?

(b) All are hearts or all are face cards?

(c) All are hearts or each of the cards has a number on it?

(d) At least one of the cards is a heart?

50. Selecting a Committee The sophomore, junior, and senior classes each send three representatives to a meeting about campus recycling. At the meeting, a committee of four is selected from among these representatives. What is the probability that

(a) The sophomores will have at least two representatives on the commitee?

(b) The sophomores will have at least one representative on the committee?

(c) The sophomores will have no representative on the committee?

51. Sampling A box contains 20 light bulbs, eight of which are 60-watt, five of which are 75-watt, and seven of which are 100-watt. A sample of three bulbs is randomly selected from this box, without replacement and without regard to order. What is the probability that

(a) At least one of the bulbs is a 60-watt bulb?

(b) All are 60-watt or all are 100-watt?

(c) Exactly one of the bulbs is a 75-watt bulb?

52. Survey Tests revealed that 30 of the 150 children in a particular school have a vision problem of some type. If five children are selected from this school, without replacement and without regard to order, what is the probability that at least one child will have a vision problem?

53. Selecting a Committee Each of the 50 states has two senators. A committee of 20 senators is selected. What is the probability that Texas will have

(a) One or more senators on the committee?

(b) No senators on this committee?

54. Family Composition In families with three children, what is the probability that

(a) The first child is a boy?

(b) The first child is a boy, or the third child is a boy?

(c) The first child is a boy, and the third child is a boy?

(d) At least one of the children is a girl?

55. Drawing Slips of Paper A box contains four slips of paper, on each of which is written the number 1, 2, 3, or 4, respectively. A slip is drawn and the number noted. The slip is not returned to the box. Then a second slip is drawn and the number noted. What is the probability that

(a) The first number is a 2, or the second number is a 4?

(b) The first number is a 2, or the sum of the two numbers drawn is six?

(c) The first number is greater than two, or the second number is less than three?

(d) The first number is at least two?

56. Matching Socks A drawer contains six blue socks and four white socks. Three of the socks are chosen at random without replacement. What is the probability of getting a matching pair (in color)?

Exercise 57 appeared on an actuarial exam.

57. A drawer contains six blue socks and four white socks. Two of the socks are chosen at random without replacement. What is the probability that the two socks are the same color?

(a) $\dfrac{1}{3}$ (b) $\dfrac{7}{15}$ (c) $\dfrac{13}{25}$ (d) $\dfrac{11}{15}$ (e) $\dfrac{19}{25}$

Exercise 58 appeared on an actuarial exam.

58. In a dice game, the player independently rolls a fair red die and a fair green die. The player wins if, and only if, the red die shows a 1, 2, or 3, or if the total on the 2 dice is 11. What is the probability the player will win?

(a) $\dfrac{7}{36}$ (b) $\dfrac{4}{9}$ (c) $\dfrac{19}{36}$ (d) $\dfrac{5}{9}$ (e) $\dfrac{29}{36}$

Problems for Exploration

1. Let A and B be any two events in a sample space. Write the event $A \cup B$ as the union of two mutually exclusive events.

2. Let A, B, and C be any three events in a sample space. Find a formula similar to the addition theorem for two events that will give $P(A \cup B \cup C)$.

3. This exercise appeared on an actuarial examination. If E and F are events for which $P[E \cup F] = 1$, then $P[E' \cup F']$ must equal

   (a) 0   (b) $P[E'] + P[F'] - P[E']P[F']$
   (c) $P[E'] + P[F']$ (d) $P[E'] + P[F'] - 1$
   (e) 1

4. A coin is tossed until, for the first time, either two heads or two tails appear in succession. What is the probability that this experiment will end before the sixth toss?

5. The Birthday Problem

   (a) Find the probability that in a group of eight people, no two of them will have the same birthday (month and day but not year).

   (b) Find the probability that in a group of eight people, at least two will have the same birthday. Then find the odds for such an event.

   (c) Repeat Question 5(a) and (b) for a group of 18 people.

   The following table shows the probability that among a group of n people, two or more will have the same birthday. Note that, when the size of the group becomes 24 or larger, the chances for at least two having the same birthday becomes increasingly favorable.

Size of group	Probability	Size of group	Probability
5	.027	24	.538
10	.117	25	.569
15	.253	30	.706
20	.411	40	.891
22	.476	50	.970
23	.507	60	.994

SECTION 7.2   Conditional Probability

Suppose that someone rolls a single die, out of your sight, and tells you it came up an even number. You are then asked, "What is the probability that a 2 has been rolled?" The answer to the question is certainly affected by the information known to you; namely, the die is known to be one of three outcomes—a 2, a 4, or a 6. This, in effect, has reduced the sample space to $S^* = \{2, 4, 6\}$, from which the question is now easily answered a "$\frac{1}{3}$." The mathematical notation for this question is

$$P(\text{a two comes up} \mid \text{an even number has been rolled}) = \frac{1}{3},$$

where the *vertical bar* is read "given that" and the event to the right of the bar is the condition under which the question on the left of the bar is to be answered—hence, the name **conditional probability.** In generic terms, $P(A \mid B)$ is read "the probability of A, given that B has already occurred."

In many cases conditional probability can be computed in a straightforward way by considering a **reduced sample space.** The next few examples will give specifics.

EXAMPLE 1
Computing Conditional Probability

If two cards are randomly drawn, in succession, without replacement, from a deck of 52 cards, what is the probability that the second card is a spade, given that the first card was a spade?

Solution   In conditional probability language, the problem becomes P (second card is a spade | first card was a spade). Because a spade has been removed from the deck on the first draw, the reduced sample space S^* from which the second card is to be drawn contains 51 cards: 12 spades and 39 nonspades. The probability of drawing a spade from S^* is then $\frac{12}{51}$. ■

EXAMPLE 2
Computing Conditional Probability

If two cards are randomly drawn, in succession, with replacement, from a deck of 52 cards, what is the probability that the second card is a king, given that the first card drawn was a king?

Solution In symbols, the problem may be stated as $P($second card is a king $\mid$ first card was a king$)$. We are given the first card drawn, a king, but this time, it was replaced. After the first draw, the sample space S^* looks just like the original; it still contains all 52 cards. The probability that the second card is a king is then $\frac{4}{52} = \frac{1}{13}$. ▪

EXAMPLE 3
Computing Conditional Probability

If three jellybeans are randomly drawn, in succession, and without replacement, from a bowl containing six red and four green jellybeans, what is the probability that the third jellybean drawn is red, given that the first two jellybeans drawn were green?

Solution Expressed in symbols, we are to find $P($third jellybean is red $\mid$ first two jellybeans drawn were green$)$. We know that two green jellybeans have already been drawn from the bowl so that the sample space for the third draw has been reduced to S^* consisting of six red and two green jellybeans. The answer to our question, as shown in Figure 8, is $\frac{6}{8} = \frac{3}{4}$. ▪

EXAMPLE 4
Computing Conditional Probability

One card is drawn from a deck of 52 cards. What is the probability that the card is

(a) A king (K), given that the card drawn was a heart (H)?

Solution Answering $P(K \mid H)$ means considering the reduced sample space S^* consisting of 13 hearts. The probability that a king is among them is $\frac{1}{13}$.

(b) A diamond, given that the card drawn is a diamond (D) or a heart (H)?

Solution In symbols, the question asks $P(D \mid D \cup H)$. The given condition implies that we may think of the reduced sample space S^* of 26 cards, of which 13 are diamonds and 13 are hearts. Under these circumstances, the chance of the drawn card being a damond is $\frac{13}{26} = \frac{1}{2}$.

(c) A diamond, given that the card drawn is a diamond (D) and a jack (J)?

Solution We are to find $P(D \mid D \cap J)$. Note that the given condition tells us that the card drawn is the jack of diamonds, so the probability that a diamond has been drawn is 1.

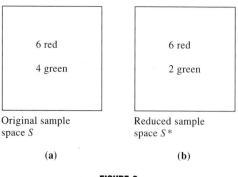

6 red

4 green

Original sample space S

(a)

6 red

2 green

Reduced sample space S^*

(b)

FIGURE 8
Computing Conditional Probability

(d) A heart, given that the card drawn was a 10 or a diamond (D)?

Solution $P(H \mid 10 \cup D)$ is found by considering the reduced sample space S^* representing the given information: four 10s along with the 12 other diamonds, making a total of 16 cards. Among these, there is only one heart; namely, the 10 of hearts. Our answer is therfore $\frac{1}{16}$. ▬▬

EXAMPLE 5
Tossing a Coin

A coin is tossed three times. Find the probability that the first two tosses resulted in heads, given that at least one tail was tossed.

Solution The visual effect of writing out the sample space for tossing a coin three times will be helpful:

$$S = \{HHH, HHT, HTH, THH, TTH, THT, HTT, TTT\}.$$

We are given that at least one tail was tossed. This means that one of the following seven outcomes making up the reduced sample space occurred:

$$S^* = \{HHT, HTH, THH, TTH, THT, HTT, TTT\}$$

Among these outcomes, only the first one listed has the first two tosses coming up heads. As a consequence, the answer is $\frac{1}{7}$. ▬▬

EXAMPLE 6
Rolling Dice

A pair of dice—one red and one green—is rolled. Find the probability that a sum of 5 was rolled, given the green die has two dots on the upper side.

Solution There are six possible outcomes that have the green die showing two dots on the upper side. They are, upon listing the red die first,

$$S^* = (1, 2), (2, 2), (3, 2), (4, 2), (5, 2), (6, 2).$$

Among these, $(3, 2)$ is the only one whose dots sum to 5. It follows that the requested probability is $\frac{1}{6}$. ▬▬

EXAMPLE 7
Pointing Toward the
Conditional Probability
Formula

Adult employment data regarding those in the work force in a small town revealed the following information:

	Employed	Unemployed
Male	500	50
Female	200	60

What is the probability that a person selected from this work force is employed, given that the person is a male?

Solution The condition "given . . . a male" reduces the sample space to the information in the top tow of the chart, of which there are a total of $500 + 50 = 550$. The question asks what proportion of *these* are employed. So we focus on the employed column but consider only the 500 (the intersection of male and employed), to get

$$P(\text{employed} \mid \text{male}) = \frac{500}{550}.$$

This answer may be viewed be in terms of sets as follows (see the data chart):

$$P(\text{employed} \mid \text{male}) = \frac{500}{550} = \frac{n(\text{employed and male})}{n(\text{male})}.$$

Now, if the numerator and denominator of the set expression on the right are both divided by the number in the sample space, $n(S)$, an expression for the conditional probability in terms of probability in the entire sample space of 810 is obtained:

$$P(\text{unemployed} \mid \text{male}) = \dfrac{\dfrac{n(\text{employed and male})}{n(S)}}{\dfrac{n(\text{male})}{n(S)}} = \dfrac{P(\text{employed and male})}{P(\text{male})}$$

$$= \dfrac{\dfrac{500}{810}}{\dfrac{550}{810}} = \dfrac{500}{550}.$$

To replay the process of Example 7 in a general setting, let A and B be two events in a sample space S, and consider $P(A \mid B)$. The given event B becomes the reduced sample space S^* (see Figure 9) so that

$$P(A \mid B) = \dfrac{n(A \cap B)}{n(B)} = \dfrac{\dfrac{n(A \cap B)}{n(S)}}{\dfrac{n(B)}{n(S)}} = \dfrac{P(A \cap B)}{P(B)},$$

provided that $P(B) \neq 0$. This justifies the **conditional probability formula** for sample spaces of equally likely outcomes, but it is also valid for all sample spaces.

Conditional Probability Formula

For any two events in a sample space, $P(A \mid B) = \dfrac{P(A \cap B)}{P(B)}$, provided that $P(B) \neq 0$.

This formula may be used to solve *any* conditional probability problem. Note that the probabilities involved are computed from the *entire original sample space*. Observe further, that the "given" event is the one that appears in the denominator of the quotient on the right.

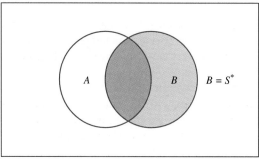

$$P(A|B) = \dfrac{P(A \cap B)}{P(B)}$$

FIGURE 9

Diagram for Conditional Probability

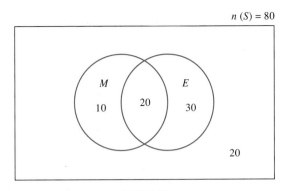

$n(S) = 80$

FIGURE 10

Class Selections at Washburn High School

EXAMPLE 8
Using the Conditional
Probability Formula

The senior class of Washburn High School has 80 students, 50 of whom are taking English, 30 are taking math, and 20 are taking both subjects.

(a) What is the probability that a senior is taking English, given that the senior is taking math?

Solution   Figure 10 will help organize the given data. Applying the conditional probability formula to this question gives $P(\text{taking English} \mid \textbf{taking math}) =$

$$\frac{P(\text{taking English and math})}{P(\textbf{taking math})} = \frac{\dfrac{20}{80}}{\dfrac{30}{80}} = \tfrac{2}{3}.$$

(b) What is the probability that a senior is taking math, given that senior is taking English?

Solution   Applying the conditional probability formula to this question gives $P(\text{taking math} \mid \textbf{taking English}) =$

$$\frac{P(\text{taking math and English})}{P(\textbf{taking English})} = \frac{\dfrac{20}{80}}{\dfrac{50}{80}} = \tfrac{2}{5}. \quad \blacksquare$$

The conditional probability formula *always* gives an expression that will solve a conditional probability problem. We can use it as a primary approach to such problems or as a foolproof option in case the reduced sample space approach becomes too complicated.

EXAMPLE 9
Drawing Slips of Paper

Three slips of paper—each with the number 1, 2, and 3, written on it, respectively—are placed in a box. The sample space for successively drawing two slips of paper, with replacement, and recording the numbers on them is as follows:

$$(1, 1) \quad (1, 2) \quad (1, 3)$$
$$(2, 1) \quad (2, 2) \quad (2, 3)$$
$$(3, 1) \quad (3, 2) \quad (3, 3)$$

Use this sample space of 9 outcomes to find the probability that the sum of the numbers on the slips is 5, given that the first number drawn was a 2.

Solution   The conditional probability formula is used first to solve this problem:

$$P(\text{sum is 5}\mid \textbf{first number is 2}) = \frac{P(\text{sum is 5 and first number is 2})}{P(\textbf{first number is 2})}.$$

$$= \frac{\dfrac{1}{9}}{\dfrac{3}{9}} = \frac{1}{3}.$$

The reduced sample space approach to this question, of course, gives the same answer: The given information leads to the reduced sample space $S^* = \{(2, 1), (2, 2), (2, 3)\}$ and among these 3 outcomes, only one has a sum of 5. Therefore, the probability that the sum is 5—given that the first number drawn was a 2—is $\frac{1}{3}$ ■

The conditional probability formula

$$P(A\mid B) = \frac{P(A \text{ and } B)}{P(B)}$$

involves three probabilities, $P(A\mid B)$, $P(A \text{ and } B)$, and $P(B)$. Anytime we know two of these probabilities, we can algebraically solve for the third. Example 10 shows how $P(A \text{ and } B)$ may be found, provided we know the other two probabilities. In the next section we will further explore this useful way to find the probability of an intersection of two events.

EXAMPLE 10

Using the Conditional Probability Formula

If $P(A\mid B) = \frac{2}{5}$ and $P(B) = \frac{7}{8}$, find $P(A \cap B)$.

Solution   The conditional probability formula

$$P(A\mid B) = \frac{P(A \cap B)}{P(B)}$$

may be solved for $P(A \cap B)$, to give

$$P(A \cap B) = P(B) \cdot P(A\mid B).$$

The given data may now be substituted to give

$$P(A \cap B) = \left(\frac{7}{8}\right)\left(\frac{2}{5}\right) = \frac{14}{40} = \frac{7}{20}. \quad ■$$

A natural way to visually display probabilities of events that occur in an order, such as a first draw, a second draw, and a third draw, is on a **probability tree diagram.** Example 11 shows how such diagrams are constructed. Even more generally, such tree diagrams can be constructed for any conditional probability problem whether there is a natural order or not. The main value of a probability tree diagram is in its visual organization of conditional probability problems and linkage to the probability of an intersection of events using the technique demonstrated in Example 10.

EXAMPLE 11
Constructing a
Probability Tree
Diagram

Suppose that two balls are randomly drawn in succession, without replacement, from a box containing five red and eight green balls. Draw and label a tree diagram that will describe the probabilities of the various outcomes.

Solution There are two generations of branches, the first representing possible outcomes on the first draw and the second the possible outcomes on the second draw. Each branch in the "second-draw" generation represents a possible outcome after the outcome of the branch to which it is attached has *already occurred from the first draw*. For instance, arriving at point, *A*, in Figure 11, from the starting point means that a red ball has been drawn on the first draw and has not been replaced; the branches attached to *A* on the right are then possibilities for the second draw *after the removal of a red ball on the first draw*. Similarly, the branches to the right of point *B* reflect possibilities for the second draw after the removal of a green ball on the first draw. Probabilities are now assigned to each branch, reflecting the outcome of the preceding branch to which it is attached, as shown in Figure 12.

The symbolism of conditional probability may now be used to relabel the tree diagram of Figure 12 into a final product as shown in Figure 13.

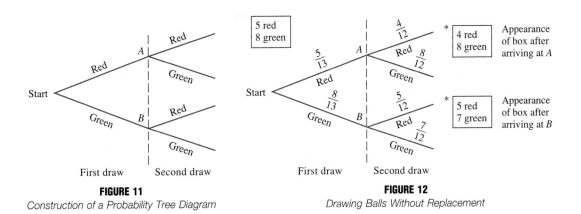

FIGURE 11
Construction of a Probability Tree Diagram

FIGURE 12
Drawing Balls Without Replacement

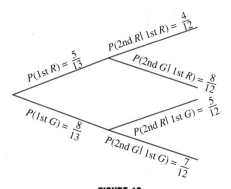

FIGURE 13
Final Apperance of Tree Diagram

In view of Example 10, the conditional probability problem

$$P(\text{2nd } R \mid \text{1st } R) = \frac{P(\text{1st } R \text{ and 2nd } R)}{P(\text{1st } R)}$$

may be algebraically solved for $P(\text{1st } R \text{ and 2nd } R)$ giving

$$P(\text{1st } R \text{ and 2nd } R) = P(\text{1st } R) \cdot P(\text{2nd } R \mid \text{1st } R).$$

Now observe that the product on the right side is precisely the product of the two probabilities along the top two branches of the tree! Thus

$$P(\text{1st } R \text{ and 2nd } R) = \left(\frac{5}{13}\right)\left(\frac{4}{12}\right) = \frac{5}{39}. \quad \blacksquare$$

EXERCISES 7.2

In each of the Exercises 1 through 8: (a) write the sample space, (b) find the requested probability by using a reduced sample space, and (c) find the requested probability by using the conditional probability formula.

1. A single die has been rolled. What is the probability that three dots show on the top side, given that an odd number was rolled?

2. A single die has been rolled. What is the probability that four dots show on the top side, given that at least a two has been rolled?

3. A single die has been rolled. What is the probability that two dots show on the top side, given that an odd number was rolled?

4. A single die has been rolled. What is the probability that at least a four has been rolled, given that an odd number was rolled?

5. A coin is tossed three times. What is the probability that exactly two tails appeared, given that at least one head appeared?

6. A coin is tossed three times. What is the probability that exactly two tails appeared, given that exactly one head appeared?

7. A coin is tossed three times. What is the probability that the second toss resulted in heads, given that exactly one head appeared?

8. A coin is tossed three times. What is the probability that heads appeared on the last two tosses, given that exactly one head appeared?

Selections   Exercises 9 through 13 refer to this experiment: Two cards are drawn in succession, without replacement, from a deck of 52 cards. What is the probability that

9. The second card was a jack, given that the first card drawn was not a jack?

10. The second card was a spade, given that the first card was a spade?

11. The second card is a king, given that the first card is a face card?

12. The second card was a club, given that the first card was a heart or a spade?

13. The second card was a nine, given that the first card was a diamond and a nine?

Selections   Exercises 14 through 18 refer to this experiment: Three cards are drawn in succession, with replacement. What is the probability that

14. The third is a spade, given that the first two cards drawn were not spades?

15. The second is a heart, given that the first and third cards drawn were hearts?

16. The third is a queen, given that the first two cards drawn were queens?

17. *Selections*   From a box containing five red and nine green marbles, two marbles are drawn in succession, with replacement. What is the probability that the second marble is

   (a) Green, given that the first was green?
   (b) Red, given that the first was red?
   (c) Red, given that the first was green?

18. *Selections*   A box contains three red and seven green marbles. If three marbles are drawn in succession, without replacement, what is the probability that

   (a) The third marble is red, given that the first two marbles drawn were red?
   (b) The third marble is red, given the first two marbles were green?
   (c) The third marble is green, given that one red and one green marble have been drawn.
   (d) All three are red?

Exercises 19 through 24 refer to these data:

During November, a store selling TV sets recorded these brands in inventory:

	With remote	With no remote
NuView	30	8
VuTech	40	12

According to these data, what is the probability that a TV set sold during November

19. Was a NuView brand, given it had a remote control?

20. Had a remote control, given it was a NuView brand?

21. Had no remote control, given it was a VuTech brand?

22. Was a VuTech brand, given it had no remote control?

23. Was a NuView brand?

24. Did not have a remote control?

Grade Distribution   *Exercises 25 through 29 refer to the grade distribution in a particular finite math class populated by freshmen and sophomores only:*

	A	B	C	D	F
Freshmen	2	5	8	3	2
Sophomores	3	4	8	2	3

If a person is randomly selected from this class, what is the probability that the person

25. Is a freshman?

26. Is a freshman and made an A?

27. Is a freshman or made an A?

28. Is a freshman, given that the student made an A?

29. Made a C, given that the student was a sophomore?

Advertisements   *Exercises 30 through 34 refer to these data: A journalism major made a study of personal ads in the classified section of newspapers from cities of 50,000 people. She found that, on the average, there were 25 such ads in each paper and among them were the following classifications:*

	Communication with lover	Looking for mate	Other matters
Placed by male	3	6	4
Placed by female	4	2	6

Based on these findings, what percentage of such ads

30. Are placed by a male, given that they communicate with a lover?

31. Communicate with a lover or seek a mate, given that they are placed by a female?

32. Are looking for a mate or other matters, given that the ad is placed by a male?

33. Communicate with a lover or seek a mate?

34. Are placed by a male?

Survey   *Exercises 35 through 39 refer to these data: One hundred adults were asked their political party affiliation and this question, "Will the European Common Market benefit the United States?" The results were as follows:*

	Yes	No	No opinion
Democrat	20	12	12
Republican	25	8	5
Independent	6	5	7

Based on this survey, what is the probability that a person

35. Will answer "yes," given that he is Democrat?

36. Will be a Republican or Independent, given that she had no opinion?

37. Will not be a Republican, given an answer of "no"?

38. Had no opinion?

39. Was an Independent?

Experiment   *Exercises 40 through 45 refer to this experiment: A pair of dice is thrown (one red and one green). What is the probability that*

40. A sum of 10 is rolled, given that at least a sum of 9 has been rolled?

41. A sum of 3 or 9 is rolled?

42. A sum of 7 is rolled, given that a sum of 3 or 9 has been rolled?

43. A sum of 3 or 4 is thrown, given that the sum of the dots showing is less than 5?

44. The red die is a 5, given that a sum of 10 has been rolled?

45. A sum of 7 is rolled, given that the green die shows 4 dots?

46. **Drawing Slips of Paper**   Four slips of paper—on each of which is written the number 1, 2, 3, or 4, respectively—are put into a box. One of the slips is drawn, and the number on it is recorded; then the slip is returned to the box. Now a second slip is drawn and the number on it noted. What is the probability that

   (a) The sum of the two numbers is five?

   (b) The sum is six, given that the first number drawn is a three?

   (c) The first number drawn is a two, given that the sum of the two numbers is four?

47. Drawing Slips of Paper   Four slips of paper—on each of which is written the number 1, 2, 3, or 4, respectively—are put into a box. Two slips are drawn in succession without replacement. What is the probability that

(a) The sum of the numbers drawn is five?

(b) The second number drawn was a three, given that the sum of the two numbers was four?

(c) The sum of the two numbers is nine, given that the first number drawn is a two?

(d) The sum of the numbers is five, given that the first number drawn is a four or a three?

Sociology   *Exercises 48 through 50 refer to three-children families. What is the probability that*

48. The first child is a boy, given that the last two are girls?

49. Two of the children are boys, given that at least one is a girl?

50. At least one is a boy, given that at least one is a girl?

Survey of Purchases   *Exercises 51 through 53 refer to these data: Fifteen percent of the farmers in an agricultural state bought a new tractor last year, 21% bought a new car, and 8% bought both a tractor and a new car. What is the probability that a farmer in this state will buy*

51. A new tractor, given that she bought a new car?

52. A new tractor or a new car?

53. Exactly one these two items, given that she bought one or the other?

Experiment   *Exercises 54 through 56 refer to this experiment: A new drug is administered to 100 people. Six of those people reported a rise in blood pressure, five reported loss of sleep, and two reported both of these side effects. According to these statistics, what is the probability that a person taking this drug*

54. Will experience a rise in blood pressure, given a report of lost sleep?

55. Will experience loss of sleep, given a reported rise in blood pressure?

56. Will experience neither of these side effects?

Use the conditional probability formula

$$P(A|B) = \frac{P(A \cap B)}{P(B)}$$

to solve each of the Exercises 57 to 62.

57. Find $P(A|B)$ if $P(A \cap B) = .3$ and $P(B) = .7$.

58. Find $P(B)$ if $P(A|B) = .8$ and $P(A \cap B) = \frac{2}{3}$.

59. Find $P(A \cap B)$ if $P(A|B) = \frac{4}{5}$ and $P(B) = \frac{5}{7}$.

60. Find $P(A|B)$ if $P(A \cap B) = .3$ and $P(B') = .2$.

61. Find $P(A|B)$ if $P(A) = .4$, $P(B) = .6$, and $P(A \cup B) = .8$.

62. Find $P(A|B)$ if $P(A) = .5$, $P(B') = .7$, and $P(A \cup B) = .7$.

In Exercises 63 through 68, draw a tree diagram, and label each branch with the correct probability.

63. A sample of two refrigerators is selected in succession, without replacement, from among six good ones and four defective ones.

64. A sample of two marbles is selected in succession, without replacement, from a box containing three red, four white, and two green marbles.

65. A sample of three marbles is selected in succession, without replacement, from a box containing eight red and three blue marbles.

66. A sample of two cards is drawn in succession, without replacement, from a deck of 52 cards; predict whether each card is a club.

67. A sample of three cards is selected in succession, with replacement; predict whether each card is a spade.

68. A sample of three balls is selected, with replacement, from a box containing four red, five green, and seven white balls.

The following exercise appeared on an actuarial exam.

69. A card hand selected from a standard card deck consists of two kings, a queen, a jack, and a 10. Three additional cards are selected at random and without replacement from the remaining cards in the deck. What is the probability that the enlarged hand contains at least three kings?

(a) $\dfrac{3}{1081}$ (b) $\dfrac{132}{1081}$ (c) $\dfrac{135}{1081}$ (d) $\dfrac{264}{1081}$

(e) $\dfrac{267}{1081}$

Excercises 70 through 73 refer to this experiment: Four-letter code words are to be made using the letters of the alphabet, with repetition of letters allowed. What is the probability that such a code

70. Has exactly one A in it?

71. Has exactly two A's, given that it has at least one A in it?

72. Has exactly one A, given that it has at least one A in it?

73. Has at least two A's, given that it has at least one A in it?

Problems for Exploration

1. A single die is rolled five times. Given that at least one face shows a three among the top five faces, what is the probability that two or more threes showed on the top faces?

2. Suppose that a card deck contains 40 cards, of which 10 are one color, 10 another color, 10 of still another color, and 10 of yet another color. Four cards are to be drawn. Would it be a good wager that all four cards will be of different colors? Explain your answer.

3. Find an expression similar to the addition theorem for this probability: $P(A \cup B \mid H)$

4. Three dice are rolled. If no two dice show the same face, what is the probability that one of the dice shows six dots?

5. Two cards are drawn in succession, without replacement, from a standard deck of 52 cards. What is the probability that
   (a) The second card was a spade, given that the first card was a spade or a diamond?
   (b) The second card was a spade, given that the first card was a spade or an ace?

6. If A and B are not mutually exclusive, and A and C are not mutually exclusive, but B and C are mutually exclusive, find the formula for $P(A \mid B \cup C)$.

7. A box contains three red and seven green balls. For each trial, a ball is selected at random, its color noted, and then this ball along with two new additional balls of the same color are put back into the box. What is the probability that a red ball is selected in each of the first three trials?

| SECTION 7.3 | **Multiplication Rules for Probability: Independent Events** |

The focus of this section is on finding the probability of the intersection of events. Examples 10 and 11 of the previous section set the stage for our work; the conditional probability formula is used to show that the probability of an intersection (the connective "and") of events may be found by multiplying probabilities. Events that occur in a natural order offer a natural entry point into these ideas.

EXAMPLE 1
The Intersection of Two
Successive Events

Suppose two calculators are to be randomly selected, in succession, without replacement, from a box that contains four defective and nine good calculators; after each selection the calculator is checked to see whether it is good or defective. What is the probability that the first calculator selected is good *and* the second calculator selected is defective?

Solution We are to find

$$P(\text{1st good } and \text{ second defective}) = P(\text{1st good} \cap \text{2nd defective}).$$

We reason like this: First, we can find the probability that the first selection will be good because there are *four defective and nine good calculators in the box*. The answer is

$$P(\text{1st } G) = \frac{9}{13}.$$

Next, we can find the probability that the second calculator is defective, given that the first calculator selected was good because the box would have *four defective and eight good* calculators in it. The answer is

$$P(\text{2nd } D \mid \text{1st } G) = \frac{4}{12}.$$

The conditional probability formula may now be used to rewrite this last conditional expression as:

$$P(\text{2nd } D \mid \text{1st } G) = \frac{P(\text{1st } G \text{ and 2nd } D)}{P(\text{1st } G)}.$$

The probability in the numerator, $P(\text{1st } G \text{ and 2nd } D)$, is precisely the one we seek! Therefore, upon solving algebraically for that probability we get the product

$$P(\text{1st } G \text{ and 2nd } D) = P(\text{1st } G) \cdot P(\text{2nd } D \mid \text{1st } G).$$

Since we know the two probabilities on the right, our problem is solved:

$$P(\text{1st } G \text{ and 2nd } D) = \left(\frac{9}{13}\right) \cdot \left(\frac{4}{12}\right) = \frac{3}{13}.$$

This conclusion tells us that "and" (intersection) means we should multiply: Multiply the probability of the first selection by the probability of the second selection, *conditioned by what happened on the first selection.*

A tree diagram as shown in Figure 14 is a natural way to view events such as the ones we are considering. Notice how the probability result just obtained implies that *"and" means to multiply the probability numbers along the specified branches of a tree.* ■

The conditional probability formula guides us to the probability of any intersection, whether the events occur in a natural order or not. Its general statement

$$P(A \mid B) = \frac{P(A \cap B)}{P(B)}, \qquad P(B) \neq 0$$

concerns *any* two events A and B where information about B is given and the probability of A is to be found. Upon solving algebraically for $P(A \cap B)$, we have the multiplication formula

$$P(A \cap B) = P(B) \cdot P(A \mid B).$$

If the roles of A and B are reversed, this formula would read

$$P(B \cap A) = P(A) \cdot P(B \mid A)$$

where information about A is given and the probability of B is to be found. But because $A \cap B = B \cap A$, the left sides of these two product expressions are equal. It follows that

$$P(A \cap B) = P(B) \cdot P(A \mid B) = P(A) \cdot P(B \mid A)$$

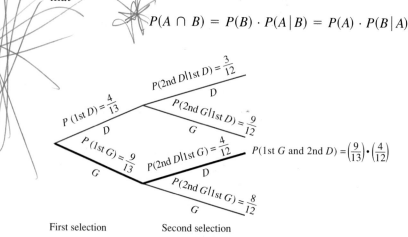

First selection   Second selection

FIGURE 14
The Probability of Selecting Defective (D) Versus Good (G) Calculators

which is a **general multiplication rule for the intersection of two events** to be used with either of the two events in the "given" slot. Notice the ordering of A and B in this rule:

Same event (either A or B)

$$P(A \cap B) = P(\quad)P(\quad | \quad)$$

↑
Other
event (either B or A)

Here are two examples of how this rule may be applied:

$$P(\text{1st good and 2nd defective}) = P(\text{1st good}) \cdot P(\text{2nd defective} \mid \text{1st good})$$
$$= P(\text{2nd defective}) \cdot P(\text{1st good} \mid \text{2nd defective})$$

Both expressions are equally valid for finding the requested intersection. Of course, the first expression would usually be more natural and easier to work with than the second, as we saw in Example 1. Another example, where no order is implied, is

$$P(\text{male and diabetic}) = P(\text{male}) \cdot P(\text{diabetic} \mid \text{male})$$
$$= P(\text{diabetic}) \cdot P(\text{male} \mid \text{diabetic}).$$

Again, both expressions are valid and the choice of which to use would depend upon what data was available to calculate the required probabilities on the right sides.

This multiplication rule can be extended to more than two events as stated next.

Multiplication Rules for Probability

Let A and B be *any* events in a sample space for which $P(A) \neq 0$ and $P(B) \neq 0$. Then

$$P(A \text{ and } B) = P(A \cap B) = P(A) \cdot P(B \mid A) = P(B) \cdot P(A \mid B).$$

This rule can be extended to several events.

$$P(A \cap B \cap C \cap D \ldots)$$
$$= P(A) \cdot P(B|A) \cdot P(C|A \cap B) \cdot P(D|A \cap B \cap C) \ldots.$$

Applied to events which occur in a natural order, 1st, 2nd, 3rd, and so on,

$$P(\text{1st and 2nd and 3rd} \ldots)$$
$$= P(\text{1st}) \cdot P(\text{2nd} | \text{1st has occurred}) \cdot P(\text{3rd} | \text{1st and 2nd have occurred}) \ldots$$

We show three more examples of events that occur in a natural order.

EXAMPLE 2
Drawing Cards

Two cards are to be randomly selected, in succession, without replacement, from a deck of 52 cards. What is the probability that the first card drawn will be a heart *and* the second card drawn will be a spade?

Solution   We need to find

$$P(\text{1st heart and 2nd spade}).$$

The product rule implies that

$$P(\text{1st heart and 2nd spade}) = P(\text{1st heart}) \cdot P(\text{2nd spade} \mid \text{1st heart}).$$

The probability that the first card drawn will be a heart is $\frac{13}{52}$. The probability that the second card will be a spade, given the first card was a heart, is $\frac{13}{51}$. (There are now 51 cards in the deck and all 13 spades are still present.) Therefore,

$$P(\text{1st heart and 2nd spade}) = \left(\frac{13}{52}\right)\left(\frac{13}{51}\right) = \frac{13}{204} \approx 0.0637.$$

EXAMPLE 3
Selecting Balls
From a Box

From a box containing five red balls and three green balls, four balls are to be randomly selected, in succession, without replacement. What is the probability that the first ball will be red, the second will be green, and the last two will be red?

Solution In symbols, the question asks

$$P(\text{1st } R \text{ and 2nd } G \text{ and 3rd } R \text{ and 4th } R).$$

The connective "and" implies that the multiplication rule should be used to answer this question; we write the probability of each selection, taking into account what has happened on the previous selections, then multiply them together:

$P(\text{1st } R) \cdot P(\text{2nd } G \mid \text{1st } R) \cdot P(\text{3rd } R \mid \text{1st } R \text{ and 2nd } G) \cdot P(\text{4th } R \mid \text{1st } R \text{ and 2nd } G$
and 3rd R).

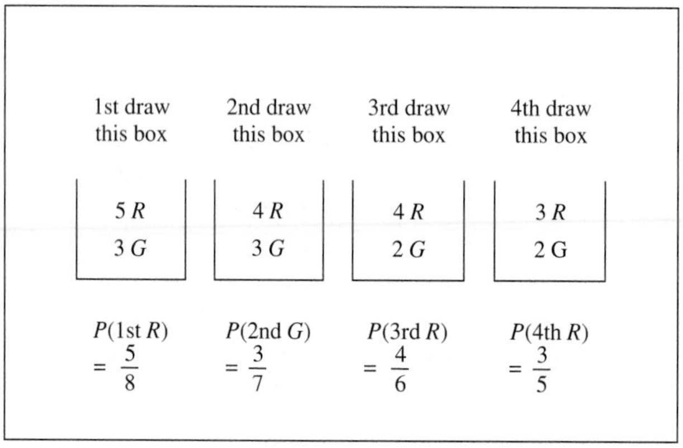

FIGURE 15
Successive Selections without Replacement

Figure 15 makes use of successive pictorial representations of these conditionals to aid in finding the probability of each selection. The answer is then the product

$$\left(\frac{5}{8}\right)\left(\frac{3}{7}\right)\left(\frac{4}{6}\right)\left(\frac{3}{5}\right) = \frac{3}{28}.$$

EXAMPLE 4
Selecting Cards

Three cards are to be randomly selected, in succession, with replacement, from a deck of 52 cards. What is the probability that the first card drawn will be a heart, the second a heart, and the third a spade?

Solution The probability that the first card drawn will be a heart is $\frac{13}{52}$. The probability that the second card will be a heart, given the first card was a heart, is again $\frac{13}{52}$ because the first card drawn was replaced in the deck. The probability that the third card will be a spade, given the first was a heart and the second was a heart, is also $\frac{13}{52}$ because, again, the second card has been returned to the deck. This means that

$$P(\text{1st heart and 2nd heart and 3rd spade})$$
$$= \left(\frac{13}{52}\right)\left(\frac{13}{52}\right)\left(\frac{13}{52}\right) = \frac{1}{64} \approx 0.0156. \quad \blacksquare$$

We now turn to an example where there is no "natural order" to the two events involved.

EXAMPLE 5
Charge Cards

Research by a department store chain revealed that 80% of the people who make purchases are women and that 75% of those women's purchases are charged on the chain's charge card. What is the probability that a person making a purchase from this chain will be a woman *and* charge the purchase on her charge card?

Solution   These data show that with probability .75 a purchase is made with a charge card, *given* that purchase was made by a woman. That is, in abbreviated form, the purchasing information tells us that $P(\text{use charge card} \mid \text{woman}) = .75$. This conditional information tells us how the product should be written in order to take advantage of the given probabilities:

$$P(\text{woman and use charge card}) = P(\text{woman}) \cdot P(\text{use charge card} \mid \text{woman})$$
$$= (.80)(.75) = .6. \quad \blacksquare$$

Example 6 shows that tree diagrams can also be used to display information about events that have no "natural" ordering. It is the conditional information that allows this to be done. Example 6 also shows that a tree diagram can be used to answer many other types of probability problems in addition to those involving the connective *and*.

EXAMPLE 6
Using a Tree Diagram

A contractor buys bags of cement from two suppliers—Smith's Lumber and X-Mart—and stores them in a warehouse. He buys 60% from Smith's and 40% from X-Mart and, upon delivery, discovers that 2% of the bags from Smith's are damaged and 4% of the bags from X-Mart are damaged.

(a) Draw a tree diagram for these data.

Solution   The tree diagram, with abbreviated notation, appears in Figure 16.

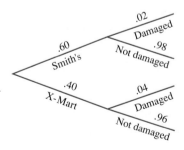

FIGURE 16
The Probability of Damage to Bags of Cement

(b) For a randomly selected bag, what is the probability it came from Smith's and was damaged?

Solution P(from Smith's and damaged) $= (.60)(.02) = .012$ from the tree diagram.

(c) For a randomly selected bag, what is the probability it was damaged, given that it came from Smith's?

Solution   The answer is read directly from the tree diagram:

$$P(\text{damaged} \mid \text{Smith's}) = .02.$$

(d) For a randomly selected bag, what is the probability it was damaged?

Solution   This would be easy if we only knew whether the bag came from Smith's or from X-Mart. Because it must come from one or the other, we interpret the problem as the union of two mutually exclusive events: P(damaged) $= P$("from Smith's and damaged" or "from X-Mart and damaged") $= P$(from Smith's and damaged) $+ P$(from X-Mart and damaged) $= P$(from Smith's) $\cdot P$(damaged $\mid$ from Smith's) $+ P$(from X-Mart) $\cdot P$(damaged $\mid$ from X-Mart) $= (.60)(.02) + (.40)(.04) = .028$. In other words, 2.8% of the bags in the warehouse are damaged.   ■

Example 6 could be extended to the purchase of bags of cement from several suppliers, $A_1, A_2, A_3, \ldots A_n$. If, as in part (d) of Example 6, we wanted to know the probability that a bag was damaged, we would write

$$P(\text{damaged}) = P(\text{"}A_1 \text{ and damaged" or "}A_2 \text{ and damaged" or "}A_3 \text{ and damaged"}$$
$$\ldots \text{ or "}A_n \text{ and damaged"})$$
$$= P(\text{from } A_1) \cdot P(\text{damaged} \mid \text{from } A_1)$$
$$+ P(\text{from } A_2) \cdot P(\text{damaged} \mid \text{from } A_2)$$
$$+ P(\text{from } A_3) \cdot P(\text{damaged} \mid \text{from } A_3)$$
$$+ \ldots + P(\text{from } A_n) \cdot P(\text{damaged} \mid \text{from } A_n).$$

We now turn our attention to the concepts of dependent and independent events. Some examples will illustrate the difference between these concepts. For example, consider the successive selection of two cards from a deck of 52 cards. If drawn *without replacement*, what is the probability that the first card drawn is a king and the second card is a queen? The answer is given by $\left(\frac{4}{52}\right)\left(\frac{4}{51}\right)$. That is, when selecting *without replacement*, the probability for the second selection definitely **depends** or is **dependent** upon what happened on the first selection. On the other hand, if the cards are drawn *with replacement*, this same probability question would be answered by $\left(\frac{4}{52}\right)\left(\frac{4}{52}\right)$. In this case, the probability for the second selection is unaffected by what happened on the first selection. Stated differently, we say that the probability for the second selection is **independent** of what happened on the first selection.

Independent Events

Two events A and B are **independent** provided the occurrence of one has no effect on any probability question asked about the other. That is, $P(A \mid B) = P(A)$ and $P(B \mid A) = P(B)$. Events that are not independent are called **dependent.**

Here are some examples to help clarify these concepts.

Independent events	Dependent events
Outcome of successive draws of cards, with replacement	Outcomes of successive draws of cards, without replacement
Outcomes of repeated tosses of a die	Outcomes of successive draws of marbles, without replacement
The main computer fails; the backup computer fails	Type of bait in mousetrap; mouse being trapped.
Outcomes of successive tosses of a coin	The weather today; the weather tomorrow

These examples show that the concepts of *independent events* and *mutually exclusive events* are not the same.

The convenience of independent events lies in the fact that the multiplication rule for two independent events A and B now becomes very simple because

$$P(A \cap B) = P(A) \cdot P(B|A) = P(A) \cdot P(B).$$

> **The Multiplication Rule for Independent Events**
>
> If A and B are independent events in a sample space, then
>
> $$P(A \cap B) = P(A) \cdot P(B).$$

The extension of the multiplication rule for independent events is particularly easy; just multiply the probabilities together!

> **The Extended Multiplication Rule for Independent Events**
>
> If $A_1, A_2, \ldots, A_n$ are independent events in a sample space, then
>
> $$P(A_1 \cap A_2 \cap \ldots \cap A_n) = P(A_1) \cdot P(A_2) \cdot \ldots \cdot P(A_n).$$

EXAMPLE 7
Probability of
Independent Events

A single die is rolled twice. What is the probability that the first roll is a four and the second roll is not a four?

Solution The independence of the two events gives

$$P(\text{1st roll a four and 2nd roll not a four})$$
$$= P(\text{1st roll a four}) \cdot P(\text{2nd roll not a four})$$
$$= \left(\frac{1}{6}\right)\left(\frac{5}{6}\right) = \frac{5}{36}. \quad \blacksquare$$

EXAMPLE 8
Probability of
Independent Events

Suppose that cataract surgery is successful 95% of the time. Of the next four people who have such surgery, what is the probability that

(a) The first three will be successful and the fourth unsuccessful?

Solution Success of the surgery on one person is assumed to have no bearing on the success or failure of the operations of another person; hence, independence is assumed. Thus, P(1st successful and 2nd successful and 3rd successful and 4th unsuccessful) $= (.95)(.95)(.95)(.05) \approx .0429$. In other words, this particular sequence of events will happen about 4.29% of the time.

(b) All four will be successful.

Solution Translating P(all four successful) into P(1st successful and 2nd successful and 3rd successful and 4th successful) means that, due to independence, the answer is found by $(.95)(.95)(.95)(.95) \approx 81.5\%$

(c) Exactly two will be successful.

Solution Among the four surgeries, the exact order of the two successful and two unsuccessful surgeries is not specified. Therefore, all possibilities of order must be considered. Using S to represent "successful" and S' to represent "unsuccessful," the surgeries might have occurred as "S and S and S' and S'''" or "S and S' and S' and S" and so on. Writing these possibilities in abbreviated form as shown next, we find that there are six mutually exclusive orderings in which these surgeries could have taken place:

P(exactly 2 successful)

$= P(SSS'S' \text{ or } SS'S'S \text{ or } S'S'SS \text{ or } SS'SS' \text{ or } S'SS'S \text{ or } S'SSS')$

$= (.95)(.95)(.05)(.05) + (.95)(.05)(.05)(.95) \ldots + (.05)(.95)(.95)(.05).$

Observing that in each of the six products there are two .95's and two .05's the answer is found to be

$$6(.95)^2(.05)^2 \approx 0.0135. \quad \blacksquare$$

Example 9 answers the question posed at the outset of this chapter.

EXAMPLE 9
Computing Probability
for Independent Events

A particular rocket assembly depends on two components, A and B. If either is working properly, the assembly will function, whereas if both fail, the assembly will not function. Assume that the functioning of either component is independent of the other, that the probability of failure of A is .002, and that the probability of failure of B is .01.

(a) What are the odds that both of these components will fail simultaneously?

Solution We first find $P(A$ fails and B fails). Because the functioning of these components is independent, the answer is $P(A$ fails$) \cdot P(B$ fails$) = (.002)(.01) = .00002$. The odds that both will fail is then

$$\frac{.00002}{1 - .00002} = \frac{.00002}{.99998} \approx .00002 \text{ or about } 1{:}50{,}000.$$

(b) What is the probability that at least one of the components is working properly?

Solution The answer is $1 - P(\text{both fail}) = 1 - .00002 = .99998$

(c) What is the probability that one will be functioning properly and one will not?

Solution This question may be rested as P ("A functioning and B failed" or "A failed and B is functioning"). The *or* separates two mutually exclusive events so that a further restatement becomes $P(A$ functioning and B failed$) + P(A$ failed and B is functioning$) = (.998)(.01) + (.002)(.99) \approx .012. \quad \blacksquare$

Finally, we show a connection between selections made in an ordered sequence and those made without regard to order where, in both cases, the selections are made *without replacement*. Consider the questions 1 and 2 that follow.

1. A four-card hand is to be dealt (selected), without replacement and without regard to order, from a deck of 52 cards. What is the probability that the hand will contain exactly three clubs?

2. A four-card hand is to be dealt (selected) in succession, without replacement, from a deck of 52 cards. What is the probability that the first three cards dealt will be clubs and the last card will not be a club?

In both cases, of course, the cards are selected or dealt in succession in *some* order. But question 1 assumes that *the order is disregarded as unimportant* and treats the hand, once dealt, as a *set*. Therefore it is a combinations problem with the answer

$$\frac{C(13,3) \cdot C(39,1)}{C(52,4)} \approx 0.0412.$$

Question 2, on the other hand, assumes *the order of selection is important* and therefore the multiplication rule for probability is the proper approach giving the answer

$$\left(\frac{13}{52}\right)\left(\frac{12}{51}\right)\left(\frac{11}{50}\right)\left(\frac{39}{49}\right) \approx 0.0103.$$

As expected, the answers to these Questions 1 and 2 are quite different.

Is there a connection at all between these two questions? The answer is "yes." Let C represent "club" and C' represent "not a club." The set of cards $\{C, C, C, C'\}$ in 1 could be described by any one of the following mutually exclusive *ordered* selections:

$$CCCC' \text{ or } CCC'C \text{ or } CC'CC \text{ or } C'CCC.$$

Therefore

$$P(\text{exactly 3 clubs}) = P(CCCC' \text{ or } CCC'C \text{ or } CC'CC \text{ or } C'CCC)$$

$$= \left(\frac{13}{52}\right)\left(\frac{12}{51}\right)\left(\frac{11}{51}\right)\left(\frac{39}{50}\right) + \left(\frac{13}{52}\right)\left(\frac{12}{51}\right)\left(\frac{39}{50}\right)\left(\frac{11}{49}\right)$$

$$+ \left(\frac{13}{52}\right)\left(\frac{39}{51}\right)\left(\frac{12}{50}\right)\left(\frac{11}{49}\right) + \left(\frac{39}{52}\right)\left(\frac{13}{51}\right)\left(\frac{12}{50}\right)\left(\frac{11}{49}\right)$$

$$= 0.0103 + 0.0103 + 0.0103 + 0.0103 = 0.0412.$$

This shows that the answer to the first question can be obtained from ordered sequences of events *if all possible orderings of the hand are taken into consideration.*

EXAMPLE 10
Selecting Marbles

A box contains three red and five green marbles. Two marbles are to be selected, without replacement and without regard to order. What is the probability that one of the marbles will be red and the other green?

Solution

(a) Since marbles are not replaced and order is disregarded, the use of combinations will give the answer:

$$P(\text{one } R, \text{ one } G) = \frac{C(3,1) \cdot C(5,1)}{C(8,2)} \approx 0.536.$$

Solution

(b) This same answer may be obtained by thinking in terms of ordered selections:

$$P(\text{one } R, \text{ one } G) = P((\text{1st } R \text{ and 2nd } G) \text{ or } (\text{1st } G \text{ and 2nd } R))$$
$$= P(\text{1st } R \text{ and 2nd } G) + P(\text{1st } G \text{ and 2nd } R)$$
$$= \left(\frac{3}{8}\right)\left(\frac{5}{7}\right) + \left(\frac{5}{8}\right)\left(\frac{3}{7}\right) = \frac{30}{56} \approx 0.536. \quad \blacksquare$$

EXAMPLE 11
Sampling

Among 20 computer chips, five are known to be defective. If a sample of three of these chips is to be selected, without replacement and without regard to order, what is the probability that all will be defective?

Solution

(a) The answer is thought of as selecting a set

$$P(\text{all 3 defective}) = \frac{C(5, 3)}{C(20, 3)} \approx 0.0087.$$

Solution

(b) The solution is thought of in terms of ordered selections

$$P(\text{all 3 defective}) = P(\text{1st } D \text{ and 2nd } D \text{ and 3rd } D)$$
$$= \left(\frac{5}{20}\right)\left(\frac{4}{19}\right)\left(\frac{3}{18}\right) = \frac{1}{114} \approx 0.0087. \quad \blacksquare$$

EXERCISES   7.3

The events in Exercises 1 through 9 are dependent.

Exercises 1 through 5 refer to this experiment: Two cards are drawn in succession without replacement. What is the probability that

1. The second was a heart, given that the first was a heart?

2. The first is an ace and the second is not an ace?

3. The first is a spade and the second is not a spade?

4. The first is a club and the second is a club?

5. Both are aces?

Exercises 6 through 9 refer to this experiment: A sample of three light bulbs is drawn in succession, without replacement, from a lot of 20 bulbs, four of which are known to be defective. What is the probability that

6. The first two are good and the last one is defective?

7. The first one is good and the last two are defective?

8. All three are good?

9. All three are defective?

$\frac{4}{20} \quad \frac{3}{19} \quad \frac{2}{18}$

The events in Exercises 10 through 18 are independent.

Exercises 10 through 12 refer to this experiment: A coin is tossed five times. What is the probability that

10. All come up heads?

11. The first three come up heads and the last two tails?

12. The last three come up heads, given that the first two were tails?

Exercises 13 through 15 refer to this experiment: A multiple choice test has five questions, and each question has four choices. If you guess at the answers, what is the probability that you will

13. Get all of them right?

14. Get the first three right and the last two wrong?

15. Get the first right, the second wrong, the third right, the fourth wrong, and the last one right?

Exercises 16 through 18 refer to these data: A baseball player has a batting average of .245. Assume that each turn at bat is not affected by the previous turns and that no walks occur on the times at bat considered. What is the probability that, for the next three times at bat, the player will

16. Get a hit all three times?

17. Get exactly two hits?

18. Get at least two hits?

In Exercises 19 through 22 assume that A and B are independent events.

19. (a) Find $P(A)$ if $P(B) = .6$ and $P(A \cap B) = .4$.
   (b) Find $P(A)$ if $P(B') = .7$ and $P(A \cap B) = .1$.

20. (a) Find $P(B)$ if $P(A) = .8$ and $P(A \cap B) = \frac{7}{16}$.
   (b) Find $P(A \cup B)$ if $P(A) = .7$ and $P(B) = .5$.

21. (a) Find $P(B)$ if $P(A') = .5$ and $P(A \cap B)' = .6$.
   (b) Find $P(A \cup B)$ if $P(A') = .4$ and $P(B) = .5$.

22. (a) Find $P(A \cap B)$ if $P(A) = \frac{1}{4}$ and $P(B) = \frac{2}{7}$.
   (b) Find $P(A' \cap B')$ if $P(A) = \frac{1}{3}$ and $P(B) = \frac{4}{5}$.

For Exercises 23 and 24, find the requested probabilities associated with these tree diagrams.

23. (a) $P(R \mid N)$
   (b) $P(M \cap S)$
   (c) $P(N \cap R)$
   (d) $P(R)$
   (e) $P(S)$

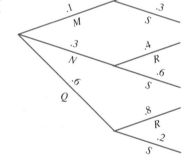

24. (a) $P(X \mid A)$
   (b) $P(Y \mid B)$
   (c) $P(B \text{ and } Y)$
   (d) $P(X)$
   (e) $P(Y)$

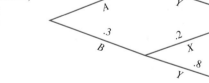

25. Three marbles are drawn, in succession and without replacement, from a box containing four red and five green marbles. Find the probability that

   (a) The first marble will be red and the last two green.
   (b) The first marble will be red, the second green, and the third red.
   (c) All three marbles will be red.

26. Four cards are drawn, in succession and without replacement, from a standard deck of 52 cards. Find the probability that

   (a) The first card will be a spade, the second a club, the third a spade, and the fourth a club.
   (b) The first two cards will be kings and the last two will be queens.
   (c) All of the cards will be hearts.

27. A box contains five red, three green, and four purple balls. Three balls are drawn, in succession with replacement from the box. Find the probability that

   (a) The first ball will be red and the next two purple.
   (b) The first ball will be red, the second green, and the third purple.
   (c) All of the balls will be green.

28. A shipment of computer chips contains 12 that are good and three that are defective. Three of these chips are selected, in succession, with replacement, and checked for a defect. Find the probability that

   (a) The first chip will be defective, the second defective, and the third good.
   (b) None of the chips will be defective.
   (c) All of the chips will be defective.

29. A single die is rolled four times. Find the probability that

   (a) The successive rolls will result in the following sequence of dots showing on the top side: 2, 2, not a 2, not a 2.
   (b) The successive rolls will result in the following sequence of dots showing on the top side: 3, 5, not a 5, 2.
   (c) The successive rolls will result in the following sequence of events: A number greater than 4, a 3, a number less than 5, a 4.

30. A single die is rolled five times. Find the probability that

   (a) The successive rolls will result in the following sequence of dots showing on the top side: Not a 3, 4, 4, not a 5, not a 5.
   (b) A 3 turns up every time.
   (c) A 3 never turns up.

31. A pair of dice is rolled twice. Find the probability that

   (a) A sum of 6 will be rolled first, and a sum of 9 will be rolled second.
   (b) A sum of 7 will be rolled both times.
   (c) A sum of 5 will be rolled first, and a sum that is not 5 will be rolled second.

32. A pair of dice is rolled three times. Find the probability that

    (a) A sum of 6 will be rolled first, a sum of 9 second, and a sum of 3 third.

    (b) A sum of 4 will be rolled all three times.

    (c) A sum of 5 will be rolled the first time, and a sum of 3 will be rolled the second and third times.

Product Effectiveness *Exercises 33 through 35 refer to these data: Suppose that it is known that flu shots are 97% effective. Suppose that two people who live in different parts of the country take flu shots. What is the probability that*

33. Both will get the flu?

34. Neither will get the flu?

35. At least one will get the flu?

Selecting a Committee *Exercises 36 through 38 refer to this experiment: A committee of three is selected at random from four juniors and five seniors. What is the probability that*

36. The first two selected are juniors and the third is a senior?

37. All three are juniors?

38. The last one selected is a junior, given that the first two selected are seniors?

Mortality Tables *Exercises 39 through 41 refer to these data: According to insurance mortality tables, the probability that a 50-year-old woman will be alive ten years from now is .91. College roommates Jane and Sue both have mothers who are 50 years old and live in different states. What is the probability that*

39. Both of their mothers will be alive ten years from now?

40. At least one of their mothers will be alive ten years from now?

41. Neither of their mothers will be alive ten years from now?

Product Reliability *Exercises 42 through 45 refer to these data: A rocket scientist estimates that a particular component in a rocket is 99.8% reliable. Three of these components are installed so that if one fails, the task of the component is automatically passed to the next component. Assume that each component works properly or fails, independently of the others. What is the probability that*

42. All three will fail?

43. The first two will fail, but the third will not?

44. At least one will function properly?

45. How many of these components should be installed so that at least one will function properly, with a probability of .99999?

Product Reliability *Exercises 46 through 48 refer to these data: A bank has a main computer and a backup computer. The main computer has reliability of .96 and the backup computer has a reliability of .90. What is the probability that*

46. Both will fail at the same time?

47. At least one will function properly?

48. One will be working properly and one will be down?

Tossing Coins *Exercises 49 through 51 refer to this experiment: A coin is tossed seven times. What is the probability that*

49. All turn up heads?

50. The last one turns up heads, given that the first six turned up heads?

51. The last three turn up heads, given that the first four have turned up tails?

Product Durability *Exercises 52 through 54 refer to these data: A particular type of light bulb has a probability of .01 of burning out in less than 800 hours.*

52. If four of these bulbs are put into four different lamps, what is the probability that all four will burn out in less than 800 hours?

53. If four of these bulbs are put into four different lamps, what is the probability that at least one will still be burning after 800 hours?

54. Now suppose that these bulbs are used in a particular security application in which it is required that at least one bulb remain lighted for 800 consecutive hours. How many bulbs are required in order to assure that the probability of success is at least .999?

 Sampling *Exercises 55 through 57 refer to these data: A sample of three TV sets is selected, in succession and without replacement, from among 12 sets, three of which are defective. What is the probability that*

55. All three of those chosen are defective?

56. The first two chosen are not defective, but the third one is defective?

57. At least one is defective?

Survey *In Exercises 58 through 60, when considering families who plan to have three children, what is the probability that*

58. The first two children will be boys and the third a girl?

59. All three will be girls?

60. There will be more boys than girls?

Survey *Exercises 61 through 63 refer to these data: Suppose that 28% of the voting-age adults consider themselves Democrats, 25% consider themselves Republican, and the rest consider themselves Independents. If ten voters are randomly contacted, what is the probability that*

61. All are Republicans?

62. None are Republicans?

63. At least one is a Republican?

Survey *Exercises 64 through 66 refer to these data: A survey in a particular state revealed that 40% of the drivers do not wear a seat belt. If three cars are to be stopped during a routine traffic check in this state, what is the probability that*

64. None of the three drivers will be wearing a seat belt?

65. At least one will be wearing a seat belt?

66. One will be wearing a seat belt and two will not?

Security *Exercises 67 and 68 refer to these data: Nuclear power plants have a threefold security system, each of which is 98% reliable, and independent of the others, to prevent unauthorized persons from entering the premises. What is the probability that an unauthorized person will*

67. Get through all three security systems?

68. Get through the first two systems, but not the third?

Sampling *Exercises 69 through 71 refer to these data: Examination of a large number of fruit flies reveals that 10% have an eye defect. If three fruit flies are selected at random, in succession and without replacement, what is the probability that*

69. All will have an eye defect?

70. None will have an eye defect?

71. Exactly one will have an eye defect?

In Exercises 72 through 76:
(a) Find the requested probability considering the selections as a set.
(b) Then use ordered sequences to find the probability.

72. Three marbles are selected, without replacement and without regard to order, from a box containing four red and five green marbles. Find the probability that
(a) Exactly two will be red.
(b) All three are red.

73. In a shipment of 20 ladies' blouses, five are known to have a flaw. If a sample of four blouses is selected from this shipment, without replacement and without regard to order, find the probability that
(a) Exactly three will have a flaw.
(b) All four will have a flaw.

74. In a shipment of 15 men's shirts, four are known to have a flaw. If a sample of four shirts is selected from this shipment, without replacement and without regard to order, find the probability that
(a) None will have a flaw.
(b) Exactly two will have a flaw.

75. Five cards are dealt, without replacement and without regard to order, from a deck of 52 cards. Find the probability that
(a) Exactly two are hearts.
(b) Exactly three are kings.
(c) All are spades.

76. Four cards are dealt, without replacement and without regard to order, from a deck of 52 cards. Find the probability that
(a) Exactly two are hearts.
(b) Exactly three are kings.
(c) All are spades.

Exercise 77 appeared on an actuarial exam.

77. A random sample of six balls is selected, with replacement, from an urn that contains ten red, five white, and five blue balls. What is the probability that the sample contains two balls of each color?
(a) $\dfrac{1}{1024}$ (b) $\dfrac{1}{646}$ (c) $\dfrac{45}{512}$
(d) $\dfrac{75}{646}$ (e) $\dfrac{45}{64}$

Exercise 78 appeared on an actuarial exam.

78. An urn contains four red balls, eight green balls, and two yellow balls. Five balls are randomly selected, with replacement, from the urn. What is the probability that one red, two green, and two yellow balls will be selected?
(a) $\dfrac{3}{512}$ (b) $\dfrac{192}{7^5}$ (c) $\dfrac{15}{512}$
(d) $\dfrac{8}{143}$ (e) $\dfrac{960}{7^5}$

Exercise 79 appeared on an actuarial exam.

79. Let S and T be independent events. Please choose (a), (b), (c), (d), or (e) to complete this equation: $P(S \cap T) = \frac{1}{10}$ and $P(S \cap T') = \frac{1}{5}$. $P[(S \cup T)'] =$
(a) $\dfrac{3}{10}$ (b) $\dfrac{11}{30}$ (c) $\dfrac{7}{15}$ (d) $\dfrac{8}{15}$ (e) $\dfrac{9}{10}$.

Exercise 80 appeared on an actuarial exam.

80. A fair coin is tossed. If a head occurs, 1 fair die is rolled; if a tail occurs, 2 fair dice are rolled. If Y is the total on the die or dice, then $P[Y = 6] =$

(a) $\dfrac{1}{9}$ (b) $\dfrac{5}{36}$ (c) $\dfrac{11}{72}$ (d) $\dfrac{1}{6}$ (e) $\dfrac{11}{36}$

Problems for Exploration

1. Prove that if A and B are independent events, then A' and B' are also independent events. (**Hint** Start with $P(A') \cdot P(B')$, and rename each factor by use of the complement theorem; then use the addition theorem, and finally one of DeMorgan's laws.)

2. Make use of question 1 to prove that if A and B are independent, then $P(A \cup B) = 1 - P(A') \cdot P(B')$.

3. Show that two mutually exclusive events cannot be independent.

This exercise is from an actuarial exam.

4. A system has two components placed in series so that the whole system fails if either of the two components fail. The second component is twice as likely to fail as the first. If the two components operate independently and if the probability that the entire system will fail is .28, then what is the probability that the first component will fail?

(a) $\dfrac{.28}{3}$ (b) .62   (c) $\dfrac{.56}{3}$ (d) .20

(e) $\sqrt{14}$

5. If A and B are independent events, decide whether A and B' are independent. Explain or justify your decision.

SECTION 7.4   ## Bayes' Theorem

Bayes' Theorem is a special application of conditional probability. An example leads to the theorem.

EXAMPLE 1

Pointing Toward Bayes' Theorem

Z-Mart, which buys men's shirts under its own private label, buys 40% of those shirts from supplier A, 50% from supplier B, and 10% from supplier C. It is found that 2% of the shirts from A have flaws, 3% from B have flaws, and 5% of those from C have flaws. A probability tree diagram representing these purchases and flaw rates, is shown in Figure 17. The first-generation branches are labeled with probabilities representing the random selection of a company. The second-generation of branches are labeled with conditional probabilities that a randomly selected shirt will have a flaw or no flaw, depending on which company produced the shirt. If one of these shirts is bought from Z-Mart, what is the probability that

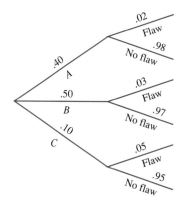

FIGURE 17
*The Probability of Flaws
in Men's Shirts*

(a) It has a flaw, given that it came from *A*?

Solution Knowing that the shirt came from *A* means that we have already traversed the *A* branch and now look for the probability, which is .02, along the attached *flaw* branch. Therefore, $P(\text{flaw}\,|\,\text{from }A) = .02$. Note that the answer was found by reading the tree diagram from left to right.

(b) It has a flaw?

Solution The solution is

$$P(\text{"from }A\text{ and a flaw" or "from }B\text{ and a flaw" or "from }C\text{ and a flaw"})$$
$$= P(\text{from }A) \cdot P(\text{a flaw}\,|\,\text{from }A) + P(\text{from }B) \cdot P(\text{a flaw}\,|\,\text{from }B)$$
$$+ P(\text{from }C) \cdot P(\text{a flaw}\,|\,\text{from }C)$$
$$= (.40)(.02) + (.50)(.03) + (.10)(.05) = 0.028.$$

(c) It came from *A*, given that it has a flaw?

Solution This time the given event *flaw* is on the right, and we are confronted with somehow trying to read back to the left to reach the event *A*. This is the *reverse* probability to which Bayes referred and which, from the viewpoint of a tree diagram, will always be the clue that the problem is a Bayes' theorem problem. How do we find the solution? Just use the conditional probability formula and read from the tree:

$$P(A\,|\,\textbf{flaw}) = \frac{P(A\text{ and flaw})}{P(\textbf{flaw})}$$

$$= \frac{(.40)(.02)}{(.40)(.02) + (.50)(.03) + (.10)(.05)} \approx 0.286. \ \blacksquare$$

Example 1(c) may be viewed in terms of three mutually exclusive events—*A*, *B*, and *C*—the union of which makes up the entire sample space, with the event *F* (flaws) intersecting each of these events; then a probability question is asked about one of the events *A*, *B*, or *C*, given that *F* has already occurred. A Venn diagram description would appear as shown in Figure 18. As noted in Example 1, quantities such as $P(A\,|\,F)$, $P(B\,|\,F)$, and $P(C\,|\,F)$ may be expressed by the conditional probability formula as

$$\frac{P(A\text{ and }F)}{P(F)}, \qquad \frac{P(B\text{ and }F)}{P(F)}, \qquad \text{and} \qquad \frac{P(C\text{ and }F)}{P(F)}.$$

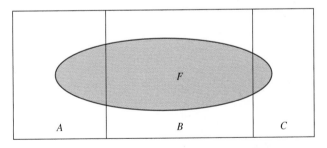

FIGURE 18
The Diagram for Bayes' Theorem

Furthermore, the denominator $P(F)$ may be rewritten as a sum using mutually exclusive events, by tracing through the tree diagram of Figure 17, $P(A$ and $F)$ + $P(B$ and $F)$ + $P(C$ and $F)$. Bayes then used the multiplication rules to rewrite each of these terms, as well as the numerator, as shown next.

> ### Bayes' Theorem:
>
> Let A, B, and C be mutually exclusive events, the union of which is the entire sample space S. Let F be any event with a probability that is not zero. Then
>
> $$P(A \mid F) = \frac{P(A \text{ and } F)}{P(A \text{ and } F) + P(B \text{ and } F) + P(C \text{ and } F)}$$
>
> $$= \frac{P(A) \cdot P(F \mid A)}{P(A) \cdot P(F \mid A) + P(B) \cdot P(F \mid B) + P(C) \cdot P(F \mid C)}$$
>
> A similar formula for $P(B \mid F)$ and $P(C \mid F)$ can be found by replacing A with, respectively, B and C. This formula applies to any finite number of mutually exclusive sets whose union is S.

As our development has suggested, a tree diagram is very useful when using Bayes' theorem.

EXAMPLE 2
Tree Diagrams and
Bayes' Theorem

Two marbles are randomly drawn, in succession and with replacement, from a box containing two red and five white marbles. The tree diagram has this appearance:

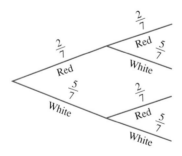

FIGURE 19
Drawing Marbles with Replacement

What is the probability that

(a) The second marble is red?

Solution The answer is $(\frac{2}{7})(\frac{2}{7}) + (\frac{5}{7})(\frac{2}{7}) \approx .286$.

(b) The second marble is red, given that the first marble is white?

Solution After locating the given event "first marble white," we continue reading naturally from *left to right* to arrive at the event "second marble red" and the answer $\frac{2}{7}$.

(c) The first marble is white, given that the second marble is red?

Solution This time the given event "second marble red" is located on the right (in two different places) of the tree diagram, and we are being asked to read from *right to left*, to locate the event "first marble white." This is a Bayes' theorem problem, and we write

$$P(\text{1st white} \mid \text{second red}) = \frac{P(\text{1st white and 2nd red})}{P(\text{2nd red})}$$

$$= \frac{\left(\frac{5}{7}\right)\left(\frac{2}{7}\right)}{\left(\frac{2}{7}\right)\left(\frac{2}{7}\right) + \left(\frac{5}{7}\right)\left(\frac{2}{7}\right)} \approx .714. \quad ■$$

EXAMPLE 3
Tree Diagrams and
Bayes' Theorem

A department store made a study of the personal checks it received for payment of goods. It discovered that 40% of the checks with insufficient funds had the wrong date on them, while only 2% of all good checks had the wrong date on them. It also found that 0.5% of all checks received had insufficient funds to cover them. The tree diagram for this situation appears in Figure 20. If a clerk in this store receives a personal check from a customer, what is the probability that

(a) It has insufficient funds and a wrong date?

Solution The answer is $(.005)(.40) = .002$.

(b) It has the wrong date, given that it has insufficient funds?

Solution $P(\text{wrong date} \mid \text{insufficient funds}) = .40$.

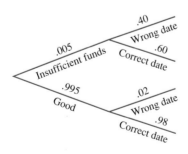

FIGURE 20
Personal Checks

(c) It has insufficient funds, given that it has the wrong date?

Solution   This is a Bayes' theorem problem.

$$P(\text{insufficient funds} \mid \text{wrong date})$$

$$= \frac{P(\text{insufficient funds and wrong date})}{P(\text{wrong date})}$$

$$= \frac{(.005)(.40)}{(.005)(.40) + (.995)(.02)} \approx .091. \quad ■$$

The examples in this section show that a tree diagram is very helpful in identifying a Bayes' theorem problem and in finding its solution. We suggest the use of a tree diagram whenever practical.

EXERCISES   7.4

In Exercises 1 through 5, use the following tree diagram to determine each probability.

1. $P(A \text{ and } D)$. 12
2. $P(D)$.
3. $P(E \mid B)$. 9
4. $P(B \mid E)$
5. $P(A \mid D)$. 6

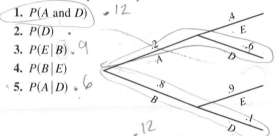

In Exercises 6 through 11, use the following tree diagram to determine each probability.

6. $P(E \mid A)$
7. $P(A \mid E)$ 4/37
8. $P(E)$
9. $P(C \mid D)$ 12/63
10. $P(D \mid C)$
11. $P(B \mid E)$

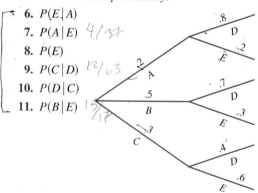

Every remaining exercise can be modeled with a tree diagram. Draw one for each set of exercises, and use it to answer the questions. Some of the exercises involve Bayes' theorem.

Sampling   *Exercises 12 through 14 refer to this experiment: Two balls are drawn, in succession and with replacement, from a box containing two red, five blue, and seven purple balls. What is the probability that*

12. The first ball is blue, and the second ball is purple?
13. The second ball is blue, given that the first is purple?
14. The first ball is blue, given that the second ball is purple?

Sampling   *Exercises 15 through 17 refer to this experiment: Three balls are drawn in succession, without replacement, from a box containing three red and six green balls. What is the probability that*

15. All three balls are red?
16. The first two balls are red and the last one is green?
17. The third ball is green?

Sampling   *Exercises 18 through 20 refer to this experiment: A box contains three slips of paper, on each of which is written the number 1, 2, or 3 respectively. Two slips of paper are drawn, in succession and with replacement, from the box. What is the probability that*

18. The second number drawn is a 3, given that the first number drawn is a 2?
19. The first number is a 3, and the second number is a 1?
20. The first number is a 2, given that the second number is a three?

Sampling Exercises 21 through 24 refer to this experiment: One box, A, contains three red and four green balls, while a second box, B, contains eight red and two green balls. A box is selected at random and a ball is drawn. What is the probability that

21. The ball is red, given that it came from box *A*?

22. The ball is red and came from box *B*?

23. The ball is red?

24. The ball was drawn from box *B*, given that it is red?

Survey Exercises 25 through 27 refer to these data: Thirty-five percent of the students at State Tech are female, while 55% of the students at State University are female. If one of these universities is selected at random (assume a 50–50 chance for each), and a student from that university is selected at random, what is the probability that the student will be

25. Female?

26. Female, given that the student comes from State Tech?

27. From State Tech, given that the student is female?

Survey Exercises 28 through 31 refer to these data: It is known that about 5% of all men and 0.25% of all women are color blind. Assume that half of the population are men and half women. If a person is selected at random, what is the probability that the person selected is

28. Female and color blind?

29. Color blind, given that the person is male?

30. Color blind?

31. Male, given that the person is color blind?

Tossing a Biased Coin Exercises 32 through 34 refer to this experiment: A coin is biased to show 60% heads and 40% tails. The coin is tossed twice. What is the probability that

32. The first toss lands heads up, and the second toss lands tails up?

33. The second toss comes up heads?

34. The first toss comes up heads, given that the second toss landed tails up?

Selection Exercises 35 through 37 refer to these data: The human resources director estimates from experience that for a particular job opening, 80% of the applicants are qualified. To help in the selection process, a test has been designed so that a qualified applicant has a probability of .85 of passing the test, while an unqualified applicant has only a probability of .05 of passing the test. If an applicant is selected at random, what is the probability that the person

35. Is qualified and will pass the test? What are the odds for this event?

36. Can pass the test, given that the applicant is unqualified for the job?

37. Is unqualified for the job, given a passing score on the test?

Selection Exercises 38 through 40 refer to these data: An auto insurance company classifies its drivers as good, medium, or bad risks. Of their policyholders, 50% are considered good risks, 35% medium risks, and 15% bad risks. Company records indicate that the probability of a driver from these three risk categories having an accident during a given year is .02, .05, and .10, respectively. What is the probability that one of their insured drivers

38. Is a good risk and will have an accident during the next year?

39. Will have an accident during the next year? What are the odds for this event?

40. Will be rated a bad risk, given that during the next year, the driver has an accident?

Product Effectiveness Exercises 41 through 43 refer to these data: During a given winter, 23% of the population received a flu shot and, of those, 1% got the flu. For those who did not receive the flu shot, 12% got the flu. What is the probability that a person

41. Did not get the flu, given that she had received the flu shot? What are the odds for this event?

42. Did not receive the flu shot and got the flu?

43. Did not receive the shot, given that he had the flu?

Test Reliability Exercises 44 through 46 refer to these data: Records indicate that 2% of the population has a certain kind of cancer. A medical test has been devised to help detect this cancer. If a person does have this type of cancer, the test will detect it 98% of the time. However, 3% of the time the test will indicate that a person has this cancer when, in fact, he or she does not. For persons using this test, what is the probability that

44. The person has this type of cancer, and the test indicates he has this type of cancer?

45. The person has this type of cancer, given the test indicates she has this type of cancer?

46. The person does not have this type of cancer, given a positive test result for this type of cancer?

Product Reliability Exercises 47 through 50 refer to these data: A manufacturer makes two models of microwave ovens, a standard model and a deluxe model, with 60% of production allotted to the standard model. About 2% of the standard models and 1% of the deluxe models have some kind of defect. If one of these ovens is selected at random, what is the probability that

47. It is a deluxe model and has a defect?

48. It is defective, given that it is a deluxe model?

49. It is a deluxe model, given that it is defective?

50. It is defective? What percentage of the microwave ovens are defective?

Product Reliability   *Exercises 51 through 54 refer to these data: An auto mechanic knows that 1% of the new fuel pumps installed fail within one year, but that about 5% of the rebuilt ones fail within one year. A mechanic selects a fuel pump from an inventory of 20, not realizing that five of the 20 are rebuilt fuel pumps. What is the probability that*

51. It was rebuilt and will fail within a year? What are the odds for this event?

52. It will fail within a year, given that it was rebuilt?

53. It was rebuilt, given that it failed during the year?

54. Any fuel pump in the inventory will fail within a year? What percentage of the inventory will fail within a year?

Product Reliability   *Exercises 55 through 59 refer to these data: Z-Mart buys ladies' slacks from four suppliers; 30% from Best Buy, 20% from Crimson Lady, 40% from Dandy Stride, and 10% from Easy Fit. It finds that the defective rates in shipments received are 4%, 5%, 3%, and 6%, respectively. What percentage of these slacks*

55. Has a defect?

56. Has a defect, given that they came from Best Buy or Crimson Lady?

57. Came from Dandy Stride and does not have a defect?

58. Came from Easy Fit, given that they have a defect?

59. Came from Crimson Lady, given that they have a defect?

Salary Structure   *Exercises 60 through 62 refer to these data: In a particular firm, 5% of the men and 3% of the women have salaries that exceed $50,000. Sixty percent of the firm's employees are men, and 40% are women. For persons working for this firm, what is the probability that*

60. The employee is a woman and has a salary that exceeds $50,000?

61. The employee is a man, give that the salary exceeds $50,000?

62. The employee's salary exceeds $50,000? What percentage of the firm's employees earn more than $50,000?

63. *Sampling*   A box contains three balls numbered 1 through 3. One ball is drawn from the box and laid aside. Then a second ball is drawn. What is the probability that the second ball is numbered 2?

64. *Sampling*   A box contains four balls, numbered 1 through 4. One ball is drawn from the box and laid aside. Then a second ball is drawn. What is the probability that the second ball drawn is numbered 3?

65. *Sampling*   A box contains five red and six green balls. Two balls are drawn in succession and laid aside. Then a third ball is drawn from the remaining balls in the box. What is the probability that the third ball drawn is red?

Exercise 66 appeared on an actuarial exam.

66. An urn contains four balls numbered 0 through 3. One ball is selected at random, removed from the urn, and not replaced. All balls with *nonzero* numbers less than that of the selected ball are also removed from the urn. Then a second ball is selected at random from those remaining in the urn. What is the probability that the second ball selected is numbered 3?

    (a) $\dfrac{1}{4}$ **(b)** $\dfrac{7}{24}$ **(c)** $\dfrac{1}{3}$ **(d)** $\dfrac{11}{24}$ **(e)** $\dfrac{13}{24}$

Problems for Exploration

Use $A_1, A_2, \ldots, A_n$ to represent mutually exclusive events whose union is the entire sample space, and use F to be any nonempty set in that sample space.

1. Write Bayes' theorem, using this general notation.

2. For Bayes' theorem to be valid, is it necessary that the intersections of F with each of the sets $A_1, A_2, \ldots, A_n$ not be empty?

3. If $F \cap A_2 = \emptyset$, what is $P(A_2 | F)$?

4. If $F \cap A_2 = \emptyset$, write the general form for Bayes's theorem that will find $P(A_1 | F)$ in simplest terms.

The following exercise appeared on an actuarial exam.

5. Three boxes are numbered 1, 2, and 3. For $k = 1$, 2, 3, box k contains k blue marbles and $5 - k$ red marbles. In a two-step experiment, a box is selected, and two marbles are drawn from it without replacement. If the probability of selecting box k is proportional to k, what is the probability that the two marbles drawn have different colors?

    (a) $\dfrac{17}{60}$ **(b)** $\dfrac{34}{75}$ **(c)** $\dfrac{1}{2}$ **(d)** $\dfrac{8}{15}$ **(e)** $\dfrac{17}{30}$

SECTION **7.5** ## Binomial Experiments; A Guide to Probability

Experiments classified as **binomial** were first studied by James Bernoulli in the late 1600s and for that reason are sometimes also known as **Bernoulli experiments.** The term *binomial* comes from the fact that these experiments can be classified as having exactly two outcomes. For an experiment to qualify as binomial (or Bernoulli), several conditions, already familiar to us, must be met.

> **HISTORICAL NOTE**
> **James Bernoulli (1654–1705)**
>
> James (Jacob, Jacques) Bernoulli was a Swiss mathematician who taught at the University of Basel. The Bernoulli distribution and the Bernoulli theorem of probability and statistics are among his many notable contributions to mathematics. The mathematical careers of James and his brother John Bernoulli form the connecting link between the mathematics of the 17th and 18th centuries.

Conditions for a Binomial (Bernoulli) Experiment

1. *Experiment repeated n times.* The experiment is to be repeated a finite number of times, say n, under the same conditions.

2. *Independence.* Each time the experiment is performed, the outcome has no effect on the outcome of any other trial of the experiment.

3. *There are two possible outcomes—success or failure.* Each time the experiment is performed, the outcome can be classified in exactly two ways; (1) success (S), with a probability of p; or (2) failure (F), with a probability of $q = 1 - p$. The probability of success never changes from trial to trial.

The concepts of repeated trials and independence are already familiar to us, so the only point of concern might be regarding the third condition, where the outcomes must be reduced to a success (S) or a failure (F).

The **central question** in a **binomial experiment** is to **find** the **probability of r successes from** among n **trials;** the outcome labeled "success" will be determined by the question asked, and all other outcomes will be lumped together and labeled "failure." Some examples clarify the terminology. The solutions follow later.

EXAMPLE 1
Determining Success and Failure for a Binomial Experiment

A single die is rolled 10 times, and we want to know the probability that exactly three 5s will be rolled. The experiment is repeated $n = 10$ times, there are to be $r = 3$ successes, and the outcome of each trial is independent of the others.

Because the question asks about rolling fives, we call "roll a 5" a success (S) and "roll a 1, 2, 3, 4, or 6" a failure (F). It follows that $p = P(S) = \frac{1}{6}$ and $q = P(F) = \frac{5}{6}$. ■

EXAMPLE 2
Determining Success
and Failure for a
Binomial Experiment

A coin is tossed 20 times, and we are interested in knowing the probability that exactly 12 heads will turn up. This experiment is repeated $n = 20$ times (or trials) and the outcomes of the trials are independent. Because the question asks about heads, success will denote the event "a head lands up" and failure will denote the event "a tail lands up." In this case we have

$$p = P(S) = \frac{1}{2}, \qquad q = P(F)\frac{1}{2}, \qquad \text{and} \qquad r = 12. \quad \blacksquare$$

EXAMPLE 3
Determining Success
and Failure for a
Binomial Experiment

Suppose that it is known that kidney transplants are successful 80% of the time. Of the next 12 people who receive such transplants, we are interested in knowing the probability that at least 10 transplants will be successful. This experiment is repeated $n = 12$ times, and we consider the success or failure of the transplant for any one person to be independent of the results of the others. Here, S represents the event "the transplant will be successful" and F represents the event "the transplant will not be successful." It follows that $p = P(S) = .80$, $q = P(F) = .20$, and r is 10, 11, or 12. $\quad \blacksquare$

EXAMPLE 4
Determining Success
and Failure for a
Binomial Experiment

Tests show that 4% of the hair dryers made by the Tight-Curl Company have a defect. Suppose that 50 of this company's hair dryers are ordered by a discount store, and we are interested in the probability that exactly one of them will be defective. In this case, $n = 50$, and because we are interested in *defects,* success will denote the event "the dryer is defective," and failure will denote the event "the dryer is not defective." We therefore have

$$P(S) = .04, \qquad P(F) = .96, \qquad \text{and} \qquad r = 1. \quad \blacksquare$$

In the first three of these examples, the labeling of success and failure was rather natural, but in Example 4, it seemed a bit unnatural. If desired, such questions can be restated so that they seem more comfortable. For instance, in Example 4, we could restate the question as, "What is the probability that exactly 49 of the dryers are not defective?" This will reverse the designations of success and failure (and their probabilities). Either interpretation will lead to the same probability answer.

Just how do we go about answering probability questions for binomial experiments such as those asked in Examples 1 through 4? Two examples will lead us to a useful formula that will give us the probabilities for any binomial experiment.

EXAMPLE 5
How Probability in a
Binomial Experiment
is Calculated

In a poll of probable voters in an upcoming gubernatorial election, candidate A gets 30% of the votes. Assume this percentage holds for the actual voting population. If four probable voters are asked about their choice of candidates, what is the probability that exactly three of them will favor candidate A?

Solution   We are interested in candidate A and therefore let S (success) denote "vote for A" and F (failure) denote "Vote for another candidate." Then, according to the data given, $P(S) = p = .30$ and $P(F) = q = .70$, the number of trials, n, is 4 and the number of successes we seek is $r = 3$.

We assume that how 1 person answers the poll is unaffected by how another person answers the poll, hence the trials are independent. The importance of independence is that the probability of any sequence of S's and F's joined by the connective "and" is found by *multiplying the individual probabilities in the sequence together*. Thus, to get the probability that exactly 3 will vote for $A(S$'s), we have these mutually exclusive *ordered possibilities* of "vote for A" (S) and "will not vote for A" (F) among the 4 persons polled:

P(exactly 3 will vote for A)

$\quad = P(SSSF$ or $SSFS$ or $SFSS$ or $FSSS)$

$\quad = (.30)(.30)(.30)(.70) + (.30)(.30)(.70)(.30) + (.30)(.70)(.30)(.30)$

$\qquad\qquad\qquad + (.70)(.30)(.30)(.30)$

$\quad = 4(.30)^3(.70) = 0.1134.$

Notice that the product in each term is the same, $(.30)^3(.70)$, because there are always 3 S's and 1 F. Therefore, if the number of terms is known, then the desired probability may be found by multiplying this number, 4, by $(.30)^3(.70)$ as we have just done. ■

EXAMPLE 6
How the Probability of a Binomial Experiment Is Calculated

A single die is rolled five times. What is the probability that exactly three of the rolls show two dots?

Solution Here, the number of trials is $n = 5$, S represents "roll a two" with $P(S) = p = \frac{1}{6}$, F represents "roll something not a two" with $P(F) = q = \frac{5}{6}$, and the number of desired successes is $r = 3$. There are several sequences of the five rolls that meet the condition of three successes and, because the outcome on each roll of the die is independent of the others, the probability of each sequence is found by multiplying the probabilities of the various outcomes together.

Sequence satisfying three successes	Probability of sequence occurring
$SFFSS$	$(\frac{1}{6})(\frac{5}{6})(\frac{5}{6})(\frac{1}{6})(\frac{1}{6}) = (\frac{1}{6})^3(\frac{5}{6})^2$
$FSSFS$	$(\frac{5}{6})(\frac{1}{6})(\frac{1}{6})(\frac{5}{6})(\frac{1}{6}) = (\frac{1}{6})^3(\frac{5}{6})^2$
$SFSSF$	$(\frac{1}{6})(\frac{5}{6})(\frac{1}{6})(\frac{1}{6})(\frac{5}{6}) = (\frac{1}{6})^3(\frac{5}{6})^2$
. . . .	. . . .

Notice that the probability is the same every time we have a sequence of throws involving three S's and two F's—namely, $(\frac{1}{6})^3(\frac{5}{6})^2$. Notice also that the sequences are *mutually exclusive* events. That is, no two of them can happen simultaneously. It follows that we can just add all of the probabilities in the right-hand column to answer one question. Better yet, if we only knew how many such sequences were on the list, we could just multiply that number by $(\frac{1}{6})^3(\frac{5}{6})^2$ to answer our question. The number of such sequences is the number of different-appearing arrangements in a row of the five letters three S's and two F's, where the S's are considered indistinguishable and the F's are considered indistinguishable. That number is

$$\frac{5!}{3! \cdot 2!} = C(5, 3) = 10.$$

Therefore, the answer to our question is

$$10\left(\frac{1}{6}\right)^3\left(\frac{5}{6}\right)^2 \approx 0.0322.$$

In symbol form, the answer is $C(5, 3)(\frac{1}{6})^3(\frac{5}{6})^2$. ■

In terms of n, p, q, and r, the final calculation for Example 6 becomes

$$C(n, r) \cdot p^r \cdot q^{n-r}.$$

Other examples show that this general pattern of thought prevails every time, leading to this formula.

The General Formula for Binomial Experiments

For a binomial experiment with n trials, $P(S) = p$ and $P(F) = q = 1 - p$, the probability of exactly r successes is $C(n, r)p^r q^{n-r}$.

EXAMPLE 7

Using the General Formula for Binomial Experiments

Tests show that about 4% of the people who take a particular drug are subject to side effects. Of 20 people taking this drug, what is the probability that exactly 5 of them will experience side effects?

Solution   We assume that one person having side effects will have no bearing on another person having side effects from the drug. With independence thus assumed, we may take the outcomes to be S: "will have side effects," with a probability of .04 and F: "will not have side effects," with a probability of .96. With $n = 20$ and $r = 5$, the binomial-experiment formula gives $C(20, 5)(.04)^5(.96)^{15} \approx .00086$. ∎

EXAMPLE 8

Using the General Formula for Binomial Experiments

A particular company polled a large number of potential customers and found that 25% of them would buy the company's product. If a salesperson calls on 10 potential customers, what is the probability that

(a) Exactly eight will buy the product?

Solution   Assuming that the customers are independent in their buying decisions, we have $n = 10$, S represents "will buy the product," F represents "will not buy the product," $p = .25$, $q = .75$, and $r = 8$. Therefore, our answer is

$$C(10, 8)(.25)^8(.75)^2 \approx .00039.$$

(b) Exactly three will buy the product?

Solution   The answer is $C(10, 3)(.25)^3(.75)^7 \doteq .250$.

(c) At least eight will buy the product?

Solution   The event "at least 8 will buy the product" is, as usual, rewritten in an equivalent statement using mutually exclusive events:

$$P(\text{at least 8 will buy}) = P(\text{exactly 8 will buy}) + P(\text{exactly 9 will buy})$$
$$+ \ P(\text{exactly 10 will buy}) = C(10, 8)(.25)^8(.75)^2 + C(10, 9)(.25)^9(.75)^1$$
$$+ \ C(10, 10)(.25)^{10}(.75)^0 \approx .00039 + .00003 + .00000 = .00042.$$

(d) At least one will buy the product?

Solution   The complement theorem will give the easiest translation:

$$P(\text{at least one will buy}) = 1 - P(\text{none will buy})$$
$$= 1 - (.75)^{10} \approx 0.944. ∎$$

Finally, we explore the connection between binomial experiments and expected value. Suppose that the probability a driver in a certain city will be wearing a seat belt is .70. Considering success as "wearing a seat belt" and failure as "not wearing a seat belt," and assuming that drivers are independent in their wearing of a belt, then the probability that r drivers among n will be wearing seat belts is found by the binomial formula $C(n, r)p^r q^{n-r}$. If 4 drivers in this city are checked, then the number of those wearing seat belts, 0, 1, 2, 3, or 4 may be considered to be values of a random variable X and a probability distribution for this random variable may be constructed as shown next. Since the conditions for a binomial experiment are met, the probability distribution is called a **binomial probability distribution,** or a **binomial distribution.**

Number wearing seat belts	Probability
0	$C(4, 0)(.70)^0(.30)^4 = .0081$
1	$C(4, 1)(.70)^1(.30)^3 = .0756$
2	$C(4, 2)(.70)^2(.30)^2 = .2646$
3	$C(4, 3)(.70)^3(.30)^1 = .4116$
4	$C(4, 4)(.70)^4(.30)^0 = .2401$
	Sum: 1

The expected value of this random variable may be found using the probability distribution table just constructed:

$$E(X) = 0(.0081) + 1(.0756) + 2(.2646) + 3(.4116) + 4(.2401) = 2.8.$$

This means that if the experiment of recording the number of drivers wearing seat belts among 4 drivers is repeated over and over for a long period of time, the average number of those wearing seat belts would be 2.8. But for binomial probability distributions, the expected value can be found much easier than the above calculation. Reason this way: If the probability of wearing a seat belt is .70 and 4 drivers are checked, then we would *expect* that $4(.70) = 2.8$ drivers would be wearing seat belts. If 800 drivers are checked, we would *expect* that $800(.70) = 56$ drivers would be wearing seat belts, and so on. These expectations can be mathematically proved in the case of binomial distributions:

$$E(X) = np$$

where n is the number of trials of the experiment and p is the probability of success.

Near the end of the exercises for this section, you will find probability problems of all types that we have studied in Chapters 6 and 7. The following guide to the various types and how their solutions are obtained is presented as a helpful review.

Guide to Probability

Outcomes that are equally likely: Write out a sample space; use counting techniques to determine the ratio of the number of outcomes in the event to those in the entire sample space.

Outcomes with unequal probability: Write out a sample space; Add the probabilities of the outcomes in an event to find the probability of that event.

Venn Diagrams: The unique wording alerts us to draw such a diagram.

The union of events: Union or "or" means to add the probabilities of each event, but don't double-count the probability of any outcome. If the events are mutually exclusive, just add their probabilities.

The intersection of events: Intersection or "and" means to multiply the probabilities of each event, but always take into account the events that already have occurred. If the events are independent, just multiply their probabilities.

Conditional probability: Reduce the sample space dictated by the "given" information, or use the conditional probability formula.

Bayes' Theorem: A special application of conditional probability. Use the conditional probability formula, and a tree diagram to aid in finding the probabilities involved.

Binomial probability: An application of independent events. Determine the events "success" and "failure," then multiply the number of sequences consisting of *S*'s and *F*'s by the probability of each sequence.

EXERCISES 7.5

The first 55 exercises involve binomial experiments. The remainder involve a random selection of all types of probability problems that we have studied.

A calculator will be helpful in obtaining a final numerical answer in exercises involving binomial experiments. Setting them up ready for calculation will be a good learning experience, even if you do not calculate to a number.

Tossing a Biased Coin   *Exercises 1 through 3 refer to this experiment: A coin found to be biased 55% for heads and 45% for tails is tossed six times. What is the probability of getting*

1. Exactly four heads?

2. Exactly two heads?

3. At least five heads?

 Success Rate   *Exercises 4 through 6 refer to these data: A baseball player is batting .335. During his next five times at bat, what is the probability that he will*

4. Get exactly three hits?

5. Get no hits?

6. Get at least one hit?

 Product Reliability   *Exercises 7 through 9 refer to these data: A popular brand of light bulbs has a 1% defect rate. If Dr. Meek buys a six-pack of such bulbs, what is the probability that*

7. Exactly one will be defective?

8. Four of the bulbs will be good and two will be defective?

9. All six will be good?

Politics   *Exercises 10 through 12 refer to these data: In a particular city, 38% of the voters consider themselves Republican. If ten voters are randomly called during an upcoming election campaign, what is the probability that*

10. Exactly six will be Republican?

11. Exactly eight will be of other political persuasions?

12. At least eight will be Republican?

Procedure Effectiveness   *Exercises 13 through 15 refer to these data: In a particular medical facility, cataract surgery is considered to be 98% successful. Of the next eight patients having such surgery, what percentage of the time*

13. Will exactly six be successful?

14. Will at least six be successful?

15. Will all eight be successful?

Product Effectiveness   *Exercises 16 through 18 refer to these data: The probability of multiple births occurring for women using a particular type of fertility drug is .20. Among 12 pregnant women who have taken this drug, what percentage of the time*

16. Will exactly ten have multiple births?

17. Will none have multiple births?

18. Will at least one have multiple births?

Elections *Exercises 19 through 21 refer to these data: A city bond issue is favored by 56% of the voters polled. Assuming this to be a representative sample, what is the probability that among five voters,*

19. Exactly four will favor the bond issue?

20. A majority of the five will favor the bond issue?

21. All five will favor the bond issue?

Family Planning *Exercises 22 through 24 refer to this experiment: What is the probability that in a family who wishes to have eight children,*

22. All will be girls?

23. Exactly four will be boys?

24. At least one will be a boy?

Rolling Dice *Exercises 25 through 28 refer to this experiment: A pair of dice is rolled four times. What is the probability that*

25. A sum of 7 is rolled every time?

26. At least one sum of 7 is rolled?

27. No sums of 7 are rolled?

28. No sum of 3 or 4 is rolled?

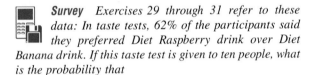

Survey *Exercises 29 through 31 refer to these data: In taste tests, 62% of the participants said they preferred Diet Raspberry drink over Diet Banana drink. If this taste test is given to ten people, what is the probability that*

29. All ten will prefer Diet Raspberry?

30. None will prefer Diet Raspberry?

31. At least one will prefer Diet Banana?

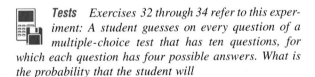

Tests *Exercises 32 through 34 refer to this experiment: A student guesses on every question of a multiple-choice test that has ten questions, for which each question has four possible answers. What is the probability that the student will*

32. Get exactly eight of the questions right?

33. Get at least eight of the questions right?

34. Get none of the questions right?

Quiz *Exercises 35 through 37 refer to this experiment: A true-false quiz in English has ten questions. If a student in the class guesses at the answer to every question, what is the probability that*

35. Exactly five are answered correctly?

36. Exactly two are answered correctly?

37. At least two are answered correctly?

Sampling *Exercises 38 through 40 refer to this experiment: Suppose that five cards are drawn in succession, with replacement, from a deck of 52 cards. What is the probability that*

38. Exactly four are clubs?

39. All five are clubs?

40. At least one is a club?

Product Reliability *Exercises 41 and 42 refer to these data: Z-Mart buys hand-held mirrors in packages of 100 from a supplier. On the average, they find one broken mirror in each package. If an employee randomly selects 12 mirrors from one of these packages, what is the probability that*

41. None will be broken?

42. At least one will be broken?

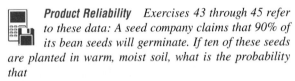

Product Reliability *Exercises 43 through 45 refer to these data: A seed company claims that 90% of its bean seeds will germinate. If ten of these seeds are planted in warm, moist soil, what is the probability that*

43. Exactly nine of them will germinate?

44. At least nine of them will germinate?

45. At least seven of them will germinate?

Safety *Exercises 46 through 48 refer to these data: Springfield has a seat-belt law on its books. A newspaper reporter did a survey of drivers in this city and found that 48% of the drivers actually wore seat belts. Based on this information, if ten motorists in this city were stopped by the police and checked for seat-belt violations, what is the probability that*

46. All ten will be wearing their seat belts?

47. Exactly eight of the ten will be wearing their seat belts?

48. None of the ten will be wearing their seat belts?

Delivery *Exercises 49 and 50 refer to these data: An overnight delivery service claims they deliver messages on time 92% of the time. A firm has a message that absolutely positively must be delivered the next morning. To make sure that the message will be delivered on time, the firm sends the same message in four separate envelopes through this delivery service. What is the probability that*

49. None of the four envelopes will be delivered on time the next moring?

50. At least one of the envelopes will be delivered on time the next morning?

51. *Success Rate* A salesperson has a success rate of 60% with the customers he calls on.

    (a) Make a binomial distribution table for the possible number of successful calls on five customers.

    (b) Calculate in two ways the expected number of successes for the five calls made.

    (c) What is the expected number of successes if the salesperson calls on 200 customers? On 800 customers?

52. *Product Reliability* A seed company claims that 90% of its tomato seeds will germinate.

    (a) Make a binomial distribution table for the possible number of germinations when four seeds are planted.

    (b) Calculate in two ways the expected number of germinations for the four seeds.

    (c) What is the expected number of germinations if 300 seeds are planted? If 1500 seeds are planted?

53. *Success Rate* A field-goal kicker is 95% accurate within the 30-yard range.

    (a) Make a binomial distribution table for the possible number of field goals made in three attempts within the 30-yard range.

    (b) Calculate in two ways the expected number of field goals made in three attempts within the 30-yard range.

    (c) What is the expected number of field goals made if 50 attempts are made? If 200 attempts are made?

54. *Success Rate* Medical records indicate that kidney transplants are 92% successful.

    (a) Make a binomial distribution table for the number of possible successful transplants among the next four patients.

    (b) Calculate in two ways the expected number of successful transplants in the next four patients.

    (c) Find the expected number of successful transplants in the next 300 and 600 patients.

Exercise 55 was given on an actuarial exam.

55. A pair of dice is tossed 10 times in succession. What is the probability of observing no 7s and no 11s in any of the 10 tosses?

    (a) $\left(\dfrac{28}{36}\right)^{10}$ **(b)** $\left(\dfrac{30}{36}\right)^{10}\left(\dfrac{34}{36}\right)^{10}$

    (c) $\left[1 - \left(\dfrac{6}{36}\right)\left(\dfrac{2}{36}\right)\right]^{10}$ **(d)** $1 - \left(\dfrac{8}{36}\right)^{10}$

    (e) $\left[1 - \left(\dfrac{6}{36}\right)^{10}\right]\left[1 - \left(\dfrac{2}{36}\right)^{10}\right]$

✷ *The remainder of the exercises randomly involve all of the types of probability we have studied.*

56. Find the probability that a five-card hand dealt without replacement will contain all spades.

57. A sample of three computers is selected, without replacement and without regard to order, from among 50, two of which are known to be defective. What is the probability that two of the computers are defective and one is not?

58. Two marbles are selected in succession and without replacement, from a box containing five red and three green marbles. What is the probability that the second marble is red, given that the first marble was green?

59. A student takes a history quiz and guesses on every question. The quiz has five questions, and each question has three possible answers. What is the probability that he will get at least three of the five answers correct?

60. One card is drawn from a deck of 52 cards. What is the probability that the card will be a queen or a club?

61. About 6% of the general public has diabetes. A particular test for this condition shows positive results 97% of the time when a person has this condition and also indicates a positive reading 2% of the time when a person does not have this condition. What is the probability that a person would actually have a diabetes condition if she tested positive?

62. If five cards are drawn in succession and without replacement, from a deck of 52 cards, what is the probability that all will be hearts or all will be spades?

63. Mortality tables show a probability of $\frac{1}{5}$ that John will live at least 20 more years and a probability of $\frac{4}{9}$ that Jane will live at least 20 more years. What is the probability that both will live at least 20 more years? That at least one of the two will be alive 20 years from now?

64. Two dice are rolled. What is the probability that one shows an even number and one an odd number?

65. Two cards are drawn in succession and without replacement, from a deck of 52 cards. What is the probability that one is an ace and one is a queen?

66. Two marbles are selected, without replacement, from a box containing three red and two green marbles. What is the probability that one is red and one is green?

67. Three boys and three girls are to be seated in a row. What is the probability that boys and girls will be in alternate seats?

68. Sixty percent of the toasters in a warehouse come from the Hot-Slice Co., and of those toasters, 3% are defective; 40% come from the Warm Morning Co., and of those, 5% are defective. What percentage of these toasters are defective?

69. If three cards are selected in succession, with replacement, from a deck of 52 cards, what is the probability that all are clubs?

70. A safety engineer claims that 15% of all automobile accidents are due to mechanical failure. Assuming this to be correct, what is the probability that of the next eight automobile accidents at least one will be due to mechanical failure?

71. If $P(A) = .6$, $P(B) = .5$, and $P(A \cup B) = .8$, what is $P(A \cap B)$?

72. A card hand selected from a deck of 52 cards contains the king of clubs, the queen of hearts, the eight of hearts, and the three of spades. Two more cards are drawn in succession, without replacement. What is the probability that at least one of these cards will be a diamond?

73. At State University, 35% of the freshmen failed mathematics, 20% failed English, and 10% failed both mathematics and English. What is the probability that a freshman failed exactly one of these two courses?

74. If five cards are dealt, without replacement, from a well-shuffled deck of 52 cards, what is the probability that exactly two will be hearts?

75. What is the probability that the last two digits of a four-digit telephone number are even?

Problems for Exploration

1. Consider a binomial experiment with n trials in which $P(S) = p$ and $P(F) = q = 1 - p$. If you calculate the probability of r successes, show that interchanging the concepts of success and failure will give the same probability.

2. Prove that $C(n, 0) + C(n, 1) + C(n, 2) + \ldots + C(n, n) = 2^n$.

Question 3 appeared on an actuarial exam.

3. In a large population of people, 50% are married, 20% are divorced, and 30% are single (never married). In a random sample of 4 people, what is the probability that exactly 3 are married?

(a) $\binom{4}{3}(.5)^3(.2)(.3)$ (b) $\binom{4}{3}(.5)^4$ (c) $(.5)^3$

(d) $(.5)^4$ (e) $\dfrac{\binom{4}{3}}{4!}$

CONCEPTS

Probabilities involving combinations of events can be determined by using the set operations of *union, intersection,* or *complement.* Pairs of events that *cannot occur simultaneously* are called *mutually exclusive. Conditional probability* refers to the likelihood of an event happening, *given that another event has occurred;* if the occurrence of this event has no effect on the probability of the other, the events are said to be *independent. Tree* and *Venn diagrams* both can be helpful visual tools for *comprehensively* and *logically* displaying pertinent probabilities for an experiment. The *addition theorem,* the *complementary theorem,* the *conditional probability formula, Bayes' theorem,* and the *general formula for binomial experiments* are all important in calculating particular types of probability outcomes.

COMPUTATION/CALCULATION

For Exercises 1 through 6, use the tree diagram to determine the requested probabilities.

1. $P(W \mid B)$

2. $P(C \text{ and } Y)$

3. $P(W)$

4. $P(A \mid W)$

5. $P(B \text{ or } Y)$

6. $P(C \mid Y)$

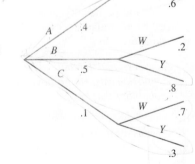

Events A and B have $P(A) = .5$ and $P(A \cup B) = .8$.

7. Find (B') if A and B are mutually exclusive.

8. Find $P(B)$ if $P(A \cap B) = .3$.

9. Find $P(B)$ if A and B are independent.

10. Find $P(A \cap B)$ if $P(B) = .7$.

11. Find $P(A|B)$ if $P(B) = .7$.

12. Find $P(B|A)$ if $P(A' \cup B') = .9$.

Events C and D are mutually exclusive, with $P(C) = \frac{3}{7}$ and $P(D) = \frac{4}{7}$. Another event E has $P(E|C) = \frac{2}{3}$ and $P(E|D) = \frac{3}{4}$.

13. Find $P(E)$.   **14.** Find $P(C|E)$.   **15.** Find $P(D|E)$.

CONNECTIONS

A tennis player has a 60% success rate on her first serves. When she does miss her first serve, 90% of her second serves succeed.

16. Find the probability that she double-faults (misses both serves).

17. If a serve is good, find the probability that it was a first serve.

A hazard forecaster predicts that for a particular ski slope in Utah, the probability of an avalanche triggering naturally during the next 10 hours is $\frac{1}{100}$, but if a skier is on it, the probability is $\frac{1}{40}$. If the slope releases, the probability that the road below it has traffic on it is $\frac{1}{25}$.

18. What is the probability of a traffic accident caused by a naturally triggered avalanche?

19. What is the probability of traffic being on the road below the slope, given that an avalanche was triggered naturally?

20. If a skier is on the slope, what is the probability that an avalanche occurs and that there is traffic on the road below the slope?

21. If a skier is on the slope, what is the probability of an avalanche occurring or of traffic being on the road below?

22. What is the probability of a traffic accident being caused by an avalanche if the odds that a skier is on the slope are 3 to 2?

 A professional football player has a 58% pass-completion rate. For his next ten passes, what is the probability that he

23. Completes exactly five passes?

24. Completes the first five passes and misses the next five passes?

25. Has more completions than incompletions or interceptions?

26. Has at least one completion?

COMMUNICATION

27. Explain what it means for two events to be mutually exclusive.

28. Describe in as many ways as you can how you can determine that two events E and F are independent.

29. Differentiate between $P(A)$ and $P(A|B)$, using Venn diagrams and verbal descriptions.

COOPERATIVE LEARNING

30. Is it possible to have events A and B that are both mutually exclusive and independent at the same time? Explain.

31. A tropical hotel in Honolulu has 31 floors. If 12 people are randomly assigned to rooms on any of the 31 floors of this hotel, find the probability that

   (a) Two or more of the 12 people are assigned to rooms on the same floor.

   (b) One or more of the other 11 people are assigned to the same floor as Ms. Tomacheck.

CRITICAL THINKING

32. Analyze the following situation, and decide who is correct. Then write a paragraph, perhaps using empirical or theoretical data, that supports your solution.

In the September 9, 1990 "Ask Marilyn" column of *Parade Magazine,* Marilyn vos Savant was asked this question: "Suppose you're on a game show, and you're given a choice of three doors. Behind one door is a car; behind the others, goats. You pick a door—say, No. 1—and the host, who knows what's behind the doors, opens another door—say, No. 3—which has a goat. He then says to you, 'Do you want to pick door No. 2?' Is it to your advantage to switch your choice?"

Marilyn said that you should switch, because the first door has a $\frac{1}{3}$ chance of winning but the second door has a $\frac{2}{3}$ chance. Numerous people wrote her, including many Ph.D. mathematicians, saying she was wrong. One wrote:

"You blew it, and you blew it big! I'll explain: After the host reveals a goat, you now have a one-in-two chance of being correct. Whether you change your answer or not, the odds are the same. There is enough mathematical illiteracy in this country, and we don't need the world's highest IQ propagating more. Shame!"

Who is right? Why?

COMPUTER/CALCULATOR

33. Use a computer or calculator program to simulate the roll of a pair of dice 360 times. Read the number of sums of 2, 3, 4, . . . , 12 obtained and compute the relative frequency of each of these outcomes. Compare these numbers to the theoretical probabilities of each sum.

Cumulative Review

34. Write an equation for the line that contains the points $(5, -2)$ and $(-1, -10)$.

35. Solve the given system of equations using any appropriate method:

$$3x - y + 2z = 19$$
$$-x + 3y + 10z = -1$$
$$2x + 4y + 5z = 7$$

36. Sketch and shade a Venn diagram, using three overlapping sets A, B, and C, that represents the region given by

$$(B' \cap C' \cap A) \cup (B \cap C \cap A').$$

37. Determine the matrix product AB given

$$A = \begin{bmatrix} 1 & 0 & 3 & 2 \\ 2 & -5 & 1 & 0 \end{bmatrix} \text{ and } B = \begin{bmatrix} 3 & -1 & 0 \\ -2 & 1 & 2 \\ 1 & 0 & -1 \\ 2 & -1 & 0 \end{bmatrix}.$$

38. Write the negation of the sentence "Some cats are not quick."

CHAPTER 7 SAMPLE TEST ITEMS

Three boys and four girls are randomly seated in a row. What is the probability that

1. A girl will be seated in the middle seat, and a boy will be seated in the right end seat?

2. A girl will be seated in the middle seat, or a boy will be seated in the right end seat?

3. Girls and boys are seated alternately, or a girl is seated in the right end seat?

4. A girl will not be seated in the middle seat, or the girls will not be seated all together?

The finite math students at State University are categorized as shown.

	Freshman	Sophomore	Junior	Senior
Male	64	37	11	8
Female	86	43	9	2

Find the probability that a finite math student at State University is

5. Female

6. A freshman

7. Female and a freshman

8. Female or a freshman

9. Female, given that the student is a freshman

10. Senior, given that the student is male

11. Not a male or not a freshman

12. Find the probability that a three-card hand dealt from a standard deck of 52 cards has exactly two queens.

One card is drawn from a standard deck:

13. Find the probability that the card is a 9, given that it is a 10.

14. Find the probability that the card is a 4 or a club.

The probability that an earthquake occurs is .002. The probability that you get an A in finite math is .2.

15. What is the probability that an earthquake occurs and you get an A in finite math?

16. What is the probability that an earthquake occurs or you get an A in finite math?

17. Three six-sided dice are rolled. Find the probability that at least one die shows six dots.

A survey of supermarket shoppers revealed that 47% bought bread, 22% bought crackers, and 61% bought bread or crackers. Find the probability that a person from the survey bought

18. Bread only

19. Bread, given that the person did not buy crackers

Consider families of three children. What is the probability that

20. Exactly two are girls?

21. Exactly two are girls, given that the third child is a girl?

A chef's school is 60% male and 40% female. Seventy percent of the males and 90% of the females like eating crab legs for dinner. What is the probability that a member of this chef's school

22. Is male or likes eating crab legs for dinner?

23. Is female, given that the member likes eating crab legs for dinner?

 A special variety of beans sprout 90% of the time. If 20 beans are planted, find the probability that

24. Exactly 18 sprout

25. At least 18 sprout

Career Uses of Mathematics

Having earned a bachelor's degree in business administration, I sort of backed into the insurance field. As an insurance agent, I have as my main function selling insurance policies and then following up on the policies to ensure that they accomplish what they are supposed to.

Mathematics is fundamental to the insurance industry, so much so that we have our own special mathematicians, who are called actuaries. Actuaries use the theory of probability in life insurance through the use of a mathematical model known as a mortality table. Mortality is defined as the probability that an individual will die within one year. The results of the actuarial mathematics are condensed into a premium ratebook for the sales force. Thus, an individual's life insurance premiums are based in part on a mathematical formula designed to calculate that individual's mortality. Insurance premiums are also affected by expenses and investments, both of which require ongoing mathematical refinements.

Ken W. Merrit
American General Life Insurance Company

8 *Statistics*

An automaker guarantees its particular type of automatic transmission for 60,000 miles. Tests have shown that such transmissions have an average life of 90,000 miles with standard deviation of 15,000 miles. If the lives of these transmissions are normally distributed, what percentage of cars will be returned to the company for transmission work while still under warranty? (See Example 6, Section 8.4.)

CHAPTER PREVIEW

Statistics is the science of collecting, organizing, and analyzing data. Many of the basic techniques used in such analyses are discussed in this chapter. These include grouping and displaying data by histograms, finding various kinds of averages, calculating the standard deviation, and using the "bell shaped" normal curve. Applications are many, ranging from the analysis of surveys taken to determine how many respondents own a certain article to the analysis of failure to satisfy a guarantee.

CALCULATOR/COMPUTER OPPORTUNITIES FOR THIS CHAPTER

On the MasterIt disk, Stat and Normal are among the support programs that reinforce the concepts of this chapter. Stat will display both the frequency and cumulative histograms, and it will calculate the median, and population and sample standard deviations. After checking your z-score entry, the program Normal will perform normal-curve calculations of the area to the left of, right of, or between given z-scores, In addition, it graphically displays the area found. Contents of the screen may be printed for a record.

Any calculator with keys labeled $\bar{x}$, σ_n, and σ_{n-1} will be of immense help in this chapter for calculating arithmetic means and standard deviations. A supplemental program for a graphing calculator will do normal curve calculations to the right or left of any z-score and show the shaded portion of the graph whose area has been found.

SECTION 8.1 ## Organizing Data: Frequency Distributions

Statistics is concerned with the collection and analysis of data and with making predictions from these data. Data are collected in many ways, often in the form of numerical quantities that measure specific characteristics. Each of these numerical quantities is called an **observation.** For example, data collected about a retail business might include observations of gross sales, net profit, inventory, earnings per share, and so forth. Data collected for medical studies might include observations of blood pressure, height, weight, and so on.

In statistical terminology, the set of things from which data are collected is called the **population.** Sometimes, we collect the entire population, in which case, we simply arrange and summarize these data as factual matter. You might think of the grade distribution of a recent test or the "stats" on a recent basketball game as examples of this type of statistics, known as **descriptive statistics.**

Another major activity of statistics is known as **inferential statistics.** It deals with the methods of collecting data from a subset of the population, called a **sample.** Based on the information from the sample, we draw conclusions about the entire population. Sampling is used when it is impossible, impractical, or too expensive to collect data on the entire population. Inferential statistics then determines how to select the samples and, with the use of probability, how accurately the sample portrays the entire population. A good example of inferential statistics is selecting a sample of voters to predict an election.

For whatever purpose data are collected, its usefulness may be enhanced by some specialized arrangements and organization. One way this may be done is by sorting the data into **classes** and then displaying the classes in a **frequency distribution table.** Sometimes, the data will fall naturally into classifications with little or no effort, such as the data in the following graph, Figure 1, and in Example 1.

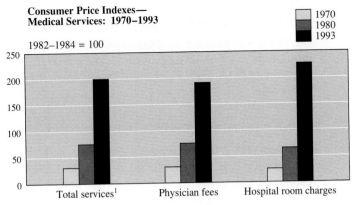

FIGURE 1
This Graph Organizes Data about Medical Services in the United States.

EXAMPLE 1
A Typical Survey

Thirty people in a shopping mall were asked how many telephones were in their home. The responses, in the order received, were as follows:

$$
\begin{array}{ccccc}
2 & 1 & 2 & 2 & 4 \\
3 & 2 & 1 & 2 & 1 \\
3 & 3 & 2 & 3 & 3 \\
1 & 1 & 0 & 2 & 2 \\
2 & 1 & 2 & 2 & 4 \\
4 & 3 & 4 & 2 & 2
\end{array}
$$

Because the subject of this survey is the number of telephones in each household, those numbers—ranging from 0 to 4—form the classes of interest. The **frequency distribution table** (see Table 1) for this data set is completed by tallying the number of 0s, 1s, 2s, 3s, and 4s, and then counting the total. Now we understand the data on phone ownership more clearly. We can get other statistical information on this survey by computing the **relative frequency** and the **cummulative frequency.**

Class: Number of telephones	Tally	Frequency
0	|	1
1	⊮⊮ |	6
2	⊮⊮ ⊮⊮ |||	13
3	⊮⊮ |	6
4						4
		Total: 30

TABLE 1
Frequency Distribution Table

Relative Frequency and Cumulative Frequency

The **relative frequency** of a class is found by dividing the number in that class by the total of all frequencies. This is also the expected probability that an individual fact (called a "datum") will fall within that class.

The **cumulative frequency** of a class is the sum of the frequency of that class and the frequencies of all the preceding classes.

Table 2 shows the relative frequency and the cumulative frequency for the telephone ownership data. Any or all of the information in Table 2 can be visually displayed in a **histogram** like those shown in Figure 2. ■■■

Number of telephones	Frequency	Relative frequency	Cumulative frequency
0	1	$\frac{1}{30}$	1
1	6	$\frac{6}{30}$	7
2	13	$\frac{13}{30}$	20
3	6	$\frac{6}{30}$	26
4	4	$\frac{4}{30}$	30
	Total: 30	Sum: $\frac{30}{30} = 1$	

TABLE 2

One obvious use of histograms is to show the nature of particular data in an immediately visual way (see Figures 2 and 3). Note that in constructing each of the histograms in Figure 2, a break was indicated in the horizontal axis so that the first bar does not overlap the vertical axis. Similar breaks can be used on either axis to show expansion or contraction of the axis for a better visual picture. Notice also that the frequency and the relative frequency histograms have the same shape. A moment's reflection will convince us that this will always be the case: The *frequency* histogram shows how frequently the collected data falls within a given class, while the *relative*

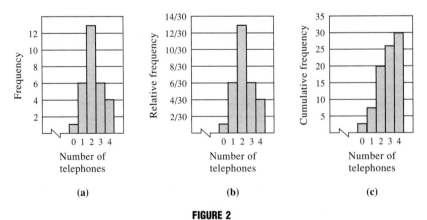

(a)   **(b)**   **(c)**

FIGURE 2
Histograms: (a) Frequency Histogram; (b) Relative Frequency Histogram;
(c) Cumulative Frequency Histogram

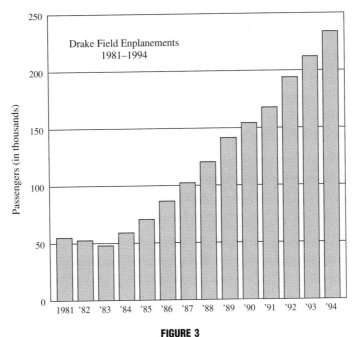

FIGURE 3

Some Data Might Be Lost in a Visual Display. Although It Shows the Airport Has Increased Its Service, the Graph Does Not Indicate the Actual Number of Flights per Year.

frequency approximates the probability that a given observation in the population will fall into that class. One item of interest later in this chapter is that if each bar is constructed to have a *width of one*, then the *area* of each bar is precisely the same as the estimated probability associated with that bar in a relative frequency histogram.

Data collected in laboratory experiments, the price of an item at various stores, and the lifespan of a flashlight battery are examples of data that usually do not fall into nice, neat classes, as did the survey on telephone ownership. In such cases, a slightly different method is helpful for grouping the data.

EXAMPLE 2
The Use of Class
Intervals

The person in charge of a college computer room observed a random sample of 25 students who came to the room and used the computers. She recorded the time (in hours) that each student spent on a computer at one sitting during a particular day. The results were as follows:

2.0	3.25	1.53	1.56	0.3
1.0	0.2	1.3	2.58	1.0
0.6	1.1	0.8	1.2	1.5
1.5	1.58	3.52	0.75	0.75
2.56	2.52	1.25	2.2	1.4

How do we analyze such an array of numbers? This time, we decide to construct **class intervals** of equal width and then count the frequency of data values that fall within each interval. The number of intervals to be used is somewhat arbitrary, but a rule of thumb is to use at least 5 and not more than 10 or 12 intervals unless an unusually large amount of data has been collected. Also, we try to select class intervals in such a way that every class contains at least one observation. To start the process, we find the smallest and largest numbers in the data, 0.2 and 3.52 hours in this case. We now decide on a starting point for the first interval (which need not be 0.2) and a width for each interval, using units that are easily understood by viewers. For instance, starting the first interval at 0.1 hours and making the interval width 17 minutes = 0.283 hours will not make an effective presentation. In this case, the easiest starting point is 0, and time intervals of 15 or 30 minutes would seem appropriate. If interval widths of 15 minutes = 0.25 hours are used, we would need at least x intervals where $0 + 0.25x = 3.52$, or $x = 14.08$. This means that 15 intervals would be needed to cover the entire data spread. If interval widths of 30 minutes = 0.5 hours are used, and starting again at 0, then at least x intervals where $0 + 0.5x = 3.52$, or $x = 7.04$, would be needed. Because we prefer not to have a fractional interval, 8 intervals are needed. Once the number of intervals has been decided, the last decision to be made before grouping these observations into the intervals is how to handle an observation that happens to precisely coincide with a number serving as the dividing point between two adjacent intervals. Because we would *not* list an observation in two different intervals, we must *arbitrarily* decide to include the dividing point in one interval and not the other. For two adjacent intervals, we choose to include the dividing point in the rightmost interval and exclude that point from the leftmost interval. (See Table 3.) The construction of the frequency and relative-frequency histograms for these data is shown in Figure 4 with the class-interval endpoints interpreted as in Table 3. ◼

Class (time : hours)	Tally	Frequency	Relative frequency	Cumulative frequency
0 up to .5	||	2	$\frac{2}{25}$	2
.5 up to 1.0	||||	4	$\frac{4}{25}$	6
1.0 up to 1.5	⫪⫪ ||	7	$\frac{7}{25}$	13
1.5 up to 2.0	⫪⫪	5	$\frac{5}{25}$	18
2.0 up to 2.5	||	2	$\frac{2}{25}$	20
2.5 up to 3.0	|||	3	$\frac{3}{25}$	23
3.0 up to 3.5	|	1	$\frac{1}{25}$	24
3.5 up to 4.0	|	1	$\frac{1}{25}$	25
	Total:	25	Sum: 1	

TABLE 3

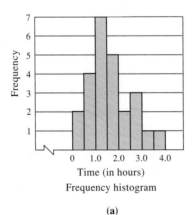

Frequency histogram

(a)

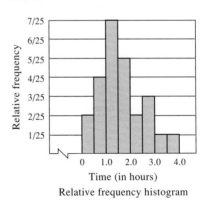

Relative frequency histogram

(b)

FIGURE 4

Histograms Showing Computer Usage

Sometimes class intervals are used to get a clearer picture of integer data values that are widely varied. Example 3 shows such a case.

EXAMPLE 3
Using Class Intervals
with Integer Data

Thirty contributions to a charitable cause were recorded as follows:

50	65	50	35	75	40
25	75	50	60	80	60
60	90	100	65	100	50
40	65	25	50	30	75
50	100	50	125	100	25

The smallest amount is $25 and the largest is $125. One way to arrange these data would be to start at $25 and use intervals of length $25 each. This means that we would need at least x intervals where $25 + 25x = 125$ or $x = 4$. However, because 25 and 125 are both included in the data and intervals are not to overlap, 5 intervals are actually needed. The data can then be tabulated as shown in Table 4. Notice that each interval contains its left endpoint, but *not* its right endpoint. The resulting frequency and cumulative-frequency histograms are shown in Figure 5. ■

Class (dollars)	Frequency	Relative frequency	Cumulative frequency
$25 up to 50	7	$\frac{7}{30}$	7
50 up to 75	13	$\frac{13}{30}$	20
75 up to 100	5	$\frac{5}{30}$	25
100 up to 125	4	$\frac{4}{30}$	29
125 up to 150	1	$\frac{1}{30}$	30
	Total : 30	Sum : 1	

TABLE 4

Can you look at Figure 5(a) and tell the size of most gifts? How are the size of the gifts distributed? Do you see how Figure 5(a), relates to Figure 5(b), as far as the growth of larger gifts is concened? These questions again point out one reason for histograms: To give an instant visual analysis of statistical data. Another very important reason for histograms in the statistical sense is to learn the shape of the probability

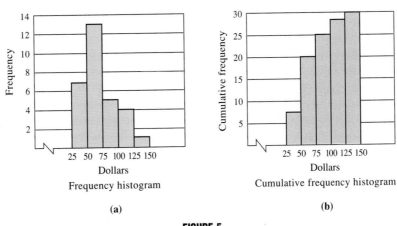

Frequency histogram
(a)

Cumulative frequency histogram
(b)

FIGURE 5
Histograms Showing Charitable Giving

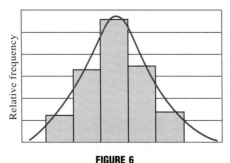

FIGURE 6
A Bell-Shaped Distribution

distribution of the data collected. As noted after Figure 3, if the bars have width one, then area can be equated to probability. (Otherwise, the area is proportional to the probability.) This suggests that we might be able to replace the histogram with a curve given by a mathematical formula and then interpret probability as an area between the curve and the *x*-axis. This is indeed the case for data whose histograms have some specific shapes and characteristics. The *bell-shaped curve* or **normal probability distribution** is undoubtedly the most famous of these special probability distributions, and it is examined in detail in Sections 8.4 and 8.5. Figure 6 shows collected data that are normal. Some statistical terminology for various distributions is shown in Figure 7.

The use of frequency tables and frequency histograms give us a better picture of the data, but we lose some of the identity of the original data. It is possible, however, to determine the shape of the probability distribution and still retain all of the data collected. Figure 8 shows one way of doing this, using the data and class intervals of Example 3. (This technique applies to noninteger data also.) The **histogram table** shown in Figure 8 shows the data to be skewed to the right, as did the original frequency and relative frequency histograms.

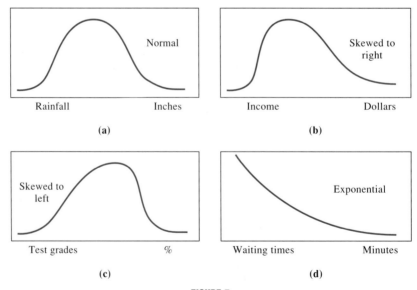

FIGURE 7
Some Types of Distributions

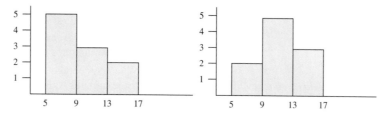

40	65			
40	65			
35	65			
30	60			
25	60			
25	60			
25	50	90		
	50	80	100	
	50	75	100	
	50	75	100	
	50	75	100	125
25–50	50–75	75–100	100–125	125–150

FIGURE 8
Histogram Table

ABUSES OF STATISTICS: Telling the Story the Way You *Want* It Told.   Two very different conclusions may often be obtained from the same data: Display the observations 6, 7, 9, 9, 9, 10, 11, 13, 15, and 16 in a histogram. The difference lies in whether the data at the right end of each class interval are or are not included in that interval. The first shows a dramatic decline while the second shows that the data are concentrated more in the 9–13 range and trailing off to the left and right. Which is right? They are both right, but a person with a vested interest in the conclusion one might draw from such a display might choose one over the other.

Similarly, exaggerating growth or decline trends may be accomplished by judicious scaling of the vertical axis. Suppose the heights of the bars decline, for instance. Then when an inch represents 20 units on the vertical axis, the ratios of the heights of adjacent bars will *appear* much smaller than if an inch represents 5 units. The former will give the appearance of a much slower decline than the latter. Sometimes, the scaling on the vertical axes is omitted altogether. Beware of these!

CAUTION   Do not accept conclusions based on graphical representation of data without asking questions!

EXERCISES   8.1

In Exercises 1 through 8, expand the given table to include the relative frequency and the cumulative frequency. Then draw frequency, relative-frequency, and cumulative-frequency histograms, and tell whether the data appear to be normally distributed, skewed to the left, skewed to the right, or none of these.

1.

Class	Frequency
0 up to 3	4
3 up to 6	8
6 up to 9	12
9 up to 12	7
12 up to 15	4

2.

Class	Frequency
0	3
1	5
2	9
3	12
4	8
5	2

3.

Class	Frequency
3	9
4	12
5	4
6	3
7	2

4.

Class	Frequency
3 up to 5	2
5 up to 7	5
7 up to 9	9
9 up to 11	4
11 up to 13	7
13 up to 15	4
15 up to 17	2

5.

Class	Frequency
150 up to 180	30
180 up to 210	50
210 up to 240	80
240 up to 270	60
270 up to 300	40

6.

Class	Frequency
200 up to 250	6
250 up to 300	20
300 up to 350	50
350 up to 400	45
400 up to 450	40
450 up to 500	25
500 up to 550	10

7.

Class	Frequency
0	2
1	3
2	8
3	5
4	2

8.

Class	Frequency
0	2
1	3
2	3
3	3
4	5
5	9

Exercises 9 through 13 refer to this frequency histogram, which shows points scored on a 12-point psychology quiz.

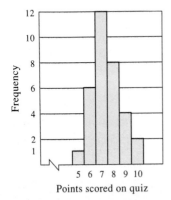

Points scored on quiz

9. What is the probability that a student scored 7 on the quiz?

10. What is the probability that a student scored 8 or more on the quiz?

11. What is the probability that a student scored less than 7 on the quiz?

12. Construct a cumulative-frequency histogram for these data.

13. What are the odds that a student will score 8 or more on the quiz?

Exercises 14 through 17 refer to this frequency histogram, which depicts the time (in minutes) people spent waiting in line at tellers' windows in a bank. Interpret the time intervals at the bottom as including the time at the left endpoint and going up to, but not including, the right endpoint.

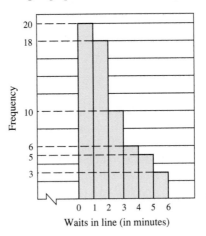

Waits in line (in minutes)

14. How many people spent three or more minutes waiting in line?

15. What is the probability that a person will spend at least two minutes waiting in line?

16. What percentage of those observed spent less than three minutes waiting in line?

17. What are the odds that a person will wait less than two minutes in line?

 Battery Length of Life *Exercises 18 through 20 refer to this experiment. Fifty flashlight batteries were tested for length of life, with these results (in hours) as shown here:*

13.5	14.6	13.2	15.0	14.5	12.4	12.0
12.6	12.8	11.1	11.6	11.5	10.6	10.2
13.0	13.5	13.2	13.8	13.6	14.0	12.1
14.2	14.5	15.2	11.6	12.8	12.5	14.2
13.9	13.5	12.6	12.3	12.9	13.6	12.2
13.5	13.7	13.3	15.4	14.6	14.8	13.4
15.4	14.8	14.6	13.0	13.5	15.8	15.6
14.4						

18. Construct the frequency, relative-frequency, and cumulative-frequency distribution tables by organizing the data into six classes of equal width.

19. Construct the frequency, relative-frequency and cumulative-frequency histograms. Do these data appear to be normally distributed or skewed?

20. About what percentage of the batteries have a life of 12 or more hours?

Plant Experiment *Exercises 21 through 26 refer to this experiment. A plant experimenter broke open a sample of 50 bean pods and counted the number of beans in each pod. The results obtained were as follows:*

```
5  4  1  5  2  0  4  5  2  4
2  3  4  4  4  6  4  3  5  4
0  2  3  2  3  5  4  4  4  6
1  3  2  4  4  3  1  5  3  5
3  3  3  1  3  4  2  3  4  3
```

21. Make a frequency and relative-frequency table for these data.

22. Construct a frequency and relative-frequency histogram for these data.

23. Arrange these data into a histogram table similar to that of Figure 8 of this section. In this arrangement, as well as in the frequency-distribution histogram, do these data appear to be normally distributed?

24. Use the relative-frequency histogram to compute the percentage of the pods that have four or more beans in them.

25. What is the probability that a pod will have two or fewer beans in it?

26. What are the odds that a pod will have three or more beans in it?

Household Survey *Exercises 27 through 32 refer to these data: A random sample of 28 households were questioned about the number of can openers they possessed. The responses were as follows:*

```
2  2  2  4  2  3  0
1  3  1  1  1  3  3
3  1  2  2  1  2  2
2  3  3  1  2  2  1
```

27. Make a frequency and relative-frequency table for these data.

28. Construct a frequency and a relative-frequency histogram for these data. Do you think these data are normally distributed or skewed?

29. Arrange these data in a histogram table similar to the one in Figure 8 of this section.

30. What percentage of homes have exactly two can openers, according to this survey?

31. What is the probability that a home will have at least two can openers?

32. What are the odds that a home will have three or more can openers?

Take-Home Pay *Exercises 33 through 35 refer to these data: The weekly take-home pay (in dollars) for 20 employees of Z-Mart were as follows:*

```
172.63   187.60   87.42    187.21   207.35
148.23   129.36   133.71   152.37   165.70
173.19   153.60   105.51   165.70   158.26
161.87   190.35   201.40   162.50   148.23
```

33. Organize these data into a frequency-distribution table that will adequately show the take-home pay for these employees.

34. Construct the frequency and relative-frequency histograms for these data.

35. Using the same class intervals as in Excercise 33, display the salaries in a histogram table similar to that in Figure 8 of this section.

Mail Delivery Times *Exercises 36 through 39 refer to these data: The time (in minutes) it took a particular mail carrier to make each of 30 deliveries was recorded as follows:*

```
1.3   5.1   7.2   4.2   7.0   7.1
2.6   1.4   2.5   3.8   2.4   3.7
3.1   3.1   3.6   6.5   4.2   4.2
1.6   6.2   6.9   4.5   6.3   3.1
2.0   1.1   5.1   2.6   5.0   4.6
```

36. Make a frequency and a relative-frequency table for these data.

37. Display your distribution in a frequency and a relative-frequency histogram.

38. Construct a cumulative-frequency histogram for these data.

39. Using the class intervals of Excecise 36, display these data in a histogram table similar to the one in Figure 8 of this section. Do these delivery times appear to be normally distributed or skewed?

Fleet Purchases *Exercises 40 through 42 refer to these data: The Speedy Cab Co. purchased 24 new cars and recorded the number of months that each car was driven before being replaced. The records were as follows:*

```
8   10   12    9   14    3
5   12   14   15    3    8
9   10   11    4   13   10
6    9   14   12    9    6
```

40. Make a frequency and a relative-frequency table for these data.

41. Using the class intervals of Exercise 40, make a histogram table similar to the one in Figure 8 of this section for these data. Do these data appear to be normally distributed? Skewed? Neither?

42. From the relative-frequency table of Exercise 40, approximate the probability that the cab company will keep a car ten months or longer.

Weekly Salaries   *Exercises 43 through 45 make use of these data: The weekly salaries of 20 persons in a particular department of a manufacturing plant are as follows:*

$300	$500	$475	$325	$225
$175	$275	$375	$425	$425
$305	$180	$400	$525	$385
$500	$292	$305	$390	$475

43. Construct a frequency-distribution table and accompanying frequency histogram to portray as accurately as you can the preceding data.

44. Make a frequency-distribution table and an accompanying frequency histogram to make the salaries appear as low as possible.

45. Make a frequency-distribution table and accompanying frequency-histogram to make the salaries appear as high as possible.

Problems for Exploration

Despite a lack of training in how and where to take a sample that would best represent the entire population, do your best to select your own random sample for the following exercises.

Select your own random sample of 15 students on your campus and collect and analyze data, using the ideas of this section that would give a picture of the information asked for in each of the Exercises 1 through 4.

1. The heights of women on your campus

2. The number of hours students sleep at night

3. How much money each student is carrying

4. How many books each student carries on campus

Select your own random sample of 35 students on your campus and do an analysis using the ideas of this section that would give a picture of the information asked for in each of the Exercises 5 through 8.

5. The heights of men on your campus

6. Grade point averages for students

7. How many hours a student listens to the radio each week

8. How much money students earn on summer jobs

9. Randomly select a sample of 20 freshmen and 20 seniors from your college. Ask how many hours they watched television last week, and analyze the information gathered.

SECTION 8.2   **Measures of Central Tendency**

We often hear statements such as "The **average** weekly salary is $254" or "The **median** price of a house in the United States is $98,750." Such statements use a single number that is considered to be the best representative of all observations under consideration. Such numbers, in some sense, measure the center of these observations; hence the term **measure of central tendency** is often applied to them.

Several specific measures of central tendency and the merits of each are discussed in this section. It should be noted that a particular measure may *not* be appropriate for a specific application and, at other times, it might be a matter of preference as to which measure is used. With this in mind, never accept one of these measures unless you know exactly what kind it is, and if you are giving a measure, use one that is defensible as being appropriate for the application.

The easiest measure of central tendency to compute is the **mode.** The mode is the observation that is recorded most often in the set of data.

The Mode

The **mode** of a set of observations is the observation that occurs most frequently.

EXAMPLE 1
Calculating the Mode

A set of observations may have one or more modes. In this set;

$$30, 70, 80, 30, 70, 30,$$

the mode is 30 because 30 is the most frequently recorded observation. However, for these data,

$$8, 6, 8, 10, 26, 17, 26,$$

there are two modes: namely, 8 and 26. These data are called **bimodal.** A set of observations may have no mode at all. The data set

$$6, 9, 11, 14, 81$$

has no mode because no observation occurs more frequently than the others. ▬

The mode is very easy to find and represents the data in its own special way. However, it is not very stable: In the first set of observations in Example 1, if one of the 30s is replaced by a 70, the mode suddenly becomes 70 rather than 30. Nonetheless, if you are interested in the **most common value** among the observations, then the mode is that value.

Another measure of central tendency is the **median.** The use of the median as a measure of central tendency is widespread in statistical analysis. (See Figure 9.)

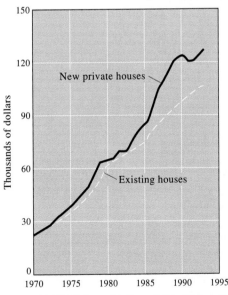

Median Sale Price of Single-Family Homes

SOURCE: U.S. Bureau of the Census, *Statistical Abstract of the United States*, 1994.

FIGURE 9
This Graph Uses the Median to Compare Sale Prices for Single-Family Homes over a 23-Year Period.

The Median

To find the **median** of a given set of data, first arrange the observations in order from the smallest to the largest (or largest to smallest). Then, if there is an odd number of observations, the median is the middle observation. If there is an even number of observations, the median is one-half the sum of the two middle observations. Thus for n observations, count from the smallest to the $\dfrac{n+1}{2}$ position to locate the median.

EXAMPLE 2
Finding the Median

(a) Find the median of these observations: 8, 3, 7, 12, 3.

Solution   The first step is to arrange all of these observations, including *both* 3s, from smallest to largest in numerical value:

$$3, 3, 7, 8, 12.$$

Because there are an odd number ($n = 5$) of observations, there is a middle observation, and that observation is 7. Therefore, the median is 7.

   Notice that if the observation 12 is replaced with 5280 in these data, the median would still be 7:

$$3, 3, 7, 8, 5280.$$

The median is not affected by extremely large or extremely small observations in collected data. This is one reason the median is a favored statistical average.

(b) Find the median of these observations: 0, 5, 0, 9, 6, 2.

Solution   Again, we begin by arranging these data from smallest to largest in numerical value:

$$0, 0, 2, 5, 6, 9.$$

This time there are an even number ($n = 6$) of observations. Hence, there is no "middle" observation, but rather two middle observations, 2 and 5. According to the definition, the median is

$$\frac{2+5}{2} = \frac{7}{2} = 3.5.$$

This example shows that the median need not be one of the given observations. ▬

Figure 10 shows data already arranged in order from the smallest to the largest and the median in each case. Once the median is found, there will *always* be an equal number of observations below and above it. This is the idea of "center" or "measure of central tendency" for the median. The median is *not* affected by *extreme* observations, but rather is determined by its *position*. The *median* is an *appropriate average* to use *if the data are spread over a large range.*

EXAMPLE 3
The Mode and Median
of Data

A Girl Scout cookie sale record had these entries in the "boxes sold" column:

$$2, 4, 4, 2, 8, 6, 12, 2, 2, 2, 6.$$

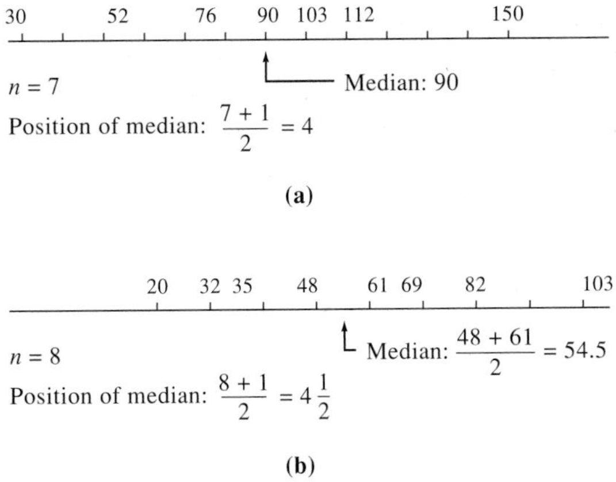

```
30      52      76    90  103 112           150
```

$n = 7$

Position of median: $\dfrac{7 + 1}{2} = 4$

Median: 90

(a)

```
        20    32 35    48    61 69    82        103
```

$n = 8$

Position of median: $\dfrac{8 + 1}{2} = 4\dfrac{1}{2}$

Median: $\dfrac{48 + 61}{2} = 54.5$

(b)

FIGURE 10
Calculating Medians

When arranged from lowest to highest, the number of boxes sold appears as

$$2, 2, 2, 2, 2, \mathbf{4}, 4, 6, 6, 8, 12.$$

The mode is 2 while the median is 4. Even though the most common order (the mode) is for two boxes, the median number of boxes sold is four. Some might view the median as the better single number to represent the "central measure" of the number of boxes sold per customer. ■

Perhaps the most common measure of central tendency is the **arithmetic mean** which is sometimes shortened to just the **mean.** This measure is the one most of us think of when we use the term **"average."**

The Arithmetic Mean

The **arithmetic mean** or **mean** of a set of n observations, $x_1, x_2, \ldots, x_n$ is the sum of all the observations divided by n: $\dfrac{x_1 + x_2 + \ldots + x_n}{n}$. If the observations are from a sample of a larger population, the mean is called the **sample mean** and is denoted by $\bar{x}$. If the observations are from the entire population, the mean is called the **population mean** and is denoted by μ (μ is the lowercase Greek letter mu).

Because the word "average" is so universally associated with the calculation of the mean, we agree that hereafter when the word average is used, it refers to the mean.

EXAMPLE 4
Comparing the Mean with the Median

A small firm makes hammers for the army. The firm's employees are the owner, a plant superintendent, and four workers. Their respective annual salaries are: $60,000, $40,000, $15,000, $15,000, $15,000, $15,000. The owner boasts that the average salary at the plant is over $26,000 per year. What kind of average is he using? The mean salary is

$$\mu = \frac{60{,}000 + 40{,}000 + 15{,}000 + 15{,}000 + 15{,}000 + 15{,}000}{6} = \$26{,}666.67,$$

so this must be the one he used. If you were considering a job with this firm, would this average give a realistic salary picture? Probably not. The median, $15,000, best represents the company's salaries and this just happens to also be the mode for this data. ▪

As shown in Example 4, the mean has the disadvantage of being affected by extreme values. Therefore it may not be too meaningful to represent the data by the mean when the data set has extreme values. To illustrate this point, consider the definition of average by a wise Fayetteville sage who claims that if you have one foot in a fire and the other in a bucket of ice water, then on the average, you are comfortable!

If you purchase three 2-liter bottles of a popular soft drink on sale for 99 cents per bottle and a week later purchase eight 2-liter bottles at the regular price of $1.79, you might be tempted to claim that the average (mean) price per bottle is $\dfrac{\$(.99 + 1.79)}{2} = \$1.39$. The correct average, however, is

$$\frac{\text{total dollars spent}}{\text{number of bottles bought}} = \frac{.99(3) + 1.79(8)}{3 + 8} \approx \$1.57.$$

Example 5 will show how to think of such an average in a "weighted" sense.

EXAMPLE 5
Comparing the Weighted Mean with Other Measures

Suppose a product is bought at three times, at prices and quantities as follows:

Purchase price per unit	Number of units purchased	Relative frequency
$2.00	500	$\frac{500}{800}$
$3.00	200	$\frac{200}{800}$
$5.00	100	$\frac{100}{800}$

Questions with very different answers come to mind about such purchases.

(a) What is the median price per unit?

Solution Think of arranging the observations from smallest to largest.

$$\underbrace{2, 2, 2, \ldots}_{500}, 2 \; , \underbrace{3, 3, 3, \ldots}_{200}, 3 \; , \underbrace{5, 5, 5, \ldots}_{100}, 5$$

There will be 500 2s, followed by 200 3s, followed by 100 5s. There are $n = 800$ observations, so the median will be located in position

$$\frac{800 + 1}{2} = 400.5$$

among these observations. This is clearly midway between two 2s, so the median price per unit is $2.00.

(b) What is the *average* (mean) *purchase price per unit?*

Solution This question is answered by finding the total dollars spent for the purchase and then dividing by the number of units purchased. The dollars spent is given by

$$\$2(500) + \$3(200) + \$5(100) = \$2100$$

while the total number of units purchased is

$$500 + 200 + 100 = 800.$$

Therefore, the mean price paid per unit is

$$\frac{\$2(500) + \$3(200) + \$5(100)}{800} = \frac{\$2100}{800} \approx \$2.63.$$

The left side of this calculation may be rewritten as

$$\$2\left[\frac{500}{800}\right] + \$3\left[\frac{200}{800}\right] + \$5\left[\frac{100}{800}\right] \approx \$2.63$$

to show each of the dollar amounts "weighted" with its proportional share (relative frequency, probability) of the total number of purchases. For this reason, such an average is sometimes given the name **weighted mean** or **weighted average.** We also recognize this computation as an **expected value** calculation. In this case the random variable values are the unit purchase prices and the expected value is the expected unit price or "average" unit price in the sense of the mean.

(c) What is the *average* (mean) *of the purchase prices?*

Solution   The answer is $\dfrac{2 + 3 + 5}{3} = \$3.33.$ ■

Example 5 serves as a guidepost to finding an average of grouped data. The basic lesson is this: *When in doubt, write it out.* By this we mean make a sketch of the data such as was done in part (a). First, this will give visual help in locating the middle or two middle numbers when finding the median. Secondly, this sketch shows how the mean, as calculated in part (b), may be viewed either as a traditional mean or as a weighted mean (expected value). Once these concepts are fully understood, such averages can be found by direct computations from the table displaying the groups.

The Weighted Mean

If $x_1, x_2, \ldots, x_n$ are n observations with respective relative frequencies p_1, $p_2, \ldots, p_n$, then the **weighted mean,** or expected value, is given by

$x_1 p_1 + x_2 p_2 + \ldots + x_n p_n.$

NOTE   A **mean** of n observations may be thought of as a **weighted mean** or expected value: Each observations is assigned the weight (relative frequency) $\dfrac{1}{n}$.

The weighted mean (hence the mean) may be interpreted geometrically as the balance point or the center of gravity of the system, using the observations as points on the number line and the relative frequencies as the respective weights at each point. Such a system is shown in Figure 11 for the data in Example 4. This visually shows

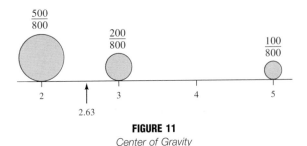

FIGURE 11
Center of Gravity

the measure of central tendency for the weighted mean, or expected value, to be the center of gravity.

If presented with a frequency histogram whose original observations were no longer available, is it possible to compute the mean of the data? The answer is "no," but we can calculate a reasonably good estimate by *assuming* that each observation lies at the midpoint of the interval in which it lies.

EXAMPLE 6
Estimating the Mean
of Grouped Data

Consider these test scores on a recent calculus test:

Scores	Frequency
91 to 99	3
82 to 90	7
73 to 81	12
64 to 72	5
55 to 63	2
Total:	29

Estimate the mean test score.

Solution   The accepted method to estimate the mean is to assume, for lack of more accurate information, that each test score lies at the midpoint of its respective class interval. Therefore, we consider the test scores to be

3 scores of **95**   (95 is midway between 91 and 100)

7 scores of **86**   (86 is midway between 82 and 90)

12 scores of **77**

5 scores of **68**

2 scores of **59**.

Because there are 29 test scores, the mean (actually the weighted mean) of the scores is

$$\frac{3(95) + 7(86) + 12(77) + 5(68) + 2(59)}{29} \approx 78.24.$$ ∎

> **Estimating the Mean of Grouped Data**
>
> If data are grouped within intervals having midpoints $m_1, m_2, \ldots, m_n$ with corresponding frequencies $f_1, f_2, \ldots, f_n$, then the **estimated mean of the data** is given by $\frac{m_1 f_1 + m_2 f_2 + \ldots m_n f_n}{f_1 + f_2 + \ldots + f_n}$.

The median for the test scores in Example 6 can also be estimated by considering, as in estimating the mean, each score to lie at the midpoint of its respective interval. In this case, the estimate of the median test score is 77.

ABUSES OF STATISTICS: What Does an Average Really Tell You?   A growth company has 30 million shares of stock outstanding. Two people own 20 million of these shares and the other 10 million shares are owned by another 50,000 people. The company states that their average stockholder owns 600 shares of company stock. This is correct (using the arithmetic mean as the method of averaging), but it portrays an inflated voting power among the stockholders. Excluding the two major stockholders, the average number of shares per stockholder is 200. The former average overstates the voting power of most stockholders by a factor of three!

If you were asked to find the average earnings of those in your father's college graduating class, you would probably do so by sending them a survey asking about such statistics. Then, based on those returned, an average could be computed. However, did you have the addresses of *all* classmates, or just those who were prominent, highly successful classmates? Who responded? Surely not everyone! Which method of averaging did you use? What does your average really mean?

CAUTION   Do not accept an average without asking questions!

EXERCISES   8.2

In Exercises 1 through 6, compute the following averages:
(a) The mode (if any)
(b) The median
(c) The mean

1. The population is 6, 9, 26, 18, 43.

2. The population is 17, 12, 29, 17, 20, 16, 20, 20.

3. The sample is 225, 102, 96, 173, 102, 112.

4. The sample is 101, 86, 137, 96, 120.

5. The poupulation is −5, −6, 12, 8, −8, 6, 0, −12, 3, 5, −3.

6. The sample is −22, 3, 7, 8, −15, 0, 2, −10, 5, 0.

In Exercises 7 through 14, decide which average—the mean, the mode, or the median—best describes the data.

7. Salaries of $20,000, $23,000, $21,000, $60,000

8. Salaries of $30,000, $60,000, $65,000, $63,000, $80,000

9. Number of boxes of Girl Scout cookies sold per scout: 23, 18, 25, 15, 80, 20, 19

10. Test scores of 76, 83, 92, 75, 83, 98, 65, 73, 82, 56, 43, 78, 80

11. Ladies' shoe sizes of 5, 7, 7, 6, 7, 8, 7, 7, 6, 7, 7

12. Grade point averages for six students of 3.51, 2.78, 3.01, 2.60, 2.95, 3.26

13. Ages of students: 18, 19, 18, 20, 22, 19, 30, 20, 42, 19, 20

14. Men's suit sizes of 40, 38, 40, 42, 44, 40, 40, 46, 40, 38

In Exercises 15 through 19, find the center of gravity (also interpreted as weighted mean or expected value).

15.

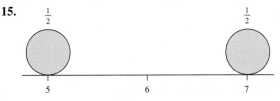

16.

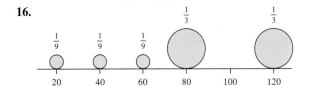

17.

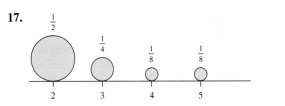

18.

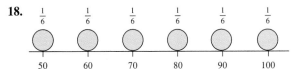

19.

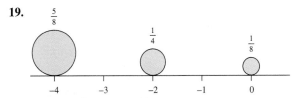

20. *Advertising*   An apartment owner has four two-bedroom units that rent for $500 per month and two efficiency units that rent for $200 per month. He advertises that his average apartment rents for $400 per month. Which average is being used? Is this truthful advertising? Does it present an accurate picture of the apartments to potential renters?

21. *Donations*   The director of a charitable cause reports that gifts received on a particular day were $20, $50, $30, $250, $50, $25, $500, $50, $25, $25, $30, $50, $25, $50, $25, and $25 and claims that the average gift exceeds $75. What average is being used? What average do you think best describes the giving for that day?

Sample *Exercises 22 through 24 refer to these data: A psychology department developed a math-anxiety test, which has 1000 points possible. They gave the test to a sample of ten students, with these results: 360, 240, 250, 300, 400, 990, 370, 210, 570, and 380.*

22. Compute the mean for the entire sample.

23. Compute the median for the entire sample. Which best describes the scores for the entire sample, the mean or the median?

24. Delete the two exceptionally high scores in the sample, and compute the mean and median of the remaining eight scores. Compare with the mean and median found using all ten test scores.

Survey *Exercises 25 and 26 refer to these data: A college journalism department has professors with the following number of years of teaching experience: 4, 12, 6, 8, 12, and 9.*

25. Find the mean and median for this population.

26. Suppose that the department hires a new professor with 23 years of teaching experience. Now find the mean and median for the years of teaching experience now possessed by the staff, and compare with the calculations of Exercise 25.

Sales *Exercises 27 through 30 refer to these data: A student organization bought 1000 replicas of their school mascot for $4.00 each to sell at an upcoming football game. They sold 700 for $6.00 each at the game and discounted the rest to a wholesaler for $3.50 each.*

27. What was the mean price for which they sold all 1000 of their mascots?

28. What was the median selling price for all 1000?

29. What was the mean of the two selling prices?

30. What was the mean profit per mascot?

Prices *Exercises 31 through 34 refer to these data: Two people bought soft drinks for a party. Maria bought 20 six-packs for $1.79 each and Bradley bought 15 six-packs for $1.59 each. They sold the drinks for 40 cents per can.*

31. What was the mean purchase price per six-pack?

32. What was the median purchase price per six-pack?

33. What was the mean of the two purchase prices?

34. What is the mean profit per six-pack?

Exercises 35 and 36 refer to these data: The ABC Fitness Company recently bought three treadmill models from a wholesaler. The prices and number of each model purchased are displayed in the table at the top of the next column.

Purchase price per unit	Number of units purchased
$150	50
$260	30
$400	10

35. Find the median purchase price per unit.

36. Find the mean purchase price per unit.

Sales *Exercises 37 through 39 refer to these data: A manufacturer's 1991 sales record for hand-held calculators follows:*

Model	Unit price	Number of units sold
KU-1	$ 15	2000
KU-2	$ 20	1500
KU-20	$ 50	1000
KU-102	$ 75	800
KU-150	$120	500

37. Find the mean price per unit at which the calculators were sold.

38. Find the median selling price per unit.

39. Find the mean of the unit prices.

Scores *Exercises 40 through 42 refer to these data: Scores on a 10 point quiz in a history course were as follows:*

Score	Frequency
10	5
9	7
8	12
7	9
6	2
5	2

40. Find the mean of the scores 5, 6, 7, 8, 9, and 10.

41. Find the mean quiz score.

42. Find the median quiz score.

Survey *Exercises 43 through 45 refer to these data: A survey of 20 households asked how many automobiles were owned by the household. The results were as follows:*

Number of automobiles	Frequency
0	1
1	3
2	12
3	2
4	2

43. Find the mean number of automobiles per household, as represented by this survey.

44. Find the mean of the numbers 0, 1, 2, 3, and 4.

45. Find the median number of automobiles per household.

Survey *Exercises 46 through 49 refer to these data: A seed company is testing a new variety of garden peas. They had a worker break open 28 pea pods and count the number of peas in each pod. The worker's record is as follows:*

Number of peas in pod	Frequency
1	3
2	4
3	5
4	8
5	6
6	2

46. Find the mean number of peas per pod.

47. Find the median number of peas per pod.

48. What is the mean of the numbers 1, 2, 3, 4, 5, and 6?

49. *Survey* Thirty students were asked how much money was in their bank accounts at the moment. Their responses were

Amount in account	Frequency
$ 0 to 50	5
$ 50 to 100	16
$100 to 200	5
$200 to 300	4

Estimate the mean amount in these accounts.

50. *Salary Ranges* A company has 20 people on its payroll in the following salary ranges:

Salary	Frequency
$15,000 to 20,000	2
$20,000 to 25,000	6
$25,000 to 30,000	8
$30,000 to 50,000	4

Estimate the mean salary for a person employed by the company.

51. Estimate the mean and median for the data displayed in this histogram.

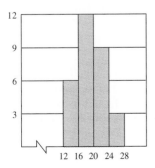

52. Estimate the mean and median for the data displayed in this histogram.

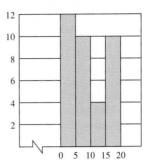

53. A group of senior citizens consisting of eight people in their 60s, five in their 70s, and eight in their 80s went on a sightseeing trip. Estimate the mean age of these people.

54. On a recent journalism test, five students scored in the 90s, eight in the 80s, fifteen in the 70s, four in the 60s, and three in the 50s. Estimate the mean test score.

55. Suppose that a student has one-hour test scores of 78, 92, and 84 and a final exam score of 85. The professor announces that the final will be weighted as two one-hour exams. What is the student's class average?

56. Latonya has a homework average of 92; one-hour test scores of 86, 90, and 89; and a final exam score of 86. The professor announces that the homework will be weighted as one-fourth of a one-hour test score and the final will be weighted as three one-hour test scores. What is Latonya's class average?

57. Suppose that you have one-hour test scores of 88, 79, and 91 thus far in a course with only the final exam left. The final is weighted as two one-hour exams, and you need a 90 average (mean) for an "A" in the course. What is the minimum score you can make on the final and still make an A?

58. Suppose that you have one-hour test scores of 92, 90, and 84, a homework average of 90%, and a quiz average of 86%. If the homework is weighted as one-third of a one-hour test, the quizzes as one-half on a one-hour test, and the final as two one-hour tests, what score would be needed on the final to have a 92% average (weighted mean)?

Problems for Exploration

1. Averages other than the ones we have discussed are possible. For the nonzero observations $x_1, x_2, \ldots x_n$, the **geometric mean** is $(x_1 \cdot x_2 \cdot \ldots \cdot x_n)^{1/n}$. Find the geometric mean of 4 and 9 and then of 5, 9, and 16. Compare with the (arithmetic) mean and median of each set of observations.

2. Suppose that you drive 100 miles on a dry interstate highway at an average speed of 50 miles per hour and return in a blinding snowstorm at an average speed of 20 miles per hour (mph). What was your average speed, in mph, for the round trip? (The answer is not 35 mph.) (The appropriate average here is called the **harmonic mean.**)

3. Ask 20 students how much time, to the nearest hour, they studied last week. Analyze these data by finding the arithmetic mean, the median, and the mode, and then constructing a histogram to give a visual picture of your data.

4. The display of data in a **box-and-whisker** plot is done like this: Arrange the data from the smallest observation s to largest observation ℓ along a line. Then find the median, m. Then find the median m_1 of the data from s to m and the median m_2 of the data from m to ℓ. Draw a box as indicated next, and "whisker out" to the extremes of the data. Such a display, like the histogram, visually shows how the data are shaped. The five numbers, s, m_1, m, m_2, and ℓ are called **Tukey's five-number summary.** For each of the following sets of data, find Tukey's five-number summary, and construct the box-and-whisker plot.

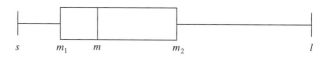

(a) 15, 3, 22, 18, 5, 12, 21, 18, 19, 9, 20, 10, 16, 21, 17

(b) The consecutive positive integers from 1 through 100

(c) The integers 1, 2, 3, followed by 30 ninety-seven times

SECTION 8.3 Measuring the Dispersion of Data

Another quantity that will help to analyze data is a number that will measure how widely spread out or dispersed the observations are, either as a set or from a central point such as the mean. One of the simplest ways to get such a number is to compute the **range** of the observations. The range is the difference between the largest observation and the smallest observation. Even though it is easy to calculate, the range is not a very good measurement of spread because it tells us nothing about the other observations. Another method uses the **deviation** (difference) between the mean and each of the observations in the list of data. This method leads to a measurement called the **standard deviation.** It has the property that the more concentrated the observations are around the mean, the smaller the standard deviation, while the further the observations are from the mean, the larger the standard deviation. Although not as easy to calculate as the range, the standard deviation is considered to be a better measurement, in the sense that it involves every observation in the set of data. Example 1 shows how the standard deviation is calculated.

EXAMPLE 1
How the Standard Deviation Is Calculated

A campus group recorded the following samples of 12 delivery times of pizza from each of two local pizza parlors (see Table 5). Even though the average (mean) delivery times are the same, which pizza parlor would you prefer to order from and why?

Solution Most of us would choose Tim's because of the more uniform and, therefore, more dependable delivery times. That is, delivery times from Tim's vary or *deviate* from the sample mean $\bar{x}$ very little while the delivery times from Larry's vary a great deal. Table 5 can now be expanded (see Table 6) to calculate how the delivery times in each case deviate from the respective means. At this point some kind of "average" of the deviations in each case is needed, to measure how the delivery times vary or deviate from the mean. Notice, however, that the sum of the deviations in each case is zero. This is no accident; the sum of the deviations from the mean of *any* set of observations is always zero. The reason this happens is that the mean is the balance point or the center of gravity of the data. This implies that the sum of the distances

Tim's pizza delivery time (in minutes)	Larry's pizza delivery time (in minutes)
35	36
38	50
31	20
34	44
30	18
32	36
36	65
37	30
32	20
33	25
34	34
36	30
Mean delivery time: $\bar{x} = 34$ minutes	Mean delivery time: $\bar{x} = 34$ minutes

TABLE 5

Deviations from the sample mean, Tim's	Deviations from the sample mean, Larry's
$x - \bar{x}$	$x - \bar{x}$
$35 - 34 = 1$	$36 - 34 = 2$
$38 - 34 = 4$	$50 - 34 = 16$
$31 - 34 = -3$	$20 - 34 = -14$
$34 - 34 = 0$	$44 - 34 = 10$
$30 - 34 = -4$	$18 - 34 = -16$
$32 - 34 = -2$	$36 - 34 = 2$
$36 - 34 = 2$	$65 - 34 = 31$
$37 - 34 = 3$	$30 - 34 = -4$
$32 - 34 = -2$	$20 - 34 = -14$
$33 - 34 = -1$	$25 - 34 = -9$
$34 - 34 = 0$	$34 - 34 = 0$
$36 - 34 = 2$	$30 - 34 = -4$

TABLE 6

(deviations) of the various observations on the left of the mean is precisely the same as the sum of the distances (deviations) to the various observations on the right of the mean.

Deviations Always Sum to Zero

For any set of observations, the sum of the deviations from the mean of those observations is *always* zero.

This fact means that the average of deviations from the mean will *always* be zero. So what do the statisticians do? They *square* each of the deviations and then compute an average. (The mathematical theory is easier to work with when squaring rather than of taking their absolute values.) Table 7 shows the result of squaring each deviation

Deviations squared, Tim's	Deviations squared, Larry's
$(x - \bar{x})^2$	$(x - \bar{x})^2$
1	4
16	256
9	196
0	100
16	256
4	4
4	961
9	16
4	196
1	81
0	0
4	16
Sum: 68	Sum: 2086

TABLE 7

from the mean for both firms. Now we have numbers whose average (mean) will have significance. However, there is yet another twist to the averaging process. If we have collected *every* observation in the entire population of n observations, then we know precisely the mean for the population, all possible deviations from the mean can be calculated, so the average of the squares of the deviations can be done as usual by adding, then dividing by n. This average is called the **population variance.** However, if we have collected only a *sample* from the entire population, such as Tim's and Larry's, the calculation of the mean of these numbers is likely to be only an approximation to the true mean of the entire population. Thus the deviations from this sample mean, hence the square of those deviations, may not be reflective of the entire population. As a consequence, averaging the squares of the deviations with n (the number in the sample) risks obtaining a smaller than a true variance, thus giving a false measurement as to how near the observations lie from the mean. To avoid this possibility, statisticians average the squares of the deviations with $n - 1$, *one less* than the number of observations in the sample, to obtain a more conservative measure of the true variance. This "average" is called the **sample variance.**

The Variance

For n observations, the **sample variance** is $s^2 = \dfrac{\Sigma (x - \bar{x})^2}{n - 1}$ and the

population variance is $\sigma^2 = \dfrac{\Sigma (x - \mu)^2}{n}$ where x is any observation, $\bar{x}$ is the

sample mean, and μ is the population mean.

For our present *sample* of pizza delivery times, the deviations squared will be averaged using $n - 1 = 12 - 1 = 11$ rather than 12. The sample variance for Tim's is

$$s^2 = \frac{68}{n - 1} = \frac{68}{11} \approx 6.18 \text{ (minutes)}^2,$$

and the sample variance for Larry's is

$$s^2 = \frac{2086}{n - 1} = \frac{2086}{11} \approx 189.64 \text{ (minutes)}^2.$$

Notice that each variance is in terms of (minutes)² units. We might wonder about the utility of such a calculation, but there is a solution to this dilemma: Take the square root to bring the units back in terms of the original units of time. The result of taking the square root of the variance is known as the **standard deviation.** The sample standard deviation for Tim's is then $\sqrt{s^2} = s = \sqrt{6.18} \approx 2.49$ minutes, and for Larry's is $\sqrt{s^2} = s = \sqrt{189.64} \approx 13.77$ minutes. Notice how these two numbers reflect the deviations of delivery times from the mean (small deviations for Tim's and large deviations for Larry's.) ▬

The Standard Deviation

The **standard deviation** is the square root of the variance. For n observations, the **sample standard deviation** is $s = \sqrt{\dfrac{\Sigma (x - \bar{x})^2}{n - 1}}$ and the **population standard deviation** is $\sigma = \sqrt{\dfrac{\Sigma (x - \mu)^2}{n}}$ where x is any observation, $\bar{x}$ is the sample mean, and μ is the population mean.

We want to emphasize that even though we are using different notations to distinguish between a sample mean and a population mean, there is only one way to find the mean of any set of n observations: Add the observations and divide by n. Contrast this with the averaging process when finding the variance.

EXAMPLE 2
Finding the Population
Mean and Standard
Deviation

The weights of ten people who are about to undergo a controlled diet program are as follows: 237, 304, 296, 345, 322, 315, 295, 317, 352, and 365 pounds. Compute the mean and standard deviation for these weights.

Solution   The population mean is the sum of the weights divided by ten:

$$\mu = \frac{3148}{10} = 314.8 \text{ pounds.}$$

Table 8 summarizes the calculations that lead to the variance and then to the standard deviation.

x	$x - \mu$	$(x - \mu)^2$
237	−77.8	6,052.84
304	−10.8	116.64
296	−18.8	353.44
345	30.2	912.04
322	7.2	51.84
315	.2	.04
295	−19.8	392.04
317	2.2	4.84
352	37.2	1,383.84
365	50.2	2,520.04
		Sum: 11,787.60

TABLE 8

Therefore, the variance is

$$\sigma^2 = \frac{11,787.60}{10} = 1178.76$$

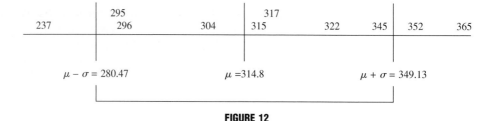

FIGURE 12

and the standard deviation is

$$\sigma = \sqrt{1178.76} \approx 34.33 \text{ lbs.}$$ ■

Knowing how many of the observations lie within one standard deviation of the mean gives another indication as to how the data points are spread from the mean. For the data in Example 2, the mean is $\mu = 314.8$ pounds, and the standard deviation is $\sigma = 34.33$ lbs. Thus, *one* standard deviation is 34.33 pounds. Weights that lie within one standard deviation of the mean are those weights that are at least as large as $\mu - \sigma = 314.8 - 34.33 = 280.47$ and not larger than $\mu + \sigma = 314.8 + 34.33 = 349.13$. As shown in Figure 12, seven of the weights lie within one standard deviation of the mean.

Even if nothing is known about the distribution of the observations, the famous Chebychev Theorem tells us something about the *probability* that an observation lies within k units of the mean.

Chebychev's Theorem

For any collection of numerical observations with mean μ and standard deviation σ, the probability that an observation x lies within k units of the mean μ is at least as large as $1 - \left(\dfrac{\sigma}{k}\right)^2$. In symbols, this theorem states that

$$P[(\mu - k) \le x \le (\mu + k)] \ge 1 - \left(\frac{\sigma}{k}\right)^2.$$

EXAMPLE 3
Using Chebychev's Theorem

Estimate the probability that a car battery will last between 57 and 63 months if it is known that such batteries have a mean life of 60 months and a standard deviation of 2 months.

Solution   According to Chebychev's theorem, we know that the probability the battery will last between 57 and 63 months is *at least* $1 - (\frac{2}{3})^2 = 1 - \frac{4}{9} = \frac{5}{9}$. ■

If observations have been grouped in a frequency distribution table and the observations are not available, we can estimate both the variance and standard deviation by assuming that each observation lies at the midpoint of their respective class intervals.

EXAMPLE 4
Estimating the Standard Deviation from a Frequency Distribution

A sample of 20 flashlight batteries was tested, and the lifetimes of the batteries were recorded as follows:

Life in hours	Frequency
8 to 10	4
10 to 12	6
12 to 14	8
14 to 16	2

Estimate the mean and standard deviation for these batteries.

Solution   We assume that the lifetime of each battery is located at the midpoint of the interval in which it lies. This implies that 4 of the batteries last for 9 hours, 6 of the batteries last for 11 hours, 8 of the batteries last for 13 hours, and 2 of the batteries last for 15 hours. Therefore, the sample mean is

$$\bar{x} = \frac{4(9) + 6(11) + 8(13) + 2(15)}{20} = \frac{236}{20} = 11.8 \text{ hours.}$$

The sample standard deviation is

$$s = \sqrt{\frac{(9 - 11.8)^2 4 + (11 - 11.8)^2 6 + (13 - 11.8)^2 8 + (15 - 11.8)^2 2}{19}}$$

$$\approx 1.88 \text{ hours.}   \blacksquare$$

Estimating the Standard Deviation from a Frequency Distribution

An estimate of the sample deviation is $s = \sqrt{\dfrac{\Sigma\,(x - \bar{x})^2 f}{n - 1}}$ and an estimate

of the population standard deviation is $\sigma = \sqrt{\dfrac{\Sigma\,(x - \mu)^2 f}{n}}$ where x

represents the midpoint of any class interval, f is the frequency of that class interval, n is the number of observations, $\bar{x}$ is the sample mean, and μ is the population mean.

Finally, we note that the variance and standard deviation of random variable values can be found using the guidelines discussed in this section with these differences: The mean is the expected value or weighted mean of the values; the squares of the deviations from the expected value are averaged as a weighted mean or expected value. Example 5 will show details.

EXAMPLE 5
Finding the Standard
Deviation of a Random
Variable

Suppose that you pay $10 to play this game. You draw 1 ball from a box containing 8 red and 3 green balls. If the ball is red, you win $5, whereas if the ball is green, you win $22. Consider your earnings to be values of a random variable X. Find your expected earnings and the standard deviation of these values from the expected value.

Solution   If the ball drawn is red, your earnings are $5 − $10 = −$5. That is, you actually lose $5. On the other hand, if the ball is green, your earnings are $22 − $10 = **$12**. Therefore, the numbers of interest are −$5 and $12, so they will become the values of our random variable.

Value of random variable:	−$5	$12
Probability distribution:	$\dfrac{8}{11}$ (red)	$\dfrac{3}{11}$ (green)

The expected value is then

$$E(X) = -5\left(\frac{8}{11}\right) + 12\left(\frac{3}{11}\right) = \frac{-40 + 36}{11} = -\frac{4}{11} \approx -36 \text{ cents.}$$

The following table sets out the steps required to find the standard deviation.

Value of random variable	Probability distribution	Deviation $x - E(X)$
-5	$\dfrac{8}{11}$	$-5 - (-.36) = -4.64$
12	$\dfrac{3}{11}$	$12 - (-.36) = 12.36$

Deviation^2 $(x - E(X))^2$	$(x - E(X))^2 P$
21.53	$(21.53)\left(\dfrac{8}{11}\right) \approx 15.66$
152.77	$(152.77)\left(\dfrac{3}{11}\right) \approx 41.67$
	Variance: 57.33

Therefore, the standard deviation is $\sigma(X) = \sqrt{57.33} \approx \$7.57$.   ■

ABUSES OF STATISTICS: Consider High and Low Data Points.   When thinking about accepting a job offer with a firm in Hartsfield, you might be impressed with the fact that the city claims that their unemployment rate over the past ten years has averaged only 4%. You might conclude that a very stable job climate exists there. However, they did not tell you that these rates ranged from a low 2% (for several of those earlier years) to a high of 14% (in recent years). Extremes do matter, and the timing of the extremes might also matter!

Suppose that you take a standardized college entrance exam and scored 146. The average score for students entering Tech University is 135, while students entering State University have an average of 140. Should you conclude that you have a better chance of getting into Tech University? No. They did not tell you that the standard deviation for such scores at Tech University is 12 and at State University is 2.5. Your score is exceptionally good when compared to other students at State University, but within just one standard deviation of the mean (average) when compared to other students at Tech University.

CAUTION   The standard deviation and extremes in the data do matter. Ask questions!

EXERCISES   8.3

In Exercises 1 through 6:
(a) Calculate the appropriate mean, $\bar{x}$ or μ
(b) Calculate the appropriate deviations, $x - \bar{x}$ or $x - \mu$ for each observation
(c) Sum the deviations
(d) Sum the deviations squared
(e) Find the appropriate variance (sample or population)
(f) Find the appropriate standard deviation (sample or population)

1. The sample: 2, 3, 4, 5, 6
2. The population: 2, 3, 4, 5, 6

3. The population: 12, 18, 20, 30, 32, 33, 80, 82, 84, 86
4. The sample: 12, 18, 20, 30, 32, 33, 80, 82, 84, 86
5. The sample: 50, 51, 52, 52.5, 53, 53.5, 54, 54.5, 55, 55.5
6. The population: 50, 51, 52, 52.5, 53, 53.5, 54, 54.5, 55, 55.5

In Exercises 7 through 10, decide, by inspection, which set of data has the largest standard deviation.

7. (a) 5, 7, 9, 10, 11, 12
   (b) 5, 7, 15, 18, 18, 20

8. (a) 50, 52, 48, 50, 51, 50, 53, 50, 49, 49
   (b) 80, 81, 80, 86, 80, 87, 80, 90, 79, 90

9. (a) 23, 25, 50, 25, 23, 25, 23, 25
   (b) 46, 46.5, 46, 46.5, 46, 45, 46, 46

10. (a) 101, 120, 102, 130, 102, 140, 102
   (b) 800, 801, 802, 803, 804, 800, 801, 802, 803, 804

11. Find the mean and standard deviation of these hourly salaries considered as the population:
   (a) 8.60, 12.30, 10.42, 9.60, 7.08
   (b) 4.30, 5.20, 9.60, 12.10, 16.80

12. Find the mean and standard deviation of these annual salaries considered as the population:
   (a) 20,000, 20,000, 25,000, 25,000
   (b) 20,000, 25,000, 12,500, 32,000

13. *Sample* The Bittle Insurance Company took a random sample of ten auto claims in Washington, D.C. The amounts were $1690.00, $823.68, $1261.18, $687.25, $972.40, $1391.52, $1560.20, $2580.12, $876.12, $920.63.
   (a) Find the mean and standard deviation of these claims.
   (b) How many of these claims are within one standard deviation of the mean?

14. *Manufacturing* A machine produces what is called a "3-inch" bolt. From a day's production run, five bolts are selected and, measured for length. The results, in inches, were 3.01, 2.97, 3.10, 3.11, and 2.98, respectively.
   (a) Find the mean and standard deviation of these lengths.
   (b) How many of these lengths are within one standard deviation of the mean?

15. *Scores* Eight graduate students took their written examination for their master's degree, with these composite scores: 88, 72, 91, 65, 73, 79, 82, 85, respectively.
   (a) Find the population mean and standard deviation of these scores.
   (b) How many of these scores lie within two standard deviations of the mean?

16. *Mileage* Seven students were awarded out-of-state scholarships from the History Department of State University. The distances from the hometowns of the students to the university were 250, 875, 1060, 780, 300, 1200, and 920, respectively.
   (a) Find the population mean and standard deviation of these mileages.
   (b) How many of these mileages lie within two standard deviations of the mean?

17. *Survey* The heart rates of eight people undertaking a physical fitness program were recorded as 72, 81, 83, 64, 72, 78, 85, and 75, respectively.
   (a) Find the population mean and the standard deviation of the heart rates.
   (b) How many of these observations lie within three standard deviations of the mean?

18. *Survey* The diastolic blood pressures of ten people about to undertake a supervised diet program are 136, 150, 129, 145, 152, 140, 160, 129, 136, and 142, respectively.
   (a) Find the population mean and standard deviation of these data.
   (b) How many of these observations lie within three standard deviations of the mean?

19. *Survey* A random sample of five January noontime temperatures in Albany was 42 degrees, 46 degrees, 61 degrees, 37 degrees, 40 degrees Fahrenheit.
   (a) Find the sample mean.
   (b) Find the variance. What units are expressed in the variance?
   (c) Find the standard deviation. What units are expressed in the standard deviation?

20. *Donations* A wealthy woman donated a bank account to each of four charities. The amount in each account, to the nearest dollar, was $28,625; $46,228; $37,258; and $63,520, respectively.
   (a) Find the population mean.
   (b) Find the variance. What units are expressed in the variance?
   (c) Find the standard deviation. What units are expressed in the standard deviation?

21. *Sample* The number of tourists visiting the Gold Nugget Saloon during the first five days in July was 236, 345, 362, 425, and 210, respectively.
   (a) Find the population mean and the median of this data.
   (b) Find the variance. What units are expressed in the variance?
   (c) Find the standard deviation. What units are expressed in the standard deviations?

22. *Sample* A radar check of the speed of a random sample of 15 cars during rush-hour traffic gave these readings: 48, 56, 50, 63, 58, 56, 60, 52, 65, 53, 60, 58, 58, 51, 57.
   (a) Find the sample mean and the median.
   (b) Find the variance. What units are expressed in the variance?
   (c) Find the standard deviation. What units are expressed in the standard deviation?

23. **Sample** A plant researcher, Jim Camp, counted the number of blooms on ten blueberry plants. The results were 86, 70, 65, 81, 72, 88, 80, 82, 81, 80.

   (a) Find the sample mean and median.
   (b) Find the variance. What units are expressed in the variance?
   (c) Find the standard deviation. What units are expressed in the standard deviation?

24. **Survey** A group dedicated to beautifying their state's highways counted the number of billboards per mile on 12 stretches of highway. Their data looked like this: 4, 8, 3, 4, 3, 5, 20, 21, 6, 3, 5, 5.

   (a) Find the sample mean and the median.
   (b) Find the variance. What units are expressed in the variance?
   (c) Find the standard deviation. What units are expressed in the standard deviation?

In Exercises 25 through 30, use Chebychev's theorem to estimate the probability of the outcome described.

25. **Life Expectancy** Estimate the probability that a particular brand of dishwasher will last between six and ten years if the average life of such dishwashers is eight years, with a standard deviation of ten months.

26. **Life Expectancy** Estimate the probability that a particular brand of hair dryer will last between three and four years if the average life of such dryers is 3.5 years, with a standard deviation of three months.

27. **Mileage** Estimate the probability that a tire will last less than 30,000 miles or more than 40,000 miles if the average life span of such tires is 35,000 miles, with a standard deviation of 2500 miles.

28. **Weights** Estimate the probability that a can of beans will weigh less than 15 ounces (oz) or more than 17 oz if such cans have an average weight of 16 oz, with a standard deviation of 1 oz.

29. Estimate the probability that an observation will always lie within two standard deviations of the mean.

30. Estimate the probability that an observation will always lie within three standard deviations of the mean.

31. **Scores** Midterm test scores for a class of 100 chemistry students were distributed as shown below:

Test scores	Frequency
50 to 60	5
60 to 70	15
70 to 80	50
80 to 90	26
90 to 100, inclusive	4

   (a) Estimate the population mean.
   (b) Estimate the population standard deviation.

32. **Life Expectancy** A sample of 20 light bulbs was tested for length of life. The results were as follows:

Length of life (hours)	Frequency
700 to 725	5
725 to 750	6
750 to 775	6
775 to 800	3

   (a) Estimate the sample mean.
   (b) Estimate the sample standard deviation.

33. **Mileage** Fifty cars of a particular make were checked for gasoline mileage, with these results:

Miles per gallon	Frequency
22 to 24	6
24 to 26	12
26 to 28	20
28 to 30	10
30 to 32	2

   (a) Estimate the sample mean.
   (b) Estimate the sample standard deviation.

34. **Survey** Twenty people have recently been admitted to a new class in the police academy in a large city. Their heights are recorded as shown in the table below:

Heights	Frequency
5′4″ to 5′6″	4
5′6″ to 5′8″	4
5′8″ to 5′10″	6
5′10″ to 6′0″	4
6′0″ to 6′2″	3

   (a) Estimate the population mean.
   (b) Estimate the population standard deviation.

35. **Game of Chance** A box contains three red and five green balls. One ball is randomly drawn from the box. If the ball is red you win $5; if the ball is green, you lose $2. Consider your earnings to be values of a random variable. Find your expected earnings and the standard deviation of those values from the expected value.

36. **Game of Chance** A box contains four red and two green marbles. Suppose you pay $10 to draw one marble from the box. If the marble is red, you win $7; if the marble is green, you win $12. Consider your earnings to be values of a random variable. Find your expected earnings and the standard deviation of those values from the expected value.

37. **Insurance** A company sells one-year term life insurance policies with a face value of $50,000 to 20-year-

Even though such curves are different for each μ and σ, they have many properties in common. Most of these properties are derived by methods of calculus. Even so, we can still make use of them without having a calculus background. An examination of the formula for any normal density function will reveal that its graph has these properties:

1. The highest point on the graph is attained when $x = \mu$, the mean. When $x = \mu$ the maximum y-coordinate is $\dfrac{1}{\sigma\sqrt{2\pi}}$.

2. The graph is always symmetrical about the line $x = \mu$, as shown in Figure 16(a). Furthermore, the curve never touches the x-axis, although it gets infinitesimally close on both ends.

In addition, calculus can be used to show that

3. The area under the curve (that is, below the curve and above the x-axis) is *always* 1, as shown in Figure 16(b).

The standard deviation σ controls the spread of the normal curve.

4. As shown in Figure 17, at the point $\mu - \sigma$, the curve stops being concave upward and becomes concave downward; at the point $\mu + \sigma$, the curve stops being concave downward and starts being concave upward.

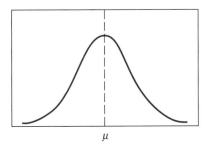

A normal curve is symmetric
about the line $x = \mu$.

(a)

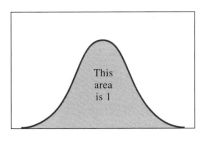

The area under any normal
curve is always 1.

(b)

FIGURE 16
Symmetry and Area Properties of a Normal Curve

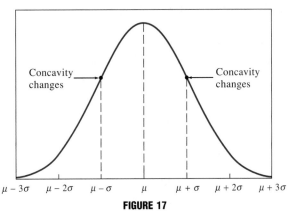

FIGURE 17
Concavity Properties of a Normal Curve

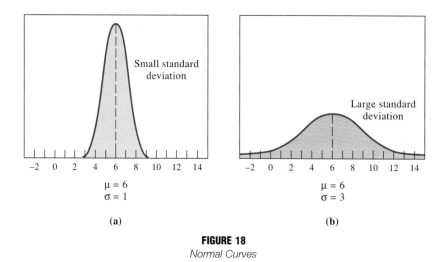

FIGURE 18
Normal Curves

It follows that small values of σ will result in tall normal curves while larger values of σ will result in flatter normal curves, as shown in Figure 18. This coincides with our knowledge of the standard deviation; small standard deviations imply that the data are clustered near the mean, whereas large standard deviations imply that the data are more spread out from the mean.

5. Regardless of the value of μ and σ, the curve for a normally distributed random variable X *always* has the properties of area coverage under that curve as shown in Figure 19. That is, approximately 68% of the area under the normal curve always lies within one standard deviation from the mean, approximately 95% of the area always lies within two standard deviations of the mean, and *nearly all*— approximately 99.7%—of the area lies within three standard deviations of the mean.

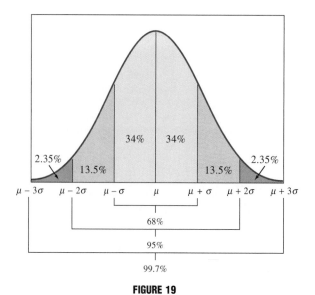

FIGURE 19
Approximate Areas under the Normal Curve

For example, if the I.Q. scores for a particular group of students were normally distributed with mean $\mu = 100$ and standard deviation $\sigma = 12$, then about 68% of the students will have I.Q. scores between 88 and 112 ($\mu - \sigma = 88$, $\mu + \sigma = 112$), about 34% of the students will have I.Q. scores between 100 and 112 ($\mu = 100$, $\mu + \sigma = 112$), and about $\frac{1}{2}(0.3\%) = .15\%$ of the students will have an I.Q. score higher than 136 ($\mu + 3\sigma = 136$).

EXAMPLE 1
Counting Standard
Deviations

A particular pizza chain requires that on a large pepperoni pizza, the weight of the pepperoni used is a normally distributed random variable with a mean of 12 oz and a standard deviation of 0.5 oz. A franchise can be revoked if a check reveals that the amount of pepperoni used is below the mean by more than three standard deviations. What is the minimum amount of pepperoni that can be put on such a pizza without violating franchise standards?

Solution   Each standard deviation is 0.5 oz, so that three standard deviations will be $3(0.5) = 1.5$ oz. Therefore, $12 - 1.5 = 10.5$ oz is the minimum amount of pepperoni that should be put on such pizzas.   ■

EXAMPLE 2
Counting Standard
Deviations

If the gasoline mileages on a particular make and model of automobile are normally distributed with a mean $\mu = 32$ mpg and a standard deviation $\sigma = 2$ mpg, what percentage of these cars get between 30 and 36 mpg?

Solution   Because *each* standard deviation is 2 mpg, it follows that 30 lies one standard deviation to the left of the mean $\mu = 32$, and 36 lies two standard deviations to the right of the mean. Using Figure 19, we find that the approximate percentage of area under the normal curve from one standard deviation to the left of the mean to two standard deviations to the right of the mean is 34% + 34% + 13.5% = 81.5%. This tells us that approximately 81.5% of such cars will attain gasoline mileage between 30 and 36 mpg. Stated differently, the approximate probability that such a car will get gasoline mileage between 30 and 36 mpg is .815.   ■

Example 2 and its connection to Figure 19 reveals the secret to finding probabilities (percentages) for normally distributed random variables: *Probability will be equated to area under the curve, and the area will depend solely upon the number of standard deviations of the value of the random variable from the mean.* Example 2 also points the way toward calculating the number of standard deviations of any observation x from the mean. As shown in Figure 20, observe that for $x = 30$, $30 = 32 + (-1)2$, which may be written symbolically as $x = \mu + (-1)\sigma$; for $x = 36$, $36 = 32 + (2)2$, which, again, may be expressed as $x = \mu + (2)\sigma$.

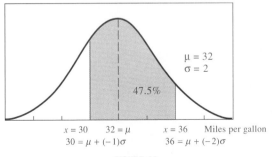

FIGURE 20
Gasoline Mileages

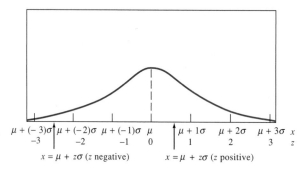

FIGURE 21
The Relationship between x and z

In general, as shown in Figure 21, for any x, $x = \mu + z\sigma$, where z is negative if x lies to the left of the mean, z is positive if x lies to the right of the mean, and $z = 0$ if x is the mean. For a given value x, how do we find the corresponding value of z? We write $x = \mu + z\sigma$ and solve for z:

$$z = \frac{x - \mu}{\sigma}.$$

This means that we can convert an area under any normal curve to an equal area under *one* normal curve, called the **standard normal curve,** as shown in Figure 22. This particular random variable is known by the distinguishing letter z, and the random variable values are called **z-scores.**

Transforming to z-scores

Let X be a normally distributed random variable with mean μ and standard deviation σ. Let x be one of its values. Then the **z-score** for x is given by

$$z = \frac{x - \mu}{\sigma}.$$

A table showing standard normal probability to four-decimal accuracy (percentage of area to the *left* of a given z-score) is provided in an appendix of this book. Some calculators have the built-in capability to find these areas, and graphics calculators may be programmed to do the same.

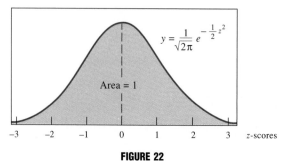

FIGURE 22
The Standard Normal Curve

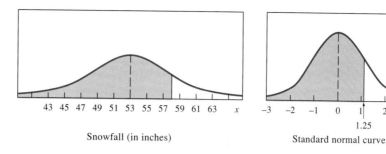

FIGURE 23

EXAMPLE 3
Using the Standard
Normal Curve

Snowfall in Plainfield is normally distributed with a mean of 53 inches and a standard deviation of 4 inches. What is the probability that snowfall in a given year will be less than 58 inches?

Solution For this example $x = 58$, $\mu = 53$, and $\sigma = 4$. Therefore $z = \frac{58-53}{4} = 1.25$. Thus, the shaded area in Figure 23(a) is the same as that of Figure 23(b). In symbols, $P(x < 58) = P(z < 1.25)$. The table for z-scores gives the shaded area as 0.8944 (to four-decimal accuracy), which is also the answer to our probability question. ∎

EXAMPLE 4
Using the Standard
Normal Curve

State University found that the GPAs for its freshmen students at the end of the fall semester were normally distributed with mean 2.74 and standard deviation 0.3. What percentage of these students had a GPA of 2.5 or more?

Solution For these data, $x = 2.5$, $\mu = 2.74$, and $\sigma = 0.3$. The z-score is

$$z = \frac{2.5 - 2.74}{.3} = -.80.$$

The equal shaded areas shown in Figure 24 give our answer. According to the table for z-scores, the area *to the left of* -0.80 is 0.2119, which means that the desired area *to the right of* -0.80 is $1 - .2119 = .7881$. It follows that 78.81% of these students have a GPA of 2.5 or more. ∎

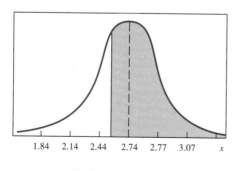
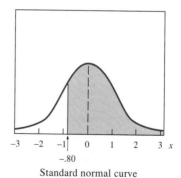

FIGURE 24

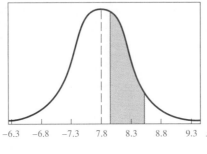
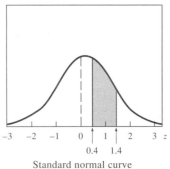

Weights of newborn infants   Standard normal curve

(a)   (b)

FIGURE 25

EXAMPLE 5
Using the Standard
Normal Curve

The weights of newborn infants in a particular hospital are normally distributed with a mean of 7.8 pounds and a standard deviation of 8 ounces. What is the probability that a baby born in this hospital will weigh between 8.0 and 8.5 pounds?

Solution   As shown in Figure 25, the desired area will be found by subtracting the area to the left of the *leftmost* z-score from the area to the left of the *rightmost* z-score. Noting that 8 ounces is 0.5 pounds, $\sigma = 0.5$ from which we find the leftmost z-score to be

$$\frac{8.0 - 7.8}{0.5} = 0.40$$

and, from the table of z-scores, the area to the left of 0.40 is 0.6554. The rightmost z-score is

$$\frac{8.5 - 7.8}{0.5} = 1.40,$$

and the area to its left is 0.9192. The desired area is then $0.9192 - .6554 = 0.2638.$

The question posed at the outset of this chapter is now answered in Example 6.

EXAMPLE 6
Using the Standard
Normal Curve

An automaker guarantees its particular type of automatic transmission for 60,000 miles. Tests have shown that such transmissions have an average life of 90,000 miles with a standard deviation of 15,000 miles, as shown in Figure 26. What percentage of cars having these transmissions will be returned to the company for transmission work while still under warranty?

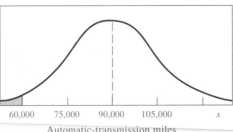
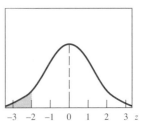

Automatic-transmission miles   Standard normal curve

(a)   (b)

FIGURE 26

Solution   The problem may be restated as asking for the probability that a transmission will last less than $x = 60{,}000$ miles. The z-score for $x = 60{,}000$ is

$$z = \frac{60{,}000 - 90{,}000}{15{,}000} = -2.$$

Thus, the percentage of the area to the left of $x = 60{,}000$ is the same as the area to the left of $z = -2$ under their respective normal curves. The area is 0.0227, so the percentage that will fail to meet the guarantee is 2.27%. ▬

Suppose that the maker of the automobiles in Example 6 wanted 1% or less of its transmissions to fail while under warranty. For how many miles should the transmissions be guaranteed? In symbols, we are asking for what value of a is $P(z \le a) = .01$. Once a is found, the final step will be to find the corresponding x value. If you use tables, you will need to find an area entry as close to .01 as possible and read the corresponding z value. In this case, the area to the left of $z = -2.32$ under the standard normal curve is 0.0102. Now use the formula $z = \dfrac{(x - \mu)}{\sigma}$ to find x: $-2.32 = \dfrac{x - 90{,}000}{15{,}000}$, or $x = 55{,}200$ miles. We therefore conclude that if the transmissions are guaranteed for 55,000 miles, about 1% of them will fail while under warranty.

ABUSES OF STATISTICS: How Normal Are Normal Distributions?   As the manager of a Z-Mart store, you purchase a large quantity of children's play watches from a firm that certifies these watches to have a mean life of one year with a standard deviation of one month. After consulting your z-score tables, you decide to guarantee these watches for nine months. A few months later, many irate mothers with children in tow come to your store and demand their money back because the watches have failed. Should you sue the firm you bought the watches from? No. You have *assumed* a normal distribution of the life of these watches when none exists because a few long-lived watches distorted the mean.

Does your professor grade on a "normal curve"? Should you question whether the class grades are normally distributed? Often, they are not.

CAUTION   Do not apply the normal distribution to data unless there is clear evidence that the data are in fact normally distributed. Ask questions!

EXERCISES   8.4

In Exercises 1 and 2, determine which normal probability curve has the larger standard deviation.

1.

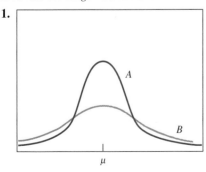

2.

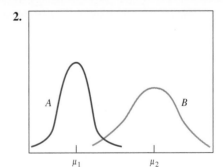

In Exercises 3 through 6, write out the normal probability density functions indicated, and sketch a graph of each on the same set of axes.

3. (a) $\mu = 0$, $\sigma = 1$
   (b) $\mu = 0$, $\sigma = 2$
   (c) $\mu = 0$, $\sigma = 3$

4. (a) $\mu = 6$, $\sigma = 2$
   (b) $\mu = 6$, $\sigma = 4$
   (c) $\mu = 6$, $\sigma = 6$

5. (a) $\mu = 4$, $\sigma = 1$
   (b) $\mu = 4$, $\sigma = 2$
   (c) $\mu = 4$, $\sigma = 3$

6. (a) $\mu = 8$, $\sigma = 0.5$
   (b) $\mu = 8$, $\sigma = 4$
   (c) $\mu = 8$, $\sigma = 7$

7. At which points on the x-axis does the normal curve with a mean of 20 and a standard deviation of 2 change concavity?

8. At which points on the x-axis does the normal curve with a mean of 60 and a standard deviation of 8 change concavity?

9. Suppose that X is a normally distributed random variable with a mean of 20 and a standard deviation of 1.5. Find the coordinate of the point on the x-axis that lies

   (a) 2 standard deviations above the mean
   (b) 3 standard deviations below the mean
   (c) 1.6 standard deviations below the mean

10. Suppose that X is a normally distributed random variable with a mean of 206.3 and a standard deviation of 16.8. Find the coordinate of the point on the x-axis that lies

    (a) 2 standard deviations below the mean
    (b) 1.6 standard deviations above the mean
    (c) 5 standard deviations above the mean

Exercises 11 through 15 refer to a normally distributed random variable X with a mean of 20 and a standard deviation of 3. Use Figure 19 to approximate the percentage of area under the curve.

11. Between $x = 20$ and $x = 23$

12. Between $x = 17$ and $x = 23$

13. To the left of $x = 20$

14. To the right of $x = 23$

15. To the left of $x = 11$

Exercises 16 through 19 refer to a normally distributed random variable X with a mean of 100 and a standard deviation of 6. Use Figure 19 to find the approximate percentage of area under the normal curve.

16. Between $x = 94$ and $x = 106$

17. To the left of $x = 88$

18. To the right of $x = 112$

19. To the left of $x = 106$

Exercises 20 through 24 refer to the normally distributed random variable Z with a mean of 0 and a standard deviation of 1. Use Figure 19 to find the approximate percentage of area under the standard normal curve.

20. Between $z = 0$ and $z = 1$

21. Between $z = -1$ and $z = 1$

22. To the left of $z = -1$

23. To the right of $z = -1$

24. To the right of $z = 2$

In the remainder of the exercises, use the appendix or appropriate calculator/computer technology to find the required area for a given z-score.

In Exercises 25 through 34, find the requested area under the standard normal curve.

25. The area to the left of $z = 2$.

26. The area to the left of $z = 1.6$.

27. The area to the left of $z = -2.3$.

28. The area to the right of $z = 2.07$.

29. The area to the right of $z = 0.03$.

30. The area between $z = 1.85$ and $z = 2.05$.

31. The area between $z = -2.1$ and $z = 0.67$.

32. The area between $z = -2.15$ and $z = -1.79$.

33. The area between $z = -6$ and $z = 0$.

34. The area between $z = -7$ and $z = 9.3$.

In Exercises 35 through 42, X is a continuous normally distributed random variable with a mean $\mu = 80$ and a standard deviation $\sigma = 5$.

(a) *Convert each probability problem involving X into a probability problem involving z.*
(b) *Sketch a graph of each problem.*
(c) *Find the indicated probability.*

35. $P(x > 90)$ 36. $P(70 < x < 85)$

37. $P(x < 62.8)$ 38. $P(68 < x < 72)$

39. $P(x < 93.67)$ 40. $P(83 < x < 93)$

41. $P(x > 76.5)$ 42. $P(80 < x < 88)$

43. **Weights** If boxes of cereal have weights that are normally distributed with a mean of 283 grams and a standard deviation of 4 grams, what should be the minimum amount in each box so the weight will not be below the mean by more than 2 standard deviations?

44. **Manufacturing** The length of what are considered "1-inch" bolts is found to be a normally distributed random variable with a mean of 1.001 inches and standard deviation of 0.002 inches. If a bolt measures more than 2 standard deviations from the mean, it is rejected as not meeting factory tolerances.

    (a) What is the minimum and maximum length a bolt must have to meet factory tolerances?
    (b) What percentage of bolts will be rejected?

45. Scores Paula scored 193 points on Professor Schein's final exam, for which the scores were normally distributed with a mean of 180 points and a standard deviation of 20 points. Meanwhile, Mary Jo scored 182 points on the same exam in Professor Feldman's section of the same class, for which the scores were normally distributed with a mean of 165 points and a standard deviation of 10 points. If both professors grade on a normal curve, who gets the higher grade, Paula or Mary Jo? (Those within one standard deviation of the mean get C's, then the B's and D's, then the A's and F's.)

46. Scores Gary scored 630 points on a particular section of a CPA examination for which the scores were normally distributed with a mean of 600 points and a standard deviation of 70 points. At a different time, Guy scored 530 points on the same section, for which the scores were normally distributed with a mean of 500 points and a standard deviation of 25 points. If both apply to the Pope Accounting Firm for a Job, who has the better chance of being hired on the basis of his excellence in relation to the norms?

47. Life Expectancy Tests of a particular brand of car battery reveal the length of life is a normally distributed random variable X with a mean of 52 months and a standard deviation of 3 months.

(a) What percentage of these batteries can be expected to last more than 55 months?

(b) What percentage of these batteries can be expected to last less than 46 months?

(c) If a store sells 200 of these batteries, how many can be expected to last less than 46 months?

(d) If these batteries are guaranteed for 46 months, what percentage of these batteries can the company expect to have returned for warranty adjustment?

48. Weights The weights of cans of corn were found to be a normally distributed random variable X with a mean of 14 oz and a standard deviation of 1 oz.

(a) What percentage of these cans weigh less than 13 oz?

(b) What percentage of these cans weigh more than 16 oz?

(c) What percentage of these cans weigh more than 20 oz?

(d) What percentage of these cans weigh between 12 oz and 16 oz?

(e) If a store has 680 cans of such corn, how many of these cans will weigh more than 16 oz?

49. Life Expectancy Light bulbs were tested for length of life; these lengths were found to be a normally dis-

tributed random variable X with a mean of 750 hours and a standard deviation of 25 hours. A sample of 10,000 such light bulbs is selected.

(a) How many of these light bulbs can be expected to last at least 775 hours?

(b) How many of these light bulbs can be expected to last less than 700 hours?

(c) How many of these light bulbs can be expected to last between 700 and 800 hours?

50. Sampling Production samples from a bolt factory indicate that a certain type of bolt produced has lengths that are a normally distributed random variable with a mean of 8 cm and a standard deviation of 0.06 cm. From a day's production of 2000 such bolts, how many would have lengths

(a) Greater than 8.1 cm?

(b) Less than 7.5 cm?

(c) Between 7.9 and 8.2 cm?

51. Sampling Pauline Pabst, a mathematics department secretary, has a large number of students come by her office each semester to complain about their professors. She decided to document the time spent listening to such complaints. After recording the times of several students, she found them to a normally distributed random variable with a mean of 4 minutes, 30 seconds, and a standard deviation of 1 minute, 45 seconds. Based on the analysis of her sample,

(a) What percentage of the students spend more than 5 minutes complaining?

(b) What is the probability that a student will spend less than 2 minutes complaining?

52. Lengths Katie measured the lengths of several carrots from her garden and decided, after analysis, that the lengths of her carrots were normally distributed with a mean length of 15.3 cm and a standard deviation of 1.2 cm. What percentage of her carrot crop will

(a) Have lengths between 15.5 and 17 cm?

(b) Have lengths more than 18 cm (to be eligible for the county fair)?

53. Schedules At the beginning of the semester, Marius clocks the time it takes to leave his apartment and arrive in the classroom for his first class. He found the times to be normally distributed with a mean of 18 minutes, 30 seconds, and a standard deviation of 2 minutes, 12 seconds.

(a) What is the probability that Marius will require more than 20 minutes to get to his first class?

(b) As the semester goes on, Marius has the habit of leaving at 7:15 A.M. for his first class, which is at 7:30 A.M. Given this, what percentage of time does Marius get to class on time?

54. *Mileage* A government agency checked the gasoline mileage of a particular make of automobile and found the mileages to be normally distributed, with a mean of 28.6 mpg and a standard deviation of 2.3 mpg. For one of these automobiles, what is the probability that the mileage will be

(a) At least 30 mpg?

(b) Between 28 and 32 mpg?

55. *Weighing* A government agency checked the weights of bags of peanuts, on which were stamped, "net weight 14 oz." The agency found that the weights on the bags checked were normally distributed, with a mean of 14.1 oz and a standard deviation of 0.2 oz. Based on this information, what is the probability that a bag of these peanuts

(a) Will weigh at least 14 oz?

(b) Will weigh between 13.8 and 14.5 oz?

56. *Billings* A particular department store has 50,000 credit card holders. During the month of January, the average billing to credit card holders (rounded to dollars) was $85, with a standard deviation of $12. Assuming that these billings are normally distributed,

(a) How many of these billings were for less than $75?

(b) How many of these billings were for more than $100?

57. *Account Balances* A particular bank presently has 100,000 checking accounts, which collectively have an average balance of $10,000, with a standard deviation of $2500. Assuming that these account balances are normally distributed,

(a) How many have balances of more than $12,000?

(b) What percentage of these accounts have balances of more than $5000?

58. *Survey* Brendan is a connoisseur of chocolate chip cookies. He decided to check the "chocolateness" of his favorite brand by buying ten packages and counting the number of chocolate chips in every cookie. After analyzing his count, he decided that the number of chocolate chips in each cookie was a normally distributed random variable, with a mean of 6.8 chips and a standard deviation of 1.7 chips. Based on these data, what is the probability that one of these cookies will

(a) Have at least eight chocolate chips in it?

(b) Have fewer than four chocolate chips in it?

59. *Survey* The ages of faculty members at State University are found to be approximately normally distributed, with a mean of 48 years, 6 months, and a standard deviation of 7 years, 2 months.

(a) The new president institutes an early retirement incentive program for all faculty over 65 years of age. What percentage of the faculty are eligible for this program?

(b) What is the probability that a faculty member is under 40 years of age?

60. *Scores* State University claims that its students' scores on the mathematics portion of the SAT are normally distributed, with a mean of 600 points and a standard deviation of 80 points. Melanie scores 700 on the test, and Brad scores 675.

(a) What is the probability that a student at State University has a score below that of Melanie's?

(b) What is the probability that a student at State University has a score below that of Brad's?

(c) Why are the probabilities in Exercise 60(a) and (b) not so radically different?

61. *Life Expectancy* A tire manufacturer claims that the life of its tires, calculated in miles, is a normally distributed random variable, with a mean of 25,000 miles and a standard deviation of 1600 miles.

(a) What percentage of these tires will last between 23,000 and 28,000 miles?

(b) If these tires are guaranteed for 22,500 miles, what percentage will be returned for warranty adjustments?

62. *Life Expectancy* The length of life of trouble-free Bright Screen TV sets is found to be a normally distributed random variable, with a mean of six years, four months, and a standard deviation of six months.

(a) What is the probability that one of these TV sets will give at least eight years of trouble-free service?

(b) If these sets are guaranteed to be trouble-free for five years and a store sells 150 of them, how many of the sets do they expect to need warranty work?

63. *Life Expectancy* A particular brand of dishwasher has a life expectancy that is estimated to be normally distributed, with a mean of ten years, eight months and a standard deviation of one year, two months.

(a) If you buy one of these dishwashers, what is the probability that it will last 12 or more years?

(b) Suppose that such dishwashers are guaranteed to last eight years. Of every 250 sold, how many will fail to last through the guarantee period?

64. *Life Expectancy* Z-Mart is considering buying a large number of automobile batteries from a manufacturer. They have an independent laboratory test for the life expectancies of these batteries. The laboratory found the lives of these batteries to be normally distributed, with a mean life of 42 months and a standard deviation of 2 months.

(a) What is the probability that one of these batteries will last more than 45 months?

(b) If Z-Mart agrees to distribute these batteries and guarantees them for 36 months, what percentage will be returned for warranty adjustment?

65. Guarantee If the life, in years, of a refrigerator is normally distributed with a mean of 15 years and a standard deviation of 3 years, what should be the guarantee period if the company wants less than 4% to fail while under warranty?

66. Guarantee If the life span of a tire, calculated in miles, is normally distributed with a mean of 30,000 miles and a standard deviation of 2000 miles, what should be the guarantee period if the company wants less than 2% to fail to last through the guarantee period?

67. Guarantee Residential air conditioners of a particular make have a life span, in years, that is normally distributed with a mean of 10 years and a standard deviation of 2.5 years. For what period of time should they be guaranteed if the company wants to be sure that less than 5% will fail while under warranty?

68. Guarantee A particular brand of roof shingles has a life span, in years, that is normally distributed with a mean of 20 years and a standard deviation of 2 years. For how long should the company guarantee such shingles if they want to be sure that less than 3% will fail while under warranty?

Problems for Exploration

1. Write the normal probability density function for a mean of 16 and a standard of deviation 2. Use your calculator to compute the y-value for these values of x: 10, 12, 13, 14, 15, 16, 17, 18, 19, 20, 22. Plot these points. What is the largest y-value on the graph of this function? Which x-values have the same y-values?

2. Explain why most z-score tables do not extend beyond $z = 3$.

3. Ask the waist size of 20 of your male acquaintances, then analyze the data collected (mean, median, standard deviation, histogram). Do you think these waist sizes are normally distributed? Explain.

4. Ask the shoe size of 20 of your female acquaintances, and analyze these data. Do you think that shoe sizes are normally distributed?

5. Collect two sets of data in your major field of study: One set that, upon analysis as far as we have learned, is normal, and one set that is definitely not normal. Show graphs such as histograms to be convincing.

6. If you were presented with a table that gave areas to the left of *positive* z-scores only, how would you use such a table to find the area under the standard normal curve between $z = -1.34$ and $z = -.97$? The area under the standard normal curve between $z = -1.34$ and $z = 0$?

SECTION 8.5

The Normal Approximation to the Binomial Distribution

Sometimes, a random variable has a distribution that is so close to being normal that its probabilities can be approximated by the normal curve. One important distribution for which this occurs is the binomial distribution. We have already seen in Section 7.5 that for a sequence of n Bernoulli trials, the number of successes that occur may be thought of as a discrete random variable that assumes one of the values 0, 1, 2, 3, . . . , n. We already know from our study of probability that the probability of exactly k successes from among n trials is $C(n, k)p^k q^{n-k}$ where p is the probability of a success and $q = 1 - p$ is the probability of a failure. The probability distribution of the number of successes is called a **binomial distribution.** Example 1 offers a reminder about binomial distributions.

EXAMPLE 1
A Binomial Distribution

A coin is tossed, and the occurrence of a head is considered a success. This experiment is repeated four times, and the number of heads is considered a random variable X. The binomial distribution for this sequence of Bernoulli trials is shown in the following table.

Value of random variable (number of heads)	Probability
0	$C(4, 0)(\frac{1}{2})^0 (\frac{1}{2})^4 = \frac{1}{16}$
1	$C(4, 1)(\frac{1}{2})^1 (\frac{1}{2})^3 = \frac{4}{16}$
2	$C(4, 2)(\frac{1}{2})^2 (\frac{1}{2})^2 = \frac{6}{16}$
3	$C(4, 3)(\frac{1}{2})^3 (\frac{1}{2})^1 = \frac{4}{16}$
4	$C(4, 4)(\frac{1}{2})^4 (\frac{1}{2})^0 = \frac{1}{16}$

The probability distribution histogram will look like this:

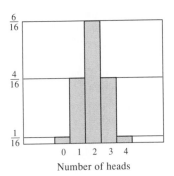

FIGURE 27
*Probability Distribution for Four
Tosses of a Coin*

The expected number of heads in four tosses is $E(X) = 0(\frac{1}{16}) + 1(\frac{4}{16}) + 2(\frac{6}{16}) + 3(\frac{4}{16}) + 4(\frac{1}{16}) = 2$. This same answer could have been obtained by multiplying the number of trials, four, by the probability of success, $\frac{1}{2}$, to obtain $E(X) = np = 4(\frac{1}{2}) = 2$ as we had discovered in Section 7.5. ▬

To calculate binomial probabilities when n is rather small is not too difficult. However, if n is large, binomial probability problems can be quite laborious to calculate. For instance, suppose that a coin is tossed 800 times and the number of heads counted. If we wanted to know the probability of getting at least 500 heads, then we would have to calculate 301 terms:

$$C(800, 500)\left(\frac{1}{2}\right)^{500}\left(\frac{1}{2}\right)^{300} + C(800, 501)\left(\frac{1}{2}\right)^{501}\left(\frac{1}{2}\right)^{299} + \ldots$$

$$+ C(800, 800)\left(\frac{1}{2}\right)^{800}\left(\frac{1}{2}\right)^{0}.$$

There must be a better way! There is. Two facts about binomial distributions allow us to approximate such binomial probability by use of a normal curve. First, the mean, or expected value, and the standard deviation for *any* binomial distribution may be found without the labor of other distributions. The formulas are shown next.

> **The Mean and Standard Deviation of a Binomial Distribution**
> If a binomial experiment consists of n Bernoulli trials, each with probability p for a success and probability $q = 1 - p$ for a failure, and the number of successes is considered a random variable, then the *mean* is $\mu = np$, the *variance* is $\sigma^2 = npq$, and the *standard deviation* is $\sigma = \sqrt{npq}$.

The second useful fact about binomial distributions has to do with the shape of the probability distribution. The probability histogram shown in Figure 27 appears to be bell-shaped. This is not an isolated incident. A theorem called the central limit theorem implies that for fixed p, the larger n becomes, the better the normal curve approximates the area in the histogram of a binomial distribution, as shown in Figure 28.

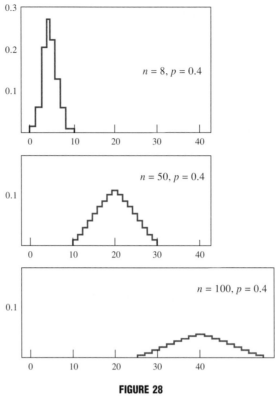

FIGURE 28
*Probability Distributions for a Binomial Experiment
with Various Values of n and p*

Approximating a Binomial Distribution

Consider any sequence of n Bernoulli trials, each with probability p for a success and probability q for a failure. The number of successes is considered to be a random variable. Then, if both $np > 5$ and $nq > 5$, we can be assured that the normal probability distribution with mean $\mu = np$ and standard deviation $\sigma = \sqrt{npq}$ will be a good approximation to the resulting binomial distribution. As n gets larger, the approximation gets better.

If either np or nq is 5 or less, the exactness of the approximation depends on the value of p. A convincing way to view the fact that a normal curve will approximate a binomial distribution under the foregoing conditions is to construct several such histograms. Some examples are shown next.

EXAMPLE 2
Approximating a
Binomial Distribution

A seed company claims that 80% of its tomato seeds will germinate. If 200 of these seeds are planted, what is the probability that at least 170 will germinate?

Solution   We interpret this experiment as a repeated Bernoulli trial, where *success* is taken to mean that the seed germinates. Then the probability of a success is .8, the probability of a failure is .2, and $n = 200$.

The normally distributed random variable X will approximate the resulting

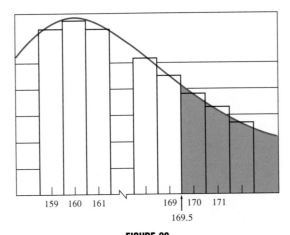

FIGURE 29

*Area Corresponding to "At Least
170 Seeds Will Germinate"*

distribution, with mean $\mu = np = 200(.8) = 160$ and standard deviation $\sigma = \sqrt{npq} = \sqrt{200(.8)(.2)} \approx 5.6569$. A normal curve may be used to approximate the probability distribution histogram for this experiment and, in particular, may be used to estimate the sum of areas in the 170th bar onward because this will answer the question asked. Because the normal curve is a probability distribution for a *continuous* random variable and we are using it to estimate probability for a *discrete* random variable, a "**continuity correction**" is needed, as shown in Figure 29. That is, with the continuous random variable value of $x = 169.5$, the area given by $P(x > 169.5)$ approximates the desired area. The z-score for $x = 169.5$ is

$$z = \frac{169.5 - 160}{5.6569} \approx 1.6794.$$

Therefore,

$$P(x > 169.5) = P(z > 1.68) = 1 - .9535 = .0465.$$

That is, if 200 of these seeds are planted, the probability that 170 or more will germinate is .0465. ▪

EXAMPLE 3

Approximating a
Binomial Distribution

A sack of 100 pennies is emptied on the table. What is the probability that 40 or fewer of them will turn up heads?

Solution   This problem may be thought of as tossing a penny 100 times and counting the number of times that heads turn up. Thus, heads is considered a success, with a probability of $\frac{1}{2}$, and tails is a failure, also with a probability of $\frac{1}{2}$. Because $n = 100$, the mean is

$$np = 100(\tfrac{1}{2}) = 50,$$

and the standard deviation is

$$\sqrt{npq} = \sqrt{100(\tfrac{1}{2})(\tfrac{1}{2})} = 5.$$

We think of a normal curve approximating the area in the 101 bars of the probability histogram for this problem and focus our attention around the bar that represents the probability of 40 heads. Because the question asks about 40 or fewer heads, the area

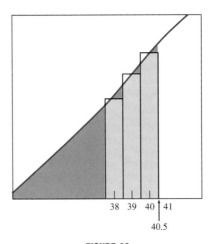

FIGURE 30
*Area Corresponding to "Forty or Fewer
Pennies Will Turn up Heads"*

under the normal curve from 40.5 leftward will approximate the desired area, as shown in Figure 30. The z-score for 40.5 is

$$z = \frac{40.5 - 50}{5} = -1.9,$$

so that for the continuous random variable X, $P(x < 40.5) = P(z < -1.9) = .0287$. Had the question asked for probability that *fewer* than 40 pennies turned up heads, the fortieth bar would *not* be included in the desired area, so that, for the continuous normal random variable X, $P(x < 39.5) = P(z < -2.1) = .0179$ is the sought-after approximation. ■

EXAMPLE 4
Approximating a
Binomial Distribution

A manufacturing process produces defective parts at the rate of 3%. In a shipment of 500 of these parts, what is the probability that fewer than 10 will be defective?

Solution   Because defective parts are of interest, consider the selection of a defective part a success this time. In so doing, $n = 500$, $p = .03$, and $q = .97$. The mean is $\mu = 500(.03) = 15$, and the standard deviation is $\sigma = \sqrt{500(.03)(.97)} \approx 3.8144$. Because the question asked about *fewer* than 10 defective parts, the area of the bar at 9, together with those leftward, will be approximated by the continuous normal random variable to the left of $x = 9.5$, as shown in Figure 31. The z-score is

$$z = \frac{9.5 - 15}{3.8144} \approx -1.4419,$$

so that $P(x < 9.5) = P(z < -1.44) = .0749$. ■

ABUSES OF STATISTICS: Data Might Depend on Events.   A plant manager looked down the list of inspectors and selected one with an accuracy rate of 98% and another with an accuracy rate of 97% to inspect gloves being made at the plant. Each pair of gloves was to be inspected by one inspector and then again by the other. The plant manager expected that a faulty pair of gloves getting by both inspectors has probability $(.02)(.03) = .0006$. The manager was dismayed to find that on some days, a much higher rate of faulty gloves made their way out of the plant. What went wrong? The

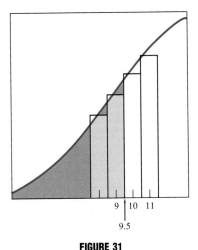

FIGURE 31
*Area Corresponding to "Fewer than
Ten Parts Will be Defective"*

manager assumed independence of inspectors when the calculation of .0006 was made, but the two became romantically linked, and independence went out the window!

CAUTION   Calculations based on independence of events will not be valid when some degree of dependence exists. Ask questions!

EXERCISES   8.5

The table for z-scores in an appendix or appropriate technology will be needed for these exercises.

Germination Rates   *Exercises 1 through 5 refer to this binomial experiment. Suppose that lettuce seeds have a 60% germination rate and that four of these seeds are planted. Consider the germination of a seed to be a success, and let X be the random variable denoting the number of seeds that germinate.*

1. Make a binomial probability distribution for the random variable *X*.

2. Construct a histogram for this distribution.

3. Compute the mean and standard deviation for this distribution, each in two different ways.

4. Use the histogram to find the probability of three or fewer seeds germinating.

5. Use the normal curve to estimate the probability of three or fewer seeds germinating. Explain the difference between this answer and the one obtained in Exercise 4.

 Defective Products   *Exercises 6 through 10 refer to this binomial experiment: A manufacturer of computer disks claims that only 4% of its disks are shipped with a defect. Suppose that you buy five of these disks. Let X be a random variable denoting the number of defective disks.*

6. Make a binomial probability distribution for this random variable *X*.

7. Construct a probability histogram for this distribution.

8. Compute the mean and the standard deviation for this distribution, each in two ways.

9. Use the histogram to compute the probability that three or fewer of the disks will be defective.

10. Use the normal curve to approximate the probability that three or fewer of the disks will be defective. Explain the difference between this answer and the one found in Exercise 9.

Performance Evaluation *Exercises 11 through 15 refer to this experiment: A basketball player is a 90% successful free-throw shooter. The player is to shoot four shots in a local sports promotion. Let X be the random variable denoting the number of shots made.*

11. Make a binomial probability distribution for the random variable X.

12. Construct a probability histogram for this random variable.

13. Compute the mean and the standard deviation for this random variable, each in two different ways.

14. Use the histogram to find the probability that three or fewer of the four shots will be made.

15. Use the normal curve to estimate the probability of making three or fewer shots. Explain the difference between this result and the one in Exercise 14.

Effectiveness of Procedure *Exercises 16 through 20 refer to this experiment: A particular medical procedure is 95% effective. This procedure is administered to six people, and the number of successes is considered to be a random variable.*

16. Make a binomial probability distribution for this random variable.

17. Construct a probability histogram for this random variable.

18. Compute the mean and the standard deviation for this distribution, each in two ways.

19. Use the histogram to find the probability of four or more successes, using this procedure.

20. Use the normal curve to estimate the probability of four or more successes, using this procedure. Explain the difference between this answer and the one found in Exercise 19.

For the remaining exercises, use the normal curve to approximate the desired probability.

21. **Survey** The probability that a male graduate student will be married is .25. From a random sample of 100 male graduate students, what is the probability that

    (a) Thirty or more will be married?
    (b) Fewer than 20 will be married?

22. **Tossing a Biased Coin** A coin is biased in such a way that the probability of heads showing on a single toss is .8. If the coin is tossed 300 times, what is the probability that

    (a) There are at least 250 heads?
    (b) There will be between 230 and 250 heads, not to include 230 or 250?

23. **Tests** A 120-question true-false test is taken, using guesswork as the only rationale for answering the questions. What is the probability that the person taking this test will have

    (a) More than 70 questions answered correctly?
    (b) Between 50 and 60 questions answered correctly, not to include 50 or 60?

24. **Survey** The probability of an adverse reaction to a flu shot is .023. If the shot is given to 500 people, what is the probability that

    (a) Between 15 and 20 will have an adverse reaction, not to include 15 or 20?
    (b) Fewer than 8 will have an adverse reaction?

25. **Survey** A survey by Washington Regional Medical Center concluded that 75% of its patients pay at least a portion of their bill with health insurance. Of the next 150 patients admitted to this hospital, what is the probability

    (a) That exactly 80% of them will pay their bill by at least partly using health insurance?
    (b) That exactly 110 of them will pay their bill using health insurance for at least part of the bill?

26. **Characteristics** A particular characteristic is present in wheat plants about 80% of the time. An experimenter has a large plot of wheat plants. If 300 of these plants are randomly selected, what is the probability that

    (a) At least 250 will have this characteristic?
    (b) Between 230 and 260 will have this characteristic, not to include 230 or 260?

27. **Manufacturing** A bolt factory estimates that about 6% of its bolts are defective in some way. Of the next 600 bolts produced, what is the probability that

    (a) Fewer than 30 will have a defect?
    (b) More than 25% of these will have a defect?

28. **Tossing a Thumbtack** If the probability that a thumbtack will land with its point up is .65, what is the probability that if the tack is dropped 1000 times,

    (a) It will land with the point up at least 635 times?
    (b) It will land with the point up between 640 and 670 times (*not* to include 640 or 670)?

29. **Scores** A college football field-goal kicker has a record of making 90% of his kicks within the 30-yard line. Of the next 20 kicks within that range, what is the probability that he will make

    (a) Fewer than 15 of them?
    (b) Exactly 17 of them?

30. **Rolling a Die** A die is rolled 800 times. What is the probability

    (a) That 130 or more 5s will come up?
    (b) That less than 140 5s will come up?

31. **Survey** A psychological study on personality disorders found that a particular disorder occurred in about 10% of the subjects studied. Based on this study, if 500 people are selected at random, what is the probability that

    (a) Fewer than 45 of them will have this disorder?

    (b) More than 52 of them will have this disorder?

32. **Survey** A journalism major was assigned the project of checking what percentage of all articles that appear in newspapers are about violent crimes. After checking several newspapers, she concluded that about 22% of the articles in these papers concerned violent crime. From a sample of 400 newspaper stories, what is the probability that

    (a) At least 90 of these articles will be about violent crime?

    (b) Fewer than 75 of these articles will be about violent crime?

Problems for Exploration

1. Explain in your own words, using graphs and perhaps an example, why the normal curve generally does not do a good job of approximating the binomial distribution for small values of n.

2. Roll a single die 100 times, and record the outcomes. Now repeat this experiment four times. From these data, compute the relative frequency of getting a five between 15 and 19 times (inclusive). Next, treat this experiment as a binomial experiment, and compute the probability that a five will turn up between 30 and 35 times (inclusive). Finally, use the normal curve to approximate this probability. Compare and discuss your results.

CHAPTER 8 REFLECTIONS

CONCEPTS

The realm of *statistics* concerns the collection, analysis, and presentation of data, usually in numerical form. The set of things from which the data are collected is called the *population*. When it is impractical or impossible to use the entire population, subsets called *samples* are collected to draw conclusions about the entire population. Sorting, arranging, or organizing data can be accomplished by using *class intervals, frequency-distribution tables,* or *histograms*. Some distributions have shapes that are normal or *skewed to the right or the left*. *Averages,* which are single numbers representing the entire set, include the *mode,* the *median,* the *arithmetic mean,* and the *weighted mean* or the *expected value*. The *standard deviation* indicates the variability of the data around the mean.

    Random variables assign numbers to each outcome of an experiment, and their related probability distributions. Normal random variables are analyzed using the *standard normal curve*. Probabilities for *binomial distributions* (Bernoulli trials) can be approximated using the normal curve.

COMPUTATION/CALCULATION

A sample of five students on campus were asked how much money (to the nearest dollar) they were carrying. The results were $2, $28, $13, $2, and $57. For this distribution, find the

1. Mode
2. Median
3. Arithmetic mean
4. Standard deviation

For the standard normal curve, find the area

5. To the left of $z = 1.3$
6. To the right of $z = -1.9$
7. Between $z = -2.4$ and $z = 0.6$

CONNECTIONS

A sample of 20 calculator batteries were tested for length of life. The results were as shown in the the table.

8. Estimate the mean.
9. Estimate the standard deviation.

Length of life (months)	Frequency
6 to 9	2
9 to 12	6
12 to 15	8
15 to 18	3
18 to 21	1

The weights of bags of Kay's Potato Chips were found to be a normally distributed random variable with a mean of 6 oz and a standard deviation of $\frac{1}{2}$ oz.

10. What percentage of these bags weigh less than $5\frac{1}{2}$ oz?
11. What is the probability that a randomly selected bag weighs between $5\frac{3}{4}$ oz and $6\frac{1}{4}$ oz?
12. If a store has 500 bags of Kay's chips in stock, about how many of these will weigh more than 7 oz?

13. A group of four women and five men from Possum Grape go to a math teachers' convention in Memphis. If a delegation of three is selected to attend the banquet, what is the expected number of women in the delegation?

A new drug cures 60% of the patients to whom it is administered. The drug is given to 150 patients.

14. What is the expected number of patients who will be cured?

15. What is the standard deviation of this distribution?

16. Use the standard normal curve to approximate the probability that at least 35 but fewer than 45 of these patients are cured.

COMMUNICATION

17. Explain what the standard deviation is and why it is useful.

18. What does it mean to say, "The expected number of women on a Senate committee is 0.63"?

19. Define *random variable* in your own words.

COOPERATIVE LEARNING

20. A collection of observations has a range from 10 to 125, with a median of 80.
  (a) Make a sketch of a frequency distribution that could fit this collection of data.
  (b) Construct two data sets, one with $n = 8$ and one with $n = 13$, that fit the given properties.

CRITICAL THINKING

21. Give reasons for and against both the *arithmetic mean* and the *median* as being the "best" average. Support your positions with examples.

COMPUTER/CALCULATOR

22. Use the computer program Stat to display a histogram using 5, 6, 7, 8, 9, and 10 groupings of these data: 3.7, 4.5, 6.7, 8.2, 4.6, 4.5, 5.8, 6.2, 8.1, 6.8, 6.8, 7.3, 4.7, 3.6, 7.7, 6.8, 7.1, 7.1, 6.9, 6.5, 6.4, 5.5, 3.1, 4.5 From each of these groupings of data, use a calculator to approximate the mean and the standard deviation. Compare with the actual mean and standard deviation given by the computer program.

Cumulative Review

23. Write the equation of a vertical line that contains the point $(-5, 7)$.

24. Solve this matrix equation:

$$\begin{bmatrix} 1 & 5 & -1 \\ 1 & 0 & 2 \\ 1 & 1 & 1 \end{bmatrix} \begin{bmatrix} x \\ y \\ z \end{bmatrix} = \begin{bmatrix} 4 \\ 1 \\ 0 \end{bmatrix}.$$

25. Graph the feasible region determined by the given constraints.

$$\begin{aligned} x + y &\leq 10 \\ -2x + y &\leq 1 \\ 7x + 3y &\geq 21 \\ -x + 2y &\geq 0 \end{aligned}$$

26. Find all the corner points of the feasible region determined by the given constraints.

27. In how many ways can four couples sit in a row of eight seats at a movie theater, if the men and women must alternate but no couple sits together?

CHAPTER **8** SAMPLE TEST ITEMS

A survey of the firms in the Industrial Park revealed the given information about their number of secretaries.

Number of secretaries	Frequency	Relative frequency
1	2	
3	3	
5	5	

1. Complete the relative frequency (probability) column.

2. What is the average (arithmetic mean) number of secretaries per firm?

3. What is the median number of secretaries per firm?

4. If a secretary from these first firms is selected at random, what is the probability that he or she works for a firm with three or fewer secretaries?

A random sample of four students revealed the following number of books they were carrying: 1, 8, 2, 1. For these data, calculate the following statistics for Questions 5 through 8.

5. The median

6. The mode

7. The arithmetic mean

8. The standard deviation

9. The daily salaries of the Goldberg Corporation employees are 90, 80, 110, 90, 100, 97, and 70 (in dollars). These salaries have a mean of 91 and a standard deviation of 13.2. How many of these salaries lie within one standard deviation of the mean?

10. Complete the probability distribution table for this experiment. Three cards are drawn (without replacement) from a standard deck of 52 cards, and the number of hearts is counted.

X	p
0	
1	
2	
3	

11. Make a probability distributioin for counting the number of girls in a family of five children.

12. Summarize these given data using appropriate statistics:

    76, 82, 61, 54, 28, 98, 97, 78, 70, 82

13. A business manager is faced with a decision to expand her business or leave things as they are. By expanding, she feels that there is a .6 probability of making an additional $25,000 profit, while there is a .4 probability of losing $12,000, through market saturation. What is the expected profit (or loss)?

14. Suppose that the lifetimes of a particular type of light bulb are normally distributed with $\mu = 1200$ hours, and $\sigma = 160$ hours. Find the probability that a light bulb will burn out in less than 1100 hours.

15. Assume that SAT verbal scores for a freshman class at a university are normally distributed with a mean of 520 and a standard deviation of 75. The top 10% of the students are placed in the honors program for English. What is the lowest score for admittance into the honors program?

A washing machine manufacturer finds that 2% of her washing machines break down within the first year. For Questions 16 through 18, use the normal curve to approximate the probability that out of a lot of 2500 washers,

16. At least 50, but fewer than 60, washers break down within one year.

17. More than 62 washers break down within one year.

18. Forty or fewer washers break down within one year.

19. You pay $10 to draw 1 card from a standard deck of 52 cards. If the card is a king, you win $15; if the card is not a king, you win $3. What are your expected earnings from this game?

Career Uses of Mathematics

I am currently a District Manager for Procter & Gamble with the responsibility of developing our Health & Beauty Aids business with our largest U.S. customer—Wal-Mart, Inc. A major part of my time is spent analyzing and organizing data to define and determine probabilities of success for marketing ideas that will develop our mutual business.

The business development area focuses on our "sales controllables" (pricing, distribution, shelving, and merchandising). We have data for each of these strategic areas of our business and have developed mathematical models for determining optimal price points and profitability, share of distribution and space allocation, and incremental merchandising volume for each brand's sizes.

I find that developing and implementing revolutionary business ideas is the most gratifying aspect of my position. Since our customers and the industry are changing constantly, we have to develop ideas for the future that will set the pace for our changing environment. Mathematical analysis and applications are critical in developing these future plans.

Ramona Kent

Ramona Kent, Procter & Gamble

Markov Chains

The Aurora planning commission found that each year 92% of those commuting by mass transportation continued to do so the next year, while 8% switched to private auto. Of those commuting by private auto, 95% continued to do so the next year, while 5% switched to mass transportation. If this rate of switching continues for a number of years, how will the percentage of commuters eventually be split between the users of mass transportation and private auto? (See Example 2, Section 9.2.)

CHAPTER PREVIEW

Markov chains provide a natural application of matrix algebra to answer probability questions about repeated trials of some types of experiments. The exact nature of these experiments is set forth at the beginning of this chapter, thus demonstrating the reason for the name "Markov chains." After introducing the transition matrix to represent Markov chains in the first section, the long-term effects of repeated trials of such experiments are explored in the form of a regular Markov chain. Finally, Markov chains that have absorbing states are studied.

CALCULATOR/COMPUTER OPPORTUNITIES FOR THIS CHAPTER

Because matrix multiplications abound in this chapter, a calculator with matrix capabilities will prove to be a valuable aid. The computer program Matrix on the disk MasterIt will also do such multiplications. Either of these tools will also be very helpful in finding the inverse of some matrices involved in absorbing Markov chains.

SECTION 9.1 **Combining Matrices with Probability: The Translation Matrix**

We have already studied repeated trials of experiments that lead to binomial (Bernoulli) distributions. In these cases, each trial had only two outcomes, and each trial was independent of other trials. We now want to generalize these conditions by considering repeated trials of an experiment in which each trial has a finite number of outcomes. Then, as the experiment is performed repeatedly, a sequence of outcomes is produced. We want to impose a special condition on those outcomes: Except for the first trial, the probability that a given outcome occurs on a particular trial depends only on which outcome occured in the previous trial. The fact that an outcome depends on the previous outcome suggests that such a sequence of outcomes may be thought of as a *chain*, in the sense that each trial is *linked* to the one preceding it. Such chains were first studied by the Russian mathematician Andrei Andreyevich Markov; hence, the name "Markov chain" is applied to a problem meeting these conditions. These conditions, along with some of the terminology are summarized next.

Conditions for a Markov Chain

1. The experiment is repeated a finite number of times. (We do not consider cases of infinite repetitions in this text.)

2. Each trial has the *same sample space*. The sample space consists of a finite number of outcomes, called **states,** and the probabilities assigned to them are called **transition probabilities.**

3. On any trial except the first, the probability that a given state is entered depends only on the state previously occupied. This is known as the **Markov Property.**

As an example of these ideas, consider the data collected by the Aurora Planning Commission on the habits of commuters, described in the opening of this chapter: Each year, about 8% of those using mass transportation (*M*) will change to using a private automobile (*A*), while 92% will continue to use mass transportation. Furthermore, 5% of those using the automobile (*A*) will change to mass transportation (*M*), while 95% will continue using the automobile. In terms of probabilities, we assume that these percents of change remain constant year after year (within limits, of course) and describe the fixed **transition probabilities** for events in the sample space:

Event	Probability
M to A	.08
M to M	.92
A to M	.05
A to A	.95

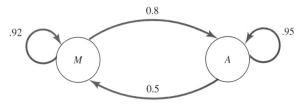

FIGURE 1
Transition Diagram

These events and corresponding probabilities can be visually displayed in a **transition diagram,** as shown in Figure 1 above. A transition diagram shows the transition probabilities from any given year (**current state**) to the following year (**next state**).

A tree diagram (see Figure 2) will also visually portray these probabilities as a transition from a current state to the next state.

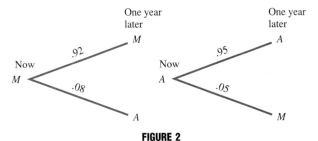

FIGURE 2
Trees Showing Transition Probabilities

EXAMPLE 1
State Probabilities Two
Years Hence

Find the probability that a person who uses mass transportation now (*current state*) will still be using mass transportation two years from now (*two states hence*).

Solution   A tree diagram will help us analyze this problem, as shown in Figure 3, and the transition diagram, as shown in Figure 1, will show how to label the branches. From looking at the diagram shown in Figure 3, we can see that there are two mutually exclusive paths in the tree that begin with mass transportation (*M*) and end

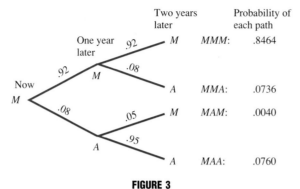

FIGURE 3
Two-stage Tree Diagram

with mass transportation (*M*) two years later: *MMM*, with a probability of .8464 and *MAM,* with a probability of .0040. Therefore, the probability that we start with *M* and end with *M* two years (two transitions, trials, steps) later is

$$P(MMM \text{ or } MAM) = P(MMM) + P(MAM) = .8464 + .0040 = .8504.$$

What is the probability that a person who now uses mass transportation will be using the automobile two years hence? Again the tree diagram gives the answer:

$$P(MMA \text{ or } MAA) = P(MMA) + P(MAA) = .0736 + .076 = .1496. \quad \blacksquare$$

EXAMPLE 2
State Probabilities Two Years Hence

What is the probability that a person using the automobile now will be using mass transportation two years from now?

Solution   A tree diagram beginning with *A* (the automobile) as the present state, as shown in Figure 4, will help solve this problem. Two paths begin with *A* and end with *M* two years later: *AMM*, with a probability of .046, and *AAM* with a probability of .0475. Consequently,

$$P(AMM \text{ or } AAM) = P(AMM) + P(AAM) = .046 + .0475 = .0935.$$

Similarly, the probability that a person using the automobile now will still be using the automobile two years later is

$$P(AMA \text{ or } AAA) = P(AMA) + P(AAA) = .004 + .9025 = .9065. \quad \blacksquare$$

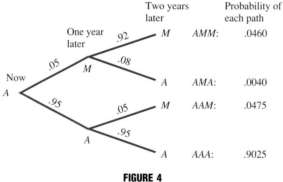

FIGURE 4
Two-stage Tree Diagram

The transition probabilities shown in Figure 1 may also be recorded in a **transition matrix,** which shows all information about the sample space:

$$\begin{array}{c} \text{Next year} \\ \text{(Next state)} \\ \begin{array}{cc} M & \quad A \end{array} \end{array}$$

$$\begin{array}{c} \text{Now } M \\ \text{(Current state) } A \end{array} \begin{bmatrix} .92 & .08 \\ .05 & .95 \end{bmatrix} = T$$

Be sure that you understand how the entries in T relate to the labels on the left and top of this matrix. (Compare with the sample space and Figure 1.) Now suppose that we compute T^2. The reason for computing it will be apparent if we put in all of the steps.

$$
\begin{aligned}
T^2 &= \begin{bmatrix} .92 & .08 \\ .05 & .95 \end{bmatrix}\begin{bmatrix} .92 & .08 \\ .05 & .95 \end{bmatrix} \\
&= \begin{bmatrix} (.92)(.92) + (.08)(.05) & (.92)(.08) + (.08)(.95) \\ (.05)(.92) + (.95)(.05) & (.05)(.08) + (.95)(.95) \end{bmatrix} \\
&= \begin{bmatrix} .8504 & .1496 \\ .0935 & .9065 \end{bmatrix}.
\end{aligned}
$$

Do you recall seeing these numbers before? Compare these entries with the results obtained in Examples 1 and 2, and you will see that this matrix gives complete information on the probability of going from one state to another in two years (two trials or steps):

$$\begin{array}{c} \text{Two years hence} \\ \begin{array}{cc} M & \quad A \end{array} \end{array}$$

$$T^2 = \begin{array}{c} \text{Now } M \\ A \end{array} \begin{bmatrix} .8504 & .1496 \\ .0935 & .9065 \end{bmatrix}$$

In other words, T^2 describes the state probabilities two years (transitions, trials, steps) later. If T is cubed we get

$$T^3 = \begin{bmatrix} .790 & .210 \\ .131 & .869 \end{bmatrix}.$$

Can you interpret these entries? How would you interpret the entries in T^4?

In general, suppose there are k states, $S_1, S_2, \ldots, S_k$. If we start in one of these states, say S_i, then after n transitions we will be in state S_j with probability equal to the number that lies in the ith row and jth column of T^n.

The Powers of a Transition Matrix

Let T^n be the nth power of a transition matrix T. If the process (or chain) starts in state S_i, then after n repetitions of the experiment (transitions) it will be in state S_j with probability equal to the number in the ith row and jth column of T^n.

For example, the entry 0.210 in the first-row, second-column position ($i = 1, j = 2$) of T^3 tells us that if we start in state $S_1 = S_i = M$, then after three transitions we will be in state $S_2 = S_j = A$ with probability 0.210.

Notice that the entries in each row of T^n, $n = 1, 2, \ldots n$, give the respective probabilities of the states (or outcomes) in that row after n transitions. This means that the probabilities in each row must sum to 1. In general, any matrix for which each row sums to 1 is called a **stochastic** matrix.

Suppose that the Aurora Planning Commission determines that this year, 25% of the commuters commute by mass transportation and 75% commute by private automobile. As noted earlier, every year commuters switch from using mass transportation to private automobile, and vice versa. Such changes are described by the transition probability matrix T. It would be interesting to know what percentage of commuters will use the mass transportation (and therefore the private automobile) next year and two years from now. To answer such questions, we first take the percentages 25% and 75%, change each to a probability and then put them into a 1×2 matrix, called a **state vector:**

$$P_0 = [.25 \quad .75].$$

Now perform the matrix multiplication $P_0 T$, and carefully note what each entry in the result means:

$$P_0 T = [.25 \quad .75] \quad \begin{matrix} M \\ A \end{matrix} \begin{bmatrix} \overset{M}{.92} & \overset{A}{.08} \\ .05 & .95 \end{bmatrix}$$

$$= [\underset{\text{M to M}}{(.25)(.92)} + \underset{\text{A to M}}{(.75)(.05)} \quad \underset{\text{M to A}}{(.25)(.08)} + \underset{\text{A to A}}{(.75)(.95)}]$$

$$= [\underset{\substack{\text{Total} \\ \text{to } M}}{.27} \quad \underset{\substack{\text{Total} \\ \text{to } A}}{.73}].$$

That is, one year later, 27% of the commuters will use the mass transportation, and 73% will use the private automobile.

If we now take the state vector $P_0 T$ at the end of one year and multiply on the right by T, the result should show the distribution for the choice of transportation of the commuters at the end of two years:

$$(P_0 T)T = P_0 T^2 = [.25 \quad .75] \begin{bmatrix} .8504 & .1496 \\ .0935 & .9065 \end{bmatrix} = [.283 \quad .717].$$

In other words, 28.3% will be commuting by mass transportation and 71.7% by automobile.

The preceding calculations show that there are two ways in which the distribution for the choice of transportation of the commuters may be found through n stages with matrix algebra. One way is to multiply the initial state vector by T to create a new state vector at the end of the first year, then multiply this state vector by T to create a new state vector at the end of the second year, etc.:

$$P_0, \ P_0 T, \ (P_0 T)T, \ [(P_0 T)T]T, \ldots$$

Because each state vector, except the initial one, is obtained by multiplying the previously obtained state vector on the right by T, we could restate the distributions at each stage as

$$P_0, \ P_1 = P_0 T, \ P_2 = P_1 T, \ P_3 = P_2 T, \ P_4 = P_3 T, \ldots,$$
$$P_n = P_{n-1} T \text{ after } n \text{ transitions.}$$

The second, and equivalent way, is to use powers of the transition matrix:

$$P_0, \ P_0 T, \ P_0 T^2, \ P_0 T^3, \ldots, \ P_0 T^n.$$

EXAMPLE 3
Brand Loyalty

A marketing researcher studied brand loyalty to three brands of popular laundry detergents. She found that 97% of the people who bought brand B still bought that brand a year later, while 2% had switched to brand C and 1% had switched to brand D. Of those who bought brand C, 95% still bought that brand after a year, while 2% had switched to B and 3% had switched to D. Of those who bought D, 93% still bought that brand a year later, while 4% had switched to B and 3% had switched to C.

(a) Find the transition matrix for this survey.

Solution

$$T = \begin{array}{c} \\ B \\ C \\ D \end{array} \begin{array}{ccc} B & C & D \\ \left[\begin{array}{ccc} .97 & .02 & .01 \\ .02 & .95 & .03 \\ .04 & .03 & .93 \end{array}\right] \end{array}$$

(b) What percentage of those buying brand B now will be buying brand D two years from now?

Solution   We first find T^2, keeping the labels as before:

$$T^2 = \begin{array}{c} \\ B \\ C \\ D \end{array} \begin{array}{ccc} B & C & D \\ \left[\begin{array}{ccc} .942 & .039 & .020 \\ .040 & .904 & .057 \\ .077 & .057 & .866 \end{array}\right] \end{array}.$$

The answer to the question is located in the first-row, third-column position of T^2: .020, or about 2%.

(c) If the market shares among these three brands this year were divided 38%, 42%, and 20%, respectively, what will the market shares be one year from now?

Solution   Let $P_0 = [.38 \quad .42 \quad .20]$ be the initial state vector showing the distribution of market share among these three brands. Then one year later the market shares will be

$$P_0 T = [.38 \quad .42 \quad .20] \left[\begin{array}{ccc} .97 & .02 & .01 \\ .02 & .95 & .03 \\ .04 & .03 & .93 \end{array}\right] = [.385 \quad .413 \quad .202]$$

Hence, about 39% of the households will be buying brand B, 41% will be buying brand C, and about 20% will be buying brand D. ▪

EXAMPLE 4
More About the
Transition Matrix

For some Markov chains, such as the one shown in Figure 5, it is impossible to pass from one state to another, or even from a given state back to that state. The probability assigned to such events would, of course, be zero.

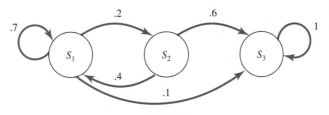

FIGURE 5
Transition Diagram

The transition matrix for such a Markov chain is

$$T = \begin{array}{c} \\ S_1 \\ S_2 \\ S_3 \end{array} \begin{array}{ccc} S_1 & S_2 & S_3 \\ \begin{bmatrix} .7 & .2 & .1 \\ .4 & 0 & .6 \\ 0 & 0 & 1.0 \end{bmatrix} \end{array}. \quad \blacksquare$$

EXAMPLE 5
More About the
Transition Matrix

The transition matrix

$$T = \begin{array}{c} \\ S_1 \\ S_2 \end{array} \begin{array}{cc} S_1 & S_2 \\ \begin{bmatrix} .3 & .7 \\ 0 & 1 \end{bmatrix} \end{array}$$

shows that it is possible to pass from state 1 to state 2 and from state 2 to state 2, but once in state 2, it is impossible to leave. ■

EXERCISES   9.1

In Exercises 1 through 6, find the missing probabilities between the various states shown, and construct the transition matrix.

1.

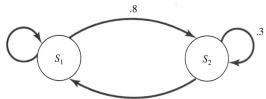

2.

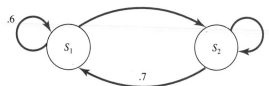

3.

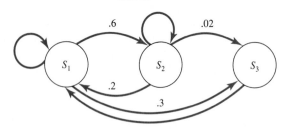

4.

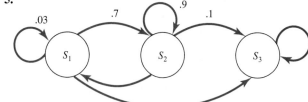

5.

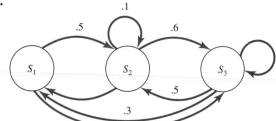

6.

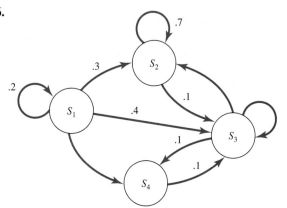

In Exercises 7 through 10, construct the transition diagram for each of the given matrices.

7. $\begin{bmatrix} .2 & .8 \\ .6 & .4 \end{bmatrix}$

8. $\begin{bmatrix} 0 & 1 \\ .4 & .6 \end{bmatrix}$

9. $\begin{bmatrix} .2 & .4 & .4 \\ 0 & .3 & .7 \\ .9 & 0 & .1 \end{bmatrix}$

10. $\begin{bmatrix} 1 & 0 & 0 \\ .4 & .6 & 0 \\ .02 & .8 & .18 \end{bmatrix}$

In Exercises 11 through 15, construct the transition diagram and the transition matrix that represents the given data.

11. Eighty percent of the people who buy a foreign make of automobile buy one on their next purchase of automobile, while 20% switch to an American make. Ninety percent of the people who buy an American make buy one on their next purchase of automobile, while 10% switch to a foreign make.

12. If it rains today, there is a 60% chance of rain tomorrow; however, if it does not rain today, there is an 80% chance that it will not rain tomorrow.

13. Sixty percent of the people in one generation who have a particular psychological characteristic will pass that characteristic on to the next generation. Fifty percent of the people in one generation who do not have this characteristic will pass it on to the next generation.

14. Eighty-five percent of those who voted Democratic in the last election will vote Democratic in the next election, 10% will vote Republican, and 5% will vote Independent. Ninety percent of those who voted Republican in the last election will do so in the next election, while 7% will vote Democratic, and 3% will vote Independent. Fifty-eight percent of those who voted Independent in the last election will do so in the next election, 26% will vote Democratic, and 16% will vote Republican.

15. Ninety percent of the subscribers to the *Evening Star* renew their subscription each year, while 6% change to the *Morning Tribune* and 4% change to the *Gazette.* Seventy percent of the subscribers to the *Morning Tribune* renew their subscription each year, while 20% change to the *Evening Star* and 10% change to the *Gazette.* Seventy-five percent of the subscribers to the *Gazette* renew their subscriptions each year, while 16% change to the *Evening Star* and 9% change to the *Morning Tribune.*

In Exercises 16 through 20, decide which of the matrices are transition matrices. For those that are not, tell why not.

16. $\begin{bmatrix} .6 & .4 \\ .3 & .7 \end{bmatrix}$

17. $\begin{bmatrix} .1 & .9 \\ .3 & .8 \end{bmatrix}$

18. $\begin{bmatrix} .05 & .95 \\ .85 & .15 \end{bmatrix}$

19. $\begin{bmatrix} .2 & .4 & .4 \\ 0 & .8 & .2 \\ .5 & .3 & .2 \end{bmatrix}$

20. $\begin{bmatrix} .4 & 0 & .6 \\ 0 & .9 & .1 \\ .5 & .3 & .4 \end{bmatrix}$

21. For the transition matrix

$$T = \begin{bmatrix} .6 & .4 \\ .2 & .8 \end{bmatrix}$$

and the state vector $[.7 \quad .3]$, find the state vector

(a) One transition later
(b) Two transitions later

22. For the transition matrix

$$T = \begin{bmatrix} .02 & .98 \\ .3 & .7 \end{bmatrix}$$

and the state vector $[.4 \quad .6]$, find the state vector

(a) Two transitions later
(b) Three transitions later

23. For the transition matrix

$$T = \begin{bmatrix} .3 & .2 & .5 \\ .4 & 0 & .6 \\ .1 & .8 & .1 \end{bmatrix}$$

and the state vector $[.1 \quad .4 \quad .5]$, find the state vector

(a) One transition later
(b) Two transitions later

24. *Customer Preferences* Suppose that people who own NumberKrunch computers for home use will purchase another NumberKrunch with a probability of .80 and will switch to a QuickDigit computer with a probability of .20. Those who own a QuickDigit will purchase another with a probability of .60 and will switch to a NumberKrunch with a probability of .40.

(a) Construct the transition matrix for this two-state Markov chain.
(b) Use a tree diagram to find the probability that a person now using a NumberKrunch computer will have switched to a QuickDigit computer two purchases later.
(c) Find the probability asked for in Exercise 24(b) by using a matrix multiplication.
(d) Find the probability that if a person now has a NumberKrunch computer, two computer purchases later they will also buy a NumberKrunch computer.

25. *Reliability* A space probe contains equipment that will diagnose and correct misalignment of the probe as it travels through space. If the probe is aligned one minute, then 60% of the time it is still aligned one minute later. On the other hand, if it is misaligned one minute, then 40% of the time it is misaligned the next minute.

   (a) Construct the transition diagram and the transition matrix for these data.
   (b) Use a tree diagram to find the probability that if the probe is aligned now, it will still be aligned two minutes from now.
   (c) Use a matrix multiplication to find the probability asked for in Exercise 25(b).
   (d) Find the probability that if the probe is not aligned now, then it still will not be aligned three minutes from now.

26. *Survey* A study of freshmen, sophomores, and juniors at State University revealed that each year, 10% of the engineering students change to liberal arts, 5% change to business, and the rest remain in engineering. Three percent of the liberal arts students change to engineering, 15% change to business, and the rest stay in liberal arts. Five percent of the business students change to engineering, 8% change to liberal arts, and the rest remain in business.

   (a) Construct the transition diagram and transition matrix for these data.
   (b) What percentage of the students now in engineering will change to business three years from now?
   (c) What percentage of the students now in liberal arts will still be in liberal arts three years from now?

27. *Customer Preferences* The records of an automobile leasing company show the following information about preference for two-door sedans (2-d), four-door sedans (4-d), and station wagons (SW):

Next Choice

		2-d	4-d	SW
	2-d	70%	20%	10%
Choice	4-d	5%	90%	5%
Now	SW	10%	30%	60%

   (a) Construct the transition diagram and transition matrix for these data.
   (b) Construct a tree diagram to find the probability that a person leasing a two-door sedan now will be leasing a station wagon three choices later.
   (c) Find the probability asked for in Exercise 27(b) by matrix multiplication.

28. *Sports Scores* Gordon and Dennis have a standing tennis match every Saturday. A record of their matches shows that if Gordon wins on any given Saturday, he wins the next Saturday with a probability of .60, whereas if Dennis wins, he wins the match the following Saturday with a probability of .45. If Gordon wins the match this Saturday, what is the probability that he will win the match three Saturdays from now?

29. *Sports Scores* Records indicate that Denita's free-throw shooting can be represented by a Markov chain: If she makes a free throw, she makes the next one 80% of the time and misses 20% of the time. If she misses a free throw, she makes the next one 60% of the time and misses 40% of the time. If Denita misses her first free throw of a game, what is the probability that she will miss the third free throw of that game?

30. *Experiment* A mouse is put into a maze, the diagram of which is shown below. A transition is considered either to move to an adjacent room or to stay in the same room for a given length of time, say one minute. Assume that the mouse is twice as likely to remain in the same room as to leave, but that if it does leave, it is equally likely to go into any adjacent room. Find the transition matrix for this experiment, and find the probability that if the mouse starts in room 2, it will be in room 4 three transitions from now.

Problems for Exploration

1. For a transition matrix for a Markov chain

$$P = \begin{bmatrix} 1 & 0 \\ .3 & .7 \end{bmatrix},$$

explain what the first row means in terms of being able to go from one state to another. Does this property apply to all powers of P?

2. For a transition matrix for a Markov chain

$$P = \begin{bmatrix} 1 & 0 & 0 \\ 0 & 1 & 0 \\ .3 & .4 & .3 \end{bmatrix},$$

explain what the first two rows mean in terms of being able to go from one state to another. Does this property apply to all powers of P?

Regular Markov Chains

Some interesting and useful results are found in the long-term behavior of a Markov chain. In some cases, the powers of the transition matrix tend to get closer and closer to one fixed matrix as the number of steps in the chain increases, while in other cases, those powers continue to exhibit noticeable change. As an example of the latter, the transition matrix

$$T = \begin{bmatrix} 0 & 1 \\ 1 & 0 \end{bmatrix} \text{ has } T^2 = \begin{bmatrix} 1 & 0 \\ 0 & 1 \end{bmatrix}, T^3 = \begin{bmatrix} 0 & 1 \\ 1 & 0 \end{bmatrix} = T, T^4 = \begin{bmatrix} 1 & 0 \\ 0 & 1 \end{bmatrix} = T^2,$$

and so on, showing that the powers of T change back and forth between T and T^2 as the power increases. On the other hand, consider the transition matrix

$$T = \begin{bmatrix} .1 & .9 \\ .7 & .3 \end{bmatrix}.$$

Notice the pattern developing in each entry of the various powers of T: (For convenience, entries of T^3, and so on, are rounded to two decimals.)

$$T^2 = \begin{bmatrix} .64 & .36 \\ .28 & .72 \end{bmatrix}; \quad T^3 = \begin{bmatrix} .32 & .68 \\ .53 & .47 \end{bmatrix}; \quad T^4 = \begin{bmatrix} .51 & .49 \\ .38 & .63 \end{bmatrix};$$

$$T^5 = \begin{bmatrix} .39 & .61 \\ .47 & .53 \end{bmatrix}; \quad T^6 = \begin{bmatrix} .46 & .54 \\ .42 & .58 \end{bmatrix}; \quad T^7 = \begin{bmatrix} .42 & .58 \\ .45 & .55 \end{bmatrix};$$

$$T^8 = \begin{bmatrix} .45 & .55 \\ .43 & .57 \end{bmatrix}; \quad T^9 = \begin{bmatrix} .43 & .57 \\ .44 & .56 \end{bmatrix}; \quad T^{10} = \begin{bmatrix} .44 & .56 \\ .44 & .57 \end{bmatrix};$$

$$T^{11} = \begin{bmatrix} .44 & .57 \\ .44 & .56 \end{bmatrix}; \quad T^{12} = \begin{bmatrix} .44 & .56 \\ .44 & .56 \end{bmatrix}; \quad T^{13} = \begin{bmatrix} .44 & .56 \\ .44 & .56 \end{bmatrix}.$$

Our calculations show that powers of 13 and above applied to T will give the same matrix as T^{12} upon rounding to two decimals. This fact is described by saying that, in the long run, the powers of T tend to approach the **stable** or **limiting matrix**

$$L = \begin{bmatrix} .44 & .56 \\ .44 & .56 \end{bmatrix}.$$

A transition matrix having such a property is called **regular.** Is there a way to tell when such a phenomenon will occur? Was it just blind luck that the rows in the limiting matrix were the same? The answers to these questions are revealed next.

Regular Transition Matrices

A transition matrix T is **regular** if all entries of T^n are *positive* for some positive integer n. The Markov chain associated with T is called a **regular Markov chain.** The importance of a regular matrix T lies in the fact that as n increases, T^n always approaches some stable or limiting matrix L having identical rows.

A word of caution about round-off errors is in order when raising a matrix to higher and higher powers with a calculator. Often when a matrix has small decimal entries, increasingly higher powers of that matrix can result in errors of such magnitude that they actually exceed the original entries themselves. Without going into error analysis, a subject in its own right, be cautious of the accuracy of the last few digits on a calculator supposedly displaying several-digit accuracy.

EXAMPLE 1
Deciding When a Matrix Is Regular

(a) The matrix $T = \begin{bmatrix} .5 & .5 \\ .2 & .8 \end{bmatrix}$ is regular because the entries in T^1 are all positive.

(b) The matrix $T = \begin{bmatrix} 1 & 0 \\ 0 & 1 \end{bmatrix}$ is not regular because $T^n = T$ for every positive integer n.

(c) The matrix $T = \begin{bmatrix} .1 & .9 \\ 0 & 1 \end{bmatrix}$ is not regular because every positive integer power of T has a 0 in the second-row, first-column position.

(d) The matrix $T = \begin{bmatrix} .1 & .2 & .7 \\ .5 & 0 & .5 \\ .4 & .4 & .2 \end{bmatrix}$ is regular because $T^2 = \begin{bmatrix} .39 & .3 & .31 \\ .25 & .3 & .45 \\ .32 & .16 & .52 \end{bmatrix}$. ▆

Regular Markov chains have another important and almost amazing property. Regardless of what the initial distribution in a state vector may be, as the number of steps in the chain increases, they all tend to approach a fixed **stable** or **limiting distribution.** To see why this is so, let T be a regular matrix and let L be the limiting matrix. This means that we may select a positive integer n large enough so that, after rounding the decimal approximations for both T^n and L to a reasonable number of places, $T^n = L$. This implies that $T^n = T^{n+1} = T^{n+2} = \ldots = L$. Now if P_0 is a state vector representing some initial distribution, then for the nth stage and beyond, the following equalities hold:

$$P_0 T^n = P_0 T^{n+1} = P_0 T^{n+2} = \ldots = P_0 L. \tag{1}$$

In other words, from the nth stage and thereafter, the state vector does not change but remains stable at the constant limiting value of $P_0 L$. The stable or long-term state vector in a regular Markov chain is of great interest in applications. However, as you can imagine, it will usually require considerable calculational effort to find the limiting matrix L for computing $P_0 L$. Fortunately, there is an easier way. For simplicity, rename the stable state vector with the letter S. That is, let $P_0 L = S$. Then the first two and last terms of Equation (1) allow us to write

$$P_0 T^n = P_0 T^{n+1}$$
$$P_0 T^n = (P_0 T^n)T$$
$$P_0 L = (P_0 L)T$$
$$S = ST.$$

Furthermore, we know that the entries in S are probability numbers so that their sum must be 1. We now have the defining properties for S.

Finding the Stable State Vector

In any regular Markov chain with transition matrix T, any initial distribution given in an initial state vector P_0 eventually reaches the stable vector S. The stable state vector $S = [p_1, p_2, \ldots, p_n]$ has these properties:

$$S = ST \quad \text{and} \quad p_1 + p_2 + \cdots + p_n = 1.$$

We are now in a position to solve the problem posed at the outset of this chapter.

EXAMPLE 2
Finding a Stable
State Vector

The Aurora planning commission (Section 9.1) found that each year 92% of those commuting by mass transportation (M) continued to commute by mass transportation the next year, while 8% changed to private automobile (A). Of those commuting by private automobile, 95% continued to do so the next year, while 5% changed to mass transportation. What is the eventual choice of transportation of commuters among these two modes?

Solution   The transition matrix for this Markov chain was found (in Section 9.1) to be

$$T = \begin{array}{c} \\ M \\ A \end{array}\begin{array}{cc} M & A \\ \begin{bmatrix} .92 & .08 \\ .05 & .95 \end{bmatrix} \end{array}.$$

According to the development preceding the present Example 2, we seek a stable vector $S = \begin{bmatrix} u & v \end{bmatrix}$ such that $S = ST$ and $u + v = 1$. Specifically, we are to find u and v such that

$$\begin{bmatrix} u & v \end{bmatrix} = \begin{bmatrix} u & v \end{bmatrix}\begin{bmatrix} .92 & .08 \\ .05 & .95 \end{bmatrix} \quad \text{and} \quad u + v = 1.$$

Doing the indicated matrix multiplication, this system becomes

$$\begin{bmatrix} u & v \end{bmatrix} = \begin{bmatrix} .92u + .05v & .08u + .95v \end{bmatrix} \quad \text{and} \quad u + v = 1.$$

Equating the entries in like addresses of the matrices gives

$$u = .92u + .05v$$
$$v = .08u + .95v$$
$$u + v = 1,$$

or, upon combining like terms in the first two equations,

$$-.08u + .05v = 0$$
$$.08u - .05v = 0$$
$$u + \quad v = 1.$$

Because the first two equations are the same, the system to be solved is

$$.08u - .05v = 0$$
$$u + \quad v = 1.$$

The solution is $u = .38$ and $v = .62$. In other words, regardless of what the choice of transportation may have been at any step of the Markov process, eventually about 38% of the commuters will use mass transportation and about 62% will use private automobile. ■

EXAMPLE 3
Finding a Stable-State
Vector

The regular transition matrix

$$\begin{array}{c} \text{This} \\ \text{year} \end{array} T = \begin{array}{c} \\ B \\ C \\ D \end{array}\begin{array}{c} \begin{array}{ccc} B & \overset{\text{Next}}{\underset{\text{year}}{C}} & D \end{array} \\ \begin{bmatrix} .97 & .02 & .01 \\ .02 & .95 & .03 \\ .04 & .03 & .93 \end{bmatrix} \end{array}$$

described the changes in brand loyalty each year by shoppers who bought brands B, C, and D. (See Example 3, Section 9.1.) What is the eventual market share for each of these three brands?

Solution   We are to find the state vector $S = [u \quad v \quad w]$ for which $S = ST$ and $u + v + w = 1$. Specifically,

$$[u \quad v \quad w] = [u \quad v \quad w] \begin{bmatrix} .97 & .02 & .01 \\ .02 & .95 & .03 \\ .04 & .03 & .93 \end{bmatrix} \quad \text{and} \quad u + v + w = 1$$

are to be satisfied. The indicated matrix multiplication gives

$$[u \quad v \quad w] = [.97u + .02v + .04w \quad .02u + .95v + .03w \quad .01u + .03v + .93w]$$

so that

$$u = .97u + .02v + .04w, \, v = .02u + .95v + .03w$$

and

$$w = .01u + .03v + .93w.$$

Combining like terms shows that these three equations become

$$-.03u + .02v + .04w = 0, \, .02u - .05v + .03w = 0$$

and

$$.01u + .03v - .07w = 0.$$

It follows that u, v, and w must satisfy the system of equations:

$$-.03u + .02v + .04w = 0$$
$$.02u - .05v + .03w = 0$$
$$.01u + .03v - .07w = 0$$
$$u + \quad v + \quad w = 1.$$

This system has the solution (to two-decimal accuracy) $u \approx .48$, $v \approx .31$, and $w \approx .20$, which means that among these three brands, in the long run, brand B will have about 48% of the market, brand C will have about 31% of the market, and brand D will have about 20% of the market. ▬

We stated earlier in this section that the limiting matrix L obtained when a regular transition matrix is raised to successively higher powers has identical rows. This was done after raising

$$T = \begin{bmatrix} .1 & .9 \\ .7 & .3 \end{bmatrix}$$

to the 13th power before seeing that

$$L = \begin{bmatrix} .44 & .56 \\ .44 & .56 \end{bmatrix}.$$

For the matrix

$$T = \begin{bmatrix} .92 & .08 \\ .05 & .95 \end{bmatrix}$$

of Example 2, which shows how commuters changed their mode of commuting, it takes well over 100 powers of T to arrive at

$$L = \begin{bmatrix} .38 & .62 \\ .38 & .62 \end{bmatrix}$$

Looking back at Example 2, do you see a relationship between the rows of L and the stable-state vector S found there? What do you think the stable-state vector for the matrix

$$T = \begin{bmatrix} .1 & .9 \\ .7 & .3 \end{bmatrix}$$

will be? If you answered $S = [.44 \quad .56]$, you are correct! Mathematically, we can prove that for a regular transition matrix T, each row of the limiting matrix L is precisely the stable-state vector S. So, if we can find S, we can find L upon deciding the desired decimal accuracy. When constructing L from S, we do not know how many steps it takes to reach the stable condition.

Limit Matrix for Regular Markov Chains

Let T be the transition matrix for a regular Markov chain. As the positive integer n increases, the matrix T^n approaches a limit matrix L each of whose rows is the stable-state vector for the chain.

EXERCISES   9.2

In Exercises 1 through 6, decide which of the transition matrices are regular.

1. $\begin{bmatrix} .5 & .5 \\ .9 & .1 \end{bmatrix}$ **2.** $\begin{bmatrix} 0 & 1 \\ .4 & .6 \end{bmatrix}$

3. $\begin{bmatrix} 0 & .5 & .5 \\ .2 & .4 & .4 \\ .8 & .1 & .1 \end{bmatrix}$ **4.** $\begin{bmatrix} 0 & .1 & .9 \\ .2 & 0 & .8 \\ .4 & .6 & 0 \end{bmatrix}$

5. $\begin{bmatrix} 1 & 0 & 0 \\ .5 & .5 & 0 \\ .2 & .3 & .5 \end{bmatrix}$ **6.** $\begin{bmatrix} 0 & .4 & .6 \\ 1 & 0 & 0 \\ 0 & 0 & 1 \end{bmatrix}$

For each of the regular matrices in Exercises 7 through 12, find the limiting matrix L by calculating the powers of the matrix.

7. $\begin{bmatrix} .3 & .7 \\ .5 & .5 \end{bmatrix}$ **8.** $\begin{bmatrix} .6 & .4 \\ .3 & .7 \end{bmatrix}$ **9.** $\begin{bmatrix} .9 & .1 \\ .2 & .8 \end{bmatrix}$

10. $\begin{bmatrix} 0 & 1 \\ .9 & 1 \end{bmatrix}$ **11.** $\begin{bmatrix} .6 & .4 \\ 1 & 0 \end{bmatrix}$

12. $\begin{bmatrix} .3 & .4 & .3 \\ .4 & .4 & .2 \\ .1 & .6 & .3 \end{bmatrix}$

For each of the regular matrices in Exercises 13 through 21, find the stable-state vector S and, from that, find the limiting matrix L.

13. $\begin{bmatrix} .4 & .6 \\ .2 & .8 \end{bmatrix}$ **14.** $\begin{bmatrix} \frac{2}{5} & \frac{3}{5} \\ \frac{2}{3} & \frac{1}{3} \end{bmatrix}$ **15.** $\begin{bmatrix} \frac{7}{8} & \frac{1}{8} \\ \frac{1}{6} & \frac{5}{6} \end{bmatrix}$

16. $\begin{bmatrix} 0 & 1 \\ .8 & .2 \end{bmatrix}$ **17.** $\begin{bmatrix} \frac{1}{7} & \frac{6}{7} \\ \frac{9}{10} & \frac{1}{10} \end{bmatrix}$ **18.** $\begin{bmatrix} .33 & .67 \\ .08 & .92 \end{bmatrix}$

19. $\begin{bmatrix} \frac{1}{5} & \frac{2}{5} & \frac{2}{5} \\ \frac{1}{2} & \frac{1}{4} & \frac{1}{4} \\ \frac{1}{3} & \frac{1}{6} & \frac{1}{2} \end{bmatrix}$ **20.** $\begin{bmatrix} 0 & .4 & .6 \\ .2 & .6 & .2 \\ .8 & .1 & .1 \end{bmatrix}$

21. $\begin{bmatrix} 0 & .9 & .1 \\ .8 & 0 & .2 \\ .3 & .3 & .4 \end{bmatrix}$

22. Donations Records kept by the United Way of Springdale indicate that 90% of those who contribute one year will also contribute the following year, while the other 10% will not contribute the following year. On the other hand, 12% of those who do not contribute one year will contribute the following year, while 88% still do not contribute the following year.

(a) Construct the transition diagram and the transition matrix representing these contribution rates.

(b) If 75% of the citizens are now contributing to the United Way each year, in the long run, what percentage of the citizens will and will not be contributing to the United Way each year?

23. Subscriptions Circulation statistics in Charlotte show that the *Evening Star* keeps 65% of its subscribers and loses 35% to the *Tribune* each year. On the other hand, the *Tribune* keeps 80% of its subscribers and loses 20% to the *Evening Star* each year.

(a) Construct the transition diagram and transition matrix representing these statistics.

(b) In the long run, what percentage of the subscribers will each paper have?

24. Brand Loyalty Surveys indicate that among people who buy an American-built car, 85% will buy American on their next purchase, and, among those who buy a foreign make, 80% will buy a foreign make on their next purchase. Assuming that this trend holds, what will the distribution of buyers between American and foreign cars eventually become?

25. Advertising A retail store with one major competitor is considering an intensive advertising campaign. The designers of the ad campaign claim that each week, 20% of the competitor's customers will switch to their store, while about 5% of their customers will be annoyed by the ad campaign and switch to the competitor.

(a) If the store now has 70% of the market and the competitor has 30%, what will be the long-term gain in market share if the ad campaign continues for several weeks?

(b) How many weeks of the ad campaign will be needed before the market share stabilizes?

26. Political Parties In Mt. Vernon, the voting population remains rather constant at 12,000. From election to election, the changes among Democrats, Republicans, and Independents are as shown in this transition matrix:

		Next election		
		D	R	I
This election	D	.80	.15	.05
	R	.08	.90	.02
	I	.20	.15	.65

In the long run, how many voters will each party have?

27. Social Strata Sociological studies indicate that from generation to generation, there is some linkage between the social stratum of the parent and that of the offspring. Assume that the population is divided into three classes—low, middle, and high—with the probability of moving from one class to another in the next generation given by this transition matrix:

		Next generation		
		L	M	H
This generation	L	.65	.23	.12
	M	.08	.80	.12
	H	.03	.37	.60

(a) If these probabilities hold true generation after generation, what is the long-term percentage of people in each class?

(b) What is the long-term probability of a person going from L in one generation to H in the next, as opposed to those staying in L?

28. Genetics Suppose that the transition matrix for the inheritance of a particular trait—being right handed (RH), left handed (LH), or ambidextrous (AM)—is given by this transition matrix:

		Offspring		
		RH	LH	AM
Parent	RH	.70	.30	0
	LH	.25	.50	.25
	AM	.40	.40	.20

(a) Find the stable-state vector S.

(b) Find the limiting matrix L, and interpret the results.

29. Psychology Experiment A rat placed in a small area is faced with three choices of holes to go into. In one hole (C), the rat is rewarded with a piece of cheese; in another (M), a toy mouse awaits; in the third (S), a shock awaits. Repeated trials of this experiment reveal that if the rat goes into C on one trial, on the next trial, it will go into C 80% of the time, into M 15% of the time, and into S 5% of the time. If the rat goes into M on one trial, on the next trial, it will go into C 40% of the time, into M 50% of the time, and into S 10% of the time. If the rat goes into S on one trial, then on the next trial, it will go into C 40% of the time, into M 40% of the time, and back into S 20% of the time.

(a) Construct the transition diagram and the transition matrix for this experiment.

(b) Find the stable-state vector S, and interpret the entries.

(c) Find the limiting matrix L, and interpret the entries.

30. *Insurance* An auto insurance company rates the drivers it insures as low risk (L), medium risk (M), and high risk (H). Records indicate that each year, in the low-risk category, about 95% of those drivers remain in the low-risk category, while 4% move down to M, and 1% fall to H. Of those in the medium-risk category, 8% rise to L, 90% remain in M, and 2% fall to H. Of those in the high-risk category, 1% rise to L, 15% rise to M, while 84% remain in H.

(a) Construct the transition diagram and the transition matrix for these data.

(b) In the long run, what will be the company's distribution of drivers in each category?

(c) Find the limiting matrix L, and interpret the entries.

Problems for Exploration

1. Assuming that T is a transition matrix for a regular Markov chain, and each of its entries is positive, explain why all powers of T must have entries that are all positive.

2. Let T be the transition matrix for a regular Markov chain. Show that the limiting matrix can be found by solving the matrix equation $AX = B$ where

$$A = \begin{bmatrix} 1 & 1 & \cdots & 1 \\ & T^T - I & \end{bmatrix} \quad \text{and} \quad B = \begin{bmatrix} 1 \\ 0 \\ \cdot \\ \cdot \\ \cdot \\ 0 \end{bmatrix}.$$

HINT (The exponent T means transpose.)

SECTION 9.3 **Absorbing Markov Chains**

A state in a Markov chain is said to be an **absorbing state** if, once entered, it is impossible to leave. A transition diagram for such a state is shown next in Figure 6. This diagram shows that it is possible to move back and forth between states S_1 and S_2 and that it is possible to move to S_3 from either of these states. However, once S_3 has been entered, the process can never leave this state. The transition matrix for this

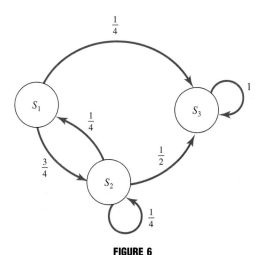

FIGURE 6
Transition Diagram Showing an Absorbing State

diagram also shows that S_3 is an absorbing state. A moment's thought convinces us that the third row of T represents an absorbing state for S_3:

$$\begin{array}{c} \\ \text{Current} \\ \text{period} \end{array} T = \begin{array}{c} \\ S_1 \\ S_2 \\ S_3 \end{array} \begin{array}{c} \overset{\text{Next period}}{\begin{array}{ccc} S_1 & S_2 & S_3 \end{array}} \\ \begin{bmatrix} 0 & \frac{3}{4} & \frac{1}{4} \\ \frac{1}{4} & \frac{1}{4} & \frac{1}{2} \\ 0 & 0 & 1 \end{bmatrix} \end{array}$$

The zeros show that it is impossible to pass from S_3 to either S_1 or S_2, and the only possibility is to pass from S_3 to S_3, with a probability of 1. If S_2 were an absorbing state, what would the second row of the transition matrix look like? The answer is "all zeros except for a 1 in the second-row, second-column entry." We now know the general pattern: S_j is an absorbing state if every entry in the jth row of the transition matrix is zero except the entry in the jth-row, jth-column position, and that entry is a 1. Just having an absorbing state is not enough to call a Markov chain an *absorbing* Markov chain, however. The definition insists that, in addition to having at least one absorbing state, it must be possible to pass from any nonabsorbing state to any absorbing state in the chain.

> ### An Absorbing Markov Chain
>
> A Markov chain is **absorbing** if:
>
> **1.** It has at least one absorbing state, and
>
> **2.** It is possible to go from any nonabsorbing state to any absorbing state in a finite number of steps.

If a Markov chain does not satisfy the conditions of an absorbing chain, then it is classified as a **nonabsorbing chain.**

EXAMPLE 1
Recognizing Absorbing Markov Chains from the Transition Matrix

Use the given transition matrix to classify the corresponding Markov chain as absorbing or nonabsorbing.

(a) $\begin{array}{c} \\ S_1 \\ S_2 \\ S_3 \end{array} \begin{array}{c} \begin{array}{ccc} S_1 & S_2 & S_3 \end{array} \\ \begin{bmatrix} .2 & .3 & .5 \\ 0 & 1 & 0 \\ .6 & .1 & .3 \end{bmatrix} \end{array}$
(b) $\begin{array}{c} \\ S_1 \\ S_2 \\ S_3 \end{array} \begin{array}{c} \begin{array}{ccc} S_1 & S_2 & S_3 \end{array} \\ \begin{bmatrix} .3 & .7 & 0 \\ .5 & .5 & 0 \\ 0 & 0 & 1 \end{bmatrix} \end{array}$
(c) $\begin{array}{c} \\ S_1 \\ S_2 \\ S_3 \\ S_4 \end{array} \begin{array}{c} \begin{array}{cccc} S_1 & S_2 & S_3 & S_4 \end{array} \\ \begin{bmatrix} .1 & .3 & .2 & .4 \\ 0 & 1 & 0 & 0 \\ .2 & .3 & 0 & 0 \\ 0 & 0 & 0 & 1 \end{bmatrix} \end{array}$

Solution   The matrix in (a) represents as absorbing Markov chain. The reason is that row 2 shows state 2 to be absorbing and rows 1 and 3 show that it is possible to pass from states 1 and 3 to state 2.

   The matrix in (b) does not represent an absorbing Markov chain. Even though state 3 is absorbing, it is not possible to pass from either state 1 or state 2 to state 3. It follows that the second condition for an absorbing Markov chain is not satisfied.

   The matrix in (c) does represent an absorbing Markov chain. States 2 and 4 are absorbing, and it is possible to pass directly from state 1 to either state 2 or state 4. It is also possible to pass directly from state 3 to state 1 and then from state 1 to both states 2 to 4.   ∎

When dealing with the transition matrix representing an absorbing Markov chain, some information can be obtained by rearranging the rows and columns so that the absorbing states appear at the top and the diagonal elements represent the transition of a state into itself. For the matrices (a) and (c) of Example 1, this new appearance would be

$$\textbf{(a)}\quad \begin{array}{c} \\ S_2 \\ S_1 \\ S_3 \end{array} \begin{array}{ccc} S_2 & S_1 & S_3 \\ \begin{bmatrix} 1 & 0 & 0 \\ .3 & .2 & .5 \\ .1 & .6 & .3 \end{bmatrix} \end{array} \qquad \textbf{(c)}\quad \begin{array}{c} \\ S_2 \\ S_4 \\ S_1 \\ S_3 \end{array} \begin{array}{cccc} S_2 & S_4 & S_1 & S_3 \\ \begin{bmatrix} 1 & 0 & 0 & 0 \\ 0 & 1 & 0 & 0 \\ .3 & .4 & .1 & .2 \\ .3 & 0 & .2 & 0 \end{bmatrix} \end{array}$$

Other examples written in this form are

$$\begin{bmatrix} 1 & 0 & 0 & 0 & 0 \\ 0 & 1 & 0 & 0 & 0 \\ .2 & .3 & .1 & .2 & .2 \\ .1 & .6 & .3 & 0 & 0 \\ .5 & .2 & .1 & .1 & .1 \end{bmatrix} \quad \text{and} \quad \begin{bmatrix} 1 & 0 & 0 & 0 & 0 \\ 0 & 1 & 0 & 0 & 0 \\ 0 & 0 & 1 & 0 & 0 \\ .1 & .4 & 0 & .3 & .2 \\ .2 & .2 & .1 & .2 & .3 \end{bmatrix}.$$

Notice the similar structure of these matrices. Each may be divided into four submatrices with the upper left submatrix being an identity I and the upper right submatrix a zero matrix $\mathbf{0}$:

$$A = \left[\begin{array}{c|c} I & \mathbf{0} \\ \hline R & Q \end{array}\right].$$

When an absorbing transition matrix T is written in the form of A above, it is said to be in **canonical form.** Notice that I and Q are *always* square matrices, although they need not be of the same order. If A is $n \times n$ and I is the $k \times k$ identity matrix, then Q is an $(n - k) \times (n - k)$ matrix.

EXAMPLE 2
Finding the Canonical Form

Find the canonical form of this absorbing matrix:

$$\begin{array}{c} \\ S_1 \\ S_2 \\ S_3 \\ S_4 \end{array} \begin{array}{cccc} S_1 & S_2 & S_3 & S_4 \\ \begin{bmatrix} .2 & .3 & .1 & .4 \\ 0 & 1 & 0 & 0 \\ .6 & .1 & .1 & .2 \\ 0 & 0 & 0 & 1 \end{bmatrix} \end{array}$$

Solution Keeping track of the states is easy if we do row and column interchanges. Interchanging the first and last rows puts the two absorbing states at the top of the matrix and gives it this appearance:

$$\begin{array}{c} & & \text{Next period} \\ & & \begin{array}{cccc} S_1 & S_2 & S_3 & S_4 \end{array} \\ \begin{array}{c} \\ \text{Current} \\ \text{period} \end{array} \begin{array}{c} S_4 \\ S_2 \\ S_3 \\ S_1 \end{array} & \begin{bmatrix} 0 & 0 & 0 & 1 \\ 0 & 1 & 0 & 0 \\ .6 & .1 & .1 & .2 \\ .2 & .3 & .1 & .4 \end{bmatrix} \end{array}$$

Now, if we interchange the first and fourth columns, the canonical form is achieved, and the matrix looks like this:

$$
\begin{array}{c}
\\
s_4 \\
s_2 \\
s_3 \\
s_1
\end{array}
\begin{array}{cccc}
s_4 & s_2 & s_3 & s_1 \\
\end{array}
\left[
\begin{array}{cccc}
1 & 0 & 0 & 0 \\
0 & 1 & 0 & 0 \\
.2 & .1 & .1 & .6 \\
.4 & .3 & .1 & .2
\end{array}
\right]
$$

It follows that

$$
R = \begin{bmatrix} .2 & .1 \\ .4 & .3 \end{bmatrix}
$$

and

$$
Q = \begin{bmatrix} .1 & .6 \\ .1 & .2 \end{bmatrix}.
$$

Note that the last two rows or columns could be interchanged, and the canonical form would still be intact. Of course, the position of the states would change accordingly. The rule to follow is this: Once all interchanges have been completed, the labels for the rows must be in the same order as those for the columns. ▬

In any absorbing Markov chain, the process will eventually pass from any *initial state* to some *absorbing state*. How many stages or steps are required to pass from a nonabsorbing state to an absorbing state and with what probability? These questions are answered using the **fundamental matrix.**

The Fundamental Matrix

For a transition matrix of an absorbing Markov chain written in the canonical form $A = \begin{bmatrix} I & 0 \\ \hline R & Q \end{bmatrix}$, the **fundamental matrix** is $F = (I - Q)^{-1}$.

(The order of the identity matrix I used in formulating F must match the order of Q and not necessarily the order of I found in A.)

EXAMPLE 3
Computing the Fundamental Matrix

Compute the fundamental matrix from the transition matrix of Example 2.

Solution After putting the original transition matrix of Example 2 in canonical form, we find that $Q = \begin{bmatrix} .1 & .6 \\ .1 & .2 \end{bmatrix}$. Therefore,

$$
\begin{aligned}
F = (I - Q)^{-1} &= \left(\begin{bmatrix} 1 & 0 \\ 0 & 1 \end{bmatrix} - \begin{bmatrix} .1 & .6 \\ .1 & .2 \end{bmatrix} \right)^{-1} \\
&= \begin{bmatrix} .9 & -.6 \\ -.1 & .8 \end{bmatrix}^{-1} = \begin{bmatrix} \frac{40}{33} & \frac{30}{33} \\ \frac{5}{33} & \frac{45}{33} \end{bmatrix}. \quad \blacksquare
\end{aligned}
$$

The importance of F is twofold, as is revealed in the next theorem.

> ### The Importance of the Fundamental Matrix
>
> **1.** A Markov chain with absorbing states always has a limiting matrix of this form:
>
> $$\left[\begin{array}{c|c} I & 0 \\ \hline FR & 0 \end{array}\right],$$
>
> and the entry in the ith row and jth column of FR gives the probability of passing from the ith nonabsorbing state to the jth absorbing state.
>
> **2.** The sum of the entries in the ith row of F gives the expected number of stages or steps that the chain will need before the ith nonabsorbing state will reach an absorbing state.

EXAMPLE 4
Using the Fundamental Matrix

From Example 2, we found that, with the states intact,

$$R = \begin{array}{c} \\ s_3 \\ s_1 \end{array}\begin{array}{cc} s_4 & s_2 \\ \left[\begin{array}{cc} .2 & .1 \\ .4 & .3 \end{array}\right] \end{array} \text{ and from Example 3, } F = \left[\begin{array}{cc} \frac{40}{33} & \frac{30}{33} \\ \frac{5}{33} & \frac{45}{33} \end{array}\right].$$

Therefore,

$$FR = \left[\begin{array}{cc} \frac{40}{33} & \frac{30}{33} \\ \frac{5}{33} & \frac{45}{33} \end{array}\right]\left[\begin{array}{cc} .2 & .1 \\ .4 & .3 \end{array}\right] = \begin{array}{c} \\ s_3 \\ s_1 \end{array}\begin{array}{cc} s_4 & s_2 \\ \left[\begin{array}{cc} \frac{20}{33} & \frac{13}{33} \\ \frac{19}{33} & \frac{14}{33} \end{array}\right] \end{array}.$$

Here is how the entries in the last matrix are interpreted: The probability of eventually passing from the nonabsorbing state S_3 to the absorbing state S_4 is $\frac{20}{33}$. The probability of eventually passing from the nonabsorbing state S_3 to the absorbing state S_2 is $\frac{13}{33}$. The probability of eventually passing from the nonabsorbing state S_1 to the absorbing state S_4 is $\frac{19}{33}$. The probability of eventually passing from the nonabsorbing state S_1 to the absorbing state S_2 is $\frac{14}{33}$.

Part (2) of the theorem immediately preceding Example 4 tells us that if we start in state S_3, the expected number (in the sense given in Section 7.4 of Chapter 7) of steps needed to pass to an absorbing state is the sum of the entries in the first row of F—namely, $\frac{40}{33} + \frac{30}{33} = \frac{70}{33} \approx 2.12$ or 3 steps (rounding up). If we start in state S_1, the expected number of steps needed to pass to an absorbing state is similarly found by the sum of the entries in the second row, $\frac{5}{33} + \frac{45}{33} = \frac{50}{33} \approx 1.52$ or 2 steps. ∎

EXAMPLE 5
A Rat in a Maze

Suppose that a rat is placed in room 1 of the following maze. Once in a room, the rat is just as likely to stay in that room as it is to enter any other room that has at least one opening. What is the expected number of visits to room 3 before it is trapped (enters the absorbing state)?

Solution There is only one absorbing state (T), while states 1, 2, and 3 are nonabsorbing. The transition matrix for this maze is given by

$$
\begin{array}{c}
\\
\\
\text{Current} \\
\text{room}
\end{array}
\begin{array}{c}
\\
1 \\
2 \\
3 \\
T
\end{array}
\overset{\displaystyle\begin{array}{cccc} \text{Next room} \\ \begin{array}{cccc} 1 & 2 & 3 & T \end{array} \end{array}}{
\begin{bmatrix}
\frac{1}{2} & \frac{1}{4} & \frac{1}{4} & 0 \\
\frac{1}{2} & \frac{1}{2} & 0 & 0 \\
\frac{1}{4} & 0 & \frac{1}{2} & \frac{1}{4} \\
0 & 0 & 0 & 1
\end{bmatrix}
}.
$$

The canonical form has this appearance:

$$
\begin{array}{c}
T \\
1 \\
2 \\
3
\end{array}
\overset{\begin{array}{cccc} T & 1 & 2 & 3 \end{array}}{
\begin{bmatrix}
1 & 0 & 0 & 0 \\
0 & \frac{1}{2} & \frac{1}{4} & \frac{1}{4} \\
0 & \frac{1}{2} & \frac{1}{2} & 0 \\
\frac{1}{4} & \frac{1}{4} & 0 & \frac{1}{2}
\end{bmatrix}
}.
$$

This shows that $R = \begin{bmatrix} 0 \\ 0 \\ \frac{1}{4} \end{bmatrix}$ and $Q = \begin{bmatrix} \frac{1}{2} & \frac{1}{4} & \frac{1}{4} \\ \frac{1}{2} & \frac{1}{2} & 0 \\ \frac{1}{4} & 0 & \frac{1}{2} \end{bmatrix}$, so that

$$
F = (I - Q)^{-1} = \left(\begin{bmatrix} 1 & 0 & 0 \\ 0 & 1 & 0 \\ 0 & 0 & 1 \end{bmatrix} - \begin{bmatrix} \frac{1}{2} & \frac{1}{4} & \frac{1}{4} \\ \frac{1}{2} & \frac{1}{2} & 0 \\ \frac{1}{4} & 0 & \frac{1}{2} \end{bmatrix} \right)^{-1}
$$

$$
= \begin{bmatrix} \frac{1}{2} & -\frac{1}{4} & -\frac{1}{4} \\ -\frac{1}{2} & \frac{1}{2} & 0 \\ -\frac{1}{4} & 0 & \frac{1}{2} \end{bmatrix}^{-1} = \begin{array}{c} 1 \\ 2 \\ 3 \end{array} \overset{\begin{array}{ccc} 1 & 2 & 3 \end{array}}{\begin{bmatrix} 8 & 4 & 4 \\ 8 & 6 & 4 \\ 4 & 2 & 4 \end{bmatrix}}.
$$

The matrix F tells us that the expected number of visits to room 3 before entering the trap is $4 + 2 + 4 = 10$. It is interesting to note that

$$
FR = \begin{bmatrix} 8 & 4 & 4 \\ 8 & 6 & 4 \\ 4 & 2 & 4 \end{bmatrix} \begin{bmatrix} 0 \\ 0 \\ \frac{1}{4} \end{bmatrix} = \begin{array}{c} 1 \\ 2 \\ 3 \end{array} \overset{T}{\begin{bmatrix} 1 \\ 1 \\ 1 \end{bmatrix}},
$$

which means that, regardless of which room the rat visits, it will ultimately end in the trap. ■

EXERCISES 9.3

Use a calculator or a computer to find the inverse in these exercises.

In Exercises 1 through 8, decide which of the transition matrices are absorbing and which are nonabsorbing. For those that are absorbing, give the absorbing states, and then write the matrix in canonical form, labeling each row and column in terms of the original state.

1. $\begin{bmatrix} .6 & .2 & .2 \\ 0 & 1 & 0 \\ 0 & .2 & .8 \end{bmatrix}$ **2.** $\begin{bmatrix} 0 & 0 & 1 \\ .3 & .7 & 0 \\ .2 & .5 & .3 \end{bmatrix}$

3. $\begin{bmatrix} 1 & 0 & 0 \\ .5 & .2 & .3 \\ .1 & .1 & .8 \end{bmatrix}$ **4.** $\begin{bmatrix} 1 & 0 & 0 \\ .6 & 0 & .4 \\ 0 & 1 & 0 \end{bmatrix}$

5. $\begin{bmatrix} .3 & .6 & 0 & .1 \\ 0 & 1 & 0 & 0 \\ .2 & .3 & .1 & .4 \\ 0 & 0 & 0 & 1 \end{bmatrix}$ **6.** $\begin{bmatrix} 1 & 0 & 0 & 0 \\ .5 & 0 & .5 & 0 \\ 0 & 1 & 0 & 0 \\ .6 & 0 & 0 & .4 \end{bmatrix}$

7. $\begin{bmatrix} .3 & .5 & .2 & 0 \\ 0 & 1 & 0 & 0 \\ 0 & 0 & 1 & 0 \\ 0 & 0 & 0 & 1 \end{bmatrix}$ **8.** $\begin{bmatrix} 0 & .2 & .5 & .3 \\ .1 & .1 & .6 & .2 \\ 0 & 1 & 0 & 0 \\ 0 & 0 & 1 & 0 \end{bmatrix}$

Each of the transition matrices in Exercises 9 through 14 represents an absorbing Markov chain.
(a) Write each in canonical form, and label the rows and columns in terms of the original states.
(b) Find the fundamental matrix of each.
(c) Determine the probability of passing from each non-absorbing state to a particular absorbing state for each matrix.

9. $\begin{bmatrix} 1 & 0 & 0 \\ 0 & 1 & 0 \\ .6 & .2 & .2 \end{bmatrix}$ **10.** $\begin{bmatrix} .2 & .3 & .5 \\ 0 & 1 & 0 \\ .1 & .6 & .3 \end{bmatrix}$

11. $\begin{bmatrix} .2 & .3 & .1 & .4 \\ 0 & 1 & 0 & 0 \\ 0 & 0 & 1 & 0 \\ .1 & 0 & .3 & .6 \end{bmatrix}$ **12.** $\begin{bmatrix} 1 & 0 & 0 & 0 \\ .3 & .4 & .2 & .1 \\ 0 & 0 & 1 & 0 \\ .1 & .5 & .2 & .2 \end{bmatrix}$

13. $\begin{bmatrix} 1 & 0 & 0 & 0 \\ .2 & .3 & .4 & .1 \\ 0 & 1 & 0 & 0 \\ 0 & 0 & 0 & 1 \end{bmatrix}$ **14.** $\begin{bmatrix} .2 & .5 & .2 & .1 \\ 0 & 1 & 0 & 0 \\ .1 & .1 & .8 & 0 \\ .3 & 0 & .7 & 0 \end{bmatrix}$

Each transition matrix in Exercises 15 through 20 represents an absorbing Markov chain. For each, calculate the expected number of trials it takes to pass from each non-absorbing state to an absorbing state.

15. $\begin{bmatrix} 1 & 0 & 0 \\ .2 & .5 & .3 \\ .1 & .1 & .8 \end{bmatrix}$ **16.** $\begin{bmatrix} .3 & .2 & .5 \\ 0 & 1 & 0 \\ .3 & 0 & .7 \end{bmatrix}$

17. $\begin{bmatrix} 1 & 0 & 0 & 0 \\ .3 & .5 & .1 & .1 \\ .2 & .3 & 0 & .5 \\ 0 & 0 & 0 & 1 \end{bmatrix}$

18.  $\begin{bmatrix} .3 & .6 & 0 & .1 \\ 0 & 1 & 0 & 0 \\ .2 & 0 & .3 & .5 \\ 0 & 0 & 0 & 1 \end{bmatrix}$

19. $\begin{bmatrix} \frac{1}{2} & 0 & \frac{1}{2} & 0 \\ 0 & 1 & 0 & 0 \\ 0 & 0 & 1 & 0 \\ \frac{1}{4} & \frac{1}{4} & \frac{1}{4} & \frac{1}{4} \end{bmatrix}$ **20.** $\begin{bmatrix} \frac{1}{4} & \frac{1}{4} & \frac{1}{4} & \frac{1}{4} \\ \frac{3}{4} & 0 & \frac{1}{4} & 0 \\ 0 & 0 & 1 & 0 \\ \frac{1}{2} & \frac{1}{4} & \frac{1}{4} & 0 \end{bmatrix}$

Psychology   *Exercises 21 through 26 refer to the maze shown. A mouse is placed in either room A or B, where it is twice as likely to move from A to B as it is to remain in A and twice as likely to move from B to A as from B to the trap, and it never stays in B.*

A	B	Trap

21. Find the transition matrix for this maze, where each state will represent one of the three rooms.

22. Put the transition matrix in canonical form, and label the rearranged states.

23. If the mouse was first placed in room *A*, what is the probability that it will eventually go to the absorbing state trap?

24. If the mouse was first placed in room *B*, what is the probability that it will eventually go to the absorbing state?

25. If the mouse was first placed in room *A*, what is the expected number of rooms visited by the mouse before going into the trap?

26. If the mouse was first placed in room *B*, what is the expected number of rooms visited by the mouse before going into the absorbing state?

Practical Application   *Exercises 27 through 31 refer to this situation: A rat lives in a farmer's barn and at night comes out to forage for its favorite foods—corn and lunch scraps left by the farmer—or to find a fatally spring-loaded mouse trap or poisoned bait left by the farmer. If the rat had lunch scraps one night, there is an 80% chance of returning for lunch scraps the next night, a 10% chance of finding the corn, and a 10% chance of going to the poisoned bait. If the rat had corn one night, there is a 50% chance of returning the next night for corn, a 30% chance of finding lunch scraps, and an equal chance of being trapped or poisoned. If the rat does not find corn or lunch scraps, it is equally likely to find the trap or the poisoned bait.*

27. Construct the transition matrix for this problem, and put it into canonical form.

28. If the rat is currently nibbling on corn, what is the expected number of days before being trapped or poisoned?

29. If the rat is currently eating lunch scraps, what is the expected number of days before being trapped or poisoned?

30. If the rat is currently eating lunch scraps, what is the probability that the rat will end up being poisoned?

31. If the rat is currently eating corn, what is the probability that the rat will end up caught in the trap?

Gambler's Ruin *Exercises 32 through 36 refer to this classic gambling problem. A game is played in which the probability of winning is $\frac{1}{3}$ and the probability of losing is $\frac{2}{3}$. The player bets \$1 each time the game is played, and if he wins, he gets his \$1 bet back plus another \$1. The game is played until the player either goes broke or wins \$3.*

32. Given that the player starts with \$2, construct the transition matrix using \$0, \$1, \$2, and \$3 as the states. Then put this matrix in canonical form.

33. Given that the player starts with \$2, what is the probability that he will win \$3 and stop playing the game?

34. Having started with \$2, how many plays of the game would be expected before winning \$3 or losing the \$2?

35. Given that the player started with \$2, what is the probability that he will go broke?

36. Suppose that the player started with only \$1. What is the probability that the player wins the \$3?

Gambler's Ruin Variation *Exercises 37 through 40 refer to this variation of the gambler's ruin problem. John and Carlos are each given \$2 to play this game. A fair coin is tossed, and if it comes up heads, John gives Carlos \$1; if the coin comes up tails, Carlos gives John \$1. The game is played until one player runs out of money (in which case, the other player has \$4).*

37. Set up the transition matrix using the states \$0, \$1, \$2, \$3, and \$4.

38. What is the expected number of trials it will take to either win or lose this game?

39. Suppose that Carlos starts with \$3 and John starts with \$2. What is the probability that Carlos will win the game?

40. Repeat Exercise 38 using a coin that is biased $\frac{5}{8}$ heads and $\frac{3}{8}$ tails.

Promotion *Exercises 41 through 43 refer to this promotion problem. Some executives in the company are junior-level (J), middle-level (M), and senior-level (S). Each year 70% of those in J remain in that state, 8% leave the company (L), 2% retire (R), 20% are promoted to M, and none go to S; 80% of those in M remain in that state, 5% leave the company, 8% retire, 2% are demoted to J, and 5% are promoted to S; 72% of those in S remain in that state, 10% leave the company, 15% retire, none are demoted to J, and 3% are demoted to M.*

41. Of those who are junior-level, what percent will eventually leave the company?

42. What is the probability that a middle-level executive will eventually retire from the company?

43. What is the expected number of years a middle-level executive will be on the job before retiring or leaving the company?

Credit Accounts *Exercises 44 through 46 refer to these data amassed by the Day's Department Stores: Credit records indicate that on a monthly basis, some of their accounts are paid up, some are delinquent less than 30 days, some are delinquent from 30 up to 60 days, and some are considered uncollectable bad debts. The probabilities for such accounts are shown next:*

	Paid	<30	<60	Bad
Paid	1	0	0	0
<30	$\frac{3}{8}$	$\frac{3}{8}$	$\frac{2}{8}$	0
<60	$\frac{1}{8}$	$\frac{4}{8}$	$\frac{2}{8}$	$\frac{1}{8}$
Bad	0	0	0	1

44. If an account is now in the <30 state, what is the probability that the account will eventually end up bad?

45. If an account is now in the <30 state, what is the expected number of months before the account is either paid or declared bad?

46. If an account is now in the <60 state, what is the probability that the account will eventually be paid?

47. *Experimental Success* Each morning, a medical scientist checks the condition of a group of sick rats being used to test a new drug. Records indicate that each day, there is a probability of .4 that a rat will still be sick and a probability of .3 that it will be cured. Each morning, the scientist removes the rats that have died and those that have been cured from the group; to those still sick, she gives another dose of the drug. What percentage of these rats will eventually be cured?

48. *Consumer Preferences* Records kept on consumers who live in a very peculiar town and who purchased only detergent brands C, D, and E indicate that of those buying C this month, 50% will buy C next month, while 30% will buy D and 20% will buy E. Of those buying D this month, 30% will buy D again next month, while 20% will buy C, and 50% will buy E. Once a person buys E, he or she will never switch to either of the other two brands. For a person who bought C this month, what is the expected number of months before the person will eventually buy E?

Problems for Exploration

1. Let a be a positive number less than 1. If the transition matrix in canonical form for an absorbing Markov chain is

$$A = \begin{array}{c} \\ 1 \\ 2 \end{array} \begin{array}{cc} 1 & 2 \\ \begin{bmatrix} 1 & 0 \\ a & 1-a \end{bmatrix} \end{array},$$

find the expected number of stages an event starting in state 2 would go through before being absorbed.

2. Using the fact that matrix multiplication may be performed with submatrices in place of entries in the matrices, show that if A is a transition matrix in canonical form of an absorbing Markov chain with m absorbing states, then A^2 is also an absorbing matrix with m absorbing states.

CHAPTER 9 REFLECTIONS

CONCEPTS

Markov chains combine matrix algebra and probability in order to answer questions about repeated trials of an experiment. *Transition diagrams* and *transition matrices* are both convenient vehicles for summarizing the pertinent data relating to such situations. If some power of a transition matrix has all positive entries, the matrix is called regular, and successive powers always approach a limiting matrix having stable entries. Such *regular Markov chains* eventually reach the same stable vector, regardless of the initial distribution in a state vector. *Absorbing Markov chains* have the property that every outcome will eventually reach some absorbing state that is impossible to leave.

COMPUTATION/CALCULATION

Calculate the missing probabilities in each transition diagram, and write the transition matrix.

1.

2.

For Questions 3 and 4, given the transition matrix

$$T = \begin{bmatrix} .2 & .5 & .3 \\ .1 & .7 & .2 \\ .6 & 0 & .4 \end{bmatrix}$$

and the state vector $[.1 \quad .6 \quad .3]$, *find the state vector*

3. One transition later   **4.** Two transitions later

For Questions 5 through 7, find the limiting matrix L for each regular matrix.

5. $\begin{bmatrix} .1 & .9 \\ .7 & .3 \end{bmatrix}$ **6.** $\begin{bmatrix} 0 & 1 \\ .8 & .2 \end{bmatrix}$ **7.** $\begin{bmatrix} .5 & .5 \\ .6 & .4 \end{bmatrix}$

8. Find the fundamental matrix F for the given absorbing Markov chain.

$$\begin{bmatrix} .6 & .3 & .1 \\ 0 & 1 & 0 \\ .4 & .4 & .2 \end{bmatrix}$$

9. For the given Markov chain, find the probability of passing from each nonabsorbing state to a particular absorbing state.

$$\begin{bmatrix} .4 & .5 & .1 \\ .6 & .2 & .2 \\ 0 & 0 & 1 \end{bmatrix}$$

10. For the given absorbing Markov chain, calculate the expected number of steps required to pass from each nonabsorbing state to an absorbing state.

$$\begin{bmatrix} 1 & 0 & 0 \\ .3 & .5 & .2 \\ .8 & .2 & 0 \end{bmatrix}$$

CONNECTIONS

An archer's record indicates that her performance can be represented by a Markov chain. If she hits the bull's-eye, she hits the next one 82% of the time and misses 18% of the time. If she misses the bull'e-eye, she hits the next one 73% of the time and misses 27% of the time.

11. Given that the archer misses the bull's-eye on her first shot, find the probability that she hits the bull's-eye on her third shot.

12. Determine the archer's limiting matrix and the number of steps it takes her to reach this stable condition.

COMMUNICATION

13. Describe the conditions necessary for a repeated experiment to be a Markov chain.

14. Indicate how to tell whether a transition matrix is regular and why regular matrices are important.

15. In your own words, summarize the conditions necessary for a Markov chain to be absorbing.

COOPERATIVE LEARNING

16.
$$\begin{bmatrix} 1 & 0 & 0 \\ 0 & 1 & 0 \\ a & b & 1-(a+b) \end{bmatrix}$$

Given this transition matrix in canonical form for a Markov chain (with $a > 0$, $b > 0$, $a + b < 1$), if an event starts in state 3, what is the probability that it will be absorbed by

(a) State 2? **(b)** State 1?

(c) Starting in state 3, what is the expected number of stages before being absorbed?

CRITICAL THINKING

17. By analyzing several 2×2 and 3×3 transition matrices having at least one 0 entry, determine the conditions for such matrices to be regular, absorbing, or neither.

COMPUTER/CALCULATOR

18. To four-decimal accuracy, determine the limiting matrix L for the transition matrix

$$T = \begin{bmatrix} .5 & .2 & .3 \\ .1 & .6 & .3 \\ .3 & .5 & .2 \end{bmatrix}$$

by raising T to consecutive positive integer powers, using a graphics calculator or computer. What is the smallest power of T that gives the desired limiting matrix?

Cumulative Review

19. Write the equation of a line having x-intercept 7.2, which is parallel to the line $4x + 2y = 23$.

20. Solve this system:
$$\begin{aligned} x + 2y + z &= 4 \\ 2x + y + 3z &= 11 \\ 3x + 5y - 2z &= -2 \end{aligned}$$

21. Calculate $(A + B)^2$ for the matrices
$$A = \begin{bmatrix} 1 & 2 \\ 3 & 4 \end{bmatrix} \quad \text{and} \quad B = \begin{bmatrix} 0 & 5 \\ -2 & 1 \end{bmatrix}.$$

22. Shade the Venn diagram to represent the set $A' \cap (B' - C')$.

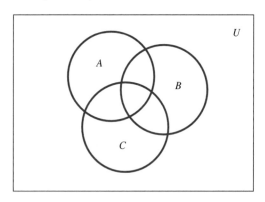

23. When a pair of dice is rolled, find the probability that the sum is not six or the product is not six.

CHAPTER **9** SAMPLE TEST ITEMS

1. Complete the transition diagram.

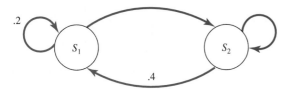

2. Find the transition matrix.

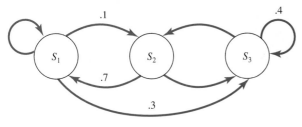

3. Complete the transition diagram.

4. Find the transition matrix.

Use each transition matrix to construct the transition diagram.

5. $\begin{bmatrix} .3 & .7 \\ .8 & .2 \end{bmatrix}$

6. $\begin{bmatrix} .5 & .2 & .3 \\ .4 & 0 & .6 \\ 0 & .1 & .9 \end{bmatrix}$

If it snows today, there is a 30% chance that it will snow tomorrow. If it does not snow today, there is a 90% chance that it will not snow tomorrow.

7. Construct the transition diagram.

8. Write the transition matrix.

For the transition matrix

$$\begin{bmatrix} .4 & .5 & .1 \\ 0 & .7 & .3 \\ .6 & .4 & 0 \end{bmatrix}$$

and the state vector $[.2 \quad .5 \quad .3]$, *find the state vector*

9. One transition later.

10. Two transitions later.

Decide whether each transition matrix is regular.

11. $\begin{bmatrix} 1 & 0 \\ .8 & .2 \end{bmatrix}$

12. $\begin{bmatrix} .1 & 0 & .9 \\ .5 & .3 & .2 \\ .6 & .4 & 0 \end{bmatrix}$

Find the limiting matrix L by calculating the powers of each regular matrix.

13. $\begin{bmatrix} .1 & .9 \\ 1 & 0 \end{bmatrix}$

14. $\begin{bmatrix} .7 & .3 \\ .4 & .6 \end{bmatrix}$

For each regular matrix, find the stable-state vector S and the limiting matrix L.

15. $\begin{bmatrix} .47 & .53 \\ .26 & .74 \end{bmatrix}$

16. $\begin{bmatrix} 0 & 1 \\ .7 & .3 \end{bmatrix}$

17. Decide whether the transition matrix

$$\begin{bmatrix} 0 & 0 & 1 \\ .6 & .4 & 0 \\ .3 & .2 & .5 \end{bmatrix}$$

is absorbing or nonabsorbing.

This transition matrix represents an absorbing Markov chain:

$$\begin{bmatrix} .4 & .3 & .3 \\ 0 & 1 & 0 \\ .1 & .7 & .2 \end{bmatrix}.$$

18. Write this matrix is canonical form, and label the rows and columns in terms of the original states.

19. Find the fundamental matrix.

20. Determine the probability of passing from each non-absorbing state to a particular absorbing state.

CHAPTER 10 · An Introduction to Game Theory

A campaign debate between Senator Smith and her opponent will focus on foreign policy and defense. The senator's managers feel that her opponent is strong on the defense issue but not on foreign policy. Senator Smith would likely gain three points when both discuss foreign policy; her opponent, five points, when both discuss defense; the senator, ten points, if she discusses foreign policy while her opponent discusses defense; and the senator, six points, if she discusses defense while her opponent discusses foreign policy. What proportion of the debate should the senator spend on each issue to gain the best advantage from the debate regardless of how her opponent divides his time? (See Example 2, Section 10.3.)

WITH SMITH . . .

PEACE HAS A CHANCE.

A vote for **Senator Smith** is a vote for foreign policy expertise.

Paid for by the Committee to Reelect Senator Smith

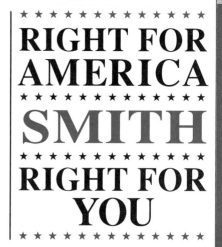

★ RIGHT FOR AMERICA ★ SMITH ★ RIGHT FOR YOU ★

CHAPTER PREVIEW

Our introduction to game theory encompasses games that are called "two-person" and "zero-sum." The mathematical analysis involves setting up the pay-off matrix and deciding from that matrix whether the game is classified as "pure strategy" or "mixed strategy." The solution to a pure strategy game may be found by inspection, while those of mixed strategy are solved by linear programming. The solution gives the best strategy for each player and the expected value of the game (in the sense of probability), indicating which player the game favors.

CALCULATOR/COMPUTER OPPORTUNITIES FOR THIS CHAPTER

A calculator with matrix capabilities or the computer program Matrix from the disk MasterIt will be useful in Section 2 for certain matrix multiplications. The computer programs LinProg and LPGraph will also be helpful in Section 3 when certain games are solved as linear programming problems.

SECTION 10.1 Strictly Determined Games

To most of us, the word *game* suggests that competitive forces are at work, each trying to devise some strategy that will allow one of the forces to win the game. Some games—such as a tennis match, a chess game, a basketball game, or negotiating teams between labor and management—are played by only two competing forces. These are known as **two-person games** (regardless of the actual number of players). On the other hand, some games may have several competing forces. An example might be the fast-food establishments in a city competing for business. In any event, there are many situations in the business world, psychology, economics, the military, and so forth, that may be considered "games" in the sense that we wish to consider in this chapter. For mathematical simplicity, we restrict our study to games involving only two competing forces—in other words, **two-person games.** In addition, and again for simplicity, we consider only games in which the amount gained by one of the players is equal to the amount lost by the opposing player. That is, at the end of the game, the sum of the earnings of two players is zero (the loss is denoted by a negative number). These are known as **zero-sum games.** Thus, all of our games are limited to those that are *two-person, zero-sum games.*

EXAMPLE 1
A Two-Person
Zero-Sum Game

Two friends, Match and Mismatch, play a game of "matching pennies." Upon a signal, each shows a penny. If the pennies turn up both heads or both tails, Match wins $1 from Mismatch; however, if only one of the coins turns up heads, Mismatch wins $1 from Match. This is a two-person game. Because any amount won by one player precisely equals the amount lost by the other, this game is also a zero-sum game.

Both Match and Mismatch have a choice to make each time they play this game: Whether to show heads or to show tails. If Match, for example, decides to show heads each time, it will not take Mismatch long to notice this and to begin to show tails each time, thereby winning $1 on every game. This would not be a good strategy for Match. Just what is the best course of action for Match?

EXAMPLE 2
A Two-Person,
Zero-Sum Game

The city of Appleton has two appliance stores: Appliance City and Appliance World. They offer discounts of 20% or 30%, depending on how aggressively they want business. If both stores offer the same discount, Appliance City gets 60% of the business. One the other hand, if Appliance City discounts more than Appliance World, then Appliance City will get 70% of the business. When Appliance World discounts more than Appliance City, the latter will get 55% of the business. This is evidently a two-person (actually two-business) game, and the sales gain of one is made at the expense of the other. ■

In this game, each business has a choice of how much to discount, and each choice leads to a different configuration of market share. For instance, if Appliance City's strategy is to discount 20% and Appliance World's strategy is to discount 30%, then Appliance City's market share would be 55% and Appliance World's share would then be 45%. From among the available courses of action, which is the best for each business?

The games of Example 1 and Example 2 demonstrate what this chapter covers: Each player has a course of action to decide, and these decisions affect the outcome of the game. Any course of action taken by a player is known as a **strategy** for that player. How then, does each player decide on a best strategy after taking into account all of the options the opposing player has available? When we find the best or **optimal strategy** for each player, we say we have "solved the game." Solving the game is our ultimate objective.

Games such as these may be mathematically analyzed if we express the **payoff** to one of the players in matrix form. Such a matrix is called the **payoff matrix.** To obtain the entries in such a matrix, we arbitrarily choose one of the players (called the **row player**) and use the rows in the matrix to represent that player's gains as positive numbers and losses as negative numbers. Of course, if no gain occurs, a zero is recorded. In the game of Example 1, suppose we call Match the row player. The resulting payoff (in dollars) matrix is

$$
\begin{array}{c}
 & \text{Mismatch} \\
 & \begin{array}{cc} H & T \end{array} \\
\text{Match} \begin{array}{c} H \\ T \end{array} &
\begin{bmatrix} 1 & -1 \\ -1 & 1 \end{bmatrix}
\end{array}
$$

For the players in Example 2, if we let Appliance City be the row player, then the payoff matrix (in percentage of market share) becomes

$$
\begin{array}{c}
 & \text{Appliance World} \\
 & \begin{array}{cc} 20\% & 30\% \end{array} \\
\begin{array}{cc} \text{Appliance} & 20\% \\ \text{City} & 30\% \end{array} &
\begin{bmatrix} .60 & .55 \\ .70 & .60 \end{bmatrix}.
\end{array}
$$

The payoff matrix for a game helps us classify the game into one of the two classifications: a **strictly determined** game and a game or **mixed strategy.** As suggested by the name, a game is strictly determined if it has a solution consisting of a *single course of action* that never varies from game to game. This solution is called a **pure strategy.** On the other hand, a game of mixed strategy has a solution involving more than one course of action, each of which is played a certain percentage of the time.

The payoff matrix for Appliance City and Appliance World shows us that a single course of action by each (a pure strategy) is the best. Obviously, Appliance City will want to choose a 30% discount because both market shares shown in the second row are larger than those in the first row. However, Appliance World is going to counter with a 30% discount to minimize that impact. This is the best strategy for both stores in the sense that Appliance City can do no worse than garner 60% of the market regardless of what Appliance World does. This is also the best Appliance World can do under the circumstances. So the term *best strategy* is not necessarily based on obtaining maximum possible payoff, but rather on doing the best a player possibly can given the available options. Notice that the entry signaling this strategy is the .60 in the second-row, second-column position. It is the smallest entry in the second row (signaling "no worse than" for the row player) and the largest in the second column (signaling the best the column player can do under the circumstances).

In terms of the payoff matrix, it is easy to tell when a game is strictly determined.

Strictly Determined Games

A game is called **strictly determined** if its payoff matrix has an entry that is the smallest element in its row and is also the largest element in its column. Such an entry is called a **saddle point** or **value** of the game. The row in which the saddle point lies is the best strategy for the row player, and the column in which the saddle point lies is the best strategy for the column player. If a game has positive value, it favors the row player; if it has a negative value, it favors the column player. If a game has zero value, it is called a **fair game.**

According to these definitions, the game in Example 1 is not strictly determined, because there is no entry in the payoff matrix that could be classified as a saddle point. This means that no single course of action, such as always showing heads, is the best strategy for either player.

The game of Example 2 is, however, a strictly determined game, or game of pure strategy, because the entry .60 in the second-row, second-column position is the smallest in the second row and the largest in the second column. Therefore, that element .60 is a saddle point or value for the game. This means that the row player, Appliance City in this case, will get 60% of the business when both businesses discount by 30%. The payoff matrix, from the location of the saddle point, also tells us that the best strategy for the row player is to pursue the single course of action (pure strategy) of discounting 30% all of the time, and for the column player to do the same. This game is now solved.

EXAMPLE 3
A Strictly Determined
Two-Person Game

The following payoff matrix comes from a two-person, zero-sum game.

$$\begin{bmatrix} 1 & -1 & -3 \\ -1 & 1 & -2 \\ 2 & 5 & -4 \end{bmatrix}$$

Determine whether this game is strictly determined. If it is, what is the saddle point and best strategy for the row and column players?

Solution   In the first row, the entry with the minimum value is -3, located in the third column, but it is not the maximum value of entries in the third column. So -3 is not a saddle point. In the second row, the entry -2 in the third column is the entry with the minimum value, and it is also the largest value in the third column and is therefore a saddle point. The third row does not have a candidate for a saddle point. It follows that the game is strictly determined and has value of -2. The row player's best strategy is row 2, while the column player's best strategy is column 3. This game favors the column player because the value of the game is -2.

A visual aid that will help decide on the location of any saddle points is circling the smallest number in each row and then boxing the largest number in each column. A number that is both circled and boxed will then be a saddle point. This scheme is used on the original matrix and is shown next.

$$\begin{bmatrix} 1 & -1 & \boxed{\enclose{circle}{-3}} \\ -1 & 1 & \boxed{\enclose{circle}{-2}} \\ \boxed{2} & \boxed{5} & \enclose{circle}{-4} \end{bmatrix}$$

EXAMPLE 4
A Strictly Determined
Two-Person Game

Graph-It has decided to bring out a new graphics calculator to compete with Tech Image's products. Graph-It is to aim its product at either the public school market only or the business market only. It will make a decision as to which market to pursue based on a financial analysis of the markets. Tech Image has the choice of putting its resources into the public school market only, the business market only, or to split its resources 50:50 among these two markets. Estimates are that if Graph-It invests its money in the public school market, then its profit will be $1 million, $8 million, or $6 million, respectively, depending on Tech Image's investments in the public school market only, the business market only, or the 50:50 split. On the other hand, if Graph-It invests its money in the business market only and Tech Image invests in public schools only, business only, or the split, then Graph-It's profit will be, respectively, $10 million, $2 million, or $5 million. Under these assumptions, what is each company's best strategy for monetary gain?

Solution   Suppose that we consider Graph-It as the row player and abbreviate public schools with P, business with B, and split with S. Then the payoff matrix will look like this:

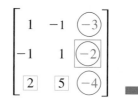

The 6 in the first-row, third-column position is the only saddle point in the payoff matrix. Consequently, the game is strictly determined with a value of $6 million favoring Graph-It. The best strategy for Graph-It is to invest its money in the public school market and for Tech Image to split its resources 50-50 between the public school and the business sectors. Notice that Graph-It could make $10 million by investing in the business market with the guarantee that Tech Image invested in the public school market. However, Graph-It cannot be sure that Tech Image will do just that. Should Tech Image also decide to put its resources into the business sector, then

Graph-It would gain only $2 million. The saddle point option tells us that regardless of what Tech Image does, if Graph-It puts its resources in the public school market, Graph-It can be assured of a $6 million profit. This again points out the true meaning of *best strategy*: The best strategy is *not necessarily based on obtaining maximum possible payoff,* but rather on *doing the best* a player possibly can *under the available options.* ■■■

EXERCISES   10.1

In Exercises 1 through 10, decide whether the game represented by the given payoff matrix is strictly determined. If so, give the value of the game and who it favors (the row or the column player), and suggest the best strategy for the row and column players.

1. $\begin{bmatrix} 3 & -1 \\ 5 & 2 \end{bmatrix}$

2. $\begin{bmatrix} 4 & 3 \\ 2 & 1 \end{bmatrix}$

3. $\begin{bmatrix} 5 & 3 \\ 2 & 7 \end{bmatrix}$

4. $\begin{bmatrix} 0 & 4 \\ -2 & 3 \end{bmatrix}$

5. $\begin{bmatrix} -3 & 3 & -5 \\ 2 & 4 & 6 \\ 1 & 2 & -2 \end{bmatrix}$

6. $\begin{bmatrix} 2 & 0 & -1 \\ 4 & -3 & 2 \\ 1 & 2 & 3 \end{bmatrix}$

7. $\begin{bmatrix} 1 & 4 & 7 \\ 0 & 2 & 4 \\ -1 & 3 & 5 \end{bmatrix}$

8. $\begin{bmatrix} 4 & -4 & -3 \\ 2 & -3 & 4 \\ 1 & -2 & 3 \end{bmatrix}$

9. $\begin{bmatrix} 2 & 1 & -2 & 4 \\ 1 & -3 & -5 & 1 \\ 3 & 2 & 4 & -2 \end{bmatrix}$

10. $\begin{bmatrix} 3 & -2 & 5 & -3 \\ 2 & 1 & 4 & -2 \\ 2 & 3 & 1 & -1 \end{bmatrix}$

11. For what values of x is this payoff matrix strictly determined?

$$\begin{bmatrix} x & 9 & 3 \\ 0 & x & -8 \\ -4 & 5 & x \end{bmatrix}.$$

12. For what values of x is this payoff matrix strictly determined?

$$\begin{bmatrix} x & 4 & 5 \\ 3 & -2 & x \\ -3 & 1 & -4 \end{bmatrix}.$$

13. *Business*   A retail store is exploring the possibility of remodeling to give a more modern look to its physical facilities. If the store remodels and the economic situation is favorable, the business will earn $100,000. However, if the store remodels and the economy goes into a recession, the business will lose $40,000. If the store is not remodeled and the economic situation is favorable, the business will lose $50,000 due to the loss of customers to other stores. If the store does not

remodel and there is a recession, the business will lose $45,000. Assume this is to be a two-person game, with the store playing against the economy. Set up the payoff matrix and decide which is the best strategy for the store: remodel or leave the facilities as they are.

14. *Politics*   The campaign managers of a political campaign are trying to decide whether to reveal a skeleton in their candidate's closet. They figure that if the skeleton is revealed and the news media are critical, the candidate will lose 20,000 votes; but if the media are not critical, she will gain 3000 sympathy votes. On the other hand, if the media learn of the skeleton first and are critical, the candidate will lose 120,000 votes; but if the media does not learn of the skeleton and it is not revealed, she will neither gain nor lose votes. Consider this a game between the candidate and the media. In the interest of getting the most votes possible in the upcoming election, should the candidate confess or try to keep the skeleton secret? Construct the payoff matrix to help you decide.

15. *Health*   Two health clinics, Nutone and Fabless, are considering offering expanded services in their town. At present, their clientele is rather stable, but they expect to divide 600 new customers with city growth and the new programs. Nutone is considering offering a sports-medicine program, a senior-citizen program, and a drug-therapy program, while Fabless is considering a sports-medicine and a senior-citizen program. Nutone figures that if it starts a sports-medicine program, it will get 250 new accounts if Fabless starts a sports-medicine program and 400 new accounts if Fabless starts a senior-citizen program. If Nutone starts a senior-citizen program, it will get 350 new accounts if Fabless starts a sports-medicine program and 300 new accounts if Fabless starts a senior-citizen program. If Nutone starts a drug-therapy program, it will get 375 new accounts if Fabless starts a sports-medicine program and 400 new accounts if Fabless starts a senior-citizen program.

(a) Construct the payoff matrix for this two-person game.

(b) Find the saddle point.

(c) What is the best strategy for each of these clinics?

16. **Marketing** Harley and Honda are considering offering new changes in their bikes. Harley is considering offering either a new body style or a new transmission, while Honda is considering changing the suspension, the engine, or the body style. Market studies show that if Harley offers a new body style, it will get 60% of the market if Honda offers a new suspension, 55% if Honda offers a new engine, and 50% if Honda offers a new body style. On the other hand, if Harley offers a new transmission, it will get 40% of the market if Honda offers a new suspension, 35% if Honda offers a new engine, and 30% if Honda offers a new body style.

    (a) Construct the payoff matrix for this two-person game.
    (b) What is the value of this game?
    (c) What are the best strategies for these two companies?

17. **Cards** John has two cards in his hand—a red 4 and a black 9—and Harry has three cards in his hand—a red 5, a black 7, and a black 8. Each simultaneously places one card on the table. If the cards are the same color, John gets the difference between the numbers from Harry; but if the cards are of different colors, Harry gets the minimum of the two numbers on the cards from John.

    (a) Construct a payoff matrix for this game.
    (b) Decide whether this game is strictly determined. If it is, what are the best strategies for both players?

18. **Cards** Sue and Jill each have three cards in their hands. Sue has an ace, a red 3, and a black 8. Jill has a red 5, a black 7, and a an ace. Each simultaneously lays a card on the table. If the cards are both aces, Sue gives Jill 1 cent; but if one plays an ace and the other does not, then the person playing the ace receives pennies equal to the number on the opponent's card. Otherwise, if both cards are the same color, Jill gives Sue pennies equal to the sum of the numbers on the cards; if the cards are not the same color, Sue gives Jill pennies equal to twice the difference of the numbers on the cards.

    (a) Construct a payoff matrix for this game.
    (b) Is this game strictly determined? If so, give the saddle point and the best strategy for both players.

19. **Locating Retail Outlets** Two companies, Computer Software and Computer Hardware, are each studying whether to locate a new store in Tallahassee or in Austin. Studies of the competitive nature of the business indicate that if Computer Software locates in Tallahassee and Computer Hardware locates in Austin, then Computer Software's annual profit will be $500,000 more than Computer Hardware's profit. If Computer Software locates in Austin and Computer Hardware locates in Tallahassee, then Computer Hardware's profit will be $250,000 greater than Computer Software's profit. If they both locate in Tallahassee, their annual profits will be equal; but if both locate in Austin, Computer Software's annual profit will be $50,000 greater than Computer Hardware's profit.

    (a) Construct the payoff matrix for this two-company game.
    (b) Decide whether this is a strictly determined game. If it is, determine the saddle point.
    (c) If strictly determined, what is the best strategy for both companies?

20. **Locating Retail Outlets** Competing dry cleaners Peter Pan and Cleanrite are each considering putting in a new branch, either in a shopping center or in the suburbs. If both select the shopping center site or the suburban site, then each will get 50% of the business. On the other hand, if one selects the shopping center and the other selects the suburban site, then the business is split 60:40 in favor of the shopping center site.

    (a) Construct the payoff matrix for this two-business game.
    (b) Find the saddle point, and determine the best strategy for each dry cleaning firm.

Problems for Exploration

1. For what values of a and b will this payoff matrix represent a strictly determined game?

$$\begin{bmatrix} a & 0 \\ 0 & b \end{bmatrix}$$

2. Show that the payoff matrix

$$\begin{bmatrix} x & y \\ x & z \end{bmatrix}$$

is strictly determined for all values of x, y, and z.

3. If a game has a payoff matrix

$$\begin{bmatrix} a & 3 \\ -1 & b \end{bmatrix},$$

where $-1 < a < b < 3$, find the saddle point(s) of this matrix, if any. Explain your answer.

4. Construct a payoff matrix that has exactly two saddle points and then make up a two-person game for which your matrix is the payoff matrix.

5. If a payoff matrix has two saddle points, can they be of different value, or must they be the same? Explain.

The Expected Value of Games with Mixed Strategies

In Section 10.1, we saw a way to find the optimal strategy for two-person strictly determined games. That optimal strategy was determined by the location of a saddle point, which, in turn, meant that the row player always selected the strategy of the row in which that point resided, and the column player always selected the strategy of the column in which that point resided. The term *strictly determined* or *pure strategy* meant that every time the game was played, the optimal strategy never deviated from the row-column selections determined by the saddle point. If both players select their optimal strategies, the outcome of the game is fixed and is called the *value of the game*. Even if one player discovers the strategy of the other, the first player will not change strategy because the selection of a different strategy will not improve her or his standing in the game. A little reflection will convince us that most games are *not* strictly determined but rather call for different strategies if the opposing player changes strategies. Such games are called games of **mixed strategy,** compared to the **pure strategy** of a strictly determined game.

A simple example of such a game is the game in Example 1 of Section 10.1, in which Match and Mismatch each selects heads or tails on a penny. Upon revealing their choices, Match gets \$1 from Mismatch if the coins match; if they do not match, Mismatch gets \$1 from Match. Considering Match as the row player, the payoff matrix is as before.

$$\text{Match} \begin{array}{c} \\ H \\ T \end{array} \overset{\begin{array}{cc} \text{Mismatch} \\ H \qquad\; T \end{array}}{\begin{bmatrix} 1 & -1 \\ -1 & 1 \end{bmatrix}}.$$

The payoff matrix reveals that this is not a strictly determined game. Further verification comes from the fact that if one of the players adopts a pure strategy, the other will soon discover it, counter, and win every time. If there is a best strategy for this game, it must involve random choices among the available options of showing heads or tails. It turns out that if any such strategy is determined, the expected value (expected earnings) to the row player may be computed. To see how this is done, suppose that Match decides to reveal heads $\frac{3}{5}$ of the time and tails $\frac{2}{5}$ of the time, while Mismatch decides to reveal heads $\frac{1}{6}$ of the time and tails $\frac{5}{6}$ of the time, each in some random order. Because these selections are independent in the probability sense, the probability of each head-tail combination is a product of probabilities as shown in the following table.

Match-Mismatch	Probability	Payoff matrix entry
Head-head	$\left(\frac{3}{5}\right)\left(\frac{1}{6}\right) = \frac{3}{30}$	1
Head-tail	$\left(\frac{3}{5}\right)\left(\frac{5}{6}\right) = \frac{15}{30}$	−1
Tail-head	$\left(\frac{2}{5}\right)\left(\frac{1}{6}\right) = \frac{2}{30}$	−1
Tail-tail	$\left(\frac{2}{3}\right)\left(\frac{5}{6}\right) = \frac{10}{30}$	1

Because the payoff matrix expresses gains for the row player, in this case Match, the expected value of Match's earnings is

$$E(X) = \left(\frac{3}{30}\right)(1) + \left(\frac{15}{30}\right)(-1) + \left(\frac{2}{30}\right)(-1) + \left(\frac{10}{30}\right)(1) = -\frac{4}{30}$$

or a loss of \$$\frac{4}{30} \approx 13$ cents per game. This shows that the game favors Mismatch, so if Mismatch stays with this strategy, Match should change in hope of gaining a better advantage.

The expected value just calculated can be obtained by putting the proportion of times (probabilities) for selecting heads and tails of the row player (Match) into a row matrix

$$P = \begin{bmatrix} \frac{3}{5} & \frac{2}{5} \end{bmatrix},$$

the probabilities of the two choices for the column player (Mismatch) into a column matrix

$$Q = \begin{bmatrix} \frac{1}{6} \\ \frac{5}{6} \end{bmatrix},$$

and letting the payoff matrix be represented by A. Then

$$E(X) = PAQ = \begin{bmatrix} \frac{3}{5} & \frac{2}{5} \end{bmatrix} \begin{bmatrix} 1 & -1 \\ -1 & 1 \end{bmatrix} \begin{bmatrix} \frac{1}{6} \\ \frac{5}{6} \end{bmatrix} = -\frac{4}{30}.$$

The preceding example may be generalized by denoting the payoff matrix as

$$A = \begin{bmatrix} a_{11} & a_{12} \\ a_{21} & a_{22} \end{bmatrix}, \quad P = \begin{bmatrix} p_1 & p_2 \end{bmatrix}, \quad \text{and} \quad Q = \begin{bmatrix} q_1 \\ q_2 \end{bmatrix},$$

where p_1, p_2, and q_1, q_2 are the respective proportions of time (probabilities) that the row and column players choose the first and second options. Note that $p_1 + p_2 = 1$ and $q_1 + q_2 = 1$. If the row player and the column player both choose the first option, with payoff a_{11} to the row player, the expected gain for the row player is $(p_1 q_1)a_{11}$, if the row player chooses option 1 and the column player chooses option 2, then the expected gain for the row player is $(p_1 q_2)a_{12}$, and so forth. The sum of such products gives the expected value for the row player. In matrix form, the expected value can be expressed as

$$E(X) = PAQ = \begin{bmatrix} p_1 & p_2 \end{bmatrix} \begin{bmatrix} a_{11} \\ a_{21} \end{bmatrix} \begin{bmatrix} a_{12} & q_1 \\ a_{22} & q_2 \end{bmatrix}$$

$$= \begin{bmatrix} p_1 a_{11} + p_2 a_{21} & p_1 a_{12} + p_2 a_{22} \end{bmatrix} \begin{bmatrix} q_1 \\ q_2 \end{bmatrix}$$

$$= (p_1 q_1)a_{11} + (p_2 q_1)a_{21} + (p_1 q_2)a_{12} + (p_2 q_2)a_{22}.$$

The Expected Value of a Two-Person, Zero-Sum Game

If the row player chooses the ith option, with probability p_i, and the column player chooses the jth option, with probability q_j, then the two together occur with probability $p_i q_j$. If the entry in the ith row and jth column of the payoff matrix is a_{ij}, then the product $(p_i q_j)a_{ij}$ is the expected return to the row player for this combination of options. The **expected value** for the game to the row player is $\Sigma (p_i q_j)a_{ij}$, which may be written in matrix form as PAQ, where P is the row matrix giving the proportion of time the row player chooses the first or the second option, respectively, A is the payoff matrix, and Q is the column matrix giving the proportion of time the column player chooses the first or the second option, respectively.

EXAMPLE 1

The Expected Value of
a Two-Person Game

Suppose that Match and Mismatch were simply tossing their coins before revealing whether heads or tails appeared. In this case, the strategy of both Match and Mismatch are randomly determined, with heads and tails appearing equally likely with probability $\frac{1}{2}$. Thus

$$P = \begin{bmatrix} \frac{1}{2} & \frac{1}{2} \end{bmatrix} \quad \text{and} \quad Q = \begin{bmatrix} \frac{1}{2} \\ \frac{1}{2} \end{bmatrix} \quad \text{so that } PAQ = \begin{bmatrix} \frac{1}{2} & \frac{1}{2} \end{bmatrix} \begin{bmatrix} 1 & -1 \\ -1 & 1 \end{bmatrix} \begin{bmatrix} \frac{1}{2} \\ \frac{1}{2} \end{bmatrix} = 0.$$

This result is not unexpected. We would expect this game to be fair even without the computation. ■

EXAMPLE 2

The Expected Value of
a Two-Person Game

Return to the game Match and Mismatch played in Section 10.1 and suppose that because an earlier decision to show heads $\frac{3}{5}$ of the time resulted in a negative expected payoff, Match decides to change strategy now and show heads $\frac{1}{5}$ of the time and tails $\frac{4}{5}$ of the time. In this case, the expected value to Match is

$$PAQ = \begin{bmatrix} \frac{1}{5} & \frac{4}{5} \end{bmatrix} \begin{bmatrix} 1 & -1 \\ -1 & 1 \end{bmatrix} \begin{bmatrix} \frac{1}{6} \\ \frac{5}{6} \end{bmatrix} = \frac{12}{30},$$

a much more desirable payoff. ■

Just guessing a proportion of time to reveal heads and tails for either Match or Mismatch is not a very logical method of proceeding if either truly wants to maximize the payoff. We have seen that the expected values can vary significantly upon changing proportions of time that heads and tails are revealed by Match, and the same would be true for Mismatch. Is there a best way to play this game, as well as other games of a similar nature? The fundamental theorem of game theory tells us that *every* game has a solution—that is, a best strategy.

The Fundamental Theorem of Game Theory

Using the notation of this section, there are optimal sets of probabilities (in matrix form) P_{opt} and Q_{opt} for the row and column players, respectively, and a number v so that the matrix product

$P_{opt}AQ \geq v$ for every strategy Q available to the column player and

$PAQ_{opt} \leq v$ for every strategy P available to the row player.

The number v is called the **value** of the game, and P_{opt} and Q_{opt} are called the **optimal strategies** for the row and column players. Every game has such a solution.

For strictly determined games, it is easy to find the P_{opt} and Q_{opt}. For instance, in Example 2 of Section 10.1, the optimal strategy for Appliance City was the second option—discounting 30%, and the same was true for the column player, Appliance World. Therefore

$$P_{opt} = \begin{bmatrix} 0 & 1 \end{bmatrix} \quad \text{and} \quad Q_{opt} = \begin{bmatrix} 0 \\ 1 \end{bmatrix}.$$

We can establish that this is indeed the case by noting the following:

$$\begin{bmatrix} 0 & 1 \end{bmatrix}\begin{bmatrix} .60 & .55 \\ .70 & .60 \end{bmatrix} Q = \begin{bmatrix} .70 & .60 \end{bmatrix} Q = \begin{bmatrix} .70 & .60 \end{bmatrix}\begin{bmatrix} q_1 \\ q_2 \end{bmatrix} = .70q_1 + .60q_2 \geq .60$$

The reason for the inequality statement follows from the fact that since $q_1 + q_2 = 1$, $q_2 = 1 - q_1$, and substitution into $.70q_1 + .60q_2$ gives

$$.70q_1 + .60(1 - q_1) = .70q_1 + .60 - .60q_1 = .10q_1 + .60.$$

Thus $.70q_1 + .60q_2 = .10q_1 + .60 \geq .60$ because the probability number q_1 must be between 0 and 1, inclusive.

Likewise,

$$P\begin{bmatrix} .60 & .55 \\ .70 & .60 \end{bmatrix}\begin{bmatrix} 0 \\ 1 \end{bmatrix} = P\begin{bmatrix} .55 \\ .60 \end{bmatrix} = \begin{bmatrix} p_1 & p_2 \end{bmatrix}\begin{bmatrix} .55 \\ .60 \end{bmatrix} = .55p_1 + .60p_2 \leq .60$$

because $p_1 + p_2 = 1$. Therefore, the fundamental theorem tells us that .60 is the expected value of the game, as we had discovered before. As was the case here, any pure-strategy row or column matrix will consist of one 1 and all other entries 0.

Even though finding P_{opt} and Q_{opt} is rather easy for pure-strategy games, it is not quite as easy for games of mixed strategy. The next section shows how this is done and completes our solution-finding efforts for game theory. It shows how the value of the game, v, may be determined with the aid of linear programming.

EXERCISES ' 10.2

In Exercises 1 through 8, find the expected value of the game whose payoff matrix is given in which the row player and the column player each choose strategies that have equal probabilities.

1. $\begin{bmatrix} 2 & -1 \\ 0 & 3 \end{bmatrix}$

2. $\begin{bmatrix} 4 & -5 \\ -6 & 8 \end{bmatrix}$

3. $\begin{bmatrix} 2 & -1 & 4 \\ 0 & 3 & -2 \\ 1 & 2 & -5 \end{bmatrix}$

4. $\begin{bmatrix} 0 & 1 & -1 \\ 1 & 1 & 1 \\ -3 & 2 & 1 \end{bmatrix}$

5. $\begin{bmatrix} -1 & 5 & 4 \\ 2 & 3 & -6 \end{bmatrix}$

6. $\begin{bmatrix} -3 & 4 \\ 2 & -3 \\ 1 & -1 \end{bmatrix}$

7. $\begin{bmatrix} -3 & 2 & 5 \\ 1 & 4 & -3 \\ -5 & 1 & 3 \end{bmatrix}$

8. $\begin{bmatrix} 2 & -3 & -5 \\ -1 & 4 & -3 \end{bmatrix}$

9. Suppose that the row player chooses strategies denoted in

$$P = \begin{bmatrix} \frac{3}{8} & \frac{5}{8} \end{bmatrix}$$

and the column player chooses strategies denoted in

$$Q = \begin{bmatrix} \frac{4}{11} \\ \frac{5}{11} \\ \frac{2}{11} \end{bmatrix}, \quad \text{where } A = \begin{bmatrix} -3 & 5 & 7 \\ 0 & -2 & -4 \end{bmatrix}.$$

Find the expected value of this game.

10. Suppose that the row player chooses strategies denoted in

$$P = \begin{bmatrix} \frac{1}{6} & \frac{2}{6} & \frac{3}{6} \end{bmatrix}$$

and the column player chooses strategies denoted in

$$Q = \begin{bmatrix} \frac{5}{8} \\ \frac{3}{8} \end{bmatrix}, \quad \text{where } A = \begin{bmatrix} 4 & -2 \\ 0 & 5 \\ -3 & 6 \end{bmatrix}.$$

Find the expected value of this game.

In Exercises 11 through 16, suppose that the row player has set a strategy as denoted in

$$P = \begin{bmatrix} \frac{3}{5} & \frac{2}{5} \end{bmatrix}$$

and that the column player decides to choose either

$$Q_1 = \begin{bmatrix} \frac{2}{3} \\ \frac{1}{3} \end{bmatrix} \quad or \quad Q_2 = \begin{bmatrix} \frac{7}{11} \\ \frac{4}{11} \end{bmatrix}.$$

Use the expected value to decide which of the column player's strategies is better for each of the payoff matrices listed.

11. $\begin{bmatrix} 0 & 2 \\ 1 & -3 \end{bmatrix}$ **12.** $\begin{bmatrix} -2 & 3 \\ 4 & -1 \end{bmatrix}$ **13.** $\begin{bmatrix} \frac{3}{4} & \frac{1}{2} \\ \frac{1}{3} & -\frac{5}{4} \end{bmatrix}$

14. $\begin{bmatrix} .2 & -.5 \\ -.8 & .7 \end{bmatrix}$ **15.** $\begin{bmatrix} -3 & 4 \\ 2 & -1 \end{bmatrix}$ **16.** $\begin{bmatrix} \frac{1}{2} & \frac{5}{6} \\ \frac{3}{5} & -\frac{2}{3} \end{bmatrix}$

In Exercises 17 through 20, each payoff matrix is for a game of pure strategy. Write the row and column matrices that represent the optimal strategy, and show that those matrices fulfill the conditions of the fundamental theorem of game theory.

17. $\begin{bmatrix} 3 & -5 \\ 4 & 6 \end{bmatrix}$ **18.** $\begin{bmatrix} -2 & -3 \\ 4 & 3 \end{bmatrix}$

19. $\begin{bmatrix} 4 & -7 \\ -3 & -8 \end{bmatrix}$ **20.** $\begin{bmatrix} -4 & 3 \\ -2 & 5 \end{bmatrix}$

21. *Digit Game* Tanya and Ilke play a game in which they simultaneously hold out either one or two fingers. If the sum of the fingers held out is even, Tanya wins 5 cents from Ilke; but if the sum of the fingers held out is odd, Ilke wins 5 cents from Tanya.

 (a) Find the payoff matrix for this game.

 (b) Find the expected value for this game if each player has a strategy of holding out one finger twice as often as two fingers. Whom does it favor?

22. *Raffle Tickets* Suppose that 1000 raffle tickets are sold for $1 each, and two prizes are to be awarded: one of $400 and one of $100. You are to be the row player, and you have two choices: to either buy a ticket or not buy a ticket. The luck of the draw is the column player, and there are three choices: win a net of $399, win a net of $99, or lose $1. Write the payoff matrix for this game, and determine your optimum strategy.

23. *A Coin Game* Suppose that Jack and Jill both have a penny and a dime in their hand. They simultaneously show a coin on the table. If the sum of the amounts is even, Jack wins an amount equal to the coin he has shown. If the sum is odd, Jill wins an amount equal to the coin she has shown.

 (a) Write the payoff matrix for this game, considering Jack as the row player.

 (b) If Jack always plays his penny and Jill always plays her dime, what is the expected value of this game?

 (c) If Jack always plays his penny and Jill plays her coins in a 50-50 split, what is the value of this game?

 (d) Which of the two strategies, the one from Exercise 23 (b) or the one from Exercise 23 (c), is best for Jill?

24. *A Card Game* Jim holds in his hand three cards—a 6, an 8, and a 10. He puts one of the cards face down on the table. Jenny then guesses which of the cards is on the table. If Jenny guesses correctly, she wins from Jim in cents the value of the card. If Jenny does not guess correctly, she gives Jim the value in cents of the card on the table.

 (a) Construct the payoff matrix for this game, considering Jim as the row player.

 (b) Suppose that Jim selects his card in an equally likely manner, but Jenny always guesses the card is an 8. What is the expected value of this game with these strategies?

 (c) Suppose that Jim selects his card so that the probability of selection is proportional to the value of the card, and Jenny guesses the 6 three-fifths of the time, the 8 two-fifths of the time, and never guesses the 10. Find the expected value of the game for these strategies.

 (d) Which of the strategies, the one in Exercise 24 (b) or the one in Exercise 24 (c), favors Jim the most?

Problems for Exploration

1. Show that if each row and column in a payoff matrix has a sum equal to zero and each option to each player is equally likely, then the game is fair.

2. Show that the best strategy for both the row and the column players is $(\frac{1}{3}, \frac{1}{3}, \frac{1}{3})$ in a game with a payoff matrix of

$$\begin{bmatrix} 0 & 1 & -1 \\ -1 & 0 & 1 \\ 1 & -1 & 0 \end{bmatrix}.$$

SECTION 10.3 ## Solving Mixed-Strategy Games

The fundamental theorem of game theory of the preceding section told us how to identify the solution to a mixed-strategy game, but not how to find that solution. For games that are nonstrictly determined, we show in this section how the optimal strategy for both the row and column player may be found by solving dual linear programming problems.

As a result of the Fundamental Theorem of Game Theory stated in the last section, the following result can be used to search for the optimal solutions.

Solution Theorem for Mixed Strategy Nonstrictly Determined Games

Let A be the payoff matrix for a nonstrictly determined game, and let R' and C' be strategies for the row and the column players, respectively. Let v be a number such that

1. $R'AC_i \geq v$ for all *pure* strategies C_i of the column player and

2. $R_j AC' \leq v$ for all *pure* strategies R_j of the row player.

Then the solution to the game is $R' = R_{\text{opt}}$, $C' = C_{\text{opt}}$, and v is the **value of the game.**

An example of how the optimal strategies of the row player can be cast in the light of a linear programming problem will be shown first.

EXAMPLE 1
The Row Player's
Optimal Strategy

Construct a linear programming problem that will determine the row player's optimal strategy for the payoff matrix

$$A = \begin{bmatrix} 3 & 1 \\ 2 & 4 \end{bmatrix}.$$

Solution Let

$$P = \begin{bmatrix} p_1 & p_2 \end{bmatrix} \quad \text{and} \quad Q = \begin{bmatrix} q_1 \\ q_2 \end{bmatrix}.$$

represent any strategies for the row and column players, respectively. Then the expected value of the game is

$$E = PAQ = \begin{bmatrix} p_1 & p_2 \end{bmatrix} \begin{bmatrix} 3 & 1 \\ 2 & 4 \end{bmatrix} \begin{bmatrix} q_1 \\ q_2 \end{bmatrix}$$

$$= \begin{bmatrix} 3p_1 + 2p_2 & p_1 + 4p_2 \end{bmatrix} \begin{bmatrix} q_1 \\ q_2 \end{bmatrix}$$

$$= (3p_1 + 2p_2)q_1 + (p_1 + 4p_2)q_2.$$

The column player must choose either Column 1 or Column 2. If Column 1 is the choice, then $q_1 = 1$ and $q_2 = 0$, and the expected value becomes

$$E = 3p_1 + 2p_2. \tag{1}$$

On the other hand, if Column 2 is the choice, then $q_1 = 0$ and $q_2 = 1$, and the expected value becomes

$$E = p_1 + 4p_2. \tag{2}$$

The goal of the row player is to find values of p_1 and p_2, where $p_1 + p_2 = 1$, which will maximize the expected value of the game. In view of Equations **(1)** and **(2)**, let v be the smaller of these two expected values. Then in either case, $E \geq v$, which means that if v is maximized, then so will E. We can also assert that $v > 0$, because all entries in the matrix A are positive. Equations (1) and (2) may now be written as

$$3p_1 + 2p_2 \geq v$$

and

$$p_1 + 4p_2 \geq v.$$

Now divide each of these inequalities by v to obtain

$$3\left(\frac{p_1}{v}\right) + 2\left(\frac{p_2}{v}\right) \geq 1$$

and

$$\left(\frac{P_1}{v}\right) + 4\left(\frac{P_2}{v}\right) \geq 1.$$

For simplicity, substitute $X = \dfrac{p_1}{v}$ and $Y = \dfrac{p_2}{v}$ to make these last two equations read

$$3X + 2Y \geq 1 \tag{3}$$

and

$$X + 4Y \geq 1. \tag{4}$$

Since $p_1 \geq 0$, $p_2 \geq 0$, we have that

$$X \geq 0 \quad \text{and} \quad Y \geq 0. \tag{5}$$

Equations (3), (4), and (5) are now recognized as the constraint system for a linear programming problem. The objective function for this problem comes from the fact that $p_1 + p_2 = 1$. Divide both sides by v to obtain

$$\frac{p_1}{v} + \frac{p_2}{v} = \frac{1}{v} \quad \text{or} \quad X + Y = \frac{1}{v}. \tag{6}$$

Keep in mind that we are to maximize v. This will be done if $\dfrac{1}{v}$ is *minimized,* or in view of line (6) if $X + Y$ is minimized. Finally, a linear programming problem to do just that can be stated from lines (6), (3), (4), and (5):

$$\text{Minimize } g = X + Y,$$
$$\text{subject to } 3X + 2Y \geq 1$$
$$X + 4Y \geq 1$$
$$X \geq 0, Y \geq 0.$$

Notice how the coefficients of X and Y in the main constraints are the *transpose* of the original payoff matrix A. Once the values of X and Y have been determined, line (6) allows us to find v as $v = \dfrac{1}{X + Y}$. The substitutions $X = \dfrac{p_1}{v}$ and $Y = \dfrac{p_2}{v}$ finally allow us to write the optimal row strategy as $p_1 = vX$ and $p_2 = vY$. ■

In view of Example 1, we next summarize the procedure for finding the optimal row strategy for 2×2 payoff matrices. This statement can be generalized in an evident way to larger payoff matrices.

How to Determine the Optimal Row Strategy

Let the payoff matrix be $\begin{bmatrix} a & b \\ c & d \end{bmatrix}$ where a, b, c, and d are positive numbers.
Solve the linear programming problem

$$\text{Minimize} \quad g = X + Y,$$
$$\text{subject to} \quad aX + cY \geq 1$$
$$bX + dY \geq 1$$
$$X \geq 0, Y \geq 0.$$

The expected value of the game is $v = \dfrac{1}{X + Y}$.

The optimal row strategy is $\begin{bmatrix} p_1 & p_2 \end{bmatrix}$ where $p_1 = vX$ and $p_2 = vY$.

Going through a similar analysis, we can arrive at this summary for finding the optimal column strategy.

How to Determine the Optimal Column Strategy

Let the payoff matrix be $\begin{bmatrix} a & b \\ c & d \end{bmatrix}$ where a, b, c, and d are positive numbers.
Solve the linear programming problem

$$\text{Maximize} \quad f = U + V,$$
$$\text{subject to} \quad aU + bV \leq 1$$
$$cU + dV \leq 1$$
$$U \geq 0, V \geq 0.$$

The expected value of the game is $v = \dfrac{1}{U + V}$.

The optimal column strategy is $\begin{bmatrix} q_1 \\ q_2 \end{bmatrix}$ where $q_1 = vU$ and $q_2 = vV$.

Notice this time that the coefficient matrix of the main constraint system is precisely the payoff matrix. Upon comparing the linear programming problems for the row and column strategies, we find that they are *dual problems*. This means that both problems can efficiently be solved simultaneously by using the standard simplex problem of the column player's optimal strategy and then using the concept of duality to read the row player's optimal strategy from the bottom of the slack variables column in the final tableau. To demonstrate these solutions, suppose we solve the linear programming problem that will determine the column player's optimal strategy using the payoff matrix of Example 1,

$$A = \begin{bmatrix} 3 & 1 \\ 2 & 4 \end{bmatrix}.$$

The entries in A are all positive. From the column player's point of view, the linear programming problem is the following standard maximum problem:

$$\text{Maximize} \quad f = U + V,$$
$$\text{subject to} \quad 3U + V \le 1$$
$$2U + 4V \le 1$$
$$U \ge 0, V \ge 0.$$

The initial tableau with the first pivot element circled is as follows:

$$\left[\begin{array}{cccc|c} 3 & 1 & 1 & 0 & 0 & 1 \\ 2 & \boxed{4} & 0 & 1 & 0 & 1 \\ \hline -1 & -1 & 0 & 0 & 1 & 0 \end{array}\right]$$

Completion of the first pivot yields this tableau:

$$\left[\begin{array}{ccccc|c} \boxed{\frac{5}{2}} & 0 & 1 & -\frac{1}{4} & 0 & \frac{3}{4} \\ \frac{1}{2} & 1 & 0 & \frac{1}{4} & 0 & \frac{1}{4} \\ \hline -\frac{1}{2} & 0 & 0 & \frac{1}{4} & 1 & \frac{1}{4} \end{array}\right]$$

Pivoting on the circled element in this matrix gives the final tableau:

$$\left[\begin{array}{ccccc|c} \boxed{1} & 0 & \frac{2}{5} & -\frac{1}{10} & 0 & \frac{3}{10} \\ 0 & 1 & -\frac{1}{5} & \frac{3}{10} & 0 & \frac{1}{10} \\ \hline 0 & 0 & \frac{1}{5} & \frac{1}{5} & 1 & \frac{2}{5} \end{array}\right]$$

From this matrix we find that $U = \frac{3}{10}$ and $V = \frac{1}{10}$ so that $U + V = \frac{4}{10} = \frac{2}{5}$. This means that the expected value for this game is $v = \frac{5}{2}$. Therefore, $q_1 = vU = (\frac{5}{2})(\frac{3}{10}) = \frac{3}{4}$ and $q_2 = vV = (\frac{5}{2})(\frac{1}{10}) = \frac{1}{4}$. The column player's optimal strategy is then given by $\begin{bmatrix} \frac{3}{4} \\ \frac{1}{4} \end{bmatrix}$.

The final tableau in the simplex algorithm calculations shows that the dual problem has solutions $X = \frac{1}{5}$ and $Y = \frac{1}{5}$, which we read from the bottom of the slack variable columns. Thus, $X + Y = \frac{2}{5}$ from which we find that again $v = \frac{5}{2}$, as we expect. In this case, $p_1 = vX = (\frac{5}{2})(\frac{1}{5}) = \frac{1}{2}$ and $p_2 = vY = (\frac{5}{2})(\frac{1}{5}) = \frac{1}{2}$. The row player's optimal strategy is then $[\frac{1}{2} \quad \frac{1}{2}]$.

If we want further justification of how this solution relates to the previous section, the maximum expected value, or value of the game, $v = \frac{5}{2}$, should be the result of the matrix multiplication PAQ. For our present example

$$PAQ = \begin{bmatrix} \frac{1}{2} & \frac{1}{2} \end{bmatrix} \begin{bmatrix} 3 & 1 \\ 2 & 4 \end{bmatrix} \begin{bmatrix} \frac{3}{4} \\ \frac{1}{4} \end{bmatrix} = \begin{bmatrix} \frac{5}{2} & \frac{5}{2} \end{bmatrix} \begin{bmatrix} \frac{3}{4} \\ \frac{1}{4} \end{bmatrix} = \frac{5}{2}.$$

The calculations we have just made with the simplex algorithm show that both the row and column player's optimal strategies can be done with one linear program. However, we should note that the row player's optimal strategy could have been computed independently of that of the column player, and it could be done in the 2×2 payoff matrix case either by the graphical method or by Crown's method applied to a minimizing problem. Similarly, the column player's optimal strategy could have been found independently of the row player's, and, in the 2×2 payoff matrix, either by the graphical method or the simplex algorithm applied to a standard maximum problem.

Finally, a comment on the need for all entries in the payoff matrix A to be positive. This assumption was needed so that v would be positive and hence we could be assured that inequalities just above Equations **(3)** and **(4)** in Example 1 would be preserved. However, we can relax that restriction. If some entries in the payoff matrix are not positive, find some positive number n such that when added to each entry in the payoff matrix A will produce a new matrix B in which all entries are positive. The use of B will produce precisely the same optimal strategies as would be found by using A. The value of the game, however, will be that found by using matrix B and then subtracting the number n. We will demonstrate this concept in the next example which also answers the question posed at the outset of this chapter.

EXAMPLE 2
Solving a Game that Is
Not Strictly Determined

A debate between Senator Smith and her opponent for reelection is almost certain to center on two issues, foreign policy (*FP*) and defense (*D*). The senator's managers feel that her opponent is strong on the defense issue but not as knowledgeable about foreign policy, and they estimate the points in various combinations of these two issues as follows:

A gain of three points for Senator Smith when both discuss foreign policy

A gain of five points for her opponent when both discuss the defense issue

A gain of ten points if Senator Smith discusses foreign policy while her opponent discusses defense

A gain of six points for Senator Smith if she discusses defense while her opponent discusses foreign policy

What fraction of the debate should Senator Smith spend on each issue to take best advantage of the point estimates?

Solution   Suppose that we consider Senator Smith the row player. The payoff matrix then has this appearance:

$$
\begin{array}{cc}
 & \begin{array}{cc} \text{Opponent} \\ FP \quad\ D \end{array} \\
\text{Senator Smith} \begin{array}{c} FP \\ D \end{array} & \begin{bmatrix} 3 & 10 \\ 6 & -5 \end{bmatrix}
\end{array}
$$

This is evidently not a strictly determined game, which means that we may apply the linear programming techniques just discussed. We begin by adding 6 to each entry in the payoff matrix to produce the matrix B with all positive entries:

$$
B = \begin{bmatrix} 9 & 16 \\ 12 & 1 \end{bmatrix}.
$$

From the column player's viewpoint, the linear programming problem would appear as the standard maximum problem shown next:

$$
\begin{aligned}
\text{Maximize} \quad & f = U + E, \\
\text{subject to} \quad & 9U + 16E \le 1 \\
& 12U + E \le 1 \\
& U \ge 0, E \ge 0.
\end{aligned}
$$

The tableaus needed to solve this problem are shown next:

$$\left[\begin{array}{ccccc|c} 9 & 16 & 1 & 0 & 0 & 1 \\ \boxed{12} & 1 & 0 & 1 & 0 & 1 \\ \hline -1 & -1 & 0 & 0 & 1 & 0 \end{array}\right]; \quad \left[\begin{array}{ccccc|c} 0 & \boxed{\frac{61}{4}} & 1 & -\frac{3}{4} & 0 & \frac{1}{4} \\ 1 & \frac{1}{12} & 0 & \frac{1}{12} & 0 & \frac{1}{12} \\ \hline 0 & -\frac{11}{12} & 0 & \frac{1}{12} & 1 & \frac{1}{12} \end{array}\right];$$

$$\left[\begin{array}{ccccc|c} 0 & 1 & \frac{4}{61} & -\frac{3}{61} & 0 & \frac{1}{61} \\ 1 & 0 & -\frac{1}{183} & \frac{16}{183} & 0 & \frac{5}{61} \\ \hline 0 & 0 & \frac{11}{183} & \frac{7}{183} & 1 & \frac{6}{61} \end{array}\right]$$

Since we are interested in Senator Smith's optimal strategy and she is considered the row player, we use the duality principle and read from the bottom of the slack variable columns the value of X and Y:

$$X = \frac{11}{183} \text{ and } Y = \frac{7}{183} \text{ so that } X + Y = \frac{18}{183} \text{ and hence } v = \frac{183}{18}.$$

Therefore, $p_1 = vX = \left(\frac{183}{18}\right)\left(\frac{11}{183}\right) = \frac{11}{18}$ and $p_2 = vY = \left(\frac{183}{18}\right)\left(\frac{7}{183}\right) = \frac{7}{18}$, which implies that Senator Smith's optimal strategy is $\left[\frac{11}{18} \quad \frac{7}{18}\right]$. That is, the senator should spend $\frac{11}{18}$ of her debate time talking about foreign policy and $\frac{7}{18}$ of the time talking about defense to gain the maximum number of points. The number of points gained by doing this is the value of the game. That value is $v = \frac{183}{18}$ minus the number 6 which was added to the original payoff matrix at the outset. Since $\frac{183}{18} \approx 10.167$, the value of the original game is $10.167 - 6 = 4.167$ or about 4.2 points. ■

The next example shows that the linear programming techniques discussed for finding optimal strategies also apply to matrices on any size.

EXAMPLE 3
Larger Payoff Matrices

Find the optimal strategies for both the row and column players for a game with this payoff matrix:

$$A = \left[\begin{array}{ccc} 2 & -2 & 3 \\ -4 & 0 & -1 \end{array}\right],$$

Solution We begin by adding the number 5 to each entry in A:

$$B = \left[\begin{array}{ccc} 7 & 3 & 8 \\ 1 & 5 & 4 \end{array}\right]$$

The standard maximum problem for the column player's optimal strategy becomes

$$\text{Maximize } f = U + E + W,$$
$$\text{subject to } \quad 7U + 3E + 8W \leq 1$$
$$U + 5E + 4W \leq 1$$
$$U \geq 0, E \geq 0, W \geq 0.$$

The tableaus leading to the solution are shown next.

$$\left[\begin{array}{ccccccc|c} \boxed{7} & 3 & 8 & 1 & 0 & 0 & 1 \\ 1 & 5 & 4 & 0 & 1 & 0 & 1 \\ \hline -1 & -1 & -1 & 0 & 0 & 1 & 0 \end{array}\right]; \qquad \left[\begin{array}{ccccccc|c} 1 & \frac{3}{7} & \frac{8}{7} & \frac{1}{7} & 0 & 0 & \frac{1}{7} \\ 0 & \boxed{\frac{32}{7}} & \frac{20}{7} & -\frac{1}{7} & 1 & 0 & \frac{6}{7} \\ \hline 0 & -\frac{4}{7} & \frac{1}{7} & \frac{1}{7} & 0 & 1 & \frac{1}{7} \end{array}\right];$$

$$\left[\begin{array}{ccccccc|c} 1 & 0 & \frac{7}{8} & \frac{5}{32} & -\frac{3}{32} & 0 & \frac{1}{16} \\ 0 & 1 & \frac{5}{8} & -\frac{1}{32} & \frac{7}{32} & 0 & \frac{3}{16} \\ \hline 0 & 0 & \frac{1}{2} & \frac{1}{8} & \frac{1}{8} & 1 & \frac{1}{4} \end{array}\right]$$

The last tableau shows that $U = \frac{1}{16}$, $E = \frac{3}{16}$, and $W = 0$. Thus $U + E + W = \frac{4}{16} = \frac{1}{4}$ from which $v = 4$. This means that the column player's optimal strategy is

$$q_1 = vU = (4)\left(\frac{1}{6}\right) = \frac{1}{4}, \qquad q_2 = vE = (4)\left(\frac{3}{16}\right) = \frac{3}{4}, \qquad \text{and}$$

$$q_3 = vU = (4)(0) = 0.$$

Using the duality principle, we find that $X = \frac{1}{8}$ and $Y = \frac{1}{8}$. If follows that the row player's optimal strategy is then

$$p_1 = eX = (4)\left(\frac{1}{8}\right) = \frac{1}{2} \qquad \text{and} \qquad p_2 = eY = (4)\left(\frac{1}{8}\right) = \frac{1}{2}.$$

The value of this game is $4 - 5 = -1$. ■

Sometimes, it is possible to reduce the size of the payoff matrix if there is a row that *dominates* another row or a column that *dominates* another column. By the term **dominate,** we mean that each entry in one of the rows (columns) is greater than or equal to the corresponding entry in the other row (column). For instance, in the matrix

$$\left[\begin{array}{cc} 4 & -3 \\ -2 & 5 \\ 3 & -6 \end{array}\right]$$

the first row dominates the third row because each element of the first row is *greater than* the corresponding element of the third row. When this happens, the row player need never play the third row in any optimal strategy because the player could always do at least as well by playing the first row. Consequently, the game may be considered to have this payoff matrix:

$$\left[\begin{array}{cc} 4 & -3 \\ -2 & 5 \end{array}\right].$$

Thus, even though we are restricting our games that are not strictly determined to ones having 2×2 payoff matrices, it may still be possible to analyze games with larger matrices if dominance is present.

EXERCISES   10.3

In Exercises 1 through 10, each matrix is the payoff matrix for a two-person, zero-sum game, some of which are strictly determined and some of which are not.

(a) *Solve each game. For those that are not strictly determined, solve by using a linear program.*

(b) *Find the value of each game.*

1. $\left[\begin{array}{cc} 2 & -3 \\ 1 & 1 \end{array}\right]$

2. $\left[\begin{array}{cc} 4 & -2 \\ -3 & 1 \end{array}\right]$

3. $\left[\begin{array}{cc} -2 & 3 \\ 4 & -5 \end{array}\right]$

4. $\left[\begin{array}{cc} 2 & 5 \\ 1 & -3 \end{array}\right]$

5. $\begin{bmatrix} 6 & 4 \\ 3 & -2 \end{bmatrix}$ **6.** $\begin{bmatrix} 4 & -2 \\ 3 & 4 \end{bmatrix}$

7. $\begin{bmatrix} -1 & 6 \\ 4 & -2 \end{bmatrix}$ **8.** $\begin{bmatrix} 0 & -2 \\ 3 & 4 \end{bmatrix}$

9. $\begin{bmatrix} 5 & -2 \\ 1 & -3 \\ -2 & 7 \end{bmatrix}$ **10.** $\begin{bmatrix} 8 & -1 & 6 \\ -3 & 0 & 2 \end{bmatrix}$

In Exercises 11 through 22, use the duality of a linear program to find the row and column player's optimal strategy and the value of the game.

11. $\begin{bmatrix} -2 & 4 \\ 3 & -1 \end{bmatrix}$ **12.** $\begin{bmatrix} -5 & 8 \\ 6 & -2 \end{bmatrix}$

13. $\begin{bmatrix} 1 & 1 \\ 4 & -3 \end{bmatrix}$ **14.** $\begin{bmatrix} \frac{7}{8} & \frac{1}{4} \\ -3 & \frac{5}{6} \end{bmatrix}$

15. $\begin{bmatrix} -2 & 6 \\ 8 & 0 \\ 6 & -1 \end{bmatrix}$ **16.** $\begin{bmatrix} 2 & 1 & -4 \\ -3 & -5 & 6 \end{bmatrix}$

17. $\begin{bmatrix} 1 & 3 & 5 \\ 6 & 2 & 1 \\ 4 & 1 & 4 \end{bmatrix}$ **18.** $\begin{bmatrix} 2 & 5 & 7 \\ 1 & 2 & 4 \\ 6 & 1 & 3 \end{bmatrix}$

19. $\begin{bmatrix} 1 & 0 & 1 \\ -1 & 1 & -1 \\ 0 & -1 & 2 \end{bmatrix}$ **20.** $\begin{bmatrix} 3 & -1 & 9 \\ 1 & 2 & -3 \\ -2 & 3 & 4 \end{bmatrix}$

21. $\begin{bmatrix} 5 & 0 & 4 \\ -4 & 1 & 2 \end{bmatrix}$ **22.** $\begin{bmatrix} 6 & 1 \\ -2 & 3 \\ 1 & 7 \end{bmatrix}$

23. *Coin Guessing*   Match and Mismatch play this game: Match conceals a coin in one of her hands. If Mismatch guesses which hand the coin is in, then Match gives Mismatch $3. If Mismatch does not guess correctly, then Mismatch gives Match $2 if the coin is in the right hand and $5 if the coin is in the left hand.

(a) What is the optimal strategy for Match?
(b) What is the optimal strategy for Mismatch?
(c) Which player does this game favor?

24. *Politics*   A polling firm believes that its livelihood will be affected by the political party in power. They must decide on one of two courses of action before the election. The relative strengths to the company, depending on which party wins the election, are shown in this payoff matrix:

$$\begin{array}{c} \\ \text{Action 1} \\ \text{Action 2} \end{array} \begin{array}{cc} \text{Dem} & \text{Rep} \\ \begin{bmatrix} 4 & -3 \\ -5 & 2 \end{bmatrix} \end{array}.$$

Determine the optimal row strategy for the polling firm, and find the value of this game.

25. *Retailing*   M-Mart and Q-Mart have stores in a local mall. They each have daily specials to attract customers. M-Mart features clothing or sporting goods, while Q-Mart features electronics or kitchenwares. The swing customers are affected by the specials in this way, according to surveys run by the mall management:

  M-Mart gains 200 of them if they feature clothing and Q-Mart features electronics, but loses 100 of them if Q-Mart features kitchenwares.

  M-Mart loses 300 of them if they feature sporting goods and Q-Mart features electronics, but it gains 175 if Q-Mart features kitchenwares.

(a) Determine the optimum strategy for each store.
(b) Which store does this game favor?

26. *Farming*   Tomato growers have two varieties of plants available, one of which does very well if the growing season is dry. The other variety does very well if the season is wet. The payoff matrix showing the relative performances under the two weather scenarios is shown next:

$$\begin{array}{c} \\ \text{Dry variety} \\ \text{Wet variety} \end{array} \begin{array}{cc} \text{Dry} & \text{Wet} \\ \begin{bmatrix} 20 & 8 \\ 10 & 20 \end{bmatrix} \end{array}.$$

Long-range weather bureau predictions indicate that the probability of a dry season is 60%; hence, Mother Nature's strategy is fixed at

$$\begin{bmatrix} .60 \\ .40 \end{bmatrix}.$$

What is the tomato growers' best strategy? What is the value of this game?

27. *A Coin Game*   Bob and John each have a number of nickels and dimes. They each choose one coin and simultaneously put them on the table. If the sum of the amounts is even, Bob takes both coins. If the sum of the amounts is odd, John takes both coins. Determine the optimal strategy for both players.

28. *A European Game*   Hans and Wolfgang play a game in which the objective is to accumulate as many stones from a previously collected pile as they can. They make a fist, then, upon a signal, hold forth either one or two fingers. If the number of fingers held forth is even, Hans gets that many stones, but if the number of fingers held forth is odd, Wolfgang gets that many stones. Find the optimal strategy for both players.

Problems for Exploration

1. Show that if a game having a 2 × 2 payoff matrix has a saddle point, then either one row dominates the other or one column dominates the other.

2. Show that if the 2 × 2 payoff matrix for a game has no inverse, then the game is fair.

CHAPTER **10** REFLECTIONS

CONCEPTS

Two-person, zero-sum games involve two competing forces, each trying to devise a winning strategy. When the optimal strategy for each player is found, the game is considered solved. A *strictly determined* or *pure-strategy game* has a single, invariant course of action as its solution, whereas a *mixed-strategy game* involves more than one course of action for its solution. In strictly determined games, the *payoff matrix* has a single saddle point or value that indicates which player is favored or that the game is fair. In mixed-strategy games, different strategies may produce different *expected values,* but optimal strategies still can always be determined.

COMPUTATION/CALCULATION

Calculate the value of each pure-strategy game.

1. $\begin{bmatrix} 10 & 9 \\ 8 & 7 \end{bmatrix}$

2. $\begin{bmatrix} 2 & 0 \\ 4 & 1 \end{bmatrix}$

3. $\begin{bmatrix} -4 & 5 & -6 \\ 2 & 3 & 5 \\ 0 & 1 & -3 \end{bmatrix}$

4. $\begin{bmatrix} 8 & 3 & 10 & 2 \\ 7 & 6 & 9 & 3 \\ 7 & 4 & 6 & 4 \end{bmatrix}$

Compute the values of x that make the payoff matrix strictly determined.

5. $\begin{bmatrix} x & 6 & 0 \\ -3 & x & -11 \\ -7 & 1 & x \end{bmatrix}$

6. $\begin{bmatrix} 2.5 & 2 & x \\ x & -1 & 1.5 \\ -2 & .5 & -1.5 \end{bmatrix}$

For each payoff matrix, A, calculate the expected value, given that each player chooses strategies having equal probability.

7. $\begin{bmatrix} 0 & -3 \\ -2 & 1 \end{bmatrix}$

8. $\begin{bmatrix} 11 & 2 \\ 1 & 15 \end{bmatrix}$

9. $\begin{bmatrix} 1 & 2 & 0 \\ 2 & 2 & 2 \\ -2 & 3 & 2 \end{bmatrix}$

10. $\begin{bmatrix} -3 & 3 & 2 \\ 0 & 1 & -8 \end{bmatrix}$

11. $\begin{bmatrix} -9 & 12 \\ 6 & -9 \\ 3 & -3 \end{bmatrix}$

12. $\begin{bmatrix} -4 & 1 & 4 \\ 0 & 3 & -4 \\ -6 & 0 & 2 \end{bmatrix}$

Solve each two-person, zero-sum game.

13. $\begin{bmatrix} 1 & -1 \\ 4 & 5 \end{bmatrix}$

14. $\begin{bmatrix} 9 & 0 & 7 \\ -2 & 1 & 3 \end{bmatrix}$

15. $\begin{bmatrix} 2 & 2 \\ 8 & -6 \end{bmatrix}$

16. $\begin{bmatrix} -1 & 3 \\ 4 & 0 \\ 3 & -1 \end{bmatrix}$

CONNECTIONS

17. Jamie and Twyla are playing a professional clay-court tennis match. The commentators estimate that Jamie will win 60% of the points when both stay at the baseline and 65% when both are at the net. When Jamie is at the baseline and Twyla at the net, Twyla wins 70% of the points. When Twyla is at the baseline and Jamie at the net, Twyla wins 55% of the points. What percentage of the time should Jamie come to the net, regardless of where her opponent plays?

COMMUNICATION

18. Describe the distinction between mixed-strategy games and pure-strategy games.

19. Indicate the significance of a saddle point for a game.

20. Explain why the term *saddle point* is appropriate for the value of a strictly determined game.

21. Describe the meaning of *expected value* for a two-person, mixed-strategy game.

22. Explain the meaning and significance of the term *dominate* with respect to a payoff matrix.

COOPERATIVE LEARNING

23. Construct a payoff matrix that has exactly *three* saddle points, and then make up a two-person, zero-sum game that fits this matrix.

CRITICAL THINKING

24. By considering several special examples, determine the conditions for the payoff matrix

$$\begin{bmatrix} a & b \\ c & d \end{bmatrix}.$$

(*a, b, c,* and *d* are all distinct) to strictly determine a two-person game.

Cumulative Review

25. Write the equation of a linear cost function if 30 items cost $400 and 55 items cost $600.

26. Complete the pivot operation on the circled element of the given augmented matrix.

$$\begin{bmatrix} 0 & -7 & 8 & 14 & 3 \\ 5 & 2 & 0 & 4 & 12 \\ 6 & 3 & -2 & \textcircled{1} & 5 \end{bmatrix}$$

27. Bus routes are available through the cities shown in the diagram, in the direction of the arrows only. Write a matrix that displays this transportation network.

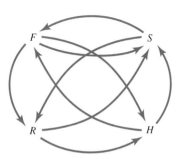

28. Make a truth table for $\sim p \lor (p \land \sim q)$.

29. Calculate the mean, median, mode, and range for the given data: 1, 3, 7, 8, 10, 10, 10.

CHAPTER **10** SAMPLE TEST ITEMS

Use the given payoff matrix to decide whether the game is strictly determined. If so, find the saddle point for the game, and find who the game favors.

1. $\begin{bmatrix} 4 & 1 \\ 6 & 3 \end{bmatrix}$ **2.** $\begin{bmatrix} 8 & 5 \\ -2 & 14 \end{bmatrix}$

3. $\begin{bmatrix} 3 & 1 & -4 \\ 0 & 5 & -2 \\ -3 & 4 & -6 \end{bmatrix}$

For what values of x is the given payoff matrix strictly determined?

4. $\begin{bmatrix} x & -2 \\ 0 & 3 \end{bmatrix}$ **5.** $\begin{bmatrix} x & -2 \\ 0 & x \end{bmatrix}$

The First and Second National Banks are each planning to open a branch on campus or at the mall. If they both select the campus or they both select the mall, they will share the business equally. However, if one selects the campus and the other the mall, the bank selecting the campus will get a 55% share of the business versus a 45% share for the other.

6. Construct the payoff matrix for this two-bank game.

7. Find the value of this game, and determine the best strategy for each bank.

For each given payoff matrix, find the expected value of the game if both players choose strategies that have equal probabilities.

8. $\begin{bmatrix} 3 & -5 \\ 0 & 8 \end{bmatrix}$ **9.** $\begin{bmatrix} -2 & 6 & 4 \\ 3 & 5 & -7 \end{bmatrix}$

10. $\begin{bmatrix} -1 & 3 & 7 \\ 2 & 3 & -2 \\ -4 & 1 & 5 \end{bmatrix}$

The row player has settled on strategy

$$P = \begin{bmatrix} \frac{1}{3} & \frac{2}{3} \end{bmatrix},$$

and the column player decides to choose either

$$Q_1 = \begin{bmatrix} \frac{1}{4} \\ \frac{3}{4} \end{bmatrix} \quad or \quad Q_2 = \begin{bmatrix} \frac{4}{13} \\ \frac{9}{13} \end{bmatrix}.$$

Use the expected value to decide whether Q_1 or Q_2 is the best strategy for the column player, given each payoff matrix.

11. $\begin{bmatrix} -1 & 2 \\ 3 & 0 \end{bmatrix}$ **12.** $\begin{bmatrix} .3 & -.6 \\ -.9 & .8 \end{bmatrix}$ **13.** $\begin{bmatrix} \frac{1}{3} & \frac{2}{5} \\ \frac{7}{8} & -\frac{1}{2} \end{bmatrix}$

Solve each two-person, zero-sum game using the given payoff matrix.

14. $\begin{bmatrix} 5 & -1 \\ 4 & 5 \end{bmatrix}$ **15.** $\begin{bmatrix} 4 & -1 \\ 3 & 3 \end{bmatrix}$ **16.** $\begin{bmatrix} 0 & -4 \\ 6 & 8 \end{bmatrix}$

17. $\begin{bmatrix} -10 & 16 \\ 12 & -4 \end{bmatrix}$ **18.** $\begin{bmatrix} 10 & -4 \\ 2 & -6 \\ -4 & 14 \end{bmatrix}$

19. $\begin{bmatrix} 9 & 0 & 7 \\ -2 & 1 & 3 \end{bmatrix}$

Mia and Elena each have a dime and a quarter. They each choose one coin and simultaneously put the selected coins on a table. If the coins match, Mia gets the product (in cents) of the two coins. If the coins do not match, Elena gets ten times the sum (in cents) of the two coins.

20. Determine the optimum strategy for each player.

21. Whom does this game favor?

Sarah has just graduated and taken her first job. She estimates that she could comfortably make payments of $200 each month toward the purchase of a new car. In addition to the money her grandfather gave her, she will need to borrow $10,000 to get the car she wants. If the bank charges 8.7% interest on the unpaid balance, how many months will be required to pay off such a loan? (See Example 5, Section 11.3.)

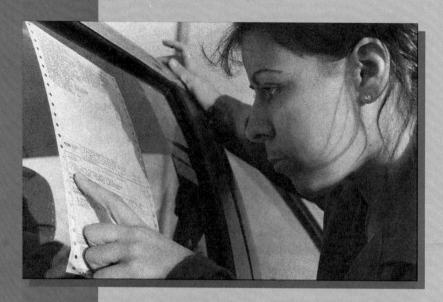

CHAPTER PREVIEW

This chapter deals with facts that everyone should know when investing or borrowing money. We begin with the concepts of simple and compound interest and how they compare. We then move to *annuities,* a term describing the accumulation of funds when periodic payments are made into an account and, once deposited, begin to earn interest immediately. The meaning of the consumer term APR (annual percentage rate) is then explored. Next, a method for calculating payments on monthly consumer loans and how to calculate the length of a loan assuming a predetermined payment level are considered. Realistic interest rates quoted by lenders are often given over a wide range of decimal percentages and, for that reason, calculators with a natural logarithm key are frequently suggested for use in this chapter.

CALCULATOR/COMPUTER OPPORTUNITIES FOR THIS CHAPTER

A calculator with a power key, y^x, a logarithm key (either base 10 or base e), and an e key is essential to doing the work of this chapter. All of the examples and answers have been worked with a calculator set to handle six decimal places (in most instances), to give greater accuracy when multiplying very small numbers by very large numbers. Inexpensive financial calculators are available, which will do all of the mathematical computations of this chapter—and more—with just the stroke of a key, after the pertinent information is entered. Graphing calculators also have these capabilities.

The computer program Finance from the disk MasterIt will do the calculations of this chapter after the data are entered.

SECTION 11.1 | ## Simple and Compound Interest

If you put money into a savings account, have a certificate of deposit at a financial institution, or buy a government bond or municipal bond, you have, in effect, agreed to loan money to that entity. To put it differently, that entity is borrowing from you. On the other hand, if you borrow money to buy a car, a stereo, or a house or to start a business, then a financial institution has agreed to loan you money. In either event, the amount of money loaned (or borrowed) is called the **principal.** The principal may be thought of as the **present value** of the money at the time it is loaned or borrowed. The person or institution that does the borrowing always agrees to pay a fee for the privilege of borrowing or using the money for a specified length of time. Such a fee is called **interest.** Interest falls into two general classifications, **simple interest** and **compound interest,** and there is a significant difference between the two.

EXAMPLE 1
Calculating Simple
Interest

If Dr. Scroggs puts $3000 into a savings account that pays 5% simple interest, how much money will be in the account at the end of one year if no withdrawals or additions are made to the account?

Solution   As usual, the interest is stated in terms of percentage of the principal *per year*. The actual interest itself is calculated as

$$I = \$3000 \times 0.05 = \$150.$$

Therefore, the amount in the account at the end of the year will be $3000 + $150 = $3150.   ▬

EXAMPLE 2
Calculating Simple
Interest

The Bank of Benton charges 16% simple interest on its MasterCard accounts. If you charged $500 on such a card and were assessed interest for one month on this amount, how much would the interest be?

Solution   The term "16% simple interest" means that the interest for borrowing the money is calculated at the rate of 16% of the $500 for an entire year. If the money is only borrowed for one month, then you should be charged only $\frac{1}{12}$ of the yearly amount. Therefore, the interest would be

$$I = \$500 \times 0.16 \times \frac{1}{12} \approx \$6.67.$$ ▬

EXAMPLE 3
Calculating Simple
Interest

Suppose that you borrowed $5000 from your grandfather for three years at the rate of 7% simple interest. How much would you owe your grandfather at the end of three years?

Solution The rate of 7% simple interest means that the borrower agrees to pay 7% of the $5000 principal *each year* as a rental fee for use of the money. Because the money is borrowed for three years, the interest will be

$$I = \$5000 \times .07 \times 3 = \$1050.$$

Therefore, the total accumulated amount to be repaid at the end of the three-year period is $A = \$5000 + \$1050 = \$6050$. ■

The total amount A that will be repaid (or accumulated by the lender) is sometimes known as the **future value** of the loan.

Simple Interest

Let P = principal (amount borrowed or loaned)
  r = interest rate *per year*
  t = length of loan measured *in years*.

Then the **simple interest** I on the loan is

$$I = Prt.$$

The **total accumulated amount** A to the person making such a loan—hence, the total amount repaid by the person borrowing the money—is

$$A = P + I = P + Prt = P(1 + rt).$$

Simple interest calculations are often used to determine the interest portion of each monthly payment made to pay back a loan. In such a case, the interest is calculated on a daily basis on the new loan balance or principal established at that time. Interest rates on savings accounts, on the other hand, are almost always given in terms of compound interest. An easy example will show how compound interest differs from simple interest. Suppose that $1000 is put into an account that pays 10% *compounded annually*. At the end of the first year, the interest will be $100, so the account balance is $1100. If the account is left untouched for a second year, the account earns 10% of the *new principal*, $1100. Therefore, the total accumulated at the end of the second year would be

$$\$1100 + (\$1100)(.10) = \$1100 + \$110 = \$1210.$$

This is the idea of **compound interest,** the *earning of interest on interest*. What is the amount accumulated at the end of the third year? In this case, the account would earn 10% interest during the third year on the new principal of $1210 established at the end of the second year. The total accumulated would then be

$$\$1210 + (\$1210)(.10) = \$1210 + \$121 = \$1331.$$

The compounding period of one year used here is only one of many commonly used periods. Following are some of the more common compounding terms advertised by financial institutions.

Interest type	Number of interest periods per year	Length of each interest period
Compounded annually	1	1 year
Compounded semiannually	2	6 months
Compounded quarterly	4	3 months
Compounded monthly	12	1 month
Compounded daily	365	1 day

What do such rates mean? Here are some examples. If rates are advertised as 6% compounded quarterly, the interest rate *per quarter* is $\frac{6\%}{4} = 1.5\%$ and this rate is applied to the total in the account at the end of each quarter. If rates are advertised as 8% compounded monthly, the interest rate *per month* is $\frac{8\%}{12} \approx .67\%$, and this rate is applied to the total in the account at the end of each month.

Now suppose that we have a particular principal P invested at the rate $i = \dfrac{r}{m}$ per interest period. At the end of the first interest period, the accumulated amount is

$$A_1 = P + Pi$$
$$= P(1 + i).$$

At the end of the second interest period, the accumulated amount is

$$A_2 = A_1 + A_1 i$$
$$= P(1 + i) + P(1 + i)i$$
$$= P(1 + i)(1 + i) = P(1 + i)^2.$$

At the end of the third interest period, the accumulated amount is

$$A_3 = A_2 + A_2 i$$
$$= P(1 + i)^2 + P(1 + i)^2 i$$
$$= P(1 + i)^2(1 + i) = P(1 + i)^3.$$

Following this, a general pattern gives the accumulated amount at the end of the nth interest period, as shown next.

The Compound Interest Formula

Let P = principal
t = number of years
m = number of interest periods per year
r = annual interest rate compounded m times per year
$i = \dfrac{r}{m}$ = interest rate *per period*
$n = mt$ = total number of interest periods.

Then the accumulated amount after n interest periods is

$$A = P\left(1 + \frac{r}{m}\right)^{mt} = P(1 + i)^n.$$

Example 4 shows the power of compounding interest.

EXAMPLE 4
Using the
Compound
Interest Formula

If $100 is invested at 8%, how much is in the account at the end of ten years if the interest is

(a) Simple?
(b) Compounded annually?
(c) Compounded semiannually?
(d) Compounded quarterly?
(e) Compounded daily?

Solution

(a) $A = P(1 + rt) = \$100.000(1 + .08(10)) = \$100.00(1 + .8)$
$= \$100.00(1.8) = \$180.00.$

(b) $A = P\left(1 + \dfrac{r}{m}\right)^{mt} = \$100.00(1 + .08)^{10} = \$100.00(1.08)^{10} \approx \$215.89.$

(c) $A = P\left(1 + \dfrac{r}{m}\right)^{mt} = \$100.00\left(1 + \dfrac{.08}{2}\right)^{2(10)} = \$100.00(1.04)^{20} \approx \$219.11.$

(d) $A = P\left(1 + \dfrac{r}{m}\right)^{mt} = \$100.00\left(1 + \dfrac{.08}{4}\right)^{4(10)} = \$100.00(1.02)^{40} \approx \$220.80.$

(e) $A = P\left(1 + \dfrac{r}{m}\right)^{mt} = \$100.00\left(1 + \dfrac{.08}{365}\right)^{365(10)} \approx \$222.53.$ ■■

NOTE   For the examples in this section and throughout this chapter, we have used a calculator for our computations. If your answers do not agree precisely, it may be due to roundoff error or the number of decimal places your calculator is set to carry, especially for numbers raised to very large powers. In all cases, accuracy is greater if you wait until the end to round rather than rounding at intermediate steps.

Example 4 shows that the greater the number of compounding periods per year, the greater the return (applied to the same principal). Could we increase indefinitely the amount earned in a ten-year period if the number of compounding periods continued to grow? The answer is yes, but there is an upper bound on the return that can be obtained this way. As the number of compounding periods increases unendingly, we approach a situation in which interest is compounded **continuously.** This means that at each instant of time, the investment grows in proportion to the amount present at that instant. The amount to which such an investment P grows after t years is given next.

Continuous Compounding

Let P = principal
r = rate of interest **compounded continuously** and stated as a decimal
t = time in years.

Then the accumulated amount after t years is

$$A = Pe^{rt}$$

where

$$e = 2.718. \ldots$$

Example 5 continues the comparisons of Example 4.

EXAMPLE 5
Compounding
Money
Continuously

If $100 is invested at 8% compounded continuously, how much will be in the account at the end of 10 years?

Solution   Substituting into the formula given above,

$$A = 100 \cdot e^{0.08(10)} = 100 \cdot e^{0.8}.$$

Now locate the key marked "e" on your calculator (equivalently, "inv" followed by "ln"), and find $100 \cdot e^{0.8} \approx \$222.55$. ▬

It may be surprising to note that after rounding to two decimals, $100 grows only slightly faster over a ten-year period at 8% when compounded continuously than when compounded daily ($222.53). Continuous compounding is better, but not that much better. The comparison between simple interest and compound interest shown by Examples 4 and 5 may also be shown graphically. Notice that simple compounding is *linear;* that is, $A = P(1 + rt) = P + Prt$ is linear in the variable t, provided P and r remain fixed. The A intercept is P and the slope of the line is Pr. On the other hand, compound interest is exponential in nature. By this we mean that the variable n is an exponent in the formula $A = (1 + i)^n$. Figure 1 shows a graphic comparison between simple and compounded interest.

Solving for some variables in finance formulas requires logarithms that we may obtain from our calculators. The solution to a commonly used equation is shown next, as a brief review of needed algebra skills.

> **An Algebra Refresher**
>
> To solve the equation $a^n = b$ for n, where a and b are positive real numbers, calculate
>
> $$n = \frac{\ln(b)}{\ln(a)} = \frac{\log(b)}{\log(a)}$$
>
> where $\ln(b)$ is the logarithm of b to the base e and $\log b$ is the logarithm of b to the base 10. The solution comes from a law of logarithms, which states that if $a^n = b$, then $n \ln(a) = \ln(b)$.

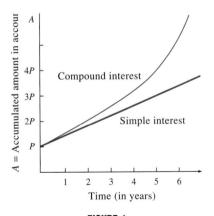

FIGURE 1
*Comparing the Growth of Principal
Between Simple and Compound Interest*

EXAMPLE 6
How Fast Will
Money Double?

How long will it take a sum of money P to double if the interest rate is 6% compounded quarterly?

Solution   The question asks how many interest periods n will it take for P to reach the accumulated amount $2P$. The compound interest formula $A = P(1 + i)^n$ then tells us that

$$2P = P\left(1 + \frac{0.06}{4}\right)^n,$$

and we are to solve for n. Upon canceling P, we have

$$2 = (1.015)^n$$

and, from the algebra refresher,

$$n = \frac{\ln(2)}{\ln(1.015)} \approx 46.56$$

interest periods, which translates into $\frac{46.56}{4} = 11.64$ years.   ■

EXAMPLE 7
Using the
Compound
Interest Formula

What sum of money should be deposited now in an account paying 8% compounded quarterly so that ten years later, the account balance will be $12,000? (Assume no further withdrawals or additions to the account.)

Solution   The number of interest periods is $n = 4 \cdot 10 = 40$, and the interest per period is $\frac{0.08}{4} = 0.02$. The compound interest formula then tells us that

$$\$12{,}000 = P(1 + 0.02)^{40}.$$
$$= P(2.2080397).$$

Solving for P gives $P \approx \$5{,}434.68$.   ■

EXAMPLE 8
Interest Needed
to Double Money

What rate of interest, compounded annually, should you receive for an amount of money P to double in ten years?

Solution   This time, we want to find the annual interest rate r for which P will become $2P$ when $n = 10$. That is, we are to solve

$$2P = P(1 + r)^{10}$$

for r. To do this, first cancel P on both sides and obtain

$$2 = (1 + r)^{10}$$

and then take the tenth root of both sides, getting

$$2^{1/10} = 1 + r$$

from which

$$r = 2^{1/10} - 1$$
$$= 2^{0.10} - 1$$
$$\approx 1.0717735 - 1$$
$$= .0717735$$

or about 7.18%.   ■

A rough estimate of the time required for money to double when compounded at a particular rate annually is given by the **rule of 72**: The time in years required for money to double when compounded at the rate of r percent annually is $\dfrac{72}{r}$. For instance, at an annual interest rate of 8%, it will take approximately $\dfrac{72}{8} = 9$ years for a sum of money to double in value.

EXAMPLE 9
The Battle
Against Inflation

If money is losing value at the rate of 5% annually because of inflation, how long will it take a dollar, now assumed to be worth 100 cents, to be worth 80 cents?

Solution Instead of increasing in value, the principal $P = \$1$ is now losing value at the rate of 5%, compounded annually. Thus after the first year, the accumulated value for an amount P is $P - .05P = P(1 - .05)$, and after the second year the accumulated value is $P(1 - .05) - P(1 - .05).05 = P(1 - .05)^2$, and so on. In particular, for $P = \$1$ we want to find n so that

$$0.80 = \$1.00(1 - 0.05)^n$$

or

$$0.80 = 0.95^n.$$

Using our algebra refresher, we find that

$$n = \frac{\ln(0.80)}{\ln(0.95)} \approx 4.35 \text{ years.} \quad \blacksquare$$

Example 9 points out that even though our investment may earn a handsome return, inflation reduces its spending power. Example 10 shows another factor affecting our investment or savings return.

EXAMPLE 10
Taxes

Suppose an investment of $5000 earns 8% compounded quarterly for a period of one year. When computing your income taxes, you find that the applicable federal rate is 28% and the applicable state rate is 7%. How much of the return on your investment do you get to keep?

Solution The return on your investment will be

$$5000\left(\frac{1 + .08}{4}\right)^4 - 5000 = \$412.16.$$

The federal tax on this amount will be

$$412.16(.28) = \$115.40$$

and the state tax will be

$$412.16(.07) = \$28.85.$$

Therefore, the total income tax due on the return in $115.40 + $28.85 = $144.25. It follows that you get to keep

$$\$412.16 - \$144.25 = \$267.91. \quad \blacksquare$$

EXERCISES 11.1

 A calculator or the computer program Finance will be helpful in these exercises.

For Exercises 1 through 10, find the amount that will be accumulated in each account under the conditions set forth.

1. The principal $2000 is accumulated with 7.5% interest, compounded monthly for 4 years.

2. The principal $5000 is accumulated with 10% interest, compounded annually for 8 years.

3. The principal $3000 is accumulated with 6% interest, compounded quarterly for 12 years.

4. The principal $7000 is accumulated with 9.5% interest, compounded annually for 15 years.

5. The principal $6000 is accumulated with simple interest of 12% for 8 months.

6. The principal P is accumulated with 6% interest, compounded quarterly for 5 years.

7. The principal $800 is accumulated for 12 years

  (a) At 7% simple interest
  (b) At 7% compounded quarterly
  (c) At 7% compounded monthly

8. The principal $4000 is accumulated for 8 years

  (a) At 9% compounded quarterly
  (b) At 9% compounded monthly
  (c) At 9% compounded daily

9. The principal $1 is accumulated for 10 years

  (a) At 12% compounded semiannually
  (b) At 12% compounded quarterly
  (c) At 12% compounded daily

10. The principal $10,000 is accumulated for 4 months

  (a) At 14% simple interest
  (b) At 14% compounded monthly
  (c) At 14% compounded daily

11. Graph on the same set of axes the accumulated value for each of the first 5 years, $5000

  (a) With simple interest of 10%
  (b) With 10% interest compounded quarterly

12. Graph on the same set of axes the accumulated value for each of the first 5 years, $10,000

  (a) With simple interest of 6%
  (b) With 6% interest compounded monthly

In Exercises 13 through 18, find the rate of interest required to achieve the conditions set forth.

13. $A = \$20,000$; $P = \$8000$; $n = 10$; interest is compounded annually

14. $A = \$12,000$; $P = \$3500$; $n = 12$; interest is compounded annually

15. $A = \$5000$; $P = \$1250$; $t = 12$ years; interest is compounded quarterly

16. $A = \$6500$; $P = \$2300$; $t = 5$ years; interest is compounded quarterly

17. $A = \$7500$; $P = \$2500$; $t = 6$ years; interest is compounded semiannually

18. $A = \$3000$; $P = \$1500$; $t = 8$ years; interest is compounded semiannually

In Exercises 19 through 24, find the number of interest periods required to achieve the conditions set forth.

19. $A = \$3500$; $P = \$1200$; interest is 8%, compounded semiannually

20. $A = \$5000$; $P = \$2400$; interest is 6%, compounded semiannually

21. $A = \$6000$; $P = \$1000$; interest is 12%, compounded quarterly

22. $A = \$30,000$; $P = \$4800$; interest is 15%, compounded quarterly

23. $A = \$50,000$; $P = \$6000$; interest is 10%, compounded annually

24. $A = \$60,000$; $P = \$10,000$; interest is 9%, compounded annually

In Exercises 25 through 30, find the principal P required to achieve the stated conditions.

25. $A = \$5000$; rate is 6%, compounded annually for a period of 10 years

26. $A = \$12,000$; rate is 9%, compounded annually for a period of 15 years

27. $A = \$30,000$; rate is 12%, compounded quarterly for a period of 10 years

28. $A = \$35,000$; rate is 6%, compounded quarterly for a period of 12 years 17,129.66

29. $A = \$3000$; rate is 7.5%, compounded monthly for a period of 10 years

30. $A = \$10,000$; rate is 9%, compounded monthly for a period of 8 years

31. *Investing* If $5000 is invested at the rate of 7%, compounded semiannually, what will be the value of the investment ten years from now, assuming no withdrawals?

32. *Investing* If $10,000 is invested at the rate of 10%, compounded quarterly, what will be the value of the investment 15 years from now assuming no withdrawals?

33. *Investing* Dr. Sekiguchi is saving money to send his children to college. How much will he need to invest now at 8% compounded quarterly if he wants the accumulated value of the investment to be $20,000 in 12 years?

34. *Investing* Ramero plans to buy a new car three years from now and, rather than borrow at that time, he plans to now invest part of a small inheritance at 7.5% compounded semiannually to cover the estimated $6000 trade-in difference. How much does he need to invest?

35. *Investing* The Smiths have received an $8000 gift from one of their parents to invest for their child's college education. They estimate that they will need $20,000 in 12 years to achieve their educational goals for their child. What interest rate compounded semi-annually would the Smiths need to achieve this goal?

36. *Investing* When she took early retirement, Nancy received $50,000 from a savings plan she had established with her company while she was still working. She plans to open her own business and work for herself for a few years before retiring completely. She estimates that she needs $75,000 to start the business she has in mind, and she plans to take three years to find the right location. What interest rate compounded quarterly would Nancy need to achieve this goal?

37. *Tripling Your Money* How many years will it take for a sum of money to triple if the interest rate is 8%, compounded quarterly?

38. *Doubling Your Money* How many years will it take for $4000 to double if the interest rate is 10%, compounded annually? Compare this with the rule of 72.

39. *Growth of Money* How many years will be required to turn $5000 into $8000 if the interest rate is 7.5%, compounded semiannually?

40. *Growth of Money* How many years will be required to turn $10,000 into $25,000 if the interest rate is 8%, compounded quarterly?

Checking Accounts *Exercises 41 and 42 refer to these regulations. Banking rules require that a specified percentage of bank funds in checking accounts be set aside as reserve and not be used for loan purposes. If a bank states that it is paying interest on available funds, they are paying interest only on the percentage of your account that is outside the reserve requirements. Assume this reserve to be 15%.*

41. Suppose that your checking account pays interest at the rate of 4.5% annually on the monthly minimum balance of available funds. If your minimum balance this month was $1200, how much interest was credited to your account?

42. Suppose that your checking account pays interest at the rate of 5% annually on the monthly minimum balance of available funds. If your minimum balance this month was $2000, how much interest was credited to your account?

43. *Cost of Living* If the cost of living is expected to increase at the rate of 4% compounded annually each year for the next 5 years, what will the cost then be of goods priced at $50 now?

44. *Cost of Living* If the cost of living is expected to increase at the rate of 6% compounded annually each year for the next 10 years, what will the cost then be of goods priced at $100 now?

45. *Inflation* If the purchasing power of the dollar is expected to decrease at the rate of 5% each year, how long will take before the purchasing power is cut in half?

46. *Inflation* If the purchasing power of the dollar is expected to decrease at the rate of 8% each year in the foreseeable future, how long will it take before the purchasing power is reduced to $\frac{3}{4}$ of its current value?

47. *Retirement* For a couple comfortably retired now on $25,000 per year, what income will they need 10 years from now to maintain the same standard of living if the inflation rate is assumed to be 5% per year?

48. *Retirement* For a couple comfortably retired now on $45,000 per year, what income will they need 8 years from now to maintain the same standard of living if the inflation rate is assumed to be 6% per year?

49. *Cars and Inflation* If a new car costs $20,000 now, and the inflation rate is assumed to be 5% per year, what will a new car cost 8 years from now?

50. *Homes and Inflation* If the median price of a new home now is $110,000, and the inflation rate is expected to be 5% per year, what will the median price of a new home be 10 years from now?

51. *The Cost of Building* The contractor building a new eight-story office building estimates that the cost of each floor will be 20% more than the floor just below it. If the cost of the first floor is $250,000, what

   (a) Will the fourth floor cost?
   (b) Will the eighth floor cost?
   (c) Will the entire building cost?

52. *Population Growth* If the population of a city has doubled in 20 years, what was the percentage rate of growth compounded annually?

53. *Rate of Return on an Investment* If Joan bought a lot for $18,000 and sold it 5 years later for $24,500, what was her percentage rate of return on this investment if it was compounded annually?

Problems for Exploration

1. Suppose that P dollars are invested at an annual rate of r percent, compounded n times per year. Find an expression that will give the number of years it will take for this amount to triple.

2. On one checkerboard, put 2 cents on the lower left square and move consecutively through all 64 squares, each time thinking of a square as a year, and each time giving yourself 100% *simple interest*. The stack of pennies on your squares will look like 2, 4, 6, 8, . . . , and the sixty-fourth square will have

$1.28. On another checkerboard, do the same thing, but give yourself 100% *compound interest*. The stack of pennies on your squares will look like 2, 4, 8, 16 . . . . In dollars, what will be the value of the stack of pennies on the sixty-fourth square?

3. Go to at least three local financial institutions, and gather information about their certificate-of-deposit rates on amounts of less than $25,000 for a two-year period. Compare the rates, and determine which is the best.

SECTION 11.2   Ordinary Annuities

An **annuity** is a sequence of equal payments, each made at equally spaced time intervals. When the interest on these payments is compounded at the same time as the payment is made, the annuity is called an **ordinary annuity.** We only consider ordinary annuities here. Examples of ordinary annuities are

    A sequence of equal annual payments into an IRA account

    A sequence of equal monthly payments to pay off a car loan

    A sequence of equal monthly payments to pay off a house loan

    A sequence of equal monthly payments into an account to be used later for a college education

When making payments into an annuity account, we are naturally interested in the total accumulated in the account after a given number of payments. This total is called the **future value** of the annuity, and it may be calculated by a formula that we develop next. Specifically, the development makes use of the compound interest formula. For example, suppose that we plan to put $500 into an account at the end of each year for five years, assuming the interest is 8% compounded annually.

```
0      1        2        3        4        5        Years
└──────┴────────┴────────┴────────┴────────┘
     $500     $500     $500     $500     $500
```

The following table shows how each of the $500 amounts earns interest compounded annually.

Payment number	Amount of payment	Number of years at 8% interest	Accumulated value
1	$500	4	$500(1 + 0.08)^4$
2	$500	3	$500(1 + 0.08)^3$
3	$500	2	$500(1 + 0.08)^2$
4	$500	1	$500(1 + 0.08)^1$
5	$500	0	$500

The future value A of the account is the sum of the entries in the last column of the table:

$$A = \$500(1.08)^4 + 500(1.08)^3 + 500(1.08)^2 + 500(1.08) + 500$$
$$= \$500[(1.08)^4 + (1.08)^3 + (1.08)^2 + (1.08) + 1].$$

Notice the pattern set out in the last line. Could you now write an expression for the future value of this annuity after the twelfth payment? If you write

$$A = \$500[(1.08)^{11} + (1.08)^{10} + (1.08)^9 + \cdots + 1]$$

you are correct! Following these guidelines, if an amount p were deposited into an account at the end of each interest-bearing period, and the interest rate was i *per period,* then the A (future value) of the annuity at the end of the nth payment would be

$$A = P[(1 + i)^{n-1} + (1 + i)^{n-2} + (1 + i)^{n-3} + \cdots + 1].$$

The sum inside the square brackets has a very special form. It is called a **geometric series,** and an algebraic formula gives its sum:

$$A = p\left[\frac{(1 + i)^n - 1}{i}\right].$$

Future Value of an Ordinary Annuity

Let p = amount of each payment made at the end of each period, $p > 0$
$\quad i$ = interest rate *per period*
$\quad n$ = number of payments.

Then the **future value** of the annuity is the sum of all payments and interest earned on those payments and is found by this formula:

$$A = p\left[\frac{(1 + i)^n - 1}{i}\right].$$

EXAMPLE 1
Calculating
Retirement
Funds

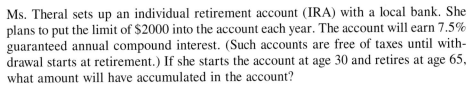

Ms. Theral sets up an individual retirement account (IRA) with a local bank. She plans to put the limit of $2000 into the account each year. The account will earn 7.5% guaranteed annual compound interest. (Such accounts are free of taxes until withdrawal starts at retirement.) If she starts the account at age 30 and retires at age 65, what amount will have accumulated in the account?

Solution This is an ordinary with $p = \$2000$, $i = .075$, and $n = 35$. The amount accumulated in the account at the end of 35 years, including the thirty-fifth payment, is given by the future value:

$$A = \$2000\frac{(1 + .075)^{35} - 1}{.075}$$
$$\approx \$2000(154.25160)$$
$$\approx \$308,505.21. \quad\blacksquare$$

Example 1 vividly shows the power of compound interest over a long period of time. Ms. Theral has actually contributed $2000 \times 35 = \$70,000$ to the account but will enjoy a retirement fund of over $300,000, even though the accumulated amount will not have the same purchasing power as her earlier contributions.

When large expenses are foreseen in the future by families or businesses, they sometimes elect to set aside in an interest-bearing account a sum of money on a regular basis to cover those expenses. For example, a family might want to set aside a certain amount each month for 16 years to have $25,000 for their child's college expenses, or a business might want to set aside a certain amount each year to accumulate $50,000 to buy a piece of equipment ten years from now. When establishing such accounts, we say they are setting up a **sinking fund.** In a *sinking fund,* we are seeking to find the amount p to deposit each period to achieve the stated financial goal in the desired length of time. We assume that the *deposits* will be made at the *end of each interest-paying period,* and we know the future value, the interest rate, and the time needed to achieve our goal. Thus, all of the quantities in

$$A = p\left[\frac{(1 + i)^n - 1}{i}\right]$$

are known except for p. Solving for p will give the amount of each payment, as shown next.

Payments to a Sinking Fund

Let A = the future value desired in the fund
$\quad i$ = interest rate *per period*
$\quad n$ = number of interest periods.

Then the amount p to be deposited at the end of each interest period in order to achieve A is

$$p = \frac{Ai}{(1 + i)^n - 1}.$$

NOTE Many books use the letter S for the sum of the terms giving the total accumulated value rather than A. In addition, because the fraction

$$\frac{(1 + i)^n - 1}{i}$$

is used so often, some books use the notation $s_{\overline{n}|i}$ for this quantity. It is read "s sub n angle i."

EXAMPLE 2
Finding the
Payments to a
Sinking Fund

Suppose that Ms. Tatum wants to save $10,000 over a period of three years to buy a fishing boat. Her bank will pay 8% compounded semiannually on an account into which she will place her payments. If the payments are made semiannually, how much should each payment be for Ms. Tatum to achieve her savings goal?

Solution This is a sinking-fund problem, with $A = \$10,000$, $i = \frac{0.08}{2} = 0.04$ per period, and $n = 3(2) = 6$. Therefore, the payments should be

$$p = \$10,000\left[\frac{0.04}{(1.04)^6 - 1}\right]$$
$$\approx \$10,000(0.1507619)$$
$$\approx \$1507.62. \quad \blacksquare$$

Suppose that an interest-bearing account already has an amount P in it, and equal payments p are made periodically, as in an ordinary annuity. What will the accumulated value of the account be after n periods? By the compound interest formula, the amount P will earn $\$P(1 + i)^n$, and, in addition, the annuity portion will earn

$$A = p\frac{(1 + i)^n - 1}{i}.$$

Therefore the total accumulated amount will be as shown next.

The Ordinary Annuity Formula

Let P = an initial amount in an account
 i = the interest rate per period
 n = the number of periods
 p = amount deposited at the end of each period.

Then the total amount accumulated in the account after n periods is

$$S = P(1 + i)^n + p\left[\frac{(1 + i)^n - 1}{i}\right].$$

This formula encompasses two of the major formulas that we have already developed: The compound interest formula (let $p = 0$) and the future value of an annuity (let $P = 0$).

EXAMPLE 3
Using the
Ordinary Annuity
Formula

Suppose that an account paying 6% compounded semiannually now has $10,000 in it and its owner plans to add $500 at the end of each six-month period for the next eight years. What will be the accumulated value of the account at the end of eight more years?

Solution The ordinary annuity formula may be used to find the accumulated value. We have $P = \$10,000$, $i = \frac{0.06}{2} = 0.03$, $n = 16$, and $p = \$500$. Therefore,

$$S = \$10,000(1 + 0.03)^{16} + \$500\frac{(1 + 0.03)^{16} - 1}{0.03}$$

$$\approx \$10,000(1.604706) + \$500(20.1568813)$$

$$\approx \$26,125.50. \quad \blacksquare\blacksquare$$

Just as an annuity has a future value, it also has a *present value*. Suppose that we have an annuity of n payments of p dollars each, where the interest rate is i per payment period. The **present value** of such an annuity is defined to be a *single deposit* of P dollars, earning the same rate of interest for n periods, that will generate the same accumulated value as the annuity. That is, we want to find P so that

$$P(1 + i)^n = p\left[\frac{(1 + i)^n - 1}{i}\right]$$

or

$$P = p\left[\frac{(1 + i)^n - 1}{i(1 + i)^n}\right] = p\left[\frac{1 - (1 - i)^{-n}}{i}\right].$$

> ### The Present Value of an Annuity
>
> For an annuity with
>
> $$p = \text{amount of payment each period}$$
> $$i = \text{interest rate per period}$$
> $$n = \text{the number of payments,}$$
>
> the **present value** is
>
> $$P = p\left[\frac{1 - (1 + i)^{-n}}{i}\right].$$

Here is an alternative way to look at the present value of an annuity: Suppose that we put P dollars into the interest-bearing account, *withdraw* p dollars each period, and redeposit it into a new account earning the same interest rate. After n periods, the new account would have the accumulated value of the annuity, which means that the original account starting with P dollars must be exhausted! In other words, the *present value of an annuity is the* **single** *deposit that will finance the annuity.*

EXAMPLE 4
Using the
Present Value of
an Annuity

Mr. and Mrs. Dunston want to put a sum of money into an account when their daughter starts college so that they could withdraw $1000 from the account each month for four years. They will make no further payments into the account once their daughter starts college. Assume that their account will earn interest at the rate of 8% compounded monthly. How much should be in the account at the start of college?

Solution This question can be restated to ask, "What is the present value of an annuity with $p = \$1000$, $i = \frac{0.08}{12}$, and $n = 48$." The present value formula gives the amount P needed as

$$P = \$1000 \frac{1 - \left(1 + \frac{.08}{12}\right)^{-48}}{\frac{.08}{12}}$$

$$\approx \$40{,}961.91. \quad \blacksquare$$

EXAMPLE 5
Small Business
Application

The ABC Appliance Center has just sold a refrigerator to a family who, after making a small down payment, agrees to make payments of $40 per month for a period of three years. Like most small businesses, they sell their loan to a bank so they can receive their money up front. If interest rates are 10% compounded annually, how much should they expect to receive from the bank for their loan?

Solution The question asks for the present value of the loan. In this case $p = \$40$, $i = \frac{.10}{12}$ because the payments are made monthly, and $n = 36$. Substituting these numbers into the present value formula gives

$$P = 40\left[\frac{1 - \left(1 + \frac{.10}{12}\right)^{-36}}{\frac{.10}{12}}\right] \approx \$1239.65. \quad \blacksquare$$

EXERCISES   11.2

In Exercises 1 through 6, find the future value of each of the annuities with the given interest rate per period.

1. $p = \$500, n = 30, i = .025$

2. $p = \$1200, n = 24, i = .04$

3. $p = \$700, n = 20, i = .01$

4. $p = \$3000, n = 40, i = .015$

5. $p = \$8000, n = 20, i = .05$

6. $p = \$10,000, n = 12, i = .03$

In Exercises 7 through 12, a sinking fund is being established, the goal of which is to reach $100,000. Calculate the payments in each case, where i is the interest rate per period.

7. $n = 20, i = .025$ 8. $n = 30, i = .01$

9. $n = 16, i = .02$ 10. $n = 18, i = .04$

11. $n = 22, i = .03$ 12. $n = 26, i = .05$

13. **Saving for the Future** Nannette plans on putting $500 in an account on her daughter's birthday, beginning with the first one and continuing through the sixteenth one. Assuming that the account pays 8% compounded annually, how much should be in the account on her daughter's sixteenth birthday?

14. **Saving for the Future** Timothy has $10,000 in an account now and plans to add $1000 to the account at the end of each quarter in the future. If the account is assumed to pay 6% compounded quarterly, how much will be in the account at the end of eight years?

15. **Investing for College** Melissa's son will enter college 16 years from now. At that time, she would like to have $20,000 available for college expenses. For this purpose, her bank will set up an account that pays 7% compounded quarterly. If she makes payments into this account at the end of each quarter, what must her payments be to achieve her goal?

16. **Business Investment** A freight-hauling firm estimates that it will need a new forklift truck in six years. They estimate that it will cost $40,000. They set up a sinking fund that pays 8% compounded semiannually, into which they will make semiannual payments to achieve their goal. Calculate the size of the payments.

17. **Investing for Retirement** Eduardo is a 40-year-old individual who plans to retire at age 65. Between now and then, $2000 is paid annually into his IRA account that is anticipated to pay 5% compounded annually. How much will be in the account upon retirement?

18. **Investing for Retirement** Jeremy invests $1000 at the end of each year in a company retirement plan in which the employer matches the employee contribu-

tion. If the plan pays 8% compounded annually and Jeremy will retire in 30 years, what will be the total accumulated value of the account?

19. **Comparing Investments** Which of the following investments earns more?

(a) At the end of each year for 20 years, $2000 is invested into an account that pays 7% compounded annually

(b) At the end of each quarter for 20 years, $1800 is invested into an account that pays 6.5% compounded quarterly

20. **Planning for Retirement** Compare the accumulated value of IRA accounts in which $2000 is invested at the end of each year into an account that earns 6% compounded annually if

(a) You start at age 25 and retire at age 65.

(b) You start at age 35 and retire at age 65.

(c) You start at age 40 and retire at age 65.

(d) You start at age 50 and retire at age 65.

21. **Planning for Retirement** Compare the accumulated value of an IRA account into which $500 is invested at the end of each quarter; the account earns 6% compounded quarterly:

(a) You start at age 25 and retire at 60.

(b) You start at age 30 and retire at 60.

(c) You start at age 35 and retire at 60.

(d) You start at age 40 and retire at 60.

22. **Planning for Retirement** If you are 30 years of age now, plan to retire at age 60, with an IRA account having a total accumulated value of $300,000, how much would you have to invest at the end of each year if the account

(a) Paid 6% compounded annually?

(b) Paid 8% compounded annually?

23. **Planning for Retirement** Yolanda wants to have enough in her retirement accounts so that upon retirement, she can withdraw $500 each month for the next 20 years. Assuming that her accounts will earn an average of 7% compounded monthly, what sum of money should she have in her accounts upon retirement?

24. **Saving** The Smiths are saving for their son's college days. They would like to be able to withdraw $800 each month from their account for five years once their son starts college. Assuming that their account will earn interest at the rate of 9% compounded monthly, what sum of money should the Smiths have in their account when the son starts college?

25. *Savings*   The Meek brothers are planning a trip around the world. They hope to work some as they go, but believe that they should have accessible $800 per month so that they can live in relative comfort for the eight months they plan to be gone. How much should they have in an account earning 6% compounded monthly when they leave so that they could withdraw the desired $800 each month for eight months?

26. *Savings*   A small business will begin paying a $5000 franchising fee beginning next year. If a fund earning 8% compounded annually is set aside from which these payments will be made, what should be the size of the fund to guarantee these annual payments for the next seven years?

27. *Lottery Earnings*   If you won a lottery that paid $5000 each year for the next 10 years and interest rates were 10% compounded annually, what is the present value of your prize?

28. *Present Value of a Loan*   You loan a friend money, and she is repaying you at the rate of $150 per month. The loan still has three years of payments left, and you would like to sell the loan and receive your money up front. If interest rates are 8% compounded annually, what is a fair price for your loan?

29. *Present Value of a Loan*   The ABC Appliance Center sells refrigerators on credit to customers. It just sold a refrigerator for payments of $65 each month for a period of three years. Like most small businesses, it sells its loan to the bank to have cash available now. If interest rates are 9% compounded annually, how much should the company expect to receive for its loan?

30. *Present Value of a Loan*   The Albright Tractor Company has just sold a tractor to a farmer who financed a portion of the purchase for payments of $1500 quar-terly for a period of four years. If interest rates are 12% compounded annually, and the company sells the loan to the bank, how much should it expect to receive?

31. Suppose that Debbie deposits $100 at the end of each month into an account that pays 5.25% compounded monthly. After three years, she puts the accumulated amount into a certificate of deposit paying 7.5% compounded quarterly for two-and-a-half years. At that time, she puts the money into a 90-day certificate of deposit paying 6.8% annually. When the certificate matures, how much money has Debbie accumulated?

32. Erin deposits $200 at the end of each month into an account that pays 6.25% compounded monthly. After two years, she puts the accumulated amount into a 90-day certificate of deposit that pays 7.5% annually. Upon maturity of the certificate, she puts the accumulated amount into a three-year certificate of deposit paying 8% compounded quarterly. When this last certificate matures, how much money will Erin have accumulated?

Problems for Exploration

1. Arrive at the present value of an annuity by interpreting the variables in the ordinary annuity formula.

2. State the formula for finding the sum of a geometric series, and use it to derive the formula for finding the future value (A) of an annuity formula.

3. Go to at least two local financial institutions, and ask them to explain their practice of buying loans from small businesses. Write a short report on what you learned.

SECTION **11.3**   **Consumer Loans and the APR**

The Truth-in-Lending Law passed by the U.S. Congress in 1969 created a common reference rate that lenders must state to protect consumers from various and sometimes confusing ways in which interest rates are quoted. That common reference rate is called the **effective interest rate** or, more commonly, the **annual percentage rate,** which is abbreviated APR. It is a statement of the interest rate that the loan earns *compounded annually*. Stated differently, the APR is the *simple interest rate* that would be needed to produce the same income as the compounding procedure over a one-year period. Any rate, annual or otherwise, can easily be converted to an APR. For instance, if an interest rate is quoted as being 6% compounded quarterly, then the compound interest principle tells us that $1 will grow to

$$\$1\left(1 + \frac{0.06}{4}\right)^4 = \$1(1.015)^4 \approx \$1(1.06136) = \$1.0614$$

in one year. Therefore, the quoted rate of 6% compounded quarterly, sometimes called the **nominal rate,** yields

$$\$1.0614 - \$1 = \$0.0614 \text{ interest}$$

for one year. This is the same yield as a simple interest rate of 6.14% applied to $1 for a year:

$$0.0614(\$1) = \$0.0614 \text{ interest}.$$

Therefore, the APR is 6.14%. This same pattern of reasoning on a $1 investment allows us to arrive at a general rule for finding the APR.

The Annual Percentage Rate (APR)

If r is the decimal equivalent of a quoted annual percentage rate, compounded n times per year, then the **effective rate of interest** or the **annual percentage rate (APR)** is given by

$$\text{APR} = \left(1 + \frac{r}{n}\right)^n - 1.$$

The intended purpose of the APR is to allow a convenient comparison of rates.

EXAMPLE 1
Calculating
the APR

Which investment earns the most?

(a) A savings account paying 6.25% compounded monthly
(b) A government bond paying 6.5% compounded semiannually
(c) A certificate of deposit paying 6.3% compounded quarterly

Solution   When each rate is converted to the APR, we will immediately know which is the better investment.

(a) $\text{APR} = \left(1 + \dfrac{0.0625}{12}\right)^{12} - 1 \approx 0.06432 \text{ or } 6.4\%.$

(b) $\text{APR} = \left(1 + \dfrac{0.065}{2}\right)^{2} - 1 \approx 0.06606 \text{ or } 6.6\%.$

(c) $\text{APR} = \left(1 + \dfrac{0.063}{4}\right)^{4} - 1 \approx 0.06450 \text{ or } 6.5\%.$

These calculations show that (b) is slightly better than either of the other two investments. ▬▬

If you bought a new car and needed to borrow $12,000 for 60 months at a 10.5% APR, do you know how much the monthly payments would be? If you borrowed $5000 at 12% APR and wanted to pay it back at the rate of $100 per month, do you know how long it would take? Of course, the auto dealers or bankers are expected to tell you the answers to these questions. They will either look up these answers in a specially prepared table or find them upon keying the pertinent information into a calculator or computer. In this section, we intend to look behind the scenes to gain some insight into the logic of how such calculations are made.

Most consumer loans (personal, automobile, home) are required to be repaid with monthly payments. In reality, however, most of us pay a little early or (more likely) a little late, and this affects the interest that is being charged to the loan. The last payment on the loan then may be a little more or a little less than the others to adjust for the interest charges. Methods of calculating interest on loans are set forth by the U.S. Comptroller of the Currency and, in some cases, by state law. One of the most common arrangements is that of paying interest on the *unpaid balance* of the loan. For instance, suppose that a $1000 personal loan is taken out at the rate of 12% simple interest applied to the monthly unpaid balance. The loan is to be repaid in six equal monthly payments of $172.55. During the first month, the borrower has use of the entire $1000 and therefore should be charged $\frac{1}{12}$ of one year's interest, or

$$\left(\frac{1}{12}\right)(0.12)(\$1000) = \$10.$$

So the first payment of $172.55 consists of $10 interest and the remainder, $162.55, is applied to the principal. The *new loan balance or unpaid balance* now is $1000 − $162.55 = $837.45. For the second payment, this unpaid balance is multiplied by the factor $(\frac{1}{12})(0.12) = 0.01$ to get the interest ($8.38) portion of the payment. The entire financial picture for this loan can be pictured in the following table. Notice that a portion of each payment is interest and the remainder is applied to the outstanding balance of the loan. The interest portion of each payment is higher in the beginning, while near the end of the payoff period, the principal portion is higher.

Payment number	Amount	Interest	Applied to principal	Unpaid balance
1	$172.55	$10.00	$162.55	$837.45
2	$172.55	$ 8.37	$164.18	$673.27
3	$172.55	$ 6.73	$165.82	$507.45
4	$172.55	$ 5.07	$167.48	$339.97
5	$172.55	$ 3.40	$169.15	$170.82
6	$172.53	$ 1.71	$170.82	$ 0.00

TABLE 1

Rather than computing interest monthly on the unpaid balance of a loan, most lending institutions charge interest on a daily basis.

EXAMPLE 2
The Interest
Portion of a
Monthly Payment

Sue has a car loan from her credit union at a rate of 10.5%, for which her payments are $235 per month. The interest is computed on a daily basis on the unpaid balance of the loan. If the loan balance after her last payment was $8647.50 and Sue makes her next payment 32 days later, how much of the $235 is paid toward interest?

Solution The interest rate per day is $\frac{0.105}{365}$. This rate is then applied to the balance of $8647.50 for 32 days, giving

$$\left(\frac{0.105}{365}\right)(\$8647.50)(32) = \$79.60. \quad \blacksquare$$

When we borrow money for a certain length of time, our immediate concern is the amount of each periodic payment. Suppose that we investigate how to arrive at the amount of each monthly payment calculated for the $1000 loan shown in Table 1. The line of thought leads us to the general formula.

First, consider the transaction from the viewpoint of the lender, who lets you borrow $1000 for six months for a rental fee of 1% compounded monthly. The lender is then entitled to an accumulated value of $1000(1 + .01)^6 = $1061.52 by the compounded interest law. Now consider the transaction from the viewpoint of the borrower. When the borrower makes a monthly payment, think of that payment as an investment in an account that earns interest at the rate of 1% compounded monthly. After making the sixth payment, the total accumulated in the account must be $1061.52, to which the lender is entitled. However, this is just an annuity where p is unknown and the future value is $1061.52.

Therefore, we must have

$$\$1061.52 = p\frac{(1 + .01)^6 - 1}{.01}$$

or

$$\$1061.52 = p(6.152015)$$

so that

$$p = \frac{\$1061.52}{6.152015} = \$172.55.$$

This thought process and the accompanying calculations serve as a model for a general formula from which monthly payments (or any other period payments) to retire a loan may be calculated. If an amount P is borrowed at i percent per period for n periods and n periodic equal payments are to be made to retire the loan, then the lender is entitled to a total accumulation of $P(1 + i)^n$ at the end of n periods. The borrower thinks of a like amount gained through an annuity of n equal payments of p dollars each, earning interest at the rate of i percent compounded at the end of each period, for a total of

$$p\frac{(1 + i)^n - 1}{i}.$$

Equating these two quantities and solving for p gives

$$P(1 + i)^n = p\frac{(1 + i)^n - 1}{i} \qquad \text{or} \qquad p = \frac{iP(1 + i)^n}{(1 + i)^n - 1}.$$

The Amount of Loan Payments Formula

Let P = amount borrowed
$\quad i$ = interest rate per payment period
$\quad n$ = number of payments to pay off the loan.

Then the amount of each payment is

$$p = \frac{iP(1 + i)^n}{(1 + i)^n - 1} = \frac{iP}{1 - (1 + i)^{-n}}.$$

EXAMPLE 3
Calculating
Payments for
a Loan

Mr. and Mrs. Callaway bought a lot for $40,000 on which they plan to build a new house in the near future. The banker required a $5000 down payment and financed the remainder at 6% compounded semiannually with equal payments to be made semiannually for eight years. What are the Callaways' semiannual payments?

Solution In this case, $P = \$35,000$, $i = \frac{0.06}{2} = 0.03$, and $n = 16$. Therefore,

$$p = \frac{0.03(\$35,000.00)}{1 - (1 + 0.03)^{-16}} \approx \$2786.38. \quad\blacksquare$$

Even though you may not know exactly how the finance company will figure the monthly payments when an APR is quoted, a very close approximation may be obtained by dividing the APR by 12 and using that as the compounding rate.

EXAMPLE 4
Estimating the
Monthly
Payments
for a Loan

Joe has just finished college; upon landing his first job, he decides to buy a new car. He thinks that after trading in his junker, he will still need to finance $13,000 at the current 10.9% APR advertised for 60 months. Find Joe's approximate monthly payment.

Solution We assume that interest will be charged monthly on the unpaid balance of the loan, so that

$$\frac{10.9\%}{12} \approx 0.908333\%$$

is to be applied to the monthly balance each month. The amount of payments formula then tells us that the monthly payments will be (approximately)

$$p = \frac{0.00908333(\$13,000)}{1 - (1 + 0.00908333)^{-60}} \approx \$282. \quad\blacksquare$$

If the amount-of-loan-payments formula is solved for $(1 + i)^n$, we get

$$(1 + i)^n = \frac{p}{p - Pi} \qquad \text{or} \qquad n = \frac{\ln\left(\dfrac{p}{p - Pi}\right)}{\ln(1 + i)}.$$

Thus, if we know p, P, and i, we can calculate the number of payments necessary to retire the loan if a solution to the preceding equation exists. (A solution exists when $p > Pi$.)

Number of Payments Needed to Retire a Loan

Let P = amount of loan
$\quad i$ = interest rate per period
$\quad p$ = amount of each payment.

Then the number of payments needed to retire the loan is

$$n = \frac{\ln\left(\dfrac{p}{p - pi}\right)}{\ln(1 + i)}$$

if n can be calculated.

This formula allows us to solve the car-payment problem posed at the outset of this chapter.

EXAMPLE 5
Calculating the
Number of
Payments
Needed to
Retire a Loan

Sarah is to repay a $10,000 loan at 8.7%, with monthly payments of $200. How many months will it take to pay off the loan?

Solution   Using the formula for calculating the number of payments, we get

$$n = \frac{\ln\left(\dfrac{200}{200 - 0.00725(10,000)}\right)}{\ln(1 + 0.00725)}$$

$$= \frac{\ln(1.568627)}{\ln(1.00725)}$$

$$\approx 62.32 \text{ months.}\quad\blacksquare$$

EXAMPLE 6
Paying Off the
Mortgage

The Cochrans obtained a loan of $87,500 to purchase a new house. The loan rate is 9.8%, with payments to be made at the end of each month for 30 years.

(a) How much are the Cochrans' monthly payments?

Solution   The solution is found by using the amount-of-loan-payments formula. In this case, $P = \$87,500$, $i = \frac{.098}{12} \approx .00816667$, and $n = 360$. The amount of each payment is found by

$$p = \frac{\$(.00816667)(87,500)}{1 - (1.00816667)^{-360}} \approx \$754.98$$

(b) Assuming that timely payments are made throughout the life of the loan, what is the total amount (principal plus interest) paid for the loan?

Solution   The total will be $754.98(360) = \$271,792.80$.

(c) After learning of the monthly payment amount, suppose that the Cochrans decided they could afford to pay an extra $50 each month toward paying off their loan. Again, assuming timely payments, what total will they now pay for their loan?

Solution   Payments of $804.98 are now being made to retire a loan of $87,500. The number of payments formula shows that

$$n = \frac{\ln\left(\dfrac{804.98}{804.98 - (87,500)(.00816667)}\right)}{\ln(1.00816667)} \approx 268.84 \text{ months.}$$

Rounding up to 269 months, we find that the total amount paid is 269($804.98) = $216,539.62. (Compare this with the amount in part (b)).

(d) After making timely payments for 20 years on this loan at the rate of $754.98 per month, how much is the unpaid balance?

Solution   After 20 years, 240 payments have been made toward retiring the loan. If we think of the remaining 120 payments as an annuity, then its present value is equal to the unpaid balance of the loan. Using the present value formula of Section 11.2, we find that

$$P = \$754.98\,\frac{1 - \left(1 + \dfrac{.098}{12}\right)^{-120}}{\dfrac{.098}{12}} \approx \$57,611.96.\quad\blacksquare$$

EXERCISES   11.3

In Exercises 1 through 8, convert the given interest rate to the APR.

1. 6.5% compounded quarterly
2. 8% compounded monthly
3. 4% compounded semiannually
4. 12% compounded daily
5. 16% compounded monthly
6. 10% compounded quarterly
7. 9% compounded daily
8. 7% compounded monthly

Rank the stated interest rates in Exercises 9 through 14.

9. (a) 5% compounded semiannually
   (b) 4.8% compounded quarterly
   (c) 4.6% compounded monthly
10. (a) 10.5% compounded quarterly
    (b) 11% compounded daily
    (c) 10.6% compounded semiannually
11. (a) 6.3% compounded monthly
    (b) 6.5% compounded quarterly
    (c) 6.8% compounded annually
12. (a) 12% compounded annually
    (b) 11.8% compounded quarterly
    (c) 11.9% compounded semiannually
13. (a) 8% compounded semiannually
    (b) 8.2% compounded annually
    (c) 6% compounded monthly
14. (a) 3% compounded quarterly
    (b) 2.5% compounded daily
    (c) 3.1% compounded semiannually
15. *Credit Card Charges* If a credit card charges 1% per month on the unpaid balance, what APR is the company charging?
16. *Credit Card Charges* If a gasoline credit card charges 0.8% per month on the unpaid balance, what APR is it charging?

In Exercises 17 through 22, the terms of a consumer loan are given.
(a) Find the monthly payment necessary to retire the loan in the prescribed time.
(b) Make a table similar to Table 1 of this section, showing the breakdown of the payment as to interest and principal reduction.

17. The loan is for $3000 at 8% simple interest on the unpaid balance and is to be repaid in five equal monthly installments.

18. The loan is for $600 at 9% simple interest on the unpaid balance and is to be repaid in four equal monthly installments.
19. The loan is for $1500 at 10.9% simple interest on the unpaid balance and is to be repaid in six equal monthly installments.
20. The loan is for $800 at 9.8% simple interest on the unpaid balance and is to be repaid in five equal monthly installments.
21. The loan is for $12,000 at 11% simple interest on the unpaid balance and is to be repaid in five equal annual installments.
22. The loan is for $15,000 at 12% simple interest on the unpaid balance and is to be repaid in six semiannual installments.

In Exercises 23 through 30, find the payment for the given length of time, amount, and interest rate.

23. $P = \$20,000$; $i = 3\%$ per month; monthly payments for 6 years
24. $P = \$18,000$; $i = 4\%$ per quarter; quarterly payments for 10 years
25. $P = \$40,000$; $i = 2\%$ per month; monthly payments for 20 years
26. $P = \$30,000$; $i = 8\%$ compounded semiannually; semiannual payments for 25 years
27. $P = \$12,000$; $i = 11.9\%$ compounded monthly; monthly payments for 50 months
28. $P = \$15,000$; $i = 0.0833\%$ per month; monthly payments for 36 months
29. $P = \$22,000$; $i = 10\%$, compounded quarterly; quarterly payments for 12 years
30. $P = \$80,000$; $i = 9.3\%$, compounded annually; annual payments for 10 years

31. Suppose that your credit union charges 9.5% simple interest on the unpaid balance of your loan for each day since the last payment. The monthly payments are $326, and you made your last payment 35 days ago, at which time the unpaid balance was $9643.68. What portion of your next payment will be interest?

32. Suppose that your house loan charges 12.3% simple interest on the unpaid balance of your loan for each day since the last payment. The last payment was made 28 days ago when the loan balance was $56,743.54. How much of the payment that you are about to make will be interest?

33. *Borrowing Money*   Suppose that you borrow $12,500 for the purchase of a new car at 9.8% APR for 48 months. What is the approximate amount of your monthly payment?

34. *Repaying a Loan*   A business borrowed $50,000 at 12% per year on the unpaid balance. The loan is to be repaid in ten equal payments, beginning at the end of the first year. Find the amount of each payment.

35. *Borrowing Money*   An appliance dealer advertises a particular brand of refrigerator for $800, which may be financed at 7.9% APR for 48 months. As usual, the payments are to be made in equal monthly installments. Find the approximate amount of each payment.

36. *Borrowing Money*   A department store advertises a stereo for $1500, which it will finance for 60 months at 8% APR. Payments are to be made monthly, and all are to be of equal amount. Find the approximate amount of each payment.

37. *Borrowing Money*   A farmer buys a new tractor for $25,000, of which his banker finances $20,000 for eight years at 11% per year on the unpaid balance. Assume the farmer repays the loan in equal annual installments. What is the amount of each payment?

38. *Borrowing Money*   Jared buys a new motorcycle from The Fastrack Motorcycle Co. for $8500. Fastrack agrees to finance 80% of the purchase at 9% APR for a period of 60 months. What is the approximate size of each of Jared's monthly payments?

39. *Borrowing Money*   The Bigger Wave Boat Co. sells George a new fishing boat for $9200. It agrees to finance 90% of the purchase at 8.5% simple interest, computed on the unpaid monthly balance, for a period of 72 months.

 (a) What is the amount of each of George's monthly payments?

 (b) Assuming timely payments of the amount found in Exercise 39(a), what is the unpaid balance of the loan after the twenty-fourth payment?

40. *Borrowing Money*   Pat and Maxine decide to buy a building lot on which they plan to build a new house. The lot costs $28,500 and the bank agrees to finance 80% of the cost at 9% compounded semiannually for a period of eight years. Payments on the loan are to be made in equal semiannual installments.

 (a) Find the amount of each payment.

 (b) Assuming timely payments of the amount found in Exercise 40(a), what is the unpaid balance after the twelfth payment?

41. *Borrowing Money*   Diane decides to buy a small farm near the edge of town for $95,000. The bank agrees to finance 70% of the cost at 10% compounded quarterly for a period of ten years. Payments on the loan are to be made in equal quarterly installments.

 (a) What is the amount of each payment?

 (b) Assuming timely payments of the amount found in Exercise 41(a), what is the unpaid balance after the twentieth payment?

Exercises 42 through 48 refer to buying a house. A couple gets financing for 90% of the $130,000 purchase price of a house at the rate of 9.5% on the monthly unpaid balance.

(a) Find the amount of the monthly payments to repay the loan only (excluding taxes and insurance).

(b) Find the total amount paid to the finance company (the monthly payment multiplied by the number of payments) for each of the following repayment periods.

42. The loan is repaid in 20 years

43. The loan is repaid in 25 years

44. The loan is repaid in 30 years

45. The loan is repaid in 35 years

46. The loan is repaid in 40 years

47. The loan is repaid in 45 years

48. The loan is repaid in 50 years

49. *Length of Loan Given Specified Payments*   Ashley is thinking about buying a new dishwasher. The model she wants costs $580, and her credit union will finance $500 of the cost at 9.8% APR. She estimates that she can afford to make monthly payments of $80 without upsetting her budget. If she can get the credit union to agree to those payments, approximately how many months will it take Ashley to pay off the loan?

50. *Length of Loan Given Specified Payments*   Chuck wants to buy a new car, but estimates that he can make payments of only $200 per month. The car he would like costs $12,500, of which the dealer will finance 80% at 11.9% APR. If the dealer agrees to be repaid in monthly payments of $200, approximately how long will it take to retire the loan?

51. *Length of Loan Given Specified Payments*   Mr. Brown buys an 80-acre addition to his farm. He plans on having his bank finance $180,000 of the purchase price. The bank is currently loaning money for such purchases at 10%, compounded annually, and Mr. Brown estimates that he can comfortably make annual payments of $15,000. If the bank will agree to these payments, how long will it take to pay off the loan?

In each of the Exercises 52 through 57, find the number of years required to pay off a $100,000 house loan at the indicated interest rates, compounded monthly, if monthly payments of $900 are made toward the loan (excluding taxes and insurance).

52. 9.5%   **53.** 9.8%   **54.** 10%

55. 10.2%   **56.** 10.5%   **57.** 11%

Problems for Exploration

1. When will the formula for finding the number of payments required to retire a loan, given a specified monthly payment, not have a solution and what does having "no solution" mean?

2. Go to two local auto dealers and investigate their rates on car loans: APR, number of months to pay off the loan, percentage of cost financed, and so on. How do they compute the amount of the monthly payment: with a calculator? a table? a computer? Do they know the formula used in the calculation? How do they compute the interest portion of each payment? Write a brief report about your findings.

CHAPTER 11 REFLECTIONS

CONCEPTS

Mortgages, car loans, savings accounts, certificates of deposit, and many other financial and consumer settings involve the lending or borrowing of money. The amount of money involved is called the *principal,* and the fee charged can be *simple interest* or *compound interest* (where interest is earned on both the principal and the interest). *Ordinary annuities* and *sinking funds* involve sequences of equal payments made at equally spaced intervals, witth interest paid at the time of the payments, and they have a predetermined goal amount or *future value*. A common reference rate, the *annual percentage rate (APR),* has been established to help consumers determine the true cost of loans and to aid them in comparing rates.

COMPUTATION/CALCULATION

Find the accumulated value if $2000 is invested for three years at 7%.

1. Simple interest
2. Compounded annually
3. Compounded quarterly
4. Compounded semiannually

Find the rate of interest if a principal of $800 accumulates to $1000 in two years, compounded

5. Monthly.   6. Semiannually.   7. Daily.

Find the number of interest periods if $1500 accumulates to $3000 at 8% compounded

8. Quarterly.   9. Daily.

Find the principal if the accumulated amount is $5000 over ten years, at 9% compounded

10. Monthly.   11. Semiannually.   12. Annually.

Find the future value of an annuity with

13. $i = 0.01, n = 18, p = \$1000.$

14. $i = 0.03, n = 40, p = \$5000.$

Calculate the payments for a sinking fund having a goal of $40,000, with

15. $i = 0.02$ and $n = 30.$ 16. $i = 0.04$ and $n = 12.$

Convert the given interest rate to the APR:

17. 8% compounded daily

18. 6% compounded quarterly

19. 7% compounded monthly

20. 10% compounded semiannually

CONNECTIONS

21. Cha Hoon Im is planning to put $125 per month into an ordinary annuity paying 7.2% interest. How much will he have available for his daughter's college education 15 years later?

22. Approximate your monthly payment if you borrow $11,000 for a new car at 10.2% APR for three years.

COMMUNICATION

23. Cite the differences between simple interest and compound interest.

24. Describe the characteristics of an ordinary annuity.

25. Indicate the purpose of a sinking fund.

26. State the purpose and meaning of the APR.

COOPERATIVE LEARNING

27. Determine the amount of the monthly payments necessary based on the best available current interest rates for you to have $1 million in a sinking fund by age 65. (First guess or estimate the amount, and then compare the estimate to your determined amount.)

CRITICAL THINKING

28. Analyze the given interest rates, and decide which gives the greatest yield:

  (a) 8% compounded annually
  (b) 7.9% compounded semiannually
  (c) 7.8% compounded quarterly
  (d) 7.7% compounded monthly
  (e) 7.6% compounded daily
  (f) 7.5% compounded continuously

COMPUTER/CALCULATOR

29. Given that $12,000 is borrowed for 60 months at 8.6% APR, estimate the monthly payments assuming the interest is charged on a

  (a) Monthly basis
  (b) Daily basis

Cumulative Reivew

30. Find the inverse of the given matrix.

$$\begin{bmatrix} 1 & 0 & 3 \\ 2 & 4 & 6 \\ 0 & -2 & 1 \end{bmatrix}$$

31. Use the simplex algorithm to

$$\text{Maximize } f = 5x + 4y,$$
$$\text{subject to } x + 2y + z \le 10$$
$$x + 3y \le 6$$
$$y + z \le 5$$
$$x \ge 0, y \ge 0, z \ge 0.$$

32. How many different arrangements are there for the letters in the word *MASSACHUSETTS*?

33. When a pair of dice is rolled, what is the probability that the sum is greater than 4, given that the sum is less than 10?

CHAPTER 11 SAMPLE TEST ITEMS

Find the accumulated value for the given conditions.

1. The principal is $1200, with simple interest of 6% for five years.

2. The principal is $3000, with 7% interest compounded semiannually for six years.

3. The prinicpal is $8000, with 9% interest compounded monthly for eight months.

4. The principal is $5000, with simple interest of 10% for eight months.

Find the rate of interest required to meet the given conditions.

5. $A = \$10,000$; $P = \$4000$; $n = 12$; the interest is compounded annually

6. $A = \$6000$; $P = \$3000$; $t = 10$ years; the interest is compounded quarterly.

Find the number of interest periods required to meet the given conditions.

7. $A = \$7000$; $P = \$2400$; the interest is 8%, compounded semiannually.

8. $A = \$25,000$; $P = \$3000$; the interest is 10%, compounded annually.

Find the principal required to achieve the given conditions.

9. $A = \$20,000$; the rate is 12%, compounded monthly for 7 years.

10. $A = \$30,000$; the rate is 6.5%, compounded quarterly for 12 years.

11. Find the number of years it will take to double your money if the interest rate is 8% compounded daily.

Find the future value of each annuity for the given interest rate per period.

12. $p = \$800$; $n = 20$; $i = 0.05$.

13. $p = \$2000$; $n = 30$; $i = 0.02$.

A sinking fund is being established with a goal of $200,000. Calculate the payments if i is the interest rate per period:

14. $n = 24$; $i = 0.03$.   **15.** $n = 20$; $i = 0.04$

16. Eli Whitney Co. sold a big-screen television to a customer for payments of $120 per month for three years. If the company sells its loan to a bank where interest rates are 8% compounded annually, how much should it receive for the loan?

In Questions 17 and 18, convert each given interest rate to the APR:

17. 7% compounded quarterly

18. 9% compounded daily

19. Find the monthly payment for a loan if $P = \$50,000$, $i = 2\%$ per month on the unpaid balance, and the payments are for 15 years.

20. Find the number of years required to pay off a $70,000 mortgage on a house at 9.4% interest rate, compounded monthly, if payments of $800 are made toward the loan.

Career Uses of Mathematics

As a junior actuary, I use math to calculate retirement benefits for the employees of our client companies. A pension plan can be thought of as a system for purchasing deferred life annuities that are payable during retirement. Many factors play a role in the calculations—an employee's age, length of employment, number of years until retirement, and current and past yearly income. These calculations are made on a yearly basis and are called a valuation. They tell the client company how much money to invest in order to provide for the current employees' income upon retirement. We use computers extensively because of the vast amount of data required.

I have always loved the problem-solving aspect of math and the feeling of accomplishment that successfully working a difficult problem gives me. In college, I majored in mathematics and learned that there was a whole world of math beyond the algebra and geometry I had taken in high school. To become an actuary, you need to have a good understanding of calculus and probability/statistics. You must also be willing to work and study hard, as there are several tests you must take when you begin your career.

Laura Tucker

Laura Tucker, Pension Actuarial Consultants

APPENDIX

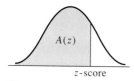

z	A(z)	z	A(z)	z	A(z)	z	A(z)
−3.00	.0013	−2.51	.0060	−2.02	.0217	−1.53	.0630
−2.99	.0014	−2.50	.0062	−2.01	.0222	−1.52	.0643
−2.98	.0014	−2.49	.0064	−2.00	.0227	−1.51	.0655
−2.97	.0015	−2.48	.0066	−2.99	.0233	−1.50	.0668
−2.96	.0015	−2.47	.0068	−1.98	.0238	−1.49	.0681
−2.95	.0016	−2.46	.0069	−1.97	.0244	−1.48	.0694
−2.94	.0016	−2.45	.0071	−1.96	.0250	−1.47	.0708
−2.93	.0017	−2.44	.0073	−1.95	.0256	−1.46	.0721
−2.92	.0017	−2.43	.0075	−1.94	.0262	−1.45	.0735
−2.91	.0018	−2.42	.0078	−1.93	.0268	−1.44	.0749
−2.90	.0019	−2.41	.0080	−1.92	.0274	−1.43	.0764
−2.89	.0019	−2.40	.0082	−1.91	.0281	−1.42	.0778
−2.88	.0020	−2.39	.0084	−1.90	.0287	−1.41	.0793
−2.87	.0020	−2.38	.0087	−1.89	.0294	−1.40	.0808
−2.86	.0021	−2.37	.0089	−1.88	.0300	−1.39	.0823
−2.85	.0022	−2.36	.0091	−1.87	.0307	−1.38	.0838
−2.84	.0023	−2.35	.0094	−1.86	.0314	−1.37	.0853
−2.83	.0023	−2.34	.0096	−1.85	.0322	−1.36	.0869
−2.82	.0024	−2.33	.0099	−1.84	.0329	−1.35	.0885
−2.81	.0025	−2.32	.0102	−1.83	.0336	−1.34	.0901
−2.80	.0026	−2.31	.0104	−1.82	.0344	−1.33	.0918
−2.79	.0026	−2.30	.0107	−1.81	.0351	−1.32	.0934
−2.78	.0027	−2.29	.0110	−1.80	.0359	−1.31	.0951
−2.77	.0028	−2.28	.0113	−1.79	.0367	−1.30	.0968
−2.76	.0029	−2.27	.0116	−1.78	.0375	−1.29	.0985
−2.75	.0030	−2.26	.0119	−1.77	.0384	−1.28	.1003
−2.74	.0031	−2.25	.0122	−1.76	.0392	−1.27	.1020
−2.73	.0032	−2.24	.0125	−1.75	.0401	−1.26	.1038
−2.72	.0033	−2.23	.0129	−1.74	.0409	−1.25	.1056
−2.71	.0034	−2.22	.0132	−1.73	.0418	−1.24	.1075
−2.70	.0035	−2.21	.0135	−1.72	.0427	−1.23	.1093
−2.69	.0036	−2.20	.0139	−1.71	.0436	−1.22	.1112
−2.68	.0037	−2.19	.0143	−1.70	.0446	−1.21	.1131
−2.67	.0038	−2.18	.0146	−1.69	.0455	−1.20	.1151
−2.66	.0039	−2.17	.0150	−1.68	.0465	−1.19	.1170
−2.65	.0040	−2.16	.0154	−1.67	.0475	−1.18	.1190
−2.64	.0041	−2.15	.0158	−1.66	.0485	−1.17	.1210
−2.63	.0043	−2.14	.0162	−1.65	.0495	−1.16	.1230
−2.62	.0044	−2.13	.0166	−1.64	.0505	−1.15	.1251
−2.61	.0045	−2.12	.0170	−1.63	.0515	−1.14	.1271
−2.60	.0047	−2.11	.0174	−1.62	.0526	−1.13	.1292
−2.59	.0048	−2.10	.0179	−1.61	.0537	−1.12	.1314
−2.58	.0049	−2.09	.0183	−1.60	.0548	−1.11	.1335
−2.57	.0051	−2.08	.0188	−1.59	.0559	−1.10	.1357
−2.56	.0052	−2.07	.0192	−1.58	.0570	−1.09	.1379
−2.55	.0054	−2.06	.0197	−1.57	.0582	−1.08	.1401
−2.54	.0055	−2.05	.0202	−1.56	.0594	−1.07	.1423
−2.53	.0057	−2.04	.0207	−1.55	.0606	−1.06	.1446
−2.52	.0059	−2.03	.0212	−1.54	.0618	−1.05	.1469

z	A(z)	z	A(z)	z	A(z)	z	A(z)
−1.04	.1492	−0.52	.3015	0.00	.5000	0.52	.6985
−1.03	.1515	−0.51	.3050	0.01	.5040	0.53	.7019
−1.02	.1539	−0.50	.3085	0.02	.5080	0.54	.7054
−1.01	.1562	−0.49	.3121	0.03	.5120	0.55	.7088
−1.00	.1587	−0.48	.3156	0.04	.5160	0.56	.7123
−0.99	.1611	−0.47	.3192	0.05	.5199	0.57	.7157
−0.98	.1635	−0.46	.3228	0.06	.5239	0.58	.7190
−0.97	.1660	−0.45	.3264	0.07	.5279	0.59	.7224
−0.96	.1685	−0.44	.3300	0.08	.5319	0.60	.7258
−0.95	.1711	−0.43	.3336	0.09	.5359	0.61	.7291
−0.94	.1736	−0.42	.3372	0.10	.5398	0.62	.7324
−0.93	.1762	−0.41	.3409	0.11	.5438	0.63	.7357
−0.92	.1788	−0.40	.3446	0.12	.5478	0.64	.7389
−0.91	.1814	−0.39	.3483	0.13	.5517	0.65	.7422
−0.90	.1841	−0.38	.3520	0.14	.5557	0.66	.7454
−0.89	.1867	−0.37	.3557	0.15	.5596	0.67	.7484
−0.88	.1894	−0.36	.3594	0.16	.5636	0.68	.7518
−0.87	.1921	−0.35	.3632	0.17	.5675	0.69	.7549
−0.86	.1949	−0.34	.3669	0.18	.5714	0.70	.7580
−0.85	.1977	−0.33	.3707	0.19	.5754	0.71	.7612
−0.84	.2004	−0.32	.3745	0.20	.5793	0.72	.7642
−0.83	.2033	−0.31	.3783	0.21	.5832	0.73	.7673
−0.82	.2061	−0.30	.3821	0.22	.5871	0.74	.7704
−0.81	.2090	−0.29	.3859	0.23	.5910	0.75	.7734
−0.80	.2119	−0.28	.3897	0.24	.5948	0.76	.7764
−0.79	.2148	−0.27	.3936	0.25	.5987	0.77	.7794
−0.78	.2177	−0.26	.3974	0.26	.6026	0.78	.7823
−0.77	.2206	−0.25	.4013	0.27	.6064	0.79	.7852
−0.76	.2236	−0.24	.4052	0.28	.6103	0.80	.7881
−0.75	.2266	−0.23	.4090	0.29	.6141	0.81	.7910
−0.74	.2296	−0.22	.4129	0.30	.6179	0.82	.7939
−0.73	.2327	−0.21	.4168	0.31	.6217	0.83	.7967
−0.72	.2358	−0.20	.4207	0.32	.6255	0.84	.7996
−0.71	.2388	−0.19	.4246	0.33	.6293	0.85	.8023
−0.70	.2420	−0.18	.4286	0.34	.6331	0.86	.8051
−0.69	.2451	−0.17	.4325	0.35	.6368	0.87	.8079
−0.68	.2482	−0.16	.4364	0.36	.6406	0.88	.8106
−0.67	.2514	−0.15	.4404	0.37	.6443	0.89	.8133
−0.66	.2546	−0.14	.4443	0.38	.6480	0.90	.8159
−0.65	.2578	−0.13	.4483	0.39	.6517	0.91	.8186
−0.64	.2611	−0.12	.4522	0.40	.6554	0.92	.8212
−0.63	.2643	−0.11	.4562	0.41	.6591	0.93	.8238
−0.62	.2676	−0.10	.4602	0.42	.6628	0.94	.8264
−0.61	.2709	−0.09	.4641	0.43	.6664	0.95	.8289
−0.60	.2742	−0.08	.4681	0.44	.6700	0.96	.8315
−0.59	.2776	−0.07	.4721	0.45	.6736	0.97	.8340
−0.58	.2810	−0.06	.4761	0.46	.6772	0.98	.8365
−0.57	.2843	−0.05	.4801	0.47	.6808	0.99	.8389
−0.56	.2877	−0.04	.4840	0.48	.6844	1.00	.8413
−0.55	.2912	−0.03	.4880	0.49	.6879	1.01	.8438
−0.54	.2946	−0.02	.4920	0.50	.6915	1.02	.8461
−0.53	.2981	−0.01	.4960	0.51	.6950	1.03	.8485

z	A(z)	z	A(z)	z	A(z)	z	A(z)
1.04	.8508	1.54	.9382	2.04	.9793	2.54	.9945
1.05	.8531	1.55	.9394	2.05	.9798	2.55	.9946
1.06	.8554	1.56	.9406	2.06	.9803	2.56	.9948
1.07	.8577	1.57	.9418	2.07	.9808	2.57	.9949
1.08	.8599	1.58	.9430	2.08	.9812	2.58	.9951
1.09	.8621	1.59	.9441	2.09	.9817	2.59	.9952
1.10	.8643	1.60	.9452	2.10	.9821	2.60	.9953
1.11	.8665	1.61	.9463	2.11	.9826	2.61	.9955
1.12	.8686	1.62	.9474	2.12	.9830	2.62	.9956
1.13	.8708	1.63	.9485	2.13	.9834	2.63	.9957
1.14	.8729	1.64	.9495	2.14	.9838	2.64	.9959
1.15	.8749	1.65	.9505	2.15	.9842	2.65	.9960
1.16	.8770	1.66	.9515	2.16	.9846	2.66	.9961
1.17	.8790	1.67	.9525	2.17	.9850	2.67	.9962
1.18	.8810	1.68	.9535	2.18	.9854	2.68	.9963
1.19	.8830	1.69	.9545	2.19	.9857	2.69	.9964
1.20	.8849	1.70	.9554	2.20	.9861	2.70	.9965
1.21	.8869	1.71	.9564	2.21	.9865	2.71	.9966
1.22	.8888	1.72	.9573	2.22	.9868	2.72	.9967
1.23	.8907	1.73	.9582	2.23	.9871	2.73	.9968
1.24	.8925	1.74	.9591	2.24	.9875	2.74	.9969
1.25	.8944	1.75	.9599	2.25	.9878	2.75	.9970
1.26	.8962	1.76	.9608	2.26	.9881	2.76	.9971
1.27	.8980	1.77	.9616	2.27	.9884	2.77	.9972
1.28	.8997	1.78	.9625	2.28	.9887	2.78	.9973
1.29	.9015	1.79	.9633	2.29	.9890	2.79	.9974
1.30	.9032	1.80	.9641	2.30	.9893	2.80	.9974
1.31	.9049	1.81	.9649	2.31	.9896	2.81	.9975
1.32	.9066	1.82	.9656	2.32	.9898	2.82	.9976
1.33	.9082	1.83	.9664	2.33	.9901	2.83	.9977
1.34	.9099	1.84	.9671	2.34	.9904	2.84	.9977
1.35	.9115	1.85	.9678	2.35	.9906	2.85	.9978
1.36	.9131	1.86	.9686	2.36	.9909	2.86	.9979
1.37	.9147	1.87	.9693	2.37	.9911	2.87	.9980
1.38	.9162	1.88	.9700	2.38	.9913	2.88	.9980
1.39	.9177	1.89	.9706	2.39	.9916	2.89	.9981
1.40	.9192	1.90	.9713	2.40	.9918	2.90	.9981
1.41	.9207	1.91	.9719	2.41	.9920	2.91	.9982
1.42	.9222	1.92	.9726	2.42	.9922	2.92	.9983
1.43	.9236	1.93	.9732	2.43	.9925	2.93	.9983
1.44	.9251	1.94	.9738	2.44	.9927	2.94	.9984
1.45	.9265	1.95	.9744	2.45	.9929	2.95	.9984
1.46	.9279	1.96	.9750	2.46	.9931	2.96	.9985
1.47	.9292	1.97	.9756	2.47	.9932	2.97	.9985
1.48	.9306	1.98	.9762	2.48	.9934	2.98	.9986
1.49	.9319	1.99	.9767	2.49	.9936	2.99	.9986
1.50	.9332	2.00	.9773	2.50	.9938	3.00	.9987
1.51	.9345	2.01	.9778	2.51	.9940		
1.52	.9357	2.02	.9783	2.52	.9941		
1.53	.9370	2.03	.9788	2.53	.9943		

ANSWERS TO SELECTED EXERCISES

TO THE STUDENT If you need further help with finite mathematics, you may want to obtain a copy of the *Student's Solutions Manual* that goes with this book. It contains solutions to all the odd-numbered exercises and all chapter test exercises. Your college bookstore either has this book or can order it for you.

The answers provided in the text, as well as supplemental material, are the ones we think most students will obtain when they work the exercises using the methods explained in the text. However, in many cases there are equivalent forms of the answer that are correct. For example, if the answer section shows $\frac{1}{4}$ and you obtain .25, then you have obtained the right answer but written in a different (but equivalent) form. Sometimes when and where you round off numbers can change an answer somewhat.

In general, if your answer does not agree with the one given in the text, see whether it can be transformed into the other form. If it can, then it is correct. If you still have doubts, talk with your instructor.

CHAPTER 1

1.1 EXERCISES (page 8)

1.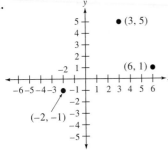

3.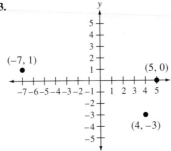

5.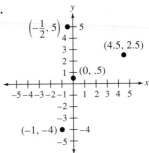

7.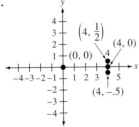

9.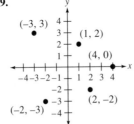

11.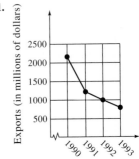

13.

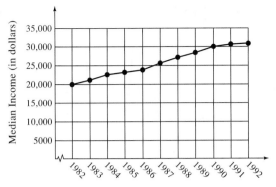

15.

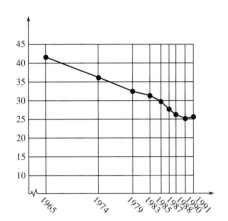

17.

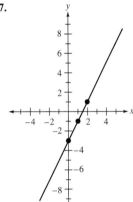

19.

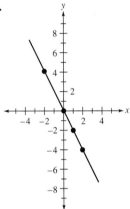

21.

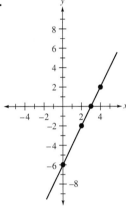

23.

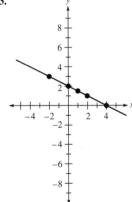

25.

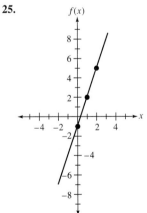

27.

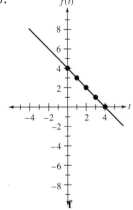

29.

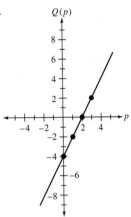

31.

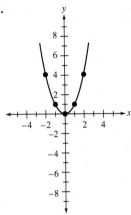

33.

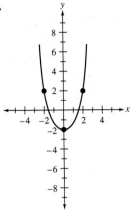

35.

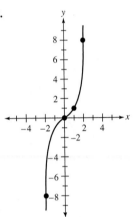

37. $0 \leq t \leq 120$

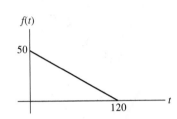

39. $0 \leq p \leq 5$

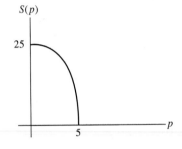

41. $0 \leq x \leq 25$

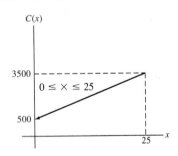

1.2 EXERCISES (page 19)

1. (a) -3 **(b)**

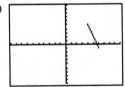

The slope is negative

3. (a) 0 **(b)**

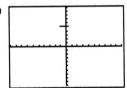

The Line segment appears horizontal. Slope 0.

5. (a) 16.8
(b)

The slope is a rather large positive number.

7. (a) Not defined.
(b)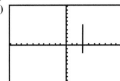

The segment is vertial.
Slope not defined.

9. (a) 1
(b)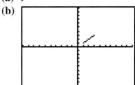

The slope is positive and near 1.

11. (a) $y = -2x + 9$
$2x + y = 9$
(b)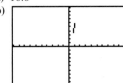

13. (a) $y = \frac{2}{3}x + 5.7$
$-2x + 3y = 17.1$
(b)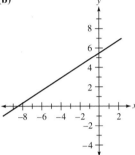

15. (a) $y = -\frac{5}{3}x + 9$
$5x + 3y = 27$
(b)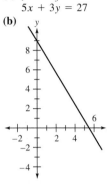

17. (a) $y = -0.87x + 1.26$
$0.87x + y = 1.26$
(b)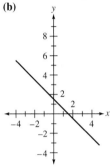

19. (a) $y = -25x + 725$
$25x + y = 725$
(b)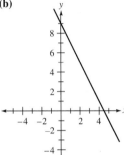

21. (a) $y = -2.69x + 4.30$
$2.69x + y = 4.30$
(b)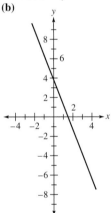

23. Vertical: $x = -\frac{1}{2}$
Horizontal: $y = 4$

25. Vertical: $x = 1.2$
Horizontal: $y = -8.6$

27. Possible answers: $4x - 7y = 8$;
$4x - 7y = 12$; $4x - 7y = 18$

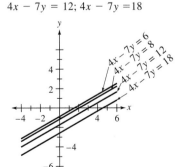

29. Possible answers: $x = 8$; $x = -3$; $x = 2$

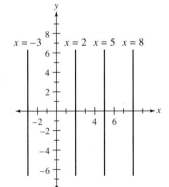

31. Possible answers: $-2x + 4y = 4$;
$-2x + 4y = 8$; $-2x + 4y = 12$

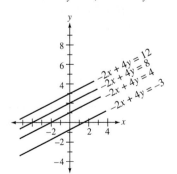

33. $y = -3x + 5.9$ **35.** $y = -\frac{2}{3}x - 3.5$ **37. (a)** Possible answers: $y = 2x - 4$; $y = 8x - 4$; $y = -4x - 4$ **(b)** Possible answers: $x + 3y = -4$; $x - 6y = -4$; $x + 5y = -4$ **39. (a)** Possible answers: $2x - 2y = 5$; $5x - 2y = 5$; $-3x - 2y = 5$
(b) Possible answers: $x + 3y = 5$; $x + 7y = 5$; $x - 7y = 5$ **41. (a)** Possible answers: $2x + 5y = 0$; $x + y = 0$; $x - y = 0$
(b) Possible answers: $x + 2y = 0$; $x + y = 0$; $x + 9y = 0$ **43. (a)** 3;   **(b)** 6;   **(c)** 15   **45. (a)** -2;   **(b)** -4;
(c) -10 **47. (a)** $-\frac{2}{5}$;   **(b)** $-\frac{4}{5}$;   **(c)** -2

49. (a) 10,000; 22,000
(b) Yes; 800
(c) The slope is the number of new subscribers per week; the y-intercept gives the number of subscribers for the first issue.
(d)

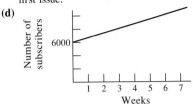

(e) 15

51. (a) 70,000; 60,000
(b) Decreasing by 2000 per day
(c) The slope gives the decrease (in thousands) of grasshoppers per day on each acre; the y-intercept gives the number of grasshoppers (in thousands) on each acre as of August 31.
(d)

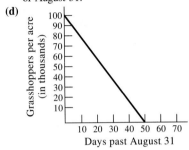

(e) 40 days

53. (a) $\frac{5}{9}$ degree   **(b)** 26.67 degrees   **(c)** $F = \frac{9}{5}C + 32$ **(d)** $\frac{9}{5}$ degree
55. (a)

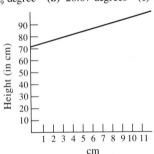

(b) 204 cm
(c) 38.75 cm

57. (a) 38,970 cases
(b) 2009
(c)

1.3 EXERCISES (page 29)

1. Not linear   **3.** Linear; $f(x) = 20x + 500$ **5.** Linear; $f(x) = 60$ **7. (a)** \$3   **(b)** \$20   **(c)** \$80   **(d)** \$4; \$3.20; \$3.10
9. (a) \$3.20   **(b)** \$1680   **(c)** \$1744   **(d)** \$87.20; \$20; \$11.60   **11. (a)** \$1.60   **(b)** \$5000   **(c)** \$5032
(d) \$251.60; \$51.60; \$26.60
13.

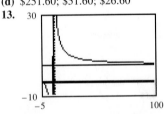

The average cost per item decreases toward the marginal cost, 8, as production increases. The graphs indicate that after 100 units, both costs are about the same.

15. $C(x) = 300x + 4000$ **17.** $C(x) = 25x + 3000$ **19.** $C(x) = 140x + 600$
21. $C(x) = 20x$ **23.** $C(x) = 44x + 240$ **25.** $y = .25x$

27. (a) $E(x) = 600x + 12,000$
(b)

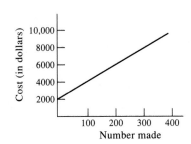

(c) The slope is the increase in students per year past 1985. The y-intercept is the number of students enrolled in 1985.
(d) 14 years

31. (a) $C(x) = 20x + 2000$

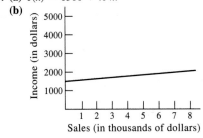

(b) $5000
(c) $20
(d) $30
(e) $24; $22; $21

35. (a) $I(x) = 1500 + .04x$
(b)

(c) $2300
(d) 4 cents
(e) $37,500

29. (a) $V(x) = 12,000 - 1000x$
(b) The slope gives the annual decline in value of the car. The y-intercept gives the value of the car when new.
(c) 7 years; 10.5 years
(d)

33. (a) $C(x) = 1500x + 25,000$
(b) $100,000
(c) $1500
(d) $2000; $1750; $1625
(e)

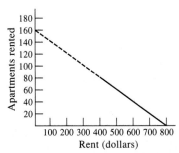

37. (a) $O(x) = -.20x + 160$

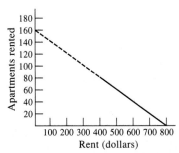

(b) Let $x = 400$; then $O(400) = 80$
Let $x = 500$; then $O(x) = 60$.
(c) 68

39. (a) $y = .45x + 2.7$ (x = number of years past 1986) **(b)** $6.75 million **(c)** In the year 1994 **(d)** No; no
41. (a) $y = .517x + 5.4$ (x = number of years past 1960) **(b)** The average increase per year (in millions of tons) of paper and paperboard recovered **(c)** The formula gives 18.842 million tons. Overstates amount recovered by over 25%; 26.08 million tons **43. (a)** 3.597 million tons **(b)** $y = 3.597x + 87.8$; 177.7 (overstates amount recovered by over 8%) **(c)** 249.7
(d) In the year 2006

45. (a) $5000

(b) $V(x) = 100,000 - 5000x$ $(0 \le x \le 20)$

(c)

Book value (in dollars) vs Years after purchase. $V(x)$ axis shows 100,000 and 50,000; x axis shows 5 10 15 20.

47. (a) $3500

(b) $V(x) = 80,000 - 3500x$ $(0 \le x \le 20)$

(c)

Book value (in dollars) vs Years after purchase. $V(x)$ axis shows 80,000, 40,000, 10,000; x axis shows 5 10 15 20.

49. (a) $20,000

(b) $V(x) = 550,000 - 20,000x$ $(0 \le x \le 25)$

(c)

Book value (in thousands of dollars) vs Years after purchase. Axis shows 100, 200, 300, 400, 500, 600; x axis shows 5 10 15 20 25.

51. (a) 40; Surplus

(b) $10; Shortage

(c) (13.33, 79.98)

(d)

Graph with $y = D(p)$ and $y = S(p)$; values 160, 79.98 on y-axis; 13.33, 26.67 on p-axis; Surplus and Shortage regions labeled.

53. (a) $D(p) = -\frac{3}{2}p + 18$ (millions of bushels)   **(b)** 10.125 million bushels   **(c)** $7.33   **(d)** 18 million bushels

1.4 EXERCISES (page 42)

1. (I) One point in common.   **3.** (II) No point in common.   **5.** (I) One point common.

7. One point in common; $x = \frac{5}{3}$, $y = -\frac{1}{3}$ **9.** All points in common; $x =$ any number, $y = -2x + 3$

11. One point in common; $x = 0$, $y = 0$ **13.** All points in common; $x =$ any number, $y = -\frac{1}{2}x$

15. No points in common.   **17.** One point in common: $x = 12$, $y = -\frac{8}{3}$ **19.** One point in common; $x = 2$, $y = 1$

21. One point in common; $x = 0$, $y = 0$ **23.** One point in common; $x = 3$, $y = 0$ **25.** (15, 40)

27. $y = \frac{3}{4}x - 2$, x any number

29. (a) 30 **(b)** (30, 1500)

(c)

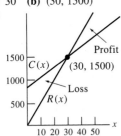

Graph with $C(x)$, $R(x)$, Profit, Loss regions; point (30, 1500); axes 1500, 1000, 500 and 10 20 30 40 50.

31. (a) 250 **(b)** (250, 5000)

(c)

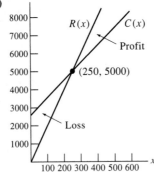

Graph with $R(x)$, $C(x)$, Profit, Loss regions; point (250, 5000); y-axis 1000–8000; x-axis 100 200 300 400 500 600.

33. (a) 70 **(b)** (70, 1050)

(c)

Graph with $R(x)$, $C(x)$, Profit, Loss regions; point (70, 1050); y-axis 500, 1000, 1500, 2000, 2500; x-axis 10 50 100.

35. (a) 12   **(b)** (12, 36)   **(c)** $P(x) = x - 12$
(d)

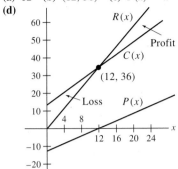

37. (a) 61.25   **(b)** (61.25, 1225)
(c) $P(x) = 8x - 490$

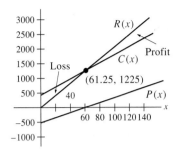

39. (a) 137   **(b)** (137, 4170)   **(c)** $P(x) = 20x - 2740$
(d)

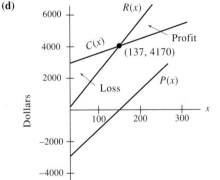

41. (a) 1382.35 or $\approx$ 1382   **(b)** \$11,680.86
43. (a) 65.32 or $\approx$ 65   **(b)** \$520.09

45. (a) $C(x) = 8.36x + 8000; R(x) = 10.80x$ **(b)** 3279
$P(x) = 2.44x - 8000$
(c)

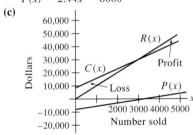

(d) Loss of \$6048; loss of \$3120

49. (a) $C(x) = 1.50x + 50,000; R(x) = 2x$
(b) $P(x) = 0.50x - 50,000$

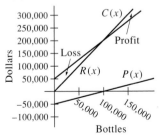

(c) Loss of \$20,000
(d) 100,000
(e) \$80,000; \$4.   \$110,000; \$2.75
(f) \$1.50

47. (a) $C(x) = .40x + 20,000$
(b) $R(x) = .50x + 8000$ **(c)** 120,000
(d)

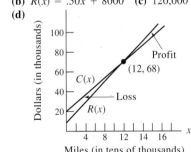

Miles (in tens of thousands)

51. (a) Computer: $C(x) = 25x + 300$,
Offset: $C(x) = 20x + 500$ (x in thousands)
(b)

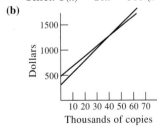

Thousands of copies

(c) 40,000 copies
(d) Computer at a cost of \$1050,
Offset at a cost of \$1900

53. (a) Brand A: $C(x) = 500$
Brand B: $C(x) = 35x + 350$

55. During the second quarter of 1997
57. In 2022

59.
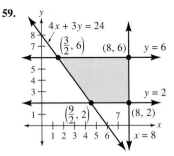

(b)
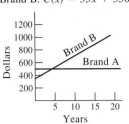

(c) Approximately 4.3 years
(d) Cost of Brand A is $ 500.
Cost of Brand B is $ 525.

61.
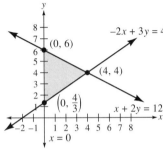

63.
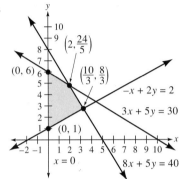

1.5 EXERCISES (page 55)

1. (a), (c)
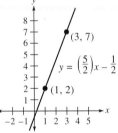

(b) $2b + 4m = 9$
$4b + 10m = 23$
$y = \frac{5}{2}x - \frac{1}{2}$
(c) The regression line is the line passing through the two points.

3. (a), (c)
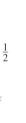

(b) $2b + m = 2$
$b + 5m = -5$
$y = -\frac{4}{3}x + \frac{5}{3}$
(c) The regression line is the line passing through the two points.

5. (a), (c)

(b) $3b + 6m = 13$
$6b + 14m = 33$
$y = \frac{7}{2}x - \frac{8}{3}$

7. (a), (c)
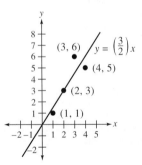

(b) $4b + 10m = 15$
$10b + 30m = 45$
$y = \frac{3}{2}x$

9. (a)
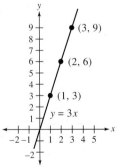

(b) $y = 3x$
(c) $r = 1$

11. (a)

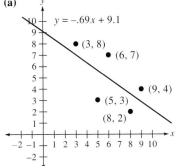

$y = -.69x + 9.1$

(3, 8) (6, 7) (9, 4) (5, 3) (8, 2)

13. (a)

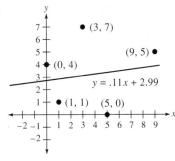

(3, 7) (9, 5) (0, 4) $y = .11x + 2.99$ (1, 1) (5, 0)

15. Rather high; possibly .9
17. Very low; possibly 0
19. Rather high; possibly .8

(b) $y = 0.11x + 2.99$ **(c)** $r = 0.14$

(b) $y = -.69x + 9.10$ **(c)** $r = -.64$

21. (a) $y = -0.47x + 48.46$

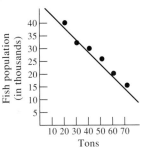

Fish population (in thousands)

10 20 30 40 50 60 70
Tons

(b) $-.99$ **(c)** 470   **(d)** 17,910   **(e)** 28.638

23. (a), (b)

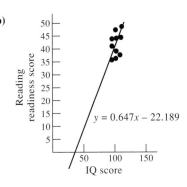

Reading readiness score

$y = 0.647x - 22.189$

50 100 150
IQ score

(c) 0.675

25. (a) $y = 11.801x - 209.382$ **(b)** 1180   **(c)** 32,166   **(d)** 38.927 in.
    $r = .981$
27. (a) Current dollars: $y = .183x + 4.933$; $r = .999$
    Constant dollars: $y = -.0575x + 5.686$; $r = -.996$
    (x = number of years after 1980)

(b)

Dollars per hour

Current dollars

Constant dollars

5 10 15
Years after 1980

(c) Current dollars in 2005: $9.51
    Constant dollars in 2005: $4.25
29. (a) $y = 12.916x + 65.203$; $r = .831$ (x = number of years after 1980)   **(b)** $388.103 million
31. (a) Waste generated: $y = 3.392x + 86.586$; $r = .99$ **(b)** No   **(c)** Waste generated: 239.23 million tons
    Materials recovered; $y = .731x + 2.847$; $r = .86$ Materials recovered: 35.74 million tons
    (x = number of years after 1960)

CHAPTER 1 REFLECTIONS (page 58)

1. x-intercept: 10; y-intercept: -6 **3.** Horizontal line: $y = 8$; Vertical line: $x = -3$ **5.** $y + 6$ **7.** $y = \frac{5}{2}x$
9. $x = 3$, $y = \frac{2}{3}$ **11.** $x =$ any number, $y = -\frac{5}{3}x + 3$ **13.** $2994   **15.** $7.90   **17.** In the year 2031
19. (100, 6500)   **21.** Slopes different, precisely one point in common; slopes equal, y-intercepts unequal, no point in common; slopes equal, y-intercepts equal, all points in common.   **23.** Data are nonlinear, using the equation too far outside the range of the data.   **25.** $x - 2y = 6$; $3x + y = 5$ has the solution $x = \frac{16}{7}$, $y = -\frac{13}{7}$. Easy to solve first equation for x, the second for y. $x + 4y = 5$; $2x - 4y = 7$ has the solution $x = 4$, $y = \frac{1}{4}$. Easy to add equations and find x.

SAMPLE TEST ITEMS FOR CHAPTER 1 (page 60)

1.

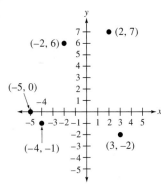

2.

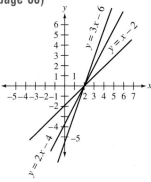

3. $y = 3.7x - 2.8$ **4.** $y = \frac{16}{5}x + 490$ **5.** $y = 250x + 2300$ **6.** $x = -2$ **7.** $m = \frac{1}{2}$; x-intercept is 9; y-intercept is $-\frac{9}{2}$
8. $y = 3.2x + 6.4$ **9.** $y = -20x + 173$ **10.** $V(x) = 5600 - 350x$ **11.** Slopes are different; one point in common
12. $x = \frac{17}{3}, y = -\frac{2}{3}$ **13.** $y =$ any number, $x = 15 - 4y$ **14.** $P(x) = 30x - 1200$; $(40, 3000)$ **15.** $(1600, 156{,}800)$
16.   **17.** $y = 2.1x + 5.5$ **18.** $r = 0.984$ **19.** 16%   **20.** Approximately 7 years

(graph: Percent return vs Years)

CHAPTER 2

2.1 EXERCISES (page 67)

1. $\begin{bmatrix} 1 & 3 & | & 5 \\ 2 & -1 & | & -4 \end{bmatrix}; \begin{bmatrix} 1 & 3 \\ 2 & -1 \end{bmatrix}$ **3.** $\begin{bmatrix} 1 & 3 & 0 & | & 0 \\ 2 & 1 & -1 & | & 0 \\ 1 & -1 & 1 & | & 0 \end{bmatrix}; \begin{bmatrix} 1 & 3 & 0 \\ 2 & 1 & -1 \\ 1 & -1 & 1 \end{bmatrix}$

5. $\begin{bmatrix} 1 & 2 & 3 & 1 & | & 5 \\ 1 & 1 & -1 & 0 & | & 3 \\ 3 & 5 & 0 & 2 & | & -2 \end{bmatrix}; \begin{bmatrix} 1 & 2 & 3 & 1 \\ 1 & 1 & -1 & 0 \\ 3 & 5 & 0 & 2 \end{bmatrix}$ **7.** $\begin{bmatrix} 5 & 3 & 2 & | & 0 \\ 1 & -2 & 0 & | & 0 \\ 3 & 1 & 1 & | & 1 \\ 1 & 0 & 4 & | & 2 \end{bmatrix}; \begin{bmatrix} 5 & 3 & 2 \\ 1 & -2 & 0 \\ 3 & 1 & 1 \\ 1 & 0 & 4 \end{bmatrix}$

9.

	Datafix	Rocktite		
Cost	30	20	4000	$30x + 20y = 4000$
Dividends	2	1	220	$2x + y = 220$
Numbers of shares	x	y		

11.

	Coal	Gas		
Electricity	3	4.5	183	$3x + 4.5y = 183$
Sulfer dioxide	60	1.0	990	$60x + y = 990$
Quantity	x	y		

13.

	Wood	Silver	Gold	
Grinder	1	$\frac{1}{2}$	3	12,000
Bonder	2	2	$\frac{5}{2}$	9600
Number of pens	x	y	z	

$x + \frac{1}{2}y + 3z = 12{,}000$

$2x + 2y + \frac{5}{2}z = 9600$

15.

	Skilled	Semiskilled	Supervisors	
Wages	12	9	15	1560
Training/safety	3	5	1	588
Production needs	1	1		140
Number of each type	x	y	z	

$12x + 9y + 15z = 1560$

$3x + 5y + z = 588$

$x + y = 140$

17.

	A	B	C	D	
Vitamin A	2	1	3	1	300
Protein	4	6	2	2	600
Batch	1	1	1	1	200
Number of lb. each	x	y	z	w	

$2x + y + 3z + w = 300$

$4x + 6y + 2z + 2w = 600$

$x + y + z + w = 200$

19.

	Soybeans	Wheat	Corn	Barley	
Profit	25	50	40	30	10,000,000
Manpower	5	6	8	3	50,000
Number of acres	x	y	z	w	

$25x + 50y + 40z + 30w = 10{,}000{,}000$

$5x + 6y + 8z + 3w = 50{,}000$

21.

	1st Shift	2nd Shift	3rd Shift	
1 & 2 overlap	1	1		10
2 & 3 overlap		1	1	8
1 & 3 equal	1		-1	0
Workers/shift	x	y	z	

$x + y = 10$

$y + z = 8$

$x - z = 0$

23.

	A	B	C	
Cost (millions)	2	$\frac{1}{2}$	1	120
Ratio	1	-2		0
Number of miles	x	y	z	

$2x + \frac{1}{2}y + z = 120$

$x - 2y = 0$

25.

	4-ply	Radial	
Total bought	1	1	600
Cost	40	50	28,500
Ratio	3	-1	0
Number bought	x	y	

$x + y = 600$

$40x + 50y = 28{,}500$

$3x - y = 0$

27.

	8% Tank	13% Tank	
Total gallons	1	1	2000
10% mix	.08	.13	200
Number of gallons	x	y	

$x + y = 2000$

$.08x + .13y = 200$

29.

	21% Solution	14% Solution	
Total liters	1	1	2
18% mix	.21	.14	0.36
Number of liters	x	y	

$x + y = 2$

$.21x + .14y = .36$

2.2 EXERCISES (page 80)

1. $x = 1, y = 1, z = 2$ **3.** $x = 2, y = 1, z = 4$ **5.** $x = -\frac{1}{2}, y = 1, z = 1$ **7.** No solution   **9.** $x = \frac{20}{7}, y = \frac{12}{7}$
11. $x = \frac{1}{7}, y = -\frac{10}{7}$ **13.** $x = 1, y = 0, z = -3$ **15.** No solution   **17.** $x = 0, y = 0, z = 0$
19. $w = \frac{56}{261}, x = -\frac{53}{174}, y = \frac{1357}{1044}, z = \frac{73}{348}$ **21.** $x = .11, y = 2.61, z = 5.31$ (two decimal accuracy)
23. $w = -5.57, x = 7.42, y = 8.49, z = 5.02$ (two decimal accuracy)   **25.** 16 tons of coal; 30,000 cubic feet of gas.
27. 3 Q6 models, 2 Q10 models, 2 Q12 models   **29.** 8 newspaper ads, 12 radios ads, 4 TV ads
31. No. The system has no solution.   **33.** 200 4-ply, 400 radial   **35.** 32 inches and 128 inches
37. 60 type A saws, 0 type B saws, and 70 type C saws   **39.** $C_A = 2.4685, C_B = 2.98, C_C = 4.1815$ (million)

2.3 EXERCISES (page 89)

1. (a) $x = \frac{7}{3} + \frac{7}{3}z$
 $y = -\frac{1}{3} + \frac{5}{3}z$
 z any number
(c) $x = \frac{7}{3}, y = -\frac{1}{3}, z = 0$
 $x = \frac{14}{3}, y = \frac{4}{3}, z = 1$
 $x = 0, y = -2, z = -1$

3. (a) $x = \frac{1}{7} + \frac{2}{7}z$
 $y = -\frac{2}{7} + \frac{3}{7}z$
 z = any number
(c) $x = \frac{1}{7}, y = -\frac{2}{7}, z = 0$
 $x = \frac{3}{7}, y = \frac{1}{7}, z = 1$
 $x = \frac{5}{7}, y = \frac{4}{7}, z = 2$

5. (a) $w = \frac{1}{5}y + z$
 $x = 2 - \frac{3}{5}y + z$
 y any number, z any number
(c) $w = 0, x = 2, y = 0, z = 0$
 $w = 1, x = 3, y = 0, z = 1$
 $w = \frac{6}{5}, x = \frac{12}{5}, y = 1, z = 1$

7. (a) $x_1 = \frac{15}{2} - 3x_4$
 $x_2 = -8 + 3x_4$
 $x_3 = \frac{51}{4} - 3x_4$
 x_4 = any number
(c) $x_1 = \frac{15}{2}, x_2 = -8, x_3 = \frac{51}{4}, x_4 = 0$
 $x_1 = \frac{9}{2}, x_2 = -5, x_3 = \frac{39}{4}, x_4 = 1$
 $x_1 = \frac{3}{2}, x_2 = -2, x_3 = \frac{27}{4}, x_4 = 2$

9. (a) $x = 3 - y - z$
 y any number
 z any number
(c) $x = 3, \quad y = 0, \quad z = 0$
 $x = -1, y = 1, \quad z = 3$
 $x = 0, \quad y = -2, z = 1$

11. (a) $x = -\frac{7}{5}z$
 $y = \frac{3}{5}z$
 z any number
(c) $x = 0, \quad y = 0, z = 0$
 $x = -7, \quad y = 3, z = 5$
 $x = -14, y = 6, z = 10$

13. (a) $x = z$
 $y = -z$
 z any number
(c) $x = 0, y = 0, \quad z = 0$
 $x = 1, y = -1, z = 1$
 $x = 2, y = -2, z = 2$

15. (a) $x = -z$
 $y = 0$
 z any number
(c) $x = 0, \quad y = 0, z = 0$
 $x = -1, y = 0, z = 1$
 $x = 3, \quad y = 0, z = -3$

17. No solution   **19.** $w = 1 - 2x + 3y - z$ **21.** No solution
 x any number
 y any number
 z any number
23. $x = 3.7097 - .7742z, y = 5.7742 - 1.5355z, z$ any number   **25.** No solution
27. $w = 2.6211 - 1.0864z, x = -.65625 + .95564z, y = -1.6638 + .60829z, z$ any number
29. $x = 2z$ **31.** $x = -z$ **33. (a)** $x = 0, y = 0, u = 10, v = 5$ **35. (a)** $x = 0, y = 0, u = -6, v = 4$
 $y = 0$ $y = -2z$ **(b)** $x = 5, y = 0$ $u = 0, \quad v = 0$ **(b)** $x = 2, y = 0, u = -4, v = 0$
 z any number z any number
37. x = funds spent on streets
 y = funds spent on urban renewal
 z = funds spent on parks
 Solution: $x = \$500,000$
 $y = 500,000 - z$
 $0 \le z \le 500,000$

39. x = number of newspaper ads
 y = number of radio ads
 z = number of television ads
 Solution: $x = 30 - 3z$
 $y = 2z$
 $0 \le z \le 10$

41. x = number of QT20 printers
 y = number of QT40 printers
 z = number of QT60 printers
 Solution: $x = 60 + z$
 $y = 95 - 2z$
 $0 \le z \le 47.5$

43. Let w be the number shipped from warehouse 1 to store 1.   Solution: $w = 2z$
Let x be the number shipped from warehouse 1 to store 2. $x = 18 - z$
Let y be the number shipped from warehouse 2 to store 1. $y = 30 - 2z$
Let z be the number shipped from warehouse 2 to store 2. $0 \leq z \leq 15$

$$\begin{aligned} w \phantom{{}+{}} + y \phantom{{}+{}} &= 30 \\ x \phantom{{}+{}} + z &= 18 \\ w \phantom{{}+{}} - 2z &= 0 \end{aligned}$$

$w = 0, x = 18, y = 30, z = 0$
$w = 20, x = 8, y = 10, z = 10$
$w = 30, x = 3, y = 0, z = 15$

CHAPTER 2 REFLECTIONS (page 91)

1. $x =$ any number, $y = 7, z = 3x$ **3.** $x = 0, y = 0, u = 12, v = 15; x = 0, y = 1, u = 10, v = 12; x = 1, y = 0, u = 11,$
$v = 14$ **5.** $x = \frac{1}{5}, y = \frac{17}{5}$ **7.** $x = \frac{23}{9} - \frac{2}{9}z, y = \frac{29}{9} + \frac{1}{9}z, z =$ any number   **9.** $x = 1, y = -2, z = 0$ **11.** $x = 2 + \frac{1}{2}z,$
$y = 2 - \frac{7}{4}z, z =$ any number
13. $\begin{aligned} 3x \phantom{{}+8y{}} - 2z &= 7 \\ -x + 8y + 5z &= 0 \\ 2y + z &= -4 \end{aligned}$ Solution: $x = 39, y = -\frac{59}{2}, z = 55$

15. \$475   **17.** The first several columns have 1s down the diagonal and zeros elsewhere. Below the last 1 the rows are all zeros; otherwise, the columns to the right of the last 1 may have any appearance.   **19.** Somewhere in the augmented matrix, a row of zeros will appear on the left of the vertical bar and a nonzero number on the right.   **21.** Any of the three types of solutions may result
23. $x = 4.327$ **25.** $y = -\frac{3}{5}x + 13$ **27.** $x = -1, y = 3$
$y = -.634$
$z = -1.572$

SAMPLE TEST ITEMS FOR CHAPTER 2 (page 93)

1. $\begin{bmatrix} 1 & 0 & 2 \\ -3 & 5 & -8 \\ 1 & 1 & 0 \end{bmatrix}$ **2.** $\begin{bmatrix} 3 & -1 & 5 \\ -5 & 17 & -1.2 \end{bmatrix}$
3. Let $x =$ number of nickels $x + y + z = 213$
$y =$ number of dimes $5x + 10y + 25z = 2115$
$z =$ number of quarters $y - 2z = 0$
4. Let $x =$ number of student tickets $x + y = 9073$
$y =$ number of adult tickets $12.5x + 16.75y = 129{,}673$
5. Let $x =$ quarts picked by worker 1 in a day $x + y = 62$
$y =$ quarts picked by worker 2 in a day $y + z = 70$
$z =$ quarts picked by worker 3 in a day $x + z = 58$

6. (c), (e)   **7.** $\begin{bmatrix} 1 & 1 & 3 & -6 \\ 0 & 2 & 4 & 10 \\ 0 & -7 & -14 & 30 \end{bmatrix}$ **8.** $x = 2, y = -3, z = 7$ **9.** $x = -1, y = 5$

10. $x = 108$ nickels, $y = 70$ dimes, $z = 35$ quarters   **11.** $x = 3, y = -1, z = 0; x = 2, y = -\frac{1}{2}, z = 2; x = 1, y = 0, z = 4$
12. $x = 3 + z, y = 2 - z, z =$ any number   **13.** No solution   **14.** $x = 5247$ student tickets, $y = 3826$ adult tickets
15. $25 + 37 + 33 = 95$

CHAPTER 3

3.1 EXERCISES (page 104)

1. $t = 3, x = 9$ **3.** x any number   **5.** $\begin{bmatrix} 0 & -10 \\ -3 & 5 \end{bmatrix}$ **7.** $\begin{bmatrix} \frac{8}{3} & \frac{2}{3} \\ \frac{1}{3} & \frac{1}{3} \end{bmatrix}$ **9.** Not possible   **11.** $\begin{bmatrix} -40 & 60 & -100 \\ 0 & 0 & 40 \end{bmatrix}$

13. $\begin{bmatrix} a + e & b + f \\ c + g & d + h \end{bmatrix} = \begin{bmatrix} e + a & f + b \\ g + c & h + d \end{bmatrix}$ **15.** (a) $\begin{array}{c} \\ P_1 \\ P_2 \end{array} \overset{\text{I-C \quad T-C \quad E-C}}{\begin{bmatrix} 30 & 50 & 90 \\ 50 & 60 & 100 \end{bmatrix}}$ (b) $\begin{array}{c} \\ \text{1-C} \\ \text{T-C} \\ \text{E-C} \end{array} \overset{P_1 \quad P_2}{\begin{bmatrix} 30 & 50 \\ 50 & 60 \\ 90 & 100 \end{bmatrix}}$

17. (a) $\begin{matrix} \text{Fr.} \\ \text{So.} \\ \text{Jr.} \\ \text{Sr.} \end{matrix} \begin{bmatrix} 300 \\ 287 \\ 250 \\ 240 \end{bmatrix}$ **(b)** $\begin{matrix} \text{Fr.} & \text{So.} & \text{Jr.} & \text{Sr.} \end{matrix}$ $[300 \quad 287 \quad 250 \quad 240]$

19. $\begin{matrix} & A & B & C & D & E \\ A \\ B \\ C \\ D \\ E \end{matrix} \begin{bmatrix} 0 & 1 & 1 & 1 & 0 \\ 0 & 0 & 0 & 1 & 0 \\ 0 & 0 & 0 & 1 & 0 \\ 0 & 0 & 0 & 0 & 1 \\ 0 & 0 & 0 & 0 & 0 \end{bmatrix}$

21. $\begin{matrix} & B & W & T & C \\ B \\ W \\ T \\ C \end{matrix} \begin{bmatrix} 0 & 1 & 1 & 0 \\ 0 & 0 & 1 & 0 \\ 0 & 0 & 0 & 1 \\ 1 & 1 & 0 & 0 \end{bmatrix}$

23. $X = 4B + \frac{20}{3}T$

25. $B = \frac{1}{6}A + \frac{1}{3}C - \frac{2}{3}D$ **27.** $T = \frac{1}{4}A - \frac{1}{2}C$ **29.** $X = \begin{bmatrix} \frac{1}{2} & -\frac{3}{2} & 1 \\ -\frac{7}{2} & -\frac{5}{2} & -\frac{5}{2} \end{bmatrix}$ **31.** $X = \begin{bmatrix} -\frac{4}{3} & -\frac{7}{3} \\ -1 & -1 \\ -\frac{2}{3} & 2 \end{bmatrix}$ **33.** Yes, $x = 3$.

35. $\begin{bmatrix} 3.6 & 1.2 \\ 1.8 & 3 \end{bmatrix}$ **37.** $\begin{bmatrix} 1 & 6 \\ 16 & 13 \end{bmatrix}$ **39. (a)** $.08A$ **(b)** $\begin{bmatrix} 2400 & 2000 \\ 4000 & 3200 \end{bmatrix}$ **(c)** $A + .08A$ **(d)** $\begin{bmatrix} 32,400 & 27,000 \\ 54,000 & 43,200 \end{bmatrix}$

41. $C + .05C = \begin{bmatrix} 2625 & 6300 \\ 5040 & 5670 \end{bmatrix}$ **43.** $(1 + .05)^3 A \approx \begin{bmatrix} 2894 & 6946 \\ 5557 & 6251 \end{bmatrix}$ **45.** $\begin{bmatrix} 14.3 & 28.6 \\ 37.5 & 87.5 \\ 66.7 & 116.7 \end{bmatrix}$ **47.** $\begin{bmatrix} 1 & 3 \\ 2 & -5 \end{bmatrix}$

49. $\begin{bmatrix} 1 & 2 & -3 & 4 \\ 0 & -1 & 2 & 1 \end{bmatrix}$

3.2 EXERCISES (page 115)

1. $\begin{bmatrix} 1 & 2 \\ 3 & -1 \end{bmatrix}\begin{bmatrix} x \\ y \end{bmatrix} = \begin{bmatrix} 7 \\ 6 \end{bmatrix}$ **3.** $\begin{bmatrix} 1 & 1 \\ 3 & -1 \\ 1 & 5 \end{bmatrix}\begin{bmatrix} x \\ y \end{bmatrix} = \begin{bmatrix} 2 \\ 6 \\ -1 \end{bmatrix}$ **5.** $\begin{aligned} 2x &= 3 \\ x + 5y &= 5 \\ 4x + 6y &= 2 \end{aligned}$ **7.** $\begin{aligned} x + z + v &= -2 \\ 2x + 3z + 5w + v &= 3 \\ -x + 2y + 2z + 2w &= 6 \end{aligned}$

9. $\begin{bmatrix} 8 & 9 \\ -3 & -6 \end{bmatrix}$ **11.** Not possible   **13.** $\begin{bmatrix} 8 & 6 & 13 & 9 \\ -3 & 3 & 3 & 15 \end{bmatrix}$ **15.** $\begin{bmatrix} 3 & 0 \\ 4 & 7 \end{bmatrix}$ **17.** $\begin{bmatrix} 3 & 0 \\ 8 & 11 \end{bmatrix}$ **19.** $\begin{bmatrix} 7 & 0 \\ 0 & 7 \end{bmatrix}$

21. $\begin{bmatrix} 10 & 0 \\ 6 & 16 \end{bmatrix}$ **23.** Not always. If $AB = BA$, the stated equality holds. In genereal $(A - B)(A + B) = A^2 + AB - BA - B^2$.

25. $(AB)C = \begin{bmatrix} 5 & 3 \\ 12 & 15 \end{bmatrix}\begin{bmatrix} 1 & 2 \\ 3 & 4 \end{bmatrix} = \begin{bmatrix} 14 & 22 \\ 57 & 84 \end{bmatrix}$; $A(BC) = \begin{bmatrix} 3 & 1 \\ 2 & 5 \end{bmatrix}\begin{bmatrix} 1 & 2 \\ 11 & 16 \end{bmatrix} = \begin{bmatrix} 14 & 22 \\ 57 & 84 \end{bmatrix}$

27. $A(B + C) = \begin{bmatrix} 3 & 1 \\ 2 & 5 \end{bmatrix}\begin{bmatrix} 2 & 2 \\ 5 & 7 \end{bmatrix} = \begin{bmatrix} 11 & 13 \\ 29 & 39 \end{bmatrix}$; $AB + AC = \begin{bmatrix} 5 & 3 \\ 12 & 15 \end{bmatrix} + \begin{bmatrix} 6 & 10 \\ 17 & 24 \end{bmatrix} = \begin{bmatrix} 11 & 13 \\ 29 & 39 \end{bmatrix}$

29. Possible answer: $A = \begin{bmatrix} 1 & 0 & 0 \\ 0 & 0 & 0 \\ 0 & 0 & 0 \end{bmatrix}$; $B = \begin{bmatrix} 0 & 0 & 0 \\ 0 & 0 & 0 \\ 1 & 3 & 0 \end{bmatrix}$; $C = \begin{bmatrix} 0 & 0 & 0 \\ 0 & 0 & 0 \\ 6 & 4 & 0 \end{bmatrix}$ **31. (a), (b), (c)** $\begin{bmatrix} 20 & 10 \\ 20 & 20 \end{bmatrix}$

33. (a), (b), (c) $\begin{bmatrix} 10x & 5x \\ 10x & 10x \end{bmatrix}$ **35.** $(A + I)BC$ **37.** $(A - I)X$ **39.** $A(B + I)$ **41.** $2(B + I)C$

43. $B = [80 \quad 260 \quad 120]$; $BA = [1480 \quad 2300 \quad 1580 \quad 2120]$ **45. (a)** $B = [80 \quad 100]$; $BA = [410 \quad 200 \quad 140]$

(b) $C = \begin{bmatrix} 5 \\ 12 \\ 6 \end{bmatrix}$; $AC = \begin{bmatrix} 25.00 \\ 32.90 \end{bmatrix}$ **47. (a)** $A = \begin{bmatrix} 30 & 50 \\ 50 & 60 \\ 90 & 100 \end{bmatrix}$ **(b)** $B = \begin{bmatrix} 5 \\ 8 \end{bmatrix}$; $AB = \begin{bmatrix} 550 \\ 730 \\ 1250 \end{bmatrix}$

(c) $C = \begin{matrix} J \\ A \end{matrix} \begin{bmatrix} 200 & 180 & 150 \\ 240 & 220 & 200 \end{bmatrix}$; $CA = \begin{bmatrix} 28,500 & 35,800 \\ 36,200 & 45,200 \end{bmatrix}$

49. $B = [.20 \quad .50]$; $C = \begin{bmatrix} 400 \\ 520 \end{bmatrix}$; $BAC = [5752]$ **51.** $[40 \quad 40 \quad 40]\begin{bmatrix} .60 & .38 & .02 \\ .20 & .60 & .20 \\ .05 & .40 & .55 \end{bmatrix} = [34 \quad 55.2 \quad 30.8]$

53. (a) $A = \begin{matrix} & A & B & C & D & E \\ A \\ B \\ C \\ D \\ E \end{matrix}\begin{bmatrix} 0 & 1 & 1 & 0 & 0 \\ 0 & 0 & 0 & 1 & 0 \\ 0 & 0 & 0 & 1 & 0 \\ 0 & 0 & 0 & 0 & 1 \\ 0 & 0 & 0 & 0 & 0 \end{bmatrix}$ **(b)** $A^2 = \begin{matrix} & A & B & C & D & E \\ A \\ B \\ C \\ D \\ E \end{matrix}\begin{bmatrix} 0 & 0 & 0 & 2 & 0 \\ 0 & 0 & 0 & 0 & 1 \\ 0 & 0 & 0 & 0 & 1 \\ 0 & 0 & 0 & 0 & 0 \\ 0 & 0 & 0 & 0 & 0 \end{bmatrix}$; A^2 shows the number of ways in which one executive reports

$$\begin{array}{c} \\ \\ \text{to another through } one \text{ intermediary.} \quad \textbf{(c) } A^3 = \end{array} \begin{array}{c} \\ A \\ B \\ C \\ D \\ E \end{array} \begin{array}{ccccc} A & B & C & D & E \\ \left[\begin{array}{ccccc} 0 & 0 & 0 & 0 & 2 \\ 0 & 0 & 0 & 0 & 0 \\ 0 & 0 & 0 & 0 & 0 \\ 0 & 0 & 0 & 0 & 0 \\ 0 & 0 & 0 & 0 & 0 \end{array}\right] \end{array} ; A^3 \text{ shows the number of ways in which one executive}$$

$$\text{reports to another through } two \text{ intermediaries.} \quad \textbf{(d) } A + A^2 + A^3 = \begin{array}{c} \\ A \\ B \\ C \\ D \\ E \end{array} \begin{array}{ccccc} A & B & C & D & E \\ \left[\begin{array}{ccccc} 0 & 1 & 1 & 2 & 2 \\ 0 & 0 & 0 & 1 & 1 \\ 0 & 0 & 0 & 1 & 1 \\ 0 & 0 & 0 & 0 & 1 \\ 0 & 0 & 0 & 0 & 0 \end{array}\right] \end{array} ; A + A^2 + A^3 \text{ shows the number of}$$

ways in which one executive may report to another through zero or more intermediaries.

3.3 EXERCISES (page 125)

1. Yes **3.** No **5.** $3x + 5y = 1$ $\quad 3x + 5y = 0$ **7.** $\begin{bmatrix} -5 & 3 \\ 2 & -1 \end{bmatrix}$ **9.** Inverse does not exist.
$\qquad\qquad\qquad\qquad 7x - y = 0; \quad 7x - y = 1$

11. Inverse does not exist. **13.** $\begin{bmatrix} 2 & \frac{3}{2} \\ 1 & \frac{1}{2} \end{bmatrix}$ **15.** $\begin{bmatrix} 0 & -.5 & 1 \\ -.5 & .25 & .5 \\ .5 & .25 & -.5 \end{bmatrix}$ **17.** No inverse exists. **19.** $\begin{bmatrix} 1 & 0 & -1 \\ 0 & -1 & 1 \\ -1 & 1 & 1 \end{bmatrix}$

21. $\begin{bmatrix} \frac{2}{3} & 0 & \frac{1}{3} & -\frac{1}{3} \\ -1 & \frac{1}{2} & 0 & 0 \\ \frac{1}{3} & 0 & -\frac{1}{3} & \frac{1}{3} \\ -\frac{4}{3} & 0 & \frac{1}{3} & \frac{2}{3} \end{bmatrix}$ **23. (a)** $\begin{bmatrix} -1 & -3 \\ -1 & 5 \end{bmatrix}$ **(b)** $\begin{bmatrix} -\frac{5}{8} & -\frac{3}{8} \\ -\frac{1}{8} & \frac{1}{8} \end{bmatrix}$ **25.** $\begin{bmatrix} 1 & 3 \\ 2 & 4 \end{bmatrix}\begin{bmatrix} x \\ y \end{bmatrix} = \begin{bmatrix} 7 \\ -3 \end{bmatrix}; \begin{bmatrix} x \\ y \end{bmatrix} = \begin{bmatrix} -\frac{37}{2} \\ \frac{17}{2} \end{bmatrix}$

27. $\begin{bmatrix} 5 & 2 \\ 4 & 2 \end{bmatrix}\begin{bmatrix} x \\ y \end{bmatrix} = \begin{bmatrix} 4 \\ -3 \end{bmatrix}; \begin{bmatrix} x \\ y \end{bmatrix} = \begin{bmatrix} 7 \\ -\frac{31}{2} \end{bmatrix}$ **29.** $\begin{bmatrix} 1 & 0 & 2 \\ 0 & 2 & 2 \\ 1 & 1 & 1 \end{bmatrix}\begin{bmatrix} x \\ y \\ z \end{bmatrix} = \begin{bmatrix} 1 \\ 3 \\ 5 \end{bmatrix}; \begin{bmatrix} x \\ y \\ z \end{bmatrix} = \begin{bmatrix} \frac{7}{2} \\ \frac{11}{4} \\ -\frac{5}{4} \end{bmatrix}$

31. The coefficient matrix has no inverse. Solution: $x = \frac{32}{5} - \frac{2}{5}z; y = 6 + 4z; z = $ any number

33. $\begin{bmatrix} \frac{1}{3} & 0 \\ 0 & 1 \end{bmatrix}$; Diagonal elements are the reciprocal of those in the original matrix.

35. $\begin{bmatrix} \frac{1}{3} & 0 & 0 \\ 0 & -1 & 0 \\ 0 & 0 & \frac{1}{5} \end{bmatrix}$; Diagonal elements are the reciprocal of those in the original matrix.

37. $\begin{bmatrix} -\frac{1}{3} & 0 & 0 & 0 \\ 0 & -\frac{1}{6} & 0 & 0 \\ 0 & 0 & \frac{1}{2} & 0 \\ 0 & 0 & 0 & \frac{1}{5} \end{bmatrix}$; Diagonal elements are the reciprocal of those in the original matrix.

39. $A = CB^{-1}$ **41.** $X = (I - A)^{-1}B$ **43.** $A = C(B + C)^{-1}$ **45.** $T = (C - B)A^{-1}$
47. $A = (C - D)(X + B)^{-1}$ **49.** $T = 3(2I + B)^{-1}C$
51. (a) $(AB)^{-1} = \begin{bmatrix} 9 & 14 \\ 5 & 8 \end{bmatrix}^{-1} = \begin{bmatrix} 4 & -7 \\ -\frac{5}{2} & \frac{9}{2} \end{bmatrix}$ **(b)** $B^{-1}A^{-1} = \begin{bmatrix} 1 & -2 \\ -\frac{1}{2} & \frac{3}{2} \end{bmatrix}\begin{bmatrix} 2 & -3 \\ -1 & 2 \end{bmatrix} = \begin{bmatrix} 4 & -7 \\ -\frac{5}{2} & \frac{9}{2} \end{bmatrix}$ **(c)** For two square
matrices of the same size and each with an inverse, the inverse of the product equals the product of the inverses in reverse order.
(d) $C^{-1}(AB)^{-1} = C^{-1}(B^{-1}A^{-1})$ **53. (a)** $(A^T)^{-1} = \begin{bmatrix} 2 & 1 \\ 3 & 2 \end{bmatrix}^{-1} = \begin{bmatrix} 2 & -1 \\ -3 & 2 \end{bmatrix}$ **(b)** $(A^{-1})^T = \begin{bmatrix} 2 & -3 \\ -1 & 2 \end{bmatrix}^T = \begin{bmatrix} 2 & -1 \\ -3 & 2 \end{bmatrix}$
(c) For a square matrix with an inverse, the inverse of the transpose equals the transpose of the inverse. **55. (a)** A must have an inverse. **(b)** No **(c)** No **(d)** Yes

3.4 EXERCISES (page 132)

1. A through Z corresponds consecutively with 1 through 26, space $= 30$, period $= 40$, exclamation point $= 50$:

(a) $C = \begin{bmatrix} 9 & 30 \\ 8 & 1 \\ 22 & 5 \\ 30 & 1 \\ 30 & 10 \\ 15 & 2 \\ 40 & 30 \end{bmatrix}$ **(b)** $CA = \begin{bmatrix} -21 & 48 \\ 7 & 17 \\ 17 & 49 \\ 29 & 61 \\ 20 & 70 \\ 13 & 32 \\ 10 & 110 \end{bmatrix}$ **(c)** $A^{-1} = \begin{bmatrix} \frac{1}{3} & -\frac{2}{3} \\ \frac{1}{3} & \frac{1}{3} \end{bmatrix}$; $(CA)A^{-1} = C$

3. Use the same letter-number correspondence as in Exercise 1.

(a) $C = \begin{bmatrix} 14 & 15 \\ 30 & 23 \\ 1 & 25 \\ 50 & 30 \end{bmatrix}$ **(b)** $CA = \begin{bmatrix} 29 & 44 \\ 53 & 76 \\ 26 & 51 \\ 80 & 110 \end{bmatrix}$ **(c)** $A^{-1} = \begin{bmatrix} 2 & -1 \\ -1 & 1 \end{bmatrix}$; $(CA)A^{-1} = C$

5. (a) $A^{-1} = \begin{bmatrix} 3 & -1 & 0 \\ -.5 & 0 & .5 \\ -2 & 1 & 0 \end{bmatrix}$ **(b)** $(CA)A^{-1} = \begin{bmatrix} 48 & 10 & 66 \\ 71 & 60 & 91 \\ 26 & 40 & 27 \\ 115 & 60 & 155 \end{bmatrix} \begin{bmatrix} 3 & -1 & 0 \\ -.5 & 0 & .5 \\ -2 & 1 & 0 \end{bmatrix} = \begin{bmatrix} 7 & 18 & 5 \\ 1 & 20 & 30 \\ 4 & 1 & 20 \\ 5 & 40 & 30 \end{bmatrix}$ Message: Great date

7. (a) $A^{-1} = \begin{bmatrix} \frac{1}{2} & 0 & 0 \\ 0 & 1 & 0 \\ 0 & 0 & \frac{1}{3} \end{bmatrix}$ **(b)** $(CA)A^{-1} = \begin{bmatrix} 14 & 12 & 45 \\ 4 & 1 & 36 \\ 60 & 23 & 3 \\ 36 & 13 & 27 \\ 28 & 7 & 90 \\ 6 & 15 & 42 \\ 40 & 9 & 42 \\ 42 & 5 & 57 \\ 80 & 30 & 90 \end{bmatrix} \begin{bmatrix} \frac{1}{2} & 0 & 0 \\ 0 & 1 & 0 \\ 0 & 0 & \frac{1}{3} \end{bmatrix} = \begin{bmatrix} 7 & 12 & 15 \\ 2 & 1 & 12 \\ 30 & 23 & 1 \\ 18 & 13 & 9 \\ 14 & 7 & 30 \\ 3 & 15 & 14 \\ 20 & 9 & 14 \\ 21 & 5 & 19 \\ 40 & 30 & 30 \end{bmatrix}$ Message: Global warming continues.

9. (a) Each unit of steel produced requires an input of .02 units of steel and .15 units of electronics. Each unit of electronics produced requires an input of .10 units of steel and .01 units of electronics.

(b) $\begin{bmatrix} \frac{825}{796} & \frac{125}{1194} \\ \frac{125}{796} & \frac{1225}{1194} \end{bmatrix}$ **(c)** $X = \begin{bmatrix} 601.97 \\ 899.29 \end{bmatrix}$ **(d)** $TX = \begin{bmatrix} 101.97 \\ 99.29 \end{bmatrix}$

11. (a) Each unit of manufacturing produced requires an input of .10 units of manufacturing and .15 of electronics. Each unit of electronics produced requires an input of .20 units of manufacturing and .05 units of electronics.

(b) $\begin{bmatrix} \frac{38}{33} & \frac{8}{33} \\ \frac{2}{11} & \frac{12}{11} \end{bmatrix}$ **(c)** $X = \begin{bmatrix} 351.52 \\ 581.82 \end{bmatrix}$ **(d)** $TX = \begin{bmatrix} 151.52 \\ 81.82 \end{bmatrix}$

13. (a) Each unit of agriculture produced requires an input of .02 units of agriculture, .20 units of manufacturing, and .15 units of electronics. Each unit of manufacturing produced requires an input of .01 units of agriculture, .10 units of manufacturing, and .08 units of electronics. Each unit of electronics produced requires an input of .01 units of agriculture, .10 units of manufacturing, and .05 units of electronics.

(b) $\begin{bmatrix} \frac{1694}{1653} & \frac{103}{8265} & \frac{20}{1653} \\ \frac{410}{1653} & \frac{1859}{1653} & \frac{200}{1653} \\ \frac{302}{1653} & \frac{799}{8265} & \frac{1760}{1653} \end{bmatrix}$ **(c)** $\begin{bmatrix} 549.06 \\ 1490.62 \\ 2317.48 \end{bmatrix}$ **(d)** $TX = \begin{bmatrix} 49.06 \\ 490.62 \\ 317.48 \end{bmatrix}$

15. (a) $P = \begin{matrix} & \begin{matrix} P_1 & P_2 & P_3 \end{matrix} \\ \begin{matrix} P_1 \\ P_2 \\ P_3 \end{matrix} & \begin{bmatrix} 0 & 2 & 3 \\ 0 & 0 & 1 \\ 0 & 0 & 0 \end{bmatrix} \end{matrix}$ **(b)** The element a_{ij} is the number of parts of type P_i that are used to assemble part P_j.

(c) $(I - P)^{-1}D = \begin{bmatrix} 1 & -2 & -3 \\ 0 & 1 & -1 \\ 0 & 0 & 1 \end{bmatrix}^{-1} \begin{bmatrix} 200 \\ 100 \\ 100 \end{bmatrix} = \begin{bmatrix} 1 & 2 & 5 \\ 0 & 1 & 1 \\ 0 & 0 & 1 \end{bmatrix} \begin{bmatrix} 200 \\ 100 \\ 100 \end{bmatrix} = \begin{bmatrix} 900 \\ 200 \\ 100 \end{bmatrix}$

17. (a) $P = \begin{array}{c} \\ P_1 \\ P_2 \\ P_3 \\ P_4 \end{array} \begin{array}{cccc} P_1 & P_2 & P_3 & P_4 \\ \end{array}$
$$\begin{array}{c} P_1 \\ P_2 \\ P_3 \\ P_4 \end{array}\begin{bmatrix} 0 & 3 & 2 & 0 \\ 0 & 0 & 0 & 1 \\ 0 & 1 & 0 & 2 \\ 0 & 0 & 0 & 0 \end{bmatrix}$$
(b) The element a_{ij} is the number of parts P_i that are used assemble the part P_j.

(c) $(I - P)^{-1}D = \begin{bmatrix} 1 & -3 & -2 & 0 \\ 0 & 1 & 0 & -1 \\ 0 & -1 & 1 & -2 \\ 0 & 0 & 0 & 1 \end{bmatrix}^{-1} \begin{bmatrix} 200 \\ 80 \\ 100 \\ 50 \end{bmatrix} = \begin{bmatrix} 1 & 5 & 2 & 9 \\ 0 & 1 & 0 & 1 \\ 0 & 1 & 1 & 3 \\ 0 & 0 & 0 & 1 \end{bmatrix}\begin{bmatrix} 200 \\ 80 \\ 100 \\ 50 \end{bmatrix} = \begin{bmatrix} 1250 \\ 130 \\ 330 \\ 50 \end{bmatrix}$

19. (a) $P = \begin{array}{c} P_1 \\ P_2 \\ P_3 \\ P_4 \end{array}\begin{array}{cccc} P_1 & P_2 & P_3 & P_4 \end{array}$
$$\begin{array}{c} P_1 \\ P_2 \\ P_3 \\ P_4 \end{array}\begin{bmatrix} 0 & 2 & 0 & 0 \\ 0 & 0 & 2 & 0 \\ 0 & 0 & 0 & 3 \\ 0 & 0 & 0 & 0 \end{bmatrix}$$
(b) The element a_{ij} is the number of parts P_i used to assemble part P_j.

(c) $(I - P)^{-1}D = \begin{bmatrix} 1 & -2 & 0 & 0 \\ 0 & 1 & -2 & 0 \\ 0 & 0 & 1 & -3 \\ 0 & 0 & 0 & 1 \end{bmatrix}^{-1}\begin{bmatrix} 250 \\ 300 \\ 150 \\ 60 \end{bmatrix} = \begin{bmatrix} 1 & 2 & 4 & 12 \\ 0 & 1 & 2 & 6 \\ 0 & 0 & 1 & 3 \\ 0 & 0 & 0 & 1 \end{bmatrix}\begin{bmatrix} 250 \\ 300 \\ 150 \\ 60 \end{bmatrix} = \begin{bmatrix} 2170 \\ 960 \\ 330 \\ 60 \end{bmatrix}$

21.
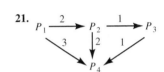

CHAPTER 3 REFLECTIONS (page 134)

1. $\begin{bmatrix} 12 & -5 \\ 9 & -1 \end{bmatrix}$ **3.** $\begin{bmatrix} 1 & 0 & 3 \\ -2 & 4 & -1 \end{bmatrix}$ **5.** $\begin{bmatrix} -4 & 2 & 1 \\ 12 & -4 & 0 \\ 3 & 1 & 3 \end{bmatrix}$ **7.** Not possible **9.** Not possible **11.** $\begin{bmatrix} \frac{1}{9} & \frac{1}{9} \\ 0 & -1 \end{bmatrix}$

13. $x = 3, w = -24, t = \frac{17}{2}, v = \frac{1}{2}$ **15.** $\begin{bmatrix} \frac{1}{3} & -\frac{1}{6} & \frac{1}{2} \\ -\frac{1}{3} & \frac{1}{6} & \frac{1}{2} \\ \frac{1}{3} & \frac{1}{3} & -1 \end{bmatrix}$ **17.** $\begin{array}{c} w \\ x \\ y \\ z \end{array}\begin{array}{cccc} w & x & y & z \end{array}$
$$\begin{array}{c} w \\ x \\ y \\ z \end{array}\begin{bmatrix} 0 & 2 & 0 & 1 \\ 0 & 0 & 1 & 0 \\ 1 & 1 & 0 & 1 \\ 1 & 0 & 0 & 0 \end{bmatrix}$$
19. $\begin{bmatrix} 9 & 41 \\ 5 & 18 \\ 7 & 51 \\ 9 & 42 \\ 7 & 35 \\ 30 & 140 \end{bmatrix}$

21. $(CA)A^{-1} = \begin{bmatrix} 9 & 41 \\ 5 & 18 \\ 7 & 51 \\ 9 & 42 \\ 7 & 35 \\ 30 & 140 \end{bmatrix}\begin{bmatrix} -3 & 1 \\ 1 & 0 \end{bmatrix} = \begin{bmatrix} 14 & 9 \\ 3 & 5 \\ 30 & 7 \\ 15 & 9 \\ 14 & 7 \\ 50 & 30 \end{bmatrix}$ **23.** Must be of the same size and have corresponding entries equal

14 9 | 3 5 | 30 7 | 15 9 | 14 7 | 50 30
N I C E   G O I N G !

25. Their product, in either order, must be the identity matrix.

27. If A is $m \times n$ and B is $n \times m$, then both AB and BA exist with AB having size $m \times m$ and BA having size $n \times n$. For $AB = BA$, A and B must be square and of the same size. Even so, the likelihood that $AB = BA$ is very small.

29. $\begin{aligned} 2a \quad\;\; - g &= 1 \\ a + d + 3g &= 0 \\ 3a \quad\;\; + 2g &= 0 \end{aligned}$ $\begin{aligned} 2b \quad\;\; - h &= 0 \\ b + e + 3h &= 1 \\ 3b \quad\;\; + 2h &= 0 \end{aligned}$ $\begin{aligned} 2c \quad\;\; - i &= 0 \\ c + f + 3i &= 0 \\ 3c \quad\;\; + 2i &= 1 \end{aligned}$

31. The inverse is $\begin{bmatrix} .833 & -.667 \\ -.167 & .333 \end{bmatrix}$

Solution: Solution: Solution: **32.** $y = 0$

$\begin{aligned} a &= \tfrac{2}{7} \\ d &= 1 \\ g &= -\tfrac{3}{7} \end{aligned}$ $\begin{aligned} b &= 0 \\ e &= 1 \\ h &= 0 \end{aligned}$ $\begin{aligned} c &= \tfrac{1}{7} \\ f &= -1 \\ i &= \tfrac{2}{7} \end{aligned}$

35. $P(x) = 20x - 2100;\ (105, 8400)$
37. $S(t) = 5t + 72,\ 0 \le t \le 5.6$

The inverse is $\begin{bmatrix} a & b & c \\ d & e & f \\ g & h & i \end{bmatrix} = \begin{bmatrix} \frac{2}{7} & 0 & \frac{1}{7} \\ 1 & 1 & -1 \\ -\frac{3}{7} & 0 & \frac{2}{7} \end{bmatrix}$

SAMPLE TEST ITEMS FOR CHAPTER 3 (page 137)

1. $y = 9, w = 0, x = 3, -3, t = 6, r = 6$

2. $\begin{bmatrix} 3 & -2 & 3 \\ -6 & -3 & -6 \\ -7 & -8 & -1 \end{bmatrix}$

3. Not possible

4. $\begin{bmatrix} 2 & -1 & 0 \\ 0 & 1 & 0 \\ 3 & 0 & 4 \end{bmatrix}$

5. $\begin{bmatrix} 30 & 36 & 42 \\ 66 & 81 & 96 \\ 102 & 126 & 150 \end{bmatrix}$

6. $\begin{bmatrix} 23 & 28 & 33 \\ 3 & 3 & 3 \\ 28 & 32 & 36 \end{bmatrix}$

7. Not possible

8. $\begin{array}{c} \\ w \\ x \\ y \end{array} \begin{array}{ccc} w & x & y \\ \begin{bmatrix} 1 & 1 & 1 \\ 0 & 1 & 1 \\ 0 & 1 & 1 \end{bmatrix} \end{array}$

9. $2x \quad + z = 4$
$3x - 2y + 5z = 17$

10. $B = [175 \quad 250]; \ BA = [4625 \quad 2850]$

11. No

12. $\begin{bmatrix} 3 & -7 \\ -2 & 5 \end{bmatrix}$

13. $\begin{bmatrix} 1 & 89 & -31 \\ 0 & -20 & 7 \\ 0 & 3 & -1 \end{bmatrix}$

14. $X = (Y - B)A^{-1}$

15. $\begin{bmatrix} x \\ y \\ z \end{bmatrix} = \begin{bmatrix} 5 \\ 13 \\ 5 \end{bmatrix}$

16. $\begin{array}{c} \\ P_1 \\ P_2 \\ P_3 \end{array} \begin{array}{ccc} P_1 & P_2 & P_3 \\ \begin{bmatrix} 0 & 1 & 2 \\ 0 & 0 & 3 \\ 0 & 0 & 0 \end{bmatrix} \end{array}$

17. The element a_{ij} is the number of parts of type P_i used to assemble part P_j.

18. $\begin{bmatrix} 1 & 1 & 5 \\ 0 & 1 & 3 \\ 0 & 0 & 1 \end{bmatrix}$

19. $\begin{bmatrix} 1800 \\ 1100 \\ 300 \end{bmatrix}$

20. 1700 of P_1, 900 of P_2, and none of P_3

CHAPTER 4

4.1 EXERCISES (page 144)

1. (a)

	Silver	Gold	Limits
Grinder	1	3	30(60)
Bonder	3	4	50(60)
Number made	x	y	
Profit	5	7	

(b) Maximize $P = 5x + 7y$,
subject to $x + 3y \leq 1800$
$3x + 4y \leq 3000$
$x \geq 0, y \geq 0$

3.

	A	B	Limits
Vitamin C	100	10	260
Vitamin D	40	80	320
Vitamin E	10	5	50
Pounds used	x	y	
Cost	20	15	

(b) Minimize $C = 20x + 15y$,
subject to $100x + 10y \geq 260$
$40x + 80y \geq 320$
$10x + 5y \geq 50$
$x \geq 0, y \geq 0$

5. Minimize $F = 2x + 3y$,
subject to $4x + 6y \geq 10$
$5x + 2y \geq 10$
$x \geq 0, y \geq 0$

7. Maximize $L = 10,000x + 2500y + 800z + 600w$,
subject to $1000x + 800y + 400z + 300w \leq 20,000$
$x \qquad\qquad\qquad \geq 5$
$\quad y \qquad\qquad \geq 3$
$x \qquad\qquad\qquad \leq 8$
$\quad y \qquad\qquad \leq 5$
$\qquad z \qquad \leq 3$
$\qquad\qquad w \leq 10$
$x \geq 0, y \geq 0, z \geq 0, w \geq 0$

9. Minimize $E = 5x + 3y + 4z$,
subject to $2x + 3y + 3z \geq 12$
$2x + y + z \geq 6$
$x \geq 0, y \geq 0, z \geq 0$

11. Maximize $P = .12x + .07y$,
subject to $x + y \leq 30,000$
$2x - y \leq 0$
$x \geq 0, y \geq 0$
(*x* and *y* in dollars; divide by 1000 to get
the number of bonds.)

13. Maximize $I = 350x + 525y$,
subject to $x + y \leq 500$
$200x + 350y \leq 25,000$
$x \geq 0, y \geq 0$

15. Minimize $C = 12,000x + 10,000y$,
     subject to $50x + 40y \geq 500$
                         $60x + 30y \geq 1000$
                    $100x + 80y \geq 1500$
                                 $x \geq 0, y \geq 0$

17. Maximize $P = 420x + 550y$,
     subject to $x + y \leq 300$
                           $x \geq 80$
                           $y \geq 120$

19. Minimize $F = 2x + 5y$,
     subject to $5x + 7y \leq 105$
                         $80x + 110y \geq 400$
                           $x - 3y \geq 0$
                                 $x \geq 0, y \geq 0$

4.2 EXERCISES (page 154)

1. (a) $3x + y = 9$
   (b) $(0, 0)$ makes inequality true;
        shade toward the origin.
   (c)

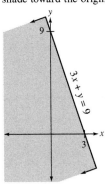

3. (a) $x + 2y = 6$
   (b) $(0, 0)$ makes inequality false;
        shade away from the origin
   (c)

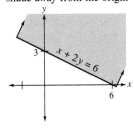

5. (a) $-2x + y = 8$
   (b) $(0, 0)$ makes inequality false;
        shade away from the origin.
   (c)

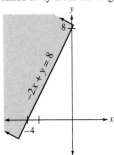

7. (a) $x = -3$
   (b) $(0, 0)$ makes inequality false;
        shade away from the origin.
   (c)

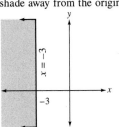

9. (a) $x = 3$
   (b) $(0, 0)$ makes inequality false;
        shade away from the origin.
   (c)

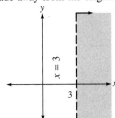

11. (a) $y = 2$
   (b) $(0, 0)$ makes inequality true;
        shade toward the origin.
   (c)

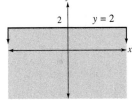

13. (a) $L:\ x + 2y = 6$
        $M:\ x = 0$
        $N:\ y = 0$
   (b)

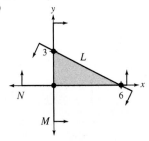

   (c) $L, M:\ (0, 3)$
        $M, N:\ (0, 0)$
        $L, N:\ (6, 0)$

15. (a) $L:\ x = y$
        $M:\ x = 2y$
        $N:\ x = 3$
   (b)

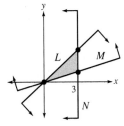

   (c) $L, M:\ (0, 0)$
        $M, N:\ (3, \frac{3}{2})$
        $L, N:\ (3, 3)$

17. (a) $L:\ \ x + 3y = 9$
        $M:\ 3x + y = 11$
        $N:\ \ x = 0$
        $O:\ \ y = 0$
   (b)

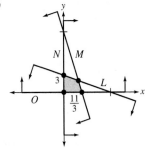

   (c) $N, L:\ (0, 3)$
        $N, O:\ (0, 0)$
        $M, O:\ (\frac{11}{3}, 0)$
        $L, M:\ (3, 2)$

19. (a) *L*: $x + y = 8$ **(b)**
     M: $2x + y = 10$
     N: $4x + y = 16$
     O: $x = 0$
     P: $y = 0$

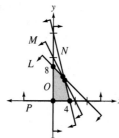

(c) *L, O*: (0, 8)
     O, P: (0, 0)
     N, P: (4, 0)
     M, N: (3, 4)
     L, M: (2, 6)

21.

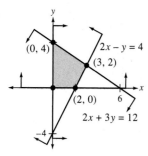

(0, 0) is in the feasible region.
(1, 3) is in the feasible region.

23.

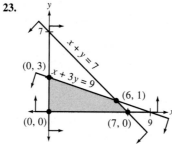

(0, 0) is in the feasible region.
(1, 3) is not in the feasible region.

25.

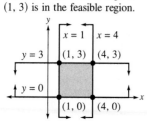

(0, 0) is not the feasible region.
(1, 3) is in the feasible region.

27.

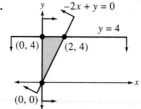

(0, 0) is in the feasible region.
(1, 3) is in the feasible region.

29.

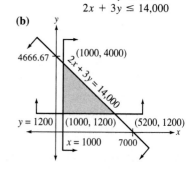

(0, 0) is not in the feasible region.
(1, 3) is not in the feasible region.

31. (a) Maximize $S = 3000x + 2500y$,
     subject to $120x + 100y \le 18{,}000$
              $x \ge 50$
              $y \ge 60$

(b)

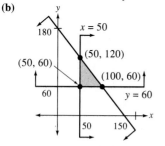

$120x + 100y = 18{,}000$

33. (a) Minimize $C = 80x + 110y$,
     subject to $30x + 40y \ge 1200$
           $90x + 50y \ge 2970$
           $x \ge 0, y \ge 0$

(b)

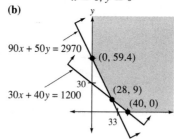

35. (a) Minimize $W = 0.2x + 0.25y$,
     subject to $x \ge 1000$
              $y \ge 1200$
           $2x + 3y \le 14{,}000$

(b)

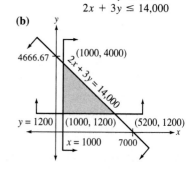

37. (a) Maximize $R = 30x + 80y$,
  subject to $x + 3y \leq 1800$
          $3x + 4y \leq 3000$
          $x \geq 0, y \geq 0$

(b)

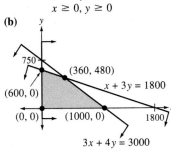

39. (a) Maximize $R = 50x + 60y$,
  subject to $10x + 20y \geq 320$
          $x - 2y \geq 0$
          $x \qquad \leq 50$
          $x \geq 0, y \geq 0$

(b)

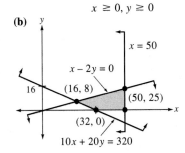

41. (a) Maximize $T = x + y$,
  subject to $30x + 35y \leq 4800$
          $20x + 45y \leq 6000$
          $x \geq 0, y \geq 0$

(b)

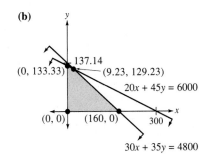

4.3 EXERCISES (page 165)

1. (a)

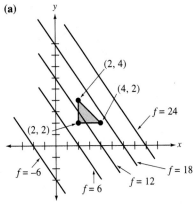

(b) The maximum of f is 16 and occurs at $(4, 2)$; the minimum is 10 and occurs at $(2, 2)$. The larger c the further upward the lines move.

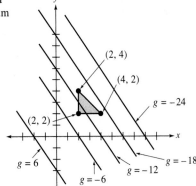

(d) The maximum of g is -10 and occurs at $(2, 2)$; the minimum of g is -16 and occurs at $(4, 2)$. The larger c the further downward the lines move.

3. (a), (b)

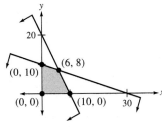

(c)

Corner points	$P = 4x + 4y$
$(0, 10)$	40
$(0, 0)$	0
$(10, 0)$	40
$(6, 8)$	56

The maximum value of P is 56 and occurs when $x = 6$ and $y = 8$.

5. (a), (b)

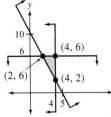

(c)

Corner points	$f = x + 2y$
(2, 6)	14
(4, 2)	8
(4, 6)	16

The maximum value of f is 16 and occurs when $x = 4$ and $y = 6$.

7. (a), (b)

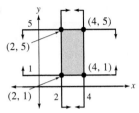

(c)

Corner points	$C = x - 2y$
(2, 5)	−8
(2, 1)	0
(4, 1)	2
(4, 5)	−6

The minimum value of C is −8 and occurs when $x = 2$ and $y = 5$.

9. (a), (b)

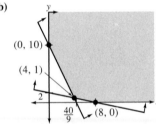

(c)

Corner points	$C = 2x + 3y$
(0, 10)	30
(4, 1)	11
(8, 0)	16

The minimum value of C is 11 and occurs when $x = 4$ and $y = 1$.

11. (a), (b)

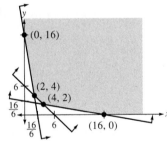

(c)

Corner points	$C = 5x + 3y$
(0, 16)	48
(2, 4)	22
(4, 2)	26
(16, 0)	80

The minimum value of C is 22 and occurs when $x = 2$ and $y = 4$.

13. (a), (b)

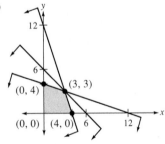

(c)

Corner points	$P = 8x + y$
(0, 4)	4
(0, 0)	0
(4, 0)	32
(3, 3)	27

The maximum value of P is 32 and occurs when $x = 4$ and $y = 0$.

15. (a), (b)

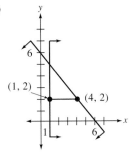

(c)

Corner points	$f = 2x - 5y$
(1, 2)	-8
(4, 2)	-2

The minimum value of f is -8 and occurs when $x = 1$ and $y = 2$.

17. (a) No **(b)** Yes, at the point (1, 2). The minimum value of f is 7.   **19.** 14 computers and 16 fax machines
21. 1200 4-ply and 2800 radials   **23.** 6 pounds of barley and 3 pounds of corn   **25.** 60 student desks and 40 secretary desks
27. Approximately 17 type A planes and 0 type B planes   **29.** 132 standard models and 168 deluxe models
31. 10 junk bonds and 20 premium bonds   **33.** 200 mice and 25 rats

4.4 EXERCISES (page 178)

1. In standard form   **3.** In standard form   **5.** Not in standard form   **7.** Not in standard form   **9.** 0   **11.** 9

13. Above the line $x + 3y = 9$.   **15.** 5   **17.** $0 \le s_2 \le 8$ **19.** $\begin{aligned} x + y + s_1 \quad\;\;\, &= 6 \\ 2x + y \quad\;\; + s_2 &= 8 \end{aligned}$ **21.** $\begin{bmatrix} 1 & 1 & 1 & 0 & 6 \\ ② & 1 & 0 & 1 & 8 \end{bmatrix}$;

a pivot on the circled element gives $\begin{bmatrix} 0 & \frac{1}{2} & 1 & -\frac{1}{2} & 2 \\ 1 & \frac{1}{2} & 0 & \frac{1}{2} & 4 \end{bmatrix}$ with basic feasible solution $x = 4$, $y = 0$, $s_1 = 2$, and $s_2 = 0$. A pivot on

the 1 in the x-column gives $\begin{bmatrix} 1 & 1 & 1 & 0 & 6 \\ 0 & -1 & -2 & 1 & -4 \end{bmatrix}$. With the nonbasic variables y and s_1 set equal to zero, $x = 6$ and $s_2 = -4$

which is not a point in the feasible region.   **23.** $\begin{bmatrix} 1 & 1 & 1 & 0 & 6 \\ ② & 1 & 0 & 1 & 8 \end{bmatrix} \rightarrow \begin{bmatrix} 0 & ⓵\frac{1}{2} & 1 & -\frac{1}{2} & 2 \\ 1 & \frac{1}{2} & 0 & \frac{1}{2} & 4 \end{bmatrix} \rightarrow \begin{bmatrix} 0 & 1 & 2 & -1 & 4 \\ 1 & 0 & -1 & 1 & 2 \end{bmatrix}$

The last matrix has basic feasible solution $x = 2$, $y = 4$, $s_1 = 0$, and $s_2 = 0$.   **25.** $\begin{aligned} x + \;\; y + s_1 \quad\;\;\, &= 9 \\ -x + 2y \quad\;\; + s_2 &= 12 \end{aligned}$

27. $\begin{bmatrix} ⓵ & 1 & 1 & 0 & 9 \\ -1 & 2 & 0 & 1 & 12 \end{bmatrix}$; a pivot on the circled element gives $\begin{bmatrix} 1 & 1 & 1 & 0 & 9 \\ 0 & 3 & 1 & 1 & 21 \end{bmatrix}$ with basic feasible solution $x = 9$, $y = 0$,

$s_1 = 0$, and $s_2 = 21$. A pivot on the -1 in the x-column gives $\begin{bmatrix} 0 & 3 & 1 & 1 & 21 \\ 1 & -2 & 0 & -1 & -12 \end{bmatrix}$. With the nonbasic variables y and s_2 set

equal to zero, $x = -12$ and $s_1 = 21$, which is not a point in the feasible region.
29. $\begin{bmatrix} ⓵ & 1 & 1 & 0 & 9 \\ -1 & 2 & 0 & 1 & 12 \end{bmatrix} \rightarrow \begin{bmatrix} 1 & 1 & 1 & 0 & 9 \\ 0 & ③ & 1 & 1 & 21 \end{bmatrix} \rightarrow \begin{bmatrix} 1 & 0 & \frac{2}{3} & -\frac{1}{3} & 2 \\ 0 & 1 & \frac{1}{3} & \frac{1}{3} & 7 \end{bmatrix}$ The last matrix has basic feasible solution $x = 2$,

$y = 7$, $s_1 = 0$, and $s_2 = 0$.   **31.** $\begin{aligned} x + y + s_1 \quad\;\;\, &= 10 \\ 2x - y \quad\;\; + s_2 &= 8 \end{aligned}$ **33.** $\begin{bmatrix} 1 & 1 & 1 & 0 & 10 \\ ② & -1 & 0 & 1 & 8 \end{bmatrix}$; a pivot on the circled element gives

$\begin{bmatrix} 0 & \frac{3}{2} & 1 & -\frac{1}{2} & 6 \\ 1 & -\frac{1}{2} & 0 & \frac{1}{2} & 4 \end{bmatrix}$ with basic feasible solution $x = 4$, $y = 0$, $s_1 = 6$, and $s_2 = 0$. A pivot on the 1 in the x-column gives

$\begin{bmatrix} 1 & 1 & 1 & 0 & 10 \\ 0 & -3 & -2 & 1 & -12 \end{bmatrix}$. With the nonbasic variables y and s_1 set equal to zero, $x = 10$ and $s_2 = -12$, which is not a point in

the feasible region.   **35.** $\begin{bmatrix} 1 & 1 & 1 & 0 & 10 \\ ② & -1 & 0 & 1 & 8 \end{bmatrix} \rightarrow \begin{bmatrix} 0 & ⓵\frac{3}{2} & 1 & -\frac{1}{2} & 6 \\ 1 & -\frac{1}{2} & 0 & \frac{1}{2} & 4 \end{bmatrix} \rightarrow \begin{bmatrix} 0 & 1 & \frac{2}{3} & -\frac{1}{3} & 4 \\ 1 & 0 & \frac{1}{3} & \frac{1}{3} & 6 \end{bmatrix}$ The last matrix has

basic feasible solution $x = 6$, $y = 4$, $s_1 = 0$, and $s_2 = 0$.   **37.** $\begin{aligned} x + \;\; y + s_1 \quad\;\;\, &= 8 \\ x + 2y \quad\;\; + s_2 &= 10 \end{aligned}$ **39.** $\begin{bmatrix} ⓵ & 1 & 1 & 0 & 8 \\ 1 & 2 & 0 & 1 & 10 \end{bmatrix}$;

a pivot on the circled element gives $\begin{bmatrix} 1 & 1 & 1 & 0 & 8 \\ 0 & 1 & -1 & 1 & 2 \end{bmatrix}$ with basic feasible solution $x = 8$, $y = 0$, $s_1 = 0$, and $s_2 = 2$.

A pivot on the other 1 in the x-column gives $\begin{bmatrix} 0 & -1 & 1 & -1 & -2 \\ 1 & 2 & 0 & 1 & 10 \end{bmatrix}$. With the nonbasic variables y and s_2 set equal to zero,

$x = 10$ and $s_1 = -2$, which is not a point in the feasible region.

41. $\begin{bmatrix} ① & 1 & 1 & 0 & | & 8 \\ 1 & 2 & 0 & 1 & | & 10 \end{bmatrix} \rightarrow \begin{bmatrix} 1 & 1 & 1 & 0 & | & 8 \\ 0 & ① & -1 & 1 & | & 2 \end{bmatrix} \rightarrow \begin{bmatrix} 1 & 0 & 2 & -1 & | & 6 \\ 0 & 1 & -1 & 1 & | & 2 \end{bmatrix}$ The last matrix has basic feasible solution $x = 6$, $y = 2$, $s_1 = 0$, and $s_2 = 0$.

43. The corner points are $(0, 8)$, $(0, 0)$, $(4, 0)$, and $(2, 6)$. Clockwise:

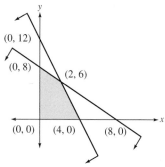

$\begin{bmatrix} 1 & ① & 1 & 0 & | & 8 \\ 3 & 1 & 0 & 1 & | & 12 \end{bmatrix} \rightarrow \begin{bmatrix} 1 & 1 & 1 & 0 & | & 8 \\ ② & 0 & -1 & 1 & | & 4 \end{bmatrix} \rightarrow \begin{bmatrix} 0 & 1 & \frac{3}{2} & -\frac{1}{2} & | & 6 \\ 1 & 0 & -\frac{1}{2} & \frac{1}{2} & | & 2 \end{bmatrix} \rightarrow \begin{bmatrix} 0 & \frac{2}{3} & 1 & -\frac{1}{3} & | & 4 \\ 1 & \frac{1}{3} & 0 & ⟨\frac{1}{3}⟩ & | & 4 \end{bmatrix} \rightarrow \begin{bmatrix} 1 & 1 & 1 & 0 & | & 8 \\ 3 & 1 & 0 & 1 & | & 12 \end{bmatrix}$

Counterclockwise:

$\begin{bmatrix} 1 & 1 & 1 & 0 & | & 8 \\ ③ & 1 & 0 & 1 & | & 12 \end{bmatrix} \rightarrow \begin{bmatrix} 0 & ⟨\frac{2}{3}⟩ & 1 & -\frac{1}{3} & | & 4 \\ 1 & \frac{1}{3} & 0 & \frac{1}{3} & | & 4 \end{bmatrix} \rightarrow \begin{bmatrix} 0 & 1 & \frac{3}{2} & -\frac{1}{2} & | & 6 \\ 1 & 0 & -\frac{1}{2} & ⟨\frac{1}{2}⟩ & | & 2 \end{bmatrix} \rightarrow \begin{bmatrix} 1 & 1 & ① & 0 & | & 8 \\ 2 & 0 & -1 & 1 & | & 4 \end{bmatrix} \rightarrow \begin{bmatrix} 1 & 1 & 1 & 0 & | & 8 \\ 3 & 1 & 0 & 1 & | & 12 \end{bmatrix}$

45. $x = 0$, $y = 2$, $s_1 = 7$, $s_2 = 0$. Yes, a corner point.   **47.** $x = 0$, $y = 5$, $s_1 = 0$, $s_2 = -2$. No, not a corner point.

49. $x = 0$, $y = 2$, $z = 1$, $s_1 = 0$, $s_2 = 0$, $s_3 = 5$. Yes, a corner point.

51. (a) $x + y \leq 6$ **(b)**
$7x + 3y \leq 21$
$x \geq 0, y \geq 0$

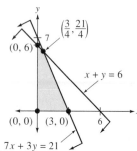

(c) $\begin{bmatrix} 1 & 1 & 1 & 0 & | & 6 \\ ⑦ & 3 & 0 & 1 & | & 21 \end{bmatrix} \rightarrow \begin{bmatrix} 0 & \frac{4}{7} & 1 & -\frac{1}{7} & | & 3 \\ 1 & \frac{3}{7} & 0 & \frac{1}{7} & | & 3 \end{bmatrix}$

(d) $\begin{bmatrix} 1 & ① & 1 & 0 & | & 6 \\ 7 & 3 & 0 & 1 & | & 21 \end{bmatrix} \rightarrow \begin{bmatrix} 1 & 1 & 1 & 0 & | & 6 \\ 4 & 0 & -3 & 1 & | & 3 \end{bmatrix}$

(e) $\begin{bmatrix} 1 & ① & 1 & 0 & | & 6 \\ 7 & 3 & 0 & 1 & | & 21 \end{bmatrix} \rightarrow \begin{bmatrix} 1 & 1 & 1 & 0 & | & 6 \\ ④ & 0 & -3 & 1 & | & 3 \end{bmatrix} \rightarrow \begin{bmatrix} 0 & 1 & \frac{7}{4} & -\frac{1}{4} & | & \frac{21}{4} \\ 1 & 0 & -\frac{3}{4} & \frac{1}{4} & | & \frac{3}{4} \end{bmatrix}$

4.5 EXERCISES (page 191)

1. (a) Yes   **(b)** $f = 20$; $x = 5$, $y = 8$, $s_1 = 0$, $s_2 = 0$ **(c)** No   **3. (a)** Yes   **(b)** $f = \frac{80}{5}$; $x = \frac{8}{5}$, $y = 0$, $s_1 = 0$, $s_2 = 16$
(c) Yes; $s_2 = 16$ or 16 units of surplus in the second constraint.   **5. (a)** No   **(b)** $f = 25$; $x = 6$, $y = 0$, $s_1 = 8$, $s_2 = 2$, $s_3 = 0$
(c) Yes; $s_1 = 8$ and $s_2 = 2$ or 8 units of surplus in the second constraint and 2 units of surplus in the third constraint.

7. (a)

(b) A: $x = 0$, $y = 7$, $s_1 = 3$, $s_2 = 0$, $f = 42$
B: $x = 0$, $y = 0$, $s_1 = 10$, $s_2 = 7$, $f = 0$
C: $x = 5$, $y = 0$, $s_1 = 0$, $s_2 = 2$, $f = 40$
D: $x = 3$, $y = 4$, $s_1 = 0$, $s_2 = 0$, $f = 48$

(c) $\begin{bmatrix} ② & 1 & 1 & 0 & 0 & | & 10 \\ 1 & 1 & 0 & 1 & 0 & | & 7 \\ \hline -8 & -6 & 0 & 0 & 1 & | & 0 \end{bmatrix} \rightarrow \begin{bmatrix} 1 & \frac{1}{2} & \frac{1}{2} & 0 & 0 & | & 5 \\ 0 & ⟨\frac{1}{2}⟩ & -\frac{1}{2} & 1 & 0 & | & 2 \\ \hline 0 & -2 & 4 & 0 & 1 & | & 40 \end{bmatrix} \rightarrow \begin{bmatrix} 1 & 0 & 1 & -1 & 0 & | & 3 \\ 0 & 1 & -1 & 2 & 0 & | & 4 \\ \hline 0 & 0 & 2 & 4 & 1 & | & 48 \end{bmatrix}$

$x = 0$, $y = 0$, $s_1 = 10$, $x = 5$, $y = 0$, $s_1 = 0$, $x = 3$, $y = 4$,
$s_2 = 7$, $f = 0$ $s_2 = 2$, $f = 40$ $s_1 = 0$, $s_2 = 0$, $f = 48$

9. (a)

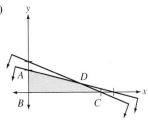

(b) A: $x = 0$, $y = 3$, $s_1 = 5$, $s_2 = 0$, $f = 12$
B: $x = 0$, $y = 0$, $s_1 = 20$, $s_2 = 12$, $f = 0$
C: $x = 10$, $y = 0$, $s_1 = 0$, $s_2 = 2$, $f = 10$
D: $x = \frac{20}{3}$, $y = \frac{4}{3}$, $s_1 = 0$, $s_2 = 0$, $f = 12$

(c)
$$\left[\begin{array}{ccccc|c} 2 & 5 & 1 & 0 & 0 & 20 \\ 1 & ④ & 0 & 1 & 0 & 12 \\ -1 & -4 & 0 & 0 & 1 & 0 \end{array}\right] \rightarrow \left[\begin{array}{ccccc|c} \frac{3}{4} & 0 & 1 & -\frac{5}{4} & 0 & 5 \\ \frac{1}{4} & 1 & 0 & \frac{1}{4} & 1 & 0 \\ 0 & 0 & 0 & 1 & 1 & 12 \end{array}\right]$$

$x = 0$, $y = 0$, $s_1 = 20$, $x = 0$, $y = 3$, $s_1 = 5$,
$s_2 = 12$, $f = 0$ $s_2 = 0$, $f = 12$

11. (a)

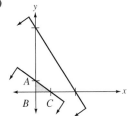

(b) A: $x = 0$, $y = 3$, $s_1 = 0$, $s_2 = 111$, $f = 3$
B: $x = 0$, $y = 0$, $s_1 = 6$, $s_2 = 120$, $f = 0$
C: $x = 6$, $y = 0$, $s_1 = 0$, $s_2 = 96$, $f = 30$

(c)
$$\left[\begin{array}{ccccc|c} ① & 2 & 1 & 0 & 0 & 6 \\ 4 & 3 & 0 & 1 & 0 & 120 \\ -5 & -1 & 0 & 0 & 1 & 0 \end{array}\right] \rightarrow \left[\begin{array}{ccccc|c} 1 & 2 & 1 & 0 & 0 & 6 \\ 0 & -5 & -4 & 1 & 0 & 96 \\ 0 & 9 & 5 & 0 & 1 & 30 \end{array}\right]$$

$x = 0$, $y = 0$, $s_1 = 6$, $x = 6$, $y = 0$, $s_1 = 0$, $s_2 = 96$,
$s_2 = 120$, $f = 0$ $f = 30$

13. (a)

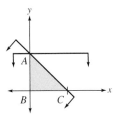

(b) A: $x = 0$, $y = 5$, $s_1 = 0$, $s_2 = 0$, $f = 5$
B: $x = 0$, $y = 0$, $s_1 = 5$, $s_2 = 5$, $f = 0$
C: $x = 5$, $y = 0$, $s_1 = 0$, $s_2 = 5$, $f = 10$

(c)
$$\left[\begin{array}{ccccc|c} ① & 1 & 1 & 0 & 0 & 5 \\ 0 & 1 & 0 & 1 & 0 & 5 \\ -2 & -1 & 0 & 0 & 1 & 0 \end{array}\right] \rightarrow \left[\begin{array}{ccccc|c} 1 & 1 & 1 & 0 & 0 & 5 \\ 0 & 1 & 0 & 1 & 0 & 5 \\ 0 & 1 & 2 & 0 & 1 & 10 \end{array}\right]$$

$x = 0$, $y = 0$, $s_1 = 5$, $x = 5$, $y = 0$, $s_1 = 0$,
$s_2 = 5$, $f = 0$ $s_2 = 5$, $f = 10$

15. (a)

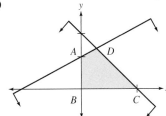

(b) A: $x = 0$, $y = 5$, $s_1 = 4$, $s_2 = 0$, $f = 15$
B: $x = 0$, $y = 0$, $s_1 = 9$, $s_2 = 10$, $f = 0$
C: $x = 9$, $y = 0$, $s_1 = 0$, $s_2 = 19$, $f = 27$
D: $x = \frac{8}{3}$, $y = \frac{19}{3}$, $s_1 = 0$, $s_2 = 0$, $f = 27$

(c)
$$\left[\begin{array}{ccccc|c} ① & 1 & 1 & 0 & 0 & 9 \\ -1 & 2 & 0 & 1 & 0 & 10 \\ -3 & -3 & 0 & 0 & 1 & 0 \end{array}\right] \rightarrow \left[\begin{array}{ccccc|c} 1 & 1 & 1 & 0 & 0 & 9 \\ 0 & 3 & 1 & 1 & 0 & 19 \\ 0 & 0 & 3 & 0 & 1 & 27 \end{array}\right]$$

$x = 0$, $y = 0$, $s_1 = 9$, $x = 9$, $y = 0$, $s_1 = 0$,
$s_2 = 10$, $f = 0$ $s_2 = 19$, $f = 27$

17. (a)

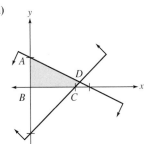

(b) A: $x = 0$, $y = 4$, $s_1 = 10$, $s_2 = 0$, $f = -4$
B: $x = 0$, $y = 0$, $s_4 = 6$, $s_2 = 8$, $f = 0$
C: $x = 6$, $y = 0$, $s_1 = 0$, $s_2 = 2$, $f = -12$
D: $x = \frac{20}{3}$, $y = \frac{2}{3}$, $s_1 = 0$, $s_2 = 0$, $f = -14$

(c)
$$\left[\begin{array}{ccccc|c} 1 & -1 & 1 & 0 & 0 & 6 \\ 1 & 2 & 0 & 1 & 0 & 8 \\ 2 & 1 & 0 & 0 & 1 & 0 \end{array}\right]$$

$x = 0$, $y = 0$, $s_1 = 6$, $s_2 = 8$,
$f = 0$

19. (a)

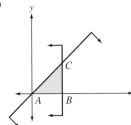

(b) A: $x = 0$, $y = 0$, $s_1 = 0$, $s_2 = 2$, $f = 0$
B: $x = 2$, $y = 0$, $s_1 = 2$, $s_2 = 0$, $f = 6$
C: $x = 2$, $y = 2$, $s_1 = 0$, $s_2 = 0$, $f = 8$

(c)
$$\left[\begin{array}{ccccc|c} -1 & 1 & 1 & 0 & 0 & 0 \\ ① & 0 & 0 & 1 & 0 & 2 \\ -3 & -1 & 0 & 0 & 1 & 0 \end{array}\right] \rightarrow \left[\begin{array}{ccccc|c} 0 & ① & 1 & 1 & 0 & 2 \\ 1 & 0 & 0 & 1 & 0 & 2 \\ 0 & -1 & 0 & 3 & 1 & 6 \end{array}\right] \rightarrow \left[\begin{array}{ccccc|c} 0 & 1 & 1 & 1 & 0 & 2 \\ 1 & 0 & 0 & 1 & 0 & 2 \\ 0 & 0 & 1 & 4 & 1 & 8 \end{array}\right]$$

$x = 0$, $y = 0$, $s_1 = 0$, $x = 2$, $y = 0$, $s_1 = 2$, $x = 2$, $y = 2$, $s_1 = 0$,
$s_2 = 2$, $f = 0$ $s_2 = 0$, $f = 6$ $s_2 = 0$, $f = 8$

21. (a)

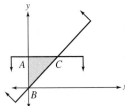

(b) $A: x = 0, y = 3, s_1 = 3, s_2 = 0, f = 6$
$B: x = 0, y = 0, s_1 = 0, s_2 = 3, f = 0$
$C: x = 3, y = 3, s_1 = 0, s_2 = 0, f = 21$

(c)
$$\left[\begin{array}{ccccc|c} ① & -1 & 1 & 0 & 0 & 0 \\ 0 & 1 & 0 & 1 & 0 & 3 \\ -5 & -2 & 0 & 0 & 1 & 0 \end{array}\right] \rightarrow \left[\begin{array}{ccccc|c} 1 & -1 & 1 & 0 & 0 & 0 \\ 0 & ① & 0 & 1 & 0 & 3 \\ 0 & -7 & 5 & 0 & 1 & 0 \end{array}\right] \rightarrow \left[\begin{array}{ccccc|c} 1 & 0 & 1 & 1 & 0 & 3 \\ 0 & 1 & 0 & 1 & 0 & 3 \\ 0 & 0 & 5 & 7 & 1 & 21 \end{array}\right]$$

$x = 0, y = 0, s_1 = 0,$ $x = 0, y = 0, s_1 = 0,$ $x = 3, y = 3, s_1 = 0,$
$s_2 = 3, f = 0$ $s_2 = 3, f = 0$ $s_2 = 0, f = 21$

23. $x = \frac{24}{7}, y = \frac{36}{7}, f = \frac{144}{7}$ **25.** $x = 0, y = 0, z = 15, f = 45$ **27.** $x = 20, y = 0, z = 0, f = 100$ **29.** $x = 7, y = \frac{3}{2},$
$z = 0, P = 17$ or $x = 0, y = \frac{17}{2}, z = 0, P = 17$ **31.** $x = 8, y = 20, z = 0, f = 76$ **33.** No solution   **35.** $x = 0,$
$y = 10, z = 0, w = 0, f = 40$ **37.** $x_1 = 1, x_2 = 0, x_3 = \frac{3}{4}, x_4 = 0, x_5 = 0, f = \frac{11}{4}$ **39.** 20 3-speed, 60 10-speed, $P = \$5200$
41. 0 T-F, 30 problem solving, total points 240   **43.** 40 brand X, 0 brand Y, number bought 40   **45.** 0 radio ads, 110
newspaper ads, 8 TV ads, people reached 100,000   **47.** 0 super deluxe, 20 deluxe, 80 standard, $P = \$1280$ **49.** 5 top star, 0
faded star, 0 local, attendance 40,000; surplus in advertising of \$1500.   **51.** 410 QP20 models, 0 QP40 models, 0 QP60 models,
20 QP100 models. Revenue: \$205,000

4.6 EXERCISES (page 206)

1. (a)

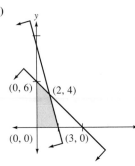

(b)
$$\left[\begin{array}{ccccc|c} 1 & 1 & 1 & 0 & 0 & 6 \\ ④ & 1 & 0 & 1 & 0 & 12 \\ -2 & -1 & 0 & 0 & 1 & 0 \end{array}\right] \rightarrow \left[\begin{array}{ccccc|c} 0 & ③/④ & 1 & -\frac{1}{4} & 0 & 3 \\ 1 & \frac{1}{4} & 0 & \frac{1}{4} & 0 & 3 \\ 0 & -\frac{1}{2} & 0 & \frac{1}{2} & 1 & 6 \end{array}\right] \rightarrow \left[\begin{array}{ccccc|c} 0 & 1 & \frac{4}{3} & -\frac{1}{3} & 0 & 4 \\ 1 & 0 & -\frac{1}{3} & \frac{1}{3} & 0 & 2 \\ 0 & 0 & \frac{2}{3} & \frac{1}{3} & 1 & 8 \end{array}\right]$$

$x = 0, y = 0, f = 0$ $x = 3, y = 0, f = 6$ $x = 2, y = 4, f = 8$

(c)

Corner	(0, 6)	(0, 0)	(3, 0)	(2, 4)
f	6	0	6	8

3. (a)

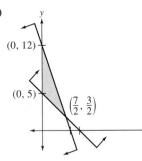

(b)
$$\left[\begin{array}{ccccc|c} -1 & -1 & 1 & 0 & 0 & -5 \\ 3 & ① & 0 & 1 & 0 & 12 \\ -4 & -2 & 0 & 0 & 1 & 0 \end{array}\right] \rightarrow \left[\begin{array}{ccccc|c} 2 & 0 & 1 & 1 & 0 & 7 \\ 3 & 1 & 0 & 1 & 0 & 12 \\ 2 & 0 & 0 & 2 & 1 & 24 \end{array}\right]$$

$x = 0, y = 0, f = 0$ $x = 0, y = 12, f = 24$

(c)

Corner	(0, 12)	(0, 5)	$\left(\frac{7}{2}, \frac{3}{2}\right)$
f	24	10	17

5. (a)

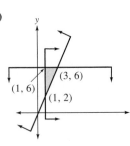

(b)
$$\left[\begin{array}{cccccc|c} ② & -1 & 1 & 0 & 0 & 0 & 0 \\ -1 & 0 & 0 & 1 & 0 & 0 & -1 \\ 0 & 1 & 0 & 0 & 1 & 0 & 6 \\ -2 & -3 & 0 & 0 & 0 & 1 & 0 \end{array}\right] \rightarrow \left[\begin{array}{cccccc|c} 1 & -\frac{1}{2} & \frac{1}{2} & 0 & 0 & 0 & 0 \\ 0 & -\frac{1}{2} & \frac{1}{2} & 1 & 0 & 0 & -1 \\ 0 & ① & 0 & 0 & 1 & 0 & 6 \\ 0 & -4 & 1 & 0 & 0 & 1 & 0 \end{array}\right] \rightarrow \left[\begin{array}{cccccc|c} 1 & 0 & \frac{1}{2} & 0 & \frac{1}{2} & 0 & 3 \\ 0 & 0 & \frac{1}{2} & 1 & \frac{1}{2} & 0 & 2 \\ 0 & 1 & 0 & 0 & 1 & 0 & 6 \\ 0 & 0 & 1 & 0 & 4 & 1 & 24 \end{array}\right]$$

$x = 0, y = 0, f = 0$ $x = 0, y = 0, f = 0$ $x = 3, y = 6, f = 24$

(c)

Corner	(1, 6)	(1, 2)	(3, 6)
f	20	8	24

7. (a)

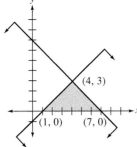

Points labeled: $(4, 3)$, $(1, 0)$, $(7, 0)$

(b)
$$\begin{bmatrix} -1 & 1 & 1 & 0 & 0 & | & -1 \\ ① & 1 & 0 & 1 & 0 & | & 7 \\ 4 & -5 & 0 & 0 & 1 & | & 0 \end{bmatrix} \rightarrow \begin{bmatrix} 0 & ② & 1 & 1 & 0 & | & 6 \\ 1 & 1 & 0 & 1 & 0 & | & 7 \\ 0 & -9 & 0 & -4 & 1 & | & -28 \end{bmatrix} \rightarrow \begin{bmatrix} 0 & 1 & \frac12 & \frac12 & 0 & | & 3 \\ 1 & 0 & -\frac12 & \frac12 & 0 & | & 4 \\ 0 & 0 & \frac92 & \frac12 & 1 & | & -1 \end{bmatrix}$$

$x = 0, y = 0, f = 0$ — $x = 7, y = 0, f = -28$ — $x = 4, y = 3, f = -1$

(c)

Corner	(1, 0)	(7, 0)	(4, 3)
f	-4	-28	-1

9. (a)

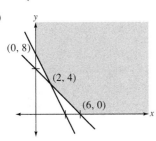

Points labeled: $(0, 8)$, $(2, 4)$, $(6, 0)$

(b)
$$\begin{bmatrix} -1 & -1 & 1 & 0 & 0 & | & -6 \\ -2 & (-1) & 0 & 1 & 0 & | & -8 \\ 1 & 3 & 0 & 0 & 1 & | & 0 \end{bmatrix} \rightarrow \begin{bmatrix} ① & 0 & 1 & -1 & 0 & | & 2 \\ 2 & 1 & 0 & -1 & 0 & | & 8 \\ -5 & 0 & 0 & 3 & 1 & | & -24 \end{bmatrix} \rightarrow$$

$x = 0, y = 0, f = 0$ — $x = 0, y = 8, f = 24$

$$\begin{bmatrix} 1 & 0 & 1 & -1 & 0 & | & 2 \\ 0 & 1 & -2 & ① & 0 & | & 4 \\ 0 & 0 & 5 & -2 & 1 & | & -14 \end{bmatrix} \rightarrow \begin{bmatrix} 1 & 1 & -1 & 0 & 0 & | & 6 \\ 0 & 1 & -2 & 1 & 0 & | & 4 \\ 0 & 2 & 1 & 0 & 1 & | & -6 \end{bmatrix}$$

$x = 2, y = 4, f = 14$ — $x = 6, y = 0, f = 6$

(c)

Corner	(0, 8)	(2, 4)	(6, 0)
f	24	14	6

11. (a)

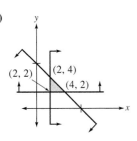

Points labeled: $(2, 2)$, $(2, 4)$, $(4, 2)$

(b)
$$\begin{bmatrix} ① & 1 & 1 & 0 & 0 & 0 & | & 6 \\ -1 & 0 & 0 & 1 & 0 & 0 & | & -2 \\ 0 & -1 & 0 & 0 & 1 & 0 & | & -2 \\ 1 & 1 & 0 & 0 & 0 & 1 & | & 0 \end{bmatrix} \rightarrow \begin{bmatrix} 1 & 1 & 1 & 0 & 0 & 0 & | & 6 \\ 0 & ① & 1 & 1 & 0 & 0 & | & 4 \\ 0 & -1 & 0 & 0 & 1 & 0 & | & -2 \\ 0 & 0 & -1 & 0 & 0 & 1 & | & -6 \end{bmatrix} \rightarrow$$

$x = 0, y = 0, f = 0$ — $x = 6, y = 0, f = 6$

$$\begin{bmatrix} 1 & 0 & 0 & -1 & 0 & 0 & | & 2 \\ 0 & 1 & 1 & 1 & 0 & 0 & | & 4 \\ 0 & 0 & ① & 1 & 1 & 0 & | & 2 \\ 0 & 0 & -1 & 0 & 0 & 1 & | & -6 \end{bmatrix} \rightarrow \begin{bmatrix} 1 & 0 & 0 & -1 & 0 & 0 & | & 2 \\ 0 & 1 & 0 & 0 & -1 & 0 & | & 2 \\ 0 & 0 & 1 & 1 & 1 & 0 & | & 2 \\ 0 & 0 & 0 & 1 & 1 & 1 & | & -4 \end{bmatrix}$$

$x = 2, y = 4, f = 6$ — $x = 2, y = 2, f = 4$

(c)

Corner	(2, 4)	(2, 2)	(4, 2)
f	6	4	6

13. (a)

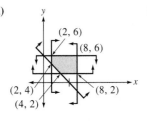

Points labeled: $(2, 6)$, $(8, 6)$, $(2, 4)$, $(4, 2)$, $(8, 2)$

(b)
$$\begin{bmatrix} -1 & -1 & 1 & 0 & 0 & 0 & 0 & 0 & | & -6 \\ -1 & 0 & 0 & 1 & 0 & 0 & 0 & 0 & | & -2 \\ 1 & 0 & 0 & 0 & 1 & 0 & 0 & 0 & | & 8 \\ 0 & -1 & 0 & 0 & 0 & 1 & 0 & 0 & | & -2 \\ 0 & ① & 0 & 0 & 0 & 0 & 1 & 0 & | & 6 \\ 1 & -2 & 0 & 0 & 0 & 0 & 0 & 1 & | & 0 \end{bmatrix} \rightarrow \begin{bmatrix} -1 & 0 & 1 & 0 & 0 & 0 & 1 & 0 & | & 0 \\ -1 & 0 & 0 & 1 & 0 & 0 & 0 & 0 & | & -2 \\ ① & 0 & 0 & 0 & 1 & 0 & 0 & 0 & | & 8 \\ 0 & 0 & 0 & 0 & 0 & 1 & 1 & 0 & | & 4 \\ 0 & 1 & 0 & 0 & 0 & 0 & 1 & 0 & | & 6 \\ 1 & 0 & 0 & 0 & 0 & 0 & 2 & 1 & | & 12 \end{bmatrix}$$

$x = 0, y = 0, C = 0$ — $x = 0, y = 6, C = -12$

$$\begin{bmatrix} 0 & 0 & 1 & 0 & 1 & 0 & 1 & 0 & | & 8 \\ 0 & 0 & 0 & 1 & ① & 0 & 0 & 0 & | & 6 \\ 1 & 0 & 0 & 0 & 1 & 0 & 0 & 0 & | & 8 \\ 0 & 0 & 0 & 0 & 0 & 1 & 1 & 0 & | & 4 \\ 0 & 1 & 0 & 0 & 0 & 0 & 1 & 0 & | & 6 \\ 0 & 0 & 0 & 0 & -1 & 0 & 2 & 1 & | & 4 \end{bmatrix} \rightarrow \begin{bmatrix} 0 & 0 & 1 & -1 & 0 & 0 & 1 & 0 & | & 2 \\ 0 & 0 & 0 & 1 & 1 & 0 & 0 & 0 & | & 6 \\ 1 & 0 & 0 & -1 & 0 & 0 & 0 & 0 & | & 2 \\ 0 & 0 & 0 & 0 & 0 & 1 & 1 & 0 & | & 4 \\ 0 & 1 & 0 & 0 & 0 & 0 & 1 & 0 & | & 6 \\ 0 & 0 & 0 & 1 & 0 & 0 & 2 & 1 & | & 10 \end{bmatrix}$$

$x = 8, y = 6, C = -4$ — $x = 2, y = 6, C = -10$

(c)

Corner	(2, 6)	(2, 4)	(4, 2)	(8, 2)	(8, 6)
C	-10	-6	0	4	-4

15. (a)

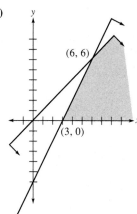

Graph with points (6, 6) and (3, 0).

(b)

$$\left[\begin{array}{ccccc|c} -1 & 1 & 1 & 0 & 0 & 0 \\ \boxed{-2} & 1 & 0 & 1 & 0 & -6 \\ 2 & 1 & 0 & 0 & 1 & -10 \end{array}\right] \rightarrow \left[\begin{array}{ccccc|c} 0 & \frac{1}{2} & 1 & -\frac{1}{2} & 0 & 3 \\ 1 & -\frac{1}{2} & 0 & -\frac{1}{2} & 0 & 3 \\ 0 & 2 & 0 & 1 & 1 & -16 \end{array}\right]$$

$x = 0, y = 0, f = 10$ $x = 3, y = 0, f = 16$

(c)

Corner	(3, 0)	(6, 6)
f	16	28

17. $x = 4, y = 4, z = 2, C = 80$ **19.** $x = 4, y = 0, z = 2, f = 24$ **21.** $x = 1, y = \frac{9}{2}, z = 0, P = 21$ **23.** $x = 0,$ $y = 3, z = 0, C = 9$ **25.** $x = 0, y = 3, z = \frac{9}{2}, f = \frac{21}{2}$ **27.** $x = \frac{3}{2}, y = 0, z = \frac{3}{2}, C = \frac{3}{2}$ **29.** $x_1 = 6, x_2 = 2, x_3 = 0,$ $f = 26$ **31.** 80 Daybrite, 220 Noglare, $R = \$154,600$ **33.** $\frac{97}{10}$ civic club speeches, 3 hours on phone, $\frac{97}{5}$ hours in shopping areas for a total of $\frac{109}{4}$ hours; or no civic club speeches, 3 hours on phone and $\frac{97}{4}$ hours in shopping areas for a total of $\frac{109}{4}$ hours. **35.** 3 doses of A, 5 doses of B, 2 doses of C for 24.5 effectiveness   **37.** 40 fax machines, 40 multimedia computers, 20 CD players for $R = \$112,000$ **39.** 1000 radial, 4200 mud and snow, 100 off-road for $R = \$152,000$ or 1000 radials, 400 mud and snow, 2950 off-road for R = \$152,000   **41.** 6 pounds of barley, 3 pounds of corn, 9 units of protein.   **43.** Logansport . . . to Lafayette 60, . . . to Marion 40, . . . to Portland 0. Lebanon . . . to Lafayette 0, . . . to Marion 40, . . . to Portland 60; total cost: $440   **45.** Mansfield . . . to Weirton 20, . . . to Delaware 8, . . . to Kenton 0. Athens . . . to Weirton 0, . . . to Delaware 7, . . . to Kenton 10; total cost: $52 or Mansfield . . . to Weirton 13, . . . to Delaware 15, to Kenton 0. Athens . . . to Weirton 7, . . . to Delaware 0, . . . to Kenton 10; total cost: $52

4.7 EXERCISES (page 219)

1. (a) $\left[\begin{array}{cc|c} 1 & 1 & 4 \\ 3 & 1 & 8 \\ 2 & 1 & 0 \end{array}\right]$ **(b)** $\left[\begin{array}{cc|c} 1 & 3 & 2 \\ 1 & 1 & 1 \\ 4 & 8 & 0 \end{array}\right]$ **(c)** Maximize $d = 4u + 8v$, subject to $u + 3v \le 2$
$u + v \le 1$
$u \ge 0, v \ge 0$

	A	B	C
Corner	(0, 8)	(2, 2)	(4, 0)
Min $f = 2x + y$	8	6	8

	P	Q	R	S
Corner	$(0, \frac{2}{3})$	(0, 0)	(1, 0)	$(\frac{1}{2}, \frac{1}{2})$
Max $d = 4u + 8v$	$\frac{16}{3}$	0	4	6

(d)

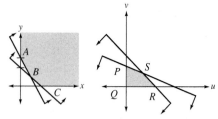

(e) $\left[\begin{array}{ccccc|c} 1 & \boxed{3} & 1 & 0 & 0 & 2 \\ 1 & 1 & 0 & 1 & 0 & 1 \\ -4 & -8 & 0 & 0 & 1 & 0 \end{array}\right] \rightarrow \left[\begin{array}{ccccc|c} \frac{1}{3} & 1 & \frac{1}{3} & 0 & 0 & \frac{2}{3} \\ \boxed{\frac{2}{3}} & 0 & -\frac{1}{3} & 1 & 0 & \frac{1}{3} \\ -\frac{4}{3} & 0 & \frac{8}{3} & 0 & 1 & \frac{16}{3} \end{array}\right] \rightarrow \left[\begin{array}{ccccc|c} 0 & 1 & \frac{1}{2} & -\frac{1}{2} & 0 & \frac{1}{2} \\ 1 & 0 & -\frac{1}{2} & \frac{3}{2} & 0 & \frac{1}{2} \\ 0 & 0 & 2 & 2 & 1 & 6 \end{array}\right]$

$u = \frac{1}{2}, v = \frac{1}{2}, d = 6; x = 2, y = 2, f = 6$

3. (a) $\begin{bmatrix} 1 & 1 & 10 \\ 2 & 1 & 12 \\ \hline 3 & 1 & 0 \end{bmatrix}$ **(b)** $\begin{bmatrix} 1 & 2 & 3 \\ 1 & 1 & 1 \\ \hline 10 & 12 & 0 \end{bmatrix}$ **(c)** Maximize $d = 10u + 12v$, **(d)**
subject to $u + 2v \leq 3$
$u + v \leq 1$
$u \geq 0, v \geq 0$

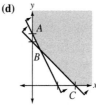

	A	B	C
Corner	$(0, 12)$	$(2, 8)$	$(10, 0)$
Min $C = 3x + y$	12	14	30

	P	Q	R
Corner	$(0, 1)$	$(0, 0)$	$(1, 0)$
Max $d = 10u + 12v$	12	0	10

(e) $\begin{bmatrix} 1 & 2 & 1 & 0 & 0 & 3 \\ 1 & ① & 0 & 1 & 0 & 1 \\ \hline -10 & -12 & 0 & 0 & 1 & 0 \end{bmatrix} \rightarrow \begin{bmatrix} -1 & 0 & 1 & -2 & 0 & 1 \\ 1 & 1 & 0 & 1 & 0 & 1 \\ \hline 2 & 0 & 0 & 12 & 1 & 12 \end{bmatrix}$ $u = 0, v = 1, d = 12$
$x = 0, y = 12, C = 12$

5. (a) $\begin{bmatrix} 4 & 1 & 18 \\ -1 & -2 & -8 \\ \hline 2 & -3 & 0 \end{bmatrix}$ **(b)** $\begin{bmatrix} 4 & -1 & 2 \\ 1 & -2 & -3 \\ \hline 18 & -8 & 0 \end{bmatrix}$

(c) Maximize $d = 18u - 8v$, **(d)**
subject to $4u - v \leq 2$
$u - 2v \leq -3$
$u \geq 0, v \geq 0$

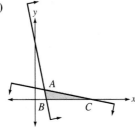

	A	B	C
Corner	$(4, 2)$	$(\frac{9}{2}, 0)$	$(8, 0)$
Min $f = 2x - 3y$	2	9	16

				Boundaries	
	P	Q	$u = \frac{1}{4}v + \frac{1}{2}$	$u = 0$
Corner	$(0, \frac{3}{2})$	$(1, 2)$		
Max $d = 18u - 8v$	-12	2	$9 - \frac{7}{2}v \rightarrow -\infty$	$-8v \rightarrow -\infty$

(e) $\begin{bmatrix} 4 & -1 & 1 & 0 & 0 & 2 \\ 1 & ⊖2 & 0 & 1 & 0 & -3 \\ \hline -18 & 8 & 0 & 0 & 1 & 0 \end{bmatrix} \rightarrow \begin{bmatrix} ⑦⁄₂ & 0 & 1 & -\frac{1}{2} & 0 & \frac{7}{2} \\ -\frac{1}{2} & 1 & 0 & -\frac{1}{2} & 0 & \frac{3}{2} \\ \hline -14 & 0 & 0 & 4 & 1 & -12 \end{bmatrix} \rightarrow \begin{bmatrix} 1 & 0 & \frac{2}{7} & -\frac{1}{7} & 0 & 1 \\ 0 & 1 & \frac{1}{7} & -\frac{4}{7} & 0 & 2 \\ \hline 0 & 0 & 4 & 2 & 1 & 2 \end{bmatrix}$ $u = 1, v = 2, d = 2$
$x = 4, y = 2, f = 2$

7. (a) $\begin{bmatrix} 1 & 1 & 8 \\ 1 & 0 & 1 \\ \hline -2 & 1 & 0 \end{bmatrix}$ **(b)** $\begin{bmatrix} 1 & 1 & -2 \\ 1 & 0 & 1 \\ \hline 8 & 1 & 0 \end{bmatrix}$

(c) Minimize $d = 8u + v$,
subject to $u + v \geq -2$
$u \qquad \geq 1$
$u \geq 0, v \geq 0$

(d)

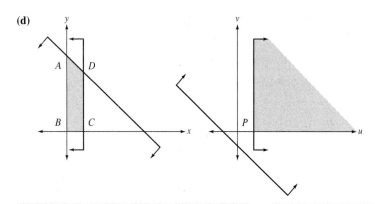

	A	B	C	D
Corner	(0, 8)	(0, 0)	(1, 0)	(1, 7)
Max $f = -2x + y$	8	0	-2	5

	P	Boundaries	
Corner	(1, 0)	$v = 0$	$u = 1$
Min $d = 8u + v$	8	$8u \to \infty$	$v + 8 \to \infty$

(e)
$$\begin{bmatrix} -1 & -1 & 1 & 0 & 0 & | & 2 \\ \boxed{-1} & 0 & 0 & -1 & 0 & | & -1 \\ 8 & 1 & 0 & 0 & 1 & | & 0 \end{bmatrix} \to \begin{bmatrix} 0 & -1 & 1 & -1 & 0 & | & 3 \\ 1 & 0 & 0 & -1 & 0 & | & 1 \\ 0 & 1 & 0 & 8 & 1 & | & -8 \end{bmatrix}$$
$u = 1, v = 0, d = 8$
$x = 0, y = 8, f = 8$

9. Minimum value of f is 8 when $x = 8$ and $y = 0$.   **11.** Minimum value of f is $\frac{52}{5}$ when $x = \frac{9}{5}$, $y = \frac{21}{5}$, and $z = \frac{2}{5}$.
13. Minimum value of f is 8 when $x = 8$, $y = 0$, and $z = 0$.   **15.** Maximum value of f is 20 when $x = 10$, $y = 0$, and $z = 0$.
17. $\frac{20}{11}$ servings of A and $\frac{5}{11}$ servings of B give 5 mg of fat or $\frac{5}{2}$ servings of A and 0 servings of B give 5 mg of fat.
19. 2 barrels of A, 2 barrels of B, and 0 barrels of C give $16   **21.** 175 cassette tapes and 75 compact discs give a cost of $1175
23. 160 standard, 50 deluxe, and 90 super deluxe give a cost of $33,700

CHAPTER 4 REFLECTIONS (page 221)

1. $10x + 3y \le 30$
    $7x + 9y \le 63$
    $x \ge 0, y \ge 0$

3.

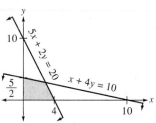

5.

Corner points	(0, 2)	(1, 1)	(3, 0)
$C = 3x + 2y$	4	5	9

Minimum of C is 4 when $x = 0$ and $y = 2$.

7. A: $x = 0, y = 5, s_1 = 0, s_2 = 2$
    B: $x = 0, y = 0, s_1 = 5, s_2 = 12$
    C: $x = 4, y = 0, s_1 = 1, s_2 = 0$
    D: $x = 2, y = 3, s_1 = 0, s_2 = 0$

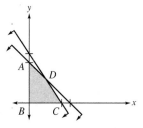

9. $\begin{bmatrix} 1 & 1 & 1 & 0 & 0 & | & 5 \\ ③ & 2 & 0 & 1 & 0 & | & 12 \\ -4 & -3 & 0 & 0 & 1 & | & 0 \end{bmatrix} \to \begin{bmatrix} 0 & ⓵ & 1 & -\frac{1}{3} & 0 & | & 1 \\ 1 & \frac{2}{3} & 0 & \frac{1}{3} & 0 & | & 4 \\ 0 & -\frac{1}{3} & 0 & \frac{4}{3} & 1 & | & 16 \end{bmatrix} \to \begin{bmatrix} 0 & 1 & 3 & -1 & 0 & | & 3 \\ 1 & 0 & -2 & 1 & 0 & | & 2 \\ 0 & 0 & 1 & 1 & 1 & | & 17 \end{bmatrix}$

$x = 2, 6 = 3, P = 17$

11. $\begin{bmatrix} ① & 4 & 1 & 1 & 0 & 0 & 0 & | & 12 \\ 0 & 1 & 4 & 0 & 1 & 0 & 0 & | & 8 \\ -4 & 0 & -1 & 0 & 0 & 1 & 0 & | & -4 \\ -2 & 1 & -1 & 0 & 0 & 0 & 1 & | & 0 \end{bmatrix} \to \begin{bmatrix} 1 & 4 & 1 & 1 & 0 & 0 & 0 & | & 12 \\ 0 & 1 & 4 & 0 & 1 & 0 & 0 & | & 8 \\ 0 & 16 & 3 & 4 & 0 & 1 & 0 & | & 44 \\ 0 & 9 & 1 & 2 & 0 & 0 & 1 & | & 24 \end{bmatrix}$

$x = 12, y = 0, z = 0, f = 24$

13. Make no sugar cookies, 120 batches of peanut butter cookies, and 75 batches of pecan sandies for a profit of $378.75.

15. Graph the boundary lines, shade in the proper feasible region, find the corner points of this region, then evaluate the objective function at each corner point.   **17.** After locating the pivot column, divide each positive number above the horizontal line into the corresponding number in the last column of the simplex tableau. The smallest nonnegative quotient then determines the pivot row.

19. Start with the first negative number in the last column of the simplex tableau. In the row that it is located, find the largest negative number to the left of the vertical bar. In that column, divide each number above the horizontal line into the corresponding number in the rightmost column. If it exists, the smallest nonnegative/positive quotient determines the pivot column. Otherwise the largest negative/negative quotient determines the pivot column.   **21.** A possibility: To make an economy stapler requires 1 minute in a bender, 4 minutes in assembly, and 5 minutes in painting. To make a deluxe stapler requires 2 minutes in a bender, 5 minutes in assembly, and 1 minute in painting. To make staplers, the company has available each hour 12 minutes for the bender, 20 minutes in assembly, and 10 minutes in painting. If each economy stapler sells for $3 and each deluxe stapler sells for $4, how many of each should be made each hour for maximum profit?   **23.** The minimum value of f is 9 when $x = 3$ and $y = 3$.

25. The minimum value of C is 9 when $x = 0$, $y = 3$, and $z = 0$.   **27.** $A^{-1} = \begin{bmatrix} 1 & -1 \\ 0 & \frac{1}{2} \end{bmatrix}$; You get an A.

29. $x = 2, y = -3, z = 6, w = 1$

SAMPLE TEST ITEMS FOR CHAPTER 4 (page 224)

1.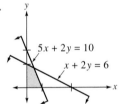

2.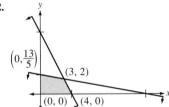

3. Let x = number of correct true-false questions.
Let y = number of correct multiple choice questions.

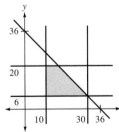

Maximize $S = 3x + 5y$
subject to $x + y \le 36$
$10 \le x \le 30$
$6 \le y \le 20$

$x = 16, y = 20$; maximum score 148

4. The maximum occurs at $(2, 10)$ and the minimum occurs at $(7, 2)$.

5.

Corner points	$(0, 2)$	$(0, 0)$	$(3, 0)$	$(\frac{171}{61}, \frac{84}{61})$
$f = 2x + 5y$	10	0	6	$\frac{762}{61}$

The maximum is $\frac{762}{61}$ when $x = \frac{171}{61}$ and $y = \frac{84}{61}$.

6.

Corner points	$(0, 6)$	$(\frac{4}{3}, 2)$	$(4, 0)$	$x = 0$	$y = 0$
$g = 4x - y$	-6	$\frac{10}{3}$	16	$-y \to -\infty$	$4x \to \infty$

There is no minimum. Along $x = 0$, $y \to -\infty$ so $g \to -\infty$.

7. $2x + y \le 8$
$x + 5y \le 10$
$x \ge 0, y \ge 0$

8. $\begin{bmatrix} ② & 1 & 1 & 0 & | & 8 \\ 1 & 5 & 0 & 1 & | & 10 \end{bmatrix} \rightarrow \begin{bmatrix} 1 & \frac{1}{2} & \frac{1}{2} & 0 & | & 4 \\ 0 & \frac{9}{2} & -\frac{1}{2} & 1 & | & 6 \end{bmatrix}$

9. No

10. $f = 35; x = 2, y = 0, s_1 = 8, s_2 = 4, s_3 = 0$

11. s_1; the surplus is 8 units. s_2; the surplus is 4 units.

12. $A: x = 0, y = 4, s_1 = 2, s_2 = 0$
$B: x = 0, y = 0, s_1 = 6, s_2 = 4$
$C: x = 6, y = 0, s_1 = 0, s_2 = 4$
$D: x = 2, y = 4, s_1 = 0, s_2 = 0$

13. $\begin{bmatrix} ① & 1 & 1 & 0 & 0 & | & 6 \\ 0 & 1 & 0 & 1 & 0 & | & 4 \\ -2 & -1 & 0 & 0 & 1 & | & 0 \end{bmatrix} \rightarrow \begin{bmatrix} 1 & 1 & 1 & 0 & 0 & | & 6 \\ 0 & 1 & 0 & 1 & 0 & | & 4 \\ 0 & 1 & 2 & 0 & 1 & | & 12 \end{bmatrix}$ Solution: $f = 12$ when $x = 6$ and $y = 0$.

14. Maximize $d = 3u + 4v + w$,
subject to $u + v + w \le 2$
$u + 2v \le 3$
$u \ge 0, v \ge 0, w \ge 0$

15.
Solution: $u = 1, v = 1, w = 0, d = 7$,
$x = 2, y = 1, f = 7$

16.

17. $\begin{bmatrix} -1 & -1 & 1 & 0 & 0 & | & -7 \\ -2 & ⊖1 & 0 & 1 & 0 & | & -10 \\ 3 & 1 & 0 & 0 & 1 & | & 0 \end{bmatrix} \rightarrow \begin{bmatrix} 1 & 0 & 1 & -1 & 0 & | & 3 \\ 2 & 1 & 0 & -1 & 0 & | & 10 \\ 1 & 0 & 0 & 1 & 1 & | & -10 \end{bmatrix}$
Solution: $x = 0, y = 10, f = 10$

18.

Corner points	(0, 10)	(3, 4)	(7, 0)	$x = 0$	$y = 0$
$f = 3x + y$	10	13	21	$y \to \infty$	$3x \to \infty$

19. $x = 0, y = 6, z = 0, f = 18$ **20.** 25 professors, 10 lecturers for a total cost of $1,150,000

CHAPTER 5

5.1 EXERCISES (page 232)

1. The student is a male or a sophomore.   **3.** The student is a male and a sophomore.   **5.** The student is not a male.
7. The blouse is red and the belt is not black.   **9.** The blouse is not red or the belt is black.   **11.** The blouse is red and the belt is black.   **13.** The die does not show an odd number and shows a five.   **15.** The die does not show an odd number or does not show a five.   **17.** $p \wedge q$ **19.** $p \wedge q$ **21.** $p \wedge \sim q$ **23.** $\sim p \vee q$

25.

p	q	$\sim q$	$p \vee \sim q$
T	T	F	T
T	F	T	T
F	T	F	F
F	F	T	T

27.

p	q	$\sim p$	$\sim q$	$\sim p \vee \sim q$
T	T	F	F	F
T	F	F	T	T
F	T	T	F	T
F	F	T	T	T

29.

p	q	$\sim p$	$\sim p \vee q$	$\sim(p \vee q)$
T	T	F	T	F
T	F	F	F	T
F	T	T	T	F
F	F	T	T	F

31.

p	q	$p \vee q$	$\sim(p \vee q)$	$\sim p$	$\sim q$	$\sim p \wedge \sim q$
T	T	T	F	F	F	F
T	F	T	F	F	T	F
F	T	T	F	T	F	F
F	F	F	T	T	T	T

33.

p	q	r	$q \vee r$	$p \wedge (q \vee r)$	$p \wedge q$	$p \wedge r$	$(p \wedge q) \vee (p \wedge r)$
T	T	T	T	T	T	T	T
T	T	F	T	T	T	F	T
T	F	T	T	T	F	T	T
T	F	F	F	F	F	F	F
F	T	T	T	F	F	F	F
F	T	F	T	F	F	F	F
F	F	T	T	F	F	F	F
F	F	F	F	F	F	F	F

35. Not equivalent   **37.** Not equivalent

39. T   **41.** T.   **43.** F.   **45.** T.   **47.** Neither of the two marbles is red.   **49.** At most one of the three students graduated with honors.   **51.** Four marbles are green.   **53.** The card is not red, and does not have a CD-player.   **55.** The blouse is blue, or the belt is not black.   **57.** Four students are taking a mathematics course and one is not, or all five are taking a mathematics course.   **59.** (Four shirts have flaws and two do not) or (five shirts have flaws and one does not) or (all six shirts have flaws).   **61.** (Three are red marbles and 23 are not) or (two are red marbles and 24 are not) or (one is a red marble and 25 are not) or (all 26 marbles are not red).

5.2 EXERCISES (page 241)

1. {0, 4, 5, 6, 7}   **3.** U   **5.** {5}   **7.** {0, 4, 5}   **9.** {2, 3}   **11.** {a, b, e, f}   **13.** {c}   **15.** ∅   **17.** {a, b, e, f}
19. A   **21.** 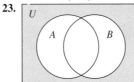   **23.**   **25.**

27. 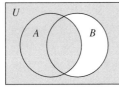   **29.** True   **31.** True   **33.** True   **35.** The number is odd and divisible by 3.
37. The number is odd or is not divisible by 3.   **39.** The number is odd, but not divisible by 3; the number is odd and not divisible by 3.   **41.** The person is at most 40 years of age.
43. The person wears glasses and is over 40 years of age.   **45.** The person wears glasses and is not over 40 years of age.

47.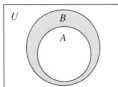

G'

49.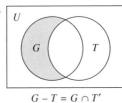

$G - T = G \cap T'$

51.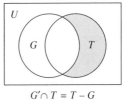

$G' \cap T = T - G$

53.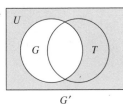

$G' \cap T' = (G \cup T)'$

55.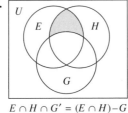

$E \cap H \cap G' = (E \cap H) - G$

57.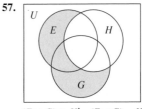

$(E \cup G) \cap H' = (E \cup G) - H$

59.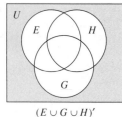

$(E \cup G \cup H)'$

61.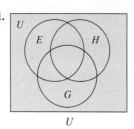

U

5.3 EXERCISES (page 248)

1.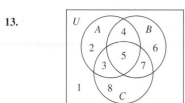

3.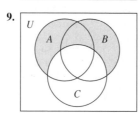

5.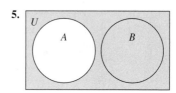

7.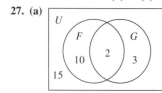

9.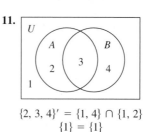

11.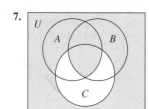

$\{2, 3, 4\}' = \{1, 4\} \cap \{1, 2\}$
$\{1\} = \{1\}$

13.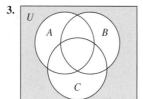

$\{6, 7, 8\} - \{4, 5, 6, 7\} = \{1, 8\} \cap \{3, 5, 7, 8\}$
$\{8\} = \{8\}$

15. Equal   **17.** Not Equal   **19.** Not Equal
21. 17   **23.** 5   **25.** 4

27. (a)

(b) $n(F \cap G') = 10$ **(c)** $n(F \cup G) = 15$ **(d)** $n(F \cup G) = 15$
(e) $n[(F - G) \cup (G - F)] = 13$ **(f)** $n((F \cup G)') = 15$
(g) $n(F \cup G) = 12 + 5 - 2 = 15$

29. (a) 59   **(b)** 41   **(c)** 21   **(d)** 7; .75%   **31. (a)** 80   **(b)** 30   **(c)** 30   **(d)** 6.67%   **33. (a)** 39   **(b)** 12   **(c)** 27   **(d)** 77
(e) 54   **35. (a)** 105   **(b)** 123   **(c)** 50   **(d)** 18   **37. (a)** 3   **(b)** 8   **(c)** 7   **39.** No.   **41. (a)** 75   **(b)** 40   **43.** A

5.4 EXERCISES (page 258)

1. 6   **3.** 11   **5.** 16; *HHHH, HHHT, HHTH, HTHH, THHH, HHTT, HTTH, TTHH, THHT, THTH, HTHT, TTTH, TTHT,*
THTT, HTTT, TTTT   **7.** 12, 33, 32, 31, 30, 23, 22; 21, 20, 13, 12, 11, 10   **9.** 3; $\{(E_1, L_1), (E_2, L_3), (E_3, L_2)\}$, $\{(E_1, L_3),$
$(E_2, L_2), (E_3, L_1)\}$, $\{(E_1, L_2), (E_2, L_1), (E_3, L_3)\}$ **11.** 12; $S_1F_1I_1, S_1F_1I_2, S_1F_2I_1, S_1F_2I_2, S_2F_1I_1, S_2F_1I_2, S_2F_2I_1, S_2F_2I_2, S_3F_1I_1,$
$S_3F_1I_2, S_3F_2I_1, S_3F_2I_2$ **13. (a)** 1,000,000,000   **(b)** 900,000,000   **(c)** 810,000,000   **15. (a)** 40,320   **(b)** 0   **(c)** 4320
17. (a) 40,320   **(b)** 5040   **19. (a)** 10,000   **(b)** 4536   **(c)** 1400   **21.** 4; 32; 128   **23.** 3; 5   **25.** 1024   **27.** 6
29. (a) 6,760,000   **(b)** 6,500,000   **(c)** 340,704   **31. (a)** 37,152   **(b)** 36,592   **(c)** 29,040   **33.** 657,720   **35.** 1560
37. 1680   **39. (a)** 6   **(b)** 6   **41. (a)** 100   **(b)** 1900   **43. (a)** 336   **(b)** 512

5.5 EXERCISES (page 266)

1. 720   **3.** 40,320   **5.** 1680   **7.** 792   **9.** 2,598,960   **11.** 2   **13.** 1680   **15.** 6   **17.** 5040   **19.** 1
21. 24   **23.** 1   **25. (a)** $P(8, 4)$ **(b)** 1680   **27. (a)** $P(7, 7)$ **(b)** 5040   **29.** 120   **31.** 336   **33.** 120   **35.** 24
37. 1260   **39.** 792   **41.** 60,060   **43.** 2520   **45.** 252

5.6 EXERCISES (page 273)

1. 15   **3.** 84   **5.** 105   **7.** 194,580   **9.** 3432   **11. (a)** Both are 10.   **(b)** Both are 126.   **(c)** Both are 56.
(d) $C(n, r) = C(n, n - r)$ **13.** $C(6, 0) = 1$; $C(6, 1) = 6$; $C(6, 2) = 15$; $C(6, 3) = 20$; $C(6, 4) = 15$; $C(6, 5) = 6$; $C(6, 6) = 1$
15. 10   **17.** 22,100   **19.** 5005   **21. (a)** 455   **(b)** 10   **(c)** 525   **23. (a)** 35   **(b)** 21   **(c)** 56   **25. (a)** 5   **(b)** 210
(c) 210   **(d)** 10   **27. (a)** 1287   **(b)** 5148   **(c)** 11,761,893   **(d)** 2574   **29. (a)** 10   **(b)** 30   **(c)** 40
31. (a) 14,950   **(b)** 1   **(c)** 69,667   **33. (a)** 50   **(b)** 990   **(c)** 1155   **35.** 330   **37. (a)** 134,044   **(b)** 20   **(c)** 1410
39. (a) 108   **(b)** 126   **(c)** 6   **(d)** 111   **41. (a)** 2   **(b)** 36

5.7 EXERCISES (page 277)

1. 19   **3.** 19,547,186,230   **5.** 55   **7.** 720   **9.** 5040   **11. (a)** 4   **(b)** 16   **(c)** 1   **(d)** 0   **(e)** 52   **(f)** 40
13. (a) 100,000   **(b)** 90,000   **15. (a)** 27   **(b)** 19   **17.** 56   **19.** 256   **21. (a)** 720   **(b)** 120   **(c)** 24   **23. (a)** 32
(b) 38   **(c)** 18   **25. (a)** 1,560,000   **(b)** 54,000   **(c)** 765,072   **27.** 5040   **29.** 210   **31. (a)** 1296   **(b)** 648
(c) 180   **33. (a)** 792   **(b)** 3003   **(c)** 80,730   **35. (a)** 6435   **(b)** 259,459,200   **37. (a)** 210   **(b)** 5040   **(c)** 10,000
39. (a) 28   **(b)** 34   **(c)** 27   **41.** 514,594,080   **43.** 256   **45. (a)** 75   **(b)** 780   **47. (a)** 12,429,382,108
(b) 7,630,529,764   **49. (a)** 252   **(b)** 30,240   **(c)** 100,000

CHAPTER 5 REFLECTIONS (page 280)

1. Yes   **3.** Some of the marbles are not green   **5.** {1}   **7.** {1, 2, 3}   **9.** {3}   **11.** {2, 5, 6}   **13.** $A \cup C - (A \cap C)$
15. 216   **17.** 7,334,887,680   **19.** 61,124,064   **21.** 1,362,649,145   **23.** Some Fords are not red and not blue.   **25.** 4
27. 1820   **29.** Two sentences are logically equivalent if they always have the same truth values.   **31.** Permutations and
combinations both involve selecting subsets of a given set with order being important for permutations but not important for
combinations.   **33.** If you assume that the order of selection *doesn't* matter, then n(at least one red) $= n$(exactly one red) $+$
n(exactly two red) $+ n$(exactly three red) $= 399$; also, n(at least one red) $= n$(any three) $- n$(none red) $= 399$. If you assume
order *is* important, then n(at least one red) $= n(RRR) + n(RRN) + n(RNR) + n(NRR) + n(RNN) + n(NRN) + n(NNR) = 2394$;
also, a tree diagram could be used to give 2394 ways as does the given proposed solution.   **35.** $y = 1.03x + 10.1$
37. $x =$ any number, $y = 3x + 2, z = 11x - 2$ **39.** $x = 1.5, y = 7.5, s_1 = 0, s_2 = 23.5, s_3 = 0, f = 34.5$

SAMPLE TEST ITEMS FOR CHAPTER 5 (page 281)

1. The first ball drawn is not red or the second ball drawn is blue.   **2.** The first ball drawn is not red and the second ball drawn
is not blue.   **3.** The first ball drawn is red and the second ball drawn is blue.

4.

p	q	$\sim p$	$\sim q$	$\sim q \wedge p$	$\sim p \vee (\sim q \wedge p)$
T	T	F	F	F	F
T	F	F	T	T	T
F	T	T	F	F	T
F	F	T	T	F	T

5. F   **6.** F   **7.** T   **8.** F   **9.** T   **10.** F   **11.** T
12. {4}   **13.** {2}   **14.** {∅, 3}

15.

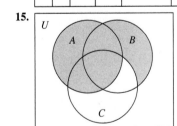

16. 0   **17.** 47   **18.** 900   **19.** 211,926   **20.** 60   **21.** 126   **22.** 70
23. 36   **24.** 384   **25.** 1023

CHAPTER 6

6.1 EXERCISES (page 292)

1. (a) $S = \{A, B, C, D\}$ **(b)** $\frac{1}{4}$ **(c)** $\frac{3}{4}$ **3. (a)** $S = \{GG, GB, BG, BB\}$ **(b)** $\frac{1}{4}$ **(c)** $\frac{3}{4}$ **(d)** $\frac{2}{4}$
5. (a) $S = \{GGG, GGB, GBG, BGG, GBB, BGB, BBG, BBB\}$ **(b)** $\frac{1}{8}$ **(c)** $\frac{7}{8}$ **(d)** $\frac{3}{8}$ **7. (a)** $S = \{LY, LM, YL, YM, ML, MY\}$
(b) $\frac{1}{6}$ **(c)** $\frac{4}{6}$ **(d)** $\frac{2}{6}$ **9. (a)** $S = \{1, 2, 3, 4\}$ **(b)** $\frac{1}{4}$ **(c)** $\frac{2}{4}$ **(d)** $\frac{1}{4}$ **11. (a)** $S = \{H1, H2, H3, H4, T1, T2, T3, T4\}$

(b) $\frac{1}{8}$ **(c)** $\frac{2}{8}$ **(d)** $\frac{2}{8}$ **13. (a)** $S = \{1, 2, 3, 4, 5, 6, 7, 8, 9\}$ **(b)** $\frac{1}{9}$ **(c)** $\frac{2}{9}$ **(d)** $\frac{5}{9}$ **15. (a)** $S = \{12, 13, 21, 23, 31, 32\}$
(b) $\frac{1}{6}$ **(c)** $\frac{2}{6}$ or $\frac{1}{3}$ **17. (a)** $\frac{6}{36}$ **(b)** $\frac{2}{36}$ **(c)** $\frac{10}{36}$ **19. (a)** 0 **(b)** $\frac{1}{36}$ **(c)** $\frac{6}{36}$ **21. (a)** 0 **(b)** $\frac{3}{216}$ **(c)** $\frac{15}{216}$ **23. (a)** 52
(b) $\frac{13}{52}$ **(c)** $\frac{4}{52}$ **25. (a)** $\approx.0005$ **(b)** $\approx.0020$ **(c)** .2215 **27. (a)** $\approx.000045$ **(b)** 0 **(c)** $\approx.603$ **29. (a)** $\approx.274$
(b) $\approx.040$ **(c)** $\approx.274$ **31. (a)** $\approx.591$ **(b)** $\approx.073$ **(c)** $\approx.00002$ **33. (a)** .6 **(b)** .1 **(c)** .1 **35. (a)** $\approx.004$
(b) $\approx.64$ **(c)** $\approx.65$ **37. (a)** $\approx.167$ **(b)** $\approx.130$ **(c)** $\approx.0001$ **39. (a)** $\approx.24$ **(b)** .09 **(c)** .0081 **41. (a)** $\approx.18$
(b) $\frac{1}{56}$ **(c)** $\approx.18$ **43. (a)** $\approx.07$ **(b)** $\approx.43$ **(c)** $\approx.0048$ **45. (a)** $\approx.024$ **(b)** $\approx.476$ **(c)** $\approx.476$

6.2 EXERCISES (page 304)

1. (a) $S = \{A, B, C\}$; $P(A) = \frac{1}{4}$; $P(B) = \frac{1}{4}$; $P(C) = \frac{1}{2}$ **(b)** $\frac{3}{4}$ **(c)** $\frac{3}{4}$ **3. (a)** $S = \{a, b, c\}$; $P(a) = \frac{1}{2}$; $P(b) = \frac{1}{4}$; $P(c) = \frac{1}{4}$ **(b)** $\frac{1}{4}$
(c) $\frac{3}{4}$ **5. (a)** $S = \{$red, green, blue$\}$; P(red) $= \frac{1}{6}$; P(green) $= \frac{1}{6}$; P(blue) $= \frac{2}{3}$ **(b)** $\frac{5}{6}$ **(c)** $\frac{5}{6}$ **7. (a)** $S = \{$success, failure$\}$;
P(success) $= \frac{1}{6}$; P(failure) $= \frac{5}{6}$ **9. (a)** $S = \{R, G, B\}$; $P(R) = \frac{6}{15}$; $P(G) = \frac{2}{15}$; $P(B) = \frac{7}{15}$ **(b)** $\frac{8}{15}$ **(c)** $\frac{9}{15}$
11. $S = \{$success, failure$\}$; $P(s) = \frac{11}{36}$; $P(f) = \frac{25}{36}$ **13.** $S = \{S, F\}$; $P(S) = \frac{1}{4}$; $P(F) = \frac{3}{4}$ **15.** $S = \{S, F\}$; $P(S) = \frac{1}{4}$; $P(F) = \frac{3}{4}$

17. (a) $S = \{0, 1, 2, 3\}$; $P(0) = \frac{1}{8}$; **19. (a)** $S = \{1, 2, 3, 4, 5, 6\}$; $P(1) = P(2) = P(3) = P(4) = P(5) = P(6) = \frac{1}{6}$
$P(1) = \frac{3}{8}$; $P(2) = \frac{3}{8}$; $P(3) = \frac{1}{8}$ **(b)** $\frac{3}{6}$ **(c)** $\frac{2}{6}$ **(d)**
(b) $\frac{1}{2}$ **(c)** $\frac{7}{8}$
(d)

21. (a) $S = \{0, 1, 2\}$; $P(0) = .3$; **23. (a)** $S = \{0, 1, 2\}$; $P(0) = .81$; $P(1) = .18$; $P(2) = .01$ **(b)** .99
$P(1) = .6$; $P(2) = .1$ **(b)** .9 **(c)**
(c)

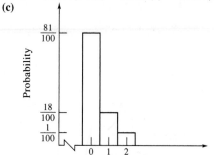

25. (a) $S = \{0, 1, 2\}$; $P(0) = .851$; $P(1) = .145$; **27. (a)** $S = \{0, 1, 2\}$; $P(0) = \frac{7}{15}$; $P(1) = \frac{7}{15}$; $P(2) = \frac{1}{15}$
$P(2) = .005$ **(b)** $\frac{8}{15}$ **(c)**
(b) .149 **(c)**

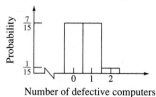

29. (a)

No. of Radios	Prob.
3	$\frac{5}{20}$
4	$\frac{8}{20}$
5	$\frac{7}{20}$

(b)

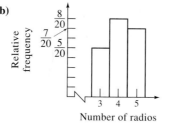

Number of radios

(c) $\frac{15}{20}$

31. (a)

No. of beans	Prob.
2	$\frac{2}{18}$
3	$\frac{5}{18}$
4	$\frac{8}{18}$
5	$\frac{3}{18}$

(b)

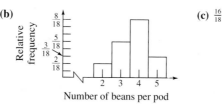

Number of beans per pod

(c) $\frac{16}{18}$

33. (a)

No. of heads	Prob.
0	.29
1	.51
2	.20

(b) .71

35. (a)

Predictions	Prob.
Recession	$\frac{64}{120}$
Slow growth	$\frac{42}{120}$
Higher growth	$\frac{14}{120}$

(b) $\frac{56}{120}$

37. (a)

Computer chips	Prob.
Usable	.92
Unusable	.08

(b) .92

39. (a) $\frac{16}{450}$ **(b)** $\frac{110}{450}$ **(c)** $\frac{166}{450}$ **(d)** $\frac{144}{450}$ **41. (a)** $\frac{182}{250}$ **(b)** $\frac{110}{250}$ **(c)** $\frac{30}{250}$ **(d)** $\frac{120}{250}$ **43. (a)** $\frac{4}{200}$ **(b)** $\frac{12}{200}$ **(c)** $\frac{20}{200}$ **(d)** $\frac{113}{200}$
45. (a) 4:48 **(b)** 13:39 **(c)** 26:26 **(d)** 39:13 **47. (a)** 11:839 **(b)** 1:5524 **(c)** 5524:1 **49. (a)** 4:16 **(b)** 14:6
(c) 16:4

6.3 EXERCISES (page 317)

1. (a) $\$-\frac{1}{2}$ **(b)** Lose \$750 **3. (a)** $\frac{2}{11}$ places forward **(b)** 8 places forward **5.** 1 **7.** 9.25 **9.** $\$-\frac{10}{14}$
11. $\$-\frac{1}{4}$ **13.** $\$\frac{2}{9}$ **15.** \$0 **17.** $\$-\frac{1}{4}$ **19.** \$48 **21. (a)** \$100 **(b)** \$200,000 **23.** \$4500 **25.** 3.6
27. $\$-\frac{1}{38}$ **29.** Business district location **31.** 3.15 **33.** \$3.99 **35.** \$ − 2.99
37. (a)

X	p
0	$\frac{10}{28}$
1	$\frac{15}{28}$
2	$\frac{3}{28}$

(b)

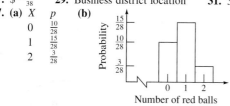

Number of red balls

(c) $\frac{3}{4}$ **39.** $\frac{34}{221}$ **41.** $\frac{5}{4}$ **43.** $\frac{1}{6};\frac{1}{3}$ **45.** $\frac{1}{2}$ **47.** 30 **49.** 7

CHAPTER 6 REFLECTIONS (page 320)

1. 312 **3.** 8:44 **5.** .252 **7.** $S = \{$success, failure$\}$ **9.** 1.5 **11.** at least 3.04 **13.** \$800 **15.** The event E
must happen. It is a certain event. Example: E is the event of rolling a sum that is less than 15 with a pair of standard dice.
17. Expected value is the average outcome you can expect in the long run.
19. The probabilities are not the same since the sets overlap and some elements get counted twice; the individual probabilities.
P(not a heart) $= \frac{39}{52}$
P(not an ace) $= \frac{48}{52}$
also sum to more than 1: One method of solution uses a Venn diagram.
The event not a heart has 3 + 36 elements, while the event not an ace
has 12 + 36 elements. Counting all the parts, but just once, gives
12 + 36 + 3 = 51 elements. So, P(not a heart or not an ace) $= \frac{51}{52}$.

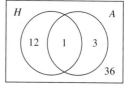

21. A linear function is $f(x) = 1.1x + 31$. The prediction for the 90s is $f(20) = 53$. **23.** $\begin{bmatrix} -5 & 0 & 18 \\ 4 & -2 & -20 \\ -6 & 0 & 25 \end{bmatrix}$

25. Three of one sex and one of the other is most likely (8 ways). The other outcomes occur in 2 and 6 ways, respectively.

SAMPLE TEST ITEMS FOR CHAPTER 6 (page 321)

1. 56 **2.** $\frac{1}{3}$ **3.** 6:10 **4.** $\frac{9}{14}$ **5.** $\frac{5}{36}$ **6.** $\frac{1}{36}$ **7.** $\frac{5}{6}$ **8.** $\approx.783$ **9.** $\approx.138$ **10.** $\approx.052$ **11.** $\approx.107$
12. $\approx.643$ **13.** **14.** $\frac{10}{32}$ **15.** $\frac{16}{32}$ **16.** 2.5 **17.** .154 **18.** 38

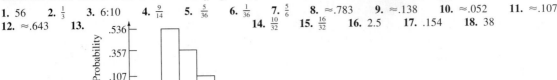

CHAPTER 7

7.1 EXERCISES (page 333)

1. (a) .8 **(b)** .5 **(c)** .5 **3. (a)** .9 **(b)** .3 **(c)** .7 **5. (a)** .3 **(b)** .2 **(c)** .7 **7. (a)** .6 **(b)** .9 **(c)** .8 **9.** .60
11. $\frac{3}{8}$ **13. (a)** $\frac{86}{180}$ **(b)** $\frac{93}{180}$ **(c)** $\frac{13}{180}$ **15. (a)** Yes **(b)** No **(c)** No **17.** $\frac{155}{165}$ **19.** $\frac{140}{165}$ **21.** $\frac{54}{100}$ **23.** $\frac{1}{10}$ **25.** $\frac{11}{20}$
27. (a) $\frac{8}{52}$ **(b)** $\frac{16}{52}$ **(c)** $\frac{1}{52}$ **(d)** $\frac{16}{52}$ **29. (a)** $\frac{4}{7}$ **(b)** $\frac{5}{7}$ **(c)** $\frac{5}{7}$ **31. (a)** $\frac{1}{5}$ **(b)** $\frac{4}{5}$ **(c)** $\frac{71}{120}$ **(d)** $\frac{3}{10}$ **33. (a)** $\frac{4}{7}$ **(b)** $\frac{5}{7}$ **(c)** $\frac{12}{35}$
(d) $\frac{11}{35}$ **35. (a)** $\frac{4}{5}$ **(b)** $\frac{1}{5}$ **(c)** $\frac{2}{5}$ **(d)** $\frac{3}{5}$ **37.** $\frac{7}{8}$ **39.** $\frac{33}{36}$ **41.** .696 **43. (a)** 0 **(b)** $\frac{6}{16}$ **(c)** $\frac{1}{4}$ **45. (a)** $\frac{10}{36}$ **(b)** $\frac{8}{36}$
(c) $\frac{11}{36}$ **47. (a)** .000990 **(b)** .367 **(c)** 0 **49. (a)** .0000257 **(b)** .0000188 **(c)** .0624 **(d)** .885 **51. (a)** .807
(b) .0798 **(c)** .461 **53. (a)** .36 **(b)** .64 **55. (a)** $\frac{5}{12}$ **(b)** $\frac{4}{12}$ **(c)** $\frac{8}{12}$ **(d)** $\frac{9}{12}$ **57.** b

7.2 EXERCISES (page 344)

1. (a) $S = \{1, 2, 3, 4, 5, 6\}$ **(b), (c)** $\frac{1}{3}$ **3. (a)** $S = \{1, 2, 3, 4, 5, 6\}$ **(b), (c)** 0
5. (a) $S = \{HHH, HHT, HTH, THH, TTH, THT, HTT, TTT\}$ **(b), (c)** $\frac{3}{7}$
7. (a) $S = \{HHH, HHT, HTH, THH, TTH, THT, HTT, TTT\}$ **(b),** **(c)** $\frac{1}{3}$
9. $\frac{4}{51}$ **11.** $\frac{1}{3}$ **13.** $\frac{1}{17}$ **15.** $\frac{1}{4}$ **17. (a)** $\frac{9}{14}$ **(b)** $\frac{5}{14}$ **(c)** $\frac{5}{14}$ **19.** $\frac{3}{7}$ **21.** $\frac{12}{52}$ **23.** $\frac{38}{90}$ **25.** $\frac{1}{2}$ **27.** $\frac{23}{40}$
29. $\frac{2}{5}$ **31.** $\frac{1}{2}$ **33.** $\frac{3}{5}$ **35.** $\frac{20}{44}$ **37.** $\frac{17}{25}$ **39.** $\frac{18}{100}$ **41.** $\frac{1}{6}$ **43.** $\frac{5}{6}$ **45.** $\frac{1}{6}$ **47. (a)** $\frac{1}{3}$ **(b)** $\frac{1}{2}$ **(c)** 0 **(d)** $\frac{1}{3}$
49. $\frac{3}{7}$ **51.** $\frac{8}{21}$ **53.** $\frac{20}{28}$ **55.** $\frac{1}{3}$ **57.** $\frac{3}{7}$ **59.** $\frac{4}{7}$ **61.** $\frac{1}{3}$
63. **65.** **67.**

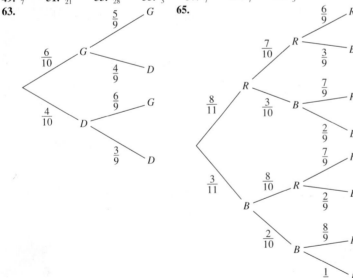

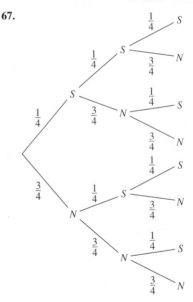

69. c **71.** .0565 **73.** .0580

7.3 EXERCISES (page 356)

1. $\frac{12}{51}$ **3.** $\frac{13}{68}$ **5.** $\frac{1}{221}$ **7.** .028 **9.** .0035 **11.** $\frac{1}{32}$ **13.** $\frac{1}{1024}$ **15.** $\frac{9}{1024}$ **17.** .136 **19.** (a) $\frac{2}{3}$ (b) $\frac{1}{3}$
21. (a) .8 (b) .8 **23.** (a) .4 (b) .03 (c) .12 (d) .67 (e) .33 **25.** (a) $\frac{10}{63}$ (b) $\frac{5}{42}$ (c) $\frac{1}{21}$ **27.** (a) $\frac{5}{108}$
(b) $\frac{5}{144}$ (c) $\frac{1}{64}$ **29.** (a) $\frac{25}{1296}$ (b) $\frac{5}{1296}$ (c) $\frac{8}{1296}$ **31.** (a) $\frac{5}{324}$ (b) $\frac{1}{36}$ (c) $\frac{8}{81}$ **33.** .0009 **35.** .0591 **37.** .0476
39. .8281 **41.** .0081 **43.** .000004 **45.** Two **47.** .996 **49.** $\frac{1}{128}$ **51.** $\frac{1}{8}$ **53.** .99999999 **55.** .0045
57. .618 **59.** $\frac{1}{8}$ **61.** .00000095 **63.** .944 **65.** .936 **67.** .000008 **69.** .001 **71.** .243 **73.** (a) .03096
(b) .001032 **75.** (a) .2743 (b) .00007388 (c) .0004952 **77.** c **79.** c

7.4 EXERCISES (page 364)

1. .12 **3.** .9 **5.** .6 **7.** $\frac{4}{37}$ **9.** $\frac{12}{63}$ **11.** $\frac{15}{37}$ **13.** $\frac{5}{14}$ **15.** $\frac{1}{84}$ **17.** $\frac{2}{3}$ **19.** $\frac{1}{9}$ **21.** $\frac{3}{7}$ **23.** $\frac{43}{70}$
25. .45 **27.** .39 **29.** .05 **31.** .95 **33.** .6 **35.** .68; 68:32 **37.** .0145 **39.** .0425; 17:383 **41.** .99; 99:1
43. .976 **45.** .4 **47.** .004 **49.** .25 **51.** .0125; 1:79 **53.** .625 **55.** .04 **57.** .388 **59.** .25
61. .714 **63.** $\frac{1}{3}$ **65.** $\frac{5}{11}$

7.5 EXERCISES (page 372)

1. .278 **3.** .164 **5.** .13 **7.** .057 **9.** .941 **11.** .142 **13.** .99% **15.** 85.1% **17.** 6.87%
19. .216 **21.** .055 **23.** .273 **25.** .00077 **27.** .48 **29.** .0084 **31.** .992 **33.** .00042 **35.** .246
37. .989 **39.** .00098 **41.** .886 **43.** .387 **45.** .987 **47.** .034 **49.** .000041

51. (a)

Number of successful calls	Probability
0	.0102
1	.0768
2	.2304
3	.3456
4	.2592
5	.0778

(b) 3
(c) 120; 480

53. (a)

Number of field goals made	Probability
0	.000125
1	.007125
2	.135375
3	.857375

(b) 2.85
(c) 47.5; 190

55. a **57.** .00245 **59.** .21 **61.** .756 **63.** $\frac{4}{45}$; $\frac{5}{9}$ **65.** .012 **67.** .1 **69.** $\frac{1}{64}$ **71.** .3 **73.** .35 **75.** .25

CHAPTER 7 REFLECTIONS (page 375)

1. .2 **3.** .33 **5.** .77 **7.** .7 **9.** .6 **11.** $\frac{4}{7}$ **13.** $\frac{5}{7}$ **15.** $\frac{3}{5}$ **17.** $\frac{5}{8}$ **19.** .04 **21.** .064 **23.** .216
25. .5822 **27.** Events are mutually exclusive if they can't happen at the same time.
29. $P(A)$ is the ratio of the number of elements in A (regions 2, 3) to the number of elements in the sample space (regions 1, 2, 3, 4). $P(A \mid B)$ is the ratio of the number of elements in A and B (region 3) to the number of elements in B (regions 3, 4).

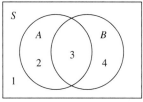

31. (a) .914 (b) .303 **33.** Answers will vary (the theoretical number of each is 10, 20, 30, 50, 60, 50, 40, 30, 20, 10, respectively. **35.** $x = 5, y = -2, z = 1$
37. $\begin{bmatrix} 10 & -3 & -3 \\ 17 & -7 & -11 \end{bmatrix}$

SAMPLE TEST ITEMS FOR CHAPTER 7 (page 377)

1. $\frac{2}{7}$ **2.** $\frac{5}{7}$ **3.** $\frac{4}{7}$ **4.** $\frac{31}{35}$ **5.** $\frac{7}{13}$ **6.** $\frac{15}{26}$ **7.** $\frac{43}{130}$ **8.** $\frac{51}{65}$ **9.** $\frac{43}{75}$ **10.** $\frac{1}{15}$ **11.** $\frac{49}{65}$ **12.** .013 **13.** 0
14. $\frac{4}{13}$ **15.** .0004 **16.** .2016 **17.** .4213 **18.** .39 **19.** .5 **20.** $\frac{3}{8}$ **21.** .5 **22.** .96 **23.** $\frac{6}{13}$ **24.** .285
25. .677

CHAPTER 8

8.1 EXERCISES (page 387)

1.

Class	Frequency	Relative frequency	Cumulative frequency
0–3	4	$\frac{4}{35}$	4
3–6	8	$\frac{8}{35}$	12
6–9	12	$\frac{12}{35}$	24
9–12	7	$\frac{7}{35}$	31
12–15	4	$\frac{4}{35}$	35

The data appears to be *normal*.

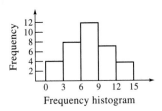

Frequency histogram

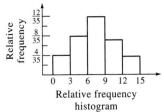

Relative frequency histogram

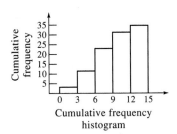

Cumulative frequency histogram

3.

Class	Frequency	Relative frequency	Cumulative frequency
3	9	$\frac{9}{30}$	9
4	12	$\frac{12}{30}$	21
5	4	$\frac{4}{30}$	25
6	3	$\frac{3}{30}$	28
7	2	$\frac{2}{30}$	30

The data appears to be *skewed to the right*.

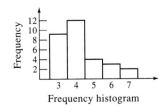

Frequency histogram

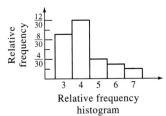

Relative frequency histogram

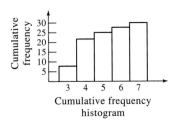

Cumulative frequency histogram

5.

Class	Frequency	Relative frequency	Cumulative frequency
150–180	30	$\frac{30}{260}$	30
180–210	50	$\frac{50}{260}$	80
210–240	80	$\frac{80}{260}$	160
240–270	60	$\frac{60}{260}$	220
270–300	40	$\frac{40}{260}$	260

The data is not normal nor skewed.

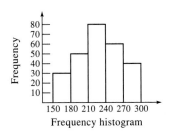

Frequency histogram

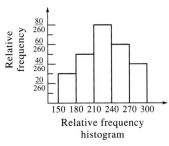

Relative frequency histogram

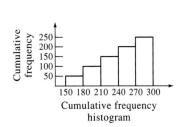

Cumulative frequency histogram

7.

Class	Frequency	Relative frequency	Cumulative frequency
0	2	$\frac{2}{20}$	2
1	3	$\frac{3}{20}$	5
2	8	$\frac{8}{20}$	13
3	5	$\frac{5}{20}$	18
4	2	$\frac{2}{20}$	20

The data is close to *normal.*

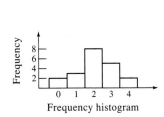

Frequency histogram

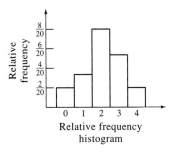

Relative frequency histogram

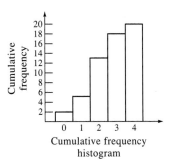

Cumulative frequency histogram

9. $\frac{12}{33}$ **11.** $\frac{7}{33}$ **13.** 14:19   **15.** $\frac{24}{62}$ **17.** 38:24

19. The data is slightly *skewed to the left.*

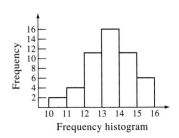

Frequency histogram

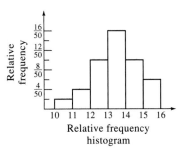

Relative frequency histogram

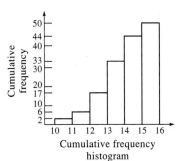

Cumulative frequency histogram

21.

Number of pods	Frequency	Relative frequency
0	2	$\frac{2}{50}$
1	4	$\frac{4}{50}$
2	7	$\frac{7}{50}$
3	13	$\frac{13}{50}$
4	15	$\frac{15}{50}$
5	7	$\frac{7}{50}$
6	2	$\frac{2}{50}$

23.

```
                        4
                        4
                  3     4
                  3     4
                  3     4
                  3     4
                  3     4
                  3     4
            2     3     4     5
            2     3     4     5
            2     3     4     5
      1     2     3     4     5
      1     2     3     4     5
0     1     2     3     4     5     6
0     1     2     3     4     5     6
0     1     2     3     4     5     6
```

The data appears to be skewed to the left rather than normal.

25. $\frac{13}{50}$

27.

Number of can openers	Frequency	Relative frequency
0	1	$\frac{1}{28}$
1	8	$\frac{8}{28}$
2	11	$\frac{11}{28}$
3	7	$\frac{7}{28}$
4	1	$\frac{1}{28}$

29.

```
            2
            2
            2
      1     2
      1     2     3
      1     2     3
      1     2     3
      1     2     3
      1     2     3
      1     2     3
      1     2     3
0     1     2     3     4
0     1     2     3     4
```

31. $\frac{19}{28}$

33.

Take-home pay	Frequency	Relative frequency
85–110	2	$\frac{2}{20}$
110–135	2	$\frac{2}{20}$
135–160	5	$\frac{5}{20}$
160–185	6	$\frac{6}{20}$
185–210	5	$\frac{5}{20}$

35.

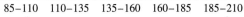

```
                            173.19
                158.26      172.63      207.35
                153.60      165.70      201.40
                152.37      165.70      190.35
105.51  133.71  148.23      162.50      187.60
 87.42  129.36  148.23      161.87      187.21
85–110  110–135  135–160    160–185     185–210
```

37.

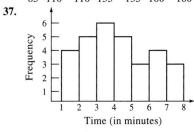

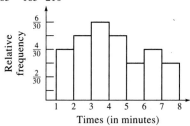

39.

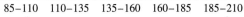

```
            3.8
      2.6   3.7   4.6
1.6   2.6   3.6   4.5               6.9
1.4   2.5   3.1   4.2   5.1   6.5   7.2
1.3   2.4   3.1   4.2   5.1   6.3   7.1
1.1   2.0   3.1   4.2   5.0   6.2   7.0
1–2   2–3   3–4   4–5   5–6   6–7   7–8
```

The data does not seem to be normal or skewed.

41.

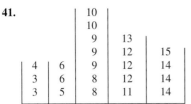

		10		
		10		
		9	13	
		9	12	15
4	6	9	12	14
3	6	8	12	14
3	5	8	11	14

2–5   5–8   8–11   11–14   14–17

The data does not seem to be normal or skewed.

43.

Salaries	Frequency
175–250	3
250–325	5
325–400	4
400–475	3
475–550	5

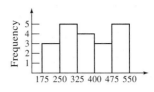

45.

Salaries	Frequency
175–225	2
225–275	1
275–325	5
325–375	1
375–425	4
425–475	2
475–525	4
525–575	1

8.2 EXERCISES (page 397)

1. (a) None   **(b)** 18   **(c)** 20.4   **3. (a)** 102   **(b)** 107   **(c)** 135   **5. (a)** None   **(b)** 0   **(c)** 0   **7.** Median   **9.** Median
11. Mean   **13.** Median or mode   **15.** 6   **17.** $2\frac{7}{8}$ **19.** -3 **21.** Mean; median   **23.** 365; median   **25.** 8.5; 8.5
27. \$5.25   **29.** \$4.75   **31.** \$1.70   **33.** \$1.69   **35.** \$150   **37.** \$39.66   **39.** \$56.00   **41.** 7.95   **43.** 2.05
45. 2   **47.** 4   **49.** \$102.50   **51.** 19.2; 18   **53.** 74.5   **55.** 84.8   **57.** 96

8.3 EXERCISES (page 406)

1.

x	$\bar{x}$	$x - \bar{x}$	$(x - \bar{x})^2$
2	4	-2	4
3	4	-1	1
4	4	0	0
5	4	1	1
6	4	2	4
		0	10

$s^2 = 2.5$ $s = 1.58$

3.

x	μ	$x - \mu$	$(x - \mu)^2$
12	47.7	-35.7	1274.49
18	47.7	-29.7	882.09
20	47.7	-27.7	767.29
30	47.7	-17.7	313.29
32	47.7	-15.7	246.49
33	47.7	-14.7	216.09
80	47.7	32.3	1043.29
82	47.7	34.3	1176.49
84	47.7	36.3	1317.69
86	47.7	38.3	1466.89
		0	8704.10

$\sigma^2 = 870.41$ $\sigma = 29.5$

5.

x	$\bar{x}$	$x - \bar{x}$	$(x - \bar{x})^2$
50	53.1	-3.1	9.61
51	53.1	-2.1	4.41
52	53.1	-1.1	1.21
52.5	53.1	$-.6$	.36
53	53.1	$-.1$	.01
53.5	53.1	.4	.16
54	53.1	.9	.81
54.5	53.1	1.4	1.96
55	53.1	1.9	3.61
55.5	53.1	2.4	5.76
		0	27.9

$s^2 = 3.1$ $s = 1.76$

7. b **9.** a **11. (a)** $\mu = 9.6$, $\sigma \approx 1.75$ **(b)** $\mu = 9.6$, $\sigma \approx 4.59$ **13. (a)** $\bar{x} = 1276.31$, $s \approx \$567.26$ **(b)** 8
15. (a) $\mu \approx 79.38$, $\sigma \approx 8.29$ **(b)** All eight of them   **17. (a)** $\mu = 76.25$, $\sigma \approx 6.48$ **(b)** All of them   **19. (a)** $\bar{x} = 45.2$
(b) $s^2 \approx 88.70$, (degrees)^2 **(c)** $s \approx 9.42$, degrees   **21. (a)** $\mu = 315.6$; 345   **(b)** $\sigma^2 = 6494.64$, (tourists)^2
(c) $\sigma \approx 80.59$, tourists   **23. (a)** $\bar{x} = 78.5$; 80.5   **(b)** $s^2 = 52.5$, (blooms)^2 **(c)** $s \approx 7.25$, blooms   **25.** $\geq .826$
27. $\leq .25$ **29.** $\geq .75$ **31. (a)** 75.42

(b) $\sigma = \sqrt{\dfrac{(54.5 - 75.42)^2 5 + (64.5 - 75.42)^2 15 + \cdots + (95 - 75.42)^2 4}{100}} \approx 8.77$

33. (a) 26.6   **(b)** $s = \sqrt{\dfrac{(23 - 26.6)^2 6 + (25 - 26.6)^2 12 + (27 - 26.6)^2 20 + (29 - 26.6)^2 10 + (31 - 26.6)^2 2}{49}} \approx 2.06$

35. $\$\frac{5}{8}$; \$3.39   **37.** \$0; \$7000

8.4 EXERCISES (page 417)

1. B

3. (a) $y = \dfrac{1}{\sqrt{2\pi}} e^{(-1/2)x^2}$

(b) $y = \dfrac{1}{2\sqrt{2\pi}} e^{(-1/2)x^2/4}$

(c) $y = \dfrac{1}{3\sqrt{2\pi}} e^{(-1/2)x^2/9}$

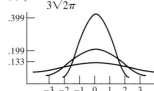

5. (a) $y = \dfrac{1}{\sqrt{2\pi}} e^{(-1/2)(x-4)^2}$

(b) $y = \dfrac{1}{2\sqrt{2\pi}} e^{(-1/2)(x-4)^2}$

(c) $y = \dfrac{1}{3\sqrt{2\pi}} e^{(-1/2)(x-4/3)^2}$

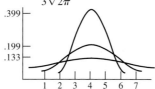

7. 18 and 22 **9.** (a) 23 (b) 15.5 (c) 17.6 **11.** 34% **13.** 50% **15.** .15% **17.** 2.5% **19.** 84% **21.** 68%
23. 84% **25.** .9773 **27.** .0107 **29.** .4880 **31.** .7305 **33.** .5
35. (a) $P(z > 2)$
(b)
(c) .0227

37. (a) $P(z < -3.44)$
(b)
(c) about 0

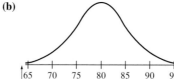

39. (a) $P(z < 2.73)$
(b)
(c) .9968

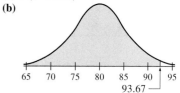

41. (a) $P(z > -.7)$
(b)
(c) .7580

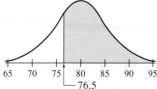

43. 275 grams **45.** Mary Jo **47.** (a) 15.87% (b) 2.27% (c) 4.54 or about 5 (d) 2.27% **49.** (a) 1587 (b) 227
(c) 9546 **51.** (a) 38.59% (b) .0764 **53.** (a) .2482 (b) 5.59% **55.** (a) .6915 (b) .9105 **57.** (a) 21,190
(b) 97.73% **59.** (a) 1.07% (b) .1170 **61.** (a) 86.44% (b) 5.94% **63.** (a) .1271 (b) About 3 **65.** Less than
9.75 years **67.** Less than 5.875 years

8.5 EXERCISES (page 426)

1.

Number of seeds	Probability
0	.0256
1	.1536
2	.3456
3	.3456
4	.1296

3. $\mu = 2.4$; $\sigma = .98$
5. .8686; np and nq and not > 5.

7.

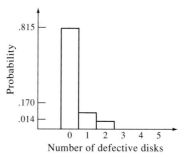

Number of defective disks

9. About 1   **11.**

Number of shots	Probability
0	.0001
1	.0036
2	.0486
3	.2916
4	.6561

13. $\mu = 3.6$; $\sigma = .6$

15. .4325; nq is very small and the distribution is far from normal.

17.

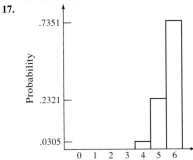

19. .9977   **21. (a)** .1492   **(b)** .1020   **23. (a)** .0274   **(b)** .4223   **25. (a)** .0279   **(b)** .0677   **27. (a)** .1314
(b) .0000   **29. (a)** .0045   **(b)** .2243   **31. (a)** .2061   **(b)** .3557

CHAPTER 8 REFLECTIONS (page 428)

1. 2   **3.** 20.4   **5.** .9032   **7.** .7176   **9.** 3.06   **11.** .3830   **13.** $1\frac{1}{3}$ **15.** 6   **17.** The standard deviation is a measure of variability and gives an indication of how the scores are spread out from the mean.
19. A random variable is a function that assigns numbers to each outcome of an experiment.   **21.** The mean is a good average because it involves every score, but it can be drastically affected by extreme scores. The median is not affected by extreme scores but depends only on the middle score(s).   **23.** $x = -5$
25.

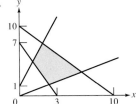

27. 144

SAMPLE TEST ITEMS FOR CHAPTER 8 (page 429)

1.

Number of secretaries	Frequency	Relative frequency
1	2	.2
3	3	.3
5	5	.5

2. 3.6   **3.** 4   **4.** .5   **5.** 1.5   **6.** 1   **7.** 3
8. 3.37   **9.** 5

10.

x	p
0	.414
1	.436
2	.138
3	.013

11.

Number of girls	Probability
0	.03125
1	.15625
2	.3125
3	.3125
4	.15625
5	.03125

12. Median = 77
Mode = 82
Mean = 72.6
$\sigma = 19.84$
$s \approx 20.91$

13. $10,200 profit   **14.** .2643   **15.** 616   **16.** .4410   **17.** .0367   **18.** .0869   **19.** -6.08

CHAPTER 9

9.1 EXERCISES (page 438)

1. From S_1 to S_2: 0.4, from S_2 to S_2: 0.3 $\begin{bmatrix} .6 & .4 \\ .7 & .3 \end{bmatrix}$ **3.** From S_1 to S_1: 0.1, from S_2 to S_2: 0.78, from S_3 to S_1: 1 $\begin{bmatrix} .1 & .6 & .3 \\ .2 & .78 & .02 \\ 1 & 0 & 0 \end{bmatrix}$

5. From S_1 to S_3: 0.27, from S_2 to S_1: 0, from S_3 to S_3: 1 $\begin{bmatrix} .03 & .7 & .27 \\ 0 & .9 & .1 \\ 0 & 0 & 1 \end{bmatrix}$

7.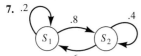

9.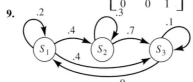

11. $\begin{array}{c} \\ F \\ A \end{array} \begin{array}{cc} F & A \\ \begin{bmatrix} .80 & .20 \\ .10 & .90 \end{bmatrix} \end{array}$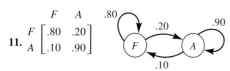

13. $\begin{array}{c} \\ P \\ P' \end{array} \begin{array}{cc} P & P' \\ \begin{bmatrix} .60 & .40 \\ .50 & .50 \end{bmatrix} \end{array}$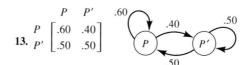

15. $\begin{array}{c} \\ E \\ M \\ G \end{array} \begin{array}{ccc} E & M & G \\ \begin{bmatrix} .90 & .06 & .04 \\ .20 & .70 & .10 \\ .16 & .09 & .75 \end{bmatrix} \end{array}$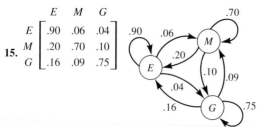

17. Not a transition matrix. Elements in the second row do not sum to 1.   **19.** Transition matrix   **21. (a)** $[.48 \quad .52]$
(b) $[.392 \quad .608]$ **23. (a)** $[.24 \quad .42 \quad .34]$ **(b)** $[.274 \quad .32 \quad .406]$

25. (a) $\begin{bmatrix} .60 & .40 \\ .60 & .40 \end{bmatrix}$

(b)

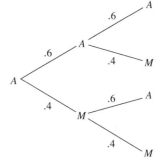

$P(AAA) + P(AMA) =$
$(.6)(.6) + (.4)(.6) = .6$

(c) $\begin{bmatrix} .6 & .4 \\ .6 & .4 \end{bmatrix}^2 = \begin{bmatrix} .6 & .4 \\ .6 & .4 \end{bmatrix}$ The answer is found in the first-row, first-column position: 0.6.   **(d)** .40

27. (a) $\begin{bmatrix} .70 & .20 & .10 \\ .05 & .90 & .05 \\ .10 & .30 & .90 \end{bmatrix}$ **(b)**

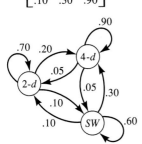

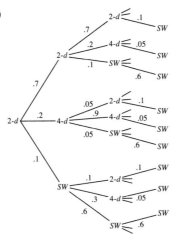

$$(.7)(.7)(.1) + (.7)(.2)(.05) + (.7)(.1)(.6)$$
$$+ (.2)(.05)(.1) + (.2)(.9)(.05) + (.2)(.05)(.6)$$
$$+ (.1)(.1)(.1) + (.1)(.3)(.05) + (.1)(.6)(.6) = .1525$$

(c) $\begin{bmatrix} .70 & .20 & .10 \\ .05 & .90 & .05 \\ .10 & .30 & .60 \end{bmatrix}^3 = \begin{bmatrix} .3885 & .459 & .1525 \\ .10925 & .7925 & .09825 \\ .1635 & .5675 & .269 \end{bmatrix}$ The answer is found in the first-row, third-column position: 0.1525.

29. 0.28

9.2 EXERCISES (page 445)

1. Regular   **3.** Regular   **5.** Not regular   **7.** $\begin{bmatrix} .3 & .7 \\ .5 & .5 \end{bmatrix}^6 = \begin{bmatrix} .417 & .583 \\ .417 & .583 \end{bmatrix}$ **9.** $\begin{bmatrix} .9 & .1 \\ .2 & .8 \end{bmatrix}^{24} = \begin{bmatrix} .667 & .333 \\ .667 & .333 \end{bmatrix}$

11. $\begin{bmatrix} .6 & .4 \\ 1 & 0 \end{bmatrix}^8 = \begin{bmatrix} .714 & .286 \\ .714 & .286 \end{bmatrix}$ **13.** $S = [.25 \quad 75], L = \begin{bmatrix} .25 & .75 \\ .25 & .75 \end{bmatrix}$ **15.** $S = [\frac{4}{7} \quad \frac{3}{7}], L = \begin{bmatrix} \frac{4}{7} & \frac{3}{7} \\ \frac{4}{7} & \frac{3}{7} \end{bmatrix}$

17. $S = [\frac{21}{41} \quad \frac{20}{41}], L = \begin{bmatrix} \frac{21}{41} & \frac{20}{41} \\ \frac{21}{41} & \frac{20}{41} \end{bmatrix}$ **19.** $S = [\frac{1}{3} \quad \frac{4}{15} \quad \frac{2}{5}], L = \begin{bmatrix} \frac{1}{3} & \frac{4}{15} & \frac{2}{5} \\ \frac{1}{3} & \frac{4}{15} & \frac{2}{5} \\ \frac{1}{3} & \frac{4}{15} & \frac{2}{5} \end{bmatrix}$

21. $S = [.388 \quad .410 \quad .201], L = \begin{bmatrix} .388 & .410 & .201 \\ .388 & .410 & .201 \\ .388 & .410 & .201 \end{bmatrix}$

23. (a)

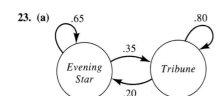

$\begin{array}{cc} & \text{E.S.} \quad \text{T} \\ \text{E.S.} & \begin{bmatrix} .65 & .35 \\ .20 & .80 \end{bmatrix} \\ \text{T} & \end{array}$ **(b)** The *Evening Star* has 36.36% and the *Tribune* has 63.64%.

25. (a) The gain will be to 80% of the market.   **(b)** 19 weeks
27. (a) 15.9% in L, 61.0% in M, and 23.1% in H.   **(b)** 23.1%; 15.9%
29. (a)

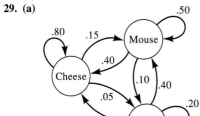

$\begin{array}{cccc} & C & M & S \\ C & \begin{bmatrix} .80 & .15 & .05 \\ .40 & .50 & .10 \\ .40 & .40 & .20 \end{bmatrix} \\ M & \\ S & \end{array}$

(b) $S = [.667 \quad .259 \quad .074]$ If the experiment is repeated many times, when placed in the small area the rat will go to C 66.7% of the time, M 25.9% of the time, and to S 7.4% of the time.

$$
\begin{array}{c}
\begin{array}{ccc} C & M & S \end{array} \\
\text{(c) } L = \begin{array}{c} C \\ M \\ S \end{array} \left[\begin{array}{ccc} .667 & .259 & .074 \\ .667 & .259 & .074 \\ .667 & .259 & .074 \end{array}\right]
\end{array}
$$
Assume the experiment has been repeated many times. If the rat goes in $C(M, S$, respectively) on one trial, then the probability that the rat will go in C, M, or S is .667, .259, and .074, respectively.

9.3 EXERCISES (page 452)

1. Absorbing; S_2;
$$
\begin{array}{c}
\begin{array}{ccc} S_2 & S_1 & S_3 \end{array} \\
\begin{array}{c} S_2 \\ S_1 \\ S_3 \end{array} \left[\begin{array}{ccc} 1 & 0 & 0 \\ .2 & .6 & .2 \\ .2 & 0 & .8 \end{array}\right]
\end{array}
$$

3. Absorbing; S_1;
$$
\begin{array}{c}
\begin{array}{ccc} S_1 & S_2 & S_3 \end{array} \\
\begin{array}{c} S_1 \\ S_2 \\ S_3 \end{array} \left[\begin{array}{ccc} 1 & 0 & 0 \\ .5 & .2 & .3 \\ .1 & .1 & .8 \end{array}\right]
\end{array}
$$

5. Absorbing; S_2, S_4;
$$
\begin{array}{c}
\begin{array}{cccc} S_4 & S_2 & S_3 & S_1 \end{array} \\
\begin{array}{c} S_4 \\ S_2 \\ S_3 \\ S_1 \end{array} \left[\begin{array}{cccc} 1 & 0 & 0 & 0 \\ 0 & 1 & 0 & 0 \\ .4 & .3 & .1 & .2 \\ .1 & .6 & 0 & .3 \end{array}\right]
\end{array}
$$

7. Not absorbing

9. (a)
$$
\begin{array}{c}
\begin{array}{ccc} S_1 & S_2 & S_3 \end{array} \\
\begin{array}{c} S_1 \\ S_2 \\ S_3 \end{array} \left[\begin{array}{ccc} 1 & 0 & 0 \\ 0 & 1 & 0 \\ .6 & .2 & .2 \end{array}\right]
\end{array}
$$
(b) $F = [1.25]$

(c) $FR = [.75 \quad .25]$. The probability that S_3 will reach S_1 is .75; the probability that S_3 will reach S_2 is .25.

11. (a)
$$
\begin{array}{c}
\begin{array}{cccc} S_3 & S_2 & S_1 & S_4 \end{array} \\
\begin{array}{c} S_3 \\ S_2 \\ S_1 \\ S_4 \end{array} \left[\begin{array}{cccc} 1 & 0 & 0 & 0 \\ 0 & 1 & 0 & 0 \\ .1 & .3 & .2 & .4 \\ .3 & 0 & .1 & .6 \end{array}\right]
\end{array}
$$
(b) $F = \begin{array}{c} \begin{array}{cc} S_1 & S_4 \end{array} \\ \begin{array}{c} S_1 \\ S_4 \end{array} \left[\begin{array}{cc} 1.43 & 1.43 \\ .36 & 1.86 \end{array}\right] \end{array}$

(c) From FR, the probability that S_1 will reach S_2 is .43, that S_1 will reach S_3 is .57, that S_4 will reach S_2 is .11, that S_4 will reach S_3 is .89.

13. (a)
$$
\begin{array}{c}
\begin{array}{cccc} S_1 & S_4 & S_3 & S_2 \end{array} \\
\begin{array}{c} S_1 \\ S_4 \\ S_3 \\ S_2 \end{array} \left[\begin{array}{cccc} 1 & 0 & 0 & 0 \\ 0 & 1 & 0 & 0 \\ 0 & 0 & 0 & 1 \\ .2 & .1 & .4 & .3 \end{array}\right]
\end{array}
$$
(b) $F = \begin{array}{c} \begin{array}{cc} S_3 & S_2 \end{array} \\ \begin{array}{c} S_3 \\ S_2 \end{array} \left[\begin{array}{cc} 2.333 & 3.333 \\ 1.333 & 3.333 \end{array}\right] \end{array}$

(c) From FR, the probability that S_3 will reach S_1 is .667, that S_3 will reach S_4 is .333, that S_2 will reach S_1 is .667, that S_2 will reach S_4 is .333.

15. S_2: 7.143; S_3: 8.572   **17.** S_3: 1.702; S_2: 2.340   **19.** S_1: 2; S_4: 2   **21.** $\left[\begin{array}{ccc} \frac{1}{3} & \frac{2}{3} & 0 \\ \frac{2}{3} & 0 & \frac{1}{3} \\ 0 & 0 & 1 \end{array}\right]$ **23.** 1   **25.** 7.5

27.
$$
\begin{array}{c}
\begin{array}{cccc} P & T & C & L \end{array} \\
\begin{array}{c} P \\ T \\ C \\ L \end{array} \left[\begin{array}{cccc} 1 & 0 & 0 & 0 \\ 0 & 1 & 0 & 0 \\ .10 & .10 & .50 & .30 \\ .10 & 0 & .10 & .80 \end{array}\right]
\end{array}
$$
29. 8.57   **31.** .286   **33.** .428   **35.** .571   **37.**
$$
\begin{array}{c}
\begin{array}{cccccc} & 0 & 1 & 2 & 3 & 4 \end{array} \\
\begin{array}{c} 0 \\ 1 \\ 2 \\ 3 \\ 4 \end{array} \left[\begin{array}{ccccc} 1 & 0 & 0 & 0 & 0 \\ \frac{1}{2} & 0 & \frac{1}{2} & 0 & 0 \\ 0 & \frac{1}{2} & 0 & \frac{1}{2} & 0 \\ 0 & 0 & \frac{1}{2} & 0 & \frac{1}{2} \\ 0 & 0 & 0 & 0 & 1 \end{array}\right]
\end{array}
$$
39. 6

41. Approximately 54%   **43.** 6.87 years   **45.** 2.9 months   **47.** 50%

CHAPTER 9 REFLECTIONS (page 455)

1. S_1 to S_2: 32; S_2 to S_2: 02; S_3 to S_2: .8 $\left[\begin{array}{ccc} .05 & .32 & .63 \\ .7 & .02 & .28 \\ .2 & .8 & 0 \end{array}\right]$ **3.** $[.26 \quad .47 \quad .27]$ **5.** $\left[\begin{array}{cc} .4375 & .5625 \\ .4375 & .5625 \end{array}\right]$ **7.** $\left[\begin{array}{cc} .545 & .455 \\ .545 & .455 \end{array}\right]$

9. For S_1 the probability is 1; for S_2 the probability is 1.   **11.** .80   **13.** Each trial must have the same sample space with a finite number of outcomes. The probability of any given outcome on a certain trial depends only on the probability of the outcome on the preceding trial (except the first).   **15.** There must be at least one absorbing state (a state from which there is no escape) and it must be possible to reach any absorbing state from any nonabsorbing state.   **17.** One zero in a 2×2 transition matrix will yield a regular matrix if, and only if, the zero is on the main diagonal. Otherwise, the matrix is absorbing. In the 3×3 case, if a matrix has two zeros in the same row or the same column not on the main diagonal, then the matrix will not be regular. If the two zeros were in the same row of a transition matrix, then the matrix would be absorbing. Zeros have no particular significance in absorbing matrices except that at least one row must contain 1 and all other entries zero.

19. $y = -2x + 14.4$ **21.** $\begin{bmatrix} 8 & 42 \\ 6 & 32 \end{bmatrix}$ **23.** $\frac{3}{4}$

SAMPLE TEST ITEMS FOR CHAPTER 9 (page 457)

1. S_1 to S_2: .8; S_2 to S_2: .6   **2.** $\begin{array}{c} \\ S_1 \\ S_2 \end{array}\begin{array}{c} S_1 \quad S_2 \\ \begin{bmatrix} .2 & .8 \\ .4 & .6 \end{bmatrix} \end{array}$ **3.** S_1 to S_1: .6; S_2 to S_3; .3; S_3 to S_2: .6   **4.** $\begin{array}{c} \\ S_1 \\ S_2 \\ S_3 \end{array}\begin{array}{c} S_1 \quad S_2 \quad S_3 \\ \begin{bmatrix} .6 & .1 & .3 \\ .7 & 0 & .3 \\ 0 & .6 & .4 \end{bmatrix} \end{array}$

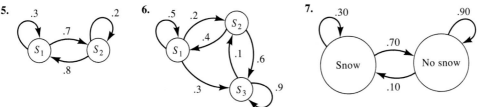

5. **6.** **7.**

8. $\begin{array}{c} \\ S \\ NS \end{array}\begin{array}{c} S \quad\; NS \\ \begin{bmatrix} .30 & .70 \\ .10 & .90 \end{bmatrix} \end{array}$ **9.** $[.26 \quad .57 \quad .17]$ **10.** $[.206 \quad .597 \quad .197]$ **11.** Not regular   **12.** Is regular

13. $\begin{bmatrix} .1 & .9 \\ 1 & 0 \end{bmatrix}^{57} = \begin{bmatrix} .53 & .47 \\ .53 & .47 \end{bmatrix}$ **14.** $\begin{bmatrix} .7 & .3 \\ .4 & .6 \end{bmatrix}^{4} = \begin{bmatrix} .57 & .43 \\ .57 & .43 \end{bmatrix}$ **15.** $[.329 \quad .671]; \begin{bmatrix} .329 & .671 \\ .329 & .671 \end{bmatrix}$

16. $[.412 \quad .588]; \begin{bmatrix} .412 & .588 \\ .412 & .588 \end{bmatrix}$ **17.** Nonabsorbing   **18.** $\begin{array}{c} \\ S_2 \\ S_1 \\ S_3 \end{array}\begin{array}{c} S_2 \quad S_2 \quad S_3 \\ \begin{bmatrix} 1. & 0 & 0 \\ .3 & .4 & .3 \\ .7 & .1 & .2 \end{bmatrix} \end{array}$ **19.** $\begin{bmatrix} 1.78 & .67 \\ 8.22 & 1.33 \end{bmatrix}$ **20.** $\begin{bmatrix} 1 \\ 1 \end{bmatrix}$

CHAPTER 10

10.1 EXERCISES (page 463)

1. Is strictly determined; 2; row player; Row player's best strategy: Row 2; Column player's best strategy: Column 2   **3.** Not strictly determined   **5.** Is strictly determined; 2; row player; Row player's best strategy: Row 2; Column player's best strategy: Column 1   **7.** Is strictly determined; 1, row player; Row player's best strategy: Row 1; Column player's best strategy: Column 1
9. Not strictly determined   **11.** The value of x must be greater than zero and less than three.

13.

	Economy	
	Favorable	Recession
Remodel	100,000	−40,000
Don't remodel	−50,000	−45,000

The saddle point is −40,000. The store should remodel.

15. **(a)**

Nutone	Fabless		
		Sports	Senior
	Sports	250	400
	Senior	350	350
	Drug	375	400

(b) The saddle point is 375.
(c) Nutone should start a drug therapy program and Fabless should start a sports medicine program.

17. **(a)**

John	Harry			
		R, 5	B, 7	B, 8
	R, 4	1	−4	−4
	B, 9	−5	2	1

(b) Not strictly determined

19. **(a)**

Computer software	Computer hardware	
	Tallahassee	Austin
Tallahassee	0	500,000
Austin	−250,000	50,000

(b) The game is strictly determined.
(c) Both companies should locate in Tallahassee.

10.2 EXERCISES (page 468)

1. 1   **3.** $\frac{4}{9}$ **5.** $\frac{7}{6}$ **7.** $\frac{5}{9}$ **9.** $-\frac{9}{88}$ **11.** Q_2 is the better column player strategy.
13. Q_2 is the better column player strategy.   **15.** Q_1 is the better column player strategy.

17. $P = [0 \quad 1], Q = \begin{bmatrix} 1 \\ 0 \end{bmatrix}; [0 \quad 1]A\begin{bmatrix} q_1 \\ q_2 \end{bmatrix} = 4q_1 + 6q_2 \geq 4; [p_1 \quad p_2]A\begin{bmatrix} 1 \\ 0 \end{bmatrix} = 3p_1 + 4p_2 \leq 4.$

19. $P = [0, 1], Q = \begin{bmatrix} 0 \\ 1 \end{bmatrix}; [1 \quad 0]A\begin{bmatrix} q_1 \\ q_2 \end{bmatrix} = 4q_1 - 7q_2 \geq -7; [p_1 \quad p_2]A\begin{bmatrix} 0 \\ 1 \end{bmatrix} = -7p_1 - 8p_2 \leq -7.$

21. (a)

	Ilke	
Tanya	1	2
1	5	−5
2	−5	5

(b) $\frac{5}{9}$; favors Tanya   **23. (a)**

	Jill	
Jack	P	D
P	1	−10
D	−10	10

(b) −10   **(c)** $-\frac{9}{2}$
(d) The one in (b).

10.3 EXERCISES (page 476)

1. (a) This game is strictly determined.   **(b)** 1   **3. (a)** Row player: $\begin{bmatrix} \frac{9}{14} & \frac{5}{14} \end{bmatrix}$ Column player: $\begin{bmatrix} \frac{4}{7} \\ \frac{3}{7} \end{bmatrix}$ **(b)** $\frac{1}{7}$ **5. (a)** This game is strictly determined.   **(b)** 4   **7. (a)** Row player: $\begin{bmatrix} \frac{6}{13} & \frac{7}{13} \end{bmatrix}$ Column player: $\begin{bmatrix} \frac{8}{13} \\ \frac{5}{13} \end{bmatrix}$ **(b)** $\frac{22}{13}$ **9. (a)** Row player: $\begin{bmatrix} \frac{9}{16} & \frac{7}{16} \end{bmatrix}$ Column player: $\begin{bmatrix} \frac{9}{16} \\ \frac{7}{16} \end{bmatrix}$ **(b)** $\frac{31}{16}$ **11.** Row player: $\begin{bmatrix} \frac{2}{5} & \frac{3}{5} \end{bmatrix}$ Column player: $\begin{bmatrix} \frac{1}{2} \\ \frac{1}{2} \end{bmatrix}; v = 1$ **13.** Row player: $[1 \quad 0]$ Column player: $\begin{bmatrix} \frac{4}{7} \\ \frac{3}{7} \end{bmatrix}; v = 1$ (Alternatively, column players should always play their second option.)

15. Row player: $\begin{bmatrix} \frac{1}{2} & \frac{1}{2} \end{bmatrix}$; Column player: $\begin{bmatrix} \frac{3}{8} \\ \frac{5}{8} \end{bmatrix}; v = 3$ **17.** Row player: $\begin{bmatrix} \frac{2}{3} & \frac{1}{3} & 0 \end{bmatrix}$; Column player: $\begin{bmatrix} \frac{1}{6} \\ \frac{5}{6} \\ 0 \end{bmatrix}; v = \frac{8}{3}$

19. Row player: $\begin{bmatrix} \frac{2}{3} & \frac{1}{3} & 0 \end{bmatrix}$; Column player: $\begin{bmatrix} \frac{1}{3} \\ \frac{2}{3} \\ 0 \end{bmatrix}; v = \frac{1}{3}$ **21.** Row player: $\begin{bmatrix} \frac{1}{2} & \frac{1}{2} \end{bmatrix}$; Column player: $\begin{bmatrix} \frac{1}{10} \\ \frac{9}{10} \\ 0 \end{bmatrix}; v = \frac{1}{2}$

23. (a) Conceal coin in right hand $\frac{8}{13}$ of the time and in the left hand $\frac{5}{13}$ of the time.   **(b)** Guess right hand $\frac{5}{13}$ of the time and left hand $\frac{8}{13}$ of the time.   **(c)** Match   **25. (a)** M-Mart should feature clothing $\frac{19}{31}$ of the time and sporting goods $\frac{12}{31}$ of the time. Q-Mart should feature electronics $\frac{11}{31}$ of the time and kitchen wares $\frac{20}{31}$ of the time.   **(b)** M-Mart   **27.** The optimal strategies are the same for both players: Play the nickel $\frac{7}{12}$ of the time and the dime $\frac{5}{12}$ of the time.

CHAPTER 10 REFLECTIONS (page 478)

1. 9   **3.** 2   **5.** Between −3 and 0   **7.** −1   **9.** $\frac{4}{3}$ **11.** 0   **13.** 4   **15.** 2   **17.** 52%   **19.** It gives the value of the game and its position determines the strategy of both players.   **21.** It is the same concept as the expected value in the probability sense. If it is positive, the game favors the row player. If it is negative, the game favors the column player. If it is zero, the game is fair.

23. $\begin{bmatrix} -1 & 0 & -3 \\ 2 & -1 & -1 \\ 3 & 3 & 3 \end{bmatrix}$ Sam and John each have a box with a red ball, a blue ball, and a green ball. Each boy simultaneously selects a ball and reveals its color to the other. Depending upon the color combination, one or the other will get to move a plastic disk a certain number of spaces around a playing board. If John's ball is red, then Sam will move 1 space if his ball is red; no move by either will occur if Sam's ball is blue, Sam will move 3 spaces if his ball is green. If John's ball is blue, then John will move 2 spaces if Sam's ball is red, and Sam will move 1 space if his ball is blue or green. If John's ball is green, then John will move 3 spaces regardless of the color of Sam's ball.

25. $y = 8x + 160$ **27.**

	F	R	S	H
F	0	1	1	1
R	0	0	1	1
S	1	1	0	0
H	1	0	1	0

29. Mean: 7   Median: 8   Mode: 10   Range: 9

SAMPLE TEST ITEMS FOR CHAPTER 10 (page 479)

1. Strictly determined; Saddle point: 3; Favors row player.   **2.** Not strictly determined   **3.** Strictly determined; Saddle point: -2. Favors column player.   **4.** None   **5.** Between -2 and 0

6. $\begin{array}{c} \\ C \\ M \end{array} \begin{array}{cc} C & M \\ \left[\begin{array}{cc} .50 & .55 \\ .45 & .50 \end{array} \right. \end{array}$ **7.** Both should open campus branches. The value is 50.   **8.** $\frac{3}{2}$ **9.** $\frac{3}{2}$ **10.** $\frac{14}{9}$ **11.** Q_1

12. Q_2 **13.** Q_1 **14.** Value: $\frac{29}{7}$; Row strategy: $\begin{bmatrix} \frac{1}{7} & \frac{6}{7} \end{bmatrix}$ Column strategy: $\begin{bmatrix} \frac{6}{7} \\ \frac{1}{7} \end{bmatrix}$ **15.** Strictly determined; Value; 3.

Both players should use their second option.   **16.** Strictly determined; Value: 6. Row player should use second option

while the column player should use the first option.   **17.** Value; 3.62; Row strategy: $\begin{bmatrix} \frac{8}{21} & \frac{13}{21} \end{bmatrix}$ Column strategy: $\begin{bmatrix} \frac{10}{21} \\ \frac{11}{21} \end{bmatrix}$.

18. Value: 3.88; Row strategy: $\begin{bmatrix} \frac{9}{16} & \frac{7}{16} \end{bmatrix}$ Column strategy $\begin{bmatrix} \frac{9}{16} \\ \frac{7}{16} \end{bmatrix}$.   **19.** Strictly determined; Value: 7. Row player should use option 1,

the column player should use option 3.   **20.** Both players should play their dime 67% of the time and their quarter 33% of the time.
21. The game favors Elena.

CHAPTER 11

11.1 EXERCISES (page 488)

(Answers rounded to two decimals.)
1. $2697.20   **3.** $6130.44   **5.** $6480.00   **7. (a)** $1472.00   **(b)** $1839.68   **(c)** $1848.58   **9. (a)** $3.21   **(b)** $3.26
(c) $3.32   **11.**

13. 9.6%   **15.** 11.72%   **17.** 19.17%   **19.** 27.29   **21.** 60.62

23. 22.25   **25.** $2791.97   **27.** $9196.71   **29.** $1420.41   **31.** $9948.94   **33.** $7730.75   **35.** 7.78%   **37.** 13.87
years   **39.** 6.38 years   **41.** $4.50   **43.** $60.83   **45.** 13.51 years   **47.** $40,722.37   **49.** $29,549.11
51. (a) $432,000   **(b)** $895,795.20   **(c)** $4,124,771.20   **53.** 6.36%

11.2 EXERCISES (page 495)

1. $21,951.35   **3.** $15,413.30   **5.** $264,527.63   **7.** $3914.71   **9.** $5365.01   **11.** $3274.74   **13.** $15,162.14
15. $171.96   **17.** $95,454.20   **19.** The better investment is in part (b).   **21. (a)** $234,660.41   **(b)** $165,644.10
(c) $114,401.52   **(d)** $76.355.43   **23.** $64,491.25   **25.** $6258.37   **27.** $30,722.84   **29.** $2044.04   **31.** $4763.48

11.3 EXERCISES (page 502)

1. 6.66%   **3.** 4.04%   **5.** 17.23%   **7.** 9.42%   **9.** High to low: **(a)** 5.06%   **(b)** 4.89%   **(c)** 4.70%   **11.** High to
low: **(c)** 6.8%   **(b)** 6.66%   **(a)** 6.49%   **13.** High to low: **(b)** 8.2%   **(a)** 8.16%   **(c)** 7.87%   **15.** 12.68%
17. (a) $612.05

(b)

Payment number	Amount	Interest	Applied to principal	Unpaid balance
1	$612.05	$20.00	$592.05	$2407.95
2	612.05	16.05	596.00	1811.95
3	612.05	12.08	599.97	1211.98
4	612.05	8.08	603.97	608.01
5	612.05	4.05	608.01	—

19. (a) $258.01

(b)

Payment number	Amount	Interest	Applied to principal	Unpaid balance
1	$258.01	$13.63	$244.38	$1255.62
2	258.01	11.41	246.60	1009.02
3	258.01	9.17	248.84	760.18
4	258.01	6.90	251.11	509.07
5	258.01	4.62	253.39	255.68
6	258.01	2.32	255.69	—

21. (a) $3246.84

(b)

Payment number	Amount	Interest	Applied to principal	Unpaid balance
1	$3246.84	$1320.00	$1926.84	$10,073.16
2	3246.84	1108.05	2138.79	7,934.37
3	3246.84	872.78	2374.06	5,560.31
4	3246.84	611.63	2635.21	2,925.10
5	3246.84	321.76	2925.10	—

23. $681.08   **25.** $806.96   **27.** $305.56   **29.** $792.13   **31.** $87.85   **33.** 315.83   **35.** $19.49
37. $3886.42   **39. (a)** $147.21 **(b)** $5972.42   **41. (a)** $2649.11 **(b)** $41,297.41   **43. (a)** $1022.23 **(b)** $306,669
45. (a) $961.29 **(b)** $403,741.80   **47. (a)** $939.54 **(b)** $507,351.60   **49.** 6.44 months   **51.** Will never be paid off
53. 24.38 years   **55.** 28.46 years   **57.** Will never be paid off

CHAPTER 11 REFLECTIONS (page 504)

1. $2420   **3.** $2462.88   **5.** 11.21%   **7.** 11.16%   **9.** 3162.83 days   **11.** $2073.21   **13.** $19,614.75
15. $986.00   **17.** 8.33%   **19.** 7.23%   **21.** $40,316.50   **23.** Simple interest is linear in nature; compound interest is not.
Simple interest does not draw interest on interest; compound interest does draw interest on interest.   **25.** To systematically
accumulate money to be used for some purpose in the future   **27.** Assume no amounts are removed from the account. Assume
age 20. Assume 6% annual interest rates. $362.85.   **29. (a)** $246.78 **(b)** $246.07   **31.** $x = 6, y = 0, z = 0, f = 30$ **33.** $\frac{4}{5}$

SAMPLE TEST ITEMS FOR CHAPTER 11 (page 505)

1. $1560   **2.** $4533.21   **3.** $8492.79   **4.** $5333.33   **5.** 7.93%   **6.** 6.99%   **7.** 27.29   **8.** 22.25
9. $8670.31   **10.** $13,838.67   **11.** 8.67 years   **12.** $26,452.76   **13.** $81,136.16   **14.** $5809.48
15. $6716.35   **16.** $3829.42   **17.** 7.19%   **18.** 9.42%   **19.** $1029.14   **20.** 12.35 years

INDEX

A

ACKNOWLEDGMENTS

LITERARY CREDITS

117: "Germany's Sky-High Costs" from *The Wall Street Journal,* August 14, 1995. Reprinted by permission of *The Wall Street Journal,* Copyright © Dow Jones & Company, Inc. All Rights Reserved Worldwide.

376: Excerpts from "Ask Marilyn" column reprinted by permission of William Morris Agency on behalf of the author. Copyright © 1990 by Marilyn vos Savant. Originally appeared in PARADE Magazine, September 9, 1990 and December 2, 1990.

381: (Figure 3): Prepared by The Northwest Arkansas Council, Fayetteville, AR.

PHOTO CREDITS

Unless otherwise acknowledged, all photographs are the property of Scott Foresman • Addison Wesley.

COVER Liaison International, © Ralph Mercer

60: Lindsay M. Hart

61: Malyszko/Stock Boston

92: Joel Gordon Photography

94: Texas Instruments

136: Texas Instruments

138: Paul L. Grafton

225: N. R. Rowan/Jeroboam Inc.

226: Roger Dollarhide/Monkmeyer Press Photo Service

282: Rosann P. Gonzales, CPA

322: Scott D. Wisner

323: NASA

378: Ken W. Merrit

379: Alan Carey/The Image Works

430: Ramona Kent

431: Milt & Joan Mann/Cameramann International, Ltd.

480: Michael Weisbrot/Stock Boston

506: Laura Tucker